2205

$erf^{-1}(0.8) = \cdot\ 6.$

AF764135

AHMAD BAZLEE MAT ZAIN
PE 7013
FALL 1992

Enhanced Oil Recovery

Enhanced Oil Recovery

Larry W. Lake

*University of Texas
at Austin*

Prentice Hall, *Englewood Cliffs, New Jersey 07632*

Library of Congress Cataloging-in-Publication Data

Lake, Larry W.
Enhanced oil recovery / Larry W. Lake.
p. cm.
Bibliography: p.
Includes index.
ISBN 0-13-281601-6
1. Secondary recovery of oil. I. Title.
TN871.L24 1989 88-19544
665.5'3—dc19 CIP

© 1989 by Prentice-Hall, Inc.
A Division of Simon & Schuster,
Englewood Cliffs, New Jersey 07632

All rights reserved. No part of this book may be
reproduced, in any form or by any means,
without permission in writing from the publisher.

Printed in the United States of America
10 9 8 7 6 5 4 3 2

ISBN 0-13-281601-6

Prentice-Hall International (UK) Limited, *London*
Prentice-Hall of Australia Pty. Limited, *Sydney*
Prentice-Hall of Canada, Inc., *Toronto*
Prentice-Hall Hispanoamericana, S.A., *Mexico*
Prentice-Hall of India Private Limited, *New Delhi*
Prentice-Hall of Japan, Inc., *Tokyo*
Simon & Schuster Asia Pte. Ltd., *Singapore*
Editora Prentice-Hall do Brasil, Ltda., *Rio de Janeiro*

Contents

Figures

Tables

Preface

During the last decades of the twentieth century, there is not and will not be an economical, abundant substitute for crude oil in the economies of industrial countries. Maintaining the supply to propel these economies requires developing additional crude reserves. For some areas, this additional development will be in the form of exploration and drilling, but for the domestic United States (and eventually for all oil-producing areas), it will very likely be sustained by applying enhanced oil recovery (EOR). Just as certainly, and somewhat dismayingly, large-scale application of EOR is not easy. It will require more people and a generally higher degree of technology to bring about substantial EOR production. The broad goal of this text is to define this technology.

But there are other goals as well: to formalize the study of EOR as an academic discipline; to illustrate the diversity of EOR and emphasize its reliance on a relatively few physical, mathematical, and chemical fundamentals; and to establish the central position of fractional flow theory as a means for understanding EOR. For these reasons, this book was formulated as a text, with exercises and an extensive reference list.

The twin pillars of this text are fractional flow theory and phase behavior. Other names for fractional flow theory are simple wave theory, coherent wave theory, and Riemann problems. Whatever the name, the most important idea is the view of flow through permeable media as the propagation of one or more waves. The text makes no fundamental advances into ways to solve fractional flow problems; however, the application to the highly nonlinear interactions, particularly as coupled through the phase behavior, brings several fairly new applications to these techniques.

The text is on a graduate level, which presumes some basic knowledge of permeable media and flow therein. Each chapter, however, contains some qualitative material and other material that could be taught with less detail and background than the general level of the rest of the chapter. The text is intended to be taught as a two-semester sequence with Chaps. 1–6 introducing the fundamentals and Chaps. 7–11 detailing the various EOR processes. The text then flows continuously through the decidedly indistinct boundary between advanced reservoir engineering and EOR. I have also had success teaching the text as Chaps. 1, 4, 5, and 7 in the first semester and Chaps. 3, 6, 8, and 9 in the second. In this mode, Chap. 2 is assigned as background reading, and Chaps. 10 and 11 are taught in a third semester.

EOR is a very diverse subject, and several items are lightly covered or omitted. The text, being oriented toward reservoir engineering, contains relatively little about the problems of producing, evaluating, and monitoring EOR projects. Moreover, the entire area of EOR simulation is discussed only in passing. On subjects directly related to EOR, the text has little on polymer gellation, microbial-enhanced oil recovery, and the cosurfactant-enhanced alkaline flooding. But I strongly feel the fundamentals in the other EOR processes will be a basis for understanding these also. Chapter 11 on thermal EOR methods is relatively light on steam stimulation and in situ combustion, which reflects my relative inexperience in these areas.

An effort of this magnitude cannot be made without the help of others. I am most grateful for the contributions of Professor Gary A. Pope, who contributed much of the basis of Chaps. 2, 8, and 9. Our alternating teaching of EOR at The University of Texas is responsible for much of the refinement of this text. I am also grateful to Professor R. S. Schechter for his continual encouragement and to his contributions to Chap. 9 and to my understanding of the method of coherence. Dr. John Cayais contributed to Sec. 9-2 with his information on the manufacturing of sulfonates.

Much of the significant proofreading and technical editing of this text has been done by students in the EOR classes. Several students were involved in this, but I give special credit to Myra Dria and Ekrem Kasap for continually pointing out glitches and suggesting improvements. Patricia Meyers provided editorial assistance. Joanna Castillo deserves considerable credit for generating many of the figures, as does my former secretary, the late Marge Lucas, for typing interminable (but still too few) revisions and my current secretary for dealing with the errata. Finally, I thank my wife Carole for putting up with the trials of writing a text and for her encouragement throughout the process.

Acknowledgments

The author gratefully acknowledges permission to use material from the following sources:

Table 1-1: From "Improved Oil Recovery Could Help Ease Energy Shortage" by Ted Geffen, *World Oil,* vol. 177, no. 5, (October 1983), copyright © Gulf Publishing Company. Used with permission. All rights reserved. *Tables 1-2 and 1-3:* From *Oil and Gas Journal* biennial surveys, © PennWell Publishing Company. *Figures 1-2, 1-3, and 1-4:* Courtesy of the American Petroleum Institute. *Figure 1-5:* From *Thermal Recovery*, Henry L. Doherty Monograph Series 7. © 1982 SPE-AIME. *Tables 1-4, 1-5, and 1-6:* Adapted from "Technical Screening Guides for the Enhanced Recovery of Oil," presented at the 58th Annual Conference and Exhibition of the SPE, San Francisco, 1983. © 1983 SPE-AIME. *Figure 1-6:* Courtesy of the National Petroleum Council. *Figure 1-7:* From Elmond L. Claridge, "Prospects for Enhanced Oil Recovery in the United States." Paper 829168, Proceedings of the 17th Intersociety Energy Conversion Engineering Conference, Los Angeles, August 1982. © 1982 IEEE. *Table 1-7:* Adapted from *Journal of Petroleum Technology*, December, 1977. © 1977 SPE-AIME.

Figure 3-2: From *Improved Oil Recovery by Surfactant and Polymer Flooding,* © 1977 by Academic Press, Inc. *Figures 3-3, 3-4, and 3-5:* Courtesy of George L. Stegemeier. *Figure 3-7:* From "Comparison and Analysis of Reservoir Rocks," presented at the 58th Annual Technical Conference and Exhibition of the SPE, San Francisco, 1983. © 1983 SPE-AIME. *Figure 3-8 and Table 3-1:* Previously published in *J. of Can. Pet. Tech.,* Vol. 15, No. 4, pp. 49–69, Oct./Dec. 1976. *Figure 3-11:* From *Reservoir Engineering Aspects,* SPE Monograph, 1971. © 1971 SPE-AIME. *Figures 3-13, 3-15, and 3-16:* From "Magnitude and Detailed Structures of Residual Oil Saturation," *SPE Journal,* Vol. 23, March-April 1983. © 1983 SPE-AIME.

Figures 3-17 and 3-18: Courtesy of the Institute for the Study of Earth and Man, Southern Methodist University Press. *Figure 3-19:* Courtesy of Dominic Camilleri. *Table 3-3:* From "Comparison and Analysis of Reservoir Rocks and Related Clays," presented at the 58th Annual Technical Conference and Exhibition of the SPE, San Francisco, 1983. © 1983 SPE-AIME. *Table 3-4:* Adapted from Egon T. Degens, *Geochemistry of Sediments: A Brief Survey,* © 1965, p. 16. Reprinted by permission of Prentice-Hall, Inc., Englewood Cliffs, New Jersey. *Figure 3-22:* From "A Study of Caustic Consumption in a High-Temperature Reservoir," 52nd Annual Technical Conference and Exhibition of the SPE, Denver, 1977. © 1977 SPE-AIME. *Figure 3-23:* Upper panels from "The Morphology of Dispersed Clays in Sandstone Reservoirs and its Effect on Sandstone Shalines," presented at the 52nd Annual Technical Conference and Exhibition of the SPE, Denver, 1977. © 1977 SPE-AIME. Lower panels from "Comparison and Analysis of Reservoir Rocks and Related Clays," presented at the 58th Annual Technical Conference and Exhibition of the SPE, San Francisco, 1983. © 1983 SPE-AIME. *Table 3-5:* Adapted from "Comparison and Analysis of Reservoir Rocks and Related Clays," presented at the 58th Annual Technical Conference and Exhibition of the SPE, San Francisco, 1983. © 1983 SPE-AIME. *Table 3-7:* From *SPE Production Engineering*, February 1988. © 1988 SPE-AIME.

Figure 4-3: From David M. Himmelblau, *Basic Principles and Calculations in Chemical Engineering,* 4th Edition, © 1982. Reprinted by permission of Prentice-Hall, Inc., Englewood Cliffs, New Jersey. *Figure 4-11:* Courtesy of the Institute for the Study of Earth and Man, Southern Methodist University Press. *Table 4-1:* Reprinted with permission from Advances in Chemistry Series No. 182, p. 48. Copyright 1979 American Chemical Society. *Table 4-2:* From "A Robust, Iterative Method of Flash Calculations Using the Soave-Redlich-Kwong or the Peng-Robinson Equations of State," presented at the 54th Annual Technical Conference and Exhibition of the SPE, Las Vegas, 1979. © 1979 SPE-AIME.

Figure 5-7: "The Effect of Capillary Pressure on Immiscible Displacements in Stratified Porous Media," presented at the 56th Annual Technical Conference and Exhibition of the SPE, San Antonio, 1981. © 1981 SPE-AIME. *Figure 5-8:* From *Transactions of the AIME,* Vol. 198, 1953. © 1953, SPE-AIME. *Figure 5-10:* From *Transactions of the AIME,* Vol. 213, 1958. © 1958 SPE-AIME. *Figure 5-11:* Courtesy of Nobuyuki Samizo. *Table 5-2 and Figure 5-13:* Courtesy of Dover Publications, Inc. *Figure 5-16:* From *SPE Journal,* Vol. 3, March 1963. © 1963 SPE-AIME. *Figure 5-17:* Courtesy of Bureau de Recherches Geologiques et Minieres. *Figure 5-18:* Courtesy of Donald J. MacAllister.

Figures 6-2, 6-3, and 6-4: From *Transactions of the AIME,* Vol. 201, April, 1954. © 1954 SPE-AIME. *Figure 6-5:* Reprinted from *Industrial and Engineering Chemistry* 61(9), p. 16. Copyright 1969 American Chemical Society. *Figures 6-7 and 6-8:* From "A Simplified Predictive Model for Micellar/Polymer Flooding," presented at the California Regional Meeting of the Society of Petroleum Engineers, San Francisco, 1982. © 1982 SPE-AIME. *Table 6-1:* Courtesy of Mary Ellen Lambert. *Figure 6-19:* From *Transactions of the AIME,* Vol. 219, 1960. © 1960 SPE-AIME.

Figure 7-1: Drawing by Joe Lindley. Courtesy of U.S. Department of Energy, Bartlesville, Oklahoma. *Figures 7-2, 7-3, and 7-4:* Courtesy of Ingersoll-Rand Company. *Figure 7-5:* From *Journal of Petroleum Technology* 18(8), August, 1966. © 1966 SPE-AIME. *Figure 7-6:* Courtesy of the U.S. Department of Energy, Bartlesville, Oklahoma. *Figure 7-7:* From *Journal of Petroleum Technology,* Vol. 33, November, 1981. © 1981 SPE-AIME. *Figure 7-8:* Courtesy U.S. Department of Energy, Bartlesville, Oklahoma. *Figure 7-9:* From "Lumped-Component Characterization of Crude Oils for Compositional Simulation," presented at the Third Joint SPE/DOE Symposium on Enhanced Oil Recovery, Tulsa, 1982. © 1982 SPE-AIME. *Figures 7-10 and 7-11:* From *Journal of Petroleum Technology,* Vol. 33, November, 1981. © 1981 SPE-AIME. *Figure 7-13:* From *Transactions of the SPE of the AIME,* Vol. 219, 1961. © 1961 SPE-AIME. *Figures 7-14 and 7-15:* From *Miscible Displacement,* SPE Monograph Series 8, 1983. © 1983 SPE-AIME. *Figure 7-18:* From "Preliminary Experimental Results of High-Pressure Nitrogen Injection for EOR Systems," presented at the 56th Annual Technical Conference and Exhibition of the SPE, San Antonio, 1981. © 1981 SPE-AIME. *Figure 7-19:* From *SPE Journal,* Vol. 19, August, 1978. © 1978 SPE-AIME. *Figures 7-20, 7-21, and 7-22:* From *Transactions of the AIME,* Vol. 234, 1965. © 1965 SPE-AIME. *Figure 7-23:* From *Journal of Petroleum Technology,* Vol. 15, March, 1963. © 1963 SPE-AIME. *Figure 7-24:* From *Journal of Petroleum Technology,* Vol. 33, November, 1981. © 1981 SPE-AIME. *Table 7-1:* From *Journal of Petroleum Technology,* Vol. 34, 1982. © 1982 SPE-AIME. *Figure 7-25:* From *SPE Journal,* Vol. 22, 1982. © 1982 SPE-AIME. *Figure 7-26:* Section a from "Measurement and Correlation of CO_2 Miscibility Pressures," presented at the Third Annual Joint SPE/DOE Symposium on Enhanced Oil Recovery, Tulsa, 1981. © 1981 SPE-AIME. Section b courtesy of U.S. Department of Energy, Bartlesville, Oklahoma. Section c from *SPE Journal*, Vol. 22, 1981. © 1981 SPE-AIME. *Figure 7-27:* From *Transactions of the SPE of the AIME,* Vol. 219, 1961. © 1961 SPE-AIME. *Figure 7-29:* From *Transactions of the AIME,* Vol. 210, 1956. © 1956 SPE-AIME. *Figure 7-30:* From *Miscible Flooding Fundamentals,* SPE Monograph Series, 1985. © 1985 SPE-AIME. *Figure 7-40:* Courtesy of U.S. Department of Energy, Bartlesville, Oklahoma. *Figure 7-41:* From "An Investigation of Phase Behavior-Macroscopic Bypassing Interaction in CO_2 Flooding," presented at the Third Annual Joint SPE/DOE Symposium on Enhanced Oil Recovery, Tulsa, 1982. © 1982 SPE-AIME. *Figure 7-43:* From *Transactions of the AIME,* Vol. 231, 1964. © 1964 SPE-AIME. *Figure 7-44:* From *SPE Journal,* Vol. 10, December, 1970. © 1970 SPE-AIME. *Figure 7-45:* From "The Effect of Microscopic Core Heterogeneity on Miscible Flood Residual Oil Saturation," presented at the 55th Annual Technical Conference and Exhibition of the SPE, Dallas, 1980. © 1980 SPE-AIME. *Figure 7-47:* From *Journal of Petroleum Technology,* Vol. 33, November, 1981. © 1981 SPE-AIME. *Figure 7-48:* From "An Investigation of Phase Behavior-Macroscopic Bypassing Interaction in CO_2 Flooding," presented at the Third Annual Joint SPE/DOE Symposium on Enhanced Oil Recovery, Tulsa, 1982. © 1982 SPE-AIME. *Figure 7-49:* From "Effects of Mobile Water on Multiple Contact Miscible Gas Displacements," presented at the 1982 Third Joint SPE/DOE Symposium on Enhanced Oil Recovery, Tulsa, 1982. © 1982 SPE-AIME. *Figure 7G:* "Slaughter Estate Unit CO_2 Pilot Reservoir Description via a Black Oil Model Waterflood History Match," presented at the Third Annual Joint SPE/DOE Symposium on Enhanced Oil Recovery, Tulsa, 1982. © 1982 SPE-AIME.

Figure 8-1: Drawing by Joe Lindley. Courtesy of U.S. Department of Energy, Bartlesville, Oklahoma. *Figure 8-2:* From "Phase Behavior Effects on the Oil Displacement Mechanisms of Broad Equivalent Weight Surfactant Systems," presented at the Second Annual Joint SPE/DOE Symposium on Enhanced Oil Recovery, Tulsa, 1981. © 1981 SPE-AIME. *Figure 8-3:* From *Improved Oil Recovery by Surfactant and Polymer Flooding*, D. C. Shah and R. S. Schechter eds. Copyright 1977 Academic Press, Inc. *Figure 8-4 and 8-5:* Courtesy of K. Tsaur. *Figure 8-6:* Courtesy of the U.S. Department of Energy, Bartlesville, Oklahoma. *Figure 8-8:* From "Preparation and Testing of Partially Hydrolized Polyacrylamide Solutions," presented at the 51st Annual Technical Conference and Exhibition of the SPE, New Orleans, 1976. © 1976 SPE-AIME. *Figure 8-9:* From *Journal of Petroleum Technology,* Vol. 23, 1971. © 1971 SPE-AIME. *Table 8-1:* From *Enhanced Oil Recovery,* © 1984 National Petroleum Council. *Figure 8-14:* From "An Economic Polymer Flood in the North Burbank Unit, Osage County, Oklahoma," presented at the 50th Annual Technical Conference and Exhibition of the SPE, Dallas, 1975. © 1975 SPE-AIME. *Table 8-2 and Figure 8L:* Courtesy of U.S. Department of Energy, Bartlesville, Oklahoma.

Figures 9-1 through 9-7: Courtesy of the Institute for the Study of Earth and Man, Southern Methodist University Press. *Table 9-1:* From *Enhanced Oil Recovery,* F. J. Fayers, ed. Copyright © 1981 Elsevier Scientific Publishing Company. *Figure 9-8:* From *SPE Journal,* Vol. 22, 1982. © 1982 SPE-AIME. *Figure 9-9:* From "Surfactant Flooding with Microemulsions formed In-Situ—Effect of Oil Characteristics," presented at the 54th Annual Technical Conference and Exhibition of the SPE, Tulsa, 1979. © 1979 SPE-AIME. *Figure 9-10:* From *Improved Oil Recovery by Surfactant and Polymer Flooding,* D. O. Shah and R. S. Schechter, eds. Copyright © 1977 Academic Press, Inc. *Figure 9-11:* From "Surfactant Flooding with Microemulsions formed In-Situ—Effect of Oil Characteristics," presented at the 54th Annual Technical Conference and Exhibition of the SPE, Tulsa, 1979. © 1979 SPE-AIME. *Figure 9-12:* Courtesy of Svein R. Engelsen. *Figure 9-13:* Courtesy of Kim Jones. *Figure 9-16:* Courtesy of the Center for Enhanced Oil and Gas Recovery, The University of Texas at Austin. *Figure 9-18:* From SPE Formation Evaluation, September, 1987. © 1987 SPE-AIME. *Figures 9-25 and 9-26:* From *SPE Journal,* Vol. 18, October, 1978. © 1978 SPE-AIME. *Figure 9-28:* Courtesy of Miguel Enrique Fernandez. *Figure 9-29:* From *SPE Journal,* Vol. 19, 1979. © 1979 SPE-AIME. *Figure 9-30:* Courtesy of U.S. Department of Energy, Bartlesville, Oklahoma. *Figure 9-31:* From *SPE Journal,* Vol. 19, 1978. © 1978 SPE-AIME. *Figure 9-32:* From *SPE Journal,* Vol. 22, 1982. © SPE-AIME. *Figure 9-33.* Courtesy of Petroleum Engineers International, Vol. 21, August, 1981. *Table 9-4:* From *SPE Journal,* Vol. 22, 1982. © 1982 SPE-AIME. *Figure 9-34:* From *Journal of Petroleum Technology,* Vol. 22, 1970. © 1970 SPE-AIME. *Figures 9-36 and 9-37:* From "A Simplified Predictive

Model for Micellar/Polymer Flooding," presented at the California Regional Meeting of the SPE, San Francisco, 1982. © 1982 SPE-AIME. *Figure 9G:* Courtesy of Svein R. Engelsen.

Figures 10-1 and 10-2: Reprinted from *Colloid and Surface Chemistry, A Self-Study Project, Part 2, Lyophobic Colloids* by J. Th. Overbeek by permission of MIT and J. Th. Overbeek. Published by MIT, Center for Advanced Engineering Study, Cambridge, MA 02139. Copyright 1972 by J. Th. Overbeek. *Figure 10-4:* From "The Rheology of Foam," presented at the 44th Annual Technical Conference and Exhibition of the SPE, Denver, 1969. © 1969 SPE-AIME. *Figure 10-5:* Courtesy of Stanford University. *Figure 10-6:* From "Effect of Foam on Permeability of Porous Media to Water," presented at the 39th Annual Technical Conference and Exhibition of the SPE, Houston, 1964. © 1964 SPE-AIME. *Figure 10-7:* From "Effect of Foam on Trapped Gas Saturation and on Permeability of Porous Media to Water," presented at the 40th Annual Technical Conference and Exhibition of the SPE, Houston, 1965. © 1965 SPE-AIME. *Figure 10-8:* Courtesy of U.S. Department of Energy, Bartlesville, Oklahoma. *Figure 10-9:* From "Mechanisms of Foam Flow in Porous Media—Apparent Viscosity and Smooth Capillaries," presented at the 58th Annual Meeting of the SPE, San Francisco, 1983. © SPE-AIME. *Figure 10-10:* Drawing by Joe Lindley. Courtesy of U.S. Department of Energy, Bartlesville, Oklahoma. *Figure 10-12:* From "Alkaline Waterflooding—A Model for Interfacial Activity of Acidic Crude/Caustic Systems," presented at the Third Symposium on Enhanced Oil Recovery of the SPE, Tulsa, 1982. © 1982 SPE-AIME. *Figure 10-13:* From *SPE Journal,* Vol. 25, October, 1985. © 1985 SPE-AIME. *Figure 10-14:* From *SPE Journal,* Vol. 22, December, 1982. © 1982 SPE-AIME. *Figure 10-15:* "Interaction of Precipitation/Dissolution Waves and Ion Exchange in Flow through Permeable Media," Bryant, S. L.; Schechter, R. S.; Lake, L. W. *AIChE Journal*, Volume 32, No. 5, p. 61 (May 1986). Reproduced by permission of the American Institute of Chemical Engineers. *Figure 10-16:* From *Journal of Petroleum Technology*, December, 1974. © 1974 SPE-AIME. *Table 10-1:* From "Alkaline Injection for Enhanced Oil Recovery—a Status Report," presented at the First Annual Joint SPE/DOE Symposium on Enhanced Oil Recovery, Tulsa, 1980. © 1980 SPE-AIME.

Figure 11-1: Courtesy of Interstate Oil Compact Commission. *Figure 11-2:* From *Thermal Recovery,* Henry L. Doherty Monograph Series 7. © 1982 SPE-AIME. *Figures 11-3 and 11-4:* Courtesy of Oil and Gas Journal, PennWell Publishing Company. *Table 11-1:* From *Steam Tables, Thermodynamic Properties of Water, Including Vapor, Liquid and Solid Phases*. Reprinted by permission of John Wiley. *Table 11-2:* Courtesy of Interstate Oil Compact Commission. *Figure 11-5:* Courtesy of Oil and Gas Journal, PennWell Publishing Company. *Table 11-3:* Courtesy of Interstate Oil Compact Commission. *Figures 11-7 and 11-8:* From *Journal of Petroleum Technology,* May, 1967. ©1967 SPE-AIME. *Figures 11-9 and 11-11:* From *Journal of Petroleum Technology,* Vol. 14, April, 1962. © 1962 SPE-AIME. *Figures 11-10 and 11-12:* From *Journal of Petroleum Technology,* Vol. 17, July, 1965. © 1965 SPE-AIME. *Figures 11-14 and 11-15:* From *Journal of Petroleum Technology,* Vol. 30, February, 1978. © 1978 SPE-AIME. *Figure 11-19:* From *Journal of Petroleum Technology,* Vol. 27, August, 1975. © 1975 SPE-AIME. *Figure 11-20:* Fig. 3.24 in Burger J., Sourieau P. and Combarnous M. "Thermal Methods of Oil Recovery," Editions Technip, Paris and Gulf Publishing Company, Houston (1985). *Figure 11-21:* From "Calculation Methods for Linear and Radial Steam Flow in Oil Reserves," presented at the 52nd Annual Technical Conference and Exhibition of the SPE, Denver, 1977. © 1977 SPE-AIME. *Table 11-5 and Figures 11-22, 11-23, and 11-24:* From *Journal of Petroleum Technology,* Vol. 27, December 1975. © 1975 SPE-AIME. *Figure 11-25:* From *Journal of Petroleum Technology,* Vol. 33, October, 1981. © SPE-AIME. *Figure 11-26:* From *Thermal Recovery,* Henry L. Doherty Monograph Series 7. © 1982 SPE-AIME. *Figure 11-27:* From *Transactions of the SPE of the AIME,* Vol. 253, October, 1972. © 1972 SPE-AIME.

1

Defining Enhanced Oil Recovery

Enhanced oil recovery (EOR) is oil recovery by the injection of materials not normally present in the reservoir. This definition embraces all modes of oil recovery processes (drive, push-pull, and well treatments) and covers many oil recovery agents. Most important, the definition does not restrict EOR to a particular phase (primary, secondary, or tertiary) in the producing life of a reservoir. Primary recovery is oil recovery by natural drive mechanisms, solution gas, water influx, gas cap drive, or gravity drainage. Secondary recovery refers to techniques, such as gas or water injection, whose purpose, in part, is to maintain reservoir pressure. Tertiary recovery is any technique applied after secondary recovery. Nearly all EOR processes have been at least fieldtested as secondary displacements. Many thermal methods are commercial in both primary or secondary modes. Much interest has been focused on tertiary EOR, but this definition does not place any such restriction.

The definition does exclude waterflooding and is intended to exclude all pressure maintenance processes. Sometimes the latter distinction is not clear since many pressure maintenance processes have displacement character. Moreover, agents such as methane in a high-pressure gas drive, or carbon dioxide in a reservoir with substantial resident CO_2, do not satisfy the definition, yet both are clearly EOR processes. Usually the EOR cases that fall outside the definition are clearly classified by the intent of the process. In this chapter, we restrict ourselves to U.S. domestic statistics.

1-1 EOR INTRODUCTION

The EOR Target

Much of the interest in EOR centers on the amount of oil it is potentially applicable to. Table 1-1 shows this target oil accounts for 278 billion barrels in the United States alone. This represents nearly 70% of the 401 billion barrels of original oil in place. If EOR could recover only 10% of this, it could more than double the proved domestic reserves.

TABLE 1-1 PRODUCTION, RESERVES, AND RESIDUAL OIL IN PLACE; U.S. ONSHORE, EXCLUDING ALASKA (FROM GEFFEN, 1973)

Category	Billions of barrels*	Percent of original oil in place
Produced	101	25.2
Proved reserves	22	5.5
EOR target	278	69.3
Total	401	100.0

*1 bbl = 0.159 m^3

The likelihood of recovering substantial additional oil by EOR lies mostly in the future. Recent production trends show less than 10% of the domestic production rate comes from EOR processes (Fig. 1-1). But this trend is growing at a significant rate. Neither this text nor Table 1-1 deals with enhanced gas recovery.

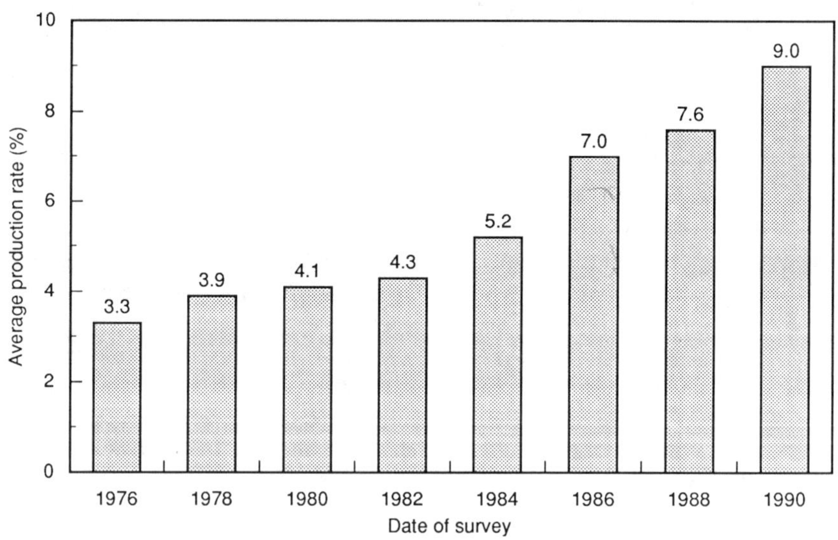

Figure 1-1 EOR oil rate as a percent of daily production (from *Oil and Gas Journal* biennial survey)

EOR Categories

With a few minor exceptions, all EOR falls distinctly into one of three categories: thermal, chemical, or solvent methods. Foam flooding, for example, could fit into all three. Each category can be divided further into individual processes (see Table 1-2), which in turn, have several variations.

Some idea of the popularity of the individual processes follows from the biennial survey of U.S. EOR activity compiled by the *Oil and Gas Journal.* These numbers tend to underestimate actual activity since they are based on voluntary surveys. The surveys do not distinguish between pilot and commercial processes.

Thermal methods, particularly steam drive and soak (combined in Table 1-2), occupy the largest share of EOR projects and have experienced growth since 1971. This density reflects the long-standing commercial success of steam flooding. All other processes have also experienced some growth, with polymer flooding and carbon dioxide flooding showing an explosive increase since 1980. Of course, the total of all EOR projects has grown steadily until oil prices declined in 1986.

TABLE 1-2 ACTIVE DOMESTIC EOR PROJECTS (FROM *OIL AND GAS JOURNAL* BIENNIAL SURVEYS)

	1971	1974	1976	1978	1980	1982	1984	1986	1988	1990
Thermal										
Steam	53	64	85	99	133	118	133	181	133	137
In situ combustion	38	19	21	16	17	21	18	17	9	8
Hot water	—	—	—	—	—	—	—	3	10	9
Total thermal	91	83	106	115	150	139	151	201	152	154
Chemical										
Micellar polymer	5	7	13	22	14	20	21	20	9	6
Polymer	14	9	14	21	22	47	106	178	111	42
Alkaline	—	2	1	3	6	18	11	8	4	2
Total chemical	19	18	28	46	42	85	138	206	124	50
Solvent										
Hydrocarbon miscible	21	12	15	15	9	12	30	26	22	23
CO_2 miscible	1	6	9	14	17	28	40	38	49	52
CO_2 immiscible	—	—	—	—	—	—	18	28	8	4
Nitrogen	—	—	—	—	—		7	9	9	9
Flue gas (miscible and immiscible)	—	—	—	—	8	10	3	3	2	3
Total solvent	22	18	24	29	34	50	98	104	90	91
Other										
Carbonated waterflood	—	—	—	—	—	—	—	1	—	—
Grand total	132	119	158	190	226	274	387	512	366	295

TABLE 1-3 DOMESTIC EOR PRODUCTION BY PROCESS TYPE (FROM *OIL AND GAS JOURNAL* BIENNIAL SURVEYS)

	Production (bbl/day or 0.159 m^3/day)					
	1980	1982	1984	1986	1988	1990
Thermal methods						
Steam	243,477	288,396	358,115	468,692	455,484	444,137
In situ combustion	12,133	10,228	6,445	10,272	6,525	6,090
Hot water				705	2,896	3,985
Total thermal	255,610	298,624	364,560	479,669	464,905	454,212
Chemical						
Micellar polymer	930	902	2,832	1,403	1,509	637
Polymer	924	2,587	10,232	15,313	20,992	11,219
Alkaline	550	580	334	185		0
Total chemical	2,404	4,069	13,398	16,901	22,501	11,856
Solvent						
Hydrocarbon miscible			14,439	33,767	25,935	55,386
CO_2 miscible	21,532	21,953	31,300	28,440	64,192	95,591
CO_2 immiscible			702	1,349	420	95
Nitrogen			7,170	18,510	19,050	22,260
Flue gas (miscible and immiscible)			29,400	26,150	40,450	17,300
Total solvent	74,807*	71,915*	83,011	108,216	150,047	190,632
Grand total	332,821	374,608	460,969	604,786	637,453	656,700

*Other solvent methods not classified separately in these years.

Project number measures activity, but oil production rate measures success. Table 1-3 shows daily oil production rates for the EOR processes. Here the preponderance of steam flooding—about 80% of total EOR production—is even more apparent. Of the remaining EOR production, about 80% is by solvent flooding.

Carefully comparing Tables 1-2 and 1-3 reveals several details. For example, polymer flooding production is quite small for the number of projects reported. This process seems to be preferred for small projects or for projects producing very little oil at initiation. The production rate per project is nearly equal for steam and carbon dioxide. This indicates the disparity between their respective production rates is caused mainly by the difference in the number of projects. The project rate for micellar-polymer and alkaline flooding reflects the marginal economics of these processes at the time of the survey.

1-2 THE NEED FOR EOR

Even given the large disparity between the commercial success of the various processes, we need all forms of EOR to maintain reserves at an acceptable level. This is because each process, even commercially successful ones, has demonstrated success on only a part of the EOR target thus far.

Reserves

Reserves are petroleum (crude and condensate) recoverable from known reservoirs under prevailing economics and technology. They are given by the following material balance equation:

$$\begin{pmatrix}\text{Present}\\ \text{reserves}\end{pmatrix} = \begin{pmatrix}\text{Past}\\ \text{reserves}\end{pmatrix} + \begin{pmatrix}\text{Additions}\\ \text{to reserves}\end{pmatrix} - \begin{pmatrix}\text{Production}\\ \text{from}\\ \text{reserves}\end{pmatrix}$$

Clearly, reserves can change with time because the last two terms can change with time.

Throughout most of the U.S. petroleum history, reserves increased (Fig. 1-2). Reserves began to decline in the late 1960s, a trend that abruptly reversed in 1971 with the addition of the approximately 10 billion barrels from the Alaskan North Slope. However, this decline immediately resumed at a higher rate, abetted by instability in non-U.S. petroleum sources. In the early 1980s, although the decline has greatly slowed, it has not vanished.

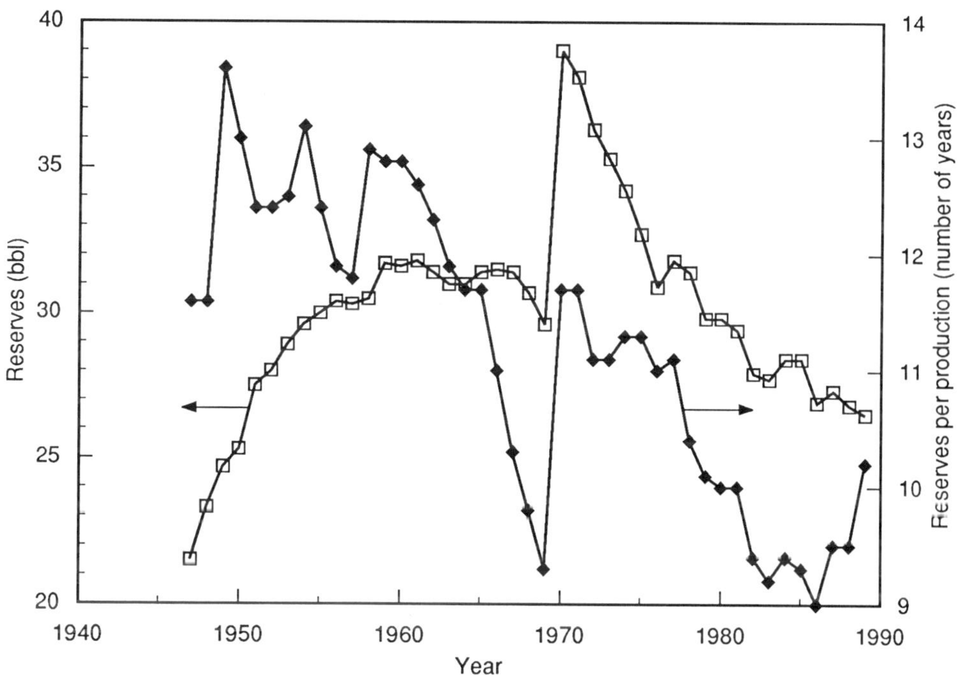

Figure 1-2 Crude reserves in the United States (from *Basic Petroleum Data Book*, 1986)

Figure 1-2 also shows the reserve-to-production ratio was relatively constant until the early 1960s and then declined beginning in 1960 through the late 1970s. The apparent stabilization beginning in 1980 is the result of the leveling out of reserves and conservation, both responses to the increase in oil price in the prior decade. Both reserves and the reserves to production ratio will fall again because of the oil price decline in 1986. EOR is one of several methods to arrest this decline.

Adding to Reserves

The four categories of adding to reserves are

1. Discovering new fields
2. Discovering new reservoirs
3. Extending reservoirs in known fields
4. Redefining reserves because of changes in economics of extraction technology

We discuss category 4 in the remainder of this text. Here we substantiate its importance by briefly discussing categories 1–3.

The question naturally arises about the probability of adding to reserves through categories 1–3. Figure 1-3 shows that whereas production is a rather

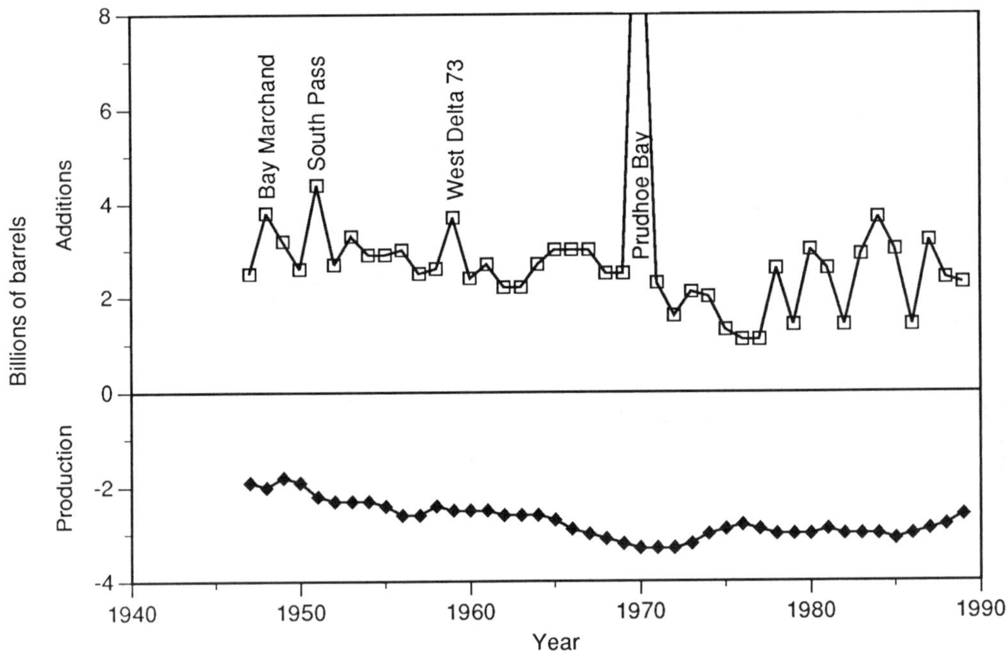

Figure 1-3 Contribution of new oil to U.S. reserves (from *Basic Petroleum Data Book*, 1986)

smoothly varying function of time, reserve additions through new oil are more sporadic. Except for years where there are discoveries of large fields, production exceeds additions even when the reserve rate is fairly constant. The reserve decline rate, therefore, can be seriously reversed only by finding large fields.

But the probability of finding large fields is declining. Figure 1-4 shows the rate of discovery of significant (1 million barrels or more) fields plotted versus the date of discovery. The figure also shows the number of exploratory wells drilled in the same year. If correlation existed between the two curves, the drilling curve should lead the discovery curve, but the reverse has happened. This may again happen in the discovery curve in the 1980s. The discovery curve terminates six years before the drilling curve because it takes this long on average to verify reserves. Both curves suggest little correlation between the number of wells drilled and the frequency of finding significant fields.

We can summarize the need for EOR, then, from Figs. 1-2 through 1-4, as follows:

1. The U.S. domestic reserves are stable but likely to decline.
2. From the viewpoint of exploration, the rate of decline can be affected only by discovering large fields.

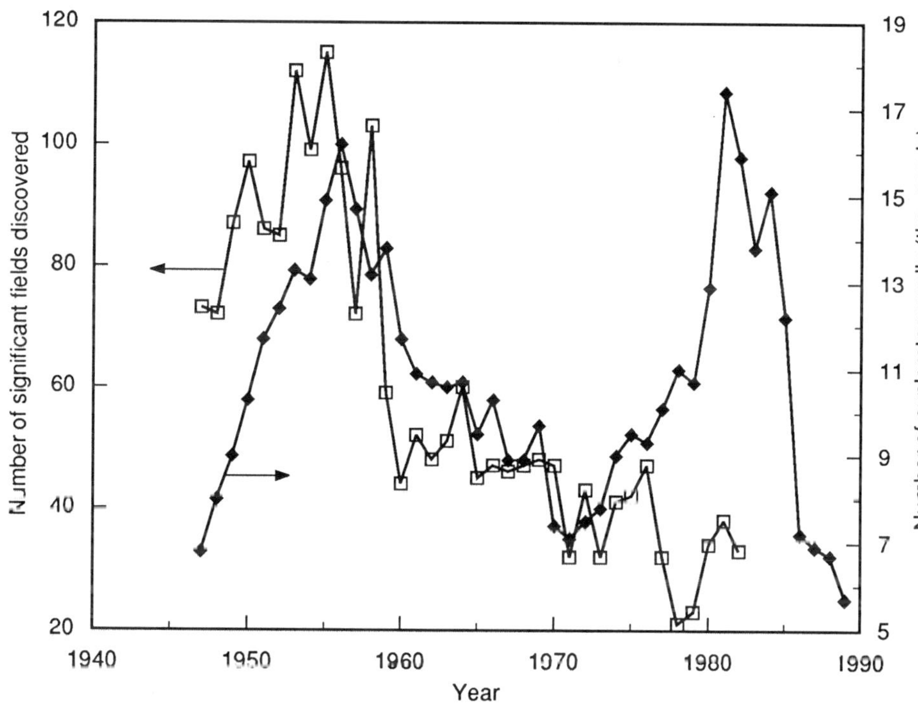

Figure 1-4 Discovery rate and drilling (from *Basic Petroleum Data Book*, 1991)

3. The rate of discovery of large fields does not correlate with the number of wells drilled.

These observations suggest techniques other than drilling and exploration are needed to replace U.S. reserves. Considering the 278 billion barrel target, these other techniques center on EOR.

1-3 INCREMENTAL OIL

A universal technical measure of the success of an EOR process is the amount of incremental oil recovered. Figure 1-5 defines incremental oil. Imagine a field, reservoir, or well whose oil rate is declining as from *A* to *B*. At *B*, an EOR project is initiated and, if successful, the rate should show a deviation from the projected decline at some time after *B*. Incremental oil is the difference between what was actually recovered, *B* to *D*, and what would have been recovered had the process not been initiated, *B* to *C*. This is the shaded region in Fig. 1-5.

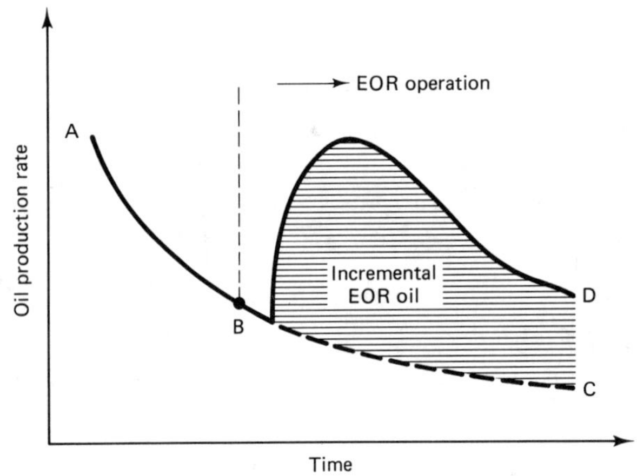

Figure 1-5 Incremental oil recovery from typical EOR response (from Prats, 1982)

As simple as the concept in Fig. 1-5 is, EOR is difficult to determine in practice. There are several reasons for this.

1. Combined production from EOR and non-EOR wells. Such production makes it difficult to ascribe the EOR-produced oil to the EOR project. Co-mingling occurs when, as is usually the case, the EOR project is phased into a field undergoing other types of recovery.
2. Oil from other sources. Usually the EOR project has experienced substantial well cleanup or other improvements. The oil produced as a result of such treatment is not easily differentiated from the EOR oil.

3. Inaccurate estimate of hypothetical decline. The curve from *B* to *C* in Fig. 1-5 must be accurately estimated. But since it did not occur, there is no way of assessing this accuracy. Techniques ranging from decline curve analysis to numerical simulation must still be tempered by good judgment.

1-4 CATEGORY COMPARISONS

We spend most of this text covering the details of EOR processes. At this point, we introduce some issues and compare performances of the three basic EOR processes. The latter is represented as typical oil recoveries (incremental oil expressed as a percent of original oil in place) and by various utilization factors. Both are based on actual experience.

Table 1-4 shows sensitivity to high salinities is common to all chemical flooding EOR. Total dissolved solids should be less than 100,000 g/m^3, and hardness should be less than 2,000 g/m^3. Chemical agents are also susceptible to loss through rock–fluid interactions. Maintaining adequate injectivity is a persistent problem. Historical oil recoveries have ranged from small to moderately large. Chemical utilization factors have meaning only when compared to the costs of the individual agents; polymer, for example, is usually three to four times as expensive (per unit mass) as surfactants.

TABLE 1-4 CHEMICAL EOR PROCESSES (ADAPTED FROM TABER AND MARTIN, 1983)

Process	Recovery mechanism	Issues	Typical recovery (%)	Typical agent utilization*
Polymer	Improves volumetric sweep by mobility reduction	Injectivity Stability High salinity	5	0.3–0.5 lb polymer per bbl oil produced
Micellar polymer	Same as polymer plus reduces capillary forces	Same as polymer plus chemical availability, retention, and high salinity	15	15–25 lb surfactant per bbl oil produced
Alkaline polymer	Same as micellar polymer plus oil solubilization and wettability alteration	Same as micellar polymer plus oil composition	5	35–45 lb chemical per bbl oil produced

*1 lb/bbl ≅ 2.86 kg/m^3

Table 1-5 shows a similar comparison for thermal processes. Recoveries are generally higher for these processes than for the chemical methods. Again, the issues are similar within a given category, centering on heat losses, override, and air pollution. Air pollution occurs because steam is usually generated by burning a portion of

TABLE 1-5 THERMAL EOR PROCESSES (ADAPTED FROM TABER AND MARTIN, 1983)

Process	Recovery mechanism	Issues	Typical recovery (%)	Typical agent utilization
Steam (drive and stimulation)	Reduces oil viscosity Vaporization of light ends	Depth Heat losses Override Pollution	50–65	0.5 bbl oil consumed per bbl oil produced
In situ combustion	Same as steam plus cracking	Same as steam plus control of combustion	10–15	10 MCF air per bbl oil produced*

*1 MCF/stb ≅ 178 SCM gas/SCM oil

the produced oil. If this burning occurs on the surface, the emission products contribute to air pollution; if the burning is in situ, production wells can be a source of pollutants.

Table 1-6 compares the solvent flooding processes. Only two groups are in this category, corresponding to whether or not the solvent develops miscibility with the oil. Oil recoveries are generally lower than micellar-polymer recoveries. The solvent utilization factors as well as the relatively low cost of the solvents have brought these processes, particularly carbon dioxide flooding, to commercial application. The distinction between a miscible and an immiscible process is slight.

TABLE 1-6 SOLVENT EOR METHODS (ADAPTED FROM TABER AND MARTIN, 1983)

Process	Recovery mechanism	Issues	Typical recovery (%)	Typical agent utilization*
Immiscible	Reduces oil viscosity Oil swelling Solution gas	Stability Override Supply	5–15	10 MCF solvent per bbl oil produced
Miscible	Same as immiscible plus development of miscible displacement	Same as immiscible	5–10	10 MCF solvent per bbl oil produced

*1 MCF/stb ≅ 178 SCM solvent/SCM oil

1-5 THE FUTURE OF EOR

Forecasting these processes in a highly volatile economic scene is risky at best. Nevertheless, predicting EOR trends does give some idea of current thinking.

Figure 1-6 gives a forecast made by the National Petroleum Council (NPC) in

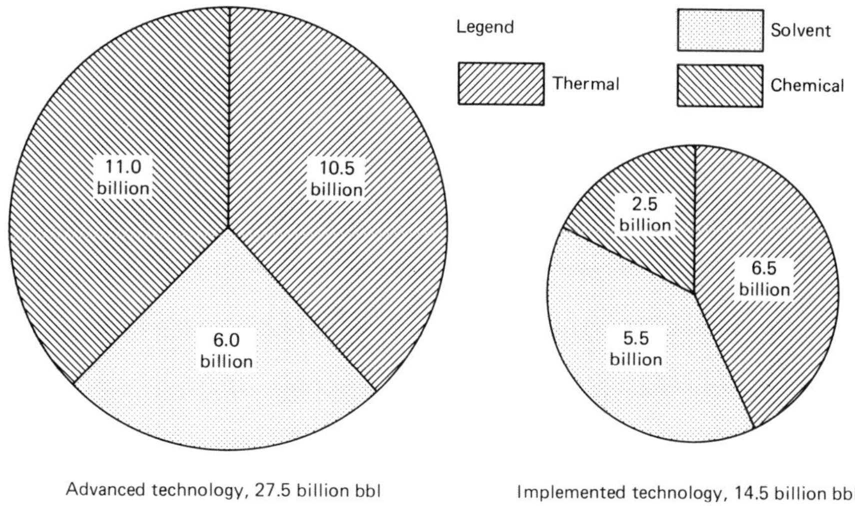

Figure 1-6 EOR projections (from National Petroleum Council, 1984)

1984. These forecasts assume a base oil price of $30 per barrel and a minimum rate of return on investment of 10%. Implemented technology, attainable without additional technical developments, shows thermal recovery has the largest potential, trailed narrowly by solvent flooding and then chemical flooding. The entire EOR projection is 14.5 billion barrels. This represents a dismayingly small 5% of the total target oil given in Table 1-1. One purpose of this text is to explain why this is so low and to indicate what must be done to improve it.

Technological advances can dramatically alter this picture. Figure 1-6 also shows an advanced technology case from the same NPC study wherein the thermal potential has nearly doubled, and the chemical has increased by more than a factor of 4. In the chemical case, where micellar-polymer potential alone represents 10 billion barrels, technological improvements can change it from the process with the least potential to that with the most. Solvent flooding, on the other hand, is believed to be fairly well developed. The 27.5 billion barrel advanced technology potential represents more than 10% of the EOR target. This quantity is larger than the proved reserves listed in Table 1-1.

To emphasize the difficulties in attaining the advanced technology case, we end this section on a sobering issue. We have discussed the EOR target as though it will remain constant. That this is not so is indicated by Fig. 1-7, a projection of the three types of reserves in Table 1-1 on into the next century. The EOR target is the amount between the top and second lines. This figure says the EOR target will actually contract rather rapidly with time, being pinched from below by actual production and from above by abandoned oil. Abandoned oil can come about by mechanical deterioration or a variety of economically driven causes. When a reservoir is actually abandoned, it is not likely that EOR can be successful because few processes

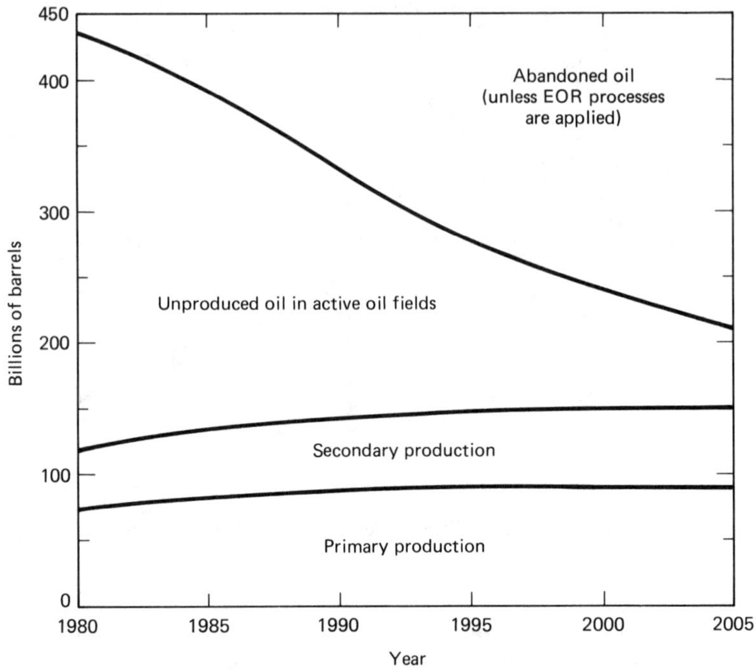

Figure 1-7 Decline of EOR target (from Claridge, 1982)

can bear the expense of drilling new wells. An advantage of EOR over exploratory drilling is that we know the location of the oil, and much of the equipment is already in place.

The shrinking target introduces a sense of urgency to the implementation of EOR. To accomplish this implementation, we must diligently seek opportunities for EOR and then pursue them to successful completion. Doing this requires substantial technical sophistication from all phases of the producing community. Providing this sophistication is one of the purposes of this text.

1-6 UNITS AND NOTATION

SI Units

The basic set of units in this text is the Systeme International (SI) system. We cannot be entirely rigorous about SI units because a great body of figures and tables has been developed in more traditional units. It is impractical to convert these; therefore, we give a list of the more important conversions in Table 1-7 and some helpful pointers in this section.

TABLE 1-7 AN ABRIDGED SI UNITS GUIDE (ADAPTED FROM CAMPBELL ET AL., 1977)

SI base quantities and units

Base quantity or dimension	SI unit	SI unit symbol (use roman type)	SPE dimensions symbol (use roman type)
Length	Meter	m	L
Mass	Kilogram	kg	m
Time	Second	s	t
Thermodynamic temperature	Kelvin	K	T
Amount of substance	Mole*	mol	

*When the mole is used, the elementary entities must be specified; they may be atoms, molecules, ions, electrons, other particles, or specified groups of such particles in petroleum work. The terms *kilogram mole*, *pound mole*, and so on are often erroneously shortened to mole.

Some common SI derived units

Quantity	Unit	SI unit symbol (use roman type)	Formula (use roman type)
Acceleration	Meter per second squared	—	m/s^2
Area	Square meter	—	m^2
Density	Kilogram per cubic meter	—	kg/m^3
Energy, work	Joule	J	$N \cdot m$
Force	Newton	N	$kg \cdot m/s^2$
Pressure	Pascal	Pa	N/m^2
Velocity	Meter per second	—	m/s
Viscosity, dynamic	Pascal-second	—	$Pa \cdot s$
Viscosity, kinematic	Square meter per second	—	m^2/s
Volume	Cubic meter	—	m^3

Selected conversion factors

To convert from	To	Multiply by
Acre (U.S. survey)	Meter² (m^2)	4.046 872 E+03
Acres	Feet² (ft^2)	4.356 000 E+04
Atmosphere (standard)	Pascal (Pa)	1.013 250 E+05
Bar	Pascal (Pa)	1.000 000 E+05
Barrel (for petroleum 42 gal)	Meter³ (m^3)	1.589 873 E−01
Barrel	Feet³ (ft^3)	5.615 E+00
British thermal unit (International Table)	Joule (J)	1.055 056 E+03
Darcy	Meter² (m^2)	9.869 232 E−13
Day (mean solar)	Second (s)	8.640 000 E+04
Dyne	Newton (N)	1.000 000 E−05
Gallon (U.S. liquid)	Meter³ (m^3)	3.785 412 E−03
Gram	Kilogram (kg)	1.000 000 E−03
Hectare	Meter² (m^2)	1.000 000 E+04
Mile (U.S. survey)	Meter (m)	1.609 347 E+03
Pound (lbm avoirdupois)	Kilogram (kg)	4.535 924 E−01
Ton (short, 2000 lbm)	Kilogram (kg)	9.071 847 E+02

1 bar = 1×10^5 Pa
1 atm = 1.01325

TABLE 1-7 CONTINUED

Selected SI unit prefixes

Factor	SI prefix	SI prefix symbol (use roman type)	Meaning (U.S.)	Meaning in other countries
10^{12}	tera	T	One trillion times	Billion
10^{9}	giga	G	One billion times	Milliard
10^{6}	mega	M	One million times	
10^{3}	kilo	k	One thousand times	
10^{2}	hecto	h	One hundred times	
10	deka	da	Ten times	
10^{-1}	deci	d	One tenth of	
10^{-2}	centi	c	One hundredth of	
10^{-3}	milli	m	One thousandth of	
10^{-6}	micro	μ	One millionth of	
10^{-9}	nano	n	One billionth of	Milliardth

1. There are several cognates, quantities having the exact or approximate numerical value, between SI and practical units. The most useful for EOR are

$$\begin{aligned} 1 \text{ cp} &= 1 \text{ mPa-s} \\ 1 \text{ dyne/cm} &= 1 \text{ mN/m} \\ 1 \text{ Btu} &\cong 1 \text{ kJ} \\ 1 \text{ Darcy} &\cong 1\ \mu\text{m}^2 \\ 1 \text{ ppm} &\cong 1 \text{ g/m}^3 \end{aligned}$$

2. Use of the unit prefixes (lower part of Table 1-7) is convenient, but it does require care. When a prefixed unit is exponentiated, the exponent applies to the prefix as well as the unit. Thus $1 \text{ km}^2 = 1(\text{km})^2 = 1(10^3 \text{ m})^2 = 1 \times 10^6 \text{m}^2$. We have already used this convention where $1\ \mu\text{m}^2 = 10^{-12} \text{ m}^2 \cong 1$ Darcy.
3. Two troublesome conversions are between pressure (147 psia $\cong$ 1 MPa) and temperature (1 K = 1.8 R). Since neither the Fahrenheit nor the Celsius scale is absolute, an additional translation is required.

$$°\text{C} = \text{K} - 273$$

and

$$°\text{F} = \text{R} - 460$$

The superscript ° is not used on absolute temperature scales.

4. The volume conversions are complicated by the interchangeable use of mass and standard volumes. Thus we have

$$0.159 \text{ m}^3 = 1 \text{ reservoir barrel, or bbl}$$

and

$$0.159 \text{ SCM} = 1 \text{ standard barrel, or stb}$$

The symbol SCM, Standard Cubic Meter, is not standard SI; it represents the amount of mass contained in one cubic meter evaluated at standard temperature and pressure.

Consistency

Maintaining unit consistency is important in all exercises. Both units and numerical values should be carried in all calculations. This ensures that the unit conversions are done correctly and indicates if the calculation procedure itself is appropriate. In maintaining consistency, three steps are required.

1. Clear all unit prefixes.
2. Reduce all units to the most primitive level necessary. For many cases, this will mean reverting to the fundamental units given in Table 1-7.
3. After calculations are complete, reincorporate the unit prefixes so that the numerical value of the result is as close to 1 as possible. Many adopt the convention that only the prefixes representing multiples of 1,000 are used.

Naming Conventions

The diversity of EOR makes it impossible to assign symbols to components without some duplication or undue complication. In the hope of minimizing the latter by adding a little of the former, Table 1-8 gives the naming conventions of phases and components used throughout this text. The nomenclature section defines other symbols.

Phases always carry the subscript j, which occupies the second position in a doubly subscripted quantity. $j = 1$ is always water, or the aqueous phase, thus freeing up the symbol w for wetting (and nw for nonwetting). The subscript s designates the solid, nonflowing phase.

A subscript i, occurring in the first position, indicates the component. Singly subscripted quantities indicate components. In general, $i = 1$ is always water; $i = 2$ is oil or hydrocarbon; and $i = 3$ refers to a displacing component, whether surfactant or light hydrocarbon. Component indices greater than 3 are used exclusively in Chaps. 8–10, the chemical flooding part of the text.

TABLE 1-8 NAMING CONVENTIONS FOR PHASES AND COMPONENTS

Phases		
j	Identity	Text locations
1	Water or aqueous	Throughout
2	Oil or oleic	Throughout
3	Gas or light hydrocarbon Microemulsion	Secs. 5-6 and 7-7 Chap. 9
s	Solid	Chaps. 2, 3, and 8–10
w	Wetting	Throughout
nw	Nonwetting	Throughout

Components		
i	Identity	Text locations
1	Water	Throughout
2	Oil or intermediate hydrocarbon	Throughout
3	Gas Light hydrocarbon Surfactant	Sec. 5-6 Sec. 7-6 Chap. 9
4	Polymer	Chaps. 8 and 9
5	Anions	Sec. 3-4 and 9-5
6	Divalents	Secs. 3-4 and 9-5
7	Divalent-surfactant component	Sec. 9-6
8	Monovalents	Secs. 3-4 and 9-5

2

Basic Equations for Fluid Flow in Permeable Media

Formulating the equations that describe flow in permeable media is a necessary first step in understanding and describing EOR processes. Each process involves at least one phase that may contain several components. Moreover, because of varying temperature, pressure, and composition, these components may mix completely in some regions of the flow domain, causing the disappearance of a phase in those regions. Atmospheric pollution and chemical and nuclear waste storage lead to similar problems.

In this chapter, we describe multiphase, multicomponent fluid flow in permeable media. Our description uses basic conservation laws and linear constitutive theory.

Our formulation differs from other sources in its generality for multiphase, multicomponent flows. For example, it contains as special cases the multicomponent, single-phase flow equations (Bear, 1972) and the three-phase, multicomponent equations (Crichlow, 1977; Peaceman, 1977; Coats, 1980). In addition, others (Todd and Chase, 1979; Fleming et al., 1981; Larson, 1979) have presented multicomponent, multiphase formulations for flow in permeable media but with more restrictive assumptions such as ideal mixing or incompressible fluids. However, many of these assumptions must be made before the equations are solved.

2-1 MASS CONSERVATION

In this section, we describe the physical nature of multiphase, multicomponent flows in permeable media and the mathematical formulation of the conservation equations.

The four most important mechanisms causing transport of chemical species in naturally occurring permeable media are viscous forces, gravity forces, dispersion (diffusion), and capillary forces. The driving forces for the first three are pressure, density, and concentration gradients, respectively. Capillary or surface forces are caused by high curvature boundaries between the various homogeneous phases. This curvature is the result of such phases being constrained by the pores of the permeable medium. Capillary forces imply differing pressures in each homogeneous fluid phase.

The Continuum Assumption

Transport of chemical species in multiple homogeneous phases occurs because of the above forces, the flow being restricted to the highly irregular flow channels within the medium. The conservation equations for each species apply at each point in the medium, including the solid phase. In principle, given constitutive relations, reaction rates, and boundary conditions, a mathematical system may be formulated for all flow channels in the medium. But owing to the extremely tortuous phase boundaries in such a system, we cannot solve species conservation equations in a local sense except for only the simplest microscopic permeable media geometry.

The practical way of avoiding this difficulty is to apply a continuum definition to a macroscopic scale so that a point within a permeable medium is associated with a representative elementary volume (REV), a volume that is large with respect to the pore dimensions of the stationary phase but small compared to the dimensions of the permeable medium. The REV is defined as a volume below which local fluctuations in some primary property of the permeable media, usually the porosity, become large (Bear, 1972). A volume-averaged form of the species conservation equations applies for each REV within the now continuous domain of the macroscopic permeable medium. (For details on volume averaging, see Bear, 1972; Gray, 1975.) The volume-averaged species conservation equations are identical to the conservation equations outside a permeable medium except for altered definitions for the accumulation, flux, and source terms. These definitions now include permeable media porosity, permeability, tortuosity, and dispersivity, all locally smooth because of the definition of the REV.

Rather than beginning with the nonpermeable media flow equations, and then volume averaging over the REV, we invoke the continuum assumption at the outset and derive the mass conservation on this basis. This approach denies many of the physical insights obtained from volume averaging, but it is far more direct.

Mass Balance

Consider then an arbitrary, fixed volume V embedded within a permeable medium through which is flowing an arbitrary number of chemical species. You must constantly be aware of the distinction between components and phases in this discussion. Let there be $i = 1, \ldots, N_C$ components, and $j = 1, \ldots, N_P$ phases. The volume V is greater than or equal to the REV but smaller than or equal to the macroscopic permeable media dimensions. As Fig. 2-1(a) shows, the surface area A of V is

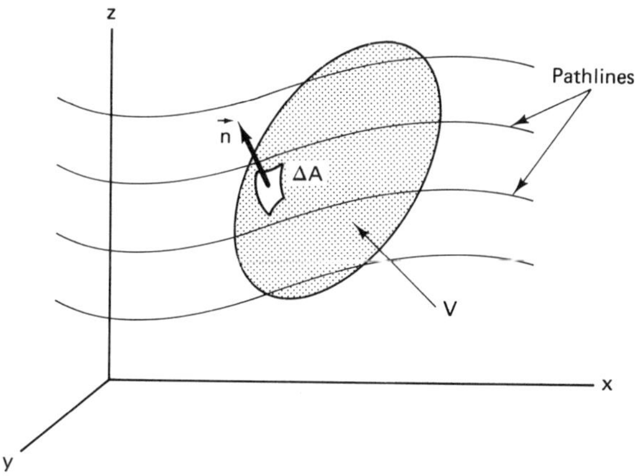

(a) Arbitrary volume in flow domain

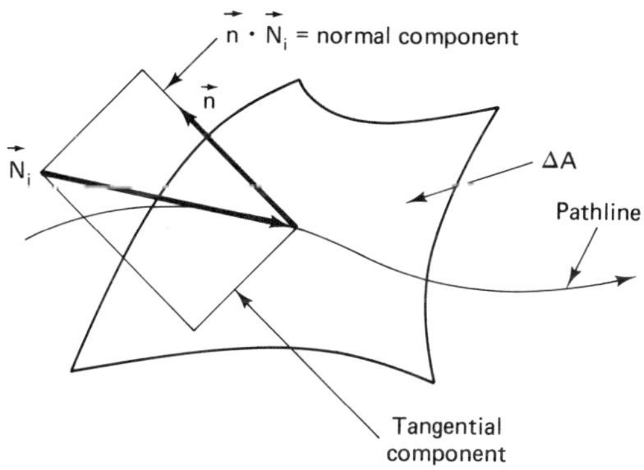

(b) Surface element detail

Figure 2-1 Geometries for conservation law derivations

made up of elemental surface areas ΔA from the center of which is pointing a unit outward normal vector $\vec{n}$. The sum of all the surface elements ΔA is the total surface area of V. This sum becomes a surface integral as the largest ΔA approaches zero.

The mass conservation equation for species i in volume V is

$$\left\{\begin{array}{c}\text{Rate of}\\ \text{accumulation}\\ \text{of } i \text{ in } V\end{array}\right\} = \left\{\begin{array}{c}\text{Rate of } i\\ \text{transported}\\ \text{into } V\end{array}\right\} - \left\{\begin{array}{c}\text{Rate of } i\\ \text{transported}\\ \text{from } V\end{array}\right\} + \left\{\begin{array}{c}\text{Rate of}\\ \text{production}\\ \text{of } i \text{ in } V\end{array}\right\}, \quad i = 1, \ldots, N_C \qquad (2.1\text{-}1)$$

This equation is the rate form of the conservation equation; an equivalent form based on cumulative flow follows from integrating Eq. (2.1-1) with respect to time (see Sec. 2-5). The first two terms on the right-hand side of Eq. (2.1-1) can be written as

$$\left\{\begin{array}{c}\text{Rate of } i\\ \text{transported}\\ \text{into } V\end{array}\right\} - \left\{\begin{array}{c}\text{Rate of } i\\ \text{transported}\\ \text{from } V\end{array}\right\} = \left\{\begin{array}{c}\text{Net rate of}\\ i \text{ transported}\\ \text{into } V\end{array}\right\}, \quad i = 1, \ldots, N_C \qquad (2.1\text{-}2)$$

With this identification, the terms from left to right in Eq. (2.1-1) are the accumulation, flux, and source, respectively. We try to give mathematical form to these terms in the following paragraphs.

The accumulation term for species i is

$$\left\{\begin{matrix}\text{Rate of}\\ \text{accumulation}\\ \text{of } i \text{ in } V\end{matrix}\right\} = \frac{\partial}{\partial t}\left\{\begin{matrix}\text{Total mass}\\ \text{of } i \text{ in } V\end{matrix}\right\} = \frac{\partial}{\partial t}\left\{\int_V W_i dV\right\} \tag{2.1-3a}$$

where W_i is the overall concentration of i in units of mass of i per unit bulk volume. The volume integral represents the sum of infinitesimal volume elements in V weighted by the overall concentration. Since V is fixed

$$\frac{\partial}{\partial t}\left\{\int_V W_i\, dV\right\} = \int_V \frac{\partial W_i}{\partial t}\, dV \tag{2.1-3b}$$

This entire development may be repeated with a time-varying V with the same result (Slattery, 1972).

The net flux term follows from considering the rate of transport across a single surface element as shown in Fig. 2-1(b). Let $\vec{N}_i$ be the flux vector of species i evaluated at the center of ΔA in units of mass of i per surface area-time. $\vec{N}_i$ may be readily decomposed into components normal and tangential to $\vec{n}$. However, only the normal component $\vec{n}\cdot\vec{N}_i$ is crossing ΔA, and the rate of transport across ΔA is

$$\left\{\begin{matrix}\text{Rate of transport}\\ \text{of } i \text{ across } \Delta A\end{matrix}\right\} = -\vec{n}\cdot\vec{N}_i\Delta A \tag{2.1-4a}$$

The minus sign arises because $\vec{n}$ and $\vec{N}_i$ have opposing directions for transport into ΔA ($\vec{n}\cdot\vec{N}_i < 0$), and this term must be positive from Eq. (2.1-1). Summing up infinitesimal surface elements yields

$$\left\{\begin{matrix}\text{Net rate}\\ \text{of transport}\\ \text{into } V\end{matrix}\right\} = -\int_A \vec{n}\cdot\vec{N}_i\, dA \tag{2.1-4b}$$

Since the surface integral is over the entire surface of V, both flow into and from V are included in Eq. (2.1-4b).

The net rate of production of i in V is

$$\left\{\begin{matrix}\text{Net rate}\\ \text{of production}\\ \text{of } i \text{ in } V\end{matrix}\right\} = -\int_V R_i dV \tag{2.1-5}$$

where R_i is the mass rate of production in units of mass of i per bulk volume-time. This term can account for both production ($R_i > 0$) and destruction ($R_i < 0$) of i, either through one or more chemical reactions or through physical sources (wells) in V.

Combining Eqs. (2.1-3) through (2.1-5) into Eq. (2.1-1) gives the following scalar equation for the conservation of i:

$$\int_V \frac{\partial W_i}{\partial t}\, dV + \int_A \vec{n} \cdot \vec{N}_i \, dA = \int_V R_i dV, \qquad i = 1, \ldots, N_C \tag{2.1-6}$$

Equation (2.1-6) is an overall balance, or "weak" form of the conservation equation.

The surface integral in Eq. (2.1-6) may be converted to a volume integral through the divergence theorem

$$\int_V \vec{\nabla} \cdot \vec{B} dV = \int_A \vec{n} \cdot \vec{B} \, dA \tag{2.1-7}$$

where B can be any scalar, vector, or tensor function of position in V (with appropriate changes in the operator definitions). The symbol $\vec{\nabla}$ is the divergence operator, a kind of generalized derivative, whose specific form depends on the coordinate system. Table 2-1 gives forms of $\vec{\nabla}$ in rectangular, cylindrical, and spherical coordinates. The function B must be single-valued in V, a requirement fortunately met by most physical problems. Finally, implicit in the representation of the surface integral of Eqs. (2.1-6) and (2.1-7) is the requirement that the integrand be evaluated on the surface A of V.

Applying the divergence theorem to Eq. (2.1-6) gives

$$\int_V \left(\frac{\partial W_i}{\partial t} + \vec{\nabla} \cdot \vec{N}_i - R_i \right) dV = 0, \qquad i = 1, \ldots, N_C \tag{2.1-8}$$

But since V is arbitrary, the integrand must be zero.

$$\frac{\partial W_i}{\partial t} + \vec{\nabla} \cdot \vec{N}_i - R_i = 0, \qquad i = 1, \ldots, N_C \tag{2.1-9}$$

Equation (2.1-9) is the differential form for the species conservation equation. It applies to any point within the macroscopic dimensions of the permeable medium. In the next section, we give specific definitions to the overall concentration W_i, flux $\vec{N}_i$, and source terms R_i.

2-2 DEFINITIONS AND CONSTITUTIVE EQUATIONS FOR ISOTHERMAL FLOW

Table 2-2 summarizes the equations needed for a complete description of isothermal, multicomponent, multiphase flow in permeable media. Column 1 in Table 2-2 gives the differential form of the equation named in column 2. Column 3 gives the number of scalar equations represented by the equation in column 1. Columns 4 and 5 give the identity and number of independent variables added to the formulation by the equation in column 1. N_D is the number of spatial dimensions ($N_D \leqq 3$). The stationary phase is a single homogeneous phase though more than one solid can exist.

TABLE 2-1 SUMMARY OF DIFFERENTIAL OPERATORS IN RECTANGULAR, CYLINDRICAL, AND SPHERICAL COORDINATES

Rectangular coordinates (x, y, z)	Cylindrical coordinates (r, θ, z)	Spherical coordinates (r, θ, ϕ)
$\vec{\nabla} \cdot \vec{B} = \frac{\partial B_x}{\partial x} + \frac{\partial B_y}{\partial y} + \frac{\partial B_z}{\partial z}$	$\vec{\nabla} \cdot \vec{B} = \frac{1}{r}\frac{\partial (rB_r)}{\partial r} + \frac{1}{r}\frac{\partial B_\theta}{\partial \theta} + \frac{\partial B_z}{\partial z}$	$\vec{\nabla} \cdot \vec{B} = \frac{1}{r^2}\frac{\partial (r^2 B_r)}{\partial r} + \frac{1}{r \sin\theta}\frac{\partial}{\partial \theta}(B_\theta \sin\theta) + \frac{1}{r \sin\theta}\frac{\partial B_\phi}{\partial \phi}$
$[\vec{\nabla} S]_x = \frac{\partial S}{\partial x}$	$[\vec{\nabla} S]_r = \frac{\partial S}{\partial r}$	$[\vec{\nabla} S]_r = \frac{\partial S}{\partial r}$
$[\vec{\nabla} S]_y = \frac{\partial S}{\partial y}$	$[\vec{\nabla} S]_\theta = \frac{1}{r}\frac{\partial S}{\partial \theta}$	$[\vec{\nabla} S]_\theta = \frac{1}{r}\frac{\partial S}{\partial \theta}$
$[\vec{\nabla} S]_z = \frac{\partial S}{\partial z}$	$[\vec{\nabla} S]_z = \frac{\partial S}{\partial z}$	$[\vec{\nabla} S]_\phi = \frac{1}{r \sin\theta}\frac{\partial S}{\partial \phi}$
$\vec{\nabla}^2 S = \frac{\partial^2 S}{\partial x^2} + \frac{\partial^2 S}{\partial y^2} + \frac{\partial^2 S}{\partial z^2}$	$\vec{\nabla}^2 S = \frac{1}{r}\frac{\partial}{\partial r}\left(r\frac{\partial S}{\partial r}\right) + \frac{1}{r^2}\frac{\partial^2 S}{\partial \theta^2} + \frac{\partial^2 S}{\partial z^2}$	$\vec{\nabla}^2 S = \frac{1}{r^2}\frac{\partial}{\partial r}\left(r^2\frac{\partial S}{\partial r}\right) + \frac{1}{r^2 \sin\theta}\frac{\partial}{\partial \theta}\left(\sin\theta\frac{\partial S}{\partial \theta}\right) + \frac{1}{r^2 \sin^2\theta}\frac{\partial^2 S}{\partial \phi^2}$

Note: B = vector function
S = scalar function

TABLE 2-2 SUMMARY OF EQUATIONS FOR ISOTHERMAL FLUID FLOW IN PERMEABLE MEDIA

Equation (1)	Name (2)	Number independent scalar* equations (3)	Dependent variables† Identity (4)	Dependent variables† Number (5)
(2.1-9) $\frac{\partial W_i}{\partial t} + \vec{\nabla} \cdot \vec{N}_i = R_i$	Species i conservation	N_C	W_i, R_i, $\vec{N}_i$	$2N_C + N_C N_D$
(2.2-1) $W_i = \phi \sum_{j=1}^{N_P} \rho_j S_j \omega_{ij} + (1 - \phi)\rho_s \omega_{is}$	Overall concentration	$N_C - 1$	ρ_j, S_j, ω_{ij}, ω_{is}	$2N_P + N_P N_C + N_C$
(2.2-2) $\vec{N}_i = \sum_{j=1}^{N_P} (\rho_j \omega_{ij} \vec{u}_j - \phi \rho_j S_j \vec{\vec{K}}_{ij} \cdot \vec{\nabla} \omega_{ij})$	Species i flux	$N_C N_D$	$\vec{u}_j$	$N_P N_D$
(2.2-3) $R_i = \phi \sum_{j=1}^{N_P} S_j r_{ij} + (1 - \phi) r_{is}$	Species i source	$N_C - 1$	r_{ij}, r_{is}	$N_P N_C + N_C$
(2.2-4) $\sum_{i=1}^{N_C} R_i = 0$	Total reaction definition	1		
(2.2-5) $\vec{u}_j = -\lambda_{rj} \vec{\vec{k}} \cdot (\vec{\nabla} P_j + \rho_j \vec{g})$	Darcy's law	$N_P N_D$	λ_{rj}, P_j	$2N_P$
(2.2-6) $\lambda_{rj} = \lambda_{rj}(S, \omega, \vec{u}_j, \vec{x})$	Relative mobility	N_P		
(2.2-7) $P_j - P_n = P_{cjn}(S, \omega, \vec{x})$	Capillary pressure definition	$N_P - 1$		

* Total independent equations $= N_D(N_P + N_C) + 2N_P N_C + 4N_P + 4N_C$

† Total dependent variables $= N_D(N_P + N_C) + 2N_P N_C + 4N_P + 4N_C$

TABLE 2-2 CONTINUED

Equation (1)	Name (2)	Number independent scalar* equations (3)	Dependent variables[†] Identity (4)	Number (5)
(2.2-8a) $\sum_{i=1}^{N_C} \omega_{ij} = 1$	Mass fraction definition	N_P		
(2.2-8b) $\sum_{i=1}^{N_C} \omega_{is} = 1$	Stationary phase mass fraction definition	1		
(2.2-9) $\sum_{j=1}^{N_P} S_j = 1$	Saturation definition	1		
(2.2-10a) $r_{ij} = r_j(\omega_{ij}, P_j)$	Homogeneous kinetic reaction rates	$(N_C - 1)N_P$		
(2.2-10b) $r_{is} = r_{is}(\omega_{is})$	Stationary phase reaction rates	$N_C - 1$		
(2.2-10c) $\sum_{i=1}^{N_C} r_{ij} = 0$	Total phase reaction definition	N_P		
(2.2-10d) $\sum_{i=1}^{N_C} r_{is} = 0$	Stationary phase total reaction rates	1		
(2.2-11a) $\omega_{ij} = \omega_{ij}(\omega_{ik})_{k \neq j}$	Equilibrium relations (or phase balances)	$N_C(N_P - 1)$		
(2.2-11b) $\omega_{is} = \omega_{is}(\omega_{ij})$	Stationary phase equilibrium relations (or phase balances)	N_C		
(2.2-12) $\rho_j = \rho_j(T, P_j)$	Equations of state	N_P		

* Total independent equations $= N_D(N_P + N_C) + 2N_PN_C + 4N_P + 4N_C$

[†] Total dependent variables $= N_D(N_P + N_C) + 2N_PN_C + 4N_P + 4N_C$

A normally subscripted quantity (for example, ω_{ij}) appearing without subscripts in Table 2-2 indicates a relationship involving, at most, all members of the subscripted set. In the listing of dependent variables, the primary media properties, such as porosity ϕ, and the permeability tensor $\vec{\vec{k}}$, are given functions of position $\vec{x}$ within the permeable media. These quantities are, strictly speaking, functions of the fluid pressure within the medium (Dake, 1978), but for pressures nondestructive to the permeable medium, this effect is generally weak. We also assume the stationary phase density ρ_s is given, as is the dispersion tensor $\vec{\vec{K}}_{ij}$, even though the latter is a function of the phase velocities and molecular diffusivities. The remaining terms in Table 2-2 are defined in the nomenclature and below.

The first four equations in Table 2-2 are the species conservation Eq. (2.1-9) and definitions for the accumulation, flux, and source terms in this equation. We take the N_C conservation equations to be the independent set of equations; the conservation of overall mass or continuity equation, which follows from summing Eq. (2.1-9) from 1 to N_C, is not listed as an independent equation (see Sec. 2-4). In solving specific problems, it may be more convenient to take the problem statement as the continuity equation and $N_C - 1$ mass conservation equations with the major component (for example, water in flow of dissolved salts in an aqueous solution) being the one omitted.

Definition of Terms

The accumulation term W_i, the overall concentration of species i, represents the sum of the species i in the N_P flowing phases plus the stationary phase (Eq. 2.2-1). The phase saturation S_j is defined as a fraction of the pore space occupied by phase j.

There are only $N_C - 1$ independent W_i since summing on i with the mass fraction definitions (Eqs. 2.2-8a and 2.2-8b) gives

$$\sum_{i=1}^{N_C} W_i = \phi \sum_{j=1}^{N_P} \rho_j S_j + (1 - \phi)\rho_s \equiv \rho(\omega_i, P) \tag{2.2-13a}$$

where ρ is the overall density of the permeable medium (total mass flowing plus stationary phases divided by the bulk volume). We can regard the overall density as a function of some local pressure P and the set of overall mass fractions defined as

$$\omega_i = \frac{W_i}{\sum_{i=1}^{N_C} W_i} \tag{2.2-13b}$$

Here ω_i is the mass of species i in all phases divided by the total mass of the permeable medium. Combining Eqs. (2.2-13a) and (2.2-13b) yields a constraint on the W_i

$$\rho(W_1, \ldots, W_{N_C}, P) = \sum_{i=1}^{N_C} W_i \tag{2.2-13c}$$

which means there are $N_C - 1$ independent W_i, not N_C. The notation on the left side of Eq. (2.2-13c) indicates that ρ is a function of two variables, the overall concentration and pressure. Equations (2.2-13a) through (2.2-13c) can be construed as a constraint on the mass fraction ω_{ij}, phase pressures P_j, and saturations S_j.

The flux $\vec{N}_i$ in Eq. (2.2-2) is the sum over all flowing phases of the flux of component i within each phase. The $\vec{N}_i$ are comprised of a convection term (determined by the superficial phase velocity $\vec{u}_j$) and a dispersion term (characterized by the dispersion tensor $\vec{\vec{K}}_{ij}$).

Dispersion has the same form as diffusion in nonpermeable media flows and, in fact, collapses to molecular diffusion in the limit of $\vec{u}_j$ small (see Chap. 5). At larger $\vec{u}_j$, the components of $\vec{\vec{K}}_{ij}$ can be many times larger than molecular diffusion since they now contain contributions from fluctuations of the velocity $\vec{u}_j$ and mass fraction ω_{ij} about their average values in the REV (Gray, 1975). Two components of $\vec{\vec{K}}_{ij}$ for homogeneous, isotropic permeable media by Bear (1972) are

$$(K_{xx})_{ij} = \frac{D_{ij}}{\tau} + \frac{\alpha_{\ell j} u_{xj}^2 + \alpha_{tj}(u_{yj}^2 + u_{zj}^2)}{\phi S_j |\vec{u}_j|} \tag{2.2-14a}$$

$$(K_{xy})_{ij} = \frac{(\alpha_{\ell j} - \alpha_{tj}) u_{xj} u_{yj}}{u S_j |\vec{u}_j|} \tag{2.2-14b}$$

where the subscript ℓ refers to the spatial coordinate in the direction parallel, or longitudinal, to bulk flow, and t is any direction perpendicular, or transverse, to ℓ. D_{ij} is the effective binary diffusion coefficient of species i in phase j (Bird et al., 1960), $\alpha_{\ell j}$ and α_{tj} are the longitudinal and transverse dispersivities, and τ is the permeable media tortuosity factor. $(K_t)_{ij}$ is positive since $(\alpha_{\ell j} - \alpha_{tj})$ is always positive.

The source term is Eq. (2.2-3), the sum over all phases of the rate of appearance of species i because of homogeneous chemical reactions within phase j (Levenspeil, 1962). Each r_{ij} could represent the sum of several reactions within the phase j if species i participates in simultaneous reactions. The sum of R_i over all species (Eq. 2.2-4) is zero since there can be no net accumulation of mass because of chemical reaction. Frequently, R_i is used to represent reactions occurring at phase boundaries even though, strictly speaking, such reactions are the consequence of flux terms evaluated at phase boundaries in the volume-averaging procedures (Gray, 1975). It is also convenient to use R_i to represent physical sources that are either specified or related to the phase pressures and saturations.

Auxiliary Relations

Equation (2.2-5) is a multiphase version of Darcy's law for flow in permeable media (Collins, 1976). The single-phase version of Darcy's law is actually a volume-averaged form of the momentum equation (Slattery, 1972; Hubbert, 1956). The form given in Eq. (2.2-5) assumes creeping flow in the permeable medium with no fluid slip at the phase boundaries. Corrections to account for non-Darcy effects appear in standard references (Collins, 1976; Bear, 1972). The potential function for the phase superficial velocity $\vec{u}_j$ is the vectorial sum $\vec{\nabla} P_j + \rho_j \vec{g}$, where P_j is the pressure within the continuous phase j. $\vec{g}$ is the gravitational vector, which is assumed constant and directed toward the earth's center. Hereafter in this text, we assume the coordinate direction parallel to $\vec{g}$ is positive upward. The gravitational vector can be written as

$$\vec{g} = -g\vec{\nabla} D_z \tag{2.2-15}$$

where g is the magnitude of the gravitational vector, and D_z is a positive distance below some horizontal reference plane. For Cartesian coordinate systems with a constant inclination with the reference plane, $\vec{\nabla}$ becomes a vector consisting of cosines of the inclination angles between the respective axis and the vertical.

The tensorial form of the permeability $\vec{\vec{k}}$ implies an anisotropic permeable medium having coordinate axes not aligned with respect to the principal axis of $\vec{\vec{k}}$. With the inclusion of $\vec{\vec{k}}$, we have now included all the primary permeable media properties, ϕ, $\vec{\vec{k}}$, $\alpha_{\ell j}$, α_{tj}, and τ, into the formulation.

The other quantity in Eq. (2.2-5) is the relative mobility λ_{rj} of phase j, defined as the quotient of the relative permeability k_{rj} and viscosity μ_j.

$$\lambda_{rj} = \frac{k_{rj}(S, \omega, \vec{x})}{\mu_j(\omega, \vec{u}_j)} \tag{2.2-16}$$

Equation (2.2-16) decomposes λ_{rj} into a rock–fluid property k_{rj} and a fluid property μ_j. k_{rj} is a function of the tendency of phase j to wet the permeable medium, of pore size distribution, and of the entire set of phase saturations (see Chap. 3). μ_j is a function of the phase composition, and, if phase j is non-Newtonian, the magnitude of the superficial velocity $\vec{u}_j$ (see Chap. 8). The relative permeabilities and viscosities k_{rj} and μ_j are usually determined experimentally to give λ_{rj}. It is slightly more general to write the $\lambda_{rj}\vec{\vec{k}}$ product in Eq. (2.2-5) as

$$\lambda_{rj}\vec{\vec{k}} = \frac{\vec{\vec{k}}_j}{\mu_j} \tag{2.2-17}$$

where $\vec{\vec{k}}_j$ is the phase permeability tensor. This form allows for anisotropic relative permeabilities.

The difference between the phase pressures of any two phases flowing in the REV is the capillary pressure defined as in Eq. (2.2-7). The capillary pressure between the phases j and n is a function of the same variables as the relative permeability (Fatt and Dykstra, 1951). That there are $N_P - 1$ independent relations follows from considering the set of all capillary pressures with j fixed, $P_{c1j}, P_{c2j}, \ldots, P_{cN_Pj}$. Ignoring the trivial case of P_{cjj}, there are clearly $N_P - 1$ capillary pressures. The capillary pressure P_{ckn} between any two other phases k and n may be expressed as a linear combination of members from the original set.

$$P_{ckn} = P_k - P_n = (P_k - P_j) + (P_j - P_n) = P_{ckj} + P_{cjn} \tag{2.2-18}$$

Hence there are only $N_P - 1$ independent capillary pressure relations, usually determined experimentally under static conditions. We discuss capillary pressure in more detail in Chap. 3.

The pressures P_j are the continuous phase pressures, not the pressures that would exist in disconnected "globules" of phase j. In the latter case, the phase pressure differences still exist but, being a reflection of the local permeable medium pore configuration, are not uniquely determined by the functions given in Eq. (2.2-7).

Equations (2.2-8a), (2.2-8b), and (2.2-9) follow from the definitions of mass fraction and phase saturation, respectively.

Equations (2.2-10a) through (2.2-10d) are definitions of the reaction rate of component i in phase j or in the stationary phase. As was true for R_i, there can be no net accumulation of mass in a phase owing to chemical reaction. Then the reaction rate terms r_{ij} and r_{is} sum to zero as indicated by Eqs. (2.2-10b) and (2.2-10c).

Local Equilibrium

Equations (2.2-11a) and (2.2-11b) are relations among the mass fractions of the N_P flowing phases and the stationary phase present in the REV. These relations arise from solving the conservation equation for each species in each phase

$$\frac{\partial W_{ij}}{\partial t} + \vec{\nabla} \cdot \vec{N}_{ij} = R_{ij} + r_{mij} \qquad (2.2\text{-}19)$$

where the second subscript on W_i, $\vec{N}_i$, and R_i refers to a single term in the sums over all phases in their original definition. The term r_{mij} expresses the rate of mass transfer of species i from or into phase j. To maintain consistency with Eq. (2.1-9), we must have $\sum_{j=1}^{N_P} r_{mij} = 0$, a relation following from the inability to accumulate mass at an interface. Since the sum of the conservation equations over all flowing phases for species i is Eq. (2.1-9), there are $N_C(N_P - 1)$ such independent phase balances. Since there are also N_C phase balances for the stationary phase, the total number of independent relations is $N_C N_P$. There are a similar number of additional unknowns, the r_{mij}, which must be independently specified.

Although the phase balance is formally correct, a much more practical approach is to assume local thermodynamic equilibrium; that is, the mass fractions of component i are related through thermodynamic equilibrium relations (Pope and Nelson, 1978). For flow through naturally occurring permeable media, the assumption of local equilibria among phases is usually adequate (Raimondi and Torcaso, 1965). Exceptions are flows at very high rates or leachant flows such as might occur in alkaline floods.

If local equilibrium applies, the number $N_C N_P$ of independent scalar equations may be derived from the phase rule (see Chap. 4). The equilibrium relations themselves are very strong functions of the particular EOR process, and much of the behavior and many of the important features of a given process can be understood from relatively simple phase equilibria considerations. In Chap. 4, we discuss phase behavior generally; we reserve more specifics for the relevant sections on solvent, chemical, and thermal flooding.

Before leaving this brief discussion of phase balances and equilibrium relations, we note the definition of local equilibrium is slightly different between permeable and nonpermeable media flows. In the former case, the compositions of the entire phases in the REV are in equilibrium; in the latter case, equilibrium is understood to apply only across the microscopic phase boundaries (Bird et al., 1960).

The final set of equations in Table 2-2 are the equations of state (Eq. 2.2-12), which relate the phase densities to its composition, temperature, and pressure. For flows in local thermodynamic equilibrium, the equilibrium relations for the flowing phases (Eq. 2.2-11), in theory, can be derived from the equation of state. This derivation is rarely done in practice (except for the simplest cases such as two-phase vapor–liquid equilibria) because of computational inefficiencies and inherent inaccuracies in even the best of the equations of state. Nevertheless, there should be a degree of internal consistency between Eqs. (2.2-11) and (2.2-12).

Continuity Equation

If we insert Eqs. (2.2-1) through (2.2-3) into (2.1-9) we arrive at

$$\frac{\partial}{\partial t}\left(\phi \sum_{j=1}^{N_P} \rho_j S_j \omega_{ij} + (1-\phi)\rho_s \omega_{is}\right) + \vec{\nabla} \cdot \left(\sum_{j=1}^{N_P} (\rho_j \omega_{ij} \vec{u}_j - \phi S_j \rho_j \vec{\vec{K}}_{ij} \cdot \vec{\nabla} \omega_{ij})\right)$$

$$= \phi \sum_{j=1}^{N_P} S_j r_{ij} + (1-\phi) r_{is}, \quad i = 1, \ldots, N_C \qquad (2.2\text{-}20)$$

We sum Eq. (2.2-20) over the N_C components to obtain the equation of continuity or conservation of total mass.

$$\frac{\partial}{\partial t}\left(\phi \sum_{j=1}^{N_P} \rho_j S_j + (1-\phi)\rho_s\right) + \vec{\nabla} \cdot \left(\sum_{j=1}^{N_P} \rho_j \vec{u}_j\right) = 0 \qquad (2.2\text{-}21)$$

Equation (2.2-21) can be written totally in terms of pressure and saturation derivatives using Eqs. (2.2-5) and (2.2-13); this equation is a form of the "pressure" equation.

2-3 ENERGY BALANCE EQUATIONS

To relax the requirement of isothermal flow on the equations of Table 2-2, we require a conservation of energy equation. This equation adds an additional dependent variable, temperature, to the formulation. A statement of the energy balance or first law of thermodynamics suitable for our purposes is

$$\left\{\begin{array}{c}\text{Rate of}\\ \text{accumulation}\\ \text{of energy in } V\end{array}\right\} = \left\{\begin{array}{c}\text{Net rate}\\ \text{of energy}\\ \text{transported into } V\end{array}\right\} + \left\{\begin{array}{c}\text{Rate of}\\ \text{energy production}\\ \text{in } V\end{array}\right\} \qquad (2.3\text{-}1)$$

where V is an arbitrary volume as in Fig. 2-1. We use the parallel between the species conservation Eq. (2.1-1) and Eq. (2.3-1) to shorten the following development.

By analogy to the procedure in Sec. 2-2, Eq. (2.3-1) can be written as

$$\int_V \frac{\partial}{\partial t}\left(\rho U + \frac{1}{2}\sum_{j=1}^{N_P} \rho_j |\vec{u}_j|^2\right) + \vec{\nabla}\cdot\vec{E}dV = \dot{W} \quad (2.3\text{-}2)$$

In Eq. (2.3-2), the term $1/2 \; \Sigma_{j=1}^{N_P} \rho_j |\vec{u}_j|^2$ represents kinetic energy per unit bulk volume. The remaining terms ρU, $\vec{E}$, and $\dot{W}$ represent energy concentration, flux, and source, respectively, to which we give specific form below. U is an overall internal energy, and ρ is the overall density given by Eq. (2.2-13a).

The source term requires considerably more elaboration than the other terms in Eq. (2.3-2). The form of the first law of thermodynamics for open systems expressed by Eq. (2.3-2) requires the $\dot{W}$ term to be composed of work components only, in the absence of external heating sources. Heats of reaction, vaporization, and solution are, of course, important in several EOR processes, but these are implicitly present in the equation in the concentration and flux terms. We consider only rate of work done against a pressure field $\dot{W}_{PV}$ and work against gravity $\dot{W}_G$ in this development. The sum $\dot{W} = \dot{W}_{PV} + \dot{W}_G$ is the rate of work done on a fluid element in the volume V.

Returning to Fig. 2-1(b), consider an element in the multiphase, multicomponent flow field crossing ΔA. Since work is the product of force times a distance, the rate of work is force times a velocity. The element crossing ΔA is, therefore, doing work $\Delta\dot{W}_{PV}$, where

$$\Delta\dot{W}_{PV} = -\sum_{j=1}^{N_P} P_j \Delta A \vec{n}\cdot\vec{u}_j \quad (2.3\text{-}3)$$

The term $P_j \Delta A \vec{n}$ is the force exerted on ΔA by the pressure in phase j. The scalar product in Eq. (2.3-3) merely expresses a more general definition of work rate when using vector forces and velocities. The negative sign in Eq. (2.3-3) is to satisfy the usual thermodynamic sign convention for work since $\Delta\dot{W}_{PV}$ must be positive for work done on a fluid element flowing into V $(\vec{n}\cdot\vec{u}_j < 0)$. The total pressure-volume work is the sum of Eq. (2.3-3) over all surface elements which, in the limit of the largest ΔA approaching zero, becomes a surface integral. Using the divergence theorem Eq. (2.1-7) on this integral gives the final form for $\dot{W}_{PV}$.

$$\dot{W}_{PV} = -\int_V \sum_{j=1}^{N_p} \vec{\nabla}\cdot(P_j \vec{u}_j)dV \quad (2.3\text{-}4)$$

To account for the gravity work, we again take a scalar product of a velocity and the gravity vector $\vec{g}$.

$$\Delta\dot{W}_G = \sum_{j=1}^{N_P} \rho_j \vec{u}_j\cdot\vec{g}\Delta V \quad (2.3\text{-}5)$$

The positive sign arises in this equation since a fluid phase flowing against gravity $(\vec{u}_j\cdot\vec{g} < 0)$ is having work done on it. Note the distinction between the elemental forms in Eqs. (2.3-3) and (2.3-5). Equation (2.3-3) is appropriate for work done against surface forces, and Eq. (2.3-5) is appropriate for work done against body forces. The rate of total work done against gravity is

$$\dot{W}_G = \int_V \sum_{j=1}^{N_P} \rho_j \vec{u}_j \cdot \vec{g} dV \tag{2.3-6}$$

from the usual limiting procedure.

The work expressions fit nicely into Eq. (2.3-2). After collecting all terms under the same volume integral and making the integrand zero because V is arbitrary, we have

$$\frac{\partial}{\partial t}\left(\rho U + \frac{1}{2}\sum_{j=1}^{N_P} \rho_j |\vec{u}_j|^2\right) + \vec{\nabla}\cdot\vec{E} + \sum_{j=1}^{N_P} \vec{\nabla}\cdot(P_j\vec{u}_j) - \sum_{j=1}^{N_P} \rho_j \vec{u}_j \cdot \vec{g} = 0 \tag{2.3-7}$$

The energy flux term is made up of convective contributions from the flowing phases, conduction, and radiation, all other forms being neglected.

$$\vec{E} = \sum_{j=1}^{N_P} \rho_j \vec{u}_j \left[U_j + \frac{1}{2}|\vec{u}_j|^2\right] + \vec{q}_c + \vec{q}_r \tag{2.3-8}$$

For brevity, we neglect radiation in the following discussion though this transport mechanism can be important in estimating heat losses from wellbores (Chap. 11). For multiphase flow, the conductive heat flux is from Fourier's law

$$\vec{q}_c = -k_{Tt}\vec{\nabla}T \tag{2.3-9}$$

where k_{Tt} is the total thermal conductivity. k_{Tt} is a complex function of the phase saturations and phase k_{Tj} and solid k_{Ts} thermal conductivities which we take to be known (see Chap. 11). The parallel between Eq. (2.3-8) and the dispersive flux term in Eq. (2.2-2) is obvious. We have also invoked the requirement of local thermal equilibrium in this definition by taking the temperature T to be the same in all phases within the REV.

Inserting definitions (2.3-8) and (2.3-9) into (2.3-7) yields

$$\frac{\partial}{\partial t}\left(\rho U + \frac{1}{2}\sum_{j=1}^{N_P} \rho_j |\vec{u}_j|^2\right) + \vec{\nabla}\cdot\left(\sum_{j=1}^{N_P} \rho_j \vec{u}_j \left[U_j + \frac{1}{2}|\vec{u}_j|^2\right]\right)$$
$$- \vec{\nabla}\cdot(k_{Tt}\vec{\nabla}T) + \sum_{j=1}^{N_P}[\vec{\nabla}\cdot(P_j\vec{u}_j) - \rho_j\vec{u}_j\cdot\vec{g}] = 0 \tag{2.3-10}$$

The first sum in the convection term and that in the pressure-volume work expression may be combined to give

$$\frac{\partial}{\partial t}\left(\rho U + \frac{1}{2}\sum_{j=1}^{N_P} \rho_j |\vec{u}_j|^2\right) + \vec{\nabla}\cdot\left(\sum_{j=1}^{N_P} \rho_j \vec{u}_j \left[H_j + \frac{1}{2}|\vec{u}_j|^2\right]\right) - \vec{\nabla}\cdot(k_{Tt}\vec{\nabla}T)$$
$$- \sum_{j=1}^{N_P} \rho_j \vec{u}_j \cdot \vec{g} = 0 \tag{2.3-11}$$

where $H_j = U_j + P_j/\rho_j$ is the enthalpy of phase j per unit mass of j. Finally, let us write the gravity vector as in Eq. (2.2-15). The last term in Eq. (2.3-11) becomes

$$\begin{aligned}\sum_{j=1}^{N_P} \rho_j \vec{u}_j \cdot \vec{g} &= -g \sum_{j=1}^{N_P} \rho_j \vec{u}_j \cdot \vec{\nabla} D_z \\ &= -g \sum_{j=1}^{N_P} \vec{\nabla} \cdot (\rho_j \vec{u}_j D_z) + g \sum_{j=1}^{N_P} D_z \vec{\nabla} \cdot (\rho_j \vec{u}_j)\end{aligned} \tag{2.3-12}$$

This when substituted into Eq. (2.3-11) gives

$$\frac{\partial}{\partial t}\left(\rho U + \frac{1}{2}\sum_{j=1}^{N_P} \rho_j |\vec{u}_j|^2\right) + \vec{\nabla} \cdot \left(\sum_{j=1}^{N_P} \rho_j \vec{u}_j \left[H_j + \frac{1}{2}|\vec{u}_j|^2 + gD_z\right]\right) - \vec{\nabla} \cdot (k_{Tt} \vec{\nabla} T)$$
$$- gD_z \vec{\nabla} \cdot \left(\sum_{j=1}^{N_P} \rho_j \vec{u}_j\right) = 0 \tag{2.3-13}$$

From Eq. (2.2-21), the last term becomes $gD_z \partial(\phi\rho)/\partial t$, which when substituted into Eq. (2.3-13), becomes the final form (Eq. 2.3-14) in Table 2-3 since gD_z is time independent. The gravity work terms are now in the equation as the more familiar potential energy.

Auxiliary Relations

Table 2-3 summarizes the equations that, together with those of Table 2-2, are needed for a complete specification of nonisothermal fluid flow problems. The first three equations we have already discussed.

The energy concentration per unit bulk volume must include internal energy contributions from all flowing phases and the solid phase

$$\rho U = \phi \sum_{j=1}^{N_P} \rho_j S_j U_j + (1 - \phi)\rho_s U_s \tag{2.3-15}$$

where U_j is the internal energy per unit mass of phase j. This definition, along with the kinetic energy term, neglects all forms of energy except internal and potential, which is included in $\dot{W}$ below.

The phase internal energies U_j and U_s and the enthalpies H_j are functions of temperature T, phase pressure P_j, and composition ω_{ij}. One form this dependency can take is Eq. (2.3-16), where the doubly subscripted internal energies (and enthalpies) are partial mass quantities. Partial mass quantities, Eq. (2.3-17), are analogies to partial molar quantities in solution thermodynamics (Denbigh, 1968). For example, the partial mass internal energy of species i in phase j is the change in U_j as ω_{ij} is changed, all other variables being held constant,

$$U_{ij} = \left(\frac{\partial U_j}{\partial \omega_{ij}}\right)_{P_j, T, \omega_{kj}, k \neq i} \tag{2.3-18}$$

TABLE 2-3 SUMMARY OF ADDITIONAL EQUATIONS FOR NONISOTHERMAL FLUID FLOW IN PERMEABLE MEDIA

Equation (1)	Name (2)	Number independent scalar equations* (3)	Dependent variables[†] Identity (4)	Dependent variables[†] Number (5)
(2.3-14) $\frac{\partial}{\partial t}\left(\rho U + \rho g D_z + \frac{1}{2}\sum_{j=1}^{N_P} \rho_j \lvert \vec{u}_j \rvert^2\right) - \vec{\nabla} \cdot \left(\sum_{j=1}^{N_P} \rho_j \vec{u}_j \left[H_j + \frac{1}{2}\lvert \vec{u}_j \rvert^2 + g D_z\right]\right) - \vec{\nabla} \cdot (k_{Tt} \vec{\nabla} T) = 0$	Energy conservation	1	U, H_j, T	$N_P + 2$
(2.3-15) $\rho U = \phi \sum_{j=1}^{N_P} \rho_j S_j U_j + (1 - \phi)\rho_s U_s$	Total internal energy	1	U_j, U_s	$N_P + 1$
(2.3-16a) $U_j = \sum_{i=1}^{N_C} \omega_{ij} U_{ij}$	Phase internal energy	N_P	U_{ij}	$N_P N_C$
(2.3-16b) $U_s = \sum_{i=1}^{N_C} \omega_{is} U_{is}$	Phase internal energy	1	U_{is}	N_C
(2.3-16c) $H_j = \sum_{i=1}^{N_C} \omega_{ij} H_{ij}$	Phase enthalpy	N_P	H_{ij}	$N_P N_C$
(2.3-17a) $U_{ij} = U_{ij}(T, P_j, \omega)$	Partial mass internal energy	$N_P N_C$		
(2.3-17b) $U_{is} = U_{is}(T, P_j, \omega)$	Partial mass internal energy	N_C		
(2.3-17c) $H_{ij} = H_{ij}(T, P_j, \omega)$	Partial mass enthalpy	$N_P N_C$		

* Total independent equations $= 2(N_P N_C) + 2N_P + N_C + 3$

[†] Total dependent variables $= 2(N_P N_C) + 2N_P + N_C + 3$

and similarly for U_{is} and H_{ij}. The partial mass properties themselves may be calculated from equations of state (Eq. 2.2-12) or empirical correlations as functions of temperature, pressure, and composition.

Equations (2.3-16) readily revert to simple forms. For example, if phase j is an ideal solution, the partial mass quantities become pure component quantities, functions of temperature and pressure only. Further, if j is an ideal gas, the partial mass quantities are functions only of temperature.

The equations presented in Tables 2-2 and 2-3 are complete, but they can only be solved with the specification of a similarly complete set of initial and boundary conditions.

2-4 SPECIAL CASES

We now consider several special cases of the general conservation equations in Tables 2-2 and 2-3. Each special case is applied in practice to describe various EOR processes occurring in permeable media fluid flow. These special cases can be accurately approximated by much simpler forms of the above general equations with fewer and simpler associated auxiliary equations and boundary conditions. We restrict our discussion to flows in local thermodynamic equilibrium.

Fractional Flow Equations

Consider one-dimensional linear flow and constant temperature, rock and fluid properties (T, ϕ, $\vec{\vec{k}}$, and ρ_j) with no sorption ($W_{is} = 0$), reaction ($R_i = 0$), or interphase mass transfer. Take the medium to have a constant dip angle α. For this case, the energy conservation equation is trivial, and Eq. (2.1-9) reduces to

$$\phi \frac{\partial S_j}{\partial t} + \frac{\partial u_j}{\partial x} = 0, \qquad j = 1, \ldots, N_P \tag{2.4-1}$$

In this equation, we have defined pseudocomponents such that $\omega_{ij} = 0$ for all i except one for which it is unity. The use of pseudocomponents is quite common in EOR descriptions since this simplification often results in greater understanding of the processes without introducing significant error.

To eliminate the need to solve for pressure, Eq. (2.4-1) is usually written in terms of a fractional flow function, which can be defined for the case of equal phase pressures ($P_{cjn} = 0$) as

$$f_j = \frac{u_j}{u} = \frac{\lambda_{rj}}{\sum_{k=1}^{N_P} \lambda_{rk}} \left[1 - \frac{kg \sin \alpha}{u} \sum_{k=1}^{N_P} \lambda_{rk}(\rho_j - \rho_k) \right] \tag{2.4-2}$$

where $u = \sum_{j=1}^{N_P} u_j$, and α is the dip angle $\tan \alpha = dD_z/dx$.

It is easily shown from Eq. (2.2-21) that under these circumstances u is a function of time only, and f_j is a function of saturation only, allowing us to write Eq. (2.4-1) in a final hyperbolic form.

$$\frac{\partial S_j}{\partial t} + \frac{u}{\phi}\frac{\partial f_j}{\partial x} = 0, j = 1, \ldots, N_P \tag{2.4-3}$$

To solve Eq. (2.4-3) for the phase saturations $S_j(x, t)$, the total volumetric fluid flux u injected at the inflow boundary and the experimentally measured fractional flow dependences of $N_P - 1$ phases (note $\Sigma_{j=1}^{N_P} f_j = 1$) are needed. Buckley and Leverett (1942) first solved this equation for two-phase flow, and the resulting estimation of waterflood oil recovery is called the Buckley-Leverett theory (see Chap. 5). Other similar cases, including three-phase flow and compositional effects such as interphase mass transfer and adsorption, have been solved in closed form (Pope, 1980). We discuss these solutions in detail in Chaps. 7–9.

Miscible Flow

The above case applies to the simultaneous flow of immiscible fluids. We now treat the isothermal case of many components flowing simultaneously in a single fluid phase. Thus only one phase flows regardless of composition, but both convection and dispersion of these components must be included. Miscible processes of interest include (1) true miscible displacement of oil by a solvent from a reservoir; (2) chromatographic processes of various sorts such as analytical chromatography, separation chromatography, ion exchange processes, and adsorption of chemicals as they percolate through soils and other naturally occurring permeable media; (3) leaching processes such as the in situ mining of uranium; and (4) chemical reaction processes of many types in fixed bed reactors.

Equation (2.1-9) for single-phase flow is

$$\frac{\partial(\phi\rho\omega_i)}{\partial t} + \frac{\partial}{\partial t}[(1 - \phi)\rho_s\omega_{is}] + \vec{\nabla}\cdot[\rho\omega_i\vec{u} - \phi\rho\vec{\vec{K}}_i\cdot\vec{\nabla}\omega_i] = R_i, i = 1, \ldots, N_C \tag{2.4-4}$$

The second subscript j is now superfluous and has been dropped. The auxiliary Eqs. (2.2-5), (2.2-6), (2.2-8), and (2.2-10) through (2.2-12) are still needed, but the others are no longer pertinent. The principal one of these, Eq. (2.2-5) or Darcy's law, has a considerably simpler form as well, namely,

$$\vec{u} = -\frac{\vec{\vec{k}}}{\mu}\cdot(\vec{\nabla}P + \rho\vec{g}) \tag{2.4-5}$$

Since the relative permeability is now constant, it is lumped with $\vec{\vec{k}}$, the saturation of the single flowing phase being unity

For miscible solvents (see Chap. 7), the sorption term, the second term in Eq. (2.4-4), can be dropped, and $R_i = 0$, giving

$$\frac{\partial(\phi\rho\omega_i)}{\partial t} + \vec{\nabla}\cdot[\rho\omega_i\vec{u} - \phi\rho\vec{\vec{K}}_i\cdot\vec{\nabla}\omega_i] = 0, \; i = 1, \ldots, N_C \qquad (2.4\text{-}6)$$

A special one-dimensional linear case of Eq. (2.4-6) is obtained when the effect of composition and pressure on density is neglected and $\vec{\vec{K}}_i$ is a constant. Letting $C_i = \omega_i\rho$ be the mass concentration of component i, it follows that

$$\phi\frac{\partial C_i}{\partial t} + u\frac{\partial C_i}{\partial x} = \phi K_{\ell i}\frac{\partial^2 C_i}{\partial x^2}, \qquad i = 1, \ldots, N_C \qquad (2.4\text{-}7)$$

where $K_{\ell i}$, the longitudinal dispersion coefficient, is now a scalar,

$$K_{\ell i} = \frac{D_i}{\tau} + \frac{\alpha_\ell |u|}{\phi} \qquad (2.4\text{-}8)$$

as a special case of the more general definition (Eq. 2.2-14a). Moreover, u is at most a function of time, depending on the boundary conditions specified. D_i is usually taken as a constant, yielding the linear convection–diffusion (CD) equation. Several closed-form solutions for simple initial and boundary conditions are available for the CD equation (see Chaps. 5 and 7).

Chromatographic Equation

Several chromatographic processes are special cases of Eq. (2.4-7). We must restore the C_{is} term that describes the accumulation of component i owing to sorption reactions, for this is the essence of a chromatographic process. These sorption reactions may be adsorption, the exchange of one ion by another on the stationary substrate, or precipitation–dissolution reactions (see Chaps. 8–10). All these processes lead to selective separation of the components as they percolate through the permeable medium. Dispersion does not alter the separation, so we neglect the second-order term, a step that results in a set of strongly coupled (via the sorption term) first-order partial differential equations

$$\phi\frac{\partial C_i}{\partial t} + (1 - \phi)\frac{\partial C_{is}}{\partial t} + u\frac{\partial C_i}{\partial x} = 0, \qquad i = 1, \ldots, N_C \qquad (2.4\text{-}9)$$

In these particular physical problems, the nature of the sorption term is usually such that the equations are hyperbolic, and the method of characteristics can be used as a solution technique.

Semimiscible Systems

In several EOR applications, a description of flow in permeable media based on strictly miscible or immiscible flow is unsatisfactory. For these situations, the equations in Table 2-2 reduce to a simpler form consistent with the known complexities of the flow behavior. As an example of this, consider the flow of N_C components in up to N_P phases in the absence of chemical reaction. Such flows are characteristic of solvent (see Chap. 7) and micellar-polymer flooding (see Chap. 9) EOR applications.

If we assume incompressible fluids, constant porosity, and ideal mixing, Eq. (2.2-20) may be divided by the respective pure component density ρ_i^o to give

$$\frac{\partial}{\partial t}\left(\phi \sum_{j=1}^{N_P} C_{ij} S_j + (1 - \phi) C_{is}\right) + \vec{\nabla} \cdot \left(\sum_{j=1}^{N_P} C_{ij} \vec{u}_j - \frac{\phi \rho_j S_j}{\rho_i^o} \vec{\vec{K}}_{ij} \cdot \vec{\nabla} \omega_{ij}\right) = 0,$$

$$i = 1, \ldots, N_C \qquad (2.4\text{-}10)$$

where $C_{ij} = \rho_j \omega_{ij} / \rho_i^o$ is now the volume fraction of component i in phase j. To write Eq. (2.4-10) completely in terms of the C_{ij}, we assume $\rho_j \vec{\nabla} \omega_{ij} = \vec{\nabla}(\rho_j \omega_{ij})$. Under these assumptions, Eq. (2.2-21) leads to

$$\vec{\nabla} \cdot \left(\sum_{j=1}^{N_P} \rho_j \vec{u}_j\right) = 0 \qquad (2.4\text{-}11)$$

where we have used the constancy of total (incompressible) mass (Eq. 2.2-13a) in eliminating the time derivative. Equation (2.4-11) can be used along with the definition of fractional flow (Eq. 2.4-2) to write Eq. (2.4-10) in one-dimensional form.

$$\frac{\partial}{\partial t}\left(\phi \sum_{j=1}^{N_P} C_{ij} S_j + (1 - \phi) C_{is}\right) + u \frac{\partial}{\partial x}\left(\sum_{j=1}^{N_P} f_j C_{ij}\right) - \frac{\partial}{\partial x}\left(\sum_{j=1}^{N_P} \phi S_j K_{\ell ij} \frac{\partial C_{ij}}{\partial x}\right) = 0,$$

$$i = 1, \ldots, N_C \qquad (2.4\text{-}12)$$

Even with the above, Eq. (2.4-12) is still fairly general and must be solved simultaneously with Darcy's law and with the definitions for relative mobility, capillary pressure, mass fractions, saturations, equations of state, and equilibria relations, (Eqs. 2.2-5, 2.2-9, 2.2-11, and 2.2-12). This form is particularly convenient because many cases of binary and ternary phase equilibria are conventionally represented as volume fractions rather than mass fractions (see Chap. 4).

Steam Flooding Equations

As a special case of nonisothermal flow, we derive the "steam" equations given by Stegemeier et al. (1977). We assume at most $N_P = 3$ phases—an aqueous phase $j = 1$, a hydrocarbon phase $j = 2$, and a gas phase $j = 3$—are present. Further, at most two unreactive, nonsorbing pseudocomponents—water and oil—are present. We restrict the hydrocarbon phase to contain only oil, and the aqueous and gaseous phases to contain only water, assumptions that eliminate volatile hydrocarbons from the equations. With these assumptions, the mass conservation equations become, for water,

$$\frac{\partial}{\partial t}[\phi(\rho_1 S_1 + \rho_3 S_3)] + \vec{\nabla} \cdot (\rho_1 \vec{u}_1 + \rho_3 \vec{u}_3) = 0 \qquad (2.4\text{-}13a)$$

and, for oil,

$$\frac{\partial}{\partial t}(\phi \rho_2 S_2) + \vec{\nabla} \cdot (\rho_2 \vec{u}_2) = 0 \tag{2.4-13b}$$

The dispersion terms are absent from these equations since the phase compositions are constant. The conservation of energy Eq. (2.3-14) becomes

$$\frac{\partial}{\partial t}[\phi(\rho_1 S_1 U_1 + \rho_2 S_2 U_2 + \rho_3 S_3 U_3) + (1 - \phi)\rho_s U_s] + \vec{\nabla} \cdot (\rho_1 H_1 \vec{u}_1 + \rho_2 H_2 \vec{u}_2 + \rho_3 H_3 \vec{u}_3) - \vec{\nabla} \cdot (k_{Tt} \nabla T) = 0 \tag{2.4-14a}$$

where kinetic and potential energy terms have been neglected. We further neglect pressure-volume work by letting the enthalpies equal internal energies and by taking porosity to be constant. The derivatives in Eq. (2.4-14a) may then be expanded to give

$$\begin{aligned}(1 - \phi)\frac{\partial(\rho_s H_s)}{\partial t} &+ \phi \rho_1 S_1 \frac{\partial H_1}{\partial t} + \phi \rho_2 S_2 \frac{\partial H_2}{\partial t} \\ &+ \phi \rho_3 S_3 \frac{\partial H_3}{\partial t} + \phi (H_3 - H_1)\frac{\partial(\rho_3 S_3)}{\partial t} \\ &+ \rho_1 \vec{u}_1 \cdot \vec{\nabla} H_1 + \rho_2 \vec{u}_2 \cdot \vec{\nabla} H_2 + (H_3 - H_1)\vec{\nabla} \cdot (\rho_3 \vec{u}_3) \\ &+ \rho_3 \vec{u}_3 \cdot \vec{\nabla} H_3 - \vec{\nabla} \cdot (k_{Tt} \vec{\nabla} T) = 0\end{aligned} \tag{2.4-14b}$$

where Eqs. (2.4-13a) and (2.4-13b) have been used to eliminate several terms. The term $(H_3 - H_1)$ equals L_V, the latent heat of vaporization of water, and we assume enthalpies are independent of pressure $dH_j = C_{pj} dT$, where C_{pj} is the specific heat of phase j. If the C_{pj} are constant, Eq. (2.4-14b) becomes

$$M_{Tt}\frac{\partial T}{\partial t} + (\rho_1 C_{p1} \vec{u}_1 + \rho_2 C_{p2} \vec{u}_2) \cdot \vec{\nabla} T - \vec{\nabla} \cdot (k_{Tt} \vec{\nabla} T) = -L_V\left\{\phi \frac{\partial(\rho_3 S_3)}{\partial t} + \vec{\nabla} \cdot (\rho_3 S_3 \vec{u}_3)\right\} \tag{2.4-14c}$$

where M_{Tt} is the overall volumetric heat capacity

$$M_{Tt} = \phi(\rho_1 C_{p1} S_1 + \rho_2 S_2 C_{p2}) + (1 - \phi)\rho_s C_{ps} \tag{2.4-15}$$

In this definition and in Eq. (2.4-14c) the terms involving the gaseous phase density ρ_3, have been neglected since gas densities are usually much smaller than liquid densities. The term on the right side of Eq. (2.4-14c) represents the production or destruction of the steam phase times the latent heat and is a source term for the energy equation. If steam disappears (condenses), the source term is positive, which causes the temperature to rise. This results in a decrease in oil viscosity, the primary recovery mechanism in thermal flooding (see Chap. 11). The latent heat, phase pressures, and temperature are related through the vapor pressure curve for water and capillary pressure relations.

2-5 OVERALL BALANCES

A common and useful way to apply the equations in the previous sections is in the form of macroscopic or overall balances (Bird et al., 1960). Rather than balances written for each point within the permeable medium, overall balances are spatially integrated forms of the differential balances that thereby apply to the entire reservoir. Since the spatial component is absent from the equations, overall balances are much simpler and far easier to integrate than differential balances. This simplification is achieved at the expense of losing spatial detail of the concentration variables; therefore, to be useful, overall balances must be supplemented with independently derived or analytical correlations.

To derive the overall mass balance for species i, we begin with the differential balance on volume V in the form Eq. (2.1-6). We then identify V with the total bulk volume V_b exclusive of the small volumes associated with a finite number of sources and sinks embedded within. However, the volume V_b is still simply connected, and the divergence theorem still applies. The boundary of V_b may also be a fluid source or sink term, as would be the case of an oil column abutting an aquifer or a free gas cap. If we assume the fluxes across the boundaries of V are normal to the cross-sectional area, Eq. (2.1-6) becomes

$$V_b \frac{d\overline{W}_i}{dt} + \dot{N}_{Pi} - \dot{N}_{Ji} = V_b \overline{R}_i, \qquad i = 1, \ldots, N_C \tag{2.5-1}$$

where the superscript bar denotes volume-averaged quantities. The terms $\dot{N}_{Pi}$ and $\dot{N}_{Ji}$ are the mass production and injection rates of species i for all the source and sink terms in V_b. These are functions of time since they are evaluated at fixed positions on V_b. $\overline{R}_i$ is the volume-averaged reaction rate term of species i and is also a function of time. Equation (2.5-1) may be integrated with respect to time

$$V_b(\overline{W}_i - \overline{W}_{iI}) = N_{Ji} - N_{Pi} + V_b \int_0^t \overline{R}_i \, dt, \qquad i = 1, \ldots, N_C \tag{2.5-2}$$

In writing Eq. (2.5-2), we have assumed the cumulative injection and production of species i at $t = 0$ is zero. In what follows, we ignore the cumulative reaction rate term.

The most common application of Eq. (2.5-2) is to calculate N_{Pi} with W_i, $\overline{W}_{iI}$, and N_{Ji} specified. In particular, $\overline{W}_i(t)$ is difficult to know without actually integrating the differential balances. This difficulty is circumvented by defining E_{Ri}, the recovery efficiency of species i, as

$$E_{Ri} = \frac{N_{Pi} - N_{Ji}}{V_b \overline{W}_{iI}} \tag{2.5-3}$$

E_{Ri} is the net amount of species i produced expressed as a fraction of the amount of species initially present. For a component injected into the reservoir, E_{Ri} is negative, but for a component to be recovered, oil or gas (which it is almost exclusively applied to), E_{Ri} is positive and lies between 0 and 1. From Eq. (2.5-2), $\overline{W}_i$ is

$$\overline{W}_i = \overline{W}_{iI}(1 - E_{Ri}) \tag{2.5-4}$$

For either Eq. (2.5-3) or (2.5-4) to be useful, E_{Ri} must be expressed independently as a function of time. This is commonly done by decomposing E_{Ri} into the displacement efficiency E_{Di} and volumetric sweep efficiency E_{Vi} of component i

$$E_{Ri} = E_{Di}E_{Vi} \tag{2.5-5a}$$

where

$$E_{Di} = \frac{\text{Amount of } i \text{ displaced}}{\text{Amount of } i \text{ contacted}} \tag{2.5-5b}$$

$$E_{Vi} = \frac{\text{Amount of } i \text{ contacted}}{\text{Amount of } i \text{ in place}} \tag{2.5-5c}$$

These quantities in turn must be specified independently: E_{Di} as a function of time and fluid viscosities, relative permeabilities, and capillary pressures (see Chap. 5) and E_{Vi} as a function of time, viscosities, well arrangements, heterogeneity, gravity, and capillary forces (see Chap. 6).

A similar procedure applied to the energy conservation Eq. (2.3-14) yields

$$V_b \frac{d}{dt}(\overline{\rho U}) + \dot{H}_P - \dot{H}_J = -\int_A \vec{q}_c \cdot \vec{n} dA = -\dot{Q} \tag{2.5-6a}$$

where kinetic and potential energy terms have been neglected, and $\dot{H}_P$ and $\dot{H}_J$ represent the rates of enthalpy production and injection into and from V. This equation, of course, is a version of the first law of thermodynamics and will be useful in calculating heat losses to wellbores (with the potential energy term restored) and the overburden and underburden of a reservoir (see Chap. 11).

The time integrated form of Eq. (2.5-6a) is

$$V_b((\overline{\rho U}) - (\overline{\rho U})_I) = \dot{H}_J - \dot{H}_P - \dot{Q} \tag{2.5-6b}$$

from which we may define a thermal efficiency $\overline{E}_{hs}$ as the ratio of thermal energy remaining in the volume V_b to the net thermal energy injected.

$$\overline{E}_{hs} = \frac{V_b((\overline{\rho U}) - (\overline{\rho U})_I)}{\dot{H}_J - \dot{H}_P} = 1 - \frac{\dot{Q}}{\dot{H}_J - \dot{H}_P} \tag{2.5-7}$$

Equation (2.5-7) is used to independently calculate $\dot{Q}$.

2-6 SUMMARY

We will use the equations introduced and developed in this chapter in the remainder of the text. Introducing all of the equations here eliminates repetitive derivation in later chapters. The compilation also emphasizes one of the main points of this text: all EOR processes are described by specializations of the same underlying conservation laws. Solving these specializations and deducing physical observations from the solutions will occupy much of the remainder of this text.

EXERCISES

2A. *Hydrostatics*. Show that for static ($\vec{u_j} = 0$) and isothermal conditions Eq. (2.2-5) reduces to

$$\frac{dP_c}{dD_z} = (\rho_1 - \rho_2)g \tag{2A-1}$$

for two-phase flow where P_c is the oil–water capillary pressure curve.

2B. *Single-Phase Flow*. Show that for the flow of a single phase ($j = 2$) in the presence of an immiscible, immobile phase ($j = 1$) the isothermal mass balance equations in one-dimensional radial coordinates reduce to

$$\frac{\phi c_t}{\lambda_{r2} k}\frac{\partial P}{\partial t} - \frac{1}{r}\frac{\partial}{\partial r}\left(r\frac{\partial P}{\partial r}\right) = 0 \tag{2B-1}$$

where

$$c_t = S_1 c_1 + S_2 c_2 + c_f \tag{2B-2}$$

$$c_j = \frac{1}{\rho_j}\left(\frac{\partial \rho_j}{\partial P}\right)_T \tag{2B-3}$$

$$c_f = \frac{1}{\phi}\left(\frac{\partial \phi}{\partial P}\right)_T \tag{2B-4}$$

Equation (2B-1), the "diffusivity" equation, has assumed that terms of the form $c_2(\partial P/\partial r)^2$ are negligible. Equation (2B-1) forms and is the basis for a large variety of well test techniques (Earlougher, 1977).

2C. *Simplified Combustion Model*. Based on two-phase ($j = 2 =$ liquid, $j = 3 =$ gas), four-component ($i = 1 =$ water, $i = 2 =$ oil $[C_nH_{2m}]$, $i = 3 = CO_2$, $i = 4 = O_2$), one-dimensional flow show that the energy conservation equations in Table 2-3 reduce to

$$M_{Tt}\frac{\partial T}{\partial t} + (\rho_2 C_{p2} u_2 + \rho_3 C_{p3} u_3)\frac{\partial T}{\partial x} - \frac{\partial T}{\partial x}\left(k_{Tt}\frac{\partial T}{\partial x}\right) = \phi S_3 \Delta H_{RXN} \tag{2C-1}$$

where ΔH_{RXN} is the heat of reaction for the gaseous phase reaction

$$(2n + m)O_2 + 2C_nH_{2m} \rightarrow 2nCO_2 + 2mH_2O \tag{2C-2}$$

$$\Delta H_{RXN} = -\sum_{i=1}^{4} H_{i0} r_{i3} \tag{2C-3}$$

Further assumptions for Eq. (2C-1) are there is only oil present in liquid phase, no sorption or dispersion, ideal solution behavior (specific heat of gaseous phase is the mass fraction $-$ weighted sum of the component specific heats), no heat of vaporization of oil ($H_{22} = H_{23}$), enthalpies and internal energies are equal, kinetic and potential energies are negligible, and solid phase density and porosity are constant.

2D. *Black Oil Equations*. The conventional representation (Peaceman, 1977) of the flow of fluids in oil and gas reservoirs is the "black oil" equations wherein up to three phases, aqueous ($j = 1$), oleic ($j = 2$), and gaseous ($j = 3$), flow simultaneously. The aqueous and gaseous phases consist of a single pseudocomponent, water ($i = 1$) and gas ($i = 3$), respectively. The oleic phase consists of oil ($i = 2$) with a dissolved gas component. Show that for isothermal flow in the absence of chemical reactions, dispersion, or sorbed components the mass balance equations of Table 2-2 reduce to, for water and oil,

$$\frac{\partial}{\partial t}\left(\frac{\phi S_j}{B_j}\right) + \vec{\nabla} \cdot \left(\frac{\vec{u}_j}{B_j}\right) = 0, \qquad j = 1, 2 \tag{2D-1}$$

and, for gas,

$$\frac{\partial}{\partial t}\left(\phi\left[\frac{S_3}{B_3} + \frac{S_2 R_s}{B_2}\right]\right) + \vec{\nabla} \cdot \left(\frac{R_s}{B_2}\vec{u}_2 + \frac{\vec{u}_3}{B_3}\right) = 0 \tag{2D-2}$$

B_1 is the water formation volume factor (volume of a given mass of water at the prevailing temperature and pressure divided by the volume of the same mass of water at standard temperature and pressure).

$$B_1 = \frac{\rho_1^s}{\rho_1} \tag{2D-3}$$

R_s is the solution gas–oil ratio (volume of dissolved gas divided by volume of oil phase, with both volumes evaluated at standard temperature and pressure).

$$R_s = \frac{\omega_{32}\rho_2^s}{\omega_{22}\rho_3^s} \tag{2D-4}$$

B_2 is the oil formation volume factor (volume of oil at prevailing temperature and pressure divided by volume of oil at standard conditions).

$$B_2 = \frac{\rho_2^s}{\omega_{22}\rho_2} \tag{2D-5}$$

B_3 is the gas formation volume factor (volume of a given mass of gas at prevailing temperature and pressure divided by volume of the same mass at standard temperature and pressure).

$$B_3 = \frac{\rho_3^s}{\rho_3} \tag{2D-6}$$

The above definitions may be introduced into the mass balances of Table 2-2 by dividing each by their respective standard densities ρ_j^s and assuming each ρ_j^s to be time independent.

Equations (2D-1) and (2D-2) are balances for water, oil, and gas in standard volumes; because of the assumptions given in their development, a standard volume is a mass quantity with units of volume.

3

Petrophysics and Petrochemistry

In Chap. 2, we saw that a complete specification of the fluid flow equations in permeable media required functions for capillary pressure, relative permeabilities, and phase behavior. In Chap. 4, we discuss, in general terms, EOR phase behavior and some of the equations necessary for its quantitative representation. We also discuss petrophysical relations in a similar fashion; we begin with the properties of immiscible phases (oil and water); go on to EOR-related quantities, such as the capillary desaturation curve; and then end with a brief discussion of petrochemistry.

In developing each petrophysical property, we follow the same basic procedure. First, we describe the property mathematically in terms of simplified physical laws, usually based on incompressible, steady-state flow in even simpler geometries. The simple geometry is intended to represent the smallest element of the permeable medium—the connected pore or the microscopic scale. Second, we modify the petrophysical property for the connected pore to account for the actual permeable medium local geometry—variable pore cross sections, lengths, nonlinear dimensions (tortuosity), and multiple connections of one pore with another. This step translates the relation from the actual permeable medium flow domain to that of the locally continuous representative elementary volume (REV) that we first discussed in Chap. 2. Unfortunately, since this step constitutes a great deal of art, we are restricted here to fairly simple idealizations about the local pore geometry.

3-1 POROSITY AND PERMEABILITY

Porosity is the ratio of void or pore volume to macroscopic or bulk volume; the rock- or solid-phase volume is the bulk volume less the pore volume. For most naturally occurring media, the porosity is between 0.10 and 0.40 although, on occasion, values outside this range have been observed. Porosity is often reported as a percent, but in calculations it should always be used as a fraction. From these typical values, the rock phase clearly occupies the largest volume in any medium.

The porosity of a permeable medium is a strong function of the variance of the local pore or grain size distribution and a weak function of the average pore size itself. For sandstones, the porosity is usually determined by the sedimentological processes the medium was originally deposited under. For limestone formations, the porosity is mainly the result of changes that took place after deposition.

The pore space as well as the porosity can be divided into an interconnected or effective porosity available to fluid flow and a disconnected porosity unavailable to fluid flow. The latter porosity is of no concern to the EOR processes discussed here; hence in the rest of this text the word *porosity* means effective porosity only (Collins, 1976). Certain EOR processes exhibit behavior whereby some of the effective porosity is shielded from the displacing agent. Such is the case for the "dead-end" pore volume to solvents (see Chap. 7) and the "inaccessible" pore volume to polymer solutions (see Chap. 8).

Permeability is also a basic permeable medium property that is as important to EOR as porosity. As we discussed in Sec. 2-2, permeability is a tensorial property that is, in general, a function of position and pressure. Usually, the pressure dependence is neglected in most calculations, but the variation with position can be quite pronounced. In fact, permeability varies spatially by three or more factors of 10 in a typical formation, whereas porosity varies by only a few percent. This is a form of reservoir heterogeneity that seriously influences the outcome of nearly all EOR displacements (see Chap. 6).

The permeability of a medium is a strong function of the local pore size and a weak function of the grain size distribution. That both porosity (strong) and permeability (weak) are functions of the grain size distribution is manifest in correlations between the two quantities.

To demonstrate the dependence of permeability on pore size and to illustrate the procedure of transforming a local model to the REV scale, we derive the Carmen-Kozeny equation. The local pore model in this case is the capillary tube, probably the most common such model in permeable media studies. Consider the single-phase, steady, laminar flow of a constant-viscosity Newtonian fluid through a horizontal capillary of radius R and length L_t as shown in Fig. 3-1(a). A force balance on an annular element of fluid yields an ordinary differential equation for the fluid velocity that can be solved, subject to radial symmetry and a no-slip condition at the tube wall, to give the volumetric flow rate q (Bird et al., 1960).

$$q = \frac{\pi R^4}{8\mu}\frac{\Delta P}{L_t} \tag{3.1-1}$$

Equation (3.1-1) is the Hagen-Poiseuille equation for laminar flow in a tube. For this equation to apply, the tube must be long enough so that the flow is free of entrance or exit effects. This condition certainly does not hold in a permeable medium pore,

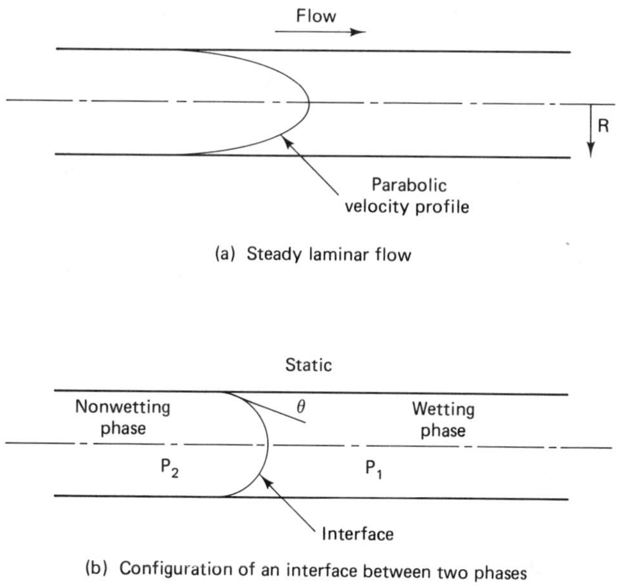

Figure 3-1 Tube flow analogues to REV conditions

but the simplicity of the equation suggests that we continue the development. The average velocity in the tube is

$$\bar{v} = \frac{q}{\pi R^2} = \frac{R^2}{8\mu}\frac{\Delta P}{L_t} \tag{3.1-2}$$

This equation is the beginning point in the transformation to the REV scale.

We wish to make the travel time of a fluid element in the capillary tube equal to that in a REV, or

$$\left(\frac{L_t}{\bar{v}}\right)_{\text{tube}} = \left(\frac{L}{v}\right)_{\text{REV}} \tag{3.1-3}$$

This equation introduces the interstitial fluid velocity v on the right-hand side, where $v = u/\phi$ from the Dupuit-Forchheimer assumption relating the interstitial v and superficial velocities u (Bear, 1972). v and u are the two most important velocity definitions for permeable media work. The superficial, or "Darcy," velocity u is the volumetric flow rate divided by the macroscopic cross-sectional area normal to flow. The interstitial, or "front," velocity is the true velocity of a fluid element as it crosses the medium macroscopic dimension. Throughout this text, we use the symbols u and v to differentiate between these velocities.

When a discrete, single-phase flow form of Darcy's law is used to eliminate v in Eq. (3.1-3), we can solve for the single-phase, one-dimensional permeability component k

$$k = \frac{R^2\phi}{8\tau} \tag{3.1-4}$$

where $\tau = (L_t/L)^2$, the squared ratio of the capillary tube length to REV length, is the tortuosity, another basic permeable medium property. Values of tortuosity for typical permeable media are readily estimated from the electrical resistivity of brine-filled media (Pirson, 1983). Tortuosity is always greater than 1 and can be greater than 10, but it is usually in the range of 2–5 for the media of interest here. The experimental best fit tortuosity for an assemblage of regularly packed spheres is a 25/12 ratio.

Even with this quantity, the value of R in a REV is difficult to visualize. To help, we invoke the concept of a hydraulic radius (Bird et al., 1960)

$$R_h = \frac{\text{Cross-sectional area open to flow}}{\text{Wetted perimeter}} \tag{3.1-5a}$$

R_h is $R/2$ for the tube geometry and may be defined for virtually any particle type with the following modification to the basic definition (Eq. 3.1-5a):

$$R_h = \frac{\text{Volume open to flow}}{\text{Wetted surface area}} \tag{3.1-5b}$$

Using the definition for porosity, this becomes

$$R_h = \frac{\phi}{a_v(1 - \phi)} \tag{3.1-6}$$

where a_v is the specific internal surface of the medium (surface area to volume), an intrinsic permeable media property. Substituting this into Eq. (3.1-4) gives

$$k = \frac{\phi^3}{2\tau(1 - \phi)^2 a_v^2} \tag{3.1-7}$$

For an assemblage of uniform spheres, a_v is

$$a_v = \frac{6}{D_p} \tag{3.1-8}$$

where D_p is the sphere or particle diameter. Combining Eqs. (3.1-4) through (3.1-7) gives the Carmen-Kozeny equation

$$k = \frac{1}{72\tau} \frac{\phi^3 D_p^2}{(1 - \phi)^2} \tag{3.1-9}$$

Equation (3.1-9) illustrates many important features about permeability. Permeability is a strong function of pore or particle size D_p and of packing through the porosity. This accounts for the low permeabilities in media that have a large clay content. Though Eq. (3.1-9) applies, strictly speaking, to assemblages of spheres, the effect of nonspherical particles does not become large until the eccentricity of the spheroids becomes large (see Exercise 3A). Experimentally, permeability is well correlated to the squared particle diameter (Fig. 3-2) for permeable media consisting of beads.

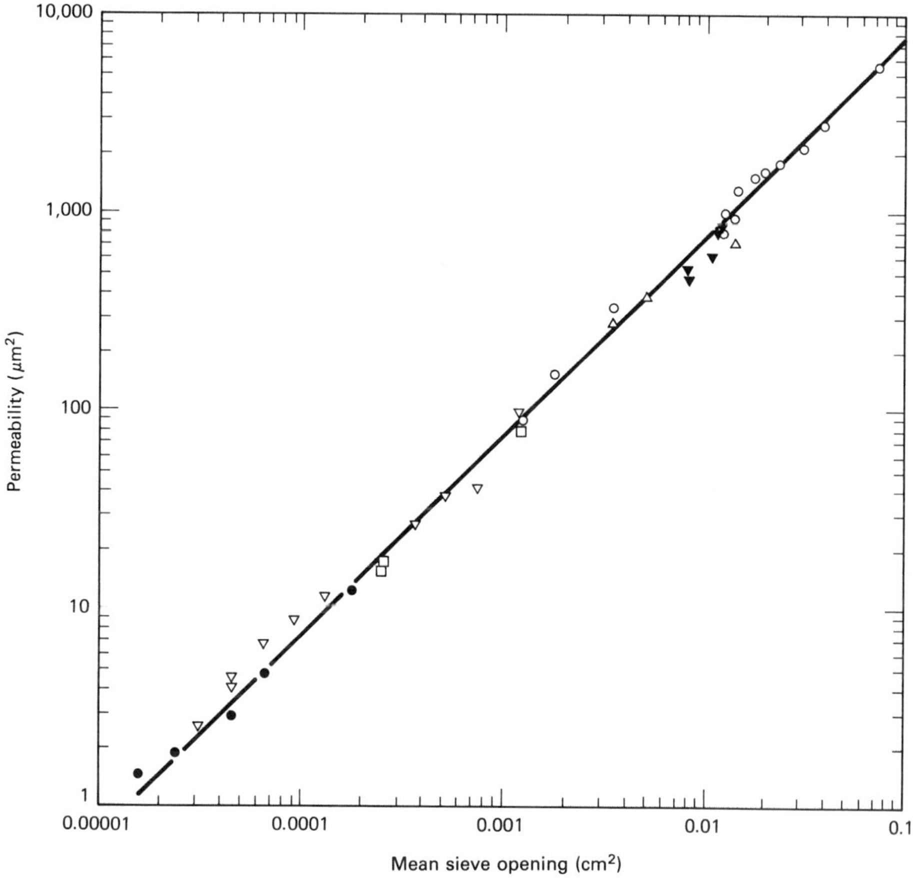

Figure 3-2 Experimental permeabilities as a function of bead size (adapted from Stegemeier et al., 1977)

The Carmen-Kozeny equation is used to make order of magnitude estimates of permeability and to estimate the pore size from a knowledge of the permeability. The latter factor is particularly relevant to EOR since the local pore size can be used to develop theoretical expressions relating to the mobility of polymer solutions. To do this we estimate the local shear rate, from an expression for the wall shear rate in a capillary tube.

$$\dot{\gamma}_{\text{wall}} = \frac{4\bar{v}}{R} \qquad (3.1\text{-}10a)$$

Equation (3.1-10a) defines an equivalent permeable media shear rate

$$\dot{\gamma}_{eq} = \frac{4v\tau^{1/2}}{R} \qquad (3.1\text{-}10b)$$

using Eq. (3.1-3) and the definition of tortuosity. Eliminating R with Eq. (3.1-4) gives

$$\dot{\gamma}_{eq} = 4v\left(\frac{\phi}{8k}\right)^{1/2} = \frac{4q}{A\sqrt{8k\phi}} \tag{3.1-11}$$

The shear rate given by Eq. (3.1-11) is useful in correlating and predicting the rheological properties of non-Newtonian fluids in permeable media flow (see Chap. 8).

Considering its simplicity, Eq. (3.1-9) does remarkably well in describing permeability. But the capillary tube model of a permeable medium is limited because of its uniform cross section and because it does not provide alternate pathways for fluid flow within each REV. (For more complicated local permeable medium models, see Dullien, 1979.) The consequence of these deficiencies is that such models cannot be made to predict relative permeabilities or trapped phase saturations without some modification. The latter effects play major roles in EOR, and we discuss them separately below. But first we discuss two-phase flow and its attendant phenomenon, capillary pressure.

3-2 CAPILLARY PRESSURE

Since interfacial forces, as manifest in capillary pressures, are easily the strongest forces within the REV in multiphase flow at typical velocities, capillary pressure is the most basic rock–fluid characteristic in multiphase flows, just as porosity and permeability are the most basic properties in single-phase flow. To discuss capillary pressure, we begin with the capillary tube concept and then proceed, through qualitative arguments, to the capillary phenomena actually observed in multiphase flow.

Returning to Fig. 3-1(b), consider a capillary tube having the same dimensions as in Fig. 3-1(a) except the tube now contains two phases, a nonwetting phase on the left and a wetting phase on the right. Phase 1 wets the tube surface because the contact angle θ, measured through this phase, is less than 90°. The boundary between the two phases is a phase boundary or interface across which one or more of the intensive fluid properties change discontinuously. This boundary is not the same as the concentration boundaries between miscible fluids across which intensive properties, if they change at all, do not change continuously.

If the phases and the interface in the tube are not flowing, a higher pressure is required in the nonwetting phase than in the wetting phase to keep the interface from moving. A static force balance across the interface in the direction parallel to the tube axis yields an expression for the nonwetting–wetting phase pressure difference.

$$P_2 - P_1 = \frac{2\,\sigma\cos\theta}{R} \equiv P_c \tag{3.2-1}$$

Equation (3.2-1) defines capillary pressure. The interface shape in Fig. 3-1(b) should not be confused with the velocity profile in Fig. 3-1(a). The latter is a static condition, whereas the former is a plot of dynamic velocity.

Equation (3.2-1), a simple form of Laplace's equation, relates the capillary pressure across an interface to the curvature of the interface R, the interfacial tension σ, and the contact angle θ. If either the interfacial tension is zero or the interface is perpendicular to the tube wall, the capillary pressure will be zero. The first condition is satisfied when the absence of interfacial tension (and, hence, the interface) renders the two adjoining phases miscible. The second condition holds only for the simple uniform tube geometry. The contact angle can take on all values between 0° and 180°; if it is greater than 90°, the wetting pattern of the two fluids is reversed, and the capillary pressure, as defined by Eq. (3.2-1), becomes negative.

In more complicated geometries, the form of the $1/R$ term in Eq. (3.2-1) is replaced with the mean curvature, a more general expression. In the following discussion, we must remember that capillary pressure is inversely proportional to a generalized interfacial curvature, which is usually dominated by the smallest local curvature of the interface.

For more complicated geometries, consider the entry of a nonwetting phase into a single pore of toroidal geometry bounded by the sphere assemblage shown in Fig. 3-3(a). Figure 3-3(b) shows the capillary pressure corresponding to the various entry positions of the interface. To force the interface into the pore, it must be compressed through the pore neck radius R_n, causing a decrease in the interface curvature and an increase in the capillary pressure as the nonwetting phase saturation in the toroid is increased from zero to point A in Fig. 3-3(b).

Once inside the pore body, which has a larger dimension R_b than the neck, the curvature increases, and the capillary pressure decreases. The decrease will continue until the interface becomes constrained by the walls on the opposite side of the pore at point B. After this point, much depends on the size of the pore exit. If the exit radius is also R_n, the capillary pressure will build back up to C. Beyond this, the interface will leave the pore, and neither the capillary pressure nor the saturation will change. However, if at point B, the capillary pressure at the pore neck is larger than in the pore body, the interface will collapse at the pore entrance rather than push on through (this is especially likely if the pore body is much larger than the pore entrance). The collapsing interface creates a disconnected globule of the nonwetting phase within the toroid. The globule conforms to the pore body to minimize its energy, and the curvature again increases, causing an abrupt decrease in the capillary pressure from points C to D. When this happens the wetting phase changes from a continuous funicular (two positive curvatures) configuration to a discontinuous pendular (one positive and one negative curvature) configuration (Stegemeier, 1976).

If the nonwetting phase saturation is again increased, the glob is forced farther into the rock–rock contacts, which manifests itself in large capillary pressure increases. The wetting phase at this point retreats to saturations approximated by monolayer coverage of the rock surfaces. Though this would seem to result in quite a small saturation, Melrose (1982) has shown that wetting phase saturations above 10% are possible at this limit. The same process takes place even if the displacing fluid has neutral wettability with respect to the solid; hence for an irregular pore geometry, $\cos\theta = 0$ does not imply zero capillary pressure.

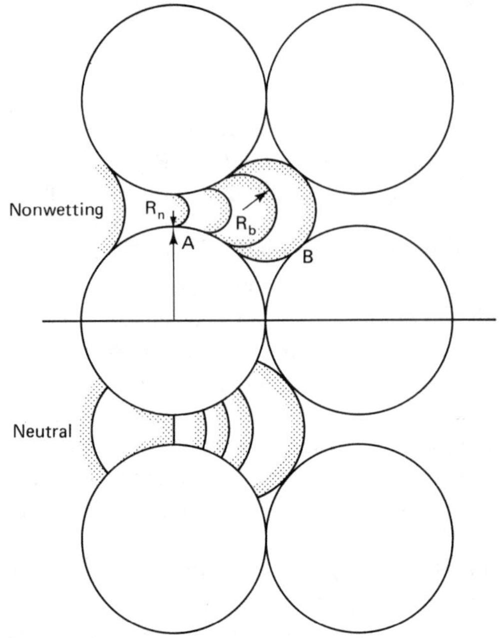

(a) Entry of nonwetting and neutral fluids in toroidal pore model

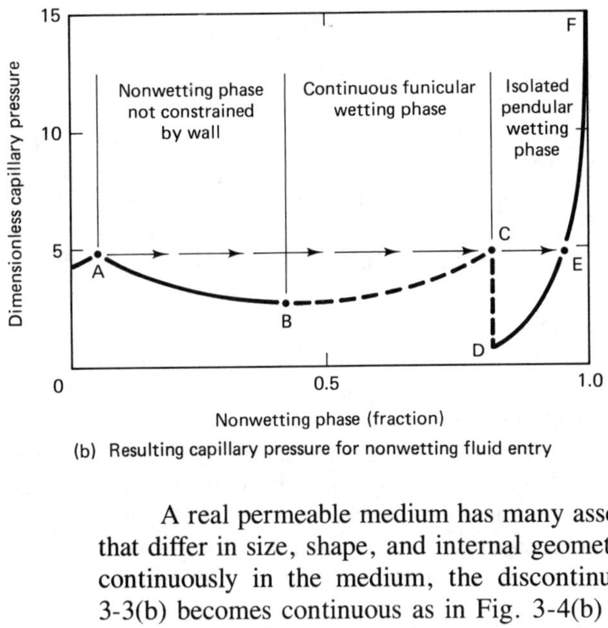

(b) Resulting capillary pressure for nonwetting fluid entry

Figure 3-3 Schematic of interface entrance into a toroidal pore (adapted from Stegemeier, 1976)

A real permeable medium has many assemblages like the toroid in Fig. 3-3(a) that differ in size, shape, and internal geometry. If these differences are distributed continuously in the medium, the discontinuous capillary pressure curve in Fig. 3-3(b) becomes continuous as in Fig. 3-4(b) (see Exercise 3B). Many of the same features are still present—the existence of an entry pressure at small nonwetting (mercury in this case) saturations and the sharp increase at large saturations. But the events corresponding to points *B–D* in Fig. 3-3(b) are absent.

Actual capillary pressure curves exhibit a sense of hysteresis, which can tell us much about the permeable medium. Consider the permeable medium idealized by the arrangement of decreasing size pores, as in Fig. 3-4(a), into which a nonwetting

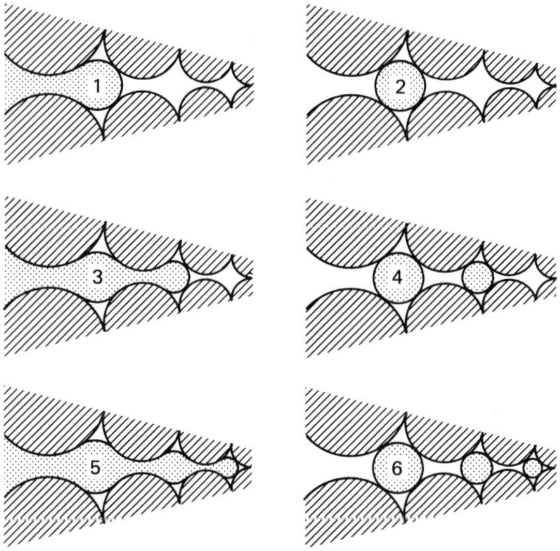

(a) Pore network cross section normal to direction of entry

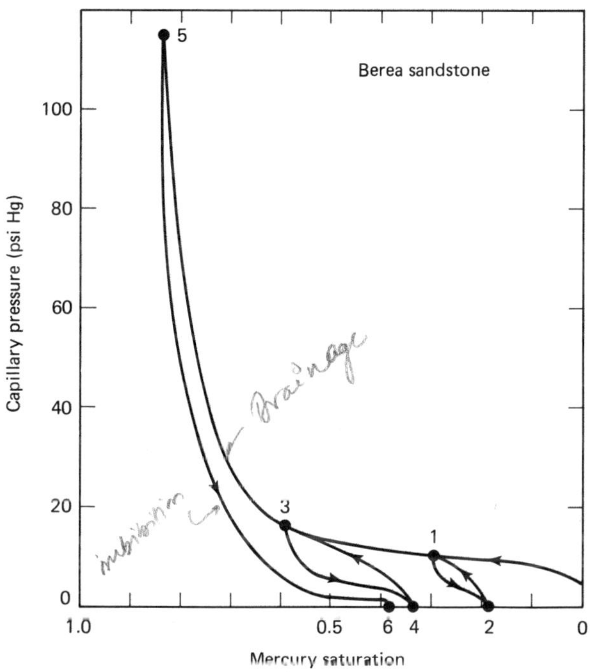

(b) Capillary pressure saturation sequence

Figure 3-4 The distribution of a nonwetting phase at various saturations (adapted from Stegemeier, 1976) phase is alternatively injected and then withdrawn. The forcing of a nonwetting phase into a pore (nonwetting saturation increasing) is a drainage process; the reverse (wetting saturation increasing) is an imbibition process. We imagine the pores have an exit for the wetting fluid somewhere on the right. Beginning at zero nonwetting saturation, we inject up to the saturation shown in condition 1 of Fig. 3-4(a). At static conditions, the pressure difference between the exit and entrance of the assemblage is the capillary pressure at that saturation. When the wetting fluid is introduced into the pore from the right, the nonwetting fluid disconnects leaving a trapped or nonflowing glob in the largest pore, condition 2. The capillary pressure curve from condition 2 to condition 1 is an imbibition curve that is different from the drainage curve because it terminates ($P_c = 0$) at a different saturation. At static condition 2, the entrance–exit pressure difference is zero since both pressures are being measured in the same wetting phase. Going from condition 2 to 3 is a second drainage process that results in an even higher nonwetting saturation, a higher capillary pressure, and a higher trapped nonwetting phase saturation, condition 4, after imbibition. At the highest capillary pressure, condition 5, all pores contain the nonwetting phase, and the post-imbibition trapped saturation is a maximum. The capillary pressure curve going from the largest nonwetting phase saturation to the largest trapped nonwetting phase saturation is the imbibition curve (curve 6). Curve 1 is the drainage curve; all other curves are designated second drainage, third drainage, etc.

Though quite simple, the representation in Fig. 3-4(a) explains many features of actual capillary pressure curves. Imbibition curves are generally different from drainage curves, but the difference shrinks at high nonwetting phase saturations where more of the originally disconnected globs are connected. The hysteresis we discuss here is *trapping* hysteresis. *Drag* hysteresis is also caused by differences in advancing and receding contact angles (see Figure 3-8). The termination of any imbibition curve is at zero capillary pressure.

The nonwetting phase residual saturation depends on the largest nonwetting phase saturation. A plot of these two quantities is the initial–residual (IR) curve. The IR curve manifests many of the same permeable media properties as do the capillary pressure curves. Figure 3-5 shows several typical IR curves. Most important, the quantity being plotted on the ordinate of capillary pressure curves is the pressure difference between the continuous nonwetting and wetting phases. When either phase exists in a disconnected form, such as conditions 2, 4, and 6 in Fig. 3-4(a), a local capillary pressure does exist, but it is not unique owing to the variable glob sizes. We use these concepts and the IR curve to estimate the capillary desaturation curve in Sec. 3-4, but first let us see how we may estimate the IR curve from hysteresis in the capillary pressure curves.

Consider the drainage and imbibition capillary pressure curves given in Fig. 3-6. The drainage curve begins at zero nonwetting phase saturation (first drainage), and the imbibition curve begins at the maximum nonwetting phase saturation possible in the experimental apparatus. Points A, B, and C represent three saturation configurations in the pores shown below the capillary pressure plot. Point B repre-

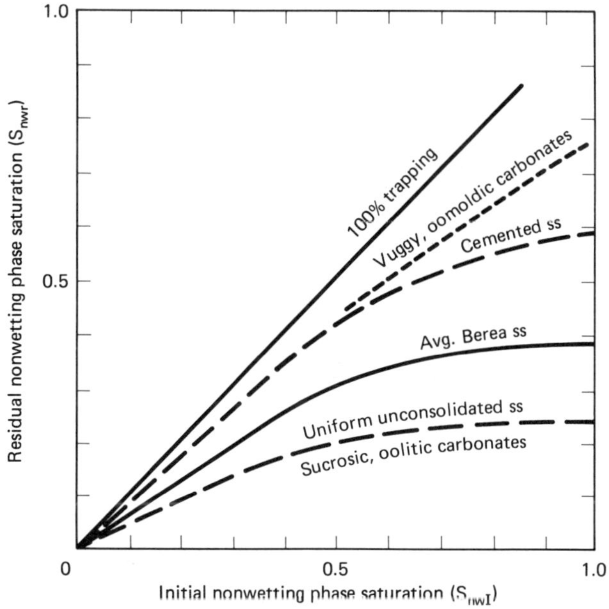

Figure 3-5 Typical IR nonwetting phase saturation curves (from Stegemeier, 1976)

sents the maximum trapped nonwetting phase saturation, and points A and C have equal capillary pressures but on different capillary pressure curves. The difference between the x coordinates of points A and C is the disconnected nonwetting phase saturation at point C. The connected nonwetting phase configuration is identical in configurations A and C; hence the trapped nonwetting phase saturation corresponding to an initial saturation at point A is the difference between the nonwetting phase saturation at point B minus the difference between the nonwetting phase saturations at points C and A (i.e., $B - [C - A]$). This procedure yields one point on the IR curve, but the whole curve may be traced by picking several points along the two P_c curves. Alternatively, if one capillary pressure curve is given along with the IR curve, the other capillary pressure curve may be estimated by the reverse procedure. A variable residual nonwetting saturation may severely affect ultimate waterflood recovery in reservoirs that have a large original transition zone (see Exercise 3C).

The capillary pressure at a given saturation is a measure of the smallest pore being entered by the nonwetting phase at that point, suggesting the curvature of the capillary pressure curve is a function of the pore size distribution. The level of the curve is determined by the mean pore size. In an effort to separate the effects of pore size and pore size distribution, Leverett (1941) proposed a nondimensional form of the drainage capillary pressure curve that should be independent of the pore size. We first replace the capillary tube radius R in Laplace's Eq. (3.2-1) with a function $R/j(S_{nw})$, where j is a dimensionless function of the nonwetting phase saturation S_{nw}. If we eliminate the hydraulic radius between the new Eq. (3.2-1) and Eq. (3.1-4), we arrive at the Leverett j-function,

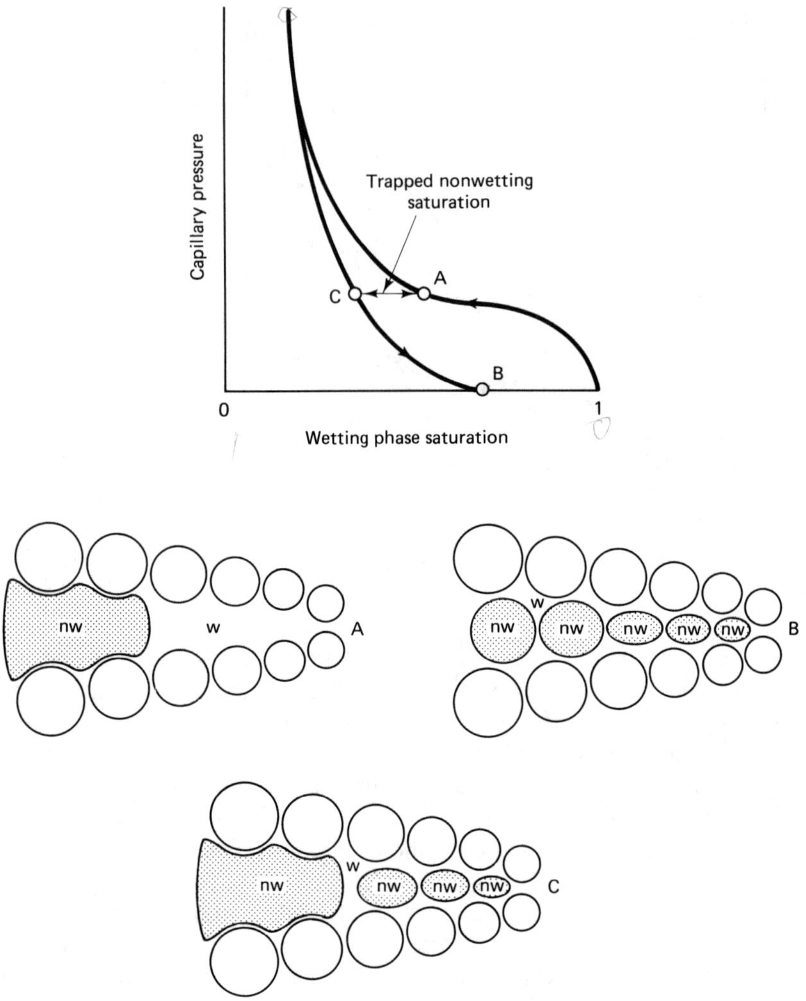

Figure 3-6 Schematic of the construction of an IR curve

$$j(S_{nw}) = \frac{P_c\sqrt{\frac{k}{\phi}}}{\sigma \cos \theta} \tag{3.2-2}$$

The numerical constants and the tortuosity have been absorbed into j in Eq. (3.2-2). As shown in the original work (1941), the j-function is independent of pore size, as was intended, but is also independent of the interfacial tension between the fluid pair used to measure P_c. j as a function of S_{nw} is presented in several standard works (Collins, 1976; Bear, 1972). The process used in deriving Eq. (3.2-2) is a form of pore size to REV scaling. (For alternate capillary pressure correlations, see Morrow, 1976.)

By a similar procedure, the first drainage capillary pressure curve may be used to calculate pore size distributions. At each capillary pressure value, Eq. (3.2-1) may be used, with $\sigma\cos\,\theta$ known, to calculate the radius R of the largest pore being entered at that nonwetting saturation. The nonwetting saturation itself is the volume fraction being occupied by pores of this size or larger. This information may be converted into a pore size percentile plot, which may then be converted into a frequency of occurrence of pores at a given R. Figure 3-7 illustrates the wide variety of pore sizes exhibited by natural media based on this procedure.

The scaling ideas in the above development are useful in estimating the capillary pressures in a rock sample of arbitrary k and ϕ from a single capillary pressure

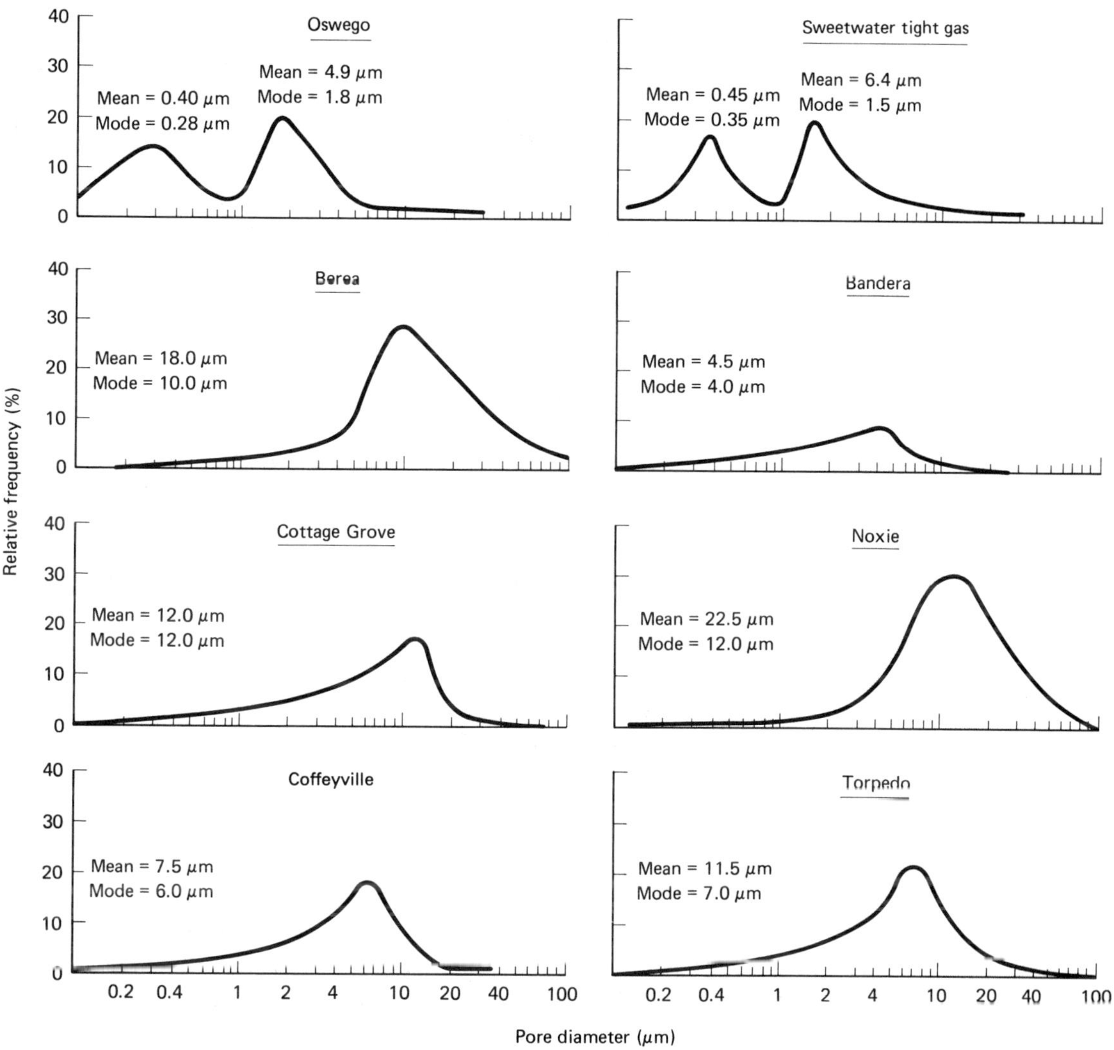

Figure 3-7 Pore size distributions of sedimentary rocks (from Crocker et al., 1983)

curve on a rock sample of known k and ϕ. Of course, this scaling assumes all samples have the same pore size distribution and tortuosity. The scaling can also be used to estimate the capillary pressure for a fluid pair different from that used to measure the capillary pressure curve. Unfortunately, the j-function was derived for a drainage capillary pressure curve, so it cannot, by itself, be used to estimate imbibition capillary pressure curves. Moreover, the effect of permeable media wettability is insufficiently represented by the contact angle (see Fig. 3-3). Even if it were, wettability is difficult to estimate.

At least three tests are commonly used to measure permeable media wettability (Anderson, 1986).

1. In the Amott test (Amott, 1959), wettability is determined by the amount of oil or water spontaneously imbibed in a core sample compared to the same values when flooded. Amott wettability values range from $+1$ for complete water wetting to -1 for complete oil wetting. This measure is the most widely cited wettability index in the literature.
2. In the U.S. Bureau of Mines test (Donaldson et al., 1969), wettability index W is the logarithm of the ratio of the areas under centrifuge-measured capillary pressure curves in both wetting phase saturation increasing and decreasing directions. W can range from $-\infty$ (oil wet) to $+\infty$ (water wet) but are characteristically between -1.5 and $+1.0$.
3. In a third test, contact angles can be measured directly on polished silica or calcite surfaces (Wagner and Leach, 1959).

As a means of estimating wettability in permeable media, none of these tests is entirely satisfactory. The Amott index and the W index can be taken in actual permeable media, but their correspondence to capillary pressure is not direct. But both of these tests are measures of aggregate rather than local wettability. The contact angle measurement is, of course, direct, but it is unclear to what extent a polished synthetic surface represents the internal surface of the permeable media. Tacit in the use of contact angle measurements is the assumption that, in determining wettability, oil–brine properties are more important than solid surface properties. Contact angle measurements do exhibit hysteresis (Fig. 3-8). Moreover, all wettability measures suffer from the disadvantage that they are not routinely measured. Figure 3-9 cross-correlates these measures.

The contact angle has been used to survey the wettability of 55 oil reservoirs with results shown in Table 3-1. Neither the total number nor either of the two major reservoir lithology classifications (sandstone and carbonate) are exclusively of one wettability. Most sandstone reservoirs tend to be water wet or intermediate wet, whereas most carbonate reservoirs tend to be intermediate wet or oil wet. The contact angle has also been used to correlate relative permeability measurements (Owens and Archer, 1971), but in actual practice, these measurements are so rare that, as we show, usually the relative permeability measurements are used to infer wettability.

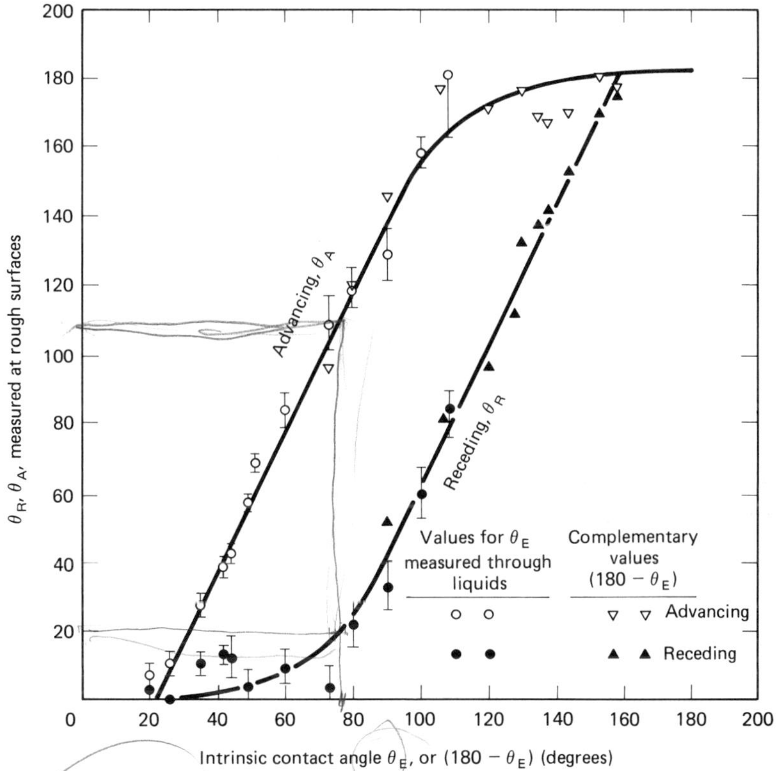

Figure 3-8 Advancing (θ_A) and receding (θ_R) contact angles observed at rough surfaces versus intrinsic contact angle (θ_E) (from Morrow, 1976)

TABLE 3-1 DISTRIBUTION OF WATER-WET, INTERMEDIATE-WET, AND OIL-WET RESERVOIRS BASED ON ADVANCING CONTACT ANGLE MEASUREMENTS ON SMOOTH MINERAL SURFACES FOR FLUIDS FROM 55 RESERVOIRS (ADAPTED FROM MORROW, 1976; BASED ON TREIBER ET AL., 1971)

Wettability class	Water wet	Intermediate wet	Oil wet
Defining contact angle range used by Treiber et al. (θ_A at smooth mineral surface)	0°–75°	75°–105°	105°–180°
No. of sandstone reservoirs	13 (43%)	2 (7%)	15 (50%)
No. of carbonate reservoirs	2 (8%)	2 (8%)	21 (84%)
Total	15 (27%)	4 (7%)	36 (66%)
Defining contact angle range from classification used in Morrow (1976)	$\theta_A < 62°$	$\theta_A > 62°$ $\theta_R > 133°$	$\theta_A > 133°$ ($\theta_R = \theta_A$)
No. of sandstone reservoirs	12 (40%)	10 (33%)	8 (27%)
No. of carbonate reservoirs	2 (8%)	16 (64%)	7 (28%)
Total	14 (26%)	26 (47%)	15 (27%)

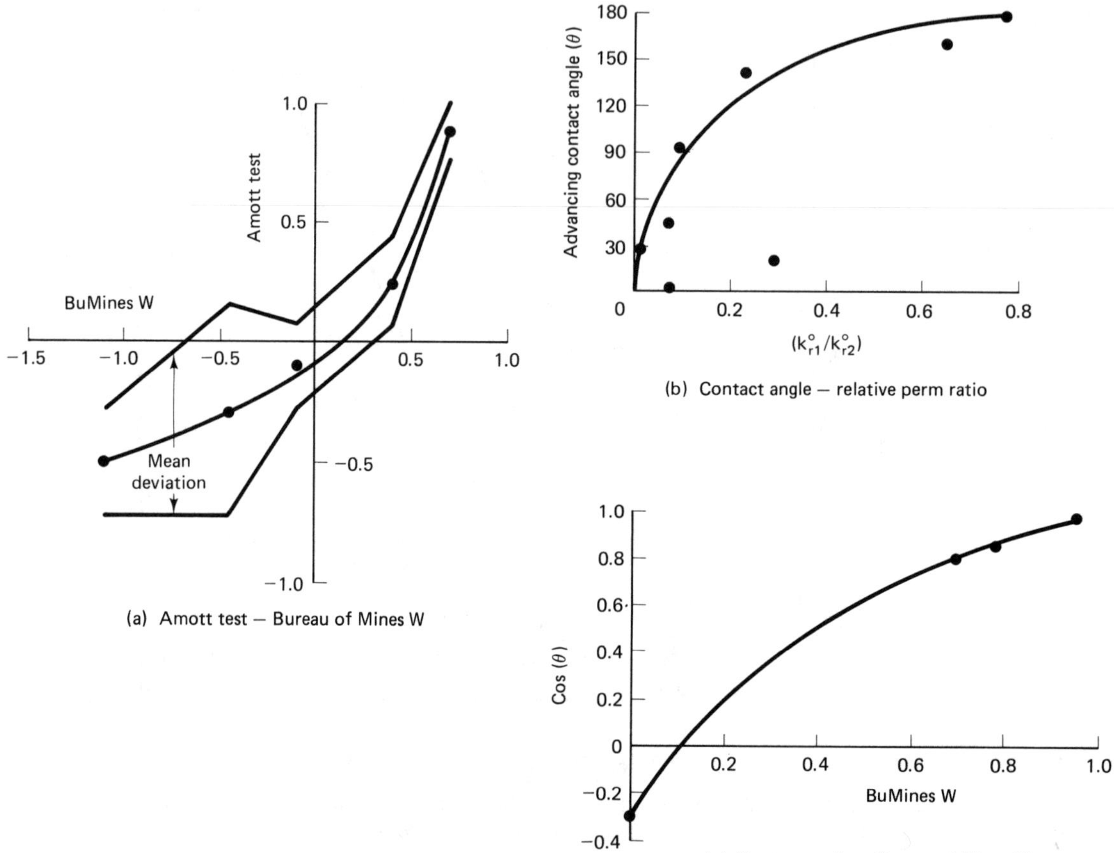

(a) Amott test – Bureau of Mines W

(b) Contact angle – relative perm ratio

(c) Contact angle – Bureau of Mines W

Figure 3-9 Correspondence between wettability tests

3-3 RELATIVE PERMEABILITY

Relative permeability curves and their associated parameters are easily the most relevant petrophysical relations for EOR. Consider the flow of several incompressible, single-component phases in a one-dimensional, linear permeable medium. If the flow is steady state—that is, the saturation of all the phases does not vary with time and position—Darcy's law (Eq. 2.2-5) may be integrated over a finite distance Δx to give

$$u_j = -\lambda_j \Delta\Phi_j/\Delta x \tag{3.3-1}$$

where λ_j is the mobility of phase j. The mobility is the "constant" of proportionality between the flux of phase u_j and the potential difference $\Delta\Phi_j = \Delta(P_j - \rho_j g D_z)$. λ_j

can be decomposed into a rock property, the absolute permeability k, a fluid property, the phase j viscosity μ_j, and a rock–fluid property, the relative permeability k_{rj}.

$$\lambda_j = k\left(\frac{k_{rj}}{\mu_j}\right) \tag{3.3-2}$$

The relative permeability is a strong function of the saturation of phase S_j. Being a rock–fluid property, the functionality between k_{rj} and S_j is also a function of rock properties (pore size distribution, for example) and wettability. It is not, in general, a strong function of fluid properties, though when certain properties (interfacial tension, for example) change drastically, relative permeability can be affected.

Alternate definitions involving the mobility and relative permeability are the relative mobility λ_{rj}

$$\lambda_{rj} = \frac{k_{rj}}{\mu_j} \tag{3.3-3a}$$

and the phase permeability k_j

$$k_j = k\,k_{rj} \tag{3.3-3b}$$

k_j is a tensorial property in three dimensions. It is important to keep the four definitions (mobility, relative mobility, phase permeability, and relative permeability) separate and clear.

Though there have been attempts to calculate relative permeabilities on theoretical grounds, by far the most common source of k_r curves has been experimental measurements. (For experimental procedures to do this, see Jones and Roszelle, 1978.)

Figure 3-10 gives schematic oil–water relative permeability curves. The relative permeability to a phase decreases as the saturation of that phase also decreases; however, the relative permeability to a phase vanishes at some point well before the phase saturation becomes zero. If the relative permeability of a phase is zero, it can no longer flow, and the saturation at this point cannot be lowered any further. Reducing the "trapped" oil saturation is one of the most important objectives of EOR (see Sec. 3-4). The trapped oil saturation is called the residual oil saturation and given the symbol S_{2r}. In later chapters, we illustrate that certain EOR fluids can lower residual oil saturation values.

It is extremely important to distinguish the residual oil saturation from the remaining oil saturation S_{2R}. The residual oil saturation is the oil remaining behind in a thoroughly waterswept region of the permeable medium; the remaining oil saturation is the oil left after a waterflood, well-swept or not. Thus $S_{2R} \geqq S_{2r}$. The trapped water saturation S_{1r} is the irreducible water saturation. It is not the connate water saturation, which is the water saturation in a reservoir before any water is injected, although many times S_{1r} is equal to the connate water saturation.

Two other important landmarks on the relative permeability curves are the endpoint relative permeabilities. These are the constant relative permeability of a

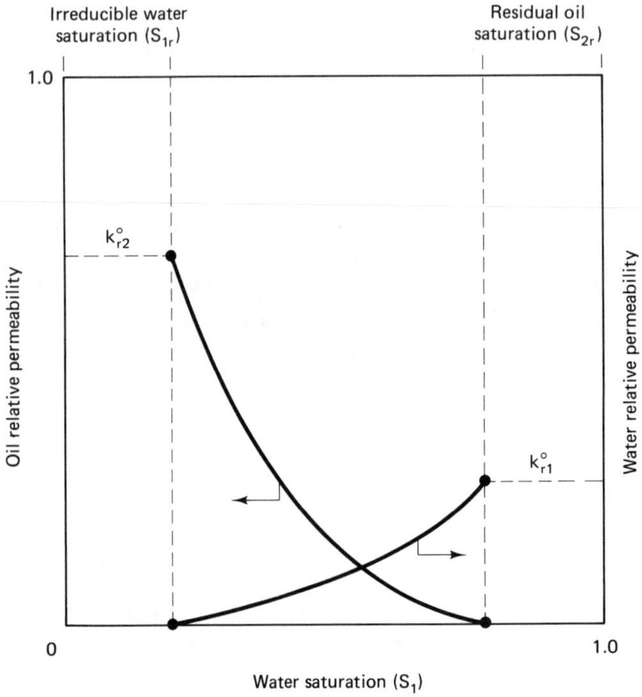

Figure 3-10 Typical water–oil relative permeabilities

phase at the other phase's residual saturation. In this text, we designate the endpoint relative permeability by a superscript *o*. The word *relative* in the name of the k_r functions indicates the phase permeability has been normalized by some quantity. As the definition (Eq. 3.3-3b) implies, we take the normalizing permeability to be the absolute permeability to some reference fluid (usually 100% air or water) though this is not always the case in the literature. This choice of normalizing factor means the endpoint permeabilities will usually be less than 1.

The endpoint values are measures of the wettability. The nonwetting phase occurs in isolated globules several pore diameters in length that occupy the center of the pores. Trapped wetting phase, on the other hand, occupies the crevices between rock grains and coats the rock surfaces. Thus we would expect the trapped nonwetting phase to be a bigger obstacle to the wetting phase than the trapped wetting phase is to the nonwetting phase. The wetting phase endpoint relative permeability will, therefore, be smaller than the nonwetting phase endpoint. The ratio of wetting to nonwetting endpoints proves to be a good qualitative measure of the wettability of the medium. The permeable medium in Fig. 3-10 is water wet since k^o_{r1} is less than k^o_{r2}. For extreme cases of preferential wetting, the endpoint relative permeability to the wetting phase can be 0.05 or less.

Others view the crossover saturation (where $k_{r2} = k_{r1}$) of the relative permeabilities as a more appropriate indicator of wettability, perhaps because it is less sensitive to the value of the residual phase saturations. Figure 3-11 illustrates both the

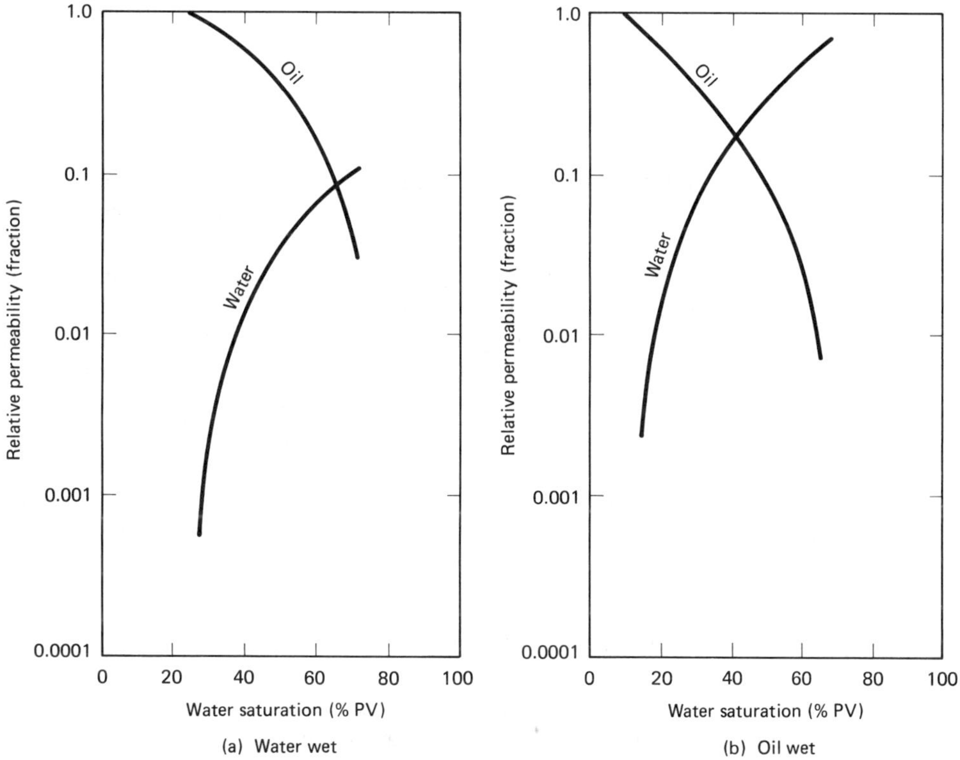

Figure 3-11 Effect of wettability on relative permeability (from Craig, 1971)

shift in the crossover point and the movement of the water endpoint relative permeability as a function of wettability. Figure 3-11 also illustrates that relative permeability can change by several factors of 10 over a normal saturation range; hence experimental curves are often presented on semilog plots as shown.

Though no general theoretical expression exists for the relative permeability function, several empirical functions for the oil–water curves are available (see, for example, Honarpour et al., 1982). When analytical expressions are needed, we use the following exponential forms for oil–water flow:

$$k_{r1} = k_{r1}^{o}\left(\frac{S_1 - S_{1r}}{1 - S_{1r} - S_{2r}}\right)^{n_1} \tag{3.3-4a}$$

$$k_{r2} = k_{r2}^{o}\left(\frac{1 - S_1 - S_{2r}}{1 - S_{1r} - S_{2r}}\right)^{n_2} \tag{3.3-4b}$$

These equations fit most experimental data and separate explicitly the relative permeability curvatures (through exponents n_1 and n_2) and the endpoints.

If capillary pressure is negligible, we may add all the phase fluxes to obtain an expression relating the total flux to the pressure gradient dP/dx in horizontal flow $u = -k\lambda_{rt}(dP/dx)$, where

$$\lambda_{rt} = \sum_{j=1}^{N_P} \lambda_{rj} \tag{3.3-5}$$

is the total relative mobility, a measure of the resistance of the medium to multiphase flow. Plots of λ_{rt} versus saturation frequently show a minimum (Fig. 9-34), meaning it is more difficult to flow multiple phases through a medium than any one of the phases alone. Even in the presence of trapped saturations, λ_{rt} is reduced because the phases cause mutual interference during flow.

Neglecting the flowing phase capillary pressures, we can also solve for the pressure gradient in the total flux expression and use it to eliminate the pressure gradient in the differential form of Eq. (3.3-1) to give, after some rearranging, the fractional flow of phase j in Eq. (2.4-2). If capillary pressure is negligible, f_j is a generally nonlinear function of saturation(s) only. This functionality forms the basis of the fractional flow analyses in Chap. 5 and the remaining chapters. Fractional flow curves are functions of phase viscosities and densities as well as of relative permeabilities.

3-4 RESIDUAL PHASE SATURATIONS

In this section, we discuss the two-phase flow behavior of residual wetting (S_{wr}) or nonwetting (S_{nwr}) saturations. These may be identified with residual oil or water saturations according to Table 3-1.

The notion of a wetting phase residual saturation is consistent with our discussion of capillary pressure. As in Fig. 3-12(a), increasing pressure gradients force ever more of the nonwetting phase into pore bodies, causing the wetting phase to retreat into the concave contacts between the rock grains and other crevices in the pore body. At very high pressures, the wetting phase approaches monolayer coverage and a low residual saturation. Because of film instability, S_{wr} is theoretically zero when P_c is infinite.

A residual nonwetting phase saturation, on the other hand, is somewhat paradoxical. After all, the nonwetting phase is repelled by the rock surfaces and, given enough contact time, all the nonwetting phase would be expelled from the medium. Repeated experimental evidence has shown this not to be the case, and, in fact, under most conditions, the S_{nwr} is as large as S_{wr}. The residual nonwetting phase is trapped in the larger pores in globules several pore diameters in length (Fig. 3-12b). Figure 3-13 shows several pore casts of these globules after waterfloods in consolidated and unconsolidated sands.

The mechanism for a residual phase saturation may be illustrated through two simplified REV-scale models. Figure 3-14 shows the double pore, or pore doublet, model, a bifurcating path in the permeable medium, and Fig. 3-15 shows three versions of the pore snap-off model, a single flow path with variable cross-sectional

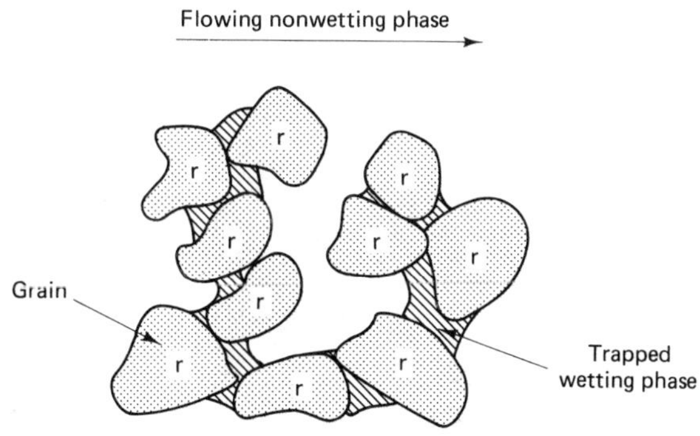

(a) Trapped wetting phase

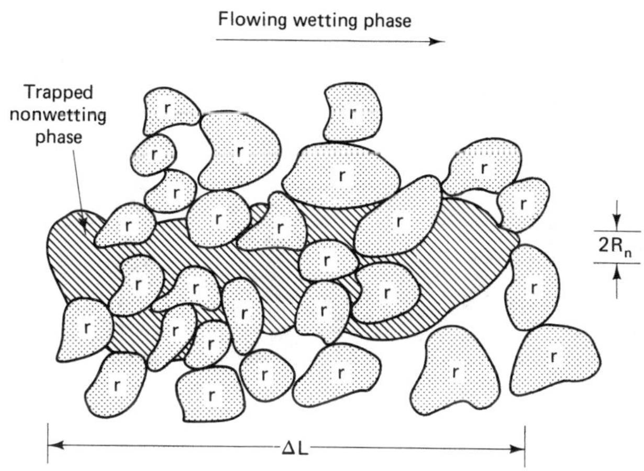

(b) Trapped nonwetting phase

Figure 3-12 Schematic illustration of trapped wetting and nonwetting phases

area. Each model contains a degree of local heterogeneity: The pore doublet model shows different radii flow paths, and the pore snap-off model shows different cross-sectional areas in each of the flow paths. This local heterogeneity is needed for there to be a residual nonwetting phase saturation. The simple capillary tube model discussed in Sec. 3-1 does not have this element and, thus, will not exhibit a nonzero S_{nwr}.

The Pore Doublet Model (Moore and Slobod, 1956)

This model assumes well-developed Poiseuille flow occurs in each path of the doublet, and the presence of the interface does not affect flow. Both assumptions would be accurate if the length of the doublet were much larger than the largest path radius

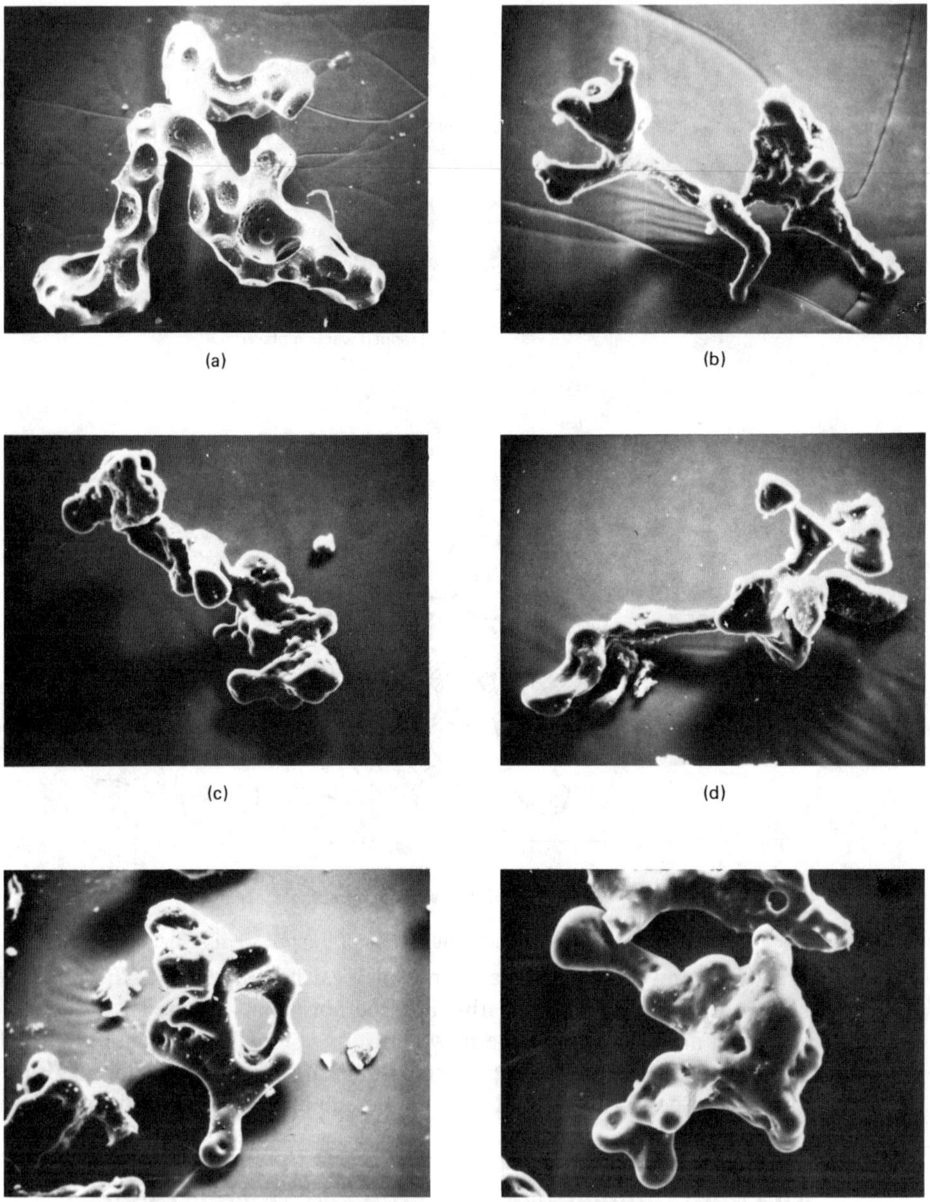

Figure 3-13 Typical large oil blobs in (a) bead packs and (b) to (f) Berea sandstone (from Chatzis et al., 1983)

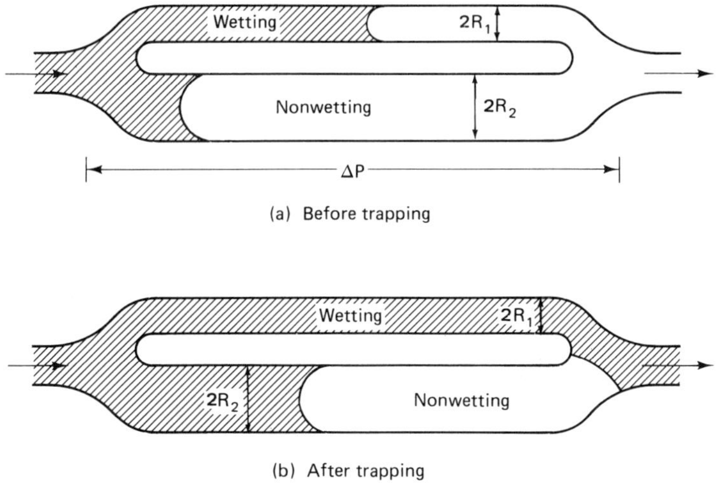

Figure 3-14 Schematic of pore doublet model

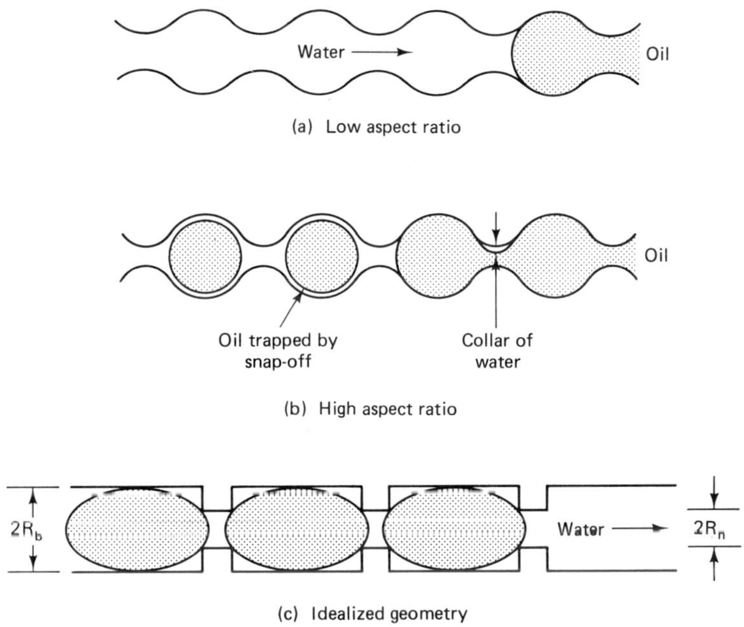

Figure 3-15 Various geometries of the pore snap-off model (Figs. 3-15(a) and (b) from Chatzis et al., 1983)

and the flow were very slow. The latter condition also permits the use of the static capillary pressure function (Eq. 3.2-1) in this flowing field. The wetting and nonwetting phases have equal viscosity in our treatment though this can be relaxed (see Exercise 3F). Most important, we assume when the wetting–nonwetting interface reaches the outflow end of the doublet in either path, it traps the resident fluid as in Fig. 3-14(b).

Based on these assumptions, the volumetric flow rate in either path is given by Eq. (3.1-1). The total volumetric flow rate through the doublet is therefore

$$q = q_1 + q_2 = \frac{\pi}{8\mu L_t}(R_1^4 \Delta P_1 + R_2^4 \Delta P_2) \tag{3.4-1}$$

and, because the paths are parallel, the driving force across each path must be equal:

$$\Delta P_1 - P_{c_1} = \Delta P_2 - P_{c_2} \tag{3.4-2}$$

In Eq. (3.4-2), the capillary pressures are positive for the imbibition process in Fig. 3-14; for a drainage process, the capillary pressures would be negative. Using these equations, we can write the volumetric flow rate in either path in terms of the total volumetric flow rate, the doublet geometry, and the interfacial tension-contact angle product from Eq. (3.2-1)

$$q_1 = \frac{q - \dfrac{\pi R_2^4 \, \sigma \cos \theta}{4\mu L_t}\left(\dfrac{1}{R_2} - \dfrac{1}{R_1}\right)}{1 + (R_2/R_1)^4} \tag{3.4-3a}$$

$$q_2 = \frac{q\left(\dfrac{R_2}{R_1}\right)^4 + \dfrac{\pi R_2^4 \, \sigma \cos \theta}{4\mu L_t}\left(\dfrac{1}{R_2} - \dfrac{1}{R_1}\right)}{1 + (R_2/R_1)^4} \tag{3.4-3b}$$

To investigate the trapping behavior of the doublet, we form the ratio of the average velocities in the paths

$$\frac{v_2}{v_1} = \frac{4N_{vc} + \left(\dfrac{1}{\beta} - 1\right)}{\dfrac{4N_{vc}}{\beta^2} - \beta^2\left(\dfrac{1}{\beta} - 1\right)} \tag{3.4-4}$$

where $\beta = R_2/R_1$ is a heterogeneity factor, and

$$N_{vc} = \left(\frac{\mu L_t q}{\pi R_1^3 \, \sigma \cos \theta}\right) \tag{3.4-5}$$

is a dimensionless ratio of viscous to capillary forces, which we henceforth call the local capillary number.

The trapping behavior of the pore doublet follows from Eq. (3.4-4) and the capillary number definition. In the limit of negligible capillary forces (large N_{vc}), the velocity in each path of the doublet is proportional to its squared radius. Hence the interface in the large-radius path will reach the outflow end before the small-radius path, and the nonwetting phase will be trapped in the small-radius path.

But if viscous forces are negligible, the small-radius path will imbibe fluid at a faster rate than is supplied at the doublet inlet. From Eqs. (3.4-2) and (3.4-3), the interface velocity in the large-radius path is negative in the fluid-starved doublet, whereas the velocity in the small-radius path is greater than that at the doublet inlet. This situation is in disagreement with the premises of the derivation: If the interface seals off the small-radius path at the doublet inlet, the flow in the small-radius tube will be zero.

Though the extreme of negligible viscous forces is hard to visualize, it is easy to imagine an intermediate case where viscous forces are small, but not negligible, compared to capillary forces. Now the doublet is no longer starved for fluid, but the interface in the small-radius path is still faster than that in the large-radius path. The nonwetting phase is trapped in the large-radius path as shown in Fig. 3-14. For typical values of the pore radii (Fig. 3-7) in Eq. (3.4-4), most flows in permeable media will be approximated by this intermediate case.

Besides explaining how a nonwetting phase can become trapped at all, the simplified behavior of the pore doublet illustrates several qualitative observations about phase trapping.

1. The nonwetting phase is trapped in large pores; the wetting phase, in small cracks and crevices.
2. Lowering capillary forces will cause a decrease in trapping. This decrease follows from simple volumetric calculations since fluids trapped in small pores will occupy a smaller volume fraction of the doublet than those in large pores.
3. There must be some local heterogeneity to cause trapping. In this case, the heterogeneity factor β must be greater than 1. Simple calculations with the pore doublet show that increasing the degree of heterogeneity increases the capillary number range over which the residual phase saturation changes.

But as a quantitative tool for estimating trapping, the pore doublet greatly overestimates the amount of residual nonwetting phase at low capillary number. At high capillary number, little evidence supports nonwetting phase trapping in the small pores. Most important, the capillary number defined by Eq. (3.4-5) is difficult to define in actual media; hence the pore doublet model is rarely used to translate to the REV scale.

Snap-Off Model

The snap-off model can readily translate to the REV scale. The exact geometry of the model (Fig. 3-15) is usually dictated by the ease with which the resulting mathematics can be solved. The sinusoidal geometry in Fig. 3-15(a) has been used by Oh and Slattery (1976) for theoretical investigation and by Chatzis et al. (1983) for experimental work. The pore snap-off model was earlier discussed by Melrose and Brandner (1974), who included the effects of contact angle hysteresis in their calculations. Later in this section, we use the idealized geometry in Fig. 3-15(c) to translate to the REV scale.

The snap-off model assumes a single-flow path of variable cross-sectional area through which is flowing a nonwetting phase. The sides of the flow path are coated with a wetting phase so that a uniquely defined local capillary pressure exists everywhere. But this capillary pressure varies with position in the flow path; it is large where the path is narrow, and small where it is wide. For certain values of the potential gradient and pore geometry, the potential gradient in the wetting phase across the path segment can be less than the capillary pressure gradient across the same segment. The external force is now insufficient to compel the nonwetting phase to enter the next pore constriction. The nonwetting phase then snaps off into globules that are localized in the pore bodies of the flow path. By this hypothesis, then, the condition for reinitializing the flow of any trapped globule is

$$\Delta\Phi_w + \Delta\rho g \, \Delta L \sin \alpha \geqq \Delta P_c \tag{3.4-6}$$

where $\Delta\Phi_w$ and ΔP_c are the wetting phase potential and capillary pressure changes across the globule. ΔL is the globule size, $\Delta\rho = \rho_w - \rho_{nw}$, and α is the angle between the globule's major axis and the horizontal axis. Equation (3.4-6) suggests a competition between external forces (viscous and gravity) and capillary forces that was also present in the pore doublet model though both models are quite different.

Of course, in any real permeable medium, local conditions approximating both the pore doublet and the snap-off model will occur. Using detailed experimental observations in consolidated cores, Chatzis et al. (1983) have determined that approximately 80% of the trapped nonwetting phase occurs in snap-off geometries, with the remaining 20% in pore doublets or in geometries that are combinations of both categories. These authors used a more elaborate classification scheme (Fig. 3-16), where the snap-off model is combined with the pore doublet model in several ways. These combinations remove many of the ad hoc assumptions about the nature of the trapping in the pore doublet model when the wetting–nonwetting interface reaches the outflow end. The theoretical treatment of the snap-off model again illustrates the basic requirements for nonwetting trapping: nonwetting trapping in large pores, the need for local heterogeneity, and strong capillary forces.

Trapping in Actual Media

We can now discuss the experimental observations of trapping in actual permeable media. The most common experimental observation is a relationship between residual nonwetting or wetting phase saturations and a local capillary number. We call this relationship the capillary desaturation curve (CDC). Figure 3-17 shows a schematic CDC curve. Typically, these curves are plots of percent residual (nonflowing) saturation for the nonwetting (S_{nwr}) or wetting (S_{wr}) phases on the y axis versus a capillary number on a logarithmic x axis. The capillary number N_{vc} is a dimension-

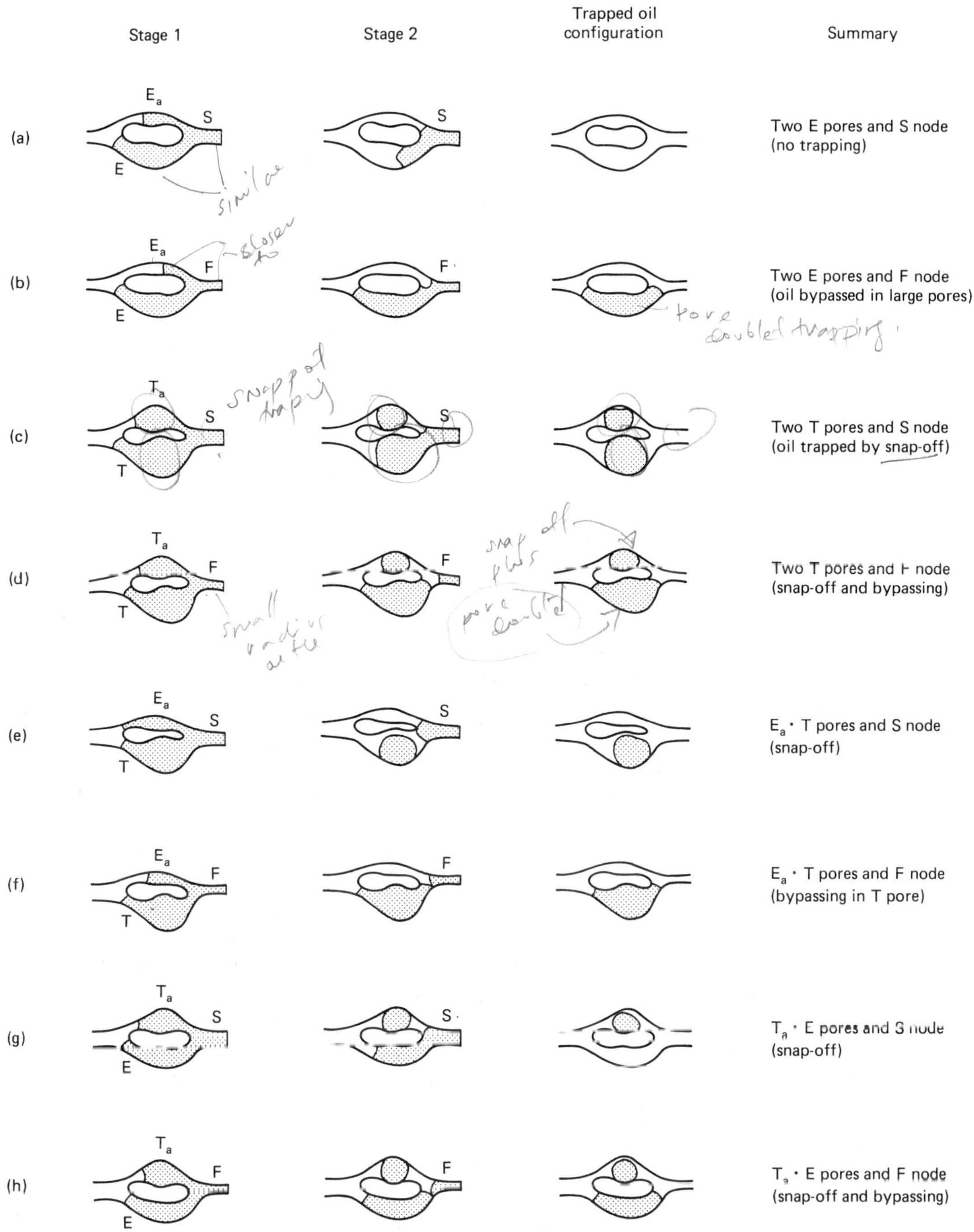

Figure 3-16 Sketches of low capillary number trapping mechanisms and configuration of residual oil in pore doublets (from Chatzis et al., 1983)

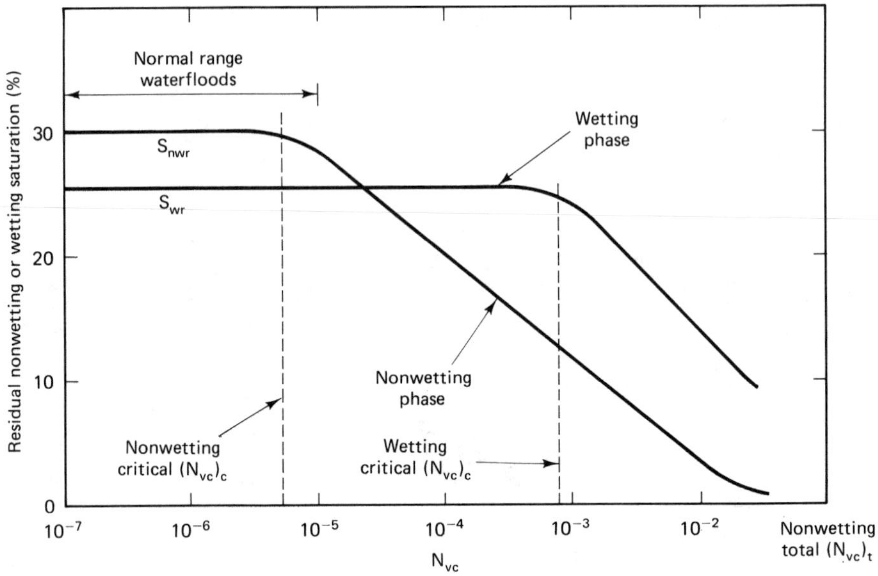

Figure 3-17 Schematic capillary desaturation curve (from Lake, 1984)

less ratio of viscous to local capillary forces, variously defined. At low N_{vc}, both S_{nwr} and S_{wr} are roughly constant at plateau values. At some N_{vc}, designated as the critical capillary number $(N_{vc})_c$, a knee in the curves occurs, and the residual saturations begin to decrease. Complete desaturation—zero residual phase saturations—occurs at the total desaturation capillary $(N_{vc})_t$ number shown in Fig. 3-17. Most waterfloods are well onto the plateau region of the CDC where, as a rule, the plateau S_{wr} is less than S_{nwr}. Frequently, the two CDC curves are normalized by their respective plateau values.

Table 3-2 summarizes the results of experimentally determined CDC curves. This list is restricted to the flow of two liquid phases in a synthetic or outcrop permeable medium; microscopic studies like those reported by Arriola et al. (1980) are not included. A more thorough review has been done by Bhuyan (1986).

No experimental data has been reported on actual reservoir permeable media; most experimental work has used pure hydrocarbons and synthetic brine. The plateau values of S_{nwr} and S_{wr} show considerable variation (there are more nonwetting phase measurements). The $(N_{vc})_c$ and $(N_{vc})_t$ for the nonwetting phase are less than the respective values for the wetting phase. For the nonwetting phase, $(N_{vc})_c$ is in the 10^{-5} to 10^{-4} range, whereas the $(N_{vc})_t$ is usually 10^{-2} to 10^{-1}. For the wetting phase, $(N_{vc})_c$ is roughly equal to the nonwetting $(N_{vc})_t$, whereas the wetting critical $(N_{vc})_c$ is 10^{-1} to 10^{0}. More precise conclusions are not warranted because of the variation in N_{vc} definitions and in the experimental conditions. In the work of du Prey (1973) for the wetting phase, as well as in a limestone sample in the work of Abrams

TABLE 3-2 SUMMARY OF EXPERIMENTAL WORK ON CAPILLARY DESATURATION CURVES

Reference	Permeable medium	Fluids	Definition N_{vc}	Nonwetting Phase: Plateau S_{nwr}	Nonwetting Phase: $(N_{vc})_c$	Nonwetting Phase: $(N_{vc})_t$	Wetting Phase: Plateau S_{wr}	Wetting Phase: $(N_{vc})_c$	Wetting Phase: $(N_{vc})_t$
Moore and Slobod (1956)	Outcrop ss	Brine-crude	$v\mu_1/\sigma \cos\theta$	0.5	10^{-7}	10^{-2}			
Taber (1969)*	Berea ss	Brine-soltrol	$v\mu_1/\sigma \cos\theta$	0.4	10^{-5}–10^{-4}	10^{-2}			
Foster (1973)	Berea ss	Brine-oil	$u\mu_1/\sigma$	0.4–0.5	10^{-5}–10^{-4}	10^{-2}–10^{-1}			
Dombrowski and Brownell (1954)	Synthetic	Distilled H_2O Pure Organics	$\dfrac{k\lvert\vec{\nabla}\Phi\rvert}{\sigma \cos\theta}$				0.075	10^{-2}	2
Du Prey (1973)	Synthetic	Water-pure hydrocarbons	$u\mu_1/\sigma$	0.2	10^{-4}	10^{-3}	None	None	0.5
Erlich et al.* (1974)	Outcrop ss	Brine-crude	$u\mu_1/\sigma$	0.3	10^{-4}				
Abrams (1975)	Outcrop ss	Brine-crude	$\dfrac{v\mu_1}{\sigma\Delta S}\cos\theta\left(\dfrac{\mu_1}{\mu_2}\right)^{0.4}$	0.3–0.4	10^{-5}–10^{-4}	10^{-2}–10^{-1}			
Gupta (1980)	Berea ss	Brine-decane	$u\mu_1/\sigma$	0.2–0.3	10^{-5}–10^{-4}	10^{-2}–10^{-1}	0.45	10^{-4}	0.03
Chatzis and Morrow (1981)	Outcrop ss	Brine-soltrol	$\dfrac{k\Delta P}{\sigma L}$ (various)	0.27–0.41	10^{-5}–10^{-4}	$\sim 10^{-3}$			

* Used surfactants or NaOH

Note: ss = sandstone

$(N_{vc})_c$ = critical capillary number

$(N_{vc})_t$ = total desaturation capillary number

(1975), the CDC knee was absent altogether, an extreme manifestation of the effect of a large pore size distribution. The range between $(N_{vc})_c$ and $(N_{vc})_t$ is considerably greater for the nonwetting phase (10^{-7} to 10^{-1}) than for the wetting phase (10^{-4} to 10^{0}).

Table 3-2 sets forth three general observations based on the CDC curve.

1. Wettability is important. The wetting phase normalized CDC curves should be two to three factors of 10 to the right of a nonwetting phase CDC curve; however, intuitively, the two CDC curves should approach each other at some intermediate wetting condition.
2. Pore size distribution is also important. The critical-total N_{vc} range should increase with increasing pore size distribution for both wetting and nonwetting phases.
3. The critical-total N_{vc} range for the nonwetting phase should be greater than for the wetting phase with, again, a continuous shift between wettability extremes.

Fig. 3-18 illustrates the last two effects, and Fig. 3-19 shows a compilation of experimental CDC data, each using a common definition of N_{vc}, in a Berea core.

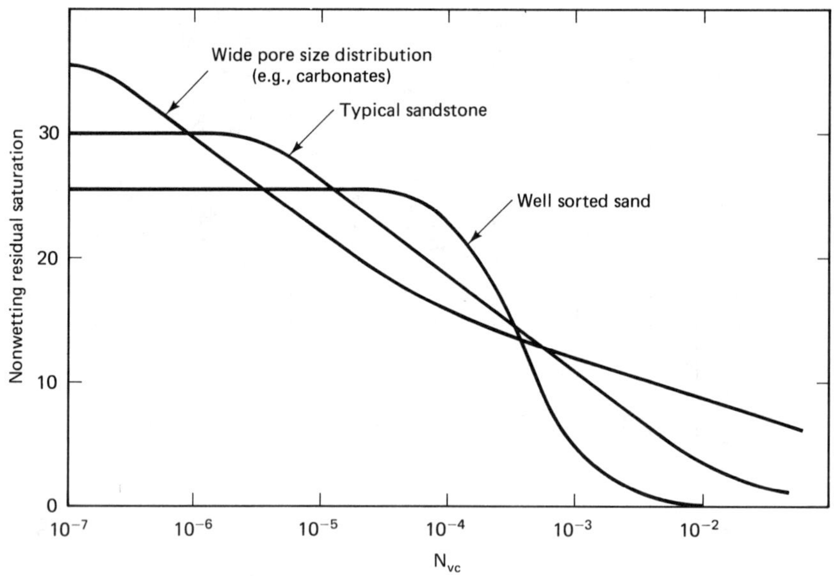

Figure 3-18 Schematic effect of pore size distribution on the CDC (from Lake, 1984)

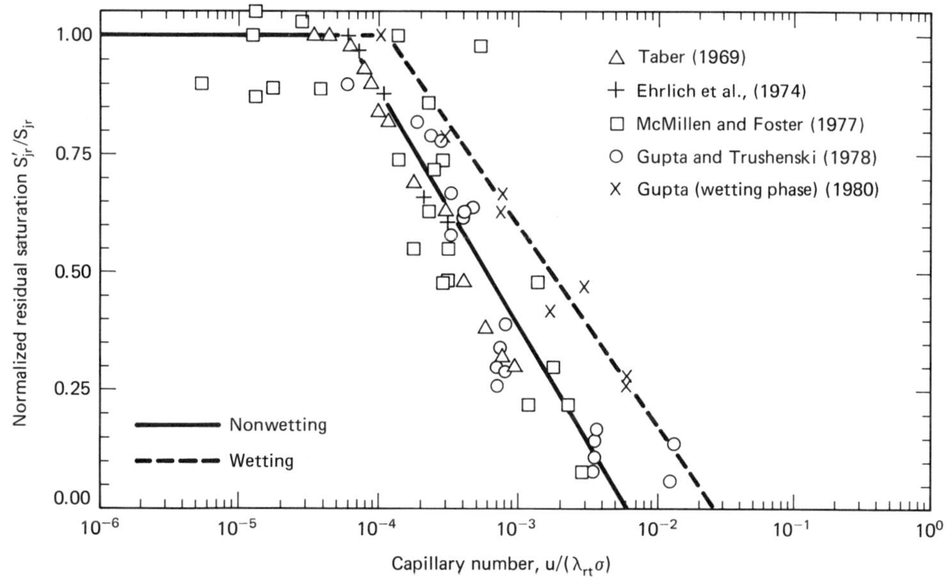

Figure 3-19 Capillary desaturation curve using a common definition (from Camilleri, 1983)

CDC Estimation

All theoretical attempts to calculate the CDC must have some means to translate microscopic physics—force balances and blob mechanics—to the REV scale. In probabilistic models, this translation is done through statistics (Larson, 1977; Mohanty and Salter, 1982). In deterministic models, the microscopic-to-REV translation follows by giving microscopic meaning to some macroscopic measurements such as capillary pressure curves or permeability (Melrose and Brandner, 1974; Oh and Slattery, 1976; Payatakes et al., 1978).

Statistical models have been successful in predicting CDC curves but, at the same time, have required calibration with experimental curves (Larson, 1977). These approaches hold best promise for nonwetting phase trapping where the disconnected nature of the trapped phase appeals to statistical treatment. The deterministic models, the method we adopt here, use conventional permeable media measurements, appear to allow greater rock specificity, and are easier to calculate.

We discuss a slightly simplified version of a deterministic theory presented by Stegemeier (1974 and 1976). This theory was, in turn, based on and associated with work by Melrose and Brandner (1974). The theory is appropriate for uniform and strongly wetted media.

Stegemeier's procedure is most appropriate for the nonwetting phase CDC. The basic idea derives from the inequality Eq. (3.4-6), which we rewrite here for horizontal flow as

$$\Delta L \, |\vec{\nabla} P_w| \geqq \Delta P_c = P_{cA} - P_{cR} \tag{3.4-7}$$

In this equation, $\Delta L \, |\vec{\nabla} P_w|$ is the projection of the pressure gradient vector in the direction of flow ($\Delta\Phi_w$ in Eq. 3.4-6), and ΔL is the length of the trapped nonwetting globule. P_{cA} and P_{cR} are the capillary pressures at the leading (advancing) and trailing (receding) ends of the globule (see Fig. 3-15 for the geometry). Using Laplace's equation, inequality Eq. (3.4-7) becomes

$$\Delta L \, |\vec{\nabla} P_w| \geqq 2\sigma\left(\frac{\cos \theta_A}{R_n} - \frac{\cos \theta_R}{R_b}\right) \tag{3.4-8}$$

Inequality Eq. (3.4-8) is a microscopic force balance that can be translated to the REV scale with the following two premises:

1. The first drainage capillary pressure function is largely determined by the pore neck radii R_n. Thus $\sigma \cos \theta$ may be eliminated between Eq. (3.2-1), with $R = R_n$, and Eq. (3.2-2) to give $R_n = 2(k/\phi)^{\frac{1}{2}}/j$. This substitution introduces the j-function into the formalism, which allows a distribution of pore necks. The pore neck being entered at a given potential depends on the nonwetting phase saturation.
2. The permeability of the medium is largely determined by the pore body radius. This premise allows R_b to be replaced with Eq. (3.1-4) with $R = R_b$. The pore body is *not* (but maybe should be) a function of saturation. These two substitutions in Eq. (3.4-8) give

$$\Delta L \, |\vec{\nabla} P_w| \geqq \sigma\left[\frac{j \cos \theta_A}{\left(\dfrac{k}{\phi}\right)^{1/2}} - \frac{\cos \theta_R}{\left(\dfrac{2\tau k}{\phi}\right)^{1/2}}\right] \tag{3.4-9}$$

It remains, then, to express ΔL on the REV scale. It is tempting to take ΔL to be equal to the pore body radius as suggested by Fig. 3-15. But experimental evidence has shown that the length of nonwetting globules is usually much longer than this (Fig. 3-13). Stegemeier (1974) derives ΔL on the basis of stability theory

$$\Delta L = \left(\frac{C\sigma}{|\vec{\nabla} P_w|}\right)^{1/2} \tag{3.4-10}$$

where C is an empirical constant. The inverse relationship between globule length and potential drop has been supported by experimental work (Arriola et al., 1980; Ng, 1978). Experimental evidence suggests a value of $C = 20$.

Eliminating ΔL between Eqs. (3.4-9) and (3.4-10) gives the final condition for mobilization of a given globule

$$N_{vc} = \frac{k \, |\vec{\nabla} P_w|}{\sigma} \geqq \frac{\phi}{C}\left[j \cos \theta_A - \frac{\cos \theta_R}{(2\tau)^{1/2}}\right]^2 \tag{3.4-11}$$

where N_{vc} here is a general version of the local capillary number, which Stegemeier calls the Dombroski-Brownell (1954) number (see Table 3-2). The CDC curve for the nonwetting phase follows from Eq. (3.4-11) if we can associate the j-function with a saturation.

The most original feature of the Stegemeier theory is the use of IR curves to estimate the capillary pressure in disconnected nonwetting phase globules. Figure 3-5 shows IR curves. Recall that the coordinates on these curves correspond to the right and left panels in Fig. 3.4(a).

Imagine a permeable medium at an irreducible wetting phase saturation as shown in condition 5 in Fig. 3-4(a). The nonwetting phase is to be displaced with a wetting phase whose potential gradient and wetting–nonwetting interfacial tension, along with the medium permeability, determine the capillary number N_{vc}. The equality in Eq. (3.4-11) will point to a precise pore radius. The nonwetting phase in pores larger than this radius will be mobilized; that in smaller pores will be trapped. Suppose the displacement mobilizes the nonwetting phase in only the largest pore. The trapped saturation will be the difference between the saturations in conditions 6 and 2. On an IR curve, this is the difference between the maximum nonwetting phase saturation and the nonwetting saturation of condition 2.

But the capillary pressure in the disconnected globule in condition 2 is the same as in condition 1, corresponding to a point on the j-function curve that determines the N_{vc} of the displacement when the equality in Eq. (3.4-11) is used. The procedure is as follows (see Fig. 3-20):

1. Pick a point on the IR curve corresponding to the maximum initial nonwetting saturation $(S_{nwI})_{\max}$. S_{nwr} at this point is the nonwetting saturation corresponding to what would be trapped if the displacement were to take place at zero capillary number, that is, spontaneous imbibition of only the wetting phase.
2. Pick another point on the IR curve at a lower S_{nwI}. The trapped nonwetting saturation is the difference between the S_{nwr} here and in step 1. The capillary pressure in the globules just mobilized corresponds to a point on the j-function curve where S_{nw} is equal to S_{nwI} on the IR curve.
3. Insert the j-value from this procedure into Eq. (3.4-11), now used as an equality, to give the N_{vc} corresponding to the residual nonwetting phase saturation. These steps generate one point on the nonwetting phase CDC curve.
4. Repeat step 2 with another S_{nwI} to generate a continuous curve.

Three other things are required in step 2. The tortuosity may be obtained from the medium's formation resistivity factor, and the constant C from matching data or from using $C = 20$ as suggested. The advancing and receding contact angles come from the correlation of Morrow (1974, 1976), reproduced in Fig. 3-8. This data requires the "intrinsic" contact angle that may come from the correspondence between the wettability measures shown in Fig. 3-9. Since the procedure is for the nonwet-

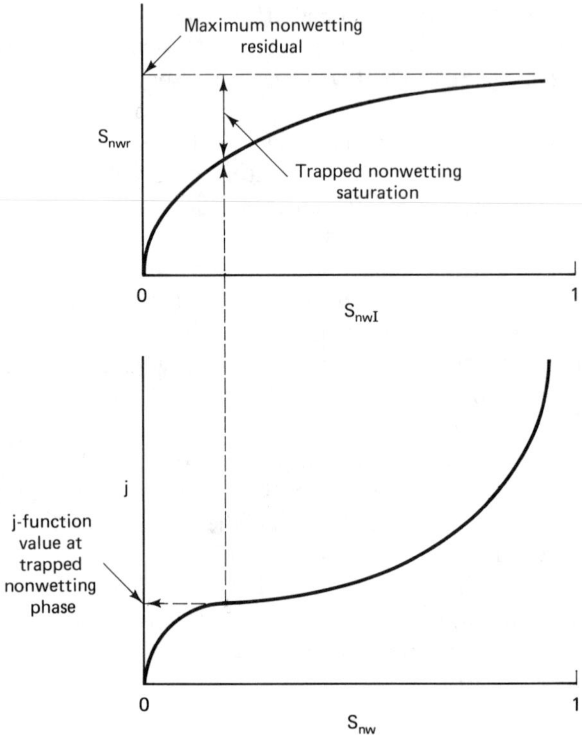

Figure 3-20 Schematic illustration of CDC construction by modified Stegemeier's method

ting phase, the intrinsic contact angle must not be greater than 90°. Figure 3-21 shows the entire nonwetting CDC curve for an Admire sand sample.

The procedure is relatively easy to use, generates reasonable results (particularly when calibrated), and reinforces the following points about the CDC curve:

1. The capillary number N_{vc} defined by Eq. (3.4-11) is the most general definition (see Table 3-2). The CDC constructed has the correct qualitative shape: The plateau in the CDC corresponds to the plateau in the IR curve.
2. Increasing N_{vc} will cause the irreducible nonwetting phase saturation to decrease. But an increase of several factors of 10 is required to bring about complete desaturation.
3. The local heterogeneity is present through the pore size distribution dependence of the *j*-function and IR curve.
4. The effect of wettability appears less important than it is in reality because the contact angles appear as differences in Eq. (3.4-11). The entire procedure is based on a specific distribution in the pores, which is determined by the wettability.
5. The original saturation of the medium has some effect through step 1 of the above procedure. This effect probably explains the dependence of the CDC on the degree of connectedness noticed by Abrams (1975).

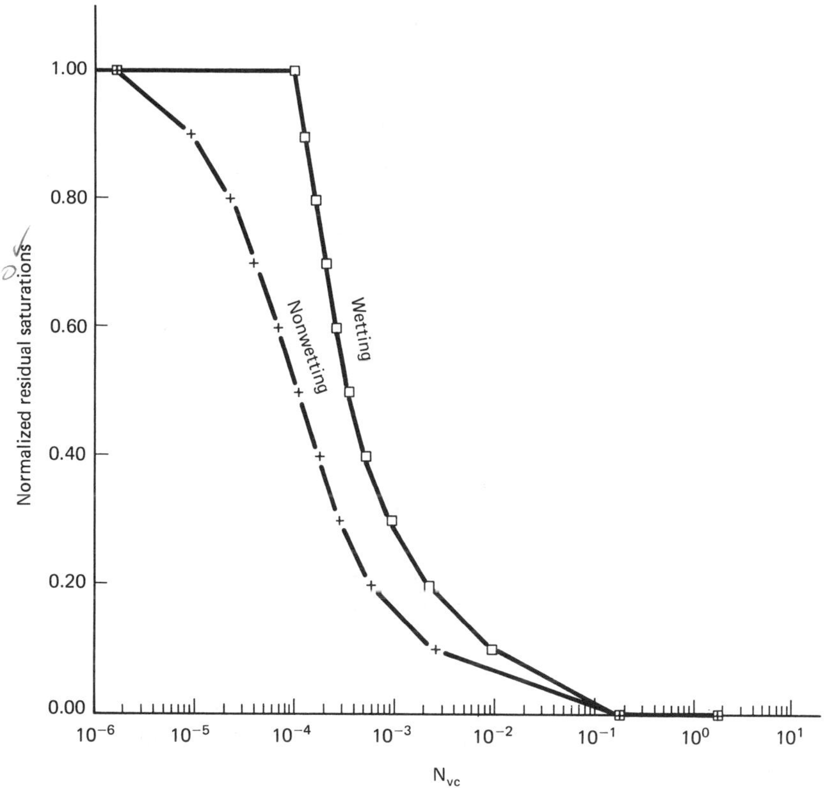

Figure 3-21 Capillary desaturation curves calculated by modified Stegemeier's method

The theory is not easily extended to wetting phase trapping. Stegemeier suggests a qualitative correction whereby the irreducible nonwetting phase saturation is multiplied by the ratio of maximum S_{nwr} to 1 minus this quantity and then plotted. This curve, truncated at the experimentally observed maximum S_{wr}, is also shown in Fig. 3-21. The wetting and nonwetting curves CDC are in qualitative agreement since it requires a higher N_{vc} for a given degree of desaturation for the wetting phase. The critical and total desaturation N_{vc}'s for the wetting phase agree poorly with those in Table 3-2.

3-5 PERMEABLE MEDIA CHEMISTRY

Several EOR processes, particularly those affected by the electrolytes in the aqueous phase, depend on the chemical makeup of the medium. We conclude this chapter by introducing a few general observations that serve as a base for subsequent chemical insights.

Species Abundance

The composition of naturally occurring permeable media is rich in the number of elements and compounds. Table 3-3 shows a comparative elemental analysis for seven sandstones and one carbonate media. The numbers in this table are a percent of the total mass reported by at least one of three methods. SEM means the value came from point counting on an image from a scanning electron micrograph. EDS is an analysis based on energy dispersed by an x-ray beam, and ICP is an inductively coupled plasma emission spectrophotometer. (For more details about this, see Crocker et al., 1983.) SEM and EDS sample the surfaces of the rock pores; ICP is a measure of the bulk chemistry. Any systematic differences between SEM/EDS and ICP, such as the persistently lower silica amounts by SEM/EDS, are reflections of surface localizations of the object mineral.

Table 3-3 indicates sandstones are about 64% to 90% silica, with the remainder distributed fairly evenly among lesser species and the clays (last three columns). Silica is important in EOR because it dissolves in aqueous solutions particularly at high temperature or at high pH. The first reaction occurs readily in thermal floods where the reaction products induce injectivity loss on precipitation. Silica minerals also have a minor anion exchange capacity at neutral or elevated pH.

The carbonate sample in Table 3-3 is only 50% to 53% calcium. This low value may be explained by a high loss on ignition in the ICP method. Calcium minerals in both sandstones and carbonates are important because they are a source of multivalent cations in solution. These cations affect the properties of polymer and micellar-polymer solutions considerably and can provide a source for pH loss in alkaline floods through hydroxide precipitation.

Clays

Clays are hydrous aluminum silicates whose molecular lattice can also contain (in decreasing prevalence) magnesium, potassium, sodium, and iron. Table 3-4 shows a summary of the most common clays. The suffix *ite* designates a clay.

Clay minerals constitute 40% of the minerals in sediments and sedimentary rocks (Weaver and Pollard, 1973). Their prevalence in commercial hydrocarbon-bearing permeable media is much less than this (see last three columns in Table 3-3), but their importance to oil recovery far exceeds their relative abundance. This importance derives from the following properties of clays: They are generally located on the pore grain surfaces, they have a large specific surface area, and they are chemically reactive. Clays affect EOR processes by influencing the medium permeability or by changing the ionic state of the resident fluids. In the following paragraphs, we give a brief exposition of clay mineral properties. (For more details, see Grim, 1968; Weaver and Pollard, 1973; or Rieke et al., 1983.)

Clays are classified by their chemical formula, crystal structure, particle size, morphology, water sensitivity, and chemical properties (see Table 3-4). The principal building block of a clay mineral is the element silicon surrounded by three oxygens in a tetrahedral structure. These tetrahedra are merged with octahedra with the

TABLE 3-3 COMPARATIVE ELEMENTAL ANALYSIS (WEIGHT PERCENT) OF ROCKS AND CLAYS (FROM CROCKER ET AL., 1983)

Rocks		SiO_2	Al_2O_3	Fe_2O_3	MgO	CaO	TiO_2	SrO	K_2O	Na_2O	Mn_2O_3	SO_3	Loss on ignition	Kaolinite	Chlorite	Illite/ Mica
Bandera sandstone	ICP	71.4	8.7	3.1	1.7	3.1	0.4	0.01	1.1	1.7			6.2			
	EDS	64.4	17.4	9.5		4.2			4.1							
	x ray	65.0	8.0	8.0										6.0	5.0	8.0
Berea sandstone	ICP	84.6	4.5	1.4	0.5	0.8	0.2	0.03	2.1	2.2	0.06		2.6			
	EDS	78.0	10.3	3.1		2.8	2.1		4.4							
	x ray	75.1	3.5	5.5	0.8	0.8			3.4					7.0	0.0	4.0
Coffeyville sandstone	ICP	81.5	7.7	3.5	0.7	0.5	1.1	0.00	1.7	3.1			2.6			
	EDS	65.1	15.4	10.5		2.7	3.8		2.4							
	x ray	70.1	10.0			6.0								4.0	4.0	6.0
Cottage Grove sandstone	ICP	84.6	4.7	1.2	0.08	0.08	0.1	0.01	0.4	2.9	0.07		1.7			
	EDS	70.4	15.9	11.2		0.6	1.1		2.8		0.4					
	x ray	75.4	5.8						5.7					6.0	1.0	6.0
Noxie sandstone	ICP	87.6	4.9	1.6	0.2	0.2	0.6	0.02	0.8	1.8	0.07		1.3			
	EDS	64.4	9.5	22.5		0.7	0.6		1.5		3.1					
	x ray	77.8	4.7						4.6					5.0	1.0	7.0
Oswego limestone	ICP	0.7	0.2	0.09	0.5	50.0		0.2	0.5	0.8		0.4	42.5			
	EDS	2.0	1.0			53.0						1.3				
	x ray	5.0				51.0								4.0	0.0	0.0
Sweetwater sandstone	ICP	88.7	4.2	0.4	0.2	0.05	0.1	0.02	0.8	1.9	0.02		1.2			
	EDS	72.4	13.3	8.1		2.9			3.3							
(tight gas sand)	x ray	90.0												0.0	2.0	8.0
Torpedo sandstone	ICP	90.5		1.9	0.2	0.2	0.5	0.8	0.8	0.2	0.2		1.6			
	EDS	72.3		11.0		0.7	1.9		2.3		2.6					
	x ray	77.0	5.1						5.0					6.0		7.0

TABLE 3-4 CLASSIFICATION OF PRINCIPAL CLAY MINERALS IN SEDIMENTS (ADAPTED FROM DEGENS, 1965)

Layers	Population of octahedral sheet	Expansion	Group	Species	Crystallochemical formula
Two sheet (1 : 1)	Dioctahedral	Nonswelling	Kaolinite	Kaolinite Dickite Nacrite	$Al_4(OH)_8[Si_4O_{10}]$
		Nonswelling and swelling	Halloysite	Halloysite Metahalloysite	$Al_4(OH)_8[Si_4O_{10}] \cdot (H_2O)_4$ $Al_4(OH)_8[Si_4O_{10}]$
	Trioctahedral	Nonswelling	7Å-chlorite (septechlorite)	Berthierine (Kaolin-Chamosite)	$(Fe^{2+},Fe^{3+},Al,Mg)_6(OH)_8[(Al,Si)_4O_{10}]$
Three sheet (2 : 1)	Dioctahedral	Swelling	Montmorillonite* (smectite)	Montmorillonite Beidellite Nontronite	$\{(Al_{2-x}Mg_xXOH)_2[Si_4O_{10}]\}^{-x}Na_x \cdot nH_2O$ $\{(Al_2(OH)_2[(Al,Si)_4O_{10}]\}^{-x}Na_x \cdot nH_2O$ $\{(Fe^{3+}_{2-x}Mg_xXOH)_2[Si_4O_{10}]\}^{-x}Na_x \cdot nH_2O$
		Nonswelling	Illite (hydromica)*	Illite-varieties	$(K_1H_3O)Al_2(H_2O,OH)_2[Al\,Si_3O_{10}]$
	Trioctahedral	Swelling	Vermiculite†	Vermiculite	$(Mg,Fe)_3(OH)_2[Al\,Si_3O_{10}]Mg \cdot (H_2O)_4$
Three sheet and one sheet (2 : 2)	Trioctahedral	Nonswelling‡	14Å-chlorite† (normal chlorite)	Chlorite-varieties	$(Al,Mg,Fe)_3(OH)_2[Al,Si)_4O_{10}]Mg_3(OH)_6$

*Also trioctahedral varieties

† Also dioctahedral varieties

‡ Swelling chlorites are rare and intermediate forms between vermiculite and chlorite

Al species in the center and OH or oxygen groups at the corners. The bond lengths of the octahedra and tetrahedra being almost identical, these geometrically regular shapes arrange themselves into planar or sheetlike structures. Hence a major classification of clays in Table 3-4 is the number of sheets in a crystal. Kaolinite is the simplest example of this structure. Obviously, such a regular structure lends itself very well to analysis by x-ray diffraction. The amount of foreign atoms (Mg, K, Fe, and Na) increases with the number of sheets.

Despite the regularity of their crystal structure, clays in permeable media usually are the smallest particles. The inference that a clay is usually any particle less than about 50 nm is imprecise: Many permeable media constituents can have particle sizes this small. But larger clay particles are not common. Because of their small size, clay particles can frequently exist as colloidal suspensions in aqueous solution.

The small particle size means clays are generally much less permeable to flow than are sandstones. Low permeability impacts on EOR in two ways. Segregated clays or *shales* are regions of very high clay content that generally aren't considered part of the reservoir. Because they lack permeability, shales are barriers to fluid flow, particularly vertical flow, and will hinder gravity segregation. Dispersed clays, on the other hand, occur distributed among the pores of permeable media. These are of more concern to EOR than segregated clays because they are more chemically reactive. Dispersed clays, though, also cause reduction of the medium permeability; in fact, the degree of clay content is a good indicator of the receptiveness of the formation to fluid injection (Fig. 3-22).

Dispersed clays have a separate morphological classification (Neasham, 1977). Figure 3-23 schematically illustrates these along with representative SEM micro-

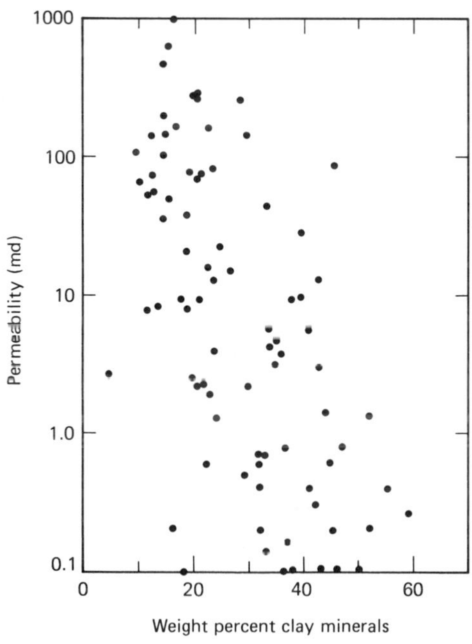

Figure 3-22 Permeability versus weight percent clay minerals (from Simlote et al., 1983)

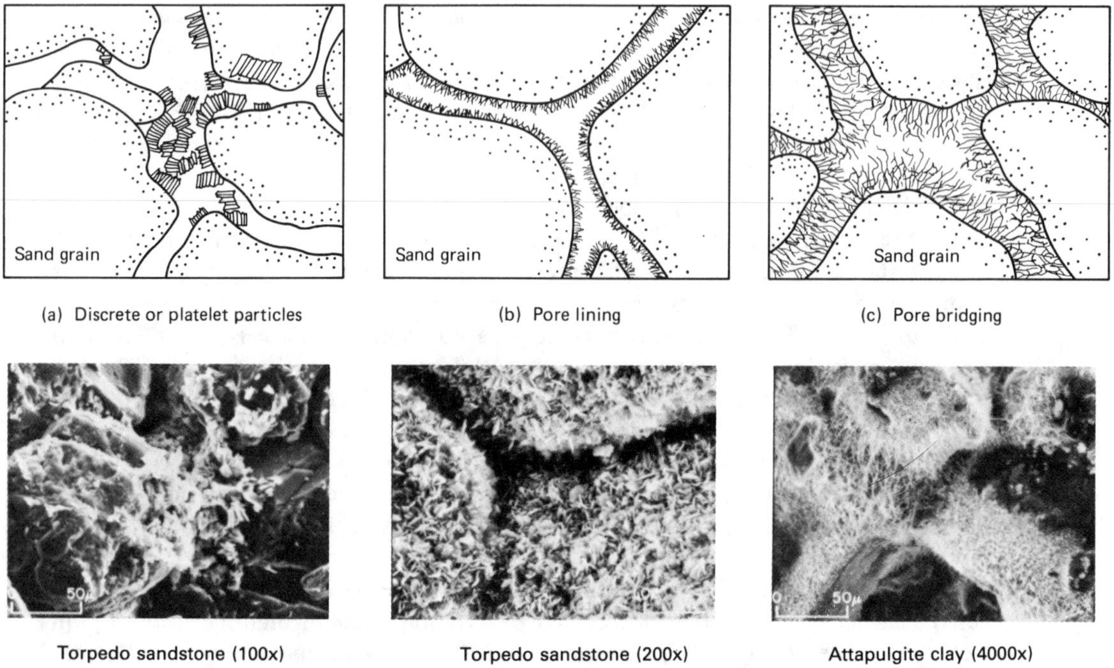

Figure 3-23 Examples of natural clays (upper panels from Nesham, 1977; lower from Crocker et al., 1982)

graphs of each type. Clays can occur as discrete platelets that are randomly arranged within the pores (Fig. 3-23a), as pore-lining clays that coat the pore walls in thin films (Fig. 3-23b), and as pore-bridging clays that exhibit filaments that extend across the pores (Fig. 3-23c). Permeable media with pore-bridging clays will have a lower permeability than what would be expected based on the sand grain size. Pore-lining clays have little effect on permeability, but clay platelets can be induced to cause permeability loss when the electrolyte balance of the fluid in the pore space is changed.

The permeability loss or clay sensitivity of media containing clay platelets is a well-documented problem in waterflooding. Many of these clays are readily swelled by very fresh water or by high concentrations of sodium cations. When swelled, they detach from the pore surface, become entrained in the flowing fluid, and then collect and bridge pore entrances farther downstream (Khilar and Fogler, 1981). The damage caused is only temporarily reversible since platelets can bridge in reverse flow. The sensitivity to fresh water is a concern in steam drives and soaks where the steam condensate is quite fresh. The sensitivity to fresh water and sodium cations affects polymer and micellar-polymer floods where efforts to remove divalent cations by preflush are common. Only swelling clays (see Table 3-4) show this effect, most prominently when the clay is in the platelet morphology.

Cation Exchange

Among the most interesting of clay characteristics is the ability of clays to exchange cations with fluids in the pore space. With diagenesis, the aluminum atoms in the simple clay structure become replaced with lower valence cations like Mg^{2+} or K^+ (see Table 3-4). This substitution imparts a deficiency of positive charge on the clay, which must be countered by cations from the fluid if the clay is to remain electrically neutral. The cation exchange capacity Q_v is a measure of the concentration of these excess negative charges. The units on Q_v are milliequivalents (meq) per unit mass of substrate. The equivalent weight of a species is the molecular weight divided by the absolute value of its charge or, basically, the amount of mass per unit charge. Q_v is expressed in this fashion to allow the concentration of any bound cation to be expressed in consistent units. The units of solution concentrations may also be expressed in milliequivalents per mass of substrate, but it is usually more convenient to work on a unit pore volume basis. The cation exchange capacity expressed per unit pore volume is Z_v, where

$$Z_v = Q_v \rho_s \left(\frac{1 - \phi}{\phi} \right) \tag{3.5-1}$$

and ρ_s is the substrate density.

The cation exchange capacity increases with degree of substitution. Since the substituted sites are in the interior of the lattice, Q_v is also a function of the morphology of the clay. Platelet clays have more edges and, therefore, more exposed sites. Following Grim (1968), typical Q_v's for montmorillonites, illites, and kaolinites are 700–1300, 200–400, 30–150 meq/kg of clay. These numbers should be compared to Q_v's for typical reservoir rocks (see Table 3-5), which are expressed in milliequivalents per kilograms of rock. Clays with large Q_v's are usually swelling clays.

The bonds between the exchange sites and the cations are chemical, but they are readily reversible. The relative ease of the replacement of one cation by another is

$$Li^+ < Na^+ < K^+ < Rb^+ < Cs^+ < Mg^{2+} < Ca^{2+} < Sr^{2+} < Ba^{2+} < H^+ \tag{3.5-2}$$

Species that have high charge densities (multivalents or small ionic radii) are more tightly held by the anionic sites. This observation suggests a possible explanation for the permeability reducing behavior of Na^+. The large Na cations disrupt the clay particles when they intrude into the structure. But only a small amount of another cation is sufficient to prevent this because most other naturally occurring cations are more tightly bound than Na^+.

Equilibrium Relations

The sequence in inequality (3.5-2) is qualitative. The actual replacement sequence depends weakly on clay type and strongly on the total composition of the fluid that is contacting the clays. More quantitative representations follow from the chemical

TABLE 3-5 PHYSICAL CHARACTERISTICS OF TYPICAL PERMEABLE MEDIA (ADAPTED FROM CROCKER ET AL., 1983)

	Porosity (fraction)	Permeability (μm^2)	Density (g/cm^3)	Surface area (m^2/g)	Clay dispersion classification	Cation exchange capacity (meq/kg · rock)
Bandera sandstone	0.174	0.012	2.18	5.50	Pore lining	11.99
Berea sandstone	0.192	0.302	2.09	0.93	Grain cementing	5.28
Coffeyville sandstone	0.228	0.062	2.09	2.85	Pore bridging	23.92
Cottage Grove sandstone	0.261	0.284	1.93	2.30	Pore bridging	17.96
Noxie sandstone	0.270	0.421	1.85	1.43	Pore lining	10.01
Oswego limestone	0.052	0.0006	2.40	0.25	Mixed in the carbonate matrix	
Sweetwater sandstone	0.052	0.0002	2.36	1.78	Discrete particles	
Torpedo sandstone	0.245	0.094	1.98	2.97	Pore bridging	29.27

equilibria. For example, the exchange reaction between a cation A of charge z_A and another B of charge z_B is given by

$$z_B A - (\text{clay})z_A + z_A B \longrightarrow z_A B - (\text{clay})z_B + z_B A \qquad (3.5\text{-}3)$$

The reaction in Eq. (3.5-3) suggests an isotherm

$$K_N = \frac{\left(\dfrac{C_{Bs}}{Z_v}\right)^{1/z_B}[C_A]^{1/z_A}}{\left(\dfrac{C_{As}}{Z_v}\right)^{1/z_A}[C_B]^{1/z_B}} \qquad (3.5\text{-}4a)$$

where K_N is the *selectivity* or *selectivity coefficient* of species A on the clay with respect to species B. In Eq. (3.5-4a), $[C_A]$ and $[C_B]$ are species concentrations in molal units, and C_{As} and C_{Bs} are in equivalents per unit pore volume (meq/cm^3 is convenient). The subscript s indicates a clay-bound species, and the equation assumes ideal behavior. For calculations, it is sometimes better to write Eq. (3.5-4a) as

$$K_{BA} = \left(\frac{Z_v}{\rho_1}\right)^{(z_A - z_B)} (K_N)^{z_A z_B} \left(\frac{z_A{}^{z_B}}{z_B{}^{z_A}}\right) = \frac{C_{Bs}^{z_A} C_A^{z_B}}{C_{As}^{z_B} C_B^{z_A}} \qquad (3.5\text{-}4b)$$

where K_{BA} is another form of the selectivity coefficient, and C_A and C_B are now in units of equivalents per unit pore volume.

In general, K_N varies with exchanging pair and the identity of the clay. For cations of interest in EOR, and clays commonly encountered in permeable media, this dependency is not great. Table 3-6 shows typical selectivities for Na^+ exchange.

TABLE 3-6 TYPICAL SELECTIVITIES (ADAPTED FROM BRUGGENWERT AND KAMPHORST, 1979)

Equation A − Clay + B ⟶	Material	Selectivity coefficient (K_N)
Na − mont. + H^+	Various	0.37–2.5
Na − mont. + NH_4^+	C. berteau maroc	4.5–6.3
Na − mont. + K^+	Bentonite	2.7–6.2
Na − kaol. + K^+	Georgia kaolinite	2.7–7.8
2Na − mont. + Ca^{+2}	Clay spur	1.9–3.5
2Na − clay + Ca^{+2}	Berea	0.3–10.5
2Na − clay + Mg^{+2}	Berea	0.2–10.0
3Na − mont. + Al^{+3}	Wyoming bentonite	2.7

Note: mont. = montmorillonite
kaol. = kaolinite

The selectivity for any other pair on this table may be obtained by eliminating the Na^+ between the two isotherms.

The cation preference expressed in inequality (3.5-2) is determined by the valences appearing in the isotherm. If we let A = monovalent and B = divalent be the only exchanging cations, electroneutrality of the clay requires

$$C_{8s} + C_{6s} = Z_v \tag{3.5-5a}$$

or

$$C_{8D} + C_{6D} = 1 \tag{3.5-5b}$$

where $C_{iD} = C_{is}/Z_v$. Further, Eq. (3.5-4b) becomes

$$K_{68} = \frac{C_{6s} r}{C_{8s}^2} \tag{3.5-6}$$

where $r = C_8^2/C_6$. The isotherm (Eq. 3.5-6) is now entirely in units of equivalents per unit volume. Using Eq. (3.5-5b) to eliminate C_{8D}, and solving for the positive root, gives

$$C_{6D} = 1 + \frac{r}{2K_{68}Z_v}\left[1 - \left(\frac{4K_{68}Z_v}{r} + 1\right)^{1/2}\right] \tag{3.5-7}$$

This equation is plotted in Fig. 3-24 (6 is Ca^{2+}), with the anion concentration C_5 as a parameter. C_5 appears in the equation because the solution must be electrically neutral

$$C_8 + C_6 = C_5 \tag{3.5-8}$$

The preference of the clay for the divalent is apparent since all curves are above the 45° line. This preference increases as the water salinity (anion concentration) decreases. We use isotherm representations like Eq. (3.5-7) to make salinity calculations when we treat chemical flooding methods in later chapters.

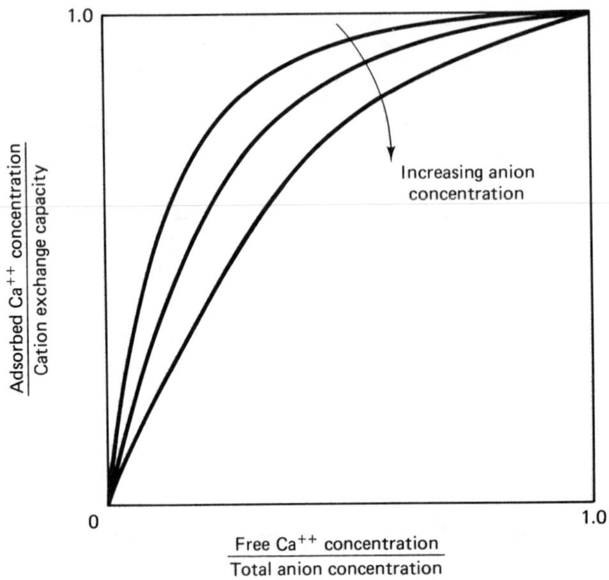

Figure 3-24 Typical isotherm for sodium–calcium exchange

Dissolution and Precipitation

Other rock–fluid interactions that affect EOR are intraaqueous reactions and dissolution–precipitation reactions. An example of the former is the combination of an anion A and cation B to form aqueous species AB.

$$A^- + B^+ \longrightarrow AB_{(aq)} \tag{3.5-9}$$

The chemical equilibrium for this reaction is

$$K_r = \frac{[C_{AB(aq)}]}{[C_A] \cdot [C_B]} \tag{3.5-10}$$

where K_r is the equilibrium constant for this ideal reaction at the temperature and pressure of interest. If the concentration of A or B exceeds a certain value, $AB_{(aq)}$ can precipitate to form $AB_{(s)}$, a solid.

$$A^- + B^+ \longrightarrow AB_{(s)} \tag{3.5-11}$$

The reverse of this is the dissolution of $AB_{(s)}$. The equilibrium relation for Eq. (3.5-11) is

$$K_r^{sp} \geqq [C_A] \cdot [C_B] \tag{3.5-12}$$

where K_r^{sp} is the solubility product of the reaction. Generally, the C's in Eqs. (3.5-10) and (3.5-12) should be the activity of the designated species. For ideal solutions—the only kind we deal with here—the activities are equal to the molal concentrations.

Equation (3.5-12) has a most interesting contrast with Eq. (3.5-10). The product concentration for the precipitation–dissolution reaction does not appear in the equilibrium expression. The state of the system must be obtained from a material balance of the individual elements, rather than the species, to be consistent with the phase rule (see Chap. 4) when each solid precipitate is considered a separate phase.

Table 3-7 gives log K_r and log K_r^{sp} for the more important reactions in permeable media. The standard enthalpies of formation in this table are to be used to approximately correct the $T_1 = 298$ K equilibrium constants and solubility products to another temperature according to

$$\log K_r \Big|_{T_2} = \frac{\Delta H_r^0}{2.303 R_g}\left\{\frac{1}{T_1} - \frac{1}{T_2}\right\} + \log K_r \Big|_{T_1} \tag{3.5-13}$$

where T_2 is the temperature of interest. The K_r's in Eq. (3.5-13) may be either equilibrium constants or solubility products.

TABLE 3-7 SELECTED SOLUBILITY DATA AT 298 K FOR AQUEOUS AND SOLID SPECIES IN NATURALLY OCCURRING PERMEABLE MEDIA (ADAPTED FROM DRIA ET AL., 1988)

Aqueous species or complexes		Log K_r	ΔH° J/kg-mole
OH^-	$= H_2O - H^+$	14.00	−133.5
$CaOH^+$	$= Ca^{2+} + H_2O - H^+$	12.70	−173.2
$Ca(OH)_2$	$= Ca^{2+} + 2H_2O - 2H^+$	27.92	−267.2
$CaCO_3$	$= Ca^{2+} + CO_3^{2-}$	−3.23	44.1
$CaHCO_3^+$	$= Ca^{2+} + CO^{2-} + H^+$	−11.23	45.0
$Ca(HCO_3)_2$	$= Ca^{+2} + 2CO_3^{2-} + 2H^+$	−20.73	66.8
HCO_3^-	$= CO_3^{2-} + H^+$	−8.84	35.5
CO_2(dissolved)	$= CO_3^{2-} - H_2O + 2H^+$	−16.68	53.8
$FeOH^+$	$= Fe^{2+} + H_2O - H^+$	6.79	−120.1
$Fe(OH)_2$	$= Fe^{2+} + 2H_2O - 2H^+$	17.60	−240.2
$FeOOH^-$	$= Fe^{2+} + 2H_2O - 3H^+$	30.52	−416.3
$Fe(OH)_3^-$	$= Fe^{2+} + 3H_2O - 3H^+$	23.03	−314.1
$H_3SiO_4^-$	$= H_4SiO_4 + OH^- - H_2O$	−4.0	
$H_3SiO_4^{2-}$	$= H_3SiO_4^- + OH^- - H_2O$	−5.0	
Solids and bases		**Log K_r^{sp}**	**ΔH_r° J/kg-mole**
$Ca(OH)_2$	$= Ca^{+2} + 2H_2O - 2H^+$	22.61	194.0
$CaCO_3$	$= Ca^{+2} + CO_3^{2-}$	−8.80	−28.0
$Fe(OH)_2$	$= Fe^{+2} + 2H_2O - 2H^+$	12.10	−219.3
$FeCO_3$	$= Fe^{+2} + CO_3^{2-}$	−10.90	−66.4
CO_2(gas)	$= CO_2^{2-} - H_2O + 2H^+$	−17.67	5.3
SiO_2(quartz)	$= H_4SiO_4 - 2H_2O$	−3.98	14.0

Note: Elements = H^+, Na^+, Ca^{2+}, Fe^{2+}, CO_3^{2-}, $C1^-$, Si^{4-}

3-6 SUMMARY

The scope of this chapter—from permeability to mineral chemistry—partly justifies the brevity of treatment. Indeed, entire books have been written on petrophysical properties (Dullien, 1979) and on aquatic chemistry (Garrels and Christ, 1965). The coverage is by no means uniform: each item was selected because it recurs in at least one later chapter. We do not cover the basics of these topics there because the EOR application is usually more advanced and many of the phenomena are important for more than one process. Most important, the introduction here and in the next chapter emphasizes that the chemistry and physics of flow in permeable media are a common base for *all* EOR processes so that proficiency in one area of a particular process inevitably supplies insights into other processes.

EXERCISES

3A. *Carmen-Kozeny Equation for Spheroids*

(a) Rederive the Carmen-Kozeny equation for a permeable medium made of oblate spheroids (ellipses rotated about their minor axes). For these shapes, the surface area A of the spheroid is

$$A = 2\pi a^2 + \pi \frac{b^2}{\epsilon} \ln\left(\frac{1+\epsilon}{1-\epsilon}\right) \tag{3A-1}$$

and the volume V is

$$V = \frac{4}{3}\pi a^2 b \tag{3A-2}$$

In these equations, a and b are the distances from the particle center to the major and minor vertices, respectively ($a > b$), and ϵ is the eccentricity ($\epsilon \le 1$) defined as

$$\epsilon = \frac{(a^2 - b^2)^{1/2}}{a} \tag{3A-3}$$

(b) Show that the permeability of the medium is a weak function of particle shape for $\epsilon < 0.5$ by plotting $k/(k)_{\epsilon=0}$ versus ϵ ($\epsilon = 0$ is a sphere). A valid comparison is possible only if the spheres and oblate spheroids have equal volume.

3B. *Capillary Pressure in a Tapered Channel.* Consider a slit of unit width that does not contain a permeable medium. The height R of the slit varies with position x between 0 and x_1 according to

$$R = R_0 + (R_1 - R_0)\left(\frac{x}{x_1}\right)^m \tag{3B-1}$$

where m is a positive constant. The channel is originally filled with a wetting fluid, and a nonwetting fluid is introduced at x_1 ($x_1 > 0$) while the wetting fluid remains in contact at $x = 0$. A dimensionless capillary pressure function P_{cD} is

$$P_{cD} = \frac{R_1 P_c}{\sigma \cos\theta} \tag{3B-2}$$

(a) Derive a set of equations that relate P_{cD} to the wetting phase saturation S_w. These equations should show that P_{cD} is a function of S_w, wettability (through θ), pore size distribution (through m), pore body to pore neck radius (R_1/R_0), and pore neck radius to particle diameter (R_0/x_1).

(b) Take the following as base case values: $\theta = 30°$, $m = 3/4$, $R_1/R_0 = 10$, and $R_0/x_1 = 0.1$. Illustrate the sensitivity to the items in part (a) by plotting five curves of P_{cD} versus S_w. Curve 1 will be for the base case values, and curves 2–5 are for each of the above parameters doubled over their base case values.

(c) From the results of part (b), state which quantities have the most effect on the capillary pressure function.

3C. *Calculating a Capillary Transition Zone*. Using the capillary pressure data in Fig. 3C,

(a) Calculate and plot water saturation versus depth profiles in a water-wet reservoir knowing that the water–oil contact ($S_1 = 1$) is 152 m deep. The water and oil densities are 0.9 and 0.7 g/cm³, respectively.

(b) Construct an IR oil saturation plot for the capillary pressure data given.

(c) Using the IR curve of part (b), plot on the graph of part (a) the residual oil saturation profile versus depth.

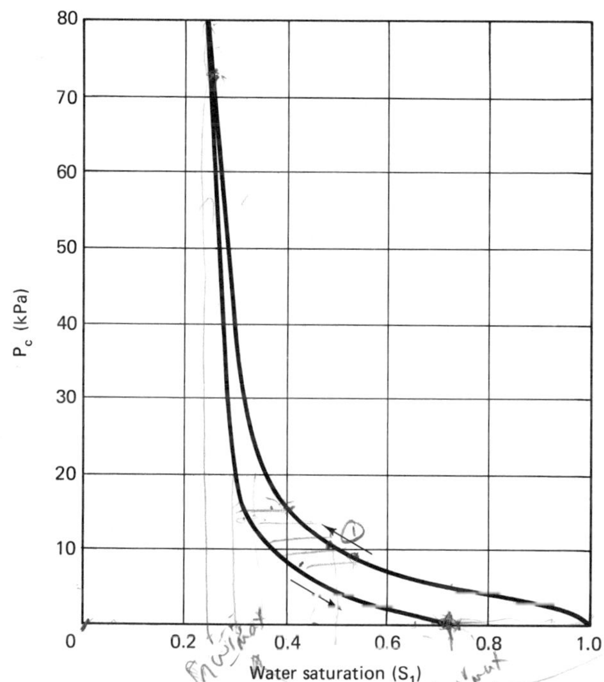

Figure 3C Water–oil capillary pressure curves

(d) If the net pay interval of the reservoir is 31 m, estimate the maximum waterflood recovery for these conditions. Compare this to the case with both the residual and initial oil saturations constant and equal to their values at the formation top.

3D. *Discontinuities in Shear Stress at Interface*. Nonequilibrium mass transfer of surfactants from bulk phases to interfaces can cause interfacial tension gradients, which in turn, can cause discontinuities in the shear stress at the interface.

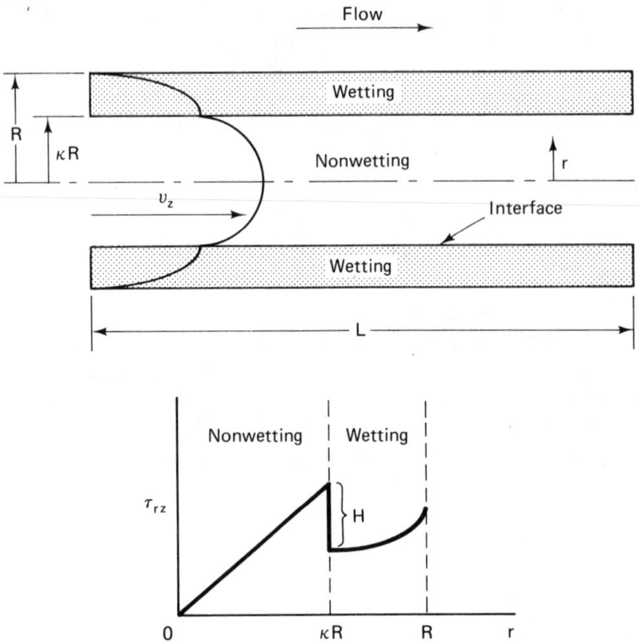

Figure 3D Simultaneous two-phase laminar flow in a tube with shear stress discontinuity at the interface

Consider the steady-state, simultaneous laminar flow of two equal viscosity, immiscible fluids in a tube, as in Fig. 3D, where $0 \leqq \kappa \leqq 1$. The wetting phase is adjacent to the tube wall, and the nonwetting phase flows in the tube core. At the interface between the two phases, there is a discontinuity H in the shear stress ($H > 0$).

(a) By making a force balance on cylindrical fluid elements, derive an expression for the shear stress τ_{rz}, including the discontinuity.

(b) If both phases are Newtonian fluids for which,

$$\tau_{rz} = -\mu \frac{dv_z}{dr} \tag{3D-1}$$

derive the local velocities and volumetric flow rates for each phase in terms of the phase viscosities, the overall pressure drop, and the tube length.

(c) Using the results of part (b) in analogy to Darcy's law, derive expressions for the wetting and nonwetting phase relative permeability as functions of phase saturations. Express the relations in terms of a capillary number of the form

$$N_{vc} = \frac{R\Delta P}{HL} \tag{3D-2}$$

Plot the relative permeability curves for each phase versus the wetting phase saturation with capillary number as a parameter.

(d) Set the relative permeability for the wetting phase equal to zero to derive an expression between the residual wetting phase and the capillary number. Plot your results in the form of a capillary desaturation curve (S_{1r} versus N_{vc}).

(e) Based on the discussion in the text, state whether you think the CDC in part (d) is qualitatively reasonable for the wetting phase. List the things in the above model that are physically unrealistic.

3E. *Equilibrium for a Trapped Globule*. Derive the equality in Eq. (3.4-6) for a static nonwetting globule under the following conditions:

(a) The globule is trapped in a horizontal channel, as in Fig. 3E(a).

(b) The globule is trapped in a tilted channel, as in Fig. 3E(b).

Take the wetting and nonwetting phases to be incompressible. The wetting phase is flowing past the trapped globule.

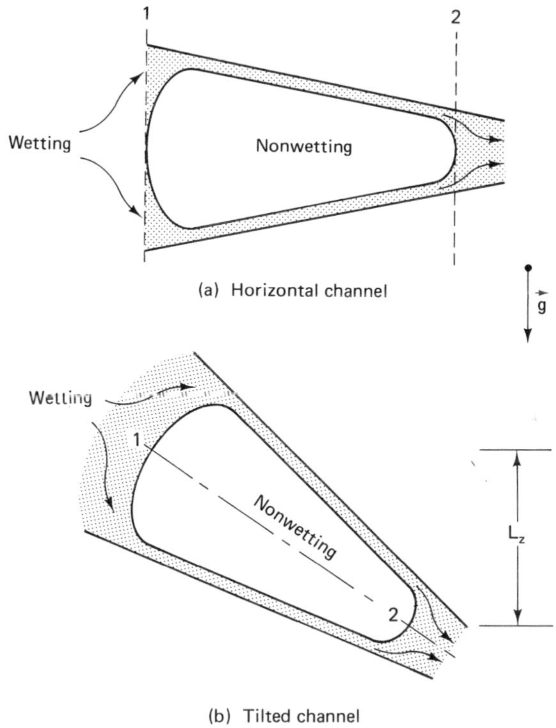

Figure 3E Schematic of trapped nonwetting phase

3F. *Capillary Desaturation in a Pore Doublet*

(a) Calculate and plot the capillary desaturation curve for a nonwetting phase based on the pore doublet model. Take the heterogeneity factor β to be 5, and the viscosities of both phases to be equal.

(b) Repeat part (a) with $\mu_{nw} = (1/2)\mu_w$.

(c) What do you conclude about the effects of viscosity on the CDC?

3G. *Calculating a Capillary Desaturation Curve*. Use the modified Stegemeier procedure (see Sec. 3-4) to calculate and plot the nonwetting CDC for the capillary pressure curves of Fig. 3C. Take k = $0.1\mu m^2$, $\phi = 0.2$, $\sigma = 30$ mN/m, the receding contact angle to be 20° (measured through the water phase), and the tortuosity to be 5. Use the IR curve of Exercise 3C.

3H. *Capillary Desaturation Curve with Gravity*

(a) Repeat the derivation given in Sec. 3-4 to develop a theoretical CDC which includes the effects of gravity. Your derivation should now contain a dimensionless ratio of gravity to capillary forces called the bond number N_b (Morrow and Chatzis, 1981). Take the characteristic length in N_b to be given by Eq. (3.4-10).

(b) Repeat Exercise 3G with the following additional data: $\Delta\rho = 0.2$ g/cm^3, and $\alpha = 45°$.

3I. *Cation Exchange Parameters*

(a) The cation exchange capacity is frequently reported in different units. If Q_v for a permeable medium is 100 meq/100 g of clay, calculate it in units of meq/100 g of medium and meq/cm^3 of pore space. The latter definition is the one given by Eq. (3.5-1). The weight percent clays in the medium is 15%, and the porosity is 22%. The density of the solid is 2.6 g/cm^3.

(b) Use the data in Table 3-6 to estimate the selectivity coefficient for calcium—magnesium exchange in a Berea medium.

3J. *Alternate Isotherms for Cation Exchange*. Equation (3.5-4) is but one of several different isotherms representing cation exchange. Another useful isotherm is the Gapon equation (Hill and Lake, 1978)

$$K_G = \frac{C_{Bs} C_A}{C_{As}\sqrt{C_B}} \tag{3J-1}$$

where K_G is the selectivity coefficient for this isotherm. We use this equation to represent sodium–calcium ($A = \mathrm{Na}^+$, $B = \mathrm{Ca}^{++}$) exchange.

(a) Invert this equation for the calcium concentration bond to the clays in the manner of Eq. (3.5-7). Take $r = C_A^2/C_B$.

(b) Show that this equation approaches consistent limits as r approaches zero and infinity.

(c) Plot the isotherm for two different solution anion concentrations in the manner of Fig. 3-24. Take $K_G = 10$.

4

Phase Behavior and Fluid Properties

The phase behavior of crude oil, water, and enhanced oil recovery fluids is a common basis of understanding the displacement mechanisms of EOR processes. Such behavior includes the two- and three-phase behavior of surfactant–brine–oil systems, the two or more phases formed in crude-oil–miscible-solvent systems, and the steam–oil–brine phases of thermal flooding. This chapter is not an exhaustive exposition of phase behavior. We concentrate on the aspects of phase behavior most pertinent to EOR. (For more complete treatments of phase behavior, see Francis, 1963; Sage and Lacey, 1939; and Standing, 1977.)

4-1 PHASE BEHAVIOR OF PURE COMPONENTS

In this section, we discuss the phase behavior of pure components in terms of pressure–temperature (P-T) and pressure–molar-volume diagrams.

Definitions

A *system* is a specified amount of material to be studied. In other chapters, the word *system* refers to the permeable medium, including the fluid with the pore space. In this chapter, the word refers only to the fluids. With this definition, a system can be described by one or more *properties*, any of several attributes of the system that can be measured. This definition implies a quantitative nature to physical properties—that is, they can be assigned a numerical value.

Properties are of two types: *extensive properties*, those dependent on the amount of mass in the system (the mass itself, volume, enthalpy, internal energy, and so on) and *intensive* properties, those independent of the amount of mass (temperature, pressure, density, specific volume, specific enthalpy, phase composition, and so on). Many times we designate intensive quantity by the modifier *specific* (quality per unit mass) or by molar (quantity per unit mole). Thermodynamic laws and physical properties are usually expressed in terms of intensive properties. The most important intensive properties in this chapter are

ρ = the density, mass per volume (or g/cm^3 in practical SI)
$\bar{V}$ = the specific volume, volume per mass (or the reciprocal of ρ)
$\bar{V}_M$ = the specific molar volume, volume per amount (or m^3/kg-mole in SI)
ρ_M = the molar density, moles per volume (or the reciprocal of $\bar{V}_M$)

Often the standard density of a fluid is given as the specific gravity, where

$$\gamma = \begin{cases} \dfrac{\rho}{\rho_{\text{water}}} & \text{for liquids} \\ \dfrac{\rho}{\rho_{\text{air}}} & \text{for gases} \end{cases} \tag{4.1-1}$$

All densities in Eq. (4.1-1) are evaluated at standard conditions of 273 K and approximately 0.1 MPa. The petroleum literature uses other standards (60°F and 14.7 psia).

In all discussions of phase behavior, it is important to understand the difference between a component and a phase. A *phase* is a homogeneous region of matter. Homogeneous means it is possible to move from any point in the region to any other without detecting a discontinuous change in a property. Such a change occurs when the point crosses an *interface*, and thereby the system consists of more than one phase. The three basic types of phases are gas, liquid, and solid, but of the last two, there can be more than one type.

A *component* is any identifiable chemical entity. This definition is broad enough to distinguish among all types of chemical isomers or even among chemical species that are different only by the substitution of a radioactively tagged element. Examples are H_2O, CH_4, C_4H_{10}, Na^+, Ca^{2+}, and CO_3^{2-}. Natural systems contain many components, and we are commonly forced to combine several components into *pseudocomponents* to facilitate phase behavior representation and subsequent calculations.

The relationship among the number of components N_C, number of phases N_P, number of chemical reactions N_R, and "degrees of freedom" N_F of the system is given by the Gibbs phase rule:

$$N_F = N_C - N_P + 2 - N_R \tag{4.1-2}$$

In many textbooks, the number of components is defined to be $N_C - N_R$, the total number of chemical species minus the chemical reactions involving them, or the number of independent species. The resulting N_F is the same. The +2 in Eq. (4.1-2) accounts for the intensive properties temperature and pressure. If one or both of these properties is specified, the equation must be reduced accordingly.

The meaning of N_C and N_P in Eq. (4.1-2) is clear enough, but N_F invariably requires some amplification. The degrees of freedom in the phase rule is the number of independent *intensive* thermodynamic variables that must be fixed to specify the thermodynamic state of all properties of the system. Intensive thermodynamic variables include phase compositions (ω_{ij} in Chap. 2) as opposed to overall compositions or volume fractions (W_i and S_j in Chap. 2), which are not thermodynamic properties. The phase rule does not itself specify the values of the N_F variables, nor does it identify the variables; it merely gives the number required.

Intuitively, we expect a relationship among the three intensive properties, temperature, pressure, and molar volume, for a pure component. Using density, molar density, or specific volume in place of the molar volume would mean no loss of generality. However, in two dimensions it is difficult to completely represent this relationship, but we can easily plot any two of these variables.

Pressure–Temperature Diagrams

Figure 4-1 shows a schematic P-T plot for a pure component. The lines on the diagram represent temperatures and pressures where phase transitions occur. These lines or phase boundaries separate the diagram into regions in which the system is

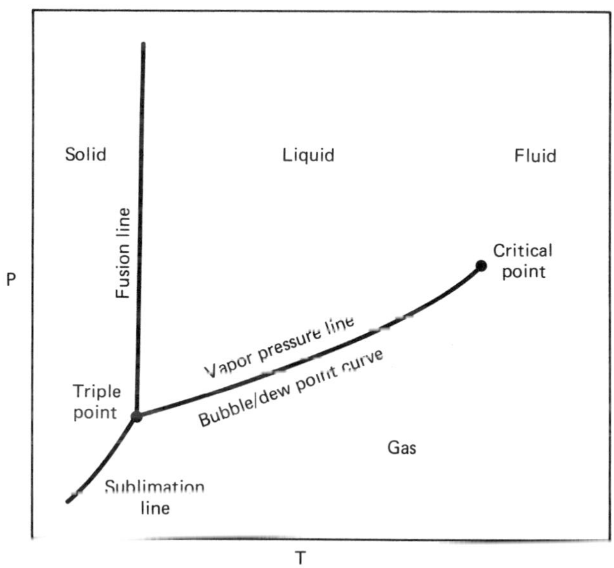

Figure 4-1 Pure component PT diagram (constant composition)

single phase. Specifically, the phase boundary separating the solid and liquid phases is the *fusion*, or melting, curve, that between the solid and gas phases is the *sublimation* curve, and that between the liquid and vapor phases is the *vapor pressure* curve. Based on our definition of a phase, a discontinuous change in system properties will occur when any phase boundary is crossed.

The phase transitions we refer to in this chapter are those of fluids in thermodynamic equilibrium. Thus it is possible for a fluid in a particular phase to momentarily exist at a P-T coordinate corresponding to another phase. But this condition is not permanent since the material would eventually convert to the appropriate stable equilibrium state.

From the phase rule we know that when two phases coexist, N_F equals 1. This can happen for a pure component only on the phase boundaries since a curve has one degree of freedom. By the same argument, three phases can coexist at only a single point in P-T space since $N_F = 0$ for this condition. This single point is a *triple point*, shown in Fig. 4-1 as the point where the three phase boundaries intersect. Other triple points such as three solids may also exist. For pure components, the phase rule says no more than three phases can form at any temperature and pressure.

Each phase boundary terminates at a *critical point*. The most interesting of these is the critical point at the termination of the vapor pressure curve. The coordinates of the critical point on a P-T plot are the critical temperature T_c and critical pressure P_c, respectively. The formal definition of P_c is the pressure above which a liquid cannot be vaporized into a gas regardless of the temperature. The definition of T_c is the temperature above which a gas cannot be condensed into a liquid regardless of the pressure. At the critical point, gas and liquid properties are identical. Obviously, the region above the critical point represents a transition from a liquid to gas state without a discontinuous change in properties. Since this region is neither clearly a liquid nor a gas, it is sometimes called the supercritical *fluid* region. The exact definition of the fluid region is arbitrary: Most texts take it to be the region to the right of the critical temperature ($T > T_c$) though it would seem that defining it to be the region to the right *and* above the critical point ($T > T_c$ and $P > P_c$) would be more consistent with the behavior of mixtures.

The behavior shown in Fig. 4-1 for a pure component is qualitatively correct though less detailed than what can be observed. There can exist, in fact, more than one triple point where solid–solid–liquid equilibria are observed. Water is a familiar example of a pure component that has this behavior. Remarkably, observations of multiple gas phases for pure components have also been reported (Schneider, 1970). Such nuances are not the concern of this text, which emphasizes gas–liquid and liquid–liquid equilibria. In fact, in all further discussions of phase behavior, we ignore triple points and solid-phase equilibria. Even with these things omitted, the P-T diagrams in this chapter are only qualitatively correct since the critical point and the vapor pressure curve vary greatly among components. Figure 7-2 shows some quantitative comparisons.

Critical phenomena do play an important role in the properties of EOR fluids. If a laboratory pressure cell contains a pure component on its vapor pressure curve

(Fig. 4-1), the pure component exists in two phases (gas and liquid) at this point, and the cell pressure is P_v. Thus the cell will contain two regions of distinctly different properties. One of these properties being the density, one phase will segregate to the top of the cell, and the other to the bottom. The phases will most likely have different light transmittance properties so that one phase, usually the upper or light phase, will be clear, whereas the other phase, the lower or heavy phase, will be translucent or dark.

We can simultaneously adjust the heat transferred to the cell so that the relative volumes of each phase remain constant, and both the temperature and pressure of the cell increase if the fluid remains on the vapor pressure curve. For most of the travel from the original point to the critical point, no change occurs in the condition of the material in the cell. But the properties of the individual phases are approaching each other. In some region near the critical point, the light phase would become darker, and the heavy phase lighter. Very near the critical point, the interface between the phases, which was sharp at the original temperature and pressure, will become blurred and may even appear to take on a finite thickness. At the critical point, these trends will continue until there is no longer a distinction between phases—that is, two phases have ceased to exist. If we continue on an extension of the vapor pressure curve, there would be a single-fluid phase and gradual changes in properties.

Pressure–Molar-Volume Diagram

A way of representing how the discontinuity in intensive properties between phases vanishes at the critical point is the pressure–molar-volume diagram. Figure 4-2 compares such a diagram with the corresponding P-T diagram. Both schematic plots show *isotherms*, changes in pressure from a high pressure P_1 to a lower pressure P_2, at four constant temperatures, T_1 through T_4.

At conditions $(P, T)_1$, the pure component is a single-phase liquid. As pressure decreases at constant temperature, the molar volume increases but only slightly since liquids are relatively incompressible. At $P = P_v\ (T_1)$, the molar volume increases discontinuously from some small value to a much larger value as the material changes from a single-liquid to a single-gas phase. Since the change takes place at constant temperature and pressure, this vaporization appears as a horizontal line in Fig. 4-2(b). Subsequent pressure lowering again causes the molar volume to increase, now at a much faster rate since the compressibility of the gas phase is much greater than that of the liquid phase. The endpoints of the horizontal segment of the pressure–molar-volume plot represent two coexisting phases in equilibrium with each other at the same temperature and pressure. The liquid and vapor phases are said to be *saturated* at $P = P_v\ (T_1)$.

At a higher temperature T_2, the behavior is qualitatively the same. The isotherm starts at a slightly higher molar volume, the vaporization at $P = P_v(T_2)$ is at a higher pressure, and the discontinuous change from saturated liquid to saturated vapor $\bar{V}_M$ is not as large as at T_1. Clearly, these trends continue as the isotherm tem-

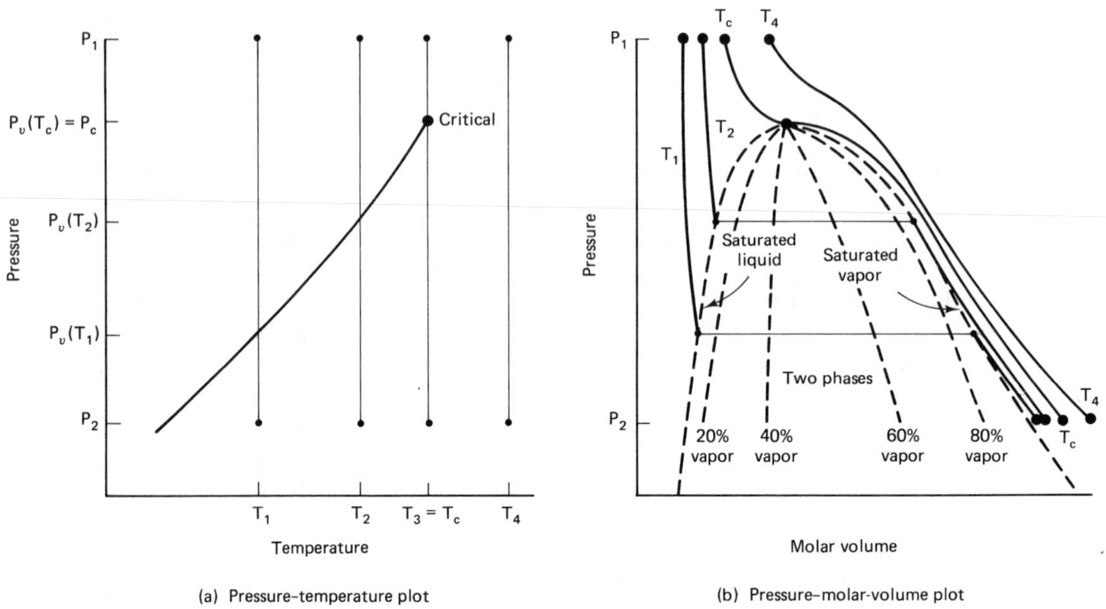

Figure 4-2 Schematic pressure–temperature and pressure–molar-volume diagrams

perature approaches $T_3 = T_c$. All isotherms on the pressure–molar-volume plot are continuously nonincreasing functions with discontinuous first derivatives at the vapor pressure line.

At the critical temperature, the two phases become identical, and the saturated liquid and gas molar volumes coincide. Since this temperature is only infinitesimally higher than one at which there would still be distinguishable liquid and gas phases, the isotherm at $T = T_c$, the critical isotherm, continuously decreases with continuous first derivatives. At the critical point $P = P_c$, the critical isotherm must have zero slope and zero curvature, or

$$\left(\frac{\partial P}{\partial \bar{V}_M}\right)_{T_c,P_c} = \left(\frac{\partial^2 P}{\partial \bar{V}_M^2}\right)_{T_c,P_c} = 0 \qquad (4.1\text{-}3)$$

These *critical constraints* follow from the physical argument given above and can also be derived by requiring a minimum in the Gibbs free energy at the critical point (Denbigh, 1968).

At isotherm temperatures above the critical temperature, $T = T_4$ in Fig. 4-2, the isotherm is monotonically decreasing with continuous first derivatives but without points of zero slope or curvature.

The endpoints of all the horizontal line segments below the critical points in the pressure–molar-volume plot define a two-phase envelope as in Fig. 4-2(b). Though rarely done, it is also possible to show lines of constant relative amounts of liquid and gas within the two-phase envelope. These *quality* lines (dotted lines in Fig. 4-2b) must converge to the critical point. The two-phase envelope on a pressure–molar-volume plot for a pure component, which projects onto a line in a P-T diagram, is not the same as the two-phase envelope on a P-T diagram for mixtures. Both Figs. 4-2(a) and 4-2(b) are merely individual planar representations of the three-dimensional relation among temperature, pressure, and molar volume. Figure 4-3 illustrates the three-dimensional character of this relation for water.

Finally, though we illustrate the phase envelope of a pure component on a pressure–molar-volume diagram, discontinuities in properties below the critical point are present in all other intensive properties except temperature and pressure (see Fig. 11-3, the pressure–enthalpy diagram for water).

4-2 PHASE BEHAVIOR OF MIXTURES

Because the purpose of EOR is to recover crude oil, an unrefined product, we need not deal with the phase behavior of pure components except as an aid to understanding mixtures. Since the phase behavior of hydrocarbon mixtures is so complex, in this and the next section, we simply compare the phase behavior of mixtures to pure components and introduce pressure–composition (P-z) and ternary diagrams.

Pressure–Temperature Diagrams

For a multicomponent mixture, $N_F > 2$ when two phases are present. Therefore, two (or more) phases can coexist in a planar region in P-T space, compared to the single component case, where two phases coexist only along a line in P-T space. Mixtures have two phases in a region, or *envelope*, in P-T space (Fig. 4-4).

Consider, along with this figure, a change in pressure from P_1 to P_5 at constant temperature T_2. The phase envelope is fixed for constant overall composition (ω_i or z_i). Since the indicated change is usually brought about by changing the volume of a pressure cell at constant composition and temperature, the process is frequently called a *constant composition* expansion.

From P_1 to P_3, the material in the cell is a single-liquid phase. At P_3, a small amount of vapor phase begins to form. The upper boundary of the phase envelope passing through this point is the *bubble point curve*, and the *y* coordinate at this point is the *bubble point pressure* at the fixed temperature. From P_3 to P_5, successively, more gas forms as the liquid phase vaporizes. This vaporization takes place over a finite pressure range in contrast to the behavior of a pure component. Continuing the constant composition expansion to pressures lower than P_5 would result in eventually reaching a pressure where the liquid phase would disappear, appearing only as drops in the cell just before this point. The pressure the liquid vanishes at is the *dew point pressure* at the fixed temperature, and the lower boundary of the phase envelope is the *dew point curve*.

For a pure component (Fig. 4-1), the dew and bubble point curves coincide.

Within the two-phase envelope, there exist quality lines that as before, indicate constant relative amounts of liquid and vapor. The composition of the liquid and gas

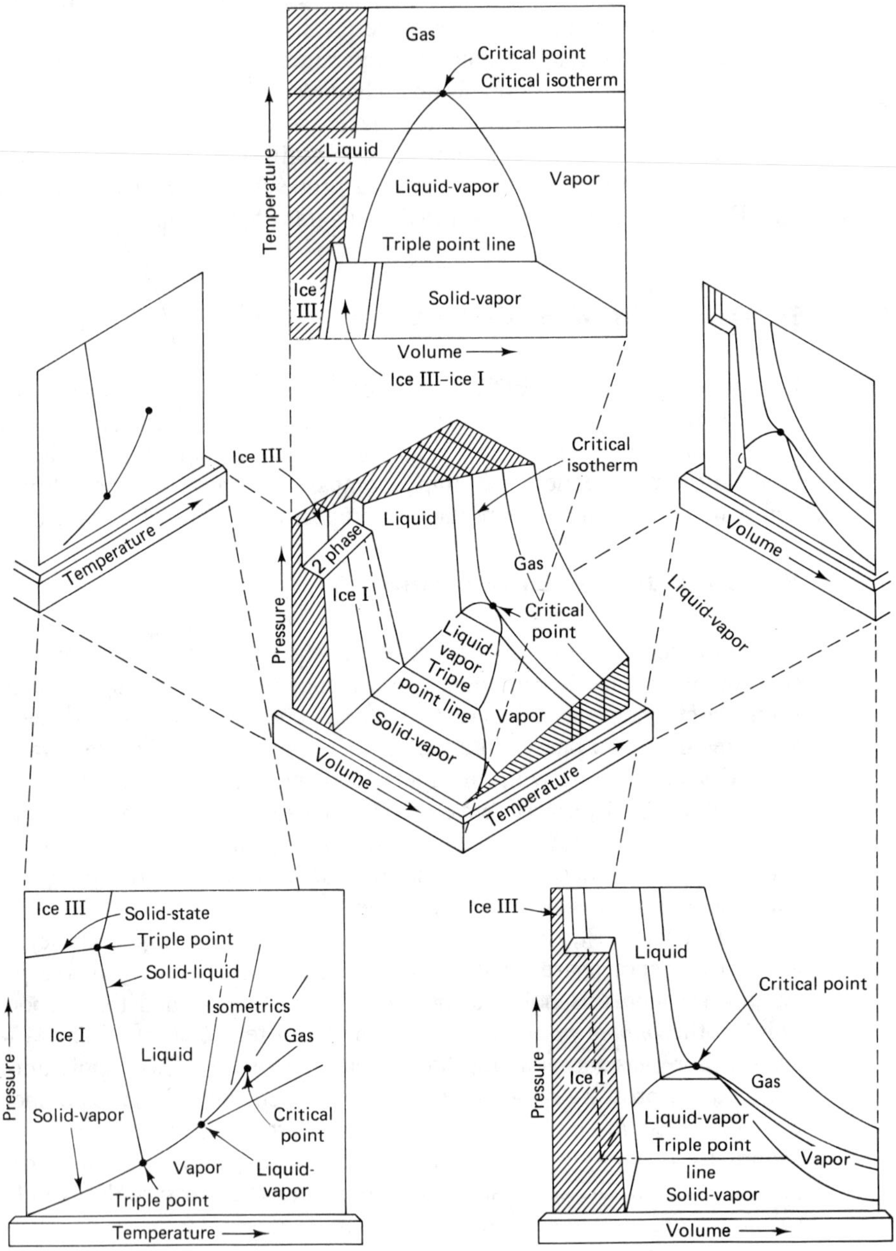

Figure 4-3 Schematic pressure-specific volume–temperature surface and projections (from Himmelblau, 1982)

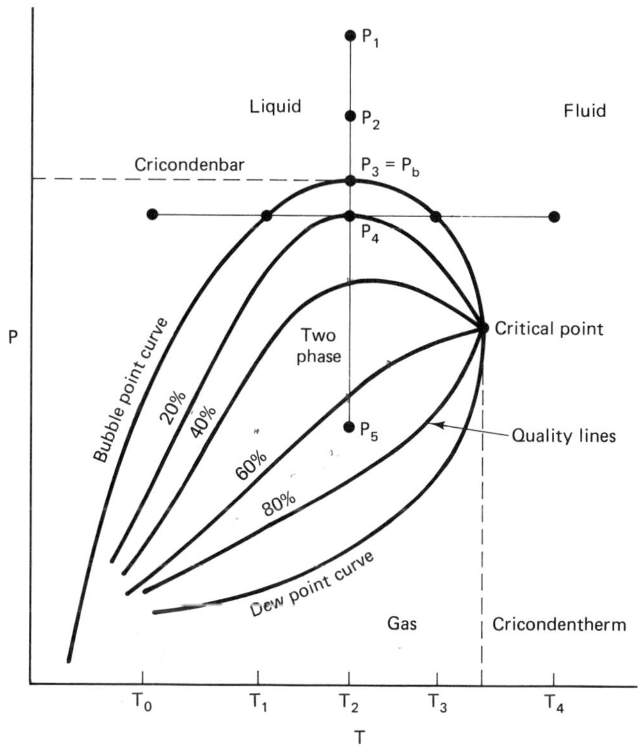

Figure 4-4 Schematic pressure–temperature diagram for hydrocarbon mixtures (constant composition)

phases is different at each point within the envelope, and both change continuously as the pressure decreases.

Phase compositions are not shown on the P-T plot. But we do know that the liquid and gas phases are saturated with respect to each other in the two-phase envelope. Hence at any T and P within the envelope, the liquid phase is at its bubble point, and the gas phase at its dew point. The quality lines converge to a common point at the critical point of the mixture though this point does not, in general, occur at extreme values of the temperature and pressure on the phase envelope boundary. The maximum pressure on the phase envelope boundary is the *cricondenbar*, the pressure above which a liquid cannot be vaporized. The maximum temperature on the phase envelope is the *cricondentherm*, the temperature above which a gas cannot be condensed. These definitions are the same as for the critical point in pure component systems; hence the best definition of the critical point for mixtures is the temperature and pressure at which the two phases become identical.

For mixtures, there exists, in general, a pressure range between the cricondenbar and P_c and between the cricondentherm and T_c where *retrograde* behavior can occur. A horizontal constant pressure line in Fig. 4-4 at $P = P_4$ begins in the liquid region at T_0 and ends in the fluid region at T_4. As temperature is increased, gas begins to form at the bubble point temperature T_1 and increases in amount from then on. But at T_2, the amount of gas begins to decrease, and the gas phase vanishes entirely at a second bubble point T_3. From T_2 to T_3, the behavior is contrary to intuition—a gas phase disappearing as temperature increases—and the phenomenon is called retrograde vaporization.

Retrograde behavior does not occur over the entire range between the two bubble point temperatures but only over the range from T_2 to T_3. By performing the above thought experiment at several pressures, one can show that retrograde behavior occurs only over a region bounded by the bubble point curve on the right and a curve connecting the points of zero slope on the quality lines on the left (McCain, 1973).

Though not possible in the P-T diagram in Fig. 4-4, retrograde phenomena are also observed for changes in pressure at constant temperature. This case, which is of more interest to a reservoir engineer, happens when the cricondentherm is larger than T_c and the constant temperature is between these extremes. This type of retrograde behavior is a prominent feature of many hydrocarbon reservoirs, but it impacts little on EOR.

We do not discuss the pressure–molar-volume behavior of hydrocarbon mixtures in detail. The main differences between the behavior of pure components and mixtures is that the discontinuous changes in $\bar{V}_M$ do not occur at constant P, and the critical point no longer occurs at the top of the two-phase region (see Exercise 4B). These differences cause interesting variations in the shape of pressure–molar-volume diagrams for mixtures but, again, are not directly relevant to EOR.

Since EOR processes are highly composition dependent, the behavior of the P-T envelope as the overall composition of the mixture changes is highly important. Consider the dilution of a crude oil M_4 with a more volatile pure component A as shown in Fig. 4-5. As the overall mole fraction of A increases, the phase envelope migrates toward the vertical axis, increasing the size of the gas region. Simultaneously, the phase envelope shrinks as it approaches the vapor pressure curve of the pure component A. There are, of course, an infinite number of mixtures (Fig. 4-5 shows only three) of the crude oil with A. Each mixture has its respective critical point in P-T space, which also migrates to the critical point of the pure component on a *critical locus*. The overall composition of a mixture at a critical point is the *critical mixture* at that temperature and pressure.

Pressure–Composition Diagrams

The phase behavior of the dilution in Fig. 4-5 on a plot of mole fraction of component A versus pressure at fixed temperature shows composition information directly. Such a plot is a *pressure composition*, or P-z, plot. The P-z plot for the sequence of mixtures in Fig. 4-5 is shown in Fig. 4-6. Since the P-T diagram in Fig. 4-5 shows only three mixtures and does not show quality lines, phase envelope boundaries are represented at relatively few points in Fig. 4-6 (see Exercise 4C).

Starting at some high pressure in Fig. 4-5 and following a line of constant temperature as pressure is reduced produces a dew point curve for mixture M_1 at

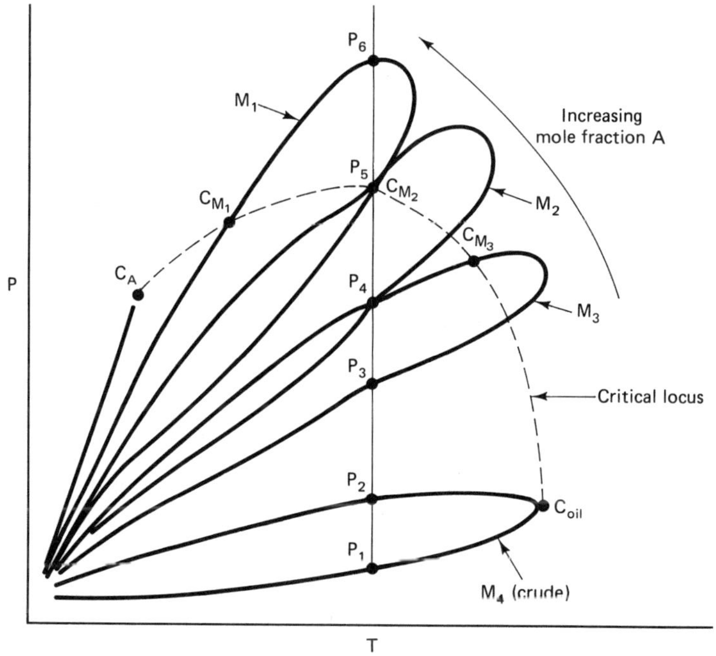

Figure 4-5 Schematic dilution of a crude oil by a more volatile pure component

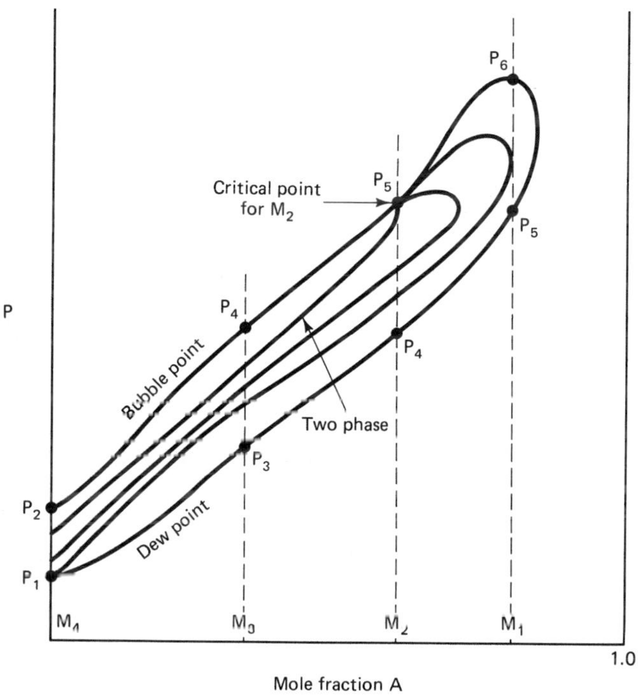

Figure 4-6 The pressure–composition plot for the dilution in Fig. 4-5

pressure P_6. Since this mixture is rich in component A, this point plots nearest the right vertical axis in Fig. 4-6 at the pressure coordinate P_6. Continuing down the constant temperature line, at P_5 the critical point for mixture M_2 is encountered (mixture M_2 is the critical composition at this temperature and pressure). But this point is also a second dew point for mixture M_1; hence P_5 plots at the same vertical coordinate for both mixtures in Fig. 4-6 but with different horizontal coordinates. At P_4 there is a bubble point for mixture M_3 and a dew point for M_2. These points again define the corresponding phase boundaries of the P-z plot in Fig. 4-6. The process continues to successively lower pressures in the same manner. Each pressure below the critical is simultaneously a bubble point and a dew point pressure for mixtures of different overall compositions. The pressures P_2 and P_1 are the bubble and dew point pressures of the undiluted crude oil. The two-phase envelope in Fig. 4-6 does not intersect the right vertical axis since the fixed temperature is above the critical temperature of the pure component A. The diagram shows the closure of the two-phase envelope as well as a few quality lines.

Since the entire P-z diagram is at constant temperature, we cannot represent the phase behavior at another temperature without showing several diagrams. More important, the composition plotted on the horizontal axis of the P-z plot is the overall composition, not either of the phase compositions. Thus horizontal lines do not connect equilibrium mixtures. Such *tie lines* do exist but are, in general, oriented on a horizontal line in a hyperspace whose coordinates are the phase compositions. However, for binary mixtures, the tie lines are in the plane of the P-z plot, and the critical point is necessarily located at the top of the two-phase region. Finally, though Fig. 4-6 is schematic, it bears qualitative similarity to the actual P-z diagrams shown in Figs. 7-10 through 7-12.

4-3 TERNARY DIAGRAMS

On a P-z plot, we sacrifice a degree of freedom (temperature) to obtain compositional information. But the diagrams can show only the composition of one component, and this representation is often insufficient for the multitude of compositions that can form in an EOR displacement. A plot that represents more composition information is the *ternary diagram*.

Definitions

Imagine a mixture, at fixed temperature and pressure, consisting of three components 1, 2, and 3. The components may be pure components. But more commonly in EOR, they are pseudocomponents, consisting of several pure components. The composition of the mixture will be a point on a plot of the mole fraction of component 3 versus that of component 2. In fact, this entire two-dimensional space is made up of points that represent the component concentrations of all possible mixtures.

We need to plot the concentrations of only two of the components since the concentration of the third may always be obtained by subtracting the sum of the mole fractions of components 2 and 3 from 1. This means all possible compositions will plot into a right triangle whose hypotenuse is a line from the 1.0 on the y axis to the 1.0 on the x axis. Though ternary diagrams are on occasion shown this way (see Fig. 7-15), they are most commonly plotted so that the right triangle is shifted to an equilateral triangle, as in Fig. 4-7.

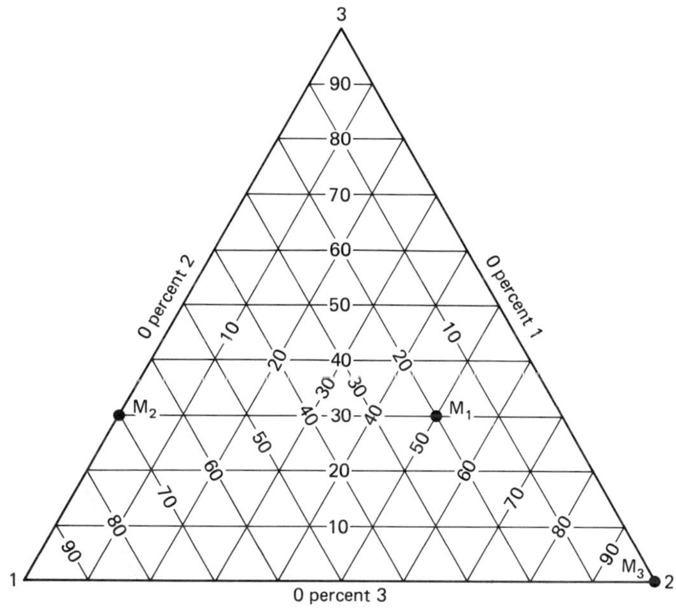

Figure 4-7 Ternary diagram

All possible ternary compositions fall on the interior of the equilateral triangle; the boundaries of the triangle represent binary mixtures (the component at the apex opposite to the particular side is absent), and the apexes represent pure components. Thus in Fig. 4-7, point M_1 is a mixture having 20%, 50%, and 30% components 1, 2, and 3, respectively; point M_2 is a binary mixture of 70% component 1 and 30% component 3, and point M_3 is 100% component 2. Representing the compositions in this manner is possible for any concentration variable (mole fraction, volume fraction, mass fraction) that sums to a constant.

Ternary diagrams are extremely useful tools in EOR because they can simultaneously represent phase and overall compositions as well as relative amounts. The correspondence of the P-T diagram to the ternary diagram in Figs. 4-8 and 4-9 compares to the P-z diagram in Figs. 4-5 and 4-6. Here we consider a ternary system consisting of components 1, 2, and 3, and consider the dilution of mixtures having constant ratios of components 2 and 3 by component 1. Each dilution represents a line corresponding to a fixed 2 : 3 ratio on the ternary in Fig. 4-9.

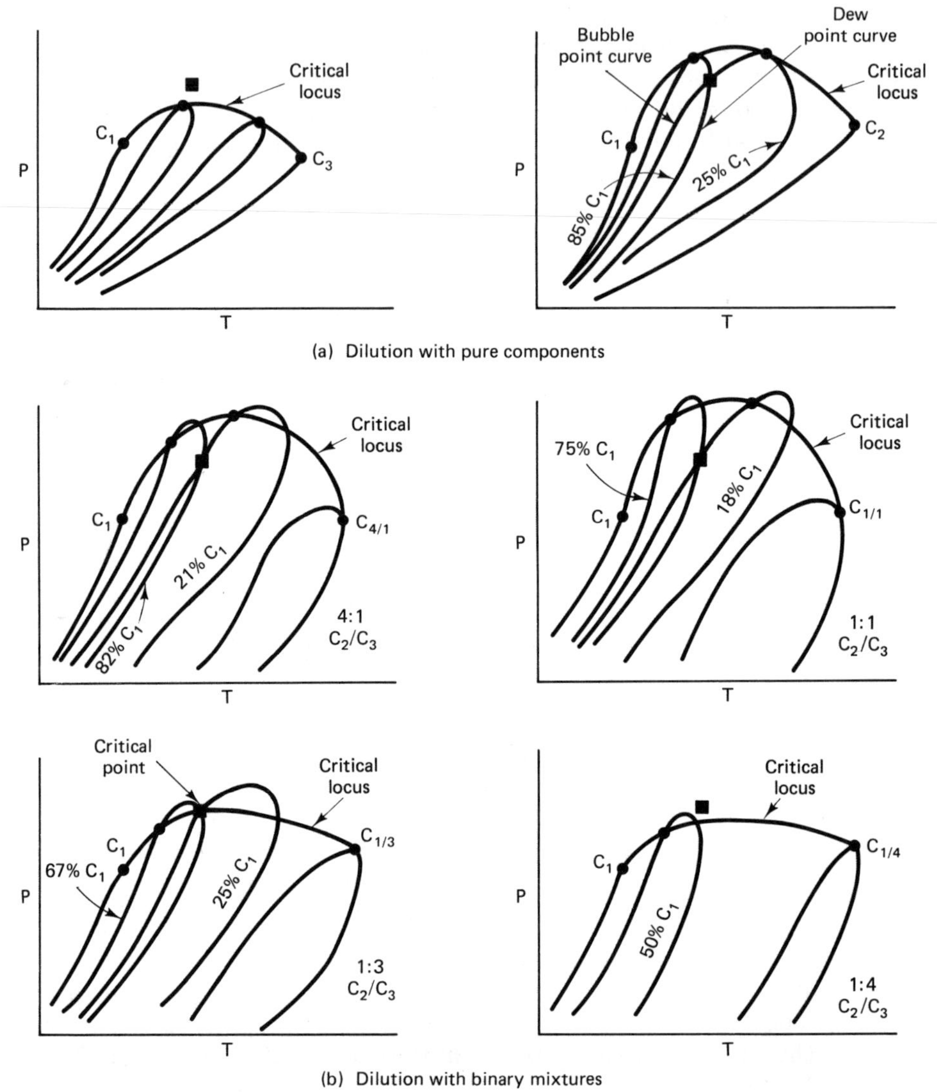

Figure 4-8 Schematic evolution of P-T diagram in three component systems

We want to follow the formation and disappearance of phases on the ternary diagram at the fixed temperature and pressure indicated by the box in Fig. 4-8. For the dilution of component 3 by component 1, the reference temperature and pressure is above the critical locus in the upper left-hand panel (Fig. 4-8a). Thus the C_1-C_3 axis of the ternary indicates no phase changes. The C_1-C_2 binary dilution in the upper right-hand panel (Fig. 4-8a) does encounter phase changes, and in fact, the reference temperature and pressure is a bubble point for a mixture of 25% C_1 and a dew point for a mixture of 85% C_1. These phase transitions are shown on the C_1-C_2 axis

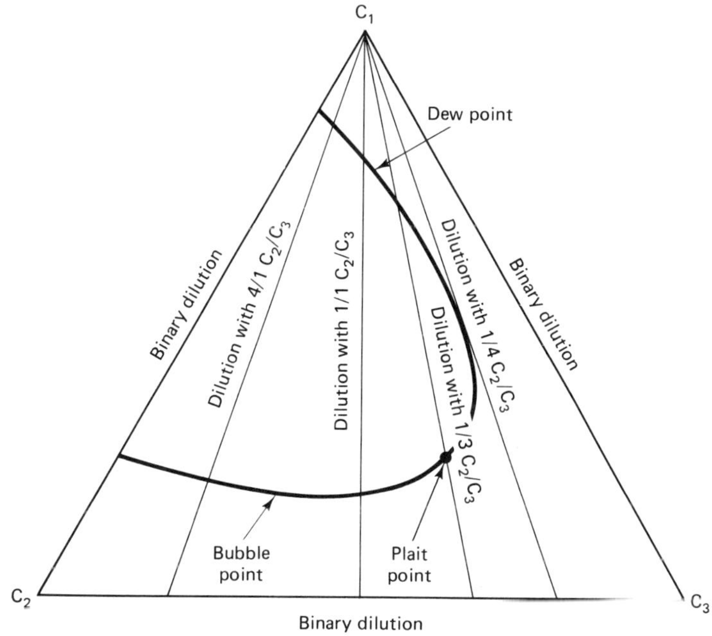

Figure 4-9 Schematic ternary diagram of dilutions in Fig. 4-8

on the ternary. The dilution indicated in the middle left-hand panel (Fig. 4-8b) shows phase transitions at 82% and 21%, respectively, which are also plotted on the ternary. For the dilution of the 1 : 3 mixture, the critical locus passes through the fixed temperature and pressure, and this composition, 25% C_1, is the critical composition of the ternary mixture. This composition is indicated on the ternary diagram in Fig. 4-9 as a *plait* point after the more common designation of the critical mixture in liquid–liquid phase equilibria. At the fixed temperature and pressure, there can exist a second phase transition—a dew point at 67% C_1—at the same temperature and pressure. After making several dilution passes through the ternary diagram, the points where there are phase transitions define a closed curve in Fig. 4-9. This curve, the *binodal* curve, separates regions of one- and two-phase behavior. Within the region enclosed by the binodal curve, two phases exist, and outside this region, all components are in a single phase.

Phase Compositions

One useful but potentially confusing feature of ternary diagrams is that it is possible to represent the composition of the phases as well as the overall composition on the same diagram. Consider an overall composition C_i on the inside of the binodal curve in Fig. 4-10

$$C_i = C_{i1} S_1 + C_{i2} S_2, \qquad i = 1, 2, 3 \tag{4.3-1}$$

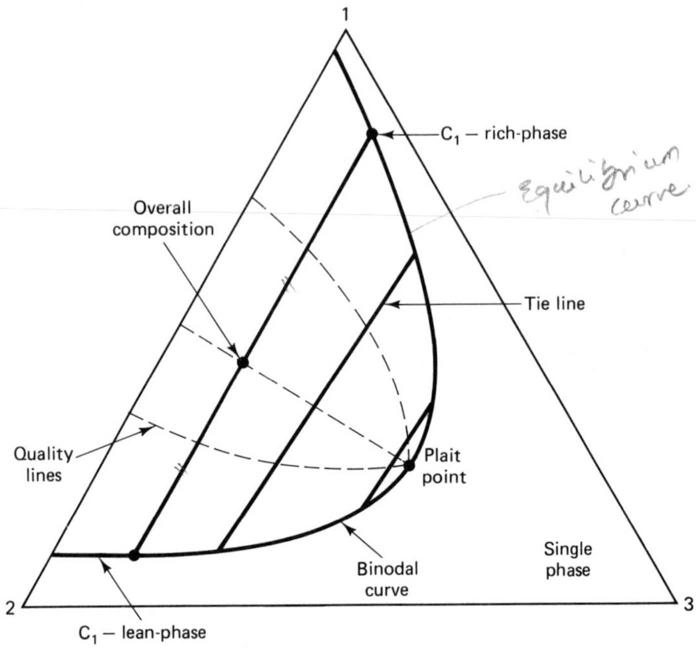

Figure 4-10 Two-phase ternary equilibria

where C_{ij} is the concentration of component i in phase j, and S_j is the relative amount of phase j. By convention, we take phase 1 to be the C_1-rich phase and phase 2 to be the C_1-lean phase. Since $S_1 + S_2 = 1$, we can eliminate S_1 from two of the equations in Eq. (4.3-1) to give

$$S_2 = \frac{C_3 - C_{31}}{C_{32} - C_{31}} = \frac{C_1 - C_{11}}{C_{12} - C_{11}} \tag{4.3-2}$$

This equation says a line through the composition of phase 1 and the overall composition has the same slope as a line passing through the composition of phase 2 and the overall composition. Both lines, therefore, are merely segments of the same straight line that passes through both phase compositions and the overall composition. The intersection of these *tie lines* with the binodal curve gives the phase compositions shown in Fig. 4-10. The entire region within the binodal curve can be filled with an infinite number of these tie lines, which must vanish as the plait point is approached since all phase compositions are equal at this point. Of course, there are no tie lines in the single-phase region.

Further, Eq. (4.3-2) implies, by a similar triangle argument, that the length of the line segment between C_i and C_{i1} divided by the length of the segment between C_{i2} and C_{i1} is the relative amount S_2. This, of course, is the well-known *lever rule*, which can also be derived for S_1. By holding S_2 constant and allowing C_i to vary, we can construct quality lines, as indicated in Fig. 4-10, which must also converge to the plait point as do the tie lines.

Tie lines are graphical representations of the equilibrium relations (Eq. 2.2-11). Assuming, for the moment, the apexes of the ternary diagram represent true components, the phase rule predicts there will be $N_F = 1$ degrees of freedom for mixtures within the binodal curve since temperature and pressure are already specified. Thus it is sufficient to specify one concentration in either phase to completely specify the state of the mixture. A single coordinate of any point on the binodal curve gives both phase compositions if the tie lines are known. This exercise does not determine the relative amounts of the phases present since these are not state variables. Nor does specifying a single coordinate of the overall concentration suffice since these, in general, do not lie on the binodal curve. Of course, it is possible to calculate the phase compositions and the relative amounts from equilibrium relations, but these must be supplemented in "flash calculations" by additional mass balance relations to give the amounts of each phase.

Three-Phase Behavior

When three phases form, there are no degrees of freedom ($N_F = 0$). The state of the system is entirely determined. It follows from this that three-phase regions are represented on ternary diagrams as smaller subtriangles embedded within the larger ternary triangle (Fig. 4-11). Since no tie lines are in three-phase regions, the apexes or *invariant* points of the subtriangle give the phase compositions of any overall composition within that subtriangle. The graphical construction indicated in Fig. 4-11 gives the relative amounts of the three phases present (see Hougen et al., 1966, and Exercise 4D).

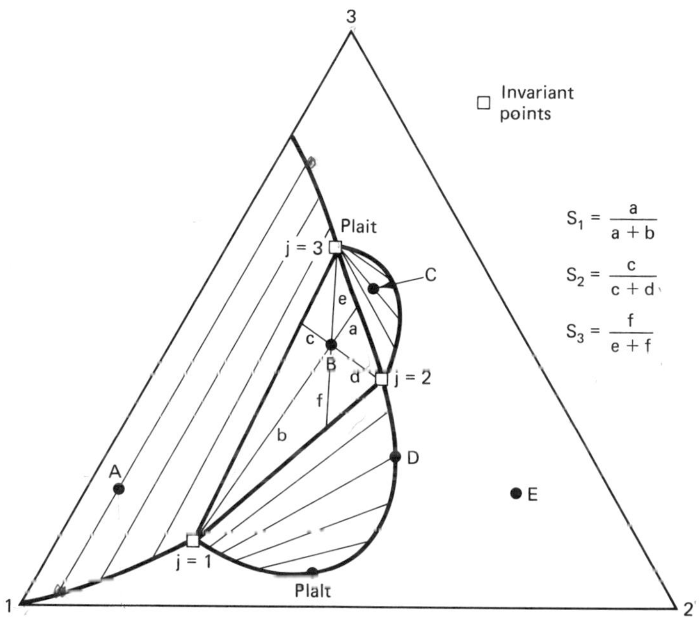

Figure 4-11 Three-phase diagram example (from Lake, 1984)

A point on a nonapex side of the subtriangle may be regarded as being simultaneously in the three-phase region or in a two-phase region; thus the subtriangle must always be bounded on a nonapex side by a two-phase region for which the side of the subtriangle is a tie line of the adjoining two-phase region. By the same argument, the apexes of the subtriangle must adjoin, at least in some nonzero region, a single-phase region. To be sure, the adjoining two-phase regions can be quite small (see Fig. 9-6).

Thus points A and C in Fig. 4-11 are two-phase mixtures, point B is three phase, and points D and E are single phase, though point D is saturated with respect to phase 1. (For more detail on the geometric and thermodynamic restrictions of ternary equilibria, see Francis, 1963.)

4-4 QUANTITATIVE REPRESENTATION OF TWO-PHASE EQUILIBRIA

Several mathematical relations describe the qualitative representations in the previous section. The most common are those based on (1) equilibrium flash vaporization ratios, (2) equations of state, and (3) a variety of empirical relations. In this section, we concentrate only on those two-phase equilibria aspects directly related to EOR. Three-phase equilibria calculations are discussed elsewhere in the literature (Mehra et al., 1980; Risnes and Dalen, 1982; Peng and Robinson, 1976) and in Chap. 9, which covers three-phase equilibria for micellar systems.

Equilibrium Flash Vaporization Ratios

If we let x_i and y_i be the mole fractions of component i in a liquid and in contact with a vapor phase, the equilibrium *flash vaporization ratio* for component i is defined by

$$K_i = \frac{y_i}{x_i}, \qquad i = 1, \ldots, N_C \tag{4.4-1}$$

This quantity is universally known as the K-value for component i.

At low pressures, the K-values are readily related to the mixture temperature and pressure. The partial pressure of component i in a low-pressure gas phase is $y_i P$ from Dalton's law of additive pressures. The partial pressure of component i in the vapor above an ideal liquid phase is $x_i P_{vi}$ from Raoult's law, where P_{vi} is the pure component vapor pressure of component i (see Figs. 4-1 and 7-2). At equilibrium for this special case, the partial pressures of component i calculated by either means must be equal; hence

$$K_i = \frac{y_i}{x_i} = \frac{P_{vi}}{P}, \qquad i = 1, \ldots, N_C \tag{4.4-2}$$

Equation (4.4-2) says at low pressures, a plot of the equilibrium K-value for a particular component at a fixed temperature will be a straight line of slope -1 on a log–log plot. Under these conditions, the K-value itself may be estimated from pure component vapor pressure data.

At higher pressures, where the assumptions behind Dalton's and Raoult's laws are inaccurate, the K-values are functions of overall composition. The additional composition information, usually based on the liquid-phase composition, can be incorporated into a *convergence pressure*, which is then correlated to the K-values. Convergence pressure correlations are usually presented in graphical form (*GPSA Data Book*, 1983) or as equations. The introduction of a composition variable directly into the K-value functions adds considerable complexity to the flash procedure.

The flash calculation proceeds as follows: Let z_i be the overall mole fraction of component i in the mixture (analogous to ω_i, the overall mass fraction in Chap. 2). Then

$$z_i = n_L x_i + n_V y_i, \qquad i = 1, \ldots, N_C \tag{4.4-3}$$

where n_L and n_V are the relative molar amounts of the liquid and gas phases, respectively. Since all quantities in Eq. (4.4-3) are relative, they are subject to the following constraints

$$\sum_{i=1}^{N_C} x_i = \sum_{i=1}^{N_C} y_i = \sum_{i=1}^{N_C} z_i = n_L + n_V = 1 \tag{4.4-4}$$

Eliminating n_L from Eq. (4.4-3) with this equation, and substituting the definition (Eq. 4.4-1) for y_i, yields the following for the liquid-phase composition:

$$x_i = \frac{z_i}{1 + (K_i - 1)n_V}, \qquad i = 1, \ldots, N_C \tag{4.4-5}$$

But these concentrations must also sum to 1.

$$\sum_{i=1}^{N_C} \frac{z_i}{1 + (K_i - 1)n_V} = 1 \tag{4.4-6a}$$

Equation (4.6-6a) is a single polynomial expression for n_V, with K_i and z_i known, that must be solved by trial and error. The equation itself is not unique since we could have eliminated n_V and x_i from Eq. (4.4-3) to give the entirely equivalent result

$$\sum_{i=1}^{N_C} \frac{z_i}{1 + \left(\dfrac{1}{K_i} - 1\right)n_L} = 1 \tag{4.4-6b}$$

The usual flash procedure is to calculate n_V or n_L by trial and error, and then use Eqs. (4.4-1) and (4.4-5) to calculate the phase concentrations.

Alternatively, Eqs. (4.4-6a) and (4.4-6b) can be used to calculate quality lines in a P-T diagram by specifying n_V or n_L and then performing trial-and-error solutions for pressure at various fixed temperatures. Two special cases of the above procedure follow directly. The bubble point curve for a mixture ($n_L = 1$) is given implicitly from Eq. (4.4-6b) as

$$1 = \sum_{i=1}^{N_C} z_i K_i \tag{4.4-7a}$$

and the dew point curve ($n_V = 1$) from Eq. (4.4-6a) as

$$1 = \sum_{i=1}^{N_C} \frac{z_i}{K_i} \tag{4.4-7b}$$

These equations suggest the necessity of doing a flash calculation. Because the K-values increase as temperature increases, a mixture of overall composition z_i at fixed temperature and pressure will be two phase only if

$$\sum_{i=1}^{N_C} z_i K_i > 1 \quad \text{and} \quad \sum_{i=1}^{N_C} \frac{z_i}{K_i} > 1 \tag{4.4-8}$$

If the first inequality in Eq. (4.4-8) is violated, the mixture is a single-phase liquid; if the second is violated, the mixture is a single-phase gas.

Equations of State

Though the K-value approach is easily the most common representation of two-phase equilibria, it suffers from a lack of generality and may result in inaccuracies particularly near the convergence pressure. In recent years, the trend has been toward equation of state (EOS) representations since these are potentially able to work near the critical point and yield internally consistent densities and molar volumes. (For more details on EOS and its underlying thermodynamic principles, see Smith and van Ness, 1975, and Denbigh, 1968.)

Pure components. An EOS is any mathematical relationship among the three intensive properties molar volume, temperature, and pressure. Usually, the relation is written in a pressure-explicit form $P = f(\overline{V}_M, T)$, and the most elementary form is the ideal gas equation

$$P = \frac{RT}{\overline{V}_M} \tag{4.4-9}$$

This equation applies only to gases at low pressure. Equation (4.4-9) can be corrected to apply to real gases by introducing a correction factor z, the *compressibility factor*

$$P = \frac{zRT}{\overline{V}_M} \tag{4.4-10}$$

The compressibility factor is itself a function of temperature and pressure that is given in many sources (see McCain, 1973, for example). Since Eq. (4.4-10) is actually a definition of the compressibility factor, the equation can also be applied to fluids and liquids though the latter is rarely done. Given the relation between z and T and P, Eq. (4.4-10) could predict volumetric behavior for all T and P.

Consider the pressure–molar-volume behavior of a pure component as shown in Fig. 4-2. Figure 4-12 also shows this type of plot with two isotherms T_1 and T_2, both below the critical temperature. Equation (4.4-9) is the equation of a hyperbola on this plot that matches the experimental isotherm well at low pressure or high molar volume. The ideal gas law fails badly in the liquid region, particularly for pressure predictions, since it predicts a zero asymptote on the molar volume axis. This is equivalent to saying the component molecules themselves have no intrinsic volume even at the highest pressure, which is, of course, a basic hypothesis in the derivation of the ideal gas law from statistical mechanics.

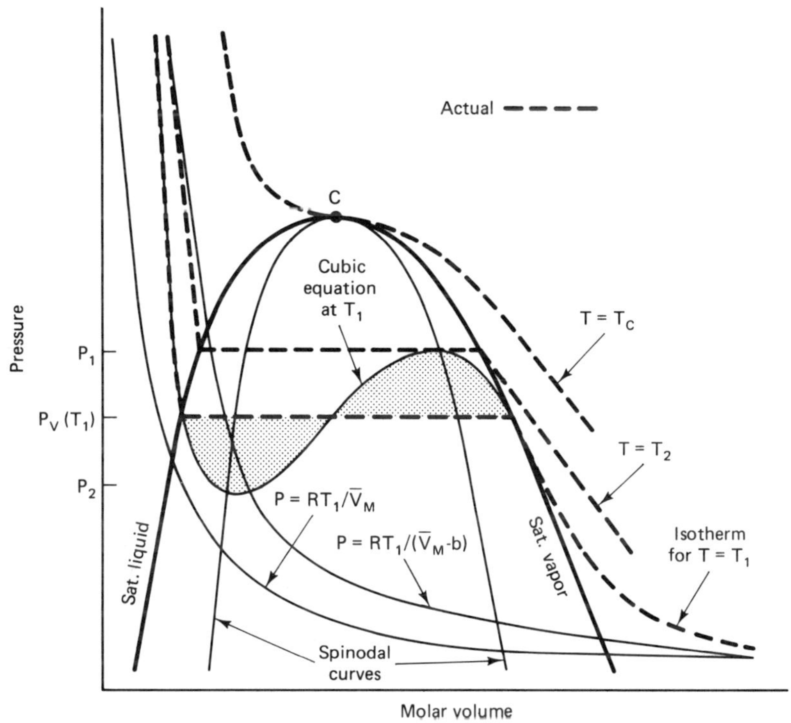

Figure 4-12 General features of cubic equations of state

To introduce a nonzero asymptote, we try an equation of the form

$$P = \frac{RT}{(\bar{V}_M - b)} \tag{4.4-11}$$

where b is now the asymptotic value of $\bar{V}_M$ as pressure increases. Figure 4-12 shows this equation can be made to match the liquid molar volumes reasonably well at high pressures. The value of b, the intrinsic molecular volume, is usually so small that Eq. (4.4-11) still provides a good estimate at low pressures.

But Eq. (4.4-11) still fails for temperature and pressure combinations that are fairly close to the pure component vapor pressure curve. To predict the molar volume up to and including the vapor pressure curve requires a function of the form

$$P = \frac{RT}{(\overline{V}_M - b)} - f(T, \overline{V}_M) \tag{4.4-12}$$

where the term $f(T, \overline{V}_M)$ is specific to the particular EOS. Equation (4.4-12) is frequently interpreted as a sum of forces, the first term being the force that will not allow the molecules to be compressed to zero volume (repulsive force), and the second being the force due to the intermolecular attraction among molecules.

A practical EOS must be accurate, internally consistent near the critical point, and relatively simple. Moreover, since we are to use it to predict vapor–liquid equilibria, it must predict both liquid and gas properties.

For pure components, there can exist two values of molar volume at a particular temperature and pressure; hence Eq. (4.4-12) must have at least two real roots at this point. Moreover, since P is a monotonically decreasing function of $\overline{V}_M$ regardless of the fluid-phase identity, f must be at least second order in $\overline{V}_M$ so that the entire function (Eq. 4.4-12) must be at least cubic in molar volume. Cubic EOS, therefore, are the simplest form that satisfy the three criteria mentioned. Though there have been more than 100 EOS proposed in the technical literature, many of which are quite complicated and have more thermodynamic rigor, we discuss only cubic EOS since these seem to be the most commonly used class of equations in EOR.

In the vicinity of the vapor pressure curve (pressures between P_1 and P_2 at temperature T_1 in Fig. 4-12), there are three real roots to the cubic EOS. The vapor pressure P_v corresponding to T_1 is the y-coordinate value that causes the shaded areas above and below P_v to have equal areas (Abbott and van Ness, 1972). For pressures above P_v, only the smallest root has physical significance and corresponds to $\overline{V}_M$ of a compressed liquid; at pressures below P_v, the largest root corresponds to $\overline{V}_M$ of a superheated vapor. At the vapor pressure, both the smallest and largest roots have physical significance corresponding to the saturated liquid and vapor molar volumes, respectively. The intermediate root has no physical significance.

As the critical point is approached, all three roots converge to the value of $\overline{V}_M$ at the critical point $\overline{V}_{Mc}$. For temperatures above T_c, cubic equations have only one real root, that of the molar volume of a fluid. For the critical isotherm itself, there is also only one real root, and the critical constraints Eq. (4.1-3) are satisfied at the critical pressure.

Within the two-phase region on the pressure–molar-volume plot, the quadratic curves defined by $(\partial P/\partial \overline{V}_M)_T = 0$ for $P < P_c$ are the *spinodal curves*. They represent the maximum degree of supersaturation with respect to the particular phase transition. Thus theoretically at least, we could lower the pressure on a single compressed liquid phase at T_1 to P_2 without changing phase. The liquid between P_v and P_2 is supersaturated with respect to the vapor phase. A phase transition must occur beyond this pressure since the partial derivative $(\partial P/\partial \overline{V}_M)_T$ is constrained to be negative on thermodynamic and physical grounds. Similarly, a vapor phase at pressure P_1 could be supercooled down to only temperature T_2 without causing a phase change, and the vapor at P_1 and T_2 is supersaturated with respect to the liquid phase. These are metastable states that will change to stable states on perturbation.

The above discussion gives the properties of any general cubic EOS. The particular form of such equations, of course, can take a wide variety of forms. Abbott (1973) gives the general form

$$P = \frac{RT}{(\overline{V}_M - b)} - \frac{\theta(\overline{V}_M - \eta)}{(\overline{V}_M - b)(\overline{V}_M^2 + \delta \overline{V}_M + \epsilon)} \tag{4.4-13}$$

where the parameters θ, η, δ, and ϵ are given in Table 4-1 for nine specifc equations of state. Equation (4.4-13) is, perhaps, not the most general form of the cubic equations available (Martin, 1979), but it does include most of the commonly accepted equations used in predicting the phase behavior of EOR fluids.

Abbott's original work (1973) contains complete references on each of the equations in Table 4-1. Thus far, only two of these equations have seen extensive use in predicting EOR phase behavior: the Soave modification (1972) of the Redlich-Kwong equation (RKS) and the Peng-Robinson (1976) equation (PR). We discuss these two equations here.

Except for the Clausius equation, all the equations in Table 4-1 are two-parameter equations. The value of these parameters may be chosen to force the equation to make internally consistent predictions in the vicinity of the critical point for pure components. Thus the values of the parameters come from enforcing the critical constraints (Eq. 4.1-3) and from evaluating the original equation at the critical point. Since there are three equations, the procedure also specifies a specific value of the critical molar volume $\overline{V}_M$ or critical z-factor z_c in addition to a and b.

It is somewhat easier, though entirely equivalent, to use the procedure of Martin and Hou (1955) to determine the parameters a and b. Expressing the RKS equation in the z-factor form will eliminate $\overline{V}_M$ between Eq. (4.4-10) and the RKS equation. By applying Descartes' rule of roots to this equation, there is either one or

TABLE 4-1 CLASSIFICATION OF SOME CUBIC EQUATIONS OF STATE (FROM ABBOTT, 1978)

Equation	θ	η	δ	ϵ
van der Waals (1873)	a	b	0	0
Berthelot (1900)	a/T	b	0	0
Clausius (1880)	a/T	b	$2c$	c^2
Redlich-Kwong (1949)	$a/T^{1/2}$	b	b	0
Wilson (1964)*	$\theta_w(T)$	b	b	0
Peng-Robinson (1976)	$\theta_{PR}(T)$	b	$2b$	$-b^2$
Lee-Erbar-Edmister (1973)	$\theta_{LEE}(T)$	(T)	b	0

*Similarly, Barner et al. (1966) and Soave (1972)

three positive and no negative real roots. The z-factor equation evaluated at the critical point must have only one real root; hence

$$(z - z_c)^3 = z^3 - 3z_c z^2 + 3z_c^2 z - z_c^3 = 0 \tag{4.4-14}$$

This equation is identically equal to the form in Table 4-2; hence equating coefficients, we immediately have $z_c = 1/3$ and

$$3z_c^2 = A - B - B^2 \tag{4.4-15a}$$

$$z_c^3 = AB \tag{4.4-15b}$$

Eliminating A from these equations gives the cubic form

$$27B^3 + 27B^2 + 9B = 1 \tag{4.4-16}$$

Moreover, using Descartes' rule, it follows that this equation has only one real positive root, which may be solved for directly to give $B = (2^{1/3} - 1)/3 = 0.08664$. Solving for A from Eq. (4.4-15b) gives $A = (9(2^{1/3} - 1))^{-1} = 0.4247$. Using the definitions for A and B gives the forms in Table 4-2 for a and b.

Clearly, the above procedure is valid for any a and b that are a function of temperature only. To match experimental vapor pressure data to subcritical temperatures, the a given by this procedure is multiplied by a factor α_i, a function of temperature that reduces to unity at the critical temperature. The factor α_i is also component specific through its dependence on the acentric factor ω_i. Acentric factors roughly express the deviation of the shape of a molecule from a sphere and are available in extensive tabulations (Reid et al., 1977).

Mixtures. The true test and practical utility of any EOS is in its prediction of mixture properties. For mixtures, many of the arguments advanced above in conjunction with Fig. 4-12 do not apply. In particular, the critical constraints are no longer satisfied at the critical point since this point is no longer at the top of the two-phase envelope.

To account for mixture behavior, the pure component parameters a_i and b_i come from various mixing rules, as shown in Table 4-2. The inclusion of the component index in Table 4-2 means the parameters used in the definitions of these quantities—T_{ci}, P_{ci}, and ω_i—are those for the pure component i.

The most general form of the mixing rules incorporates another parameter, the *binary interaction coefficient* δ_{ij} into the RKS and PR equations, which accounts for molecular interactions between two unlike molecules. By definition δ_{ij} is zero when i and j represent the same component, small when i and j represent components that do not differ greatly (for example, if i and j were both alkanes), and large when i and j represent components that have substantially different properties. Ideally, the δ_{ij} are both temperature and pressure independent (Zudkevitch and Joffe, 1970), depending only on the identities of components i and j. Though the interaction coefficients are considerably less available than acentric factors, literature tabulations are becoming more common (Yarborough, 1978; Whitson, 1982; Prausnitz et al., 1980).

TABLE 4-2 COMPARISON OF THE RKS AND PR EQUATIONS OF STATE (FROM NGHIEM AND AZIZ, 1979)

	RKS	PR
Equation	$P = \dfrac{RT}{\overline{V} - b} - \dfrac{a}{\overline{V}_M(\overline{V}_M + b)}$	$P = \dfrac{RT}{\overline{V}_M - b} - \dfrac{a}{\overline{V}_M^2 + 2b\overline{V}_M - b^2}$
z-factor	$z^3 - z^2 + z(A - B - B^2) - AB = 0$ $A = \dfrac{aP}{(RT)^2}$	$z^3 - (1 - B)z^2 + (A - 3B^2 - 2B)z - (AB - B^3 - B^2) = 0$ $B = \dfrac{bP}{RT}$
Pure component i let $a = a_i$ $b = b_i$	$a_i = \dfrac{0.42747\, R^2 T_{ci}^{2.5} \alpha_i}{P_{ci}}, \; z_c = 0.333$ $b_i = \dfrac{0.08664\, RT_{ci}}{P_{ci}}$ $m_i = 0.480 + 1.57\omega_i - 0.176\omega_i^2$	$a_i = \dfrac{0.45724\, R^2 T_{ci}^{2.5} \alpha_i}{P_{ci}}, \; z_c = 0.307$ $b_i = \dfrac{0.07780\, RT_{ci}}{P_{ci}}$ $m_i = 0.37464 + 1.54226\omega_i - 0.26992\omega_i^2$
	$\alpha_i = \left[1 + m_i\left[1 - \left(\dfrac{T}{T_{ci}}\right)^{1/2}\right]\right]^2$	
Mixture let $a = a_m$ $b = b_m$	$a_m = \sum_{ij}\sum x_i x_j a_{ij}, \; b_m = \sum_i x_i b_i$ $a_{ij} = (1 - \delta_{ij})(a_i a_j)^{1/2}$	
Pure component fugacity use z^L for f^L z^V for f^V	$\ln \dfrac{f}{P} = z - 1 - \ln(z - B) - \dfrac{A}{B} \cdot \ln\left[\dfrac{z + B}{z}\right]$	$\ln \dfrac{f}{P} = z - 1 - \ln(z - B) - \dfrac{A}{2\sqrt{2}B} \cdot \ln\left[\dfrac{z + 2.414B}{z - 0.414B}\right]$
Fugacity of component i use z^L and x_i for f_i^L z^V and y_i for f_i^V	$\ln\left(\dfrac{f_i}{Px_i}\right) = \dfrac{b_i}{b}(z - 1) - \ln(z - B) - \dfrac{A}{B}\left(\dfrac{2\sum_j x_j a_{ij}}{a} - \dfrac{b_i}{b}\right) \ln\left(\dfrac{z + B}{z}\right)$	$\ln\left(\dfrac{f_i}{x_i P}\right) = \dfrac{b_i}{b}(z - 1) - \ln(z - B) - \dfrac{A}{2\sqrt{2}B}\left(\dfrac{2\sum_j x_j a_{ij}}{a} - \dfrac{b_i}{b}\right) \ln\left(\dfrac{z + 2.414B}{z - 0.414B}\right)$

Flash calculations. To calculate vapor–liquid equilibria for mixtures from the RKS equation, an expression for the fugacity of a component i in a mixture is needed. This is most conveniently done by introducing a *fugacity coefficient* of component i defined as

$$\phi_i = \frac{f_i}{x_i P} \tag{4.4-17}$$

In Eq. (4.4-17), and all subsequent equations in this section, the composition variable may be either the liquid-phase mole fraction x_i, if calculating the fugacity coefficient of component i in the liquid phase, or the vapor-phase mole fraction y_i, if calculating the fugacity coefficient in the vapor phase. Following the arguments presented in standard texts (Smith and van Ness, 1975), the fugacity coefficient is, for a mixture,

$$\ln\phi_i = \int_{\bar{V}_M}^{\infty}\left[\left(\frac{\partial (nz)}{\partial n_i}\right)_{T,V,n} - 1\right]\frac{d\bar{V}_M}{\bar{V}_M} - \ln z \tag{4.4-18a}$$

and, for a pure component,

$$\ln\phi = \int_{\bar{V}_M}^{\infty}(z-1)\frac{d\bar{V}_M}{\bar{V}_M} + z - 1 - \ln z \tag{4.4-18b}$$

Equation (4.4-18b) is a special case of Eq. (4.4-18a) as one of the x_i becomes unity. The partial derivative in the integral of Eq. (4.4-18a) is taken at constant temperature and total volume V, where n is the total number of moles in the mixture, and n_i is the total number of moles of species i in the phase. Clearly,

$$V = n\bar{V}_M,\; n = \sum_{i=1}^{N_c} n_i, \quad \text{and} \quad x_i = n_i/n.$$

The fugacity coefficient definition (Eq. 4.4-18a) also can be written in a variety of equivalent forms (Smith and van Ness, 1975; Coats, 1980). To evaluate the integral in Eq. (4.4-18a), it is convenient to express z in an explicit form

$$z = \frac{\bar{V}_M}{\bar{V}_M - b} - \frac{a}{RT(\bar{V}_M + b)} \tag{4.4-19}$$

After multiplying Eq. (4.4-19) by n and introducing the mixing rules for a and b from Table 4-2, the resulting expression may be differentiated with respect to n_i. After some algebra, this gives

$$\left(\frac{\partial (nz)}{\partial n_i}\right)_{T,V,n} = \frac{\bar{V}_M}{\bar{V}_M - b} - \frac{\bar{V}_M b_i}{(\bar{V}_M - b_m)^2} - \frac{1}{RT}\left(\frac{\Sigma_j x_j a_{ij}}{\bar{V}_M + b_m} - \frac{ab_i}{(\bar{V}_M + b_m)^2}\right) \tag{4.4-20}$$

Equation (4.4-20) is explicit in $\bar{V}_M$ which, when substituted into Eq. (4.4-18a) and integrated, leads to the closed-form expression given in Table 4-2. Similar procedures may be used on the PR equation (see Exercise 4F).

The actual calculation of vapor–liquid equilibria follows from two general procedures based on the EOS approach. From Eqs. (4.4-1) and (4.4-17), the equilibrium K-values become

$$K_i = \frac{\phi_i^L}{\phi_i^V}, \qquad i = 1, \ldots, N_C \tag{4.4-21}$$

since the component fugacities are equal at equilibrium. Thus based on an initial estimate of the K_i, a flash calculation, as described above, will obtain the vapor and liquid compositions from Eq. (4.4-5), the K-value definition, and the K_i calculated from Eq. (4.4-21). If the beginning and initial K-value estimates agree, the calculated compositions are the correct values; if they do not agree, new values of the K_i must be estimated and the entire procedure repeated until the K-values do not change. Since the flash calculation is itself a trial-and-error procedure, this procedure is somewhat analogous to the convergence pressure approach we already described.

The second approach to calculating vapor–liquid equilibria from EOS is to directly use the equilibrium constraints. Thus the equations

$$f_i^L = f_i^V, \qquad i = 1, \ldots, N_C \tag{4.4-22a}$$

may be regarded as a set of N_C independent simultaneous nonlinear equations in either x_i or y_i (but not both since x_i and y_i are related through the K-values) that may be linearized, solved as a system of simultaneous linear equations, and iterated until the phase compositions do not change. Either way the calculation is fairly convoluted, so it is not suprising that many variations of the procedure exist (Fussell and Fussell, 1979; Mehra et al., 1980).

Equation (4.4-22a) is easily generalized to the condition for equilibrium among any number of phases N_P

$$f_{ij} = f_{ik}, \qquad i = 1, \ldots, N_C; \qquad j, k = 1, \ldots, N_P \tag{4.4-22b}$$

Empirical Representations

There are three common empirical representations of phase behavior. All are used primarily for liquid–liquid equilibria.

Hand's rule. Hand (1939) gave a fairly simple representation of two-phase equilibria that has proved useful for some EOR systems (Pope and Nelson, 1978; Young and Stephenson, 1982). The procedure is based on the empirical observation that certain ratios of equilibrium phase concentrations are straight lines on log–log or *Hand plots*.

In this section, the concentration variable C_{ij} is the volume fraction of component i (i = 1, 2, or 3) in phase j (j = 1 or 2). Using volume fractions has become conventional in the Hand representation since these are convenient in liquid–liquid equilibria.

Figure 4-13 shows the one- and two-phase regions on the ternary diagram and its correspondence to the Hand plot. The line segments AP and PB represent the binodal curve portions for phase 1 and 2, respectively, and curve CP represents the distribution curve of the indicated components between the two phases. The ratios on the distribution curve are analogous to, but entirely different from, the definitions of the K-values given above. The equilibria relations based on the Hand plot are

$$\frac{C_{3j}}{C_{2j}} = A_H\left(\frac{C_{3j}}{C_{1j}}\right)^{B_H}, \qquad j = 1, 2 \tag{4.4-23}$$

$$\frac{C_{32}}{C_{22}} = E_H\left(\frac{C_{31}}{C_{11}}\right)^{F_H} \tag{4.4-24}$$

where A_H, B_H, E_H, F_H are empirical parameters. Equation (4.4-23) represents the binodal curve, and Eq. (4.4-24) represents the distribution curve. In this form, these equations require the binodal curve to enter the corresponding apex of the ternary diagram. A simple modification overcomes this restriction (see Exercise 4G).

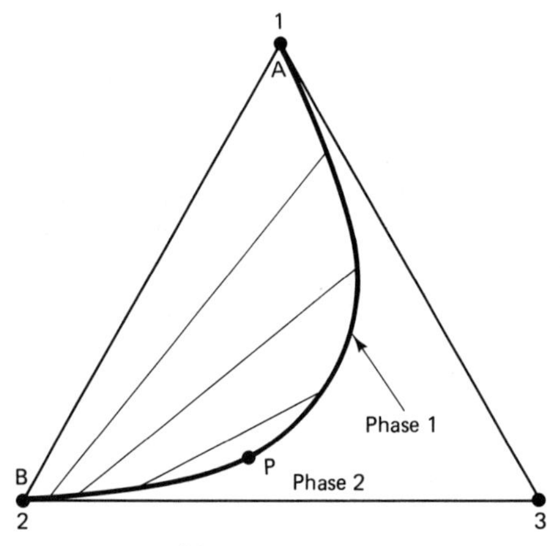

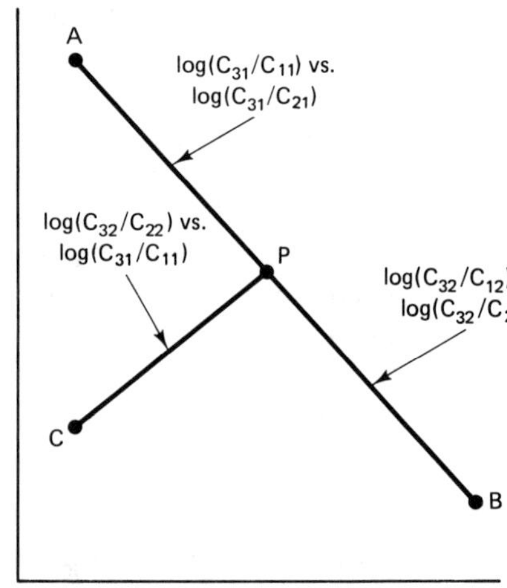

Figure 4-13 Correspondence between ternary diagram and Hand plot

Within the two-phase region of the binodal curve, there are six unknowns, the C_{ij} phase concentrations and five equations, three from Eqs. (4.4-23) and (4.4-24) and two consistency constraints

$$\sum_{i=1}^{3} C_{ij} = 1, \qquad j = 1, 2 \tag{4.4-25}$$

Thus there is $N_F = 1$ degree of freedom as required by the phase rule since temperature and pressure are fixed for ternary equilibria.

A flash calculation using the Hand procedure solves for the relative amounts of the two phases. This introduces two additional variables, S_1 and S_2, into the calculation, but there are now three additional equations, the mass balance Eq. (4.3-1) with the overall concentrations C_i known and $S_1 + S_2 = 1$. As in all the phase equilibria flash calculations, the procedure is trial and error though for certain special cases, phase concentrations follow from direct calculation. The iterative procedure is to first pick a phase concentration (say, C_{32}), calculate all the other phase concentrations from Eqs. (4.4-23) through (4.4-25), and then substitute these into the tie line Eq. (4.3-2). If this equation is satisfied, convergence has been attained; if it is not satisfied, a new C_{32} must be picked and the procedure repeated until either C_{32} does not change or Eq. (4.3-2) is satisfied.

Two other empirical representations of the distribution of components between phases are of interest: the conjugate curve and the tie line extension curve. Both require separate representations of the binodal curve, as in Eq. (4.4-23).

Conjugate curve. The conjugate curve is a curve in ternary space whose coordinates define the ends of the tie lines. Thus for phases 1 and 2, the conjugate curve would be of the form

$$C_{11} = f(C_{22}) \qquad (4.4\text{-}26)$$

Figure 4-14 shows the projections of the coordinates of this curve onto the binodal curve. The Hand distribution curve is of the form shown in Eq. (4.4-26). The conjugate curve must pass through the plait point.

Tie line extension curve. The tie line extension curve is another curve $C_3^0 = f(C_2^0)$ in ternary space that passes through the plait point, at which point it is tangent to the binodal curve (Fig. 4-15a). The two-phase tie lines are extensions of tangents from this curve through the binodal curve. Thus equations of the tie lines are given by straight lines having the equation

$$C_{3j} - f(C_2^0) = f'\bigg|_{C_2^0} (C_{2j} - C_2^0), \qquad j = 1 \text{ or } 2 \qquad (4.4\text{-}27a)$$

where $f'|_{C_2^0}$ is the slope of the tie line extension curve evaluated at the coordinate C_2^0. The tie lines follow from Eq. (4.4-27a), the equation of the extension curve, and the equation for the binodal curve.

A useful special case of the tie line extension curve occurs when all tie lines extend to a common point, as in Fig. 4-15(b). We need specify only the coordinates of this common point to define the equation for the tie lines

$$C_{3j} - C_3^0 = \eta(C_{2j} - C_2^0), \qquad j = 1 \text{ or } 2 \qquad (4.4\text{-}27b)$$

where η is the slope of the tie line. Note that if $C_3^0 > 0$, the selectivity of the components for the two phases can reverse near the base of the ternary. The representation is extremely simple because it requires only two values: any two of the coordi-

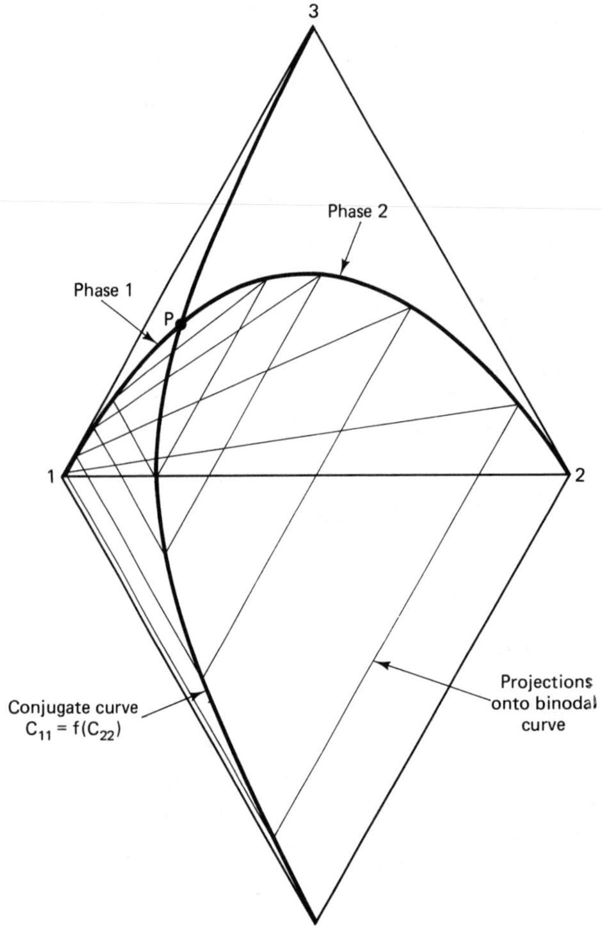

Figure 4-14 Schematic representation of a conjugate curve

nates C_i^0 or, alternatively, any of the plait point coordinates and one of the C_i^0 since the tie lines must be tangent to the binodal curve there.

This representation is far less general than either Eq. (4.4-24), (4.4-26), or (4.4-27). But experimental accuracy is often not enough to warrant more complicated equations. Moreover, the form (Eq. 4.4-27b) is extremely convenient for calculating the flow behavior of two-phase mixtures; we use it extensively in Chaps. 7 and 9.

4-5 CONCLUDING REMARKS

Multiple representations of phase behavior are clear evidence that no single method is sufficient. In most cases we find ourselves compromising between accuracy and mathematical ease in the resulting calculation. Our goal here is the exposition of the

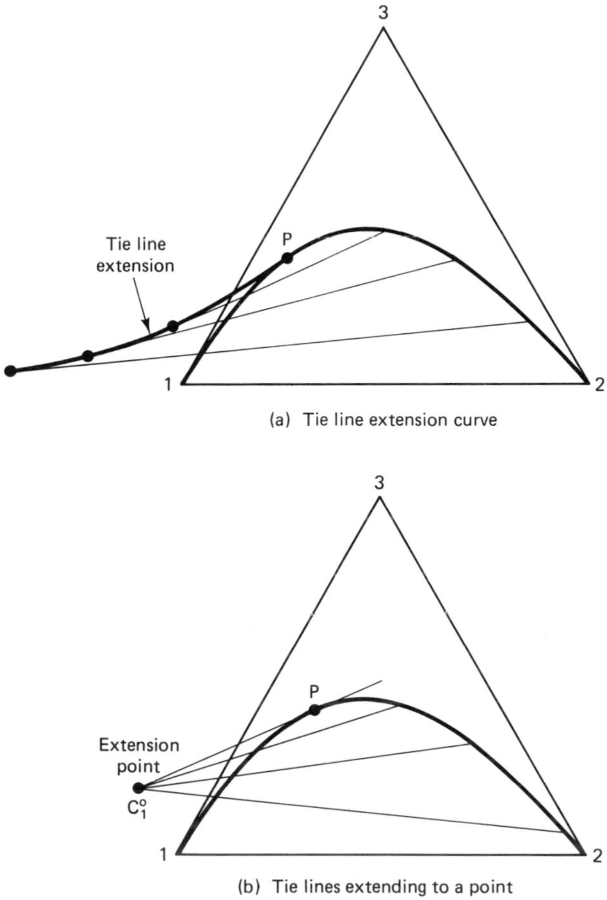

Figure 4-15 Tie line extension representation of phase behavior

underlying principles of EOR phenomena; hence we emphasize phase behavior representations which lend themselves to visual or graphical analysis in later calculations—as long as the representations themselves are qualitatively correct. The important points to grasp in this chapter, then, are graphical representations in Sec. 4-3, particularly as related to the ternary diagram; the physical meaning of tie lines and binodal curves; and the component distribution expressed by Eqs. (4.4-23), (4.4-24), and (4.4-27b).

EXERCISES

4A. *Pure Component Phase Behavior*. Sketch the following for a pure component:
- **(a)** Lines of constant pressure on a temperature–molar-volume plot
- **(b)** Lines of constant temperature on a density–pressure plot
- **(c)** Lines of constant molar volume on a temperature–pressure plot

4B. *Paths on a Pressure–Volume Plot*. Indicate the paths AA', BB', and DD', shown on the pressure–specific-volume plot in Fig. 4B, on the corresponding pressure–temperature plot.

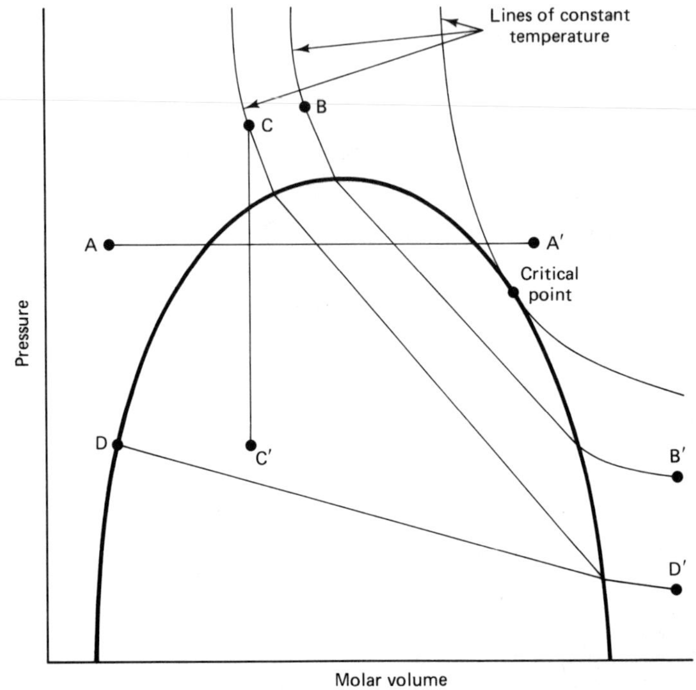

Figure 4B Pressure–specific-volume plot for Exercise 4B

4C. *Migration of P-T Envelope*. Figure 4C shows the hypothetical change in the pressure–temperature envelope of a crude oil as it is diluted with a more volatile component (CO_2). The quality lines within each envelope are in volume percent. For this data, sketch the pressure–composition diagram at 340 K and 359 K (152°F and 180°F). These temperatures are the critical temperatures for the 40% and 20% CO_2 mixtures. Include as many quality lines as possible.

4D. *Lever Rule Application*. Consider the three-component system represented in Fig. 4-11.

(a) Estimate the relative amounts of each phase present at overall compositions A, C, D, and E.

(b) Derive the expressions (indicated on the figure) for the relative amounts of each phase present at the three-phase overall composition.

(c) Estimate the relative amounts of each phase present at B.

4E. *Parameters for RKS and RP Equations of State*

(a) Derive the parameters a and b for the RKS equation using the critical constraints and the original equation given in Table 4-1.

(b) Derive the parameters a and b for the PR equation using the procedure of Martin and Hou (1955). Compare your results to Table 4-2.

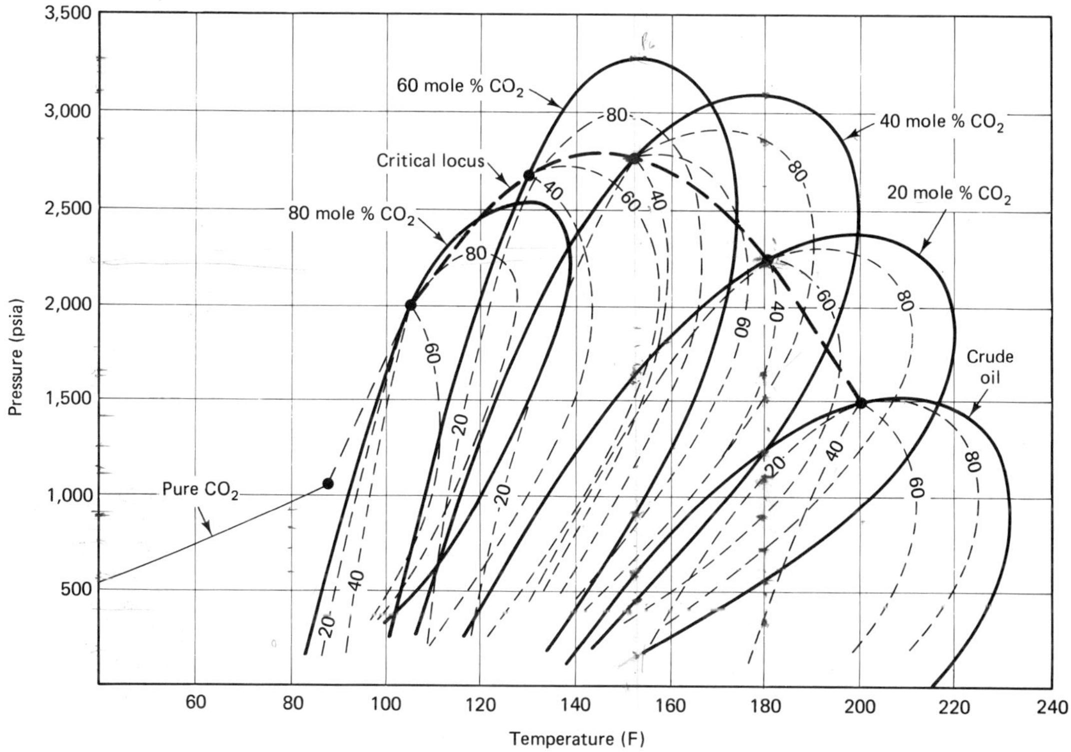

Figure 4C Change in crude oil pressure–temperature diagram with dilution by CO_2

4F. *Fugacity Coefficient from an Equation of State*

(a) Derive the expressions for the fugacity coefficient starting with Eq. (4.4-18a) for the PR equation.

(b) Show for both the PR and RKS equations that the fugacity coefficient for a mixture approaches that of a pure component as one of the x_i approaches 1.

4G. *Partially Soluble Binaries (Welch, 1982).* For cases when the partially soluble binaries on a ternary plot have some mutually soluble region, the Hand representation may be altered as

$$\frac{C'_{3j}}{C'_{2j}} = A_H \left(\frac{C'_{3j}}{C'_{1j}}\right)^{B_H}, \qquad j = 1, 2 \tag{4G-1}$$

and

$$\frac{C'_{32}}{C'_{22}} = E_H \left(\frac{C'_{31}}{C'_{11}}\right)^{F_H} \tag{4G-2}$$

where the C_{1j} are normalized concentrations

$$C'_{1j} = \frac{C_{1j} - C_{1L}}{C_{1U} - C_{1L}} \tag{4G-3}$$

$$C'_{2j} = \frac{C_{2j} - (1 - C_{1U})}{C_{1U} - C_{1L}} \tag{4G-4}$$

$$C'_{3j} = \frac{C_{3j}}{C_{1U} - C_{1L}} \tag{4G-5}$$

C_{1U} and C_{1L} are the upper and lower solubility limits of the 1-2 binary. Take $B_H = -1$ and $F_H = 1$ in the following.

(a) Derive an expression for A_H in terms of the true maximum height of the binodal curve $C_1 - C_2$. Show that the binodal curve takes value $C_{3\max}$ when $C'_1 = C'_2$ (symmetrical in normalized concentrations).

(b) Express E_H as a function of A_H and the component 1 coordinate of the plait point (C_{1p}). The A_H and E_H in parts (a) and (b) will also be a function of C_{1U} and C_{1L}.

(c) Plot the binodal curve and the two representative tie lines for $C_{1U} = 0.9$, $C_{1L} = 0.2$, $C_{3\max} = 0.5$, and $C_{1p} = 0.3$.

4H. *Using the Hand Representations.* The following data were collected from a three-component system at fixed temperature and pressure.

Phase 1		Phase 2	
Component 1	Component 2	Component 1	Component 2
0.45	0.31	0.015	0.91
0.34	0.40	0.020	0.89
0.25	0.48	0.030	0.85
0.15	0.60	0.040	0.82

The concentrations are in volume fractions.

(a) On a ternary diagram, plot as many tie lines as possible, and sketch in the binodal curve.

(b) Make a Hand plot from the data, and determine the parameters A_H, B_H, E_H, and F_H.

(c) Estimate the coordinates of the plait point from the plot in part (b).

4I. *Application of Conjugate Curve.* Consider the ternary diagram in Fig. 4I for a three-component system. The binodal curve is the solid line, and the conjugate curve the dotted line.

(a) Sketch in three representative tie lines.

(b) For the overall composition, marked as A, give the equilibrium phase compositions and the relative amounts of both phases.

(c) Plot the two-phase equilibria on a Hand plot.

(d) If the Hand equations are appropriate, determine the parameters A_H, B_H, E_H, and F_H.

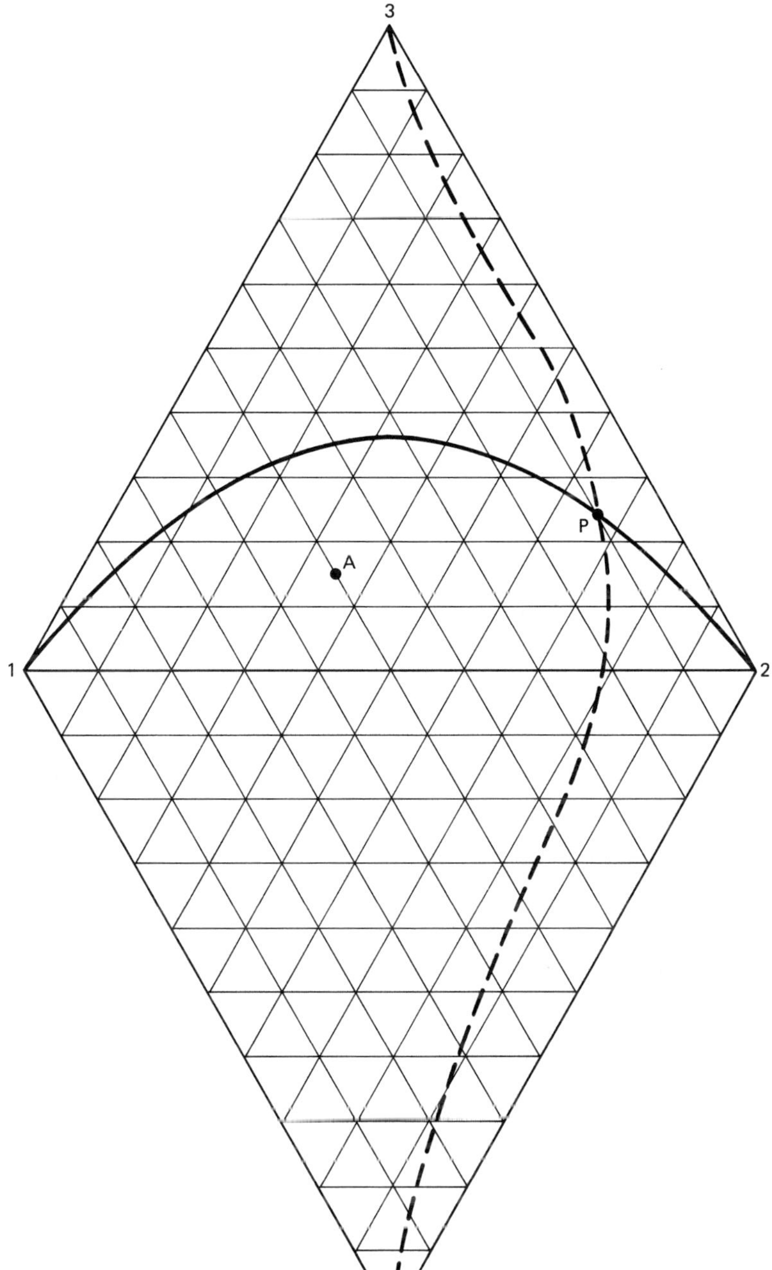

Figure 4I Diagram for Exercise 4I

5

Displacement Efficiency

The definitions for recovery, displacement, and sweep efficiencies in Eq. (2.5-5) apply to an arbitrary chemical component, but they are almost exclusively applied to oil and gas displacement. Since displacement efficiency and sweep efficiency are multiplied by each other, they are equally important to the magnitude of the recovery efficiency and, hence, the oil recovery. In Chap. 6, we discuss volumetric sweep efficiency; in this chapter, we present fundamental concepts in displacement efficiency.

For the most part, we restrict our discussion to oil displacement efficiency based on solutions to the fractional flow Eq. (2.4-3). We apply these equations to displacements in one-dimensional, homogeneous, isotropic permeable media. Thus, the results apply most realistically to displacements in laboratory floods, the traditional means of experimentally determining displacement efficiency. These results do not, of course, estimate recovery efficiency for three-dimensional, nonlinear flows without correcting for volumetric sweep efficiency and without correcting the displacement efficiency to account for differences in scale.

5-1 DEFINITIONS

If we assume constant oil density, the definition of displacement efficiency for oil becomes

$$E_D = \frac{\text{Amount of oil displaced}}{\text{Amount of oil contacted by displacing agent}} \tag{5.1-1}$$

E_D is bounded between 0 and 1. The rate at which E_D approaches 1 is strongly affected by the initial conditions, the displacing agent, and the amount of displacing agent. Fluid, rock, and fluid–rock properties also affect E_D. If the displacement is such that the displacing agent will contact all the oil initially present in the medium, the volumetric sweep efficiency will be unity, and E_D becomes the recovery efficiency E_R.

From Eq. (2.5-4) then,

$$E_D = 1 - \frac{\bar{S}_2}{\bar{S}_{2I}} \tag{5.1-2}$$

for an incompressible, single-component oil phase flowing in an incompressible permeable medium. Equation (5.1-2) says E_D is proportional to the average oil saturation in the medium. For cases where the oil may occur in more than one phase, or where more than oil can exist in the hydrocarbon phase, we must use the general definition (Eq. 2.5-5b).

5-2 IMMISCIBLE DISPLACEMENT

Virtually all of our understanding about EOR displacements begins with an understanding of the displacement of one fluid by an immiscible second fluid. The specific case of water displacing oil was first solved by Buckley and Leverett (1942) and later broadened by Welge (1952). In this section, we develop the Buckley-Leverett theory in a manner much like the original paper and several subsequent references (Collins, 1976; Craig, 1971; Dake, 1978).

For the isothermal flow of oil and water in two immiscible, incompressible phases in a one-dimensional permeable medium, the mass conservation equations of Table 2-2 reduce to

$$\phi \frac{\partial S_1}{\partial t} + u \frac{\partial f_1}{\partial x} = 0 \tag{5.2-1}$$

for flow in the positive x direction, as we discussed in Chap. 2. In this equation, f_1 is the fractional flow of water,

$$f_1 = \frac{u_1}{u} = \frac{\lambda_{r1}}{\lambda_{r1} + \lambda_{r2}} \left(1 - \frac{k\lambda_{r2} \Delta\rho g \sin\alpha}{u} \right) \tag{5.2-2}$$

in the absence of capillary pressure. In Eq. (5.2-2), α is the dip angle defined to be positive when measured in the counterclockwise direction from the horizontal, and $\Delta\rho = \rho_1 - \rho_2$ is the density difference between the water and oil phases.

The choice of S_1 as the dependent variable in Eq. (5.2-1) is largely a matter of convention; we could easily have chosen S_2 since $S_2 + S_1 = 1$, and $f_2 + f_1 = 1$. An important point is that in the absence of capillary pressure, f_1 is uniquely determined as a function of S_1 only through the relative permeability relations $\lambda_{r1} = k_{r1}/\mu_1$ and

$\lambda_{r2} = k_{r2}/\mu_2$ discussed in Sec. 3-3. In fact, since the shape of the f_1-S_1 curve proves to be the main factor in determining the character of the displacement, we digress briefly to discuss how flow conditions affect this curve.

Fractional Flow Curves

If we introduce the exponential form of the oil–water relative permeability curves (Eq. 3.3-4) into Eq. (5.2-2), we obtain

$$f_1 = \frac{1 - N_g^0(1 - S)^{n_2} \sin \alpha}{1 + \dfrac{(1 - S)^{n_2}}{M^0 S^{n_1}}} \tag{5.2-3a}$$

where

$$S = \frac{S_1 - S_{1r}}{1 - S_{2r} - S_{1r}} = \text{Reduced water saturation} \tag{5.2-3b}$$

and

$$M^0 = \frac{k_{r1}^0 \mu_2}{\mu_1 k_{r2}^0} = \text{Endpoint water–oil mobility ratio} \tag{5.2-3c}$$

$$N_g^0 = \frac{k k_{r2}^0 \Delta \rho g}{\mu_2 u} = \text{Gravity number} \tag{5.2-3d}$$

N_g^0 is the ratio of gravity to viscous pressure gradients based on the endpoint oil relative permeability. In the form of Eq. (5.2-3a), f_1 depends parametrically on M^0, N_g^0, α, and the shape of the relative permeability curves (n_1 and n_2). The f_1-S_1 curve is sensitive to all these factors, but usually M^0 and N_g^0 are most important. Figure 5-1 shows f_1-S_1 curves for various values of M^0 and $N_g^0 \sin \alpha$ with the other parameters fixed ($S_{1r} = 0.2$, $S_{2r} = 0.2$, $n_1 = n_2 = 2$). The S-shaped curves have an inflection point that varies with M^0 and $N_g^0 \sin \alpha$. The curvature of all curves generally becomes more negative as M^0 increases or $N_g^0 \sin \alpha$ decreases. The curves where f_1 is less than 0 or greater than 1 are physically correct. This circumstance indicates a flow where gravity forces are so strong that flow in the negative x direction occurs (water flows in the negative x direction for $f_1 < 0$). In Sec. 3-3, we showed that shifting the wettability of the permeable medium from water wet to oil wet caused k_{r1}^0 to increase and k_{r2}^0 to decrease. Thus for constant phase viscosities, making the medium more oil wet is qualitatively equivalent to increasing M^0. But for fixed relative permeability curves, the effect of increasing μ_1 or decreasing μ_2 is to decrease M^0.

Buckley-Leverett Solution

Returning now to Eq. (5.2-1), to calculate E_D, we seek solutions $S_1(x, t)$ subject to the initial and boundary conditions

$$S_1(x, 0) = S_{1I}, \qquad x \geqq 0 \tag{5.2-4a}$$

$$S_1(0, t) = S_{1J}, \qquad t \geqq 0 \tag{5.2-4b}$$

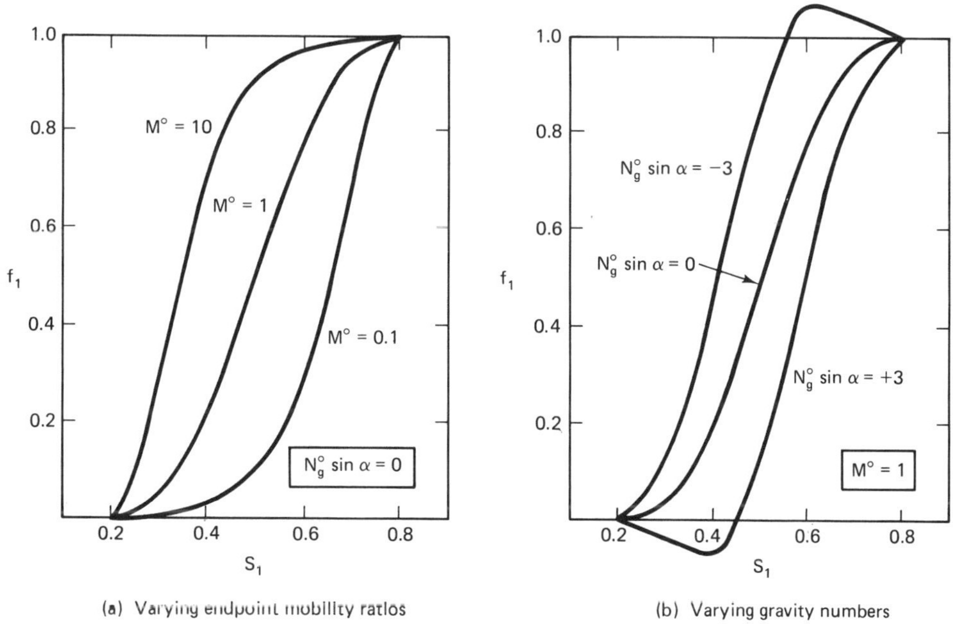

Figure 5-1 Fractional flow curves for $m = n = 2$ and $S_{1r} = S_{2r} = 0.2$

In core floods, a specified fractional flow is usually imposed on the inflow ($x = 0$) so that we may replace Eq. (5.2-4b) with

$$f_1(0, t) = f_1(S_1(0, t)) = f_{1J} = f_1(S_{1J}), \qquad t \geqq 0 \tag{5.2-4c}$$

This equation shows that f_1 is a function of x and t only through its dependence on S_1. The definition used in a given instance depends on the particular application. The conditions (Eqs. 5.2-4) also have a convenient geometrical interpretation in xt space because at the point $t = x = 0$, all values of S_1 between S_{1I} and S_{1J} exist. The Buckley-Leverett problem is usually posed with S_{1I} and S_{1J} taken to be S_{1r} and $1 - S_{2r}$, respectively.

For greater generality, we render Eqs. (5.2-1) and (5.2-4) into the following dimensionless forms:

$$\left(\frac{\partial S_1}{\partial t_D}\right) + \left(\frac{df_1}{dS_1}\right)\left(\frac{\partial S_1}{\partial x_D}\right) = 0 \tag{5.2-5a}$$

$$S_1(x_D, 0) = S_{1I}, \qquad x_D \geqq 0 \tag{5.2-5b}$$

$$S_1(0, t_D) = S_{1J}, \qquad t_D \geqq 0 \tag{5.2-5c}$$

where the dimensionless variables x_D and t_D are

$$x_D = \frac{x}{L} = \text{Dimensionless position} \tag{5.2-6a}$$

$$t_D = \int_0^t \frac{udt}{\phi L} = \text{Dimensionless time} \tag{5.2-6b}$$

L is the total macroscopic permeable medium dimension in the x direction. In these equations, u may be a function of time but not of position because of the assumption of incompressibility. Moreover, df_1/dS_1 is a total derivative since f_1 is a function of S_1 only. Introducing dimensionless variables reduces the number of parameters in the problem from four (ϕ, u, S_{1I}, and S_{1J}) in Eqs. (5.2-1) and (5.2-4) to two (S_{1I} and S_{1J}). We could further reduce the number by redefining the dependent variable S_1 (see Exercise 5A).

The dimensionless time t_D can also be expressed as

$$t_D = \int_0^t \frac{Audt}{\phi AL} = \int_0^t \frac{qdt}{V_p} \tag{5.2-7}$$

where A is the cross-sectional area of the one-dimensional medium in the direction perpendicular to the x axis, q is the volumetric flow rate, and V_p is the pore volume. t_D is the total volume of fluid injected up to time t divided by the total pore volume of the medium. In principle, V_p is well defined even for a highly irregular geometry so that t_D is a scaling variable in virtually any application. In fact, t_D is the fundamental variable used to scale from the laboratory to the field. It has been used with a wide variety of definitions for the reference volume V_p (see Table 5-1). Numerical values of t_D are frequently given as "fraction of a pore volume," or simply "pore volume"; thus it is easy to confuse with V_p, the actual pore volume, which has units of L^3 (t_D, of course, has no units).

We seek solution to Eqs. (5.2-5) in the form $S_1(x_D, t_D)$. S_1 may be written as a total differential

$$dS_1 = \left(\frac{\partial S_1}{\partial x_D}\right)_{t_D} dx_D + \left(\frac{\partial S_1}{\partial t_D}\right)_{x_D} dt_D \tag{5.2-8}$$

TABLE 5-1 TABULATION OF VARIOUS DEFINITIONS FOR DIMENSIONLESS TIME

Reference volume	Usage
Area × length × porosity	Core floods
Area × thickness × porosity	General
($A\phi h = V_p$ = total pore volume)	
V_p × volumetric sweep efficiency	
($V_p \times E_v$ = floodable pore volume = V_{PF})	Micellar polymer floods
$V_{PF} \times \Delta S_2$ = movable pore volume	Waterfloods
$V_{PF} \times S_{2I}$ = hydrocarbon pore volume ($HCPV$)	Miscible floods

Note: $t_D = \dfrac{\text{Total volume of fluid injected}}{\text{Reference volume}}\Big\}$ consistent units from which follows that the velocity v_{S_1} of a point with constant saturation S_1 in $x_D t_D$ space is

$$\left(\frac{dx_D}{dt_D}\right)_{S_1} = -\frac{(\partial S_1/\partial t_D)_{x_D}}{(\partial S_1/\partial x_D)_{t_D}} \equiv v_{S_1} \tag{5.2-9}$$

v_{S_1} is the "specific" velocity of the saturation S_1 because it has been normalized by the bulk fluid interstitial velocity u/ϕ. It is dimensionless. You can see this by converting Eq. (5.2-9) back to dimensional quantities using the definitions (Eqs. 5.2-6).

Eliminating either of the derivatives in Eq. (5.2-9) by Eq. (5.2-5a) gives

$$v_{S_1} = \frac{df_1}{dS_1} = f_1' \tag{5.2-10}$$

This equation says the specific velocity of a constant saturation S_1 is equal to the derivative of the fractional flow curve at that saturation. In dimensional form, Eq. (5.2-10) is the Buckley-Leverett equation. Since all saturations between S_{1I} and S_{1J} are initially at the origin in x_D-t_D space, and v_{S_1} is defined with S_1 constant, the position of any saturation $S_{1I} \leqq S_1 \leqq S_{1J}$ at a given t_D is

$$x_D\big|_{S_1} = \left.\frac{df_1}{dS_1}\right|_{S_1} t_D = f_1'(S_1) t_D \tag{5.2-11}$$

where we include evaluation symbols to help clarify the subsequent development. Equation (5.2-11) is the solution to the one-dimensional water-displacing-oil problem; by selecting several S_1's between S_{1I} and S_{1J}, we can construct $S_1(x_D, t_D)$. Figure 5-2(a) shows the procedure for one of the fractional flow curves of Fig. 5-1. Except for relatively simple cases (see Exercise 5E), the relation (Eq. 5.2-11) generally cannot be solved explicitly for $S_1(x_D, t_D)$.

Shock Formation

Figure 5-2(a) also shows a disconcerting tendency for an S-shaped f_1-S_1 curve to generate solutions that have three values of S_1 at the same x_D and t_D. In Fig. 5-2(b), this occurs for $0.64 < x_D < 0.94$. Of course, such triple values are nonphysical though they are entirely valid mathematically. The triple values are the result of the saturation velocity v_{S_1} increasing over some saturation region ($S_{1I} < S_1 < S_1'$ in Fig. 5-2) as S_1 changes from its initial (downstream) value to the final (upstream) value.

We eliminate the triple value region by invoking the formation of *shocks*, discontinuous changes in a physical quantity such as pressure (as in the case of sonic booms), concentration, or in this case, saturation. Shocks are characteristic features of hyperbolic equations, a class of which are the dissipation-free conservation equations. Strictly speaking, shocks are not present in nature since some dissipation (dispersion, diffusion, capillary pressure, compressibility, and thermal conductivity) is always present, which militates against their formation. When such effects are present, the shocks are smeared or spread out around the shock front position, but the position of the shock is unaltered. Despite this restriction, shocks play a central

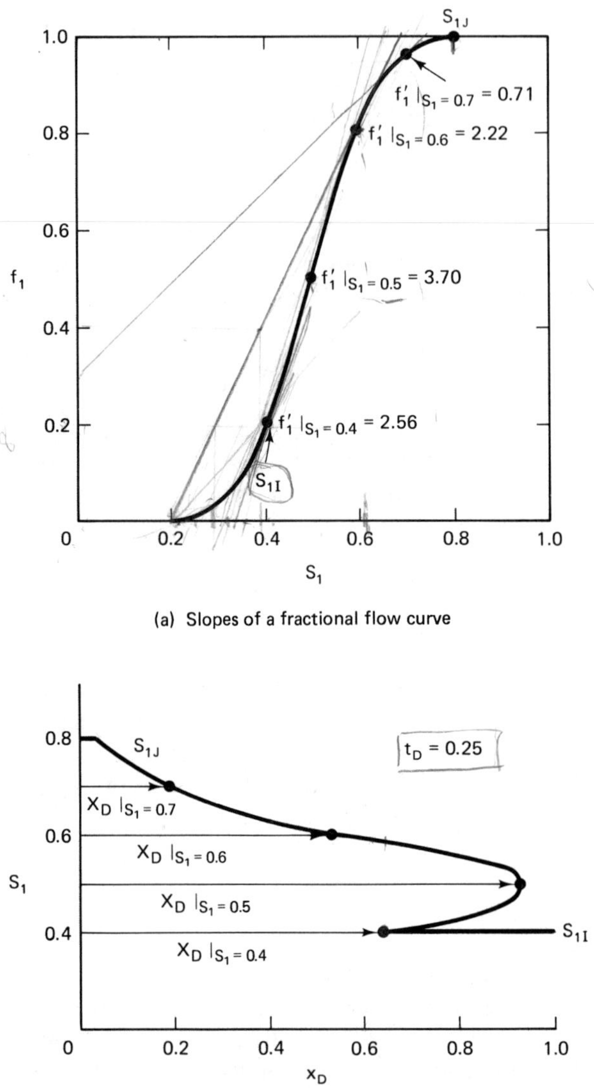

(a) Slopes of a fractional flow curve

(b) Corresponding saturation profile

Figure 5-2 Buckley-Leverett construction of $S_1(x_D, t_D)$

role in fractional flow theory, where dissipative effects are neglected, and describe many actual flows to a good approximation.

To calculate the velocity and magnitude of the shock, we recast the differential equations of this chapter into difference equations. This we do generally in Sec. 5-4; here we restrict ourselves to the water-displacing-oil problem already begun. Paradoxically, we find that calculations are considerably easier when shocks form. Figure 5-3(a) shows a water saturation shock moving from left to right. The water saturation ahead of the shock is S_1^- (downstream direction), and that behind the shock is

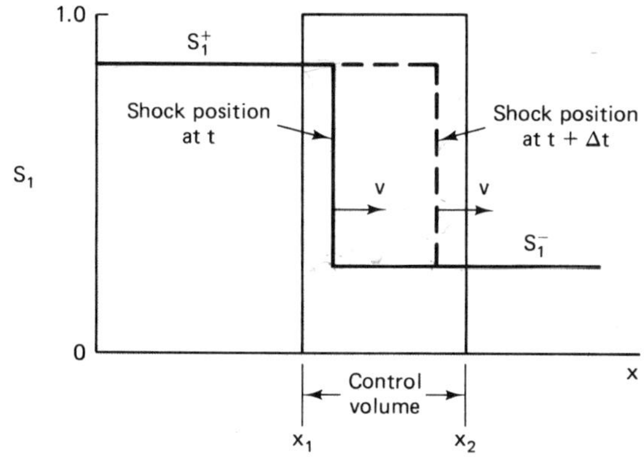

(a) Schematic of material balance around shock

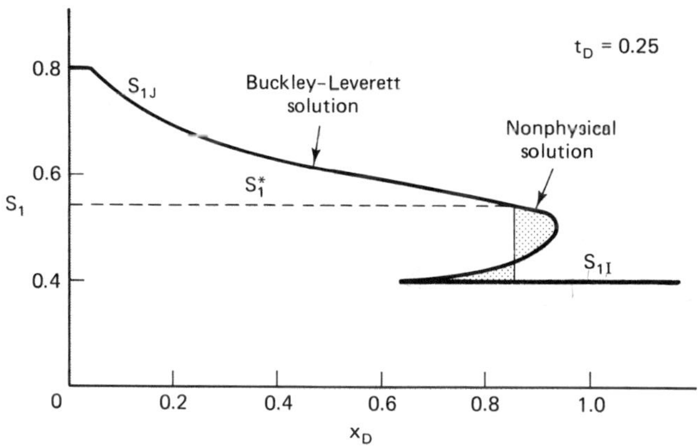

(b) Saturation profile for fractional flow curve of Fig. 5-2(a)

Figure 5-3 Water saturation profiles with shocks

S_1^+ (upstream direction). The quantity $\Delta S_1 = S_1^+ - S_1^-$ is the saturation jump across the shock. A cumulative water balance on a control volume that contains the shock in the time interval Δt is

$$\begin{pmatrix}\text{Volume water}\\ \text{present at } t+\Delta t\end{pmatrix} - \begin{pmatrix}\text{Volume water}\\ \text{present at } t\end{pmatrix} = \begin{pmatrix}\text{Volume water}\\ \text{in during } \Delta t\end{pmatrix} - \begin{pmatrix}\text{Volume water}\\ \text{out during } \Delta t\end{pmatrix}$$

$$[(v(t+\Delta t) - x_1)S_1^+ + (x_2 - v(t+\Delta t))S_1^-]A\phi$$
$$- [(vt - x_1)S_1^+ + (x_2 - vt)S_1^-]A\phi = [f_1(S_1^+) - f_1(S_1^-)]\int_t^{t+\Delta t} q\,dt$$

After some cancellation, we obtain a specific shock velocity

$$v_{\Delta S_1} = \frac{f_1(S_1^+) - f_1(S_1^-)}{S_1^+ - S_1^-} \equiv \frac{\Delta f_1}{\Delta S_1} \tag{5.2-12}$$

To incorporate shock formation into the water-displacing-oil problem, consider a saturation profile containing a triple value over some region and containing a single value elsewhere (Fig. 5-3b). In general, some saturation S_1^* will mark the end of the continuous water saturation region and the beginning of a shock. This saturation must simultaneously satisfy Eqs. (5.2-10) and (5.2-12); Eq. (5.2-10) gives velocities of S_1 greater than S_1^*, and Eq. (5.2-12) gives velocities of S_1 less than S_1^*. Equating Eqs. (5.2-10) and (5.2-12) yields the following equation for S_1^*:

$$f_1' \big|_{S_1^*} = \frac{f_1(S_1^*) - f_1(S_{1I})}{S_1^* - S_{1I}} \tag{5.2-13}$$

where we have taken $S_1^- = S_{1I}$ in Eq. (5.2-12). Equation (5.2-13) lends itself to a graphical solution since

$$f_1 - f_1(S_{1I}) = m(S_1 - S_{1I}) \tag{5.2-14}$$

is the equation of a straight line of slope m passing through the point (f_{1I}, S_{1I}) on the fractional flow plot. If $m = f_1'|_{S_1}$, then m is the slope of the fractional flow plot at S_1^*. Comparing Eq. (5.2-14) to Eq. (5.2-13), S_1^* is at the tangent to the fractional flow curve of a straight line passing through the point (f_{1I}, S_{1I}). Figure 5-4 schematically illustrates this construction. The slope of this straight line is the specific shock

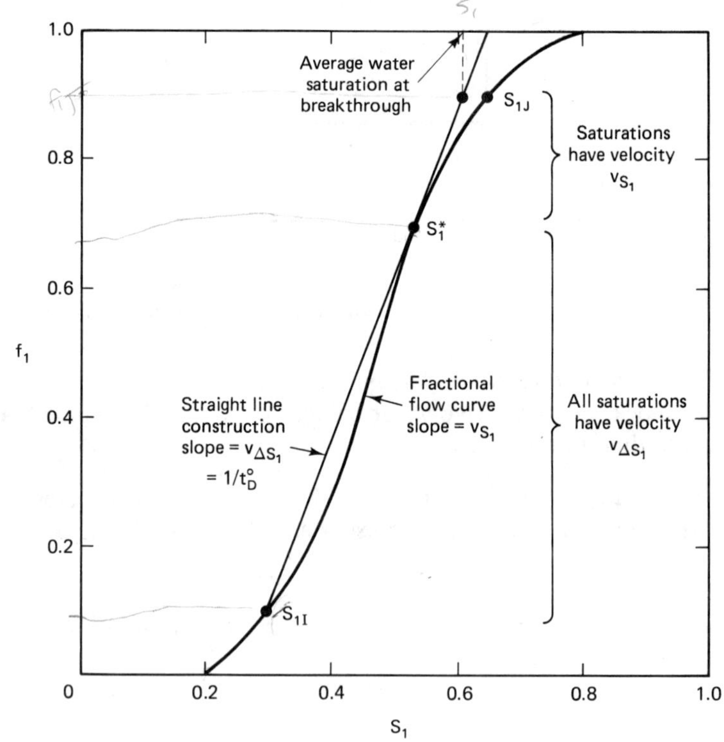

Figure 5-4 Schematic illustration of shock construction

velocity. The shock itself is a discontinuous change in saturation from S_{1I} to S_1^* at $x_D = v_{\Delta S_1} t_D$ as Fig. 5-3(b) illustrates. The saturation S_1^* is not the same as S_1' (Fig. 5-2), the saturation having the largest v_{S_1}. S_1^* is the saturation whose position requires the net area between the mathematical solution and the physical solution (shaded region in Fig. 5-3b) to be zero. This requires the shock to preserve the material balance. With this construction, all saturation velocities are monotonically (though not continuously) decreasing in the upstream direction. Figure 5-3(b) illustrates the results of the entire construction. The resulting saturation profile is sometimes called the leaky piston profile.

Wave Classification

Before further developing this theory and its applications to EOR, we define a few more terms used in subsequent discussions. These definitions are important to the interpretation of x_D-t_D plots that graphically present the solution $S_1(x_D, t_D)$.

We have been discussing how to calculate water saturation as a function of position and time for water–oil displacements. A plot of saturation, or concentration, versus time at fixed position is a saturation *history*. If the fixed position in such a plot is at the outflow end of the permeable medium, it is an *effluent* history. Plots of saturation versus position at fixed time are saturation *profiles*. Figure 5-2(b) is a water saturation profile. Changes in saturation with time and position are saturation *waves*. Thus the previous development estimates the rate of propagation of waves through a permeable medium.

An important and unifying aspect of our understanding of EOR displacements is the study and characterization of the number and types of waves they form. Depending on their spreading characters, waves may be classified into four categories.

1. A wave that becomes more diffuse on propagation is a nonsharpening, rarified, or *spreading* wave. When these waves occur, the rate of spreading is usually much larger than that caused by dissipation.
2. A wave that becomes less diffuse on propagation is a *sharpening* wave. In the absence of dissipation, these waves will become shocks even if the initial saturation profile is diffuse. When dissipation is present, these waves will asymptotically approach a *constant pattern* condition (see Sec. 5-3).
3. A wave that has both spreading and sharpening character is *mixed*. The Buckley-Leverett water saturation wave of Fig. 5-2(b) is mixed, being a sharpening wave for $S_{1I} < S_1 < S_1^*$ and a spreading wave for $S_1^* < S_1 < S_{1J}$.
4. A wave that neither spreads nor sharpens on propagation is *indifferent*. In the absence of dissipation, indifferent waves appear as shocks.

This behavior may be summarized by defining a dimensionless mixing or transition zone Δx_D. This is the fraction of the total system length that lies between arbitrary saturation limits at a given time. We take the saturation limits to be 0.1 and 0.9

of the span between the initial and injected saturations

$$\Delta x_D(t_D) = x_D\big|_{S_{0.1}} - x_D\big|_{S_{0.9}} \tag{5.2-15a}$$

where

$$S_{0.1} = 0.1(S_{1J} - S_{1I}) + S_{1I} \tag{5.2-15b}$$

$$S_{0.9} = 0.9(S_{1J} - S_{1I}) + S_{1I} \tag{5.2-15c}$$

The exact value of the limits is unimportant to the behavior of the mixing zone. The wave classification, which may be restated as Δx_D, increases with time for spreading waves, decreases for sharpening waves, and either increases or decreases for mixed waves depending on whether the shock portion of the wave exceeds the saturations used to define Δx_D. The mixing zone concept has general use in classifying mixing phenomena in a wide variety of displacements.

The final definition concerning the Buckley-Leverett development is the *time–distance diagram*. These diagrams are plots of x_D versus t_D on which appear lines of constant saturation. Figure 5-5 shows a time–distance diagram for the water–oil displacement in Figs. 5-3(b) and 5-4. The constant saturation curves are straight lines with slope given by v_{S_1} from Eq. (5.2-9). Similarly, shocks are the bold straight lines with slope given by Eq. (5.2-12). The region having varying saturation is shaded. Regions of constant saturation are adjacent to the waves and have no saturation lines. Time–distance diagrams are very convenient since they subsume both profiles and histories.

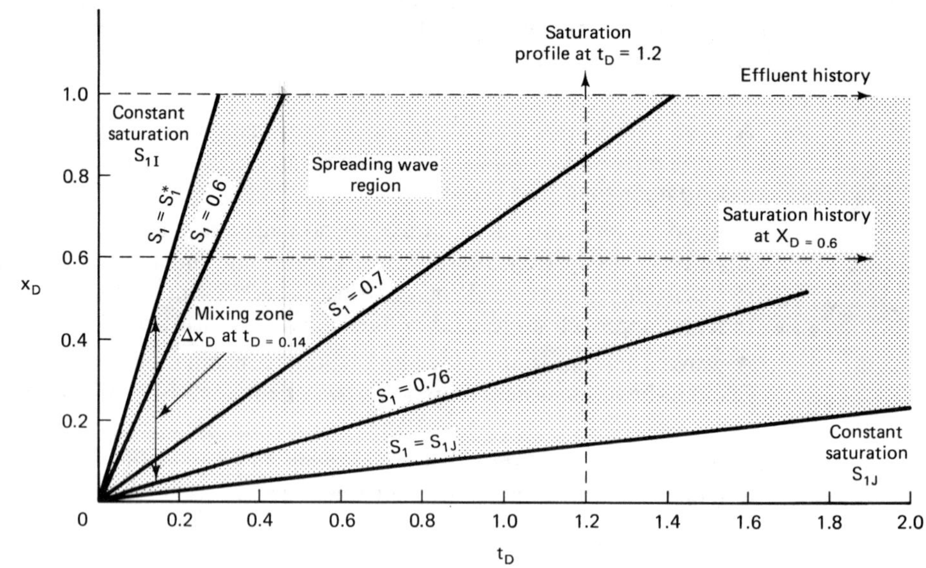

Figure 5-5 Time–distance diagram for displacement of Figs. 5-3(b) and 5-4

From the definition of effluent history, the shock portion of the water–oil displacement arrives at $x_D = 1$ when

$$t_D^0 = \frac{S_1^* - S_{1I}}{f_1^* - f_{1I}} \tag{5.2-16a}$$

from Eqs. (5.2-12) and (5.2-13). The *breakthrough time* t_D^0 is an important event in the displacement; for values $t_D > t_D^0$, we are producing some of the water being injected. The obvious inefficiency of this should suggest that we would like to conduct the displacement so that t_D^0 is as large as possible; that is, we would like to enhance the shock-forming character of the displacement. For $t_D > t_D^0$, the water saturation at the outflow end is given implicitly by

$$f_1'\big|_{x_D=1} = \frac{1}{t_D} \tag{5.2-16b}$$

from Eq. (5.2-10). In laboratory floods, it is usually more direct to measure $f_1\big|_{x_D=1}$, the water "cut," than the saturation at the effluent end. The water and oil cuts $(1 - f_1\big|_{x_D=1})$ are functions of only time from Eq. (5.2-16b).

Average Saturations

In the displacement efficiency, we must have some way to calculate average saturations since, from Eq. (5.1-2), these appear in the definition of E_D. These averages are provided by the Welge integration procedure (Welge, 1952). Consider the saturation profile in Fig. 5-3(b) at fixed t_D, and let x_{D1} be any dimensionless position at or behind the shock front position, $x_{D1} \leqq v_{\Delta S_1} t_D$. The average water saturation behind x_{D1} is

$$\hat{S}_1(t_D) = \frac{1}{x_{D1}} \int_0^{x_{D1}} S_1 \, dx_D \tag{5.2-17}$$

Equation (5.2-17) may be integrated by parts

$$\hat{S}_1 = \frac{1}{x_{D1}} \left((x_D S_1) \Big|_0^{x_{D1}} - \int_{S_{1J}}^{S_{11}} x_D \, dS_1 \right) \tag{5.2-18}$$

where $S_{11} = S_1\big|_{x_{D1}}$. Since x_{D_1} is in the spreading portion of the saturation wave, the x_D integrand may be substituted by Eq. (5.2-11)

$$\hat{S}_1 = S_{11} - \frac{1}{x_{D1}} \int_{S_{1J}}^{S_{11}} t_D f_1' \, dS_1 \tag{5.2-19}$$

which may be readily integrated (recall t_D is fixed) to

$$\hat{S}_1 = S_{11} - \frac{t_D}{x_{D1}} (f_{11} - f_{1J}) \tag{5.2-20}$$

Equation (5.2-20) relates the average water saturation behind x_{D1} to the fractional flow and saturation at that point. t_D may be replaced by Eq. (5.2-11) at this point to give

$$\hat{S}_1 = S_{11} - \frac{(f_{11} - f_{1J})}{f'_{11}} \tag{5.2-21}$$

Equation (5.2-21) is the final form of the Welge integration.

The most common use of this procedure is to let $x_{D1} = 1$ after water breakthrough ($t_D \geq t_D^0$), at which point $\hat{S}_1 = \bar{S}_1$, and f_{11} becomes the water cut. Thus the water saturation at the outflow end may be calculated from Eq. (5.2-20) as

$$S_1|_{x_D=1} = \bar{S}_1 - t_D(f_{1J} - f_1|_{x_D=1}) \tag{5.2-22}$$

If we know the water cut and average water saturation from direct measurement, simultaneously applying Eqs. (5.2-16) and (5.2-22) provides a way of estimating fractional flow curves ($f_1|_{x_D=1}$ versus $S_1|_{x_D=1}$ or f_{11} versus S_{11}) from experimental data.

The average water saturation follows from Eq. (5.2-21) for $\bar{S}_1$ with the f_1-S_1 curve known. This equation may be rearranged to give

$$f_1|_{x_D=1} - f_{1J} = f'_1|_{x_D=1}(S_1|_{x_D=1} - \bar{S}_1) \tag{5.2-23}$$

Thus $\bar{S}_1$ at any $t_D \geqq t_D^0$ is given by the extension of a straight line tangent to the fractional flow curve at $(f_1, S_1)_{x_D=1}$ to intersect with the y coordinate at $f_1 = f_{1J}$. The dimensionless time required to bring this point to $x_D = 1$ is the reciprocal slope of this line from Eq. (5.2-16). Figure 5-4 shows the graphical procedure for this. From the $\bar{S}_1$ thus determined, $\bar{S}_2 = 1 - \bar{S}_1$ may be used in the definition (Eq. 5.1-2) to calculate E_D.

The above construction and Eqs. (5.2-22) and (5.2-23) apply only to dimensionless times after breakthrough. Before breakthrough the average water saturation is

$$\bar{S}_1 = S_{1I} + t_D(f_{1J} - f_{1I}), \qquad t_D < t_D^0 \tag{5.2-24}$$

by applying the overall water material balance (Eq. 2.5-2) to this special case. Equations (5.2-22) and (5.2-24) are identical except for the value used for the effluent water cut.

We are now ready to demonstrate the effect of endpoint mobility ratio M^0, relative permeability, and $N_g^0 \sin\alpha$ on oil displacement efficiency. Figure 5-6 schematically shows the effect of these parameters for displacements with $f_{1I} = 0$ and $f_{1J} = 1$. Figure 5-6 shows, from top to bottom, plots of E_D versus t_D, water saturation profiles at various t_D, and the fractional flow curve that would give the indicated behavior. From left to right, the figures show oil displacement behavior for decreasing M^0, increasing $N_g^0 \sin\alpha$, and increasing water wetness through shifts in the relative permeability curves. Figure 5-6 represents three of the four types of waves—spreading, mixed, and sharpening. Several important conclusions follow directly from Fig. 5-6.

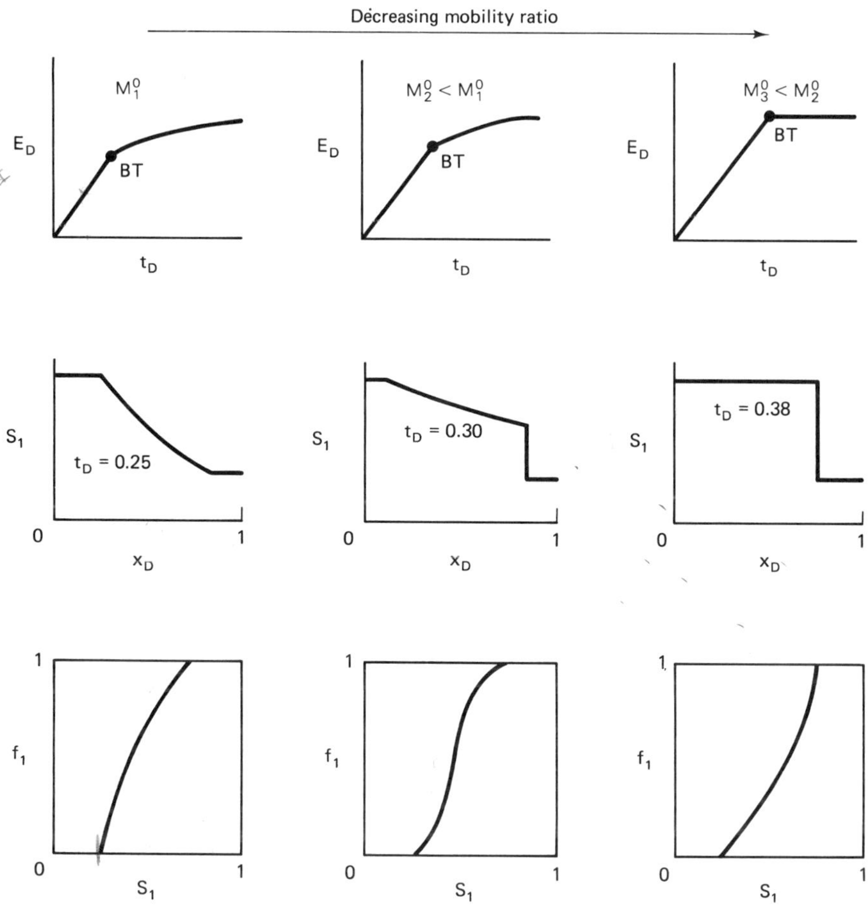

Figure 5-6 Schematic illustration of effect of mobility ratio on displacement efficiency

1. Any change that increases the size of the shock portion of the water saturation wave also increases E_D at any given t_D. These changes also delay water breakthrough and decrease the time over which the permeable medium is simultaneously producing two phases.
2. Decreasing M^0, increasing $N_g^0 \sin \alpha$, and increasing water wetness improve E_D. Of these three, M^0 is usually the only one we can have any impact on. In Chap. 6, we see that decreasing mobility ratio also increases vertical and areal sweep efficiency; hence decreasing the mobility ratio improves oil recovery in at least three ways. EOR processes that rely, partly or totally, on lowering the mobility ratio between the displacing and displaced fluids are said to be based on the *mobility ratio concept* of oil recovery. Figure 5-6 shows that when the water saturation wave becomes a complete shock, no advantage is to be gained

on E_D by further lowering M^0. Finally, there is no unique value of M^0 at which the wave changes from spreading to sharpening since the displacement is affected also by the shape of the relative permeability curves.

3. However low M^0 might be, the ultimate displacement efficiency

$$E_D^\infty = \frac{(S_{2I} - S_{2r})}{S_{2I}}$$

is limited by the presence of a residual oil saturation. EOR methods that intend to recover residual oil must rely on something other than the mobility ratio concept, such as displacing with miscible agents (see Sec. 5-5 and Chap. 7) or lowering the water–oil interfacial tension (see Chap. 9).

Besides M^0, at least two other mobility ratios are in common use. The average mobility ratio $\overline{M}$, defined as

$$\overline{M} = \frac{(\lambda_{r1} + \lambda_{r2})|_{S_1=\bar{S}_1}}{(\lambda_{r1} + \lambda_{r2})|_{S_1=S_{1I}}} \tag{5.2-25a}$$

is the ratio of total relative mobility at the average water saturation behind the shock front to the same quantity evaluated at the initial water saturation. $\overline{M}$ is commonly used to correlate the areal sweep efficiency curves (see Chap. 6). The shock front mobility ratio M_{sh} is

$$M_{sh} = \frac{(\lambda_{r1} + \lambda_{r2})|_{S_1=S_1^*}}{(\lambda_{r1} + \lambda_{r2})|_{S_1=S_{1I}}} \tag{5.2-25b}$$

M_{sh} is the quantity that controls the formation of viscous fingers. For pistonlike displacements, all three definitions are the same.

The most general definition of mobility ratio is actually the ratio of pressure gradients ahead of and behind a displacing front. The above definitions, depending on the character of the displacing front, follow from this for the case of incompressible flow (spatially independent volumetric flow rate). For compressible flows or flows of condensing fluids, the general definition is more appropriate (see Chap. 11 and Exercise 5J).

5-3 DISSIPATION IN IMMISCIBLE DISPLACEMENTS

In this section, we discuss two common dissipative effects in one-dimensional flows: capillary pressure and fluid compressibility. Both phenomena are dissipative; they cause mixing zones to grow faster than or differently from a dissipation-free flow. Both phenomena also bring additional effects.

Capillary Pressure

We do not present a closed-form solution to the water conservation equation. But we can qualitatively illustrate the effect of capillary pressure on a water–oil displacement and can give, through scaling arguments, quantitative guidelines on when it might be important. For incompressible fluids and with capillary pressure P_c included, the water material balance (Eq. 5.2-1) still applies, but the water fractional flow (Eq. 5.2-2) becomes (see Exercise 5F)

$$f_1(S_1) = \frac{\lambda_{r1}}{\lambda_{r1} + \lambda_{r2}} \left(1 - \frac{k\lambda_{r2}\Delta\rho g \sin\alpha}{u}\right) + \frac{k\lambda_{r1}(\partial P_c/\partial x)}{(1 + \lambda_{r1}/\lambda_{r2})u} \qquad (5.3\text{-}1)$$

The first term on the right side of Eq. (5.3-1) is simply the water fractional flow in the absence of capillary pressure (Eq. 5.2-2); thus many of the conclusions about displacements with $P_c = 0$, though somewhat modified, carry over to displacements with capillary pressure. The second right term in Eq. (5.3-1) is the contribution of P_c to the water fractional flow. Including the capillary pressure term causes the character of Eq. (5.2-1) to change from hyperbolic to parabolic, a general result of dissipative effects because of the spatial P_c derivative.

The capillary pressure in Eq. (5.3-1) is the phase pressure difference between two continuous oil and water phases (see Sec. 3-2). The derivative $\partial P_c/\partial x = (dP_c/dS_1) \cdot (\partial S_1/\partial x)$ has a positive sign for displacements in both oil-wet or water-wet media since dP_c/dS_1 is negative for both cases (see Fig. 3-5), and $\partial S_1/\partial x$ is also negative. Therefore, for waterfloods, capillary pressure increases the water fractional flow at a given water saturation. This augmentation is particularly important in regions having large saturation gradients, that is, around shock fronts predicted by the Buckley-Leverett theory. In an oil displacement of water, P_c causes a smaller water fractional flow since $\partial S_1/\partial x > 0$.

The effect of P_c on a one-dimensional displacement is to spread out the water saturation wave, particularly around shocks; Fig. 5-7, which illustrates how this comes about, is a simulated water saturation and pressure profile for a one-dimensional waterflood in a water-wet medium. Figure 5-7(a) shows water saturation profiles with and without capillary pressure; Fig. 5-7(b) shows the corresponding pressure profiles. Both panels are at the same t_D. The dotted phase pressures in Fig. 5-7(b) are those that would be present if the shock remained in the water saturation profile. Of course, representing shock waves with $P_c \neq 0$ is not correct, but such a portrayal presents the driving force for capillary mixing.

Ahead of the front (downstream), the difference between the oil and water phase pressures is constant and equal to the capillary pressure at S_{1I}. At the front, the phase pressures change rapidly. But behind the front (upstream), the difference between the oil and water phase pressures declines to the value at $S_1 = S_{1J}$. Compare these comments to Figs. 5-7(a) and 3-5. There is now a local pressure gradient at the shock that causes oil to flow upstream (countercurrent imbibition) and water to flow downstream faster than under the influence of viscous forces only. The resulting local mixing causes the shock to spread (Fig. 5-7a) and the pressure discontinuity to disappear. Behind the front, in the spreading portion of the water saturation wave, the effect of capillary pressure is small.

Capillary pressure will be small if the system length L is large. Consider the

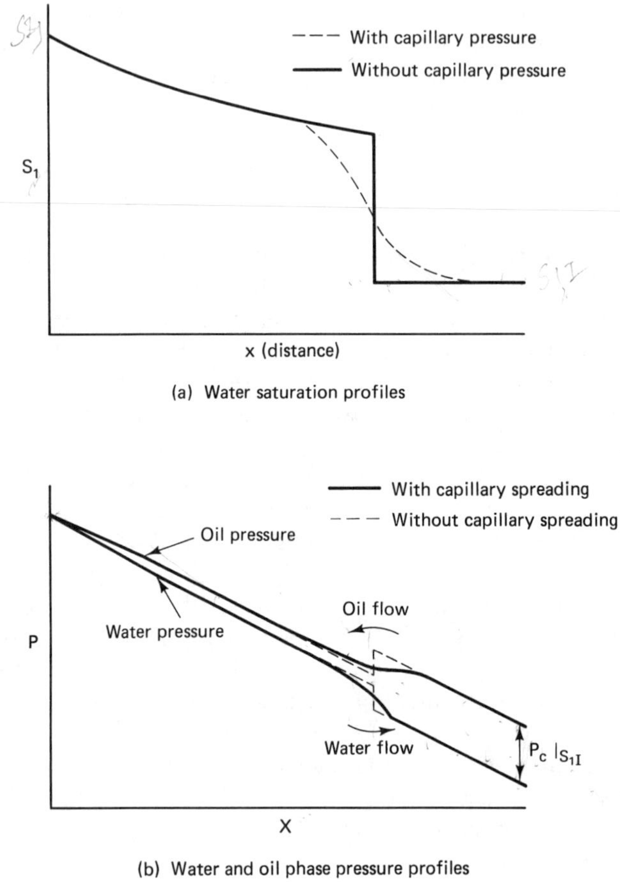

Figure 5-7 Saturation and pressure profiles under longitudinal capillary imbibition (Yokoyama, 1981)

dimensionless water conservation equation with Eq. (5.3-1) substituted and $\alpha = 0$

$$\frac{\partial S_1}{\partial t_D} + \frac{\partial}{\partial x_D}\left(\frac{1}{1 + \dfrac{\lambda_{r2}}{\lambda_{r1}}}\right) + \frac{\partial}{\partial x_D}\left(\frac{k\lambda_{r1}}{uL\left(1 + \dfrac{\lambda_{r1}}{\lambda_{r2}}\right)}\frac{\partial P_c}{\partial x_D}\right) = 0 \tag{5.3-2}$$

The last term on the left side of this equation is nonlinear in S_1 and thus difficult to estimate. Using the Leverett j-function expression (Eq. 3.2-2), we can write Eq. (5.3-2) as

$$\frac{\partial S_1}{\partial t_D} + \frac{\partial}{\partial x_D}\left(\frac{1}{1 + \dfrac{\lambda_{r2}}{\lambda_{r1}}}\right) - \frac{1}{N_{\text{RL}}}\frac{\partial}{\partial x_D}\left(g(S_1)\frac{\partial S_1}{\partial x_D}\right) = 0 \tag{5.3-3}$$

basis (C'_{is} is amount i on solid/pore volume). Thus C_i and C'_{is} are directly comparable and may be used together in later work without the need to manipulate units. The definition of overall flux is from Hirasaki (1981) and Helfferich (1981).

In principle, the fluxes F_i are functions of the C_i for $i = 1, \ldots, N_C$, and we may carry over many of the definitions, particularly those of saturation velocity, directly from Sec. 5-2. In practice, however, the relations $F_i = F_i(C_1, C_2, \ldots, C_{N_C})$ are extremely convoluted. We discuss this in more detail later, but we can give a summary of this relation here.

With C_i known, the C_{ij} and S_j may be calculated from phase equilibrium relations. The exact nature of the "flash" calculation depends on the nature of the phase behavior (see Sec. 4-4 and Chaps. 7 and 9). With the S_j and C_{ij} known, the phase relative permeabilities $k_{rj} = k_{rj}(S_j, C_{ij})$ and viscosities $\mu_j = \mu_j(C_{ij})$ may be calculated from petrophysical relations (see Sec. 3-3). From these follow the relative mobilities $\lambda_{rj} = k_{rj}/\mu_j$, which lead directly to the f_j from Eq. (2.4-2). If the phase densities are also required (if, for example, the permeable medium is not horizontal), they follow from $\rho_j = \rho_j(C_{ij})$ (Eq. 2.2-12). With the f_j and C_{ij} known F_i follows from Eq. (5.4-3c). If needed, $C'_{is} = C'_{is}(C_{ij})$ may be calculated also from the adsorption isotherm (see Chaps. 8 and 9).

Despite this complexity, we can write Eq. (5.4-2) as

$$\left(1 + \left(\frac{\partial C'_{is}}{\partial C_i}\right)_{x_D}\right)\frac{\partial C_i}{\partial t_D} + \left(\frac{\partial F_i}{\partial C_i}\right)_{t_D}\frac{\partial C_i}{\partial x_D} = 0, \qquad i = 1, \ldots, N_C \tag{5.4-4}$$

The partial derivatives $(\partial C'_{is}/\partial C_i)_{x_D}$ and $(\partial F_i/\partial C_i)_{t_D}$ in Eq. (5.4-4) follow from the chain rule. These derivatives are not the same as $(\partial C'_{is}/\partial C_j)_{C_{m \neq j}}$, which are in the definition of the total differential. The latter derivatives may be calculated directly from $C'_{is} = C'_{is}(C_{ij})$ and $F_i = F_i(C_i)$, whereas the former derivatives require knowledge of $C_i = C_i(x_D, t_D)$, which are solutions. Therefore, Eq. (5.4-4) is of little use except to allow the definition of specific concentration velocity v_{C_i}

$$v_{C_i} = \frac{(\partial F_i/\partial C_i)_{t_D}}{1 + (\partial C'_{is}/\partial C_i)_{x_D}}, \qquad i = 1, \ldots, N_C \tag{5.4-5a}$$

by analogy with Eq. (5.2-10). The definition of the specific shock velocity $v_{\Delta C_i}$ is

$$v_{\Delta C_i} = \frac{(\Delta F_i/\Delta C_i)}{1 + (\Delta C'_{is}/\Delta C_i)} \tag{5.4-5b}$$

Without additional constraints, the definitions (Eqs. 5.4-5a and 5.4-5b) impart no new information. But for the water–oil case of Sec. 5-2, they reduce to $C_i = S_1$, $F_i = f_1$, and $C'_{is} = 0$, giving

$$v_{C_i} = v_{S_1} = \left(\frac{\partial f_1}{\partial S_1}\right)_{t_D} = \frac{df_1}{dS_1} = f'_1(S_1) \tag{5.4-6}$$

The last equality is possible because f_1 is a function of S_1 only; hence $f'_1 = (\partial f_1/\partial S_1)_{t_D} = (\partial f_1/\partial S_1)_{x_D}$. Certainly for more complicated cases, this simplification where g is a positive dimensionless function of water saturation

$$g(S_1) = -\left(\frac{1}{1 + \dfrac{\lambda_{r1}}{\lambda_{r2}}}\right)\left(\frac{S_1 - S_{1r}}{1 - S_{2r} - S_{1r}}\right)^{n_1} \frac{dj}{dS_1} \tag{5.3-4}$$

and N_{RL}, the Rapoport and Leas number, is a dimensionless constant first implied by these authors (1953) to indicate when capillary pressure effects will be important.

$$N_{RL} = \left(\frac{\phi}{k}\right)^{1/2} \frac{\mu_1 u L}{k_{r1}^0 \phi \sigma_{12} \cos \theta} \tag{5.3-5}$$

Figure 5-8 is a plot of fractional oil recovery at water breakthrough versus $\mu_1 v L$ (recall $v = u/\phi$) from the experimental work of Rapoport and Leas. Since the $S_{1I} = 0$ in their cores, the vertical axis in Fig. 5-8 is the breakthrough displacement efficiency, E_D^0. As $\mu_1 v L$ increases, E_D^0 increases to a maximum of 0.58. For larger $\mu_1 v L$, E_D^0 is constant at the value predicted by the Buckley-Leverett theory.

Rapoport and Leas did not plot their results against the more general N_{RL}; however, using the given $k = 0.439\ \mu m^2$ and $\phi = 0.24$, and taking $k_{r1}^0 \sigma_{12} \cos \theta =$ 1 mN/m (typical for water-wet media), P_c will not affect a one-dimensional water–

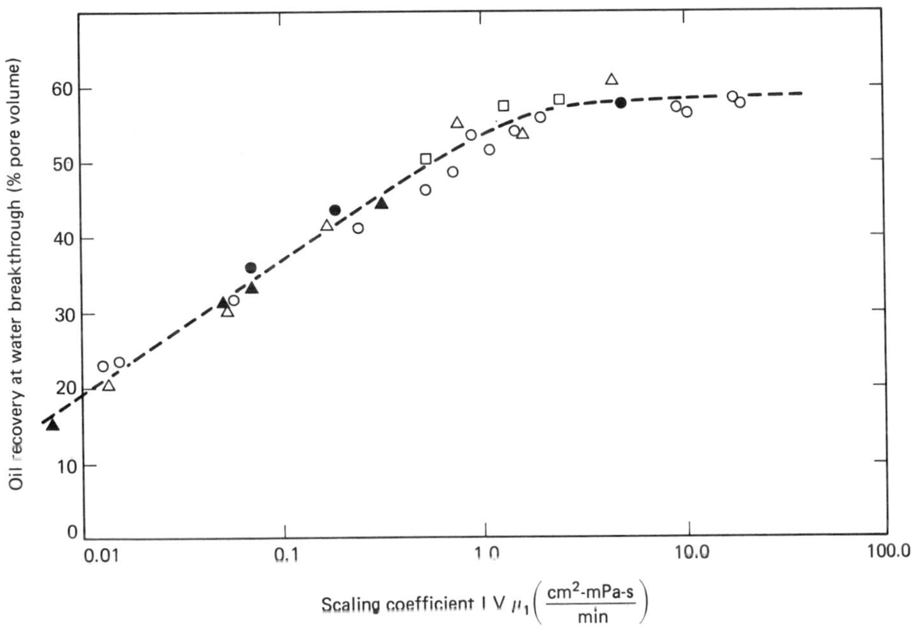

Figure 5-8 Relation between oil recovery at breakthrough and scaling coefficient in dry-filmed alumdum cores with no connate water. Different symbols represent varying core lengths and oil viscosities. (From Rapoport and Leas, 1953)

oil displacement if N_{RL} is greater than about 3. Because of the length appearing in the numerator of Eq. (5.3-5), P_c will affect the displacement front to a much greater degree in laboratory floods than in field-scale displacements because of the large disparity in L.

Of course, on a microscopic scale, capillary forces are important in determining the amount of trapped or residual oil in either laboratory or field displacements. In Sec. 3-3, we saw that the S_{2r} depended on a local viscous-to-capillary-force ratio, the capillary number N_{vc}. A common form of capillary number $N_{vc} = v\mu_1/k_{r1}^0 \sigma_{12} \cos\theta$ is embedded in the definition of N_{RL}

$$N_{RL} = \left(\frac{\phi}{k}\right)^{1/2} L N_{vc} \tag{5.3-6}$$

The factor, $L(\phi/k)^{1/2}$, is a measure of the ratio of the macroscopic permeable medium dimension to a characteristic rock dimension. Therefore, N_{vc} and N_{RL} are expressing the same physical idea—capillary-to-viscous-force ratios—but at different scales.

Recall that if N_{vc} is less than about 10^{-5}, the residual phase saturations are roughly constant. For well-sorted media, we can then put limits on N_{RL} so that capillary forces, on any scale, do not affect the displacement

$$3 < N_{RL} < 10^{-5} L \left(\frac{\phi}{k}\right)^{1/2} \tag{5.3-7}$$

(no dissipation) (constant residual saturations)

For large L, this is an extremely wide range and accounts for the common neglect of all capillary forces in one-dimensional displacement calculations. For laboratory scale, it may not be possible to satisfy both requirements.

N_{RL} may be expressed in more direct ways. From Eq. (5.3-5), we can substitute Darcy's law for water evaluated at $S_1 = 1 - S_{2r}$ for $v = u/\phi$ to obtain

$$N'_{RL} = \left(\frac{k}{\phi}\right)^{1/2} \frac{\Delta P_1}{\sigma_{12} \cos\theta} \tag{5.3-8}$$

where ΔP_1 is the pressure drop across the permeable medium measured through the water phase. The terms containing permeability and interfacial tension may be expressed in terms of the Leverett j-function to give yet another approximation to N_{RL}

$$N''_{RL} = \frac{\Delta P_1}{\Delta P_c} \tag{5.3-9}$$

where ΔP_c is the change in capillary pressure between the initial and final water saturation states. Equation (5.3-9) is a direct comparison of viscous to capillary pressure drops and is the least rigorous, but most direct, of all the measures.

For small N_{RL}, capillary pressure will cause shock waves to spread out. Though there is a parallel between dispersion in miscible displacements (see Sec. 5-5) and P_c effects in immiscible displacements, the analogy does not carry over to mixing zone growth. We show in Sec. 5-5 that dispersive mixing zones grow in proportion to the square root of time. Capillary pressure generally causes mixing zones to grow exponentially to some asymptotic limit where it proceeds, without further growth, in simple translation. How this comes about may be qualitatively explained by considering a water saturation wave that would be a shock over the entire possible saturation range, as in the right column in Fig. 5-6, where we neglected P_c effects. As we have seen, P_c effects cause such a wave to spread, but there is still a strong tendency for the wave to sharpen because of the convex-upward shape of the fractional flow curve. These two effects tend to balance each other, causing the wave to approach an asymptotic limit. The existence of such a limit further restricts the importance of capillary pressure as a mixing mechanism in one dimension. Asymptotic or "stabilized" mixing zones in one-dimensional laboratory waterfloods have been noted by several authors (Bail and Marsden, 1957).

No discussion of how capillary pressure influences a one-dimensional displacement is complete without some mention of the capillary end effect. This effect occurs when there is a discontinuity in the capillary pressure curve as, for example, when the one-dimensional permeable medium consists of two homogeneous media of differing permeabilities arranged in series. But it most commonly occurs at the end of a laboratory core where the flowing phases pass from a permeable to a nonpermeable region. The saturation behavior at the plane of discontinuity is considerably different from that predicted by the Buckley-Leverett theory.

Consider the water saturation and pressure profiles of a waterflood in a water-wet medium shown in Fig. 5-9. Capillary forces are such that they cannot be neglected. Figure 5-9(a) shows the instant that water arrives at the outflow end ($x = L$), and Fig. 5-9(b) shows some time later. On the right of the outflow end, there is no permeable medium. This region has a capillary pressure curve that is zero everywhere except at $S_1 = 0$, where all values of capillary pressure exist. The oil and water phase pressures must be continuous at $x = L$; hence the water saturation for $x > L$ is constrained to be zero because there is a nonzero phase pressure difference. This, in turn, implies water cannot flow across the outflow end of the medium until the capillary pressure just inside the system vanishes. With no production at $x = L$, but with continual water transport to the outflow end, the water saturation must build up at $x = L$ until $P_c = 0$ ($S_1 = 1 - S_{2r}$) at this plane. Hence the capillary end effect causes a delay in water production and a distortion of the water saturation at $x = L$ compared to that predicted by the Buckley-Leverett theory (Fig. 5-9b).

This delay can cause considerable error in applying the Welge integration procedure (Eq. 5.2-22). The capillary end effect has been observed experimentally by Kyte and Rapoport (1958) and in simulations by Douglas et al. (1958). Figure 5-10 reproduces data reflecting the capillary end effect.

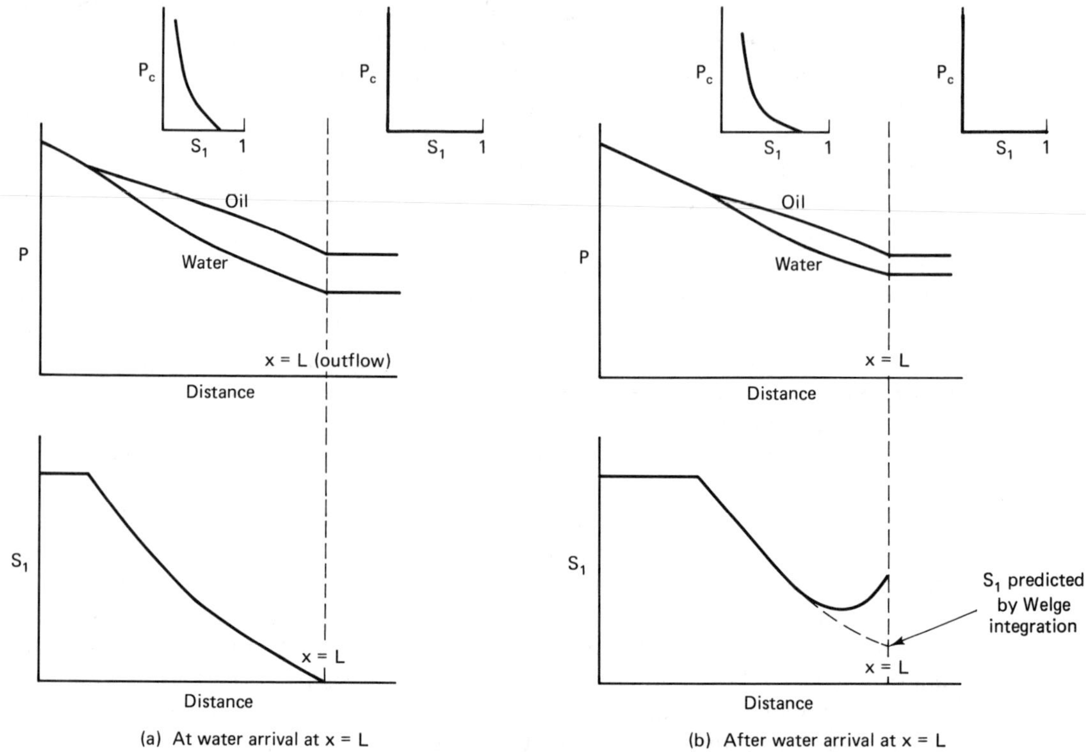

Figure 5-9 Schematic of the capillary end effect

To eliminate the capillary end effect, laboratory floods have been run at high velocities and with long lengths (both increase N_{RL}) or by placing a second permeable material at the outflow end to ensure good capillary contact.

Fluid Compressibility

A second dissipative effect is fluid compressibility. Figure 5-11 shows water saturation profiles for two waterfloods having compressible oil and incompressible water (Fig. 5-11a) and compressible water and incompressible oil (Fig. 5-11b). The completely incompressible Buckley-Leverett case is shown for comparison. These results are from computer simulations that were at constant water injection rate (Fig. 5-11a) and constant oil production rate (Fig. 5-11b). We present the results as the product of compressibility and total pressure drop ΔP (neglecting capillary forces) since this quantity determines the appropriateness of the small compressibility fluid assumption in well test analysis. For $c_j \Delta P$ products of 0.01 or less, the effect of fluid compressibility is negligible; the smearing of the shock fronts for the $c_j \Delta P = 1.25 \times 10^{-3}$ runs is because of numerical dispersion, which is an artificial dissipa-

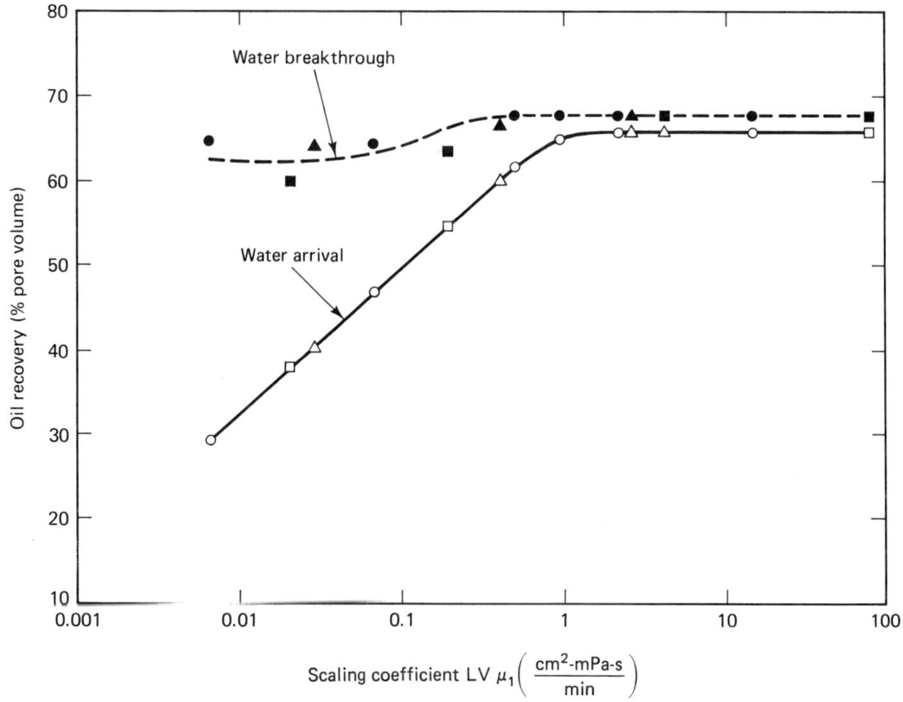

Figure 5-10 Correlation of waterflood test data in strongly water-wet alumdum cores (from Kyte and Rapoport, 1958)

tive effect. The $c_j \Delta P$ products shown in Fig. 5-11 are, of course, unrealistically high; we have selected these values merely to emphasize the effect of compressibility.

The effect of either oil or water compressibility is to spread out the Buckley-Leverett shock front in addition to the spreading caused by numerical dispersion but the effect does not become pronounced until $c_j \Delta P$ is 1 or more. However, we would expect displacements in which both fluids are compressible to experience a combined dissipative effect with greater spreading. In Fig. 5-11(a), the water saturation exceeds $1 - S_{2r}$ at the inflow end. At higher pressure, oil compression below its residual occurs. Similarly, in Fig. 5-11(b), the water saturation exceeds S_{1r} at the effluent end because, at the reduced pressure, the water will expand. These effects are characteristics of the particular conditions the runs were made under. If we had held the production pressure constant and not allowed phase saturations to decrease below their respective residuals, neither effect would be present. Still, we can see from Fig. 5-11 that the effect of compressibility is qualitatively similar to that of capillary pressure; a spreading of the shock fronts occurs but with a smaller effect on the saturation "tail."

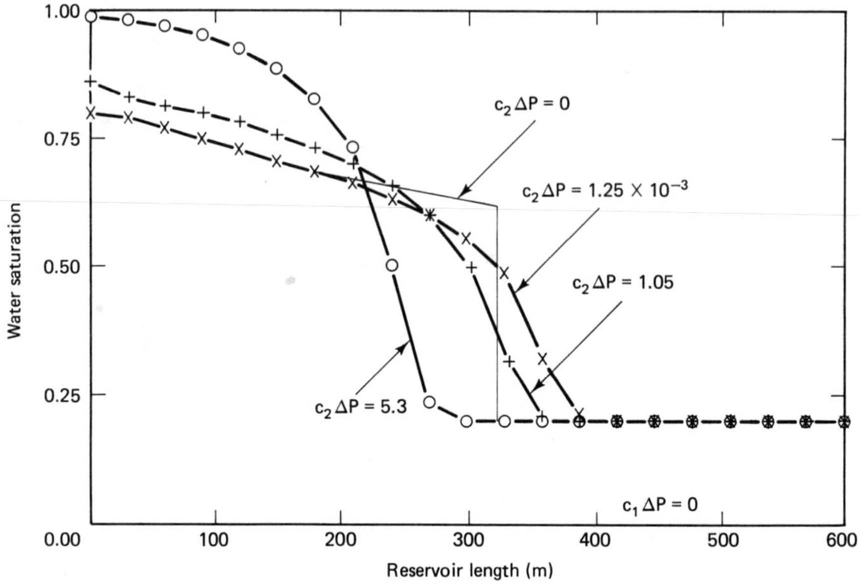

(a) Compressible oil, incompressible water

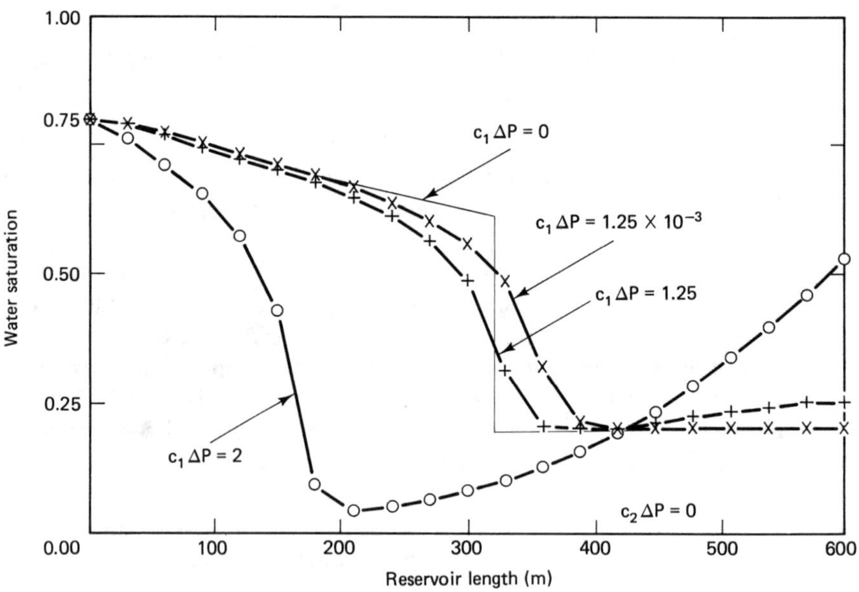

(b) Compressible water, incompressible oil

Figure 5-11 Water saturation profiles for one-dimensional water-displacing-oil floods at $t = 200$ days (adapted from Samizo, 1982)

5-4 IDEAL MISCIBLE DISPLACEMENTS

Two components are mutually miscible if they mix in all proportions without an interface forming between them. The definition is translated into the fluid flow equations by allowing a phase to be composed of several components they are mutually miscible within.

In this section, we discuss isothermal miscible displacements using fractional flow theory and with one or more phases present. Our presentation considers ideal miscible displacements with components that do not change the properties of the phases they are formed in (see Chap. 7 for more complicated displacements).

Concentration Velocities

Many of the concepts in Sec. 5-2 readily generalize to miscible displacements. We write a one-dimensional conservation equation for $i = 1, \ldots, N_C$ components as

$$\phi \frac{\partial}{\partial t}\left(\sum_{j=1}^{N_P} S_j C_{ij} + \left(\frac{1-\phi}{\phi}\right) C_{is}\right) + u \frac{\partial}{\partial x}\left(\sum_{j=1}^{N_P} f_j C_{ij}\right) = 0, \qquad i = 1, \ldots, N_C \tag{5.4-1}$$

Equation (5.4-1) is a special case of Eq. (2.4-10) with dispersion neglected. f_j is the fractional flow of phase j, given by Eq. (2.4-2) with capillary pressure neglected, and C_{ij} and C_{is} are the phase concentrations of component i in phase j and on the solid, respectively. Of course, the assumptions associated with Eq. (2.4-10)—constant porosity, incompressible fluids, and ideal mixing—also apply. In nondimensional form, Eq. (5.4-1) becomes

$$\frac{\partial}{\partial t_D}(C_i + C'_{is}) + \frac{\partial F_i}{\partial x_D} = 0, \qquad i = 1, \ldots, N_C \tag{5.4-2}$$

where

C_i = Overall fluid phase concentration of species i

$$= \sum_{j=1}^{N_P} S_j C_{ij} \tag{5.4-3a}$$

C'_{is} = Solid phase concentration of i on a pore volume basis

$$= C_{is}\left(\frac{1-\phi}{\phi}\right) \tag{5.4-3b}$$

F_i = Overall flux of species i

$$= \sum_{j=1}^{N_P} f_j C_{ij} \tag{5.4-3c}$$

The transform accomplished by Eq. (5.4-3b) changes the solid phase concentration from a solid volume basis (C_{is} is amount i on solid/volume solid) to a pore volume is not possible; still, many of the displacements of interest may be solved with the coherent or simple wave theory that we discuss in Sec. 5-5. We now discuss other particularly simple special cases of miscible displacements.

Tracers in Two-Phase Flow

The simplest case we consider is the miscible displacement in single-phase flow of component 2 by component 1. For this case, f_j and S_j are zero for all j except 1. For this particular j, f_j and S_j are unity. If component 1 does not adsorb, the concentration velocity becomes

$$v_{C_1} = 1 \tag{5.4-7}$$

from either Eq. (5.4-5a) or (5.4-5b). This seemingly trivial result has two important consequences.

1. The dimensional velocity of component 1 is equal to the bulk fluid velocity, meaning the dimensionless breakthrough time t_D^0 for component 1 is also unity. From Eq. (5.2-7), we may estimate the pore volume of the medium by knowing the cumulative fluid injected when breakthrough occurs (see Exercise 5K). Components that travel at the bulk fluid velocity are "conservative" tracers for this reason.
2. The specific concentration velocity is independent of C_1, meaning waves caused by conservative tracers are indifferent, which is generally true for ideal miscible displacements.

Most EOR displacements are only partially miscible. To illustrate a partially miscible displacement, we now consider a displacement of oil–water mixture at water saturation S_{1I} by another at a water fractional flow of $f_{1J} = f_1(S_{1J})$. We wish to distinguish between the initial and injected oil and water, so let's suppose the injected fluids contain conservative tracers. The oil-miscible tracer is completely immiscible in water, and the water-miscible tracer is similarly immiscible in oil. The process is now the displacement of an oil–water mixture by a tagged oil–water mixture. To keep this simple, we assume the tracers do not affect the fractional flow functions at all. The specific velocity of the tagged water-resident water wave is

$$v_{1'} = \frac{\partial(C_{11}f_1)}{\partial(C_{11}S_1)} = \frac{f_1}{S_1} \tag{5.4-8a}$$

from Eq. (5.4-5a), where C_{11} is the water tracer concentration. Similarly, the specific velocity of the tagged oil is

$$v_{2'} = \frac{f_2}{S_2} = \frac{1 - f_1}{1 - S_1} \tag{5.4-8b}$$

$v_{1'}$ and $v_{2'}$ are both independent of tracer concentration, hence the miscible tagged water and oil waves are indifferent. Of course, since neither of the tracers affects f_1, the saturation velocity of the water—tagged or untagged—is given by Eq. (5.2-10) or Eq. (5.2-12). The values of f_1 and S_1 in Eq. (5.4-8) are determined by the character of the oil–water wave.

Figure 5-12 illustrates some of the cases that can occur for this displacement. On each plot, the fractional flow curve is on the left, and a saturation–concentration profile is on the right. In case A, $S_{1I} = S_{1J}$ and the specific velocities are the slopes of straight lines passing through (0, 0) and $(f_1, S_1)_J$ and (1, 1) and $(f_1, S_1)_J$, respectively, from Eqs. (5.4-8a) and (5.4-8b). $v_{2'} > v_{1'}$, and the tagged oil wave leads the tagged water wave.

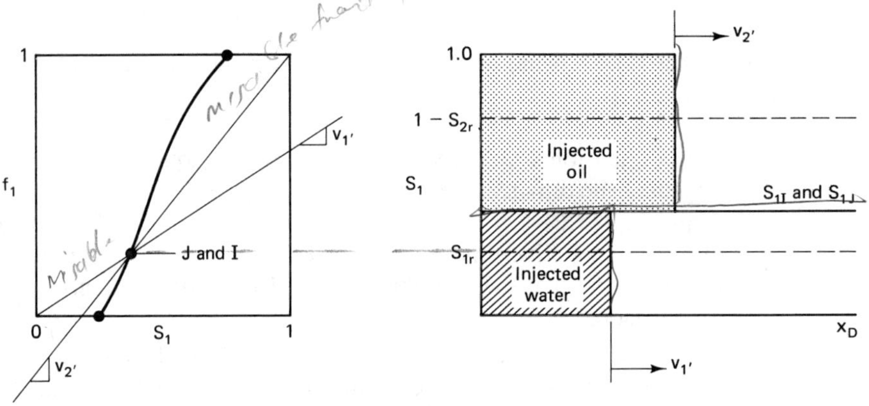

(a) Case A

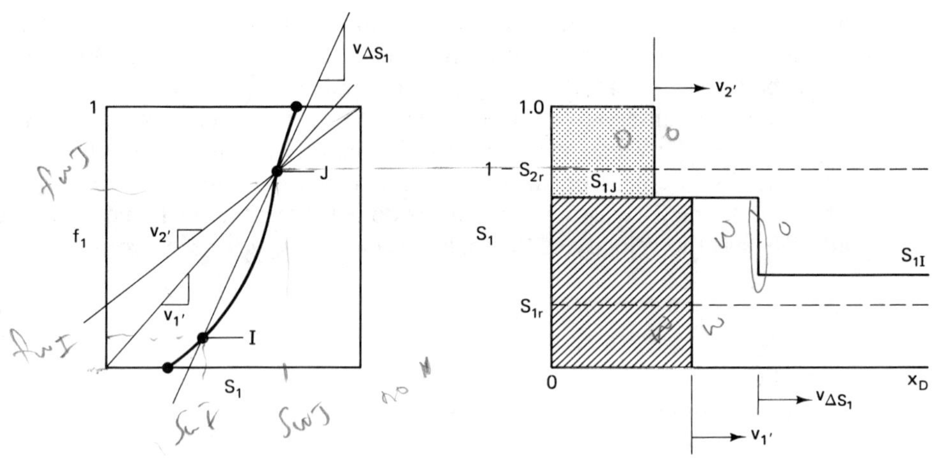

(b) Case B

Figure 5-12 Illustration of various partially miscible displacements

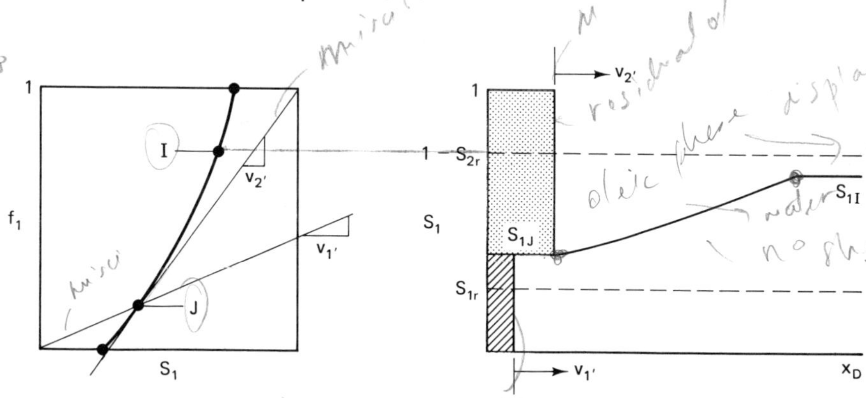

(c) Case C

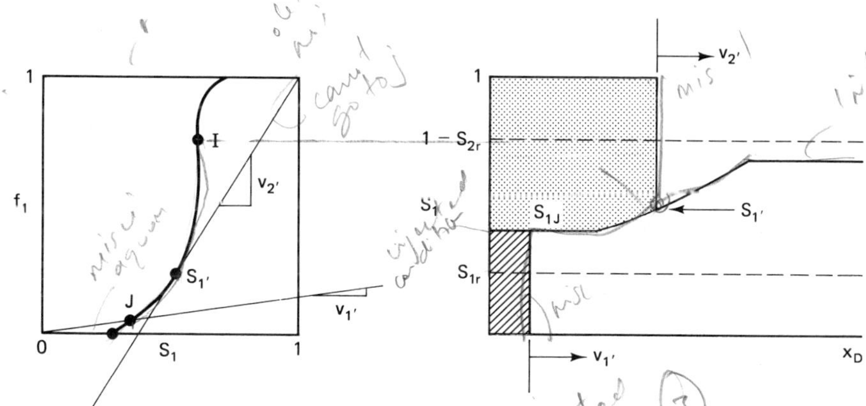

(d) Case D

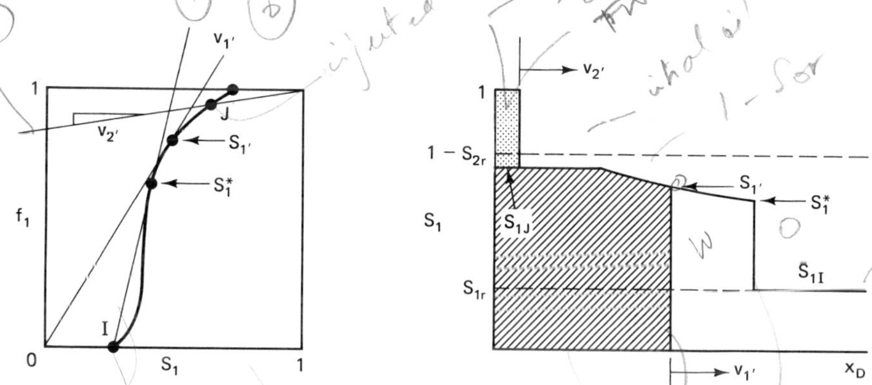

(e) Case E

Figure 5-12 Continued.

In case B, $S_{1J} > S_{1I}$, and the f_1 curve is such that the oil–water wave is a shock. Both tagged waves lag the oil–water wave. The region between the tagged water and oil–water waves contains a "bank" of resident water that will be produced before the injected water breakthrough. Breakthrough of a resident water bank in this manner has been observed experimentally (Brown, 1957) though dispersion tends to be large in such displacements (see Fig. 5-18).

Case C illustrates a spreading water–oil wave with $v_{2'} > v_{1'}$ but with all tagged concentration waves having a smaller velocity than the smallest saturation velocity at S_{1J}.

Case D is the same as case C with the fractional flow curve more convex upward. This shape causes the oil–water wave to spread more and the tagged oil front to fall somewhere in the spreading portion of the oil–water wave. The saturation, $S_{1'}$, whose velocity is the same as the tagged oil wave, is given by

$$v_{2'} = \frac{1 - f_1(S_{1'})}{1 - S_{1'}} = \left(\frac{df_1}{dS_1}\right)_{S_{1'}} \tag{5.4-9}$$

The line whose slope is $v_{2'}$ does not pass through S_{1J}, as it did in all previous cases. This is because a line through (1, 1) and $(S_1, f_1)_J$ would have a second intersection point with the fractional flow curve. The tagged oil front would then travel with two different water saturations—a physical impossibility.

Case E, the traditional Buckley-Leverett problem, is the inverse of case D where the tagged water front is now traveling in the spreading zone region. The oil–water displacement in case E is mixed, whereas in case D, it is spreading.

The important points in Fig. 5-12 are as follows:

1. As postulated, neither the tagged oil nor the tagged water causes deviation in the water–oil displacement character. When banks of resident fluids form, they do so within their respective phases.
2. One can easily imagine the tagged oil to be a hydrocarbon of less value than the oil. The tracer fronts now take on added significance since these miscible fronts are now displacing the resident oil. The resident oil, in turn, is completely displaced. Thus the ultimate E_D for these idealized displacements is 1.0. This maximum efficiency occurs without interfacial tension lowering, changes in wettability, or mobility reduction.

Of course, we have not as yet discovered a fluid that is simultaneously cheaper than and miscible with crude oil and that does not drastically change the hydrocarbon transport properties. These changes can return the ultimate displacement efficiency to something less than 1; still, the idea of displacing with miscible fluids, or those that will develop miscibility, is the central concept of Chap. 7.

5-5 DISSIPATION IN MISCIBLE DISPLACEMENTS

Because miscible waves are ideally indifferent, they are also susceptible to dissipation. By far the most prominent of the dissipative effects in miscible displacements are dispersion and viscous fingering. The latter is a two-dimensional effect, so we postpone our discussion of it to Chaps. 6 and 7. In this section, we discuss the effects of dispersion on a miscible front.

The Error Function Solution

Consider now the isothermal miscible displacement of a component by another it is completely miscible with in a one-dimensional, homogeneous permeable medium. The convection–diffusion (CD) equation (Eq. 2.4-7) describes the conservation of the displacing component with mass concentration C,

$$\phi \frac{\partial C}{\partial t} + u \frac{\partial C}{\partial x} - \phi K_\ell \frac{\partial^2 C}{\partial x^2} = 0 \tag{5.5-1}$$

Equation (5.5-1) also assumes incompressible fluid and rock, ideal mixing, and a single phase at unit saturation. The following development is valid if other phases are present (Delshad, 1981) and as long as all fractional flows and saturations are constant (see Exercise 5M). K_ℓ is the longitudinal dispersion coefficient. In dimensionless terms, Eq. (5.5-1) becomes

$$\frac{\partial C}{\partial t_D} + \frac{\partial C}{\partial x_D} - \frac{1}{N_{\text{Pe}}} \frac{\partial^2 C}{\partial x_D{}^2} = 0 \tag{5.5-2}$$

which is solved with the following boundary and initial conditions on $C(x_D, t_D)$:

$$C(x_D, 0) = C_I, \qquad x_D \geqq 0 \tag{5.5-3a}$$

$$C(x_D \to \infty, t_D) = C_I, \qquad t_D \geqq 0 \tag{5.5-3b}$$

$$C(0, t_D) = C_J, \qquad t_D \geqq 0 \tag{5.5-3c}$$

where C_I and C_J are the initial and injected compositions, respectively. In Eq. (5.5-2), N_{Pe}, the Peclet number, is defined as

$$N_{\text{Pe}} = \frac{uL}{\phi K_\ell} \tag{5.5-4}$$

which is the ratio of convective to dispersive transport. N_{Pe} is the analogue of N_{RL} for immiscible displacements as seen by comparing Eqs. (5.3-3) and (5.5-2). This displacement must take place at constant u unlike Eq. (5.2-6b). The equation and boundary conditions contain three independent parameters, C_I, C_J, and N_{Pe}, but the problem may be restated with only N_{Pe} as a parameter by defining a dimensionless concentration C_D

$$C_D = \frac{C - C_I}{C_J - C_I} \tag{5.5-5}$$

With this definition, the equation and boundary conditions become

$$\frac{\partial C_D}{\partial t_D} + \frac{\partial C_D}{\partial x_D} - \frac{1}{N_{\text{Pe}}} \frac{\partial^2 C_D}{\partial {x_D}^2} = 0 \tag{5.5-6}$$

$$C_D(x_D, 0) = 0, \qquad x_D \geqq 0 \tag{5.5-7a}$$

$$C_D(x_D \to \infty, t_D) = 0, \qquad t_D \geqq 0 \tag{5.5-7b}$$

$$C_D(x_D \to -\infty, t_D) = 1, \qquad t_D \geqq 0 \tag{5.5-7c}$$

We have replaced the original boundary condition at $x_D = 0$ (Eq. 5.5-3c) with one at $x_D \to -\infty$ (Eq. 5.5-7c). This is an approximation to simplify the following derivation of an analytic solution. The approximate solution thus obtained will be valid, strictly speaking, for large t_D or large N_{Pe} where the influence of the inlet boundary appears as though it were a great distance from the displacing front. In practice, the resulting approximate analytic solution accurately describes single-phase displacements for all but extreme cases.

The first step in deriving $C_D(x_D, t_D)$ is to transform Eqs. (5.5-6) and (5.5-7) to a moving coordinate system x_D' where $x_D' = x_D - t_D$. This may be done by regarding C_D as a function of x_D and t_D where a differential change in C_D caused by differential changes in x_D and t_D is

$$dC_D = \left(\frac{\partial C_D}{\partial t_D}\right)_{x_D} dt_D + \left(\frac{\partial C_D}{\partial x_D}\right)_{t_D} dx_D \tag{5.5-8a}$$

But regarded as a function of x_D' and t_D, dC_D is

$$dC_D = \left(\frac{\partial C_D}{\partial t_D}\right)_{x_D'} dt_D + \left(\frac{\partial C_D}{\partial x_D'}\right)_{t_D} dx_D' \tag{5.5-8b}$$

Differential changes in variables are equal regardless of the coordinate system they are viewed in. The right-hand sides of Eqs. (5.5-8a) and (5.5-8b) are therefore equal. But x_D' is a known function of x_D and t_D, from which

$$dx_D' = dx_D - dt_D \tag{5.5-9}$$

When dx_D' is replaced in the above equality, we have

$$\left[\left(\frac{\partial C_D}{\partial x_D}\right)_{t_D} - \left(\frac{\partial C_D}{\partial x_D'}\right)_{t_D}\right] dx_D + \left[\left(\frac{\partial C_D}{\partial t_D}\right)_{x_D} - \left(\frac{\partial C_D}{\partial t_D}\right)_{x_D'} + \left(\frac{\partial C_D}{\partial x_D'}\right)_{t_D}\right] dt_D = 0 \tag{5.5-10}$$

Since x_D and t_D are independent variables, dx_D and dt_D are not linearly related; hence the terms in brackets in Eq. (5.5-10) are zero, giving

$$\left(\frac{\partial C_D}{\partial x_D}\right)_{t_D} = \left(\frac{\partial C_D}{\partial x'_D}\right)_{t_D} \tag{5.5-11a}$$

$$\left(\frac{\partial C_D}{\partial t_D}\right)_{x_D} = \left(\frac{\partial C_D}{\partial t_D}\right)_{x'_D} - \left(\frac{\partial C_D}{\partial x'_D}\right)_{t_D} \tag{5.5-11b}$$

When these are substituted into Eq. (5.5-6), we have

$$\left(\frac{\partial C_D}{\partial t_D}\right)_{x'_D} - \frac{1}{N_{\text{Pe}}}\frac{\partial^2 C_D}{\partial (x'_D)^2} = 0 \tag{5.5-12}$$

and the boundary conditions retain the form of Eq. (5.5-7) thanks to the replacement of the inlet boundary condition at $x_D = 0$ with one at $x_D \to -\infty$.

Equation (5.5-12) is now the heat conduction equation whose solution may be obtained by the method of combination of variables (Bird et al., 1960). To do this, we define yet another dimensionless variable $\eta = x'_D/2\sqrt{t_D/N_{Pe}}$, with which the governing equations and boundary conditions may be transformed into

$$2\eta\frac{dC_D}{d\eta} + \frac{d^2C_D}{d\eta^2} = 0 \tag{5.5-13a}$$

$$C_D(\eta \to \infty) = 0 \tag{5.5-13b}$$

$$C_D(\eta \to -\infty) = 1 \tag{5.5-13c}$$

As required for the successful transformation of a partial to an ordinary differential equation, the conditions (Eqs. 5.5-7a and 5.5-7b) collapse into the single condition (Eq. 5.5-13b). The transformation to an ordinary differential equation is sometimes called Boltzmann's transformation. Equations (5.5-13) may be separated and integrated twice to give

$$C_D = \frac{1}{2}\left(1 - \frac{2}{\sqrt{\pi}}\int_0^{\eta} e^{-u^2}\,du\right) \tag{5.5-14}$$

The product times the integral on the right side of Eq. (5.5-14) is the error function, a widely tabulated integral (see Table 5-2 and Fig. 5-13), and abbreviated with the symbol $\text{erf}(\eta)$. By substituting the definitions for η and x'_D, we have the final form for the approximate analytic solution.

$$C_D = \frac{1}{2}\left[1 - \text{erf}\left(\frac{x_D - t_D}{2\sqrt{\frac{t_D}{N_{\text{Pe}}}}}\right)\right] = \frac{1}{2}\,\text{erfc}\left(\frac{x_D - t_D}{2\sqrt{\frac{t_D}{N_{\text{Pe}}}}}\right) \tag{5.5-15}$$

TABLE 5-2 TABULATED VALUES OF ERF (x) (FROM JAHNKE AND EMDE, 1945)

x		0	1	2	3	4	5	6	7	8	9	d
0.0	0.0	000	113	226	338	451	564	676	789	901	*013	113
1	0.1	125	236	348	459	569	680	790	900	*009	*118	111
2	0.2	227	335	443	550	657	763	869	974	*079	*183	106
3	0.3	286	389	491	593	694	794	893	992	*090	*187	100
4	0.4	284	380	475	569	662	755	847	937	*027	*117	93
5	0.5	205	292	379	465	549	633	716	798	879	959	84
6	0.6	039	117	194	270	346	420	494	566	638	708	74
7		778	847	914	981	*047	*112	*175	*238	*300	*361	65
8	0.7	421	480	538	595	651	707	761	814	867	918	56
9		969	*019	*068	*116	*163	*209	*254	*299	*342	*385	46
1.0	0.8	427	468	508	548	586	624	661	698	733	768	38
1		802	835	868	900	931	961	991	*020	*048	*076	30
2	0.9	103	130	155	181	205	229	252	275	297	319	24
3		340	361	381	400	419	438	456	473	490	507	19
4	0.95	23	39	54	69	83	97	*11	*24	*37	*49	14
5	0.96	61	73	84	95	*06	*16	*26	*36	*45	*55	10
6	0.97	63	72	80	88	96	*04	*11	*18	*25	*32	8
7	0.98	38	44	50	56	61	67	72	77	82	86	6
8		91	95	99	*03	*07	*11	*15	*18	*22	*25	4
9	0.99	28	31	34	37	39	42	44	47	49	51	3
2.0	0.995	32	52	72	91	*09	*26	*42	*58	*73	*88	17
1	0.997	02	15	28	41	53	64	75	85	95	*05	11
2	0.998	14	22	31	39	46	54	61	67	74	80	8
3		86	91	97	*02	*06	*11	*15	*20	*24	*28	5
4	0.999	31	35	38	41	44	47	50	52	55	57	3
5		59	61	63	65	67	69	71	72	74	75	2
6		76	78	79	80	81	82	83	84	85	86	1
7		87	87	88	89	89	90	91	91	92	92	1
8	0.9999	25	29	33	37	41	44	48	51	54	56	3
9		59	61	64	66	68	70	72	73	75	77	2

where erfc denotes the complementary error function. The exact analytic solution as derived by Laplace transforms is (Marle, 1981)

$$C_D = \frac{1}{2}\operatorname{erfc}\left(\frac{x_D - t_D}{2\sqrt{\dfrac{t_D}{N_{Pe}}}}\right) + \frac{e^{x_D N_{Pe}}}{2}\operatorname{erfc}\left(\frac{x_D + t_D}{2\sqrt{\dfrac{t_D}{N_{Pe}}}}\right) \tag{5.5-16}$$

The second term in Eq. (5.5-16) approaches zero exponentially as x_D and N_{Pe} grow.

Figure 5-14 shows concentration profiles of C_D versus x_D with t_D and N_{Pe} varying. As N_{Pe} increases, the concentration profile approaches the step function at $x_D = t_D$ suggested by Eq. (5.4-7). In fact, the concentration profile given by Eq.

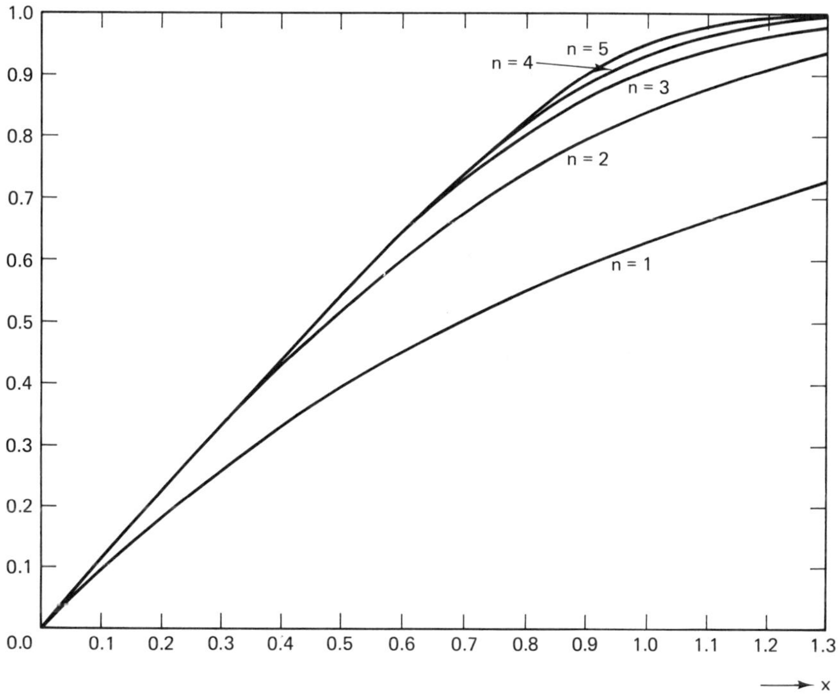

Figure 5-13 The function $E_n(x) = \frac{1}{n!} \int_0^{x^n} e^{-v}\, v^{(1/n)-1}\, dv$, where $n = 2$ is the error function (from Jahnke and Emde, 1945)

(5.5-15) is symmetric and centered on this point. The complete solution (Eq. 5.5-16) is not symmetric, but as we noted, this effect is small. Dispersion, therefore, does not affect the rate of wave propagation, but it does affect the degree of mixing in the wave.

The displacement efficiency for the displaced component is

$$E_D = \overline{C}_D = \int_0^1 C_D(x_D, t_D)\, dx =$$

$$\left(\frac{\sqrt{t_D}}{N_{Pe}}\right)\left[\text{ierfc}\left(\frac{-\sqrt{t_D N_{Pe}}}{2}\right) - \text{ierfc}\left(\frac{1-t_D}{2\sqrt{\frac{t_D}{N_{Pe}}}}\right)\right] \qquad (5.5\text{-}17)$$

where ierfc $(x) = \int_x^\infty$ erfc $(\xi)\, d\xi$ is the integral complementary error function also tabulated (Carslaw and Jaeger, 1959). Figure 5-15 plots E_D versus t_D for various N_{Pe}. E_D decreases at fixed t_D as dispersion increases. Since miscible displacements do not have residual phase saturations, E_D approaches 1 as t_D increases. Figures 5-14 and 5-15 indicate a stronger effect of N_{Pe} on concentration profiles than on displacement

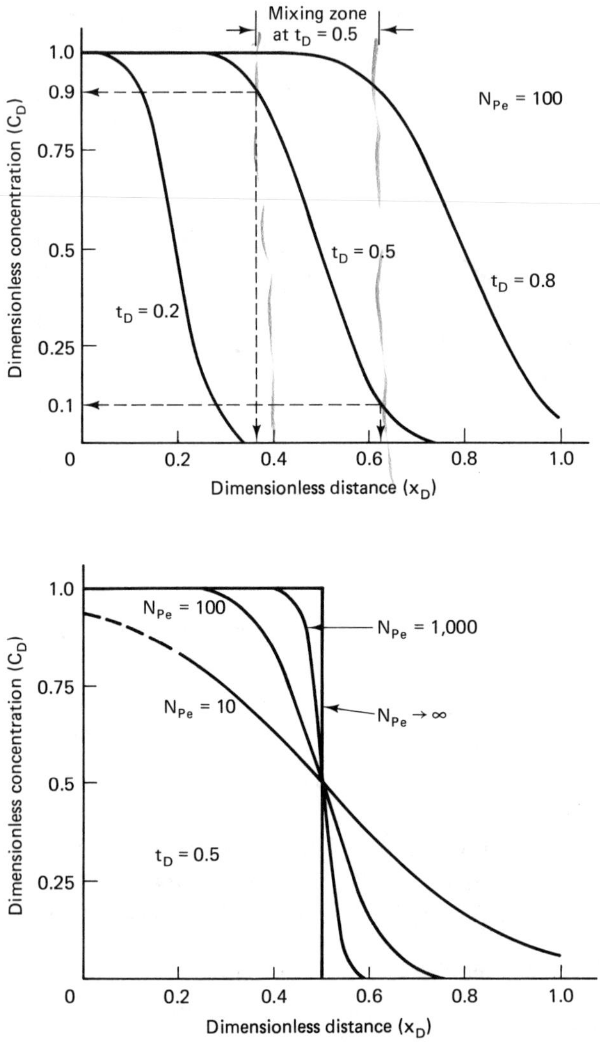

Figure 5-14 Dimensionless concentration profiles

efficiency; hence concerns about the detrimental effect of dispersion on recovery are usually limited to slugs. This topic we defer to Sec. 7-6.

The dimensionless mixing zone, the distance between the distances where $C_D = 0.1$ and $C_D = 0.9$, follows from Eq. (5.5-15),

$$\Delta x_D = x_D\big|_{C_D=0.1} - x_D\big|_{C_D=0.9} = 3.625\sqrt{\frac{t_D}{N_{Pe}}} \tag{5.5-18}$$

To arrive at this, invert Eq. (5.5-15) for $x_D\big|_{C_D=0.1}$ to yield

$$x_D\big|_{C_D=0.1} = t_D + 2\sqrt{\frac{t_D}{N_{Pe}}}\,\text{erf}^{-1}(0.8)$$

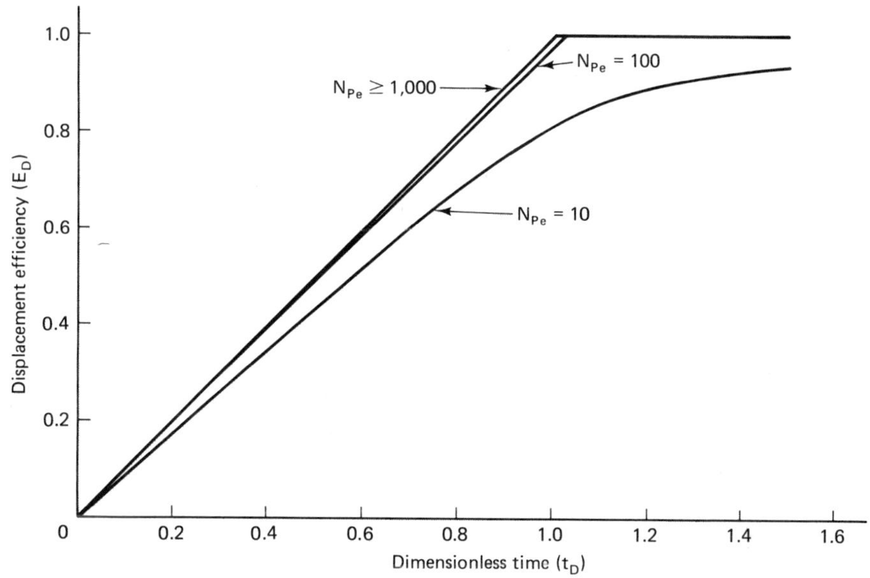

Figure 5-15 Displacement efficiency for one-dimensional miscible displacements

A similar procedure yields $x_D|_{C_D = 0.9}$, and these substituted into the definition for Δx_D give Eq. (5.5-18). Equation (5.5-18) shows that dispersive mixing zones grow in proportion to the square root of time. Immiscible mixing zones grow in proportion to time. The growth suggested by Eq. (5.5-18) is generally slower than that for an immiscible mixing zone, particularly if N_{Pe} is large. This slow growth is a partial justification for neglecting dispersion in modeling semimiscible displacements compared to fractional flow effects.

Δx_D is also useful to compare laboratory to field mixing zone lengths. An immiscible mixing zone contains no free parameters if dissipation is small. Therefore, if we conduct a laboratory immiscible flood under conditions as nearly identical to a field prototype as possible (displacement in native or restored state cores, at reservoir temperature and pressure, using actual reservoir fluids), the laboratory Δx_D will be the same as in the field.

In miscible displacements, we are generally unable to make N_{Pe} equal between the laboratory and the field. Moreover N_{Pe} is usually smaller in the laboratory; thus Δx_D usually will be larger in the laboratory than in the field. Of course, the dimensional mixing zone length, $\Delta x_D L$, will always be greater in the field because L is much greater. Why we are unable to match N_{Pe} is derived from the following discussion of dispersion coefficients.

Dispersivity

Bear (1972) suggests "hydrodynamic" dispersion is "the macroscopic outcome of the actual movements of the individual tracer particles through the pores and various physical and chemical phenomena that take place within the pores." This movement can arise from a variety of causes. In this text, *dispersion* is the mixing of two miscible fluids caused by diffusion, local velocity gradients (as between a pore wall and a pore center), locally heterogeneous streamline lengths, and mechanical mixing in pore bodies. Gravity tonguing and viscous fingering are two-dimensional effects that we discuss in Chap. 6. Here we summarize experimental findings on dispersion coefficients and some qualitative reasons for these observations.

For one-dimensional flow, the longitudinal dispersion coefficient K_ℓ is given by

$$\frac{K_\ell}{D_0} = C_1 + C_2\left(\frac{|v|D_p}{D_0}\right)^\beta \tag{5.5-19}$$

where C_1, C_2, and β are properties of the permeable medium and the flow regime. D_0 is the effective binary molecular diffusion coefficient between the miscible displacing and displaced fluids. D_p is an average particle diameter.

For very slow flows, the second term in Eq. (5.5-19) is negligible, and K_ℓ is proportional to D_0. This case is analogous to a slow displacement in a wide channel where mixing is due entirely to molecular diffusion. The constant C_1 has been found to be $1/\phi F$, where F is the electrical formation resistivity factor (Pirson, 1983) to account for the presence of the stationary phase.

For faster displacements, the second term in Eq. (5.5-19) becomes significant. Deans (1963) has shown that well-stirred tanks in series give mixing zones that can be described by dispersion coefficients proportional to velocity. Here, mixing is the result of the highly irregular flow paths in the REV, which cause fluids to mix completely as they are produced from each cell. Diffusion, of course, is negligible if the fluids are well mixed.

An alternate, two-dimensional interpretation, including diffusion in this flow regime, is the theory of Taylor (1953), whereby the flow channels are visualized as having lateral dimensions much smaller than the longitudinal dimensions. For this idealization, diffusion equalizes concentration gradients in the lateral direction giving rise to an "effective" diffusion coefficient. Mixing is now the result of transverse diffusion and variations in velocity caused by the no-slip condition at the pore wall. Taylor's theory predicts dispersion coefficients proportional to velocity squared.

Experimentally, it is found (Perkins and Johnston, 1963) that $\beta = 1$ to 1.25 in Eq. (5.5-19); hence it seems the local mixing interpretation is closer to the mark than Taylor's theory.

This local mixing flow regime is where most EOR processes will occur. In fact, if the interstitial velocity is greater than about 3 cm/day, the local mixing term in Eq. (5.5-19) dominates the first term, and we can write

$$K_\ell = \frac{D_0}{\phi F} + C_2\left(\frac{|v|D_p}{D_0}\right)^\beta D_0 \cong \alpha_\ell |v| \tag{5.5-20}$$

This does not imply that diffusion is categorically negligible in miscible flow. Several phenomena involve flow around stagnant regions (for example, dead-end pores, water blocked pores, or adjacent nonflowing zones) where diffusion rates are important even in regimes that would otherwise be well described by Eq. (5.5-20). α_ℓ in Eq. (5.5-20) is the longitudinal *dispersivity* of the permeable medium (Eq. 2.2-14), a measure of the local heterogeneity scale. Bear (1970) classifies α_ℓ as one of the fundamental properties of the medium. For the local mixing flow regime, α_ℓ is a more fundamental measure of dispersion than K_ℓ.

Figure 5-16 shows the three flow regimes from Perkins and Johnston (1963). Similar data is in Bear (1970) and in several references of Perkins and Johnston.

The form of Eq. (5.5-20) is particularly convenient as the Peclet number (Eq. 5.5-4), and the dimensionless concentration balance (Eq. 5.5-2) now become independent of velocity

$$N_{\text{Pe}} = \frac{L}{\alpha_\ell} \tag{5.5-21}$$

Therefore, the dimensionless mixing zone is directly related to α_ℓ through Eq. (5.5-18). In fact, α_ℓ/L can be crudely regarded as the dimensionless mixing zone length.

Suppose we try to design a laboratory displacement that has the same dimensionless mixing zone length as a field prototype. Then we must have

$$\left(\frac{\alpha_\ell}{L}\right)_{\text{field}} = \left(\frac{\alpha_\ell}{L}\right)_{\text{lab}} \tag{5.5-22}$$

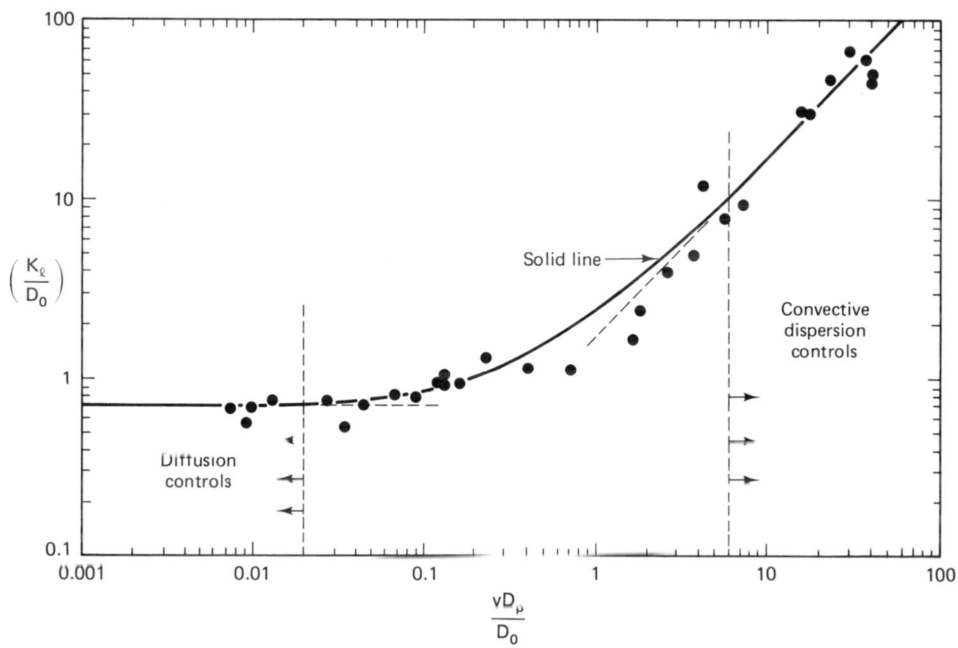

Figure 5-16 Longitudinal dispersion coefficients in permeable media flow (from Perkins and Johnston, 1963)

Equation (5.5-22) clearly cannot be satisfied if the laboratory and field dispersivities are assumed equal.

To enforce the equality in Eq. (5.5-22), we must have laboratory and field values of α_ℓ. Laboratory-measured α_ℓ's are available through correlations or experiments. They are generally a few centimeters or less depending on the core material. The estimated field-measured values of α_ℓ are far less certain. A good summary of field-measured dispersivities is shown in Fig. 5-17. This figure shows field-measured α_ℓ's for several formation types plotted against the length scale it was measured over. On the log–log scale, there is clearly considerable variation in α_ℓ at

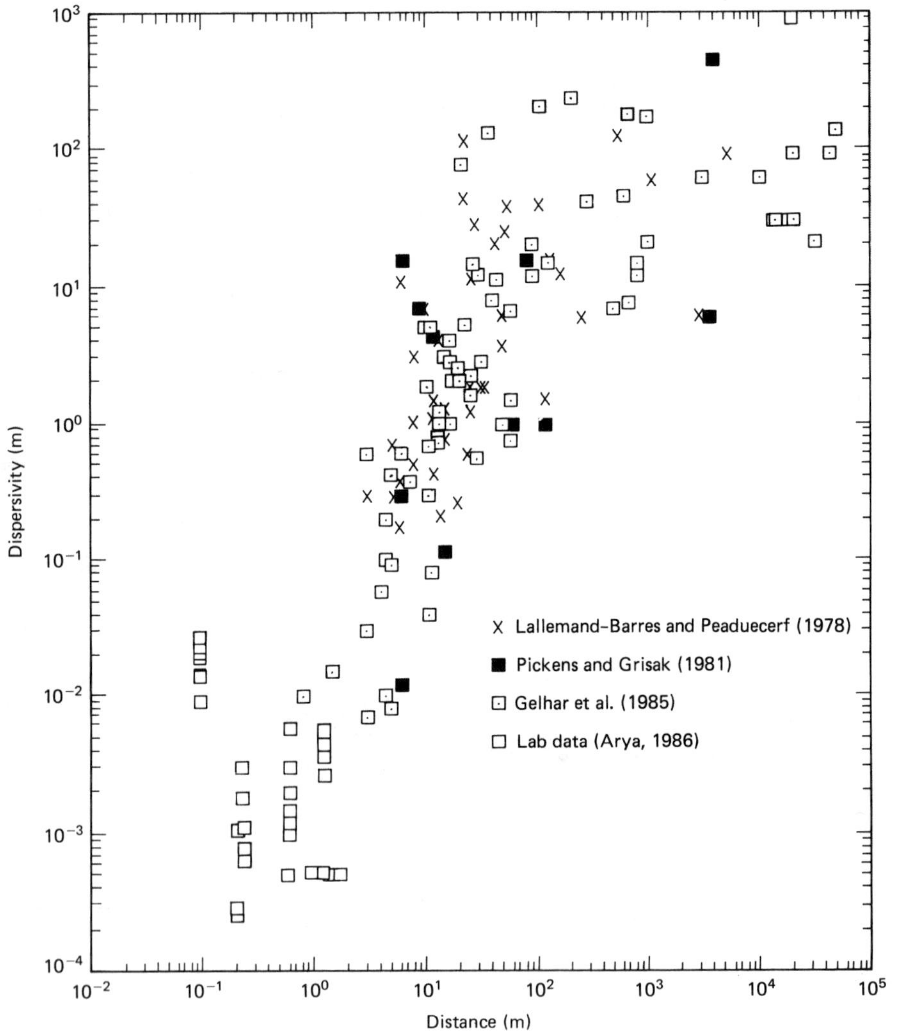

Figure 5-17 Field and laboratory measured dispersivities (from Arya et al., 1988)

the same length, and even for the same formation, even though there is little or no correlation with the latter.

Despite the scatter, there is a clear trend of increasing α_ℓ with measurement distance. We can explain this increase qualitatively by saying the scale of heterogeneity captured by a given measurement increases as the volume sampled increases. Quantitatively, the phenomenon is the subject of active research (Gelhar et al., 1979; Dagan, 1984) because of the complicated interplay between heterogeneity, local dispersion coefficients, diffusion, and other permeable media properties that combine to make α_ℓ length dependent.

Figure 5-17 points out the interesting and significant behavior of α_ℓ as system macroscopic length increases. But even on a local scale, the behavior of longitudinal dispersivity is not well known when multiple phases are flowing. Figure 5-18 gives experimental data showing how the intraphase dispersivity changes as the phase saturation changes. The data in this figure are for constant saturation flow of micellar fluids for which the more general definition of K_ℓ (Eq. 2.2-14) is appropriate. Figure 5-18 shows that aqueous phase dispersivity can increase by more than a factor of 10 as the aqueous phase saturation decreases. (This dispersivity increases as the effective heterogeneity increases, but now the "heterogeneity" must be related to the

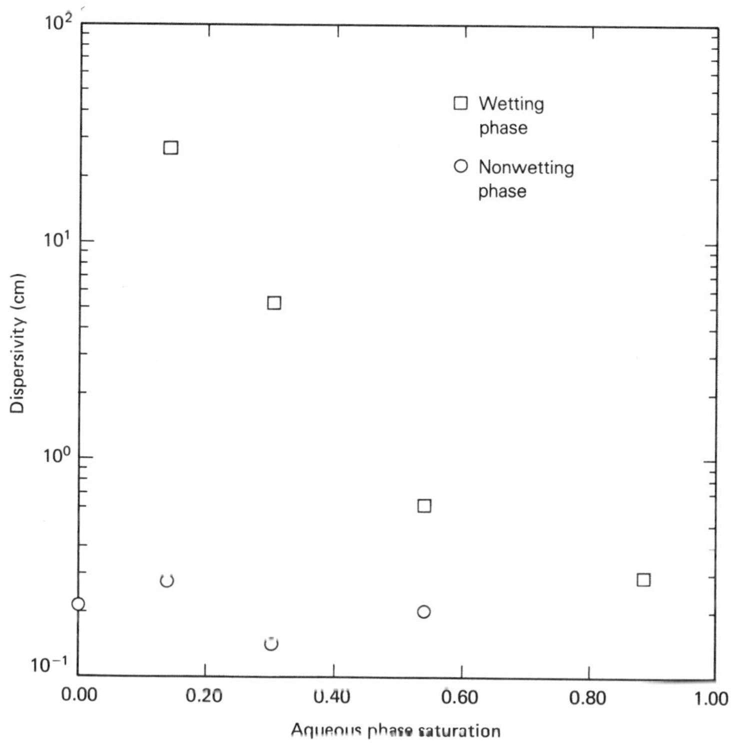

Figure 5-18 Dispersivities for constant saturation miscible flows (from MacAllister, 1982)

characteristics of the flowing fluids.) It is likely that wetting conditions play a large role in the α_ℓ increase since no such changes in α_ℓ were observed for the nonwetting phase in Fig. 5-18.

We summarize the most important points about the effects of dispersion on one-dimensional miscible flow as follows:

1. Dispersion controls the rate of mixing of two fluids but does not affect wave velocity.
2. Dispersive mixing zones can grow no faster than in proportion to the square root of time.
3. The fluid velocity of most EOR processes is such that the flow is in the local mixing flow regime where the dispersion coefficient is proportional to the interstitial velocity. The proportionality constant is the longitudinal dispersivity α_ℓ.
4. α_ℓ is a measure of the heterogeneity of the permeable medium and varies with phase saturation and the measurement scale.
5. Neglecting dispersion in field-scale displacements is not proper because dispersivity appears to increase with travel distance.

5-6 GENERALIZATION OF FRACTIONAL FLOW THEORY

In this section, we present the mathematical formalities to broaden the fractional flow theories of Secs. 5-2 and 5-4 to multiple component, multiphase flow. As in those sections, we neglect dissipative effects and restrict the equations to one-dimensional flow. Our presentation is based on a subset of the method of characteristics (MOC) solution technique known as simple wave theory, or coherence theory. (For more careful mathematical detail, see Courant and Friedrichs, 1948; Helfferich and Klein, 1970; and Jeffrey and Taniuti, 1964.)

The fundamental principle in the MOC is to solve partial differential equations (PDEs) by first converting them to a set of ordinary differential equations (ODEs) that may then be integrated simultaneously. This set of ODEs can rarely be integrated in closed form, but there is a large class of permeable media flow problems for which the integrations will appear in a general form. To illustrate these ideas, we consider first a single PDE and then pairs of PDEs in the dependent variables u and v. The theory may be generalized to more than two PDEs, but in practice, the procedures become cumbersome.

One Dependent Variable

Consider the following partial differential equation for $u(x, t)$

$$L(u) = Au_t + Bu_x + E = 0 \tag{5.6-1}$$

where A, B, and E are known functions of u, x, and t. The operator $L(u)$ is linear in the derivatives of u. The notation u_x and u_t means partial differentiation with respect to x and t holding the other variable constant. We want solutions to Eq. (5.6-1) in the form $u(x, t)$ subject to the appropriate initial and boundary conditions. In the MOC, we seek these solutions in the form $u(s)$, where s is a parameter along a curve C in x-t space such that $x = x(s)$ and $t = t(s)$. We may, therefore, write the total derivative of u with respect to s as

$$u_s = t_s u_t + x_s u_x \tag{5.6-2}$$

Equation (5.6-2) is a mixture of total derivatives u_s, t_s, and x_s and partial derivatives u_t and u_x. However, we use the same notation for both types of derivatives since the type of derivative should be clear from the usage. Comparing Eqs. (5.6-1) and (5.6-2) leads to

$$t_s = A \tag{5.6-3a}$$

$$x_s = B \tag{5.6-3b}$$

$$u_s = -E \tag{5.6-3c}$$

Equations (5.6-3), which imply the operator $L(u)$ is a directed derivative along C, are a set of three ODEs that may be integrated from an initial curve, as shown in Fig. 5-19(a), to give a characteristic curve C in xt space along which u varies as given by the integration of Eq. (5.6-3c).

The various integrations are possible only if C is nowhere tangent to the initial curve. Figure 5-19(a) schematically shows the integration of these equations for a curve C that begins at the point (x_0, t_0) on the initial curve. We could take other points on the initial curve and thereby cover the shaded *domain of dependence* in Fig. 5-19(a) defined by the characteristics through the points A and B on ends of the initial curve. If α is a parameter along the initial curve, the solution to Eqs. (5.6-3) is $t = t(s, \alpha)$, $x = x(s, \alpha)$, and $u = u(s, \alpha)$. s and α are the coordinates of a natural, generally curved, coordinate system for Eq. (5.6-1). Since α, in effect, determines which curve C passes through the point (x, t), at which the value of u is desired, the characteristics for Eq. (5.6-1) are a one-parameter (α) family of curves, and α is a label for this one-parameter family. An important observation is that at every point (x, t) in the shaded region in Fig. 5-19(a), the slope of the characteristic curve is given by

$$\frac{x_s}{t_s} = \left.\frac{dx}{dt}\right|_C = \frac{B}{A} = \sigma(u, x, t) \tag{5.6-4}$$

σ is the *characteristic direction* at a given (x, t). Equation (5.6-4) implies it will generally be unnecessary to determine $t = t(s, \alpha)$ and $x = x(s, \alpha)$ since $t = t(x, \alpha)$ will follow directly from Eqs. (5.6-4) and (5.6-3c).

Consider now a special case of Eq. (5.6-1) where $E = 0$, and A and B are functions of u only. The initial data are a curve that coincides with the x axis, where

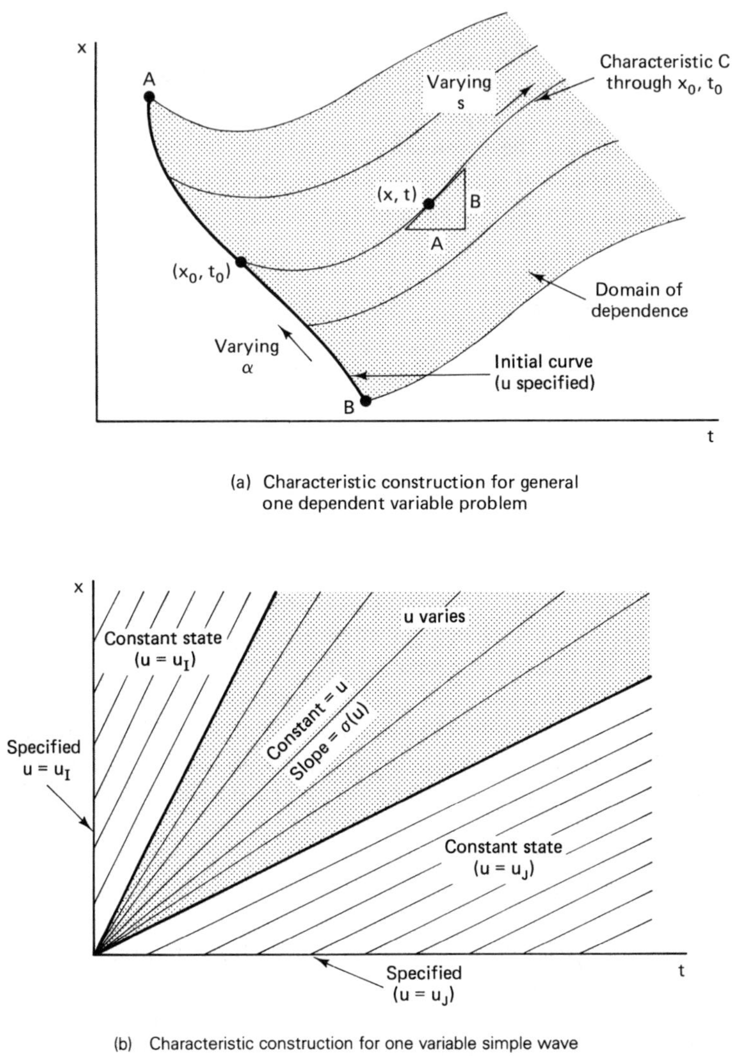

Figure 5-19 Domains of dependence for one-variable hyperbolic equations

$u = u_I$ and then coincides with the t axis, where $u = u_J$. Thus the boundary ($x = 0$) and initial ($t = 0$) data are uniform except for a step change at the origin where all values of u between u_I and u_J exist. It follows immediately from Eqs. (5.6-3c) and (5.6-4) that u is constant along the characteristics C, which are themselves straight lines. Figure 5-19(b) shows the characteristics for this case. In regions adjacent to the x and t axes, the characteristics are parallel with slopes $\sigma(u_I)$ and $\sigma(u_J)$, respectively. These regions are constant-state regions since the dependent variable u is constant therein. The shaded region in Fig. 5-19(b) is a fanlike region where σ changes continuously between the limits imposed by the constant-state regions. Each ray emanating from the origin carries a particular constant σ from the infinite numbers of u's between u_I and u_J, and each has a slope σ evaluated at that u. Therefore, the shaded region in Fig. 5-19(a) is a wave since, in any noncharacteristic direction, u is changing.

From Fig. 5-19(b), the characteristics cannot cross, but there is nothing that requires σ to decrease monotonically, as in the case shown. When σ does not decrease monotonically, a mathematically valid solution exists that leads to the formation of shock waves, u being a physical variable.

Finally, the characteristic direction σ may clearly be interpreted as a velocity (if x and t are distance and time) and written as

$$\sigma = \left.\frac{dx}{dt}\right|_C = \left.\frac{dx}{dt}\right|_u \tag{5.6-5}$$

With the appropriate forms for A, B, t, and x, Eq. (5.6-5) becomes the Buckley-Leverett equation (Eq. 5.2-10) for water displacing oil in a permeable medium as we discussed in Sec. 5-2. Note the similarity between Figs. 5-5 and 5-19(b).

Two Dependent Variables

Let us consider now a pair of PDEs in the dependent variables $u(x, t)$ and $v(x, t)$

$$L_1(u, v) = A_1 u_t + B_1 u_x + C_1 v_t + D_1 v_x + E_1 = 0 \tag{5.6-6a}$$

$$L_2(u, v) = A_2 u_t + B_2 u_x + C_2 v_t + D_2 v_x + E_2 = 0 \tag{5.6-6b}$$

Initially, we consider the most general case of the coefficients A–E being functions of x, t, u, and v. The first pair of terms in the linear operators L_1 and L_2 may be regarded as directed derivatives of u and v. From the total derivative of du and dv, there are four such directions ($A_1 u_t + B_1 u_x$, $C_1 v_t + D_1 v_x$, and so on) for each PDE. But to transform the pair to a set of ODEs, we seek a curve in (x, t) space where $u = u(s)$, $v = v(s)$, $x = x(s)$, and $t = t(s)$. We, therefore, seek a combination $L = \lambda_1 L_1 + \lambda_2 L_2$ so that L is a linear function of total derivatives u_s and v_s. As before, s is a parameter along such a curve. For solutions to the equations, the operator L must be equal to zero, hence

$$\begin{aligned} L = (A_1\lambda_1 + A_2\lambda_2)u_t + (B_1\lambda_1 + B_2\lambda_2)u_x + (C_1\lambda_1 + C_2\lambda_2)v_t \\ + (D_1\lambda_1 + D_2\lambda_2)v_x + (E_1\lambda_1 + E_2\lambda_2) = 0 \end{aligned} \tag{5.6-7}$$

For the directed derivatives of u and v to be colinear, it is necessary that

$$\frac{x_s}{t_s} = \sigma = \frac{B_1\lambda_1 + B_2\lambda_2}{A_1\lambda_1 + A_2\lambda_2} = \frac{D_1\lambda_1 + D_2\lambda_2}{C_1\lambda_1 + C_2\lambda_2} \tag{5.6-8}$$

be obtained from the total derivative for each dependent variable. The two equations in Eq. (5.6-8) may be written as

$$(A_1x_s - B_1t_s)\lambda_1 + (A_2x_s - B_2t_s)\lambda_2 = 0 \tag{5.6-9a}$$

$$(C_1x_s - D_1t_s)\lambda_1 + (C_2x_s - D_2t_s)\lambda_2 = 0 \tag{5.6-9b}$$

For nonzero λ_1 and λ_2, the determinant of the coefficient matrix must be zero; hence

$$(A_1C_2 - A_2C_1)\sigma^2 + (A_2D_1 - D_2A_1 + C_1B_2 - C_2B_1)\sigma + (B_1D_2 - D_1B_2) = 0 \tag{5.6-10}$$

where we have substituted the characteristic direction σ from Eq. (5.6-8). Immediately, it is apparent that there are, in general, two characteristic directions, not one as in the analogous expression (Eq. 5.6-4) for the one variable problem. Whether or not these directions are real for all (x, t) depends on the form of the coefficients. For permeable media flow problems, σ is real in at least some, and usually all, of the domain (x, t). This, in fact, is the definition of hyperbolic PDEs. Further, the roots in Eq. (5.6-10) are generally distinct. Let σ^+ designate the larger root and σ^- the smaller of Eq. (5.6-10). Clearly, the corresponding characteristic curves, C^+ and C^-, cover the domain of dependence in (x, t) since the slope of C^+ is everywhere larger than the slope of C^-. Figure 5-20 shows these curves. The shaded domain of dependence is bounded for the two dependent variable problem by the fast σ^+ characteristic through B and the slow σ^- characteristic through A.

Each point in the domain of dependence is on the intersection of an σ^+ and σ^- characteristic. The coordinates of a point may then be located as a distance s along a particular characteristic having label α; that is, $x = x(s, \alpha)$, and $t = t(s, \alpha)$. Alternatively, the coordinates may also be located by giving the labels of both characteristics passing through it, or $x = x(\alpha, \beta)$ and $t = t(\alpha, \beta)$, where β is now the label of the other characteristic. The notion of labels is somewhat confusing since α and β can take on the same numerical values on the initial curve; however, in the interior of the domain of influence, they are distinct.

The characteristic curves cannot be obtained, in general, unless it is known how u and v change along the characteristic directions. This may be obtained by replacing the coefficients of u_x and v_x in Eq. (5.6-7) by the numerators in Eq. (5.6-8)

$$(A_1\lambda_1 + A_2\lambda_2)u_s + (C_1\lambda_1 + C_2\lambda_2)v_s + (E_1\lambda_1 + E_2\lambda_2)t_s = 0 \tag{5.6-11a}$$

where we have rearranged with $u_s = u_xx_s + u_tt_s$, and so on. A similar procedure on u_t and v_t gives

$$(B_1\lambda_1 + B_2\lambda_2)u_s + (D_1\lambda_1 + D_2\lambda_2)v_s + (E_1\lambda_1 + E_2\lambda_2)x_s = 0 \tag{5.6-11b}$$

These equations rearranged are

$$(A_1u_s + C_1v_s + E_1x_s)\lambda_1 + (A_2u_s + C_2v_s + E_2t_s)\lambda_2 = 0 \tag{5.6-12a}$$

$$(B_1u_s + D_1v_s + E_1x_s)\lambda_1 + (B_2u_s + D_2v_s + E_2t_s)\lambda_2 = 0 \tag{5.6-12b}$$

Again, for nontrivial λ_1 and λ_2, the determinant of the coefficient matrix must be zero, and again, the characteristic equation has two real, distinct roots for u_s or v_s.

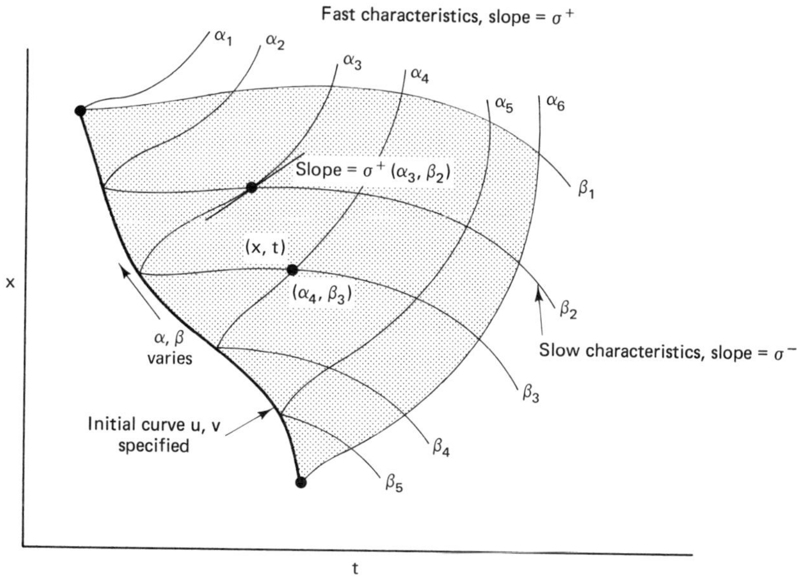

(a) Characteristics for general two-variable problem

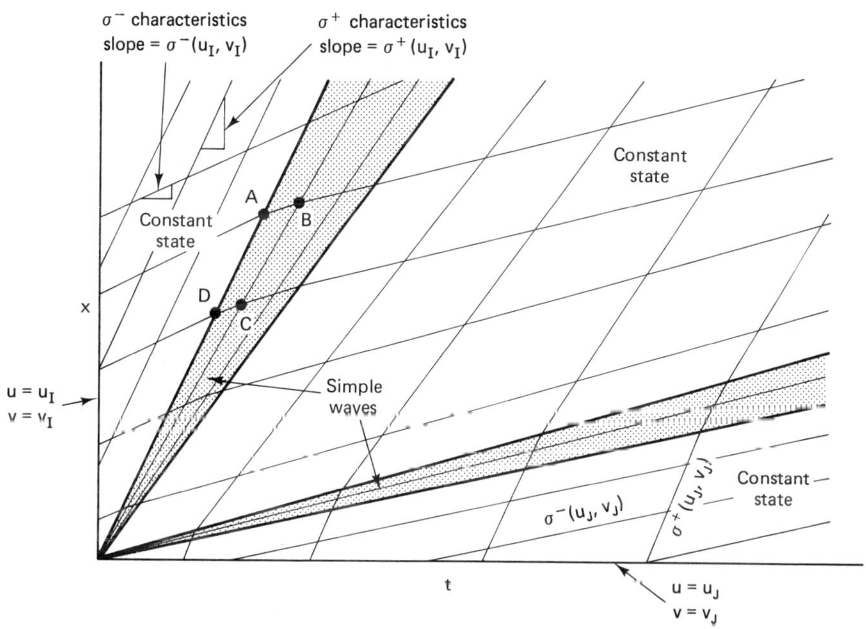

(b) Characteristics for two-variable centered simple wave

Figure 5-20 Domains of dependence for two-variable hyperbolic equations

The roots to these equations form, along with the two roots to Eq. (5.6-10), four ODEs that may be integrated simultaneously for u, v, x, and t from an initial curve.

Rather than expound further on this, let's consider the special case where E_1 and $E_2 = 0$, and the remaining coefficients are functions of u and v only. Such equations are said to be *reducible*. The characteristic directions from Eq. (5.6-10) are a known function of u and v only. Further, along each characteristic curve, there is a relation between u and v given by

$$(B_2A_1 - B_1A_2)\frac{du}{dv} = (B_1C_2 - C_1B_2) + \sigma^{\pm}(C_1A_2 - A_1C_2) \tag{5.6-13}$$

from the determinant of the singular matrix formed by Eqs. (5.6-9a) and (5.6-12a). There are three equivalent forms to this equation, but the important point is that u and v are related to each other along the $C^{\pm}$ curves in x-t space since all the coefficients in Eq. (5.6-13) are known functions of u and v. The u-v plot that contains the above solution is said to be the image, or hodograph, space and the function $u = u(v)$ for C^+ is the image curve Γ^+ or C^+, as is $u = u(v)$ for Γ^- the image curve for C^-. In this text, we call the u-v space the *composition path diagram,* and the Γ^+ and Γ^- curves the *composition paths*.

Besides the restriction to reducible PDEs, let's now consider the special case of $u = u_I$ and $v = v_I$ being specified on the x axis, and $u = u_J$ and $v = v_J$ being specified on the t axis. As before, this means that all (u, v) values between $(u, v)_I$ and $(u, v)_J$ exist at the origin. Also as before, there are regions of constant state adjacent to both axes where the characteristic directions, and hence, the labels α and β are constant. But unlike before, there are now two fan-shaped regions (Fig. 5-20b) where first the fast or α characteristics change slope, and then the β characteristics change. The regions cannot overlap, or there would be finite regions where $\sigma^+ < \sigma^-$. This fact causes the creation of a new constant-state region (u, v) between the fans that is, in general, different from either $(u, v)_I$ or $(u, v)_J$. Within the fan-shaped regions, the α and β characteristics cannot both be straight, or else these would be constant-state regions. But one of the characteristic directions must be straight in each region (σ^+ in the first shaded region in Fig. 5-20(b), and σ^- in the second). This is so because two points A and D on the boundary have the same (u, v) values since they can be regarded as being in the constant-state region. This must be true of all other rays in the fan-shaped region, for example, that passing through C and B. Otherwise, the ray would be curved (from Eq. 5.6-10) and would ultimately intersect either of the constant-state regions. It follows, then, that all points on the straight-line characteristic carry the same (u, v) value. Since $(u, v)_A = (u, v)_D$ and $(u, v)_B = (u, v)_C$, it follows that $\sigma_B^- = \sigma_C^-$ and the slope of the σ^- characteristics is the same on all the σ^+ characteristics. This means the $u = u(v)$ relationship defined by Eq. (5.6-13) is the same on any slow characteristic in the region. Thus $(du/dv)_{\sigma^-}$ and $(du/dv)_{\sigma^+}$ uniquely determine the variations in u and v in the respective fan-shaped regions. The function $u = u(v)$ is always calculated based on the curved characteristic.

The above concepts apply generally to reducible sets of PDEs in any number of N dependent variables. Stated concisely, the observations are

1. Adjacent to any constant-state region, there is a region having at least one straight line characteristic. The second region is a *simple wave* region.
2. Within a simple wave region, the dependent variables are related to each other through a set of ODEs.
3. For boundary and initial conditions that are uniform except for a step change at the origin, the entire (x, t) domain consists of alternating constant-state and simple wave regions. The simple waves in this case are *centered* simple waves.

Coherence

The information on reducible equations may be restated with more physical insight by referring to simple waves in the terminology of *coherent* waves (Helfferich and Klein, 1970). Since (u, v) is constant on a straight-line characteristic in a simple or coherent wave region, and since σ is a function of u and v only, it follows that

$$\left.\frac{dx}{dt}\right|_u = \left.\frac{dx}{dt}\right|_v \tag{5.6-14a}$$

or for $u_1, \ldots, u_N$ dependent variables

$$\left.\frac{dx}{dt}\right|_{u_1} = \left.\frac{dx}{dt}\right|_{u_2}, \ldots, = \left.\frac{dx}{dt}\right|_{u_N} \tag{5.6-14b}$$

Equation (5.6-14b) states that the velocity of constant values of the dependent variables is the same—the coherence condition. As we illustrate in Sec. 5-7, the coherence method of calculating simple waves is more direct than using MOC. Equation (5.6-14b) implies, further, that there can be no more than N waves.

5-7 APPLICATION TO THREE-PHASE FLOW

In this section, we apply the results of the coherence theory by calculating the displacement efficiencies for a three-phase water ($i = 1$), oil ($i = 2$), gas ($i = 3$) flow problem. We assume away dissipative effects—capillary pressure and pressure-dependent fluid properties—and restrict the fluids to be single pseudocomponent phases. The assumption of an incompressible gas phase is, of course, realistic only if

$$c_3 \Delta P \cong \frac{\Delta P}{P} \tag{5.7-1}$$

is small. This condition is not met in general although for flows in high permeability media $c_3\,\Delta P$ can be fairly small, particularly considering that gas viscosity is also small. But even if $c_3 \Delta P$ is large, we have seen from Sec. 5-3 that fluid compressibility causes waves to spread and does not affect wave velocity.

Subject to the above restrictions, the species conservation Eq. (5.4-1) becomes

$$\frac{\partial S_j}{\partial t_D} + \frac{\partial f_j}{\partial x_D} = 0, \qquad j = 1 \text{ or } 2 \tag{5.7-2}$$

in dimensionless form, where for a horizontal reservoir

$$f_j = \left(\frac{\lambda_{rj}}{\sum_{m=1}^{3} \lambda_{rm}} \right) \tag{5.7-3}$$

The relative mobilities in Eq. (5.7-3) are known functions of S_1 and S_2. Only two independent saturations are in this example, since $S_1 + S_2 + S_3 = 1$, which we arbitrarily take to be the water and oil saturations. Equation (5.7-3) implies that the fractional flows are known functions of S_1 and S_2.

From Eq. (5.4-5), the specific velocity of a constant saturation S_j is

$$v_{S_j} = \left(\frac{\partial f_j}{\partial S_j} \right)_{t_D}, \qquad j = 1 \text{ or } 2 \tag{5.7-4a}$$

if the wave is nonsharpening and

$$v_{\Delta S_j} = \frac{\Delta f_j}{\Delta S_j}, \qquad j = 1 \text{ or } 2 \tag{5.7-4b}$$

if the wave is a shock. We cannot take the derivative in Eq. (5.7-4a) without knowing the solution to the problem $S_j(x_D, t_D)$. The results of the previous section carry over to this problem with $A_1 = 1$, $B_1 = f_{11}$, $D_1 = f_{12}$, $B_2 = f_{21}$, $C_2 = 1$, $D_2 = f_{22}$, $A_2 = C_1 = 0$, $E_1 = E_2 = 0$. For brevity, we have adopted the convention that $f_{12} = (\partial f_1 / \partial S_2)_{S_1}$, and so on. B_1, B_2, C_1, and C_2 are known functions of S_1 and S_2—though perhaps very complicated—but we can calculate them without knowing the solution $S_1(x_D, t_D)$ and $S_2(x_D, t_D)$.

Let's now let the initial saturations in the medium be uniform at $(S_1, S_2)_I$ and impose at $x_D = 0$ the saturations $(S_1, S_2)_J$. From Sec. 5-6, we know the coherence condition applies at all points in the domain where

$$\frac{df_1}{dS_1} = \frac{df_2}{dS_2} = \sigma \tag{5.7-5}$$

from Eqs. (5.6-14b) and (5.7-4a). The derivatives in Eq. (5.7-5) are total derivatives since the coherence condition implies the existence of a relation $S_2 = S_2(S_1)$ in saturation space. We expand the derivatives in Eq. (5.7-5) and write the two equations in matrix form as

$$\begin{pmatrix} f_{11} & f_{12} \\ f_{21} & f_{22} \end{pmatrix} \begin{pmatrix} dS_1 \\ dS_2 \end{pmatrix} = \sigma \begin{pmatrix} dS_1 \\ dS_2 \end{pmatrix} \tag{5.7-6}$$

To solve for $S_2(S_1)$, we first solve this equation for the eigenvalues, $\sigma^{\pm}$

$$\sigma^{\pm} = \frac{1}{2}\{(f_{22} + f_{11}) \pm [(f_{11} - f_{22})^2 + 4f_{21}f_{12}]^{1/2}\} \tag{5.7-7}$$

Both roots to Eq. (5.7-7) are real, $\sigma^+ > \sigma^-$, and both are known functions of S_1 and S_2. Recall that the $\sigma^{\pm}$ are saturation velocities. Solving for dS_1 and dS_2 in Eq. (5.7-6) gives

$$\frac{dS_2}{dS_1} = \frac{\sigma^{\pm} - f_{11}}{f_{12}} \tag{5.7-8}$$

Equations (5.7-7) and (5.7-8) are the special cases of Eqs. (5.6-10) and (5.6-13). Equation (5.7-8) is an ordinary differential equation whose integration gives the function $S_2(S_1)$. There are two such functions corresponding to σ^+ and σ^-. The velocity of any saturation along $S_2(S_1)$ is given by σ^+ and σ^- depending on whichever is physically realistic.

The above procedure could perhaps be made clearer by addressing a particular problem. Consider an oil–gas–water mixture being displaced by water. To make the problem simple, we take the relative permeabilities to be

$$k_{rj} = \frac{S_j - S_{jr}}{1 - S_{1r} - S_{2r} - S_{3r}}, \qquad j = 1 \text{ or } 2 \tag{5.7-9}$$

and let $S_{1r} = S_{2r} = S_{3r} = 0.1$. Equation (5.7-9) is not a realistic three-phase relative permeability function (see Exercise 5N), but it is sufficient for illustration. We further take $\mu_1 = 1$ mPa-s, $\mu_2 = 5$ mPa-s, and $\mu_3 = 0.01$ mPa-s, and consider the initial conditions to be $S_{2I} = 0.45$, and $S_{1I} = 0.1$. Therefore, the medium is initially at residual water saturation with equal volumes of oil and gas. We are to displace this mixture with water, that is $S_{1J} = 0.8$ and $S_{2J} = 0.1$. This procedure corresponds to a waterflood initiated well into the primary production phase.

Figure 5-21 shows the functions $S_2(S_1)$ obtained by numerically integrating Eq. (5.7-8) with the indicated physical relations. The plot is on a triangular diagram to emphasize the relation $S_1 + S_2 + S_3 = 1$. The integration of Eq. (5.7-8) for various initial values of S_1 and S_2 produces two families of curves corresponding to σ^+ and σ^-, which are the image curves σ^+ and σ^- (light lines in Fig. 5-21) referred to previously. Since $\sigma^+ > \sigma^-$, the image curves nowhere coincide, and further, to every point in the saturation diagram, there are associated two velocities σ^+ and σ^-. The two families of curves we call the saturation *paths* after Helfferich (1981). The particular paths that pass from the initial to the injected condition are the saturation *routes* (bold lines in Fig. 5-21). Though we henceforth restrict our attention to the saturation routes, Fig. 5-21 gives a rapid visual perspective for any displacement having arbitrary initial and injected conditions.

In moving from the initial to injected conditions, there are two alternative saturation routes: (1) a σ^- segment going from the initial conditions to the upper apex of the three-phase flow region and then a σ^+ segment on the gas–water boundary to the injected condition and (2) a σ^+ segment from the initial conditions to

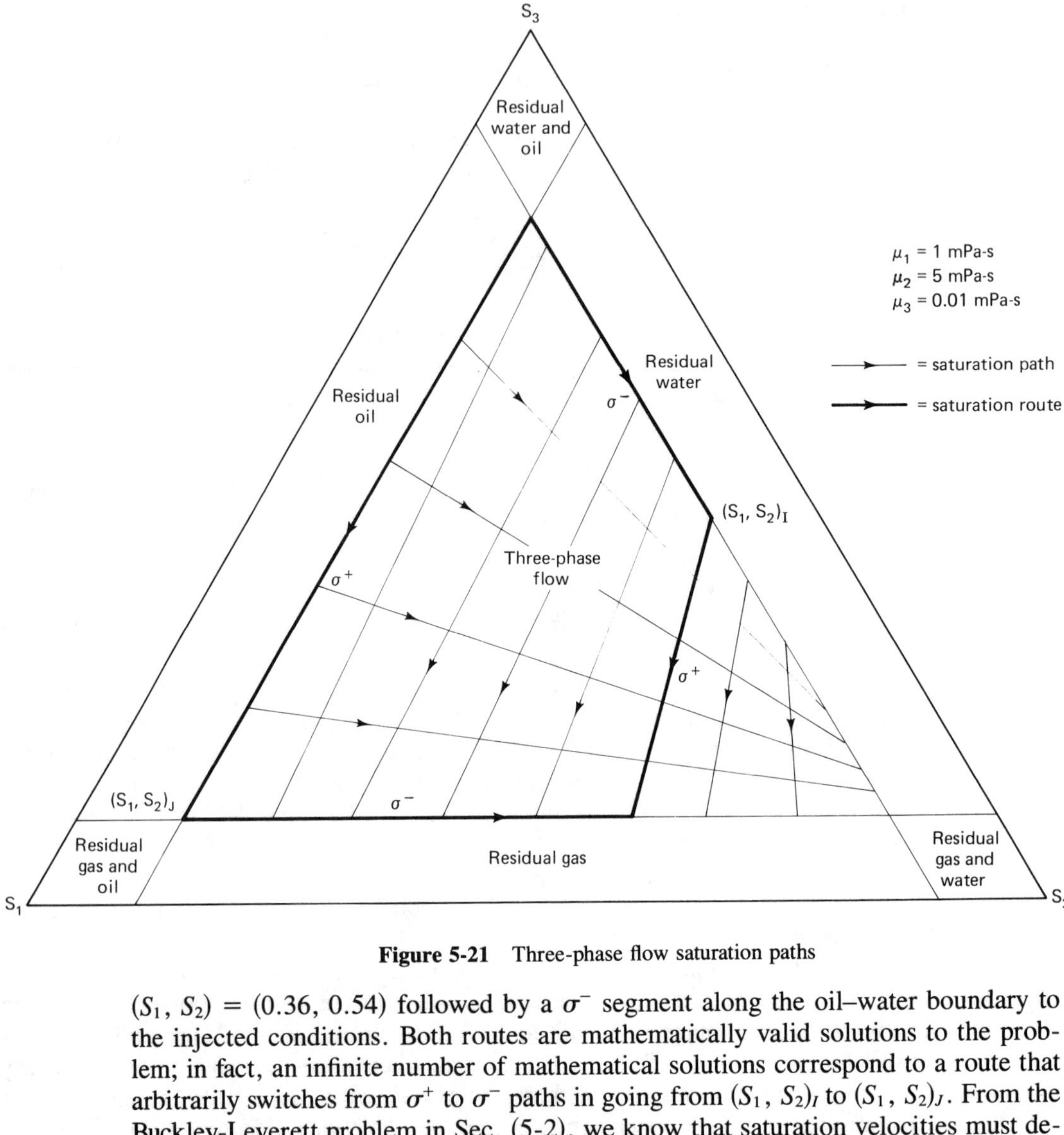

Figure 5-21 Three-phase flow saturation paths

$(S_1, S_2) = (0.36, 0.54)$ followed by a σ^- segment along the oil–water boundary to the injected conditions. Both routes are mathematically valid solutions to the problem; in fact, an infinite number of mathematical solutions correspond to a route that arbitrarily switches from σ^+ to σ^- paths in going from $(S_1, S_2)_I$ to $(S_1, S_2)_J$. From the Buckley-Leverett problem in Sec. (5-2), we know that saturation velocities must decrease monotonically (though not continuously) in the upstream direction. The only physical solution for the problem is route (2) because $\sigma^+ > \sigma^-$ forces this to be the only possible route where σ decreases monotonically from $(S_1, S_2)_I$ to $(S_1, S_2)_J$.

Within a route segment, the saturation velocities must decrease monotonically in the upstream direction also. This condition is not met on the σ^+ route segment (the arrows on the saturation routes indicate the direction of increasing saturation velocity). Such behavior indicates the wave is a shock, and we can find the shock velocity by a procedure entirely analogous to that used in Sec. (5-2). Figure 5-22(a)

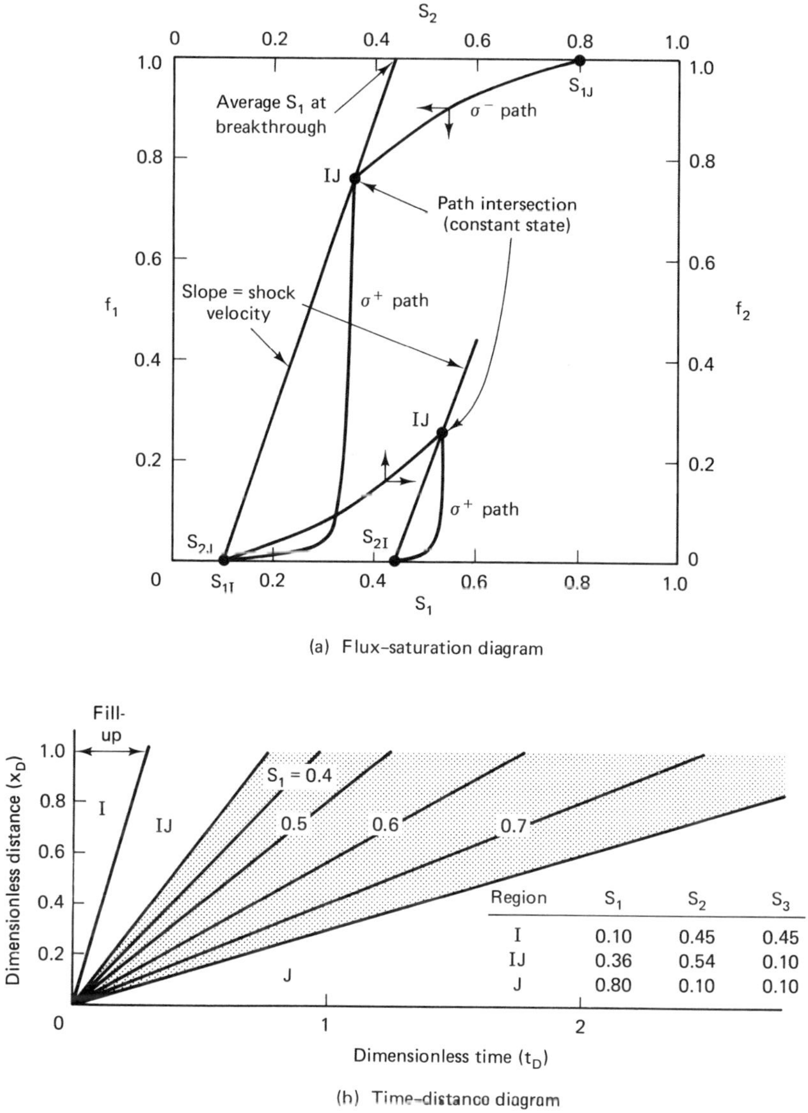

Region	S_1	S_2	S_3
I	0.10	0.45	0.45
IJ	0.36	0.54	0.10
J	0.80	0.10	0.10

Figure 5-22 Diagrams for three-phase flow example

plots the oil and water fluxes (f_1, f_2) versus (S_1, S_2) along the composition route. The shock construction is exactly as suggested in Fig. 5-4, and may be performed on either the f_1-S_1 curve or the f_2-S_2 curve. Equation (5.7-5) guarantees this equivalence. The only real difference between the three-phase and two-phase flow problems at this point is the existence of the constant-state region at IJ. The time–distance diagram for the displacement is in Fig. 5-22(b), which should be compared to Figs. 5-5 and 5-20(b).

Despite the simplified nature of the relative permeability curves used in this example, Fig. 5-22 illustrates that the most important feature of three-phase oil–gas–water flow is the extremely small gas viscosity. This viscosity causes the oil fractional flow to be small initially and to delay the appearance of an appreciable amount of oil at the outflow end until $t_D = 0.28$. This delay, or "fill-up," time is an omnipresent feature of waterfloods begun with appreciable amounts of free gas in the medium (Caudle, 1968). A fill-up period occurs because of the very large gas mobility, not as the result of gas compressibility or redissolution. The last two effects would serve to reduce the fill-up time. A second consequence of the small gas viscosity is no simultaneous three-phase flow occurs in the medium. In fact, by assuming an oil–water mixture banks up the free gas, it is possible to repeat the results in Figs. 5-21 and 5-22 with much less effort (see Exercise 5O). A final consequence of the small gas viscosity is this behavior is qualitatively accurate regardless of the relative permeability functions used.

We end this section by discussing the displacement efficiency of the three-phase flow problem. There is now a displacement efficiency for both oil and gas for which we need average saturations for the definition (Eq. 5.1-2). Considering the fractional flux-saturation curve in Fig. 5-22(a), the average saturations follow from a procedure directly analogous to the Welge procedure in Sec. 5-2.

$$\bar{S}_j = S_j\big|_{x_D=1} - t_D(f_j\big|_{x_D=1} - f_{jJ}), \qquad j = 1, 2, \text{ or } 3 \tag{5.7-10}$$

where $t_D = (df_j/dS_j)^{-1}$ is the reciprocal slope of the f_j-S_j curve evaluated at $x_D = 1$. Figure 5-22(a) shows the average water saturation at water breakthrough, and Fig. 5-23 shows the displacement efficiencies for this example. Once again E_D is limited by the residual phase saturations, oil production is delayed for a fill-up period, and

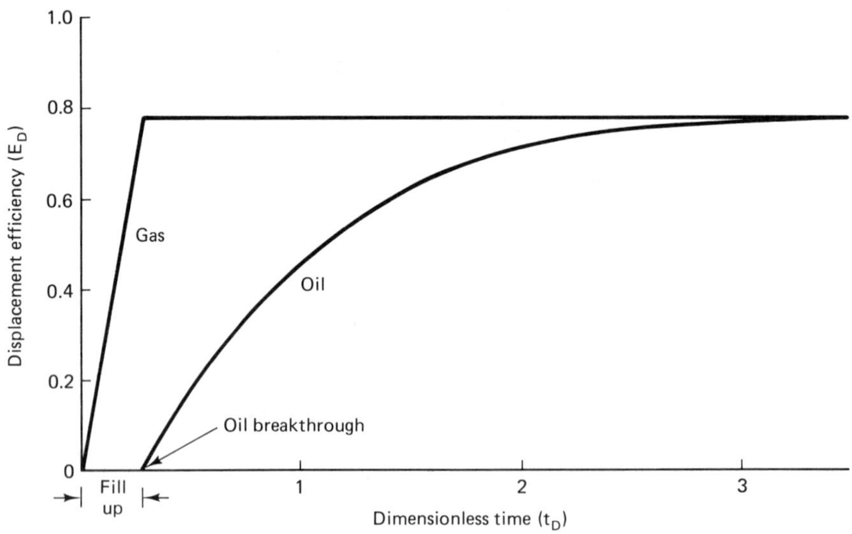

Figure 5-23 Displacement efficiencies for three-phase flow problem

the oil displacement efficiency is determined by the water–oil relative permeabilities and viscosities.

This example demonstrates the strength of simple wave theory. In later chapters, we return to these procedures for specific EOR applications.

5-8 CONCLUDING REMARKS

Any oil recovery calculation of a field-scale displacement based solely on the procedures discussed in this chapter will seriously overestimate the actual recovery: Such one-dimensional calculations neglect volumetric sweep issues which are at least as important as displacement efficiency. Nevertheless, the fractional flow calculations are important in establishing a framework for advancing our study. The items important in establishing this framework are the Buckley-Leverett theory and its generalization in Sec. 5-7, the ideas of coherent waves and their representations, and the notion of the ideal miscible displacement.

EXERCISES

5A. *Parameter-Free Statement*. Show that Eqs. (5.2-5) can be reduced to a parameter-free statement by defining and introducing a reduced saturation S_D, where

$$S_D = \frac{S_1 - S_{1I}}{S_{1J} - S_{1I}} \tag{5A-1}$$

5B. *Radial Form of Water Material Balance*

(a) Show that the one-dimensional water conservation Eq. (5.2-1) for incompressible flow in radial geometry is

$$\phi \frac{\partial S_1}{\partial t} + \frac{q}{2\pi H_t r}\left(\frac{\partial f_1}{\partial r}\right) = 0 \tag{5B-1}$$

where q is the volumetric flow rate, H_t the medium thickness, and f_1 is the same as Eq. (5.2-2).

(b) If we let $r_D = (r/R)^2$ and $t_D = \int_0^t q dt/\pi\phi H_t R^2 = \int_0^t q dt/V_p$, show that Eq. (5B-1) becomes identical to the linear Eq. (5.2-5a).

5C. *Buckley-Leverett Application*. Calculate effluent histories (water cut $f_1|_{x_D=1}$ versus t_D) for water ($\mu_1 = 1$ mPa-s) displacing oil given the following experimental data (Chang et al., 1978):

S_1	k_{r1}	k_{r2}
0.40	0.00	0.36
0.45	0.005	0.26
0.50	0.009	0.14
0.55	0.02	0.08
0.60	0.035	0.036
0.65	0.050	0.020
0.70	0.080	0.00

Use three values of oil viscosity: $\mu_2 = 1$, 5, and 50 mPa-s. For $\mu_2 = 5$ mPa-s, calculate the endpoint, shock, and average saturation mobility ratios. The dip angle is zero.

5D. *Gravity and Fractional Flow Theory*. For the exponential relative permeability functions of Eq. (3.3-4), plot water saturation profiles at $t_D = 0.3$ for dip angles of $\alpha = 0°$, 30°, and $-30°$. Additional data are $S_{1r} = S_{2r} = 0.2$, $n_1 = 1$, $n_2 = 2$, $k_{r1}^0 = 0.1$, $k_{r2}^0 = 0.8$, $\mu_1 = 1$ mPa-s, $\mu_2 = 10$ mPa-s, $k = 0.5\ \mu\text{m}^2$, $\Delta\rho = 0.2$ g/cm³, and $u = 0.6$ cm/day.

5E. *Buckley-Leverett Theory with Straight Line Relative Permeabilities*. Use straight line exponential relative permeability functions with zero residual phase saturations in the following ($n_1 = n_2 = 1$, $S_{1r} = S_{2r} = 0$ in the exponential relative permeability functions). Also take $f_{1I} = 0$ and $f_{1J} = 1$.

(a) Show that the sign of $(1 - M^0 + M^0 N_g^0 \sin\alpha)$ uniquely determines the character (spreading, indifferent, sharpening) of the water saturation wave.

(b) For the spreading wave case—$(1 - M^0 + M^0 N_g^0 \sin\alpha) < 0$—Eq. (5.2-10) may be inverted explicitly for $S_1(x_D, t_D)$. Derive this expression in terms of the quadratic formula.

(c) Use the equation in part (b) to show that for $\alpha = 0$ the water saturation function is given by

$$S_1(x_D, t_D) = \begin{cases} 0, & \dfrac{x_D}{t_D} > M^0 \\[2ex] \dfrac{\left(\dfrac{t_D M^0}{x_D}\right)^{1/2} - 1}{M^0 - 1}, & \dfrac{1}{M^0} \le \dfrac{x_D}{t_D} \le M^0 \\[2ex] 1, & \dfrac{x_D}{t_D} < \dfrac{1}{M^0} \end{cases} \tag{5E-1}$$

(d) Use Eq. (5E-1) to derive an expression for the average water saturation $\bar{S}(t_D)$ and the displacement efficiency $E_D(t_D)$.

5F. *Water Fractional Flow with Capillary Pressure*. Derive the expression for water fractional flow including capillary pressure (Eq. 5.3-1).

5G. *Analytic Relative Permeability Ratios (Ershaghi and Omoregie, 1978)*. Over intermediate water saturation ranges, the oil–water relative permeability ratio plots approximately as a straight line on a semilog scale, using

$$\frac{k_{r2}}{k_{r1}} = Ae^{-BS_1} \tag{5G-1}$$

where A and B are positive constants. Using the Buckley-Leverett theory, show that a plot of the product of oil and water cuts is a straight line with slope $1/B$ when plotted against $1/t_D$. The dip angle is zero.

5H. *Fractional Flow with Two Inflections*. For the fractional flow curve of Fig. 5H, construct plots of fractional flow versus dimensionless distance at breakthrough for saturation $S_1 = 1$ displacing $S_1 = 0$, and $S_1 = 0$ displacing $S_1 = 1$.

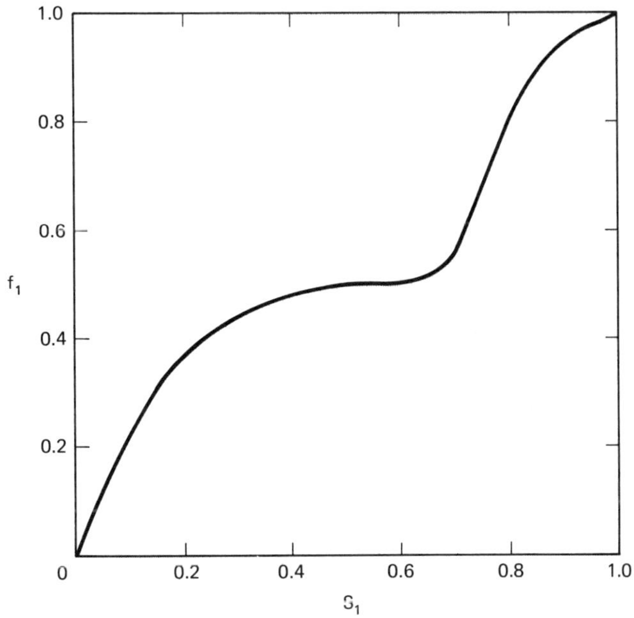

Figure 5H Fractional flow curve for Exercise 5H

5I. *The Reversibility of Dispersion and Fractional Flow*. Fluid 2 is to be partially displaced by fluid 1 in a one-dimensional permeable medium. Fluid 1 is injected until just before it is produced, and then the flow is reversed (that is, fluid 2 is *injected* at the effluent end). In all that follows, take the initial (I) condition to be 100% fluid 2 flowing and the injected (J) condition to be 100% fluid 1.

(a) Sketch two time–distance diagrams for this case using fractional flow curves like those on the extreme right and left of Fig. 5-6.

(b) If fluids 1 and 2 are completely miscible with identical viscosities and mix only by dispersion, use Eq. (5.5-18) to sketch the time–distance diagram.

(c) Based on the results of parts (a) and (b), what can you conclude about the mixing caused by fractional flow compared to that caused by dispersion?

(d) If fluids 1 and 2 are water and oil and a fractional flow curve like that on the middle panel of Fig. 5-6 applies, calculate and plot the time–distance diagram.

5J. *Mobility Ratio for Compressible Flow*. Consider the pistonlike displacement of fluid 2 by fluid 1 in the x direction. Use the general definition of mobility ratio (pressure gradient ahead of front divided by pressure gradient behind front) in the following:

(a) Show that the mobility ratio becomes the endpoint mobility ratio if the volumetric flowrate uA is not a function of x (fluids are incompressible).

(b) If the mass flux ρuA is not a function of x, on the other hand, show that the mobility ratio becomes

$$M_\nu = \frac{k^{\circ}_{r1}\,\nu_2}{k^{\circ}_{r2}\,\nu_1} \tag{5J-1}$$

where $\nu = \mu/\rho$ is the kinematic viscosity.

(c) Calculate both M° and M_ν for the following conditions: $\rho_1 = 1$ mg/cm^3, $\mu_1 = 1$ μPa-s, $\rho_2 = 0.8$ g/cm^3, $\mu_2 = 2$ mPa-s, $k^{\circ}_{r1} = 0.1$ and $k^{\circ}_{r2} = 1.0$.

5K. *Using Tracer Data*. Consider a one-dimensional permeable medium containing oil at a uniform residual saturation S_{2r}, and through which is flowing 100% water at a constant rate. At $t = 0$, a second water stream is introduced at the inlet that contains two ideal (nondispersing and nonabsorbing) tracers. Tracer 1 remains only in the water phase, but tracer 2 partitions into the residual oil phase with a partition coefficient of 2. The partition coefficient is the ratio of the concentration of tracer 2 in the oil phase to that in the water phase $K_{21}^{2} = C_{22}/C_{21}$. Tracer 1 breaks through after three hours, and tracer 2 after six hours. If the volumetric injection rate is 1 cm³/min, calculate the pore volume and S_{2r}.

5L. *Laboratory Estimation of Dispersivity*. Dispersivity may be estimated from laboratory, first-contact miscible displacements with the following development:

(a) Show from Eq. (5.5-15) that a plot of $(1 - t_D)/\sqrt{t_D}$ versus $\text{erf}^{-1}(1 - 2C_e)$ will yield a straight line with slope $2N_{Pe}^{-1/2}$. Here C_e is the effluent concentration $(C_D|_{x_D=1})$

(b) Estimate the pore volume, dispersion coefficient, and dispersivity from the following experimental data:

Volume produced (cm³)	Effluent concentration
60	0.010
65	0.015
70	0.037
80	0.066
90	0.300
100	0.502
110	0.685
120	0.820
130	0.906
140	0.988
150	0.997

The interstitial velocity is 20 cm/day, and the length is 0.5m. Note that $\text{erf}^{-1}(1 - 2x)$ is the probability axis (x axis) on probability paper.

5M. *Tracers in Two-Phase Flow*. Consider a permeable medium flowing oil and water at constant oil fractional flow (case A in Fig. 5-12). Show that if a tracer with partition coefficient defined as in Exercise 5K is introduced at $t_D = 0$, the conservation equation for the tracer concentration C in the aqueous phase is (Delshad, 1981)

$$\frac{\partial C}{\partial t_D} + \frac{\partial C}{\partial x_D} - \frac{\overline{K}}{v_T L}\frac{\partial^2 C}{\partial x_D^2} = 0 \qquad \text{(5M-1)}$$

where

$$t_D = t\frac{v_T}{L} \qquad \text{(5M-2)}$$

$$v_T = \frac{q}{A\phi}\,\frac{f_1 + K_{21} f_2}{S_1 + K_{21} S_2} \qquad \text{(5M-3)}$$

$$\overline{K} = \frac{S_1 K_{\ell 1} + K_{21} S_2 K_{\ell 2}}{S_1 + K_{21} S_2} \tag{5M-4}$$

$K_{\ell 1}$ and $K_{\ell 2}$ are the longitudinal dispersion coefficients for the tracer in the oil and water phases. Take $(q/A\phi)$ to be constant.

5N. *Three-Phase Coherence Calculation.* A more realistic three-phase relative permeability for oil, gas, and water is

$$k_{r1} = k_{r1}^0 \left(\frac{S_1 - S_{1r}}{1 - S_{1r} - S_{2r1}} \right)^{n_1} \tag{5N-1}$$

$$k_{r3} = k_{r3}^0 \left(\frac{1 - S_1 - S_2 - S_{3r}}{1 - S_{1r} - S_{3r}} \right)^{n_3} \tag{5N-2}$$

$$k_{r2} = k_{r2}^0 \left\{ \left(\frac{k_{r21}}{k_{r2}^0} + k_{r1} \right) \left(\frac{k_{r23}}{k_{r2}^0} + k_{r3} \right) - (k_{r1} + k_{r3}) \right\} \tag{5N-3}$$

where

$$k_{r21} = k_{r2}^0 \left(\frac{1 - S_1 - S_{2r1}}{1 - S_{2r1} - S_{1r}} \right)^{n_{21}} \tag{5N-4}$$

$$k_{r23} = k_{r2}^0 \left(\frac{S_2 + S_1 - (S_{2r3} + S_{1r})}{1 - (S_{2r3} + S_{1r}) - S_{3r}} \right)^{n_{23}} \tag{5N-5}$$

These are modifications of the Stone relative permeability model (1970).

In Eqs. (5N-1) through (5N-5)

n_{21} = Oil relative permeability exponent in water–oil system
n_{23} = Oil relative permeability exponent in gas–oil system
S_{2r1} = Residual oil saturation in water–oil system
S_{2r3} = Residual oil saturation in gas–oil system

Calculate and plot the following:

(a) Lines of constant k_{r1}, k_{r2}, k_{r3}, in the triangular composition space, S_1, S_2, and S_3.
(b) The composition paths and a waterflood composition route for initial saturations of 0.5, 0.3, and 0.2 for oil, gas, and water.
(c) The wave positions in a dimensionless time–distance diagram.

Use the following data:

$\mu_1 = 1$ mPa-s	$\mu_2 = 2$ mPa-s	$\mu_3 = 0.01$ mPa-s
$S_{2r1} = 0.3$	$k_{r2}^0 = 0.6$	$n_{21} = 1.5$
$S_{2r3} = 0.05$	$k_{r1}^0 = 0.3$	$n_{23} = 2$
$S_{1r} = 0.2$	$k_{r3}^0 = 0.7$	$n_1 = 3$
$S_{3r} = 0.05$	$\alpha = 0$	$n_3 = 2.5$

This problem requires a numerical solution.

5O. *Simplified Three-Phase Fractional Flow.* Rework part (c) of Exercise 5N by assuming the displacement becomes a shock wave from the initial conditions to a region of simultaneous two-phase oil–water flow followed by a wave of undetermined character to the injected conditions. The velocity of the first wave is given by

$$v_{\Delta S_1} = \frac{f_{3I}}{S_{3I} - S_{3r}} = \frac{f_{1I} - f_1^+}{S_{1I} - S_1^+} = \frac{f_{2I} - f_2^+}{S_{2I} - S_2^+} \tag{5O-1}$$

where f_1^+ and S_1^+ are the water fractional flow and saturation behind the shock. The velocity of the second wave is then given by the Buckley-Leverett construction. Plot an effluent history of oil and water cuts to demonstrate the fill-up phenomenon.

5P. *Method of Characteristics for Reducible Equations.* Consider the following pair of partial differential equations for $u(x, t)$ and $v(x, t)$

$$\frac{\partial u}{\partial t} + \frac{\partial (u^2 v)}{\partial x} = 0 \tag{5P-1}$$

$$\frac{\partial v}{\partial t} + \frac{\partial v^2}{\partial x} = 0 \tag{5P-2}$$

where both u and v are less than or equal to 1.

(a) Write these equations in the "canonical" form of Eqs. (5.6-6). Are these reducible? Why or why not?

(b) Write the coherence requirement for Eqs. (5P-1) and (5P-2). Use this to develop an expression for σ, the composition velocity along the characteristic directions.

(c) Use σ to develop an expression for $u = u(v)$ along both characteristic directions.

(d) If the boundary data are specified along a line $u = 1$ plot the "composition" path grid (u,v space) for $u < 1$ and $v < 1$.

(e) On the plot of part (d) show the "composition" route for $(u, v)_J = (0.6, 0.2)$ displacing $(u, v)_I = (1, 1)$. Treat u and v as physical variables so that the composition velocity must decrease monotonically from I to J. Plot the time (t) − distance (x) diagram for this "displacement" where $t > 0$ and $1 > x > 0$.

(f) Based on this problem and what you know about the ideal miscible displacement, discuss why the constructions in Fig. 5-12 can be done without the procedures in parts (a) through (e).

5Q. *Gravity Segregation and Fractional Flow.* Consider the homogeneous, one-dimensional permeable medium shown in Fig. 5Q for which all the fractional flow assumptions apply. Both ends of the medium are sealed. For $t < 0$, the medium contains a completely saturated water zone above a saturated oil zone ($0 < \epsilon < 1$). At $t = 0$, the more dense water is allowed to flow downward while the less dense oil flows upward. This results in a complete reversal of the oil and water zones after a sufficiently long time. Figure 5Q also shows the long-time condition of the medium.

(a) Show that there is no bulk flow ($u = 0$) at any point in the medium.

(b) Derive a water conservation equation for this special case from the general equations in Chap. 2. Give also the boundary conditions needed to solve this equation for $S_1(x, t)$.

(c) Make the equation of part (b) dimensionless by introducing appropriate scaling factors.

(d) Derive a dimensionless water flux (analogous to a fractional flow) by eliminating the water pressure gradient from the equation of part (c). The absence of bulk flow does *not* eliminate pressure gradients (Martin, 1958).

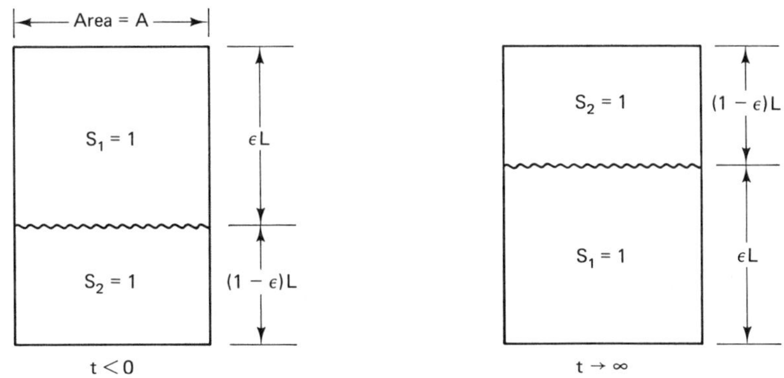

Figure 5Q Gravity segregation with fractional flow

(e) For the following values, plot the dimensionless water flux of part (d) versus water saturation.

$$k_{r1} = 0.1\, S_1^4 \qquad k_{r2} = .8(1 - S_1)^2$$
$$\mu_1 = 1 \text{ mPa-s} \qquad \mu_2 = 5 \text{ mPa-s}$$

(f) Based on the curve of part (e) and $\epsilon = 0.6$, construct the time–distance diagram showing the progress to complete gravity segregation of the water and oil zones. Estimate the dimensionless time this occurs at.

5R. *An Alternate Derivation of the Characteristic Equations.* Consider the following reducible equations for $u(x, t)$ and $v(x, t)$:

$$\begin{aligned} A_1 u_t + B_1 u_x + C_1 v_t + D_1 v_x &= 0 \\ A_2 u_t + B_2 u_x + C_2 v_t + D_2 v_x &= 0 \end{aligned} \tag{5R-1}$$

(a) Suppose that u and v are functions of the combined variable $\sigma = x/t$. Show that Eq. (5R-1) can be written as

$$\begin{pmatrix} B_1 - \sigma A_1 & D_1 - \sigma C_1 \\ B_2 - \sigma A_2 & D_2 - \sigma C_2 \end{pmatrix} \begin{pmatrix} u' \\ v' \end{pmatrix} = 0 \tag{5R-2}$$

where $u' = du/d\sigma$, and so on.

(b) For a nontrivial solution, the determinant of the coefficient matrix of Eq. (5R-2) must be zero. Show that this gives the characteristic directions given by Eq (5.6-10).

(c) Again, for a nontrivial solution, the determinant of the augmented matrix (matrix with the solution vector substituted for a column) must also be zero. Show that if we replace the second column, this operation yields the following relation between u and v:

$$\frac{du}{dv} = \frac{B_1 - \sigma^{\pm} A_1}{B_2 - \sigma^{\pm} A_2} \tag{5R-3}$$

Solutions that can be expressed in terms of (x/t) are said to be *self-similar*.

6

Volumetric Sweep Efficiency

Typical values of residual oil and connate water saturations indicate ultimate displacement efficiency should normally be between 50% and 80% of the contacted oil in a waterflood. This range is substantially higher than the 30% average recovery efficiency observed in waterfloods; it is also higher than recovery efficiency in most EOR projects (see Sec. 1-4). Of course, the reason displacement efficiency is higher than the recovery efficiency is that not all the oil is contacted by the displacing agent. This effect is present in the oil recovery Eq. (2.5-5) where the displacement efficiency is multiplied by the volumetric sweep efficiency E_V. Based on these approximate figures, the volumetric sweep efficiency is between 40% and 60% for a waterflood. For many EOR processes, it can be much smaller, and for others, effecting a large E_V is a primary design objective.

In this chapter, we provide both an overview of volumetric sweep efficiency and techniques to combine areal, vertical, and displacement sweep to arrive at a recovery efficiency. We deal almost exclusively with the immiscible water–oil displacement since this literature on recovery efficiency is well established and many of the more important features also carry over to EOR. In later chapters, we discuss the volumetric sweep efficiency of specific EOR processes. To further distinguish between volumetric and displacement sweep efficiency, we usually deal with indifferent or self-sharpening displacements in which dispersive effects are small. For these cases, the calculation techniques are equally valid whether the displacement is miscible or immiscible since there is no simultaneous flow of components.

6-1 DEFINITIONS

Based on the overall material balance of Sec. 2-5, the cumulative mass of oil recovered is

$$N_{p2} = V_b \overline{W}_{2I} E_{R2}$$

from Eq. (2.5-3) with no oil injection. We wish to convert this equation to a more standard form by the following transformations: Eliminate the recovery efficiency E_{R2} through Eq. (2.5-5a), and replace $\overline{W}_{2I}$ with $\phi(\rho_2 S_2 \omega_{22})_I$, which assumes oil is in only the liquid oleic phase. This gives

$$N_{p2} = V_b \phi (\rho_2 S_2 \omega_{22})_I E_D E_V$$

Next, eliminate $(\rho_2 \omega_{22})_I$ with the oil formation volume factor definition Eq. (2D-5), and let $V_b \phi = V_p$, the pore volume, and $N_p = N_{p2}/\rho_2^\circ$, the oil production in standard volumes. These substitutions yield

$$N_p = \frac{E_D E_V S_{2I} V_p}{B_{2I}} \tag{6.1-1}$$

In Eq. (6.1-1), E_D is the displacement sweep efficiency defined in Eq. (5.1-1), and E_V is the volumetric sweep efficiency defined as

$$E_V = \frac{\text{Volumes of oil contacted by displacing agent}}{\text{Volumes of oil originally in place}} \tag{6.1-2}$$

The term $\left(\frac{S_{2I} V_p}{B_{2I}}\right)$ represents the oil in place at the start of the displacement in standard volumes. We have also dropped the subscript $i = 2$ because all efficiencies in this chapter refer to oil recovery.

The volumetric sweep efficiency can be decomposed into the product of an areal sweep efficiency and a vertical sweep efficiency

$$E_V = E_A E_I \tag{6.1-3}$$

The definition of the areal sweep efficiency is

$$E_A = \frac{\text{Area contacted by displacing agent}}{\text{Total area}} \tag{6.1-4}$$

Figure 6-1(a) shows a schematic of a highly idealized pistonlike displacement in a four-layer areally homogeneous reservoir. Figure 6-1(b) is an areal view of Figure 6-1(a). Based on the definition of Eq. (6.1-4), E_A is the doubly cross-hatched area (at t_2) divided by the singly cross-hatched area. The vertical sweep efficiency,

$$E_I = \frac{\text{Cross-sectional area contacted by displacing agent}}{\text{Total cross sectional area}} \tag{6.1-5}$$

is also similarly defined in Fig. 6-1(a) at a particular time.

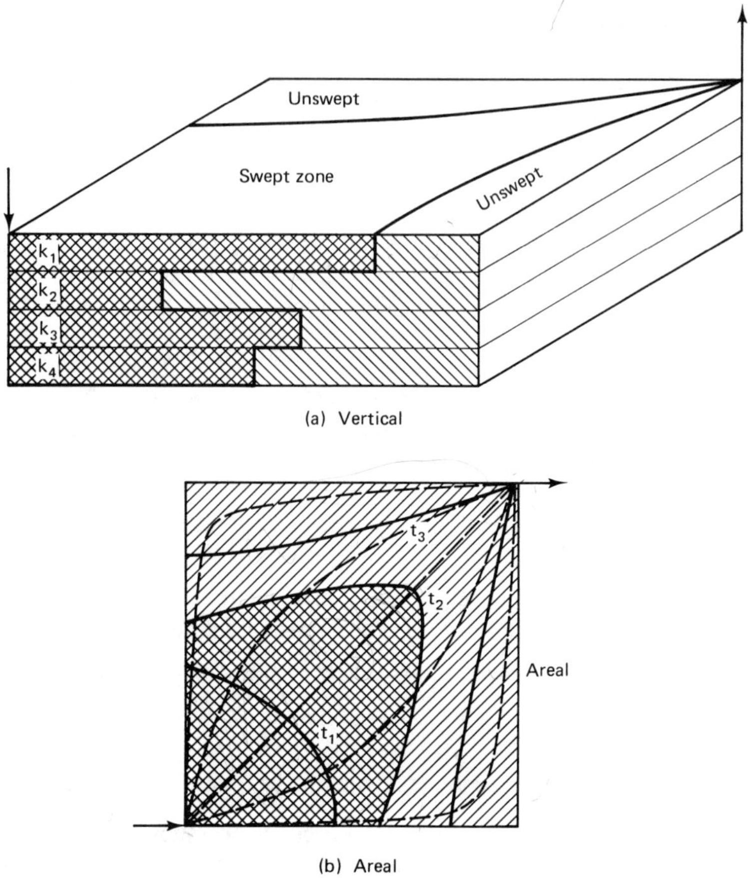

Figure 6-1 Sweep efficiency schematic

The definitions of Eqs. (6.1-3) through (6.1-5) have several subtle difficulties. Both areal and vertical sweep efficiency are ratios of areas; therefore, their product E_V must be a ratio of areas squared. This observation contradicts the definition of Eq. (6.1-2), which says E_V must be a ratio of lengths cubed. The redundant dimension in either Eq. (6.1-4) or (6.1-5) is the dimension parallel to the displacement direction. This direction is nonlinear and varies with both position and time. Thus the decomposition Eq. (6.1-3) transforms E_V into a product of two plane flows.

A second consequence of the redundant dimension in E_V is that both E_A and E_I depend on each other. Note from Fig. 6-1 that E_A depends on vertical position. Similarly, though not so obviously, E_I will be different from the cross section shown for each cross section between the injector and producer. If we restrict ourselves to cross sections defined by pathlines between the injector and producer (dotted lines in Fig. 6-1b), E_I will be the same for each cross section if it can be expressed in a dimensionless form independent of rate. But for the general case, E_I is a function of rate and will be different for each cross section. As we see in Sec. 6-7, the practical consequence of this observation is that neither the areal nor the vertical sweep efficiency in Eq. (6.1-3) can be evaluated at the same time for which the volumetric sweep efficiency is desired.

To use Eq. (6.1-1), even with the above complications, we must have independent estimates of E_A and E_I. For certain very special cases—confined displacements in areally homogeneous regular patterns with no or very good vertical communication—these are available to us through correlation (see Sec. 6-2) or calculation (see Secs. 6-3 through 6-5). When these conditions are not met, E_V must be estimated through scaled laboratory experiments or numerical simulation. In the latter case, though certainly possible to obtain sweep efficiency estimates, the oil recovery itself may be obtained directly, and Eq. (6.1-1) is unnecessary. Still, the equation provides a better understanding of sweep efficiency concepts and the factors necessary to maximize E_V than does simulation alone.

6-2 AREAL SWEEP EFFICIENCY

Though areal sweep efficiency may be determined through simulation or by analytical methods (Morel-Seytoux, 1966), the most common source of areal sweep efficiency data is from displacements in scaled physical models. Figures 6-2 through

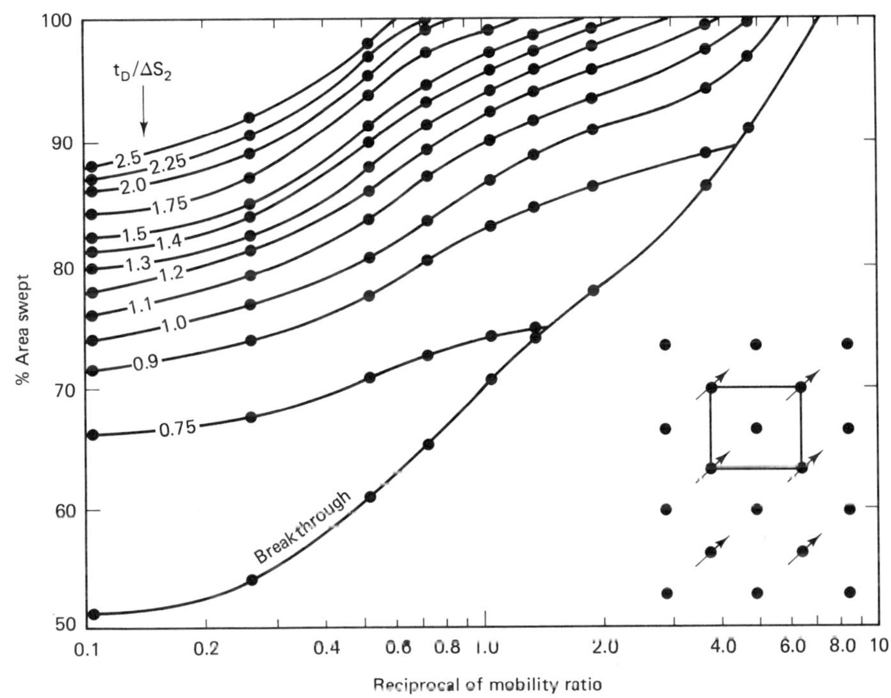

Figure 6-2 Areal sweep efficiency for a confined five-spot pattern (from Dyes et al., 1954)

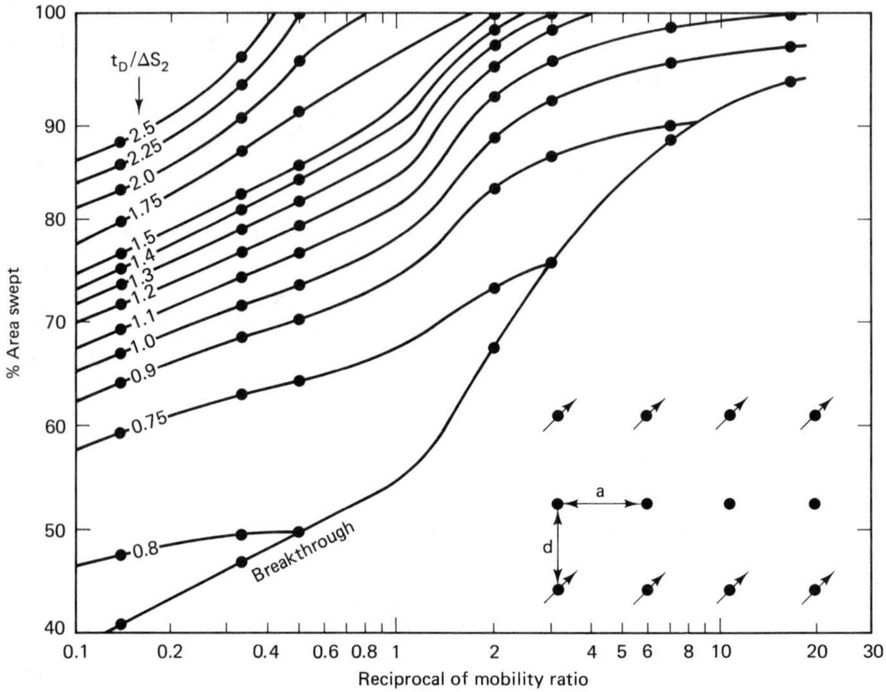

Figure 6-3 Areal sweep efficiency for a confined direct line drive pattern, $d/a = 1$ (from Dyes et al., 1954)

6-4 present three of these areal sweep "correlations" from the work of Dyes et al. (1954) for three different regular well patterns. Several more of these correlations are in the work of Craig (1971), and an extensive bibliography of areal sweep efficiency is given by Claridge (1972). For the patterns shown in the lower right-hand corner, Figs. 6-2 through 6-4 plot E_A on the y axis versus the reciprocal mobility ratio on the x axis with time as a parameter. Since mobility ratio and pattern type are fixed for a given displacement, time is actually the dependent variable. The dimensionless time in Figs. 6-2 through 6-4 is cumulative displaceable pore volumes of displacing agent injected. Since time is the dependent variable in these correlations, a more direct representation would be a plot of E_A versus dimensionless time at fixed mobility ratio and pattern type (see Exercise 6A). You should remember that these correlations are for pistonlike displacements in regular, homogeneous, confined patterns. When the well patterns are unconfined, the reference area in Eq. (6.1-4) can be much larger, and E_A smaller. Based on an extensive survey of the available correlations for spreading displacements, Craig (1971) determined that the appropriate mobility ratio for the areal sweep correlations is the average saturation mobility ratio $\overline{M}$ given by Eq. (5.2-25a).

From the correlations, E_A increases with increasing time, or throughput, and decreasing mobility ratio. At a fixed mobility ratio, E_A is equal to the displaceable

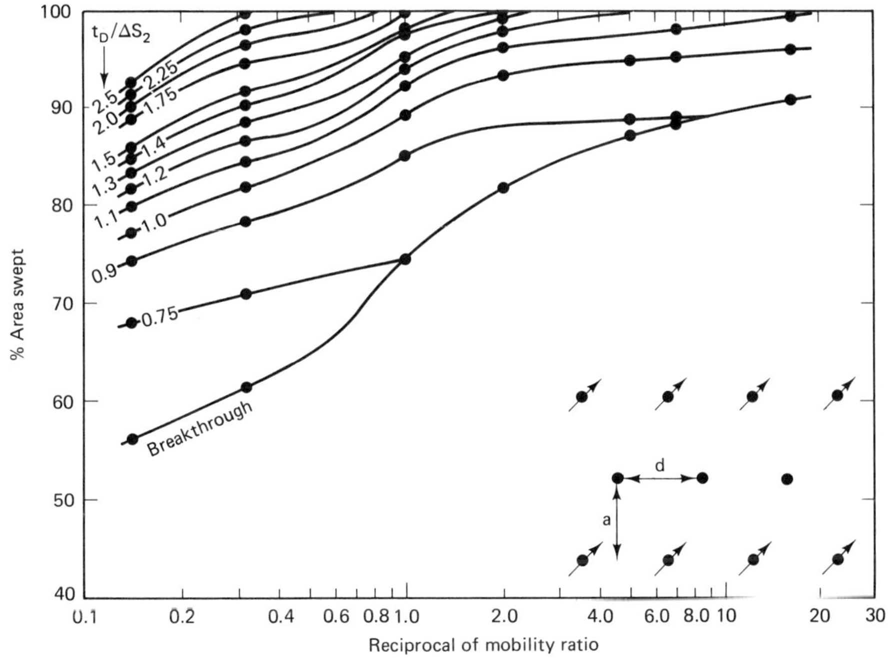

Figure 6-4 Areal sweep efficiency for a staggered line drive pattern, $d/a = 1$ (from Dyes et al., 1954)

pore volumes injected until breakthrough and then given by the indicated curves in Figs. 6-2 through 6-4 thereafter. E_A also increases as the pattern type more closely approaches linear flow, but this sensitivity is not great for the more common patterns. The decrease in E_A with increasing $\overline{M}$ is in the same direction as the change in displacement efficiency with mobility ratio discussed in Sec. 5-2; thus large mobility ratios are detrimental to both areal and displacement sweep.

6-3 MEASURES OF HETEROGENEITY

Considering the manner reservoirs are deposited in and the complex diagenetic changes that occur thereafter, it should not be surprising that no reservoir is homogeneous. This does not imply all reservoirs are dominated by their heterogeneity since in many cases, one mechanism is so strong that it completely overshadows all others. For example, gravity can be so pronounced in a high-permeability reservoir that it may be considered homogeneous to good approximation.

Nevertheless, heterogeneity is always present in reservoirs, is the most difficult feature to define, and usually has the largest effect on vertical sweep efficiency. Therefore, before we explore vertical sweep efficiency, we discuss the most common measures of heterogeneity and their limitations.

Definitions

The three principal forms of nonidealities in reservoirs are anisotropies, nonuniformities, and heterogeneities. These terms can be applied to any property but usually describe permeability, porosity, and occasionally, relative permeability. An *anisotropic* property varies with the direction of measurement and, hence, has intrinsic tensorial character (see Sec. 2-2). Following Greenkorn and Kessler (1969), the definitions of nonuniformity and heterogeneity are closely related (Fig. 6-5). A homogeneous, uniform property is represented on a frequency distribution plot as a single delta function (spike), and a heterogeneous, uniform property by a finite number of these functions. A homogeneous, nonuniform property cannot be represented by a finite number of delta functions but can be a continuous function having only one peak. A heterogeneous, nonuniform property is represented by a continuous distribution function having two or more peaks. Most laboratory displacements are homogeneous and nonuniform. Most calculation techniques assume the reservoir is uniform and heterogeneous. Actual nonuniformities are frequently "averaged" out by capillary pressure or dispersion.

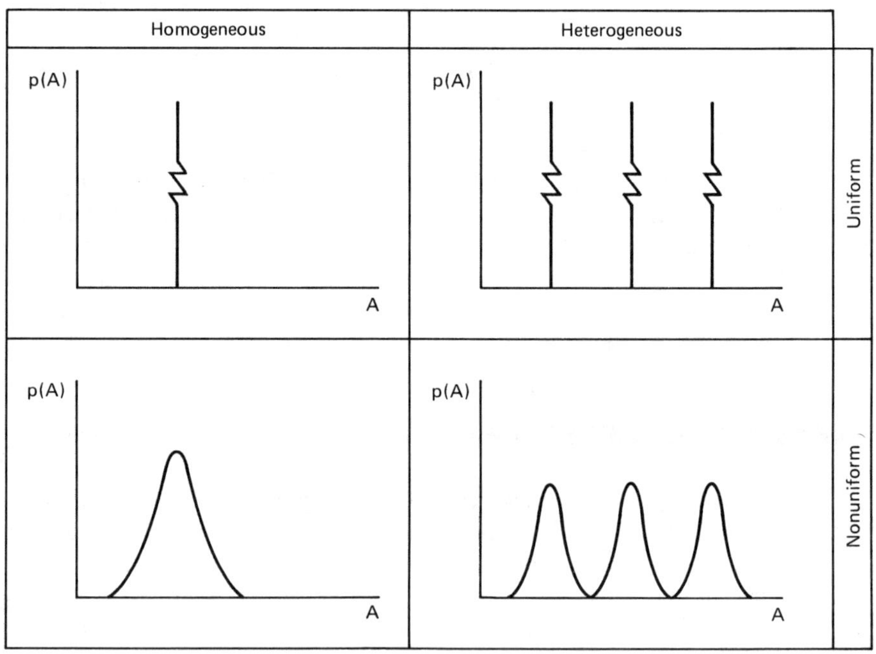

Figure 6-5 Probability distribution functions for parameter *A* (from Greenkorn and Kessler, 1965)

Flow and Storage Capacity

Since permeability can change several factors of 10 within a short distance in a reservoir, whereas porosity changes by only a few percent over the same scale, it is common to view the reservoir as homogeneous with respect to porosity and heterogeneous with respect to permeability. Although most of the traditional measures of heterogeneity adopt this convention it is not necessary and can even lead to occasional errors. In the following discussion, we include porosity variations in the definitions; the more traditional definitions can be recovered by letting porosity and thicknesses be constant.

Imagine an ensemble of N_L permeable media elements each having different permeability k_ℓ, thickness h_ℓ, and porosities ϕ_ℓ. The elements are arranged as resistances parallel to flow. From Darcy's law, the interstitial velocity of the single-phase flow of a conservative tracer is proportional to the ratio of permeability to porosity $r_\ell = k_\ell/\phi_\ell$. Thus if r_ℓ is a random variable, we can rearrange the elements in order of decreasing r_ℓ (this is equivalent to arranging in order of decreasing fluid velocity), and we can define a cumulative flow capacity at a given cross section as

$$F_n = \sum_{\ell=1}^{n} \frac{k_\ell h_\ell}{H_t \bar{k}} \tag{6.3-1a}$$

where H_t is the total thickness,

$$H_t = \sum_{\ell=1}^{N_L} h_\ell \tag{6.3-1b}$$

The average quantities are defined as

$$\bar{k} = \frac{1}{H_t} \sum_{\ell=1}^{N_L} (kh)_\ell \tag{6.3-1c}$$

and similarly for porosity. A cumulative storage capacity follows in a similar fashion

$$C_n = \sum_{\ell=1}^{n} \frac{\phi_\ell h_\ell}{H_t \bar{\phi}} \tag{6.3-1d}$$

The physical interpretation of F_n is that if the N_L elements are arranged in parallel, F_n is the fraction of total flow of velocity r_n or faster. C_n is the volume fraction of these elements. A plot of F_n versus C_n yields the curve shown in Fig. 6-6(a); if N_L becomes very large, the ensemble approaches the continuous distribution shown in Fig. 6-6(b). We designate the continuous distribution by F and C without subscripts. From the definitions F, C, and r, the slope of either curve at any C is the interstitial velocity at that point divided by the average interstitial velocity of the whole ensemble

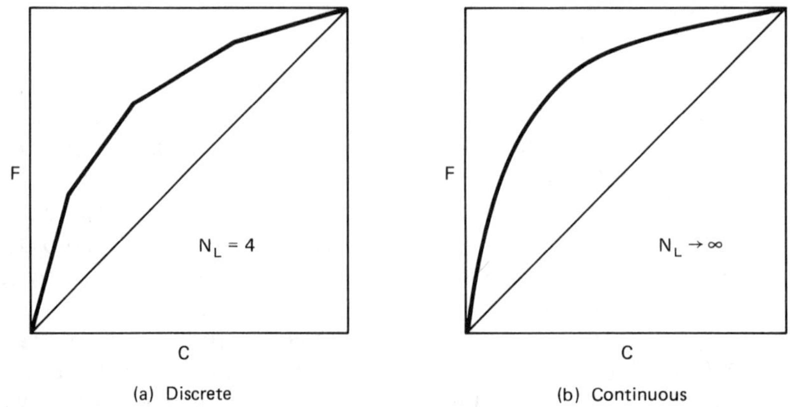

Figure 6-6 Schematic of discrete and continuous flow-storage capacity plots

$$\frac{dF}{dC} = F' = \begin{cases} \dfrac{r_n}{\bar{r}} & \text{(discrete)} \\ \dfrac{r}{\bar{r}} & \text{(continuous)} \end{cases} \tag{6.3-2}$$

Because the elements were rearranged, the slope is monotonically decreasing and, from the definitions, $F_n = C_n = 1$, when $n = N_L$.

Measures of Heterogeneity

A common measure of reservoir heterogeneity is the Lorenz coefficient L_c, defined as the area between the F-C curve and a 45° line (homogeneous F-C curve) and normalized by 0.5,

$$L_c = 2\left\{\int_0^1 F\,dC - \frac{1}{2}\right\} \tag{6.3-3}$$

for the continuous curve. The Lorenz coefficient varies between 0 (homogeneous) and 1 (infinitely heterogeneous). A second, perhaps more common, measure that lies between the same limits is the Dykstra-Parsons (1950) coefficient V_{DP}.

$$V_{\text{DP}} = \frac{(F')_{C=0.5} - (F')_{C=0.841}}{(F')_{C=0.5}} \tag{6.3-4}$$

Both L_c and V_{DP} are independent of the particular form of the k/ϕ distribution, and both rely on the rearrangement of this ratio. As originally defined, V_{DP} is actually taken from a straight line fit to the k-ϕ data plotted on a log–probability scale. This procedure introduces a nonuniqueness (two different distributions having the same V_{DP}) into V_{DP} when the data are not lognormal (Jensen and Lake, 1986). For lognormal data, Eq. (6.3-4) is unique.

To relate F to C, we assume the permeability assembly is lognormally distributed; hence the relationship between cumulative frequency Λ and r is (Aithison and Brown, 1957)

$$\Lambda = \frac{1}{2}\left[1 - \operatorname{erf}\left\{\frac{\ln\left(\frac{r}{\hat{r}}\right)}{\sqrt{2\nu_{LN}}}\right\}\right] \tag{6.3-5}$$

where $\hat{r}$ is the geometric or log-mean of the distribution, and ν_{LN} is the variance of the distribution. The relationship between $\hat{r}$ and $\bar{r}$ is given by

$$\bar{r} = \hat{r}\, e^{(\nu_{LN}/2)} \tag{6.3-6}$$

If we identify Λ with the storage capacity C and use Eqs. (6.3-2), (6.3-5), and (6.3-6), we obtain

$$C = \frac{1}{2}\left[1 - \operatorname{erf}\left\{\frac{\ln(e^{\nu_{LN}/2} F')}{\sqrt{2\nu_{LN}}}\right\}\right] \tag{6.3-7}$$

Equation (6.3-7) may be solved for F' and then integrated subject to the boundary condition $F = C = 0$,

$$F = \int_0^C \exp\left\{\frac{\nu_{LN}}{2} + \sqrt{2\nu_{LN}}\operatorname{erf}^{-1}(1 - 2\xi)\right\} d\xi \tag{6.3-8}$$

We integrate Eq. (6.3-8) numerically to give the F-C curve for fixed ν_{LN}. Figure 6-7, which uses V_{DP} instead of ν_{LN}, shows the results of such an integration where the

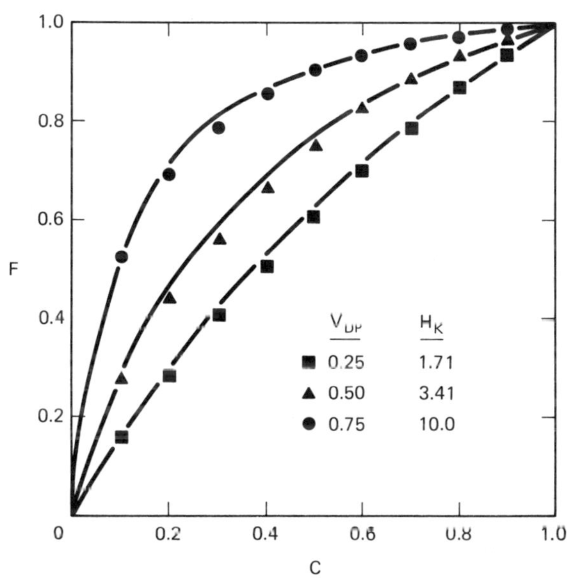

Figure 6-7 Flow-capacity–storage-capacity curves (from Paul et al., 1982)

filled points are the results of the integration. It follows from Eqs. (6.3-4) and (6.3-5) that

$$V_{DP} = 1 - e^{-\sqrt{\nu_{LN}}} \tag{6.3-9}$$

and, further, that the relationship among Lorenz and Dykstra-Parsons coefficients and ν_{LN} is

$$L_c = \text{erf}\left(\frac{\sqrt{\nu_{LN}}}{2}\right) = \text{erf}\left(\frac{-\ln(1 - V_{DP})}{2}\right) \tag{6.3-10}$$

Note from Eq. (6.3-10) that L_c and V_{DP} are bounded, whereas ν_{LN} is not.

Considering the three heterogeneity measures in Eq. (6.3-10), it must seem odd to propose a fourth, but none of the measures discussed so far directly relates to flow in permeable media. To ameliorate this, Koval (1963) proposed a heterogeneity factor H_K as a fourth measure of heterogeneity. H_K is defined by

$$H_K = \frac{1 - C}{C} \cdot \frac{F}{1 - F} \tag{6.3-11}$$

Equation (6.3-11) follows from observing the similarity between a homogeneous media fractional flow curve having straight-line relative permeabilities and zero residual phase saturations, and the points generated in Fig. 6-7. In fact, the solid lines in Fig. 6-7 are calculated from Eq. (6.3-11), with H_K adjusted to fit the calculated points. Hence there is a unique correspondence between V_{DP} and H_K, which is shown in Fig. 6-8 as the filled points. From Eqs. (6.3-8) and (6.3-11), it follows

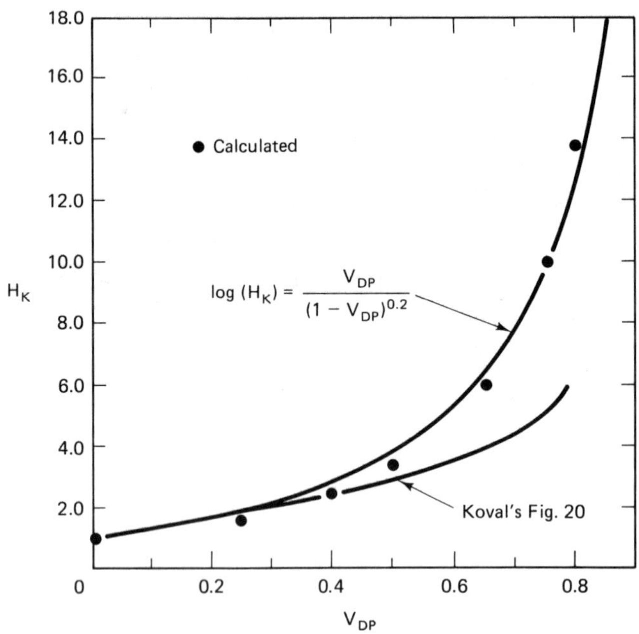

Figure 6-8 Relation between effective mobility ratio and heterogeneity (from Paul et al., 1982)

that $H_K \rightarrow \infty$ as $V_{DP} \rightarrow 1$ (infinitely heterogeneous) and $H_K \rightarrow 1$ as $V_{DP} \rightarrow 0$. Between these limits, the relation between V_{DP} and H_K is given by the following empirical fit to points in Fig. 6-8:

$$\log(H_K) = \frac{V_{DP}}{(1 - V_{DP})^{0.2}} \tag{6.3-12}$$

which is also shown in Fig. 6-8.

Using the *F-C* curve in Eq. (6.3-11), the vertical sweep efficiency of a unit mobility ratio displacement may be calculated by the one-dimensional theory in Sec. 5-2 (see Exercise 5E).

Table 6-1 shows various statistical information from several producing formations. V_{DP} varies between 0.65 and 0.89 for these formations. This rather tight range corresponds to the range in Fig. 6-7 where H_K begins to be an exceedingly strong function of V_{DP}. Since E_I decreases with increasing H_K, displacements in most reservoirs should be affected by heterogeneity. Table 6-1 also shows the areal V_{DP} based on the distribution of the average well permeabilities. In only three of the entries shown are the areal variations larger than the vertical variations. This, plus the lack of sensitivity of E_I to heterogeneous permeabilities arranged in series, partly accounts for the popularity of the stratified, or "layer-cake," model for reservoirs. We use the layer-cake model (uniform and heterogeneous) in the next two sections to calculate E_I. Finally, Table 6-1 shows the number of wells in each formation whose core plug permeabilities most closely conform to the normal, lognormal, and exponential distributions. As you can see, the wells are usually lognormal though there are significant exceptions (see Jensen et al., 1987). If the permeabilities are distributed normally, the procedure for calculating V_{DP} and L_c is still correct, but the form of the distribution function (Eq. 6.3-5) changes (see Exercise 6B).

None of the measures of heterogeneity given above are entirely satisfactory for predicting displacement performance. Since all the measures capture both heterogeneities and nonuniformities, there is a persistent, and largely ignored, question about how to use them in displacement calculations. It seems reasonable that nonuniformities would alter values of the permeable media properties such as dispersivity and capillary pressure; however, the scale of the nonuniformity is different between the field and laboratory measurements, and there is little to suggest how lab-measured properties can be changed to reflect these nonuniformities. For this reason, nonuniformities are usually ignored, and displacement calculations are based on uniformly heterogeneous permeable media models. A second reason for the inadequacies in the heterogeneity measures is that for many reservoirs it is inappropriate to treat permeability and porosity as independent variables. Correlations exist between permeability and porosity (bivariate correlations), and these variables themselves can have spatial structure (autocorrelation). When such structure does exist, the displacement response of a rearranged ensemble of layers will not be the same as that of the original distribution of layers. Determining when structure exists, and separating it from the random stochastic component, are tasks usually left to the geological interpretation in current practice.

TABLE 6-1 TYPICAL VALUES OF VERTICAL AND AREAL DYKSTRA-PARSONS COEFFICIENTS (ADAPTED FROM LAMBERT, 1981)

	Field	Formation	Number of wells studied*	Lognormal	Exponential	$\bar{k}$	$\bar{\phi}$	$(V_{DP})_{areal}$	$\bar{V}_{DP}$
1	El Dorado	Admire	262	35	42	370.14	0.2538	0.484	0.697
2	Keystone	Cardium	67	61	5	15.15	0.1063	0.752	0.653
3	Garrington	Manville B	38	35	2	5.73	0.1124	0.671	0.822
4	Madison	Bartlesville	36	10	10	29.95	0.1790	0.238	0.823
5	Pembina	Cardium	16	13	0	273.64	0.1220	0.837	0.894
		Belly River	17	15	0	12.66	0.1623	0.687	0.814
6	Hamilton Dome	Tensleep	33	11	10	98.42	0.1430	0.501	0.694
7	Rozet	Muddy	33	25	3	43.14	0.1708	0.457	0.846
8	Salt Creek	2nd Wall Creek	30	8	6	59.08	0.1843	0.495	0.851
9	Kitty	Muddy	20	19	0	11.74	0.0871	0.795	0.731
10	E. Salt Creek	2nd Wall Creek	5	3	2	35.71	0.1660	0.124	0.840
		Lakota	7	6	0	38.01	0.1540	0.424	0.899
11	Dixie West	Tradewater	16	5	3	129.13	0.1880	0.202	0.598
12	Burke Ranch	Dakota	14	6	3	23.18	0.1191	0.625	0.663
13	Oklahoma City	Prue	14	6	4	15.90	0.1368	0.473	0.683
14	Gas Draw	Muddy	14	6	0	71.61	0.1572	0.615	0.899
15	Recluse	Muddy	12	9	0	74.93	0.1437	0.591	0.855
16	W. Moorcroft	Muddy	8	6	0	201.39	0.2150	0.973	0.833
17	S. Rozet	Minnelusa	8	7	1	135.86	0.1283	0.443	0.861
18	Ute	Muddy	8	2	2	62.14	0.1790	0.752	0.758
19	Riverton Dome	Tensleep	7	2	0	2.68	0.0480	0.474	0.729
20	Carson-Hamm	Minnelusa	7	2	3	160.36	0.1624	0.465	0.722
21	N.W. Sumatra	Heath	6	0	2	124.98	0.1285	0.254	0.890
22	Pitchfork	Tensleep	5	1	4	91.54	0.1410	0.229	0.728
		Phosphoria	6	4	1	18.16	0.1430	0.544	0.833
		Total	689	297	102				

* Difference between lognormal and exponential represents number of normally distributed wells

6-4 DISPLACEMENTS WITH NO VERTICAL COMMUNICATION

In this section, we illustrate the effects of mobility ratio and heterogeneity for noncommunicating reservoirs. We treat pistonlike displacements of oil ($i = 2$) by water ($i = 1$) in uniformly heterogeneous, horizontal layer-cake models (see Exercise 6C). Further, we do not allow permeability or transmissibility in the vertical direction, a condition that could apply in actual practice if the reservoir contains impermeable and continuous shale breaks in the total interval. The reservoir now consists of an ensemble of one-dimensional elements arranged in parallel. Since there is no vertical communication, we can rearrange the layers in decreasing k/ϕ, as in Sec. 6-3. We also ignore dissipative effects to derive the noncommunication displacement model first proposed by Dykstra and Parsons (1950).

Subject to the above assumptions, the vertical sweep efficiency of the reservoir is

$$E_I = \frac{\sum_{\ell=1}^{n} (\phi h)_\ell + \sum_{\ell=n+1}^{N_L} (\phi h x_D)_\ell}{\overline{\phi} H_t} \tag{6.4-1}$$

where $x_{D\ell}$ is the dimensionless front position ($x_{f\ell}/L$) between the displacing fluid (water) and the displaced fluid (oil). The index n denotes the layer that has just broken through to the producer at a particular dimensionless time t_D

$$t_D = \frac{\int_0^t q\,dt}{H_t W L \overline{\phi}} = \frac{\sum_{\ell=1}^{N_L} (\phi h x_D)_\ell}{\overline{\phi} H_t} \tag{6.4-2}$$

where $x_{D\ell}$ for $\ell > n$ is greater than 1, W is the width of the cross section, and L is the length.

Two Layers

First, let's consider the case of a reservoir having only two layers ($N_L = 2$) as shown in Fig. 6-9 with water saturation change $\Delta S = S_{1J} - S_{1I}$. The $k/\phi \Delta S$ for the upper or fast layer is greater than that for the lower or slow layer. The front position in each layer may be determined from Darcy's law

$$\frac{dx_{f\ell}}{dt} = v_\ell = \left(\frac{k}{\phi \Delta S}\right)_\ell \lambda_{re_\ell} \frac{\Delta P}{L}, \qquad \ell = 1,\ 2 \tag{6.4-3}$$

where v_ℓ is the interstitial x velocity in layer ℓ, and λ_{re_ℓ} is the effective relative mobility in layer ℓ defined by

$$\lambda_{re_\ell} = \begin{cases} \left[\dfrac{x_{D\ell}}{\lambda_{r1}^0} + \dfrac{(1 - x_{D\ell})}{\lambda_{r2}^0}\right]^{-1} & \text{for } x_{D\ell} < 1 \\ \lambda_{r1}^0 & \text{for } x_{D\ell} > 1 \end{cases} \tag{6.4-4}$$

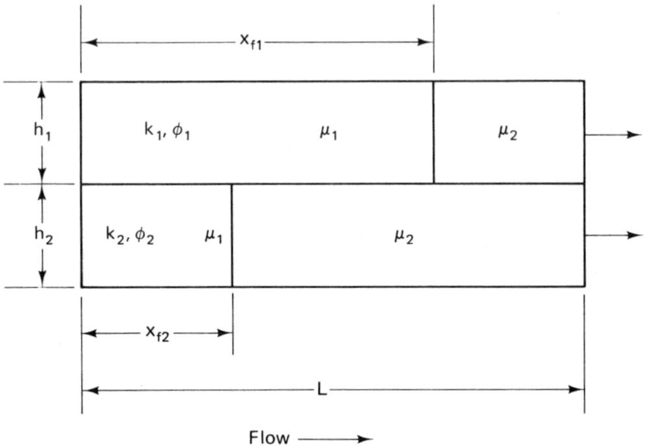

Figure 6-9 Schematic illustration of heterogeneous reservoir for Dykstra-Parsons model

Taking the ratio of the interstitial velocities in the two layers will eliminate time and the pressure drop since both layers experience the same ΔP. This equality implies communication in the wells even though there is no communication elsewhere. Because ΔP cancels, the calculation is valid whether the displacement is at constant rate or constant ΔP. Thus before breakthrough ($x_{D1} < 1$), we have

$$\frac{dx_{D1}}{dx_{D2}} = r_{12}\frac{x_{D2} + M^0(1 - x_{D2})}{x_{D1} + M^0(1 - x_{D1})} \tag{6.4-5a}$$

where r_{12}, the heterogeneity contrast ($k_1\phi_2\Delta S_2/k_2\phi_1\Delta S_1$), is greater than 1. After breakthrough ($x_{D1} > 1$), the same quantity is

$$\frac{dx_{D1}}{dx_{D2}} = r_{12}(x_{D2} + M^0(1 - x_{D2})) \tag{6.4-5b}$$

In both equations, M^0 is the endpoint mobility ratio defined in Eq. (5.2-3c). Before breakthrough x_{D1} and x_{D2} are less than 1, we can integrate Eq. (6.4-5a) subject to the boundary condition that $x_{D1} = 0$ when $x_{D2} = 0$ to give

$$\frac{1 - M^0}{2}x_{D1}^2 + M^0x_{D1} = r_{12}\left(\frac{1 - M^0}{2}x_{D2}^2 + M^0x_{D2}\right) \tag{6.4-6}$$

The front position in the lower layer at breakthrough x_{D2}^0 follows from Eq. (6.4-6) by setting $x_{D1} = 1$

$$x_{D2}^0 = \frac{\left\{(M^0)^2 + \dfrac{1 - (M^0)^2}{r_{12}}\right\}^{1/2} - M^0}{1 - M^0} \tag{6.4-7a}$$

After breakthrough, the front in the upper layer (outside the reservoir) is given by integrating Eq. (6.4-5b) with the boundary condition $x_{D2} = x_{D2}^0$ when $x_{D1} = 1$.

$$x_{D1} = 1 + r_{12}\left[\frac{1 - M^0}{2}(x_{D2}^2 - (x_{D2}^0)^2) + M^0(x_{D2} - x_{D2}^0)\right] \tag{6.4-7b}$$

The front "position" in the upper layer at complete sweepout is given by Eq. (6.4-7b) with $x_{D2} = 1$

$$x_{D1} = 1 + (r_{12} - 1)\left(\frac{1 + M^0}{2}\right) \tag{6.4-7c}$$

For fixed values of the mobility ratio, and heterogeneity contrast, E_I at a given dimensionless cumulative injection may be obtained by substituting the front positions calculated from Eqs. (6.4-6) and (6.4-7) in the definitions (Eqs. 6.4-1 and 6.4-2). Figure 6-10 shows the results of this procedure for three values of M^0 and two values of the permeability contrast.

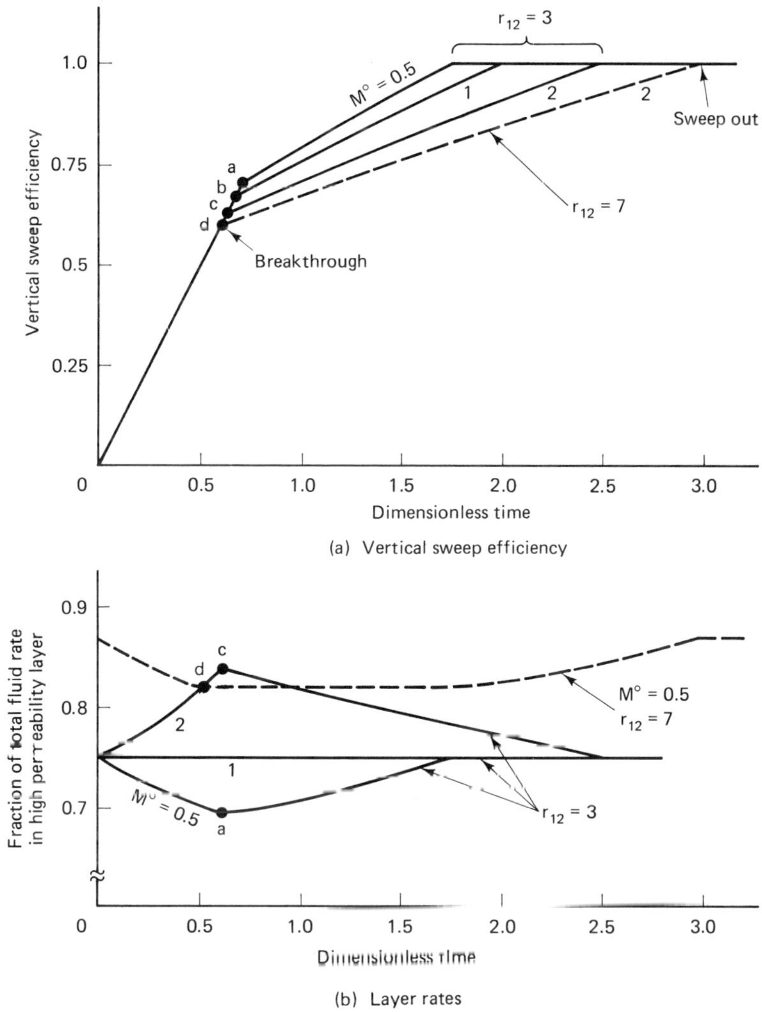

Figure 6-10 Two-layer Dykstra-Parsons calculation

For increasing permeability contrast E_I decreases (Fig. 6-10a). A decreasing M^0 improves, and an increasing M^0 worsens E_I, just as similar changes do for the areal and displacement sweep efficiencies. Fig. 6-10(b) plots the ratio of the volumetric flow rate into layer 1 to the total flow rate as a function of t_D. This follows from Eqs. (6.4-6) and (6.4-7)

$$\frac{q_1}{q_1 + q_2} = \frac{(\phi h \Delta S v)_1}{(\phi h \Delta S v)_1 + (\phi h \Delta S v)_2} = \frac{1}{1 + \dfrac{(kh)_2}{(kh)_1}\left[\dfrac{M^0 + x_{D1}(1 - M^0)}{M^0 + x_{D2}(1 - M^0)}\right]} \tag{6.4-8}$$

Equation (6.4-8) shows the reason for the changes in E_I. For $M^0 < 1$, the fast layer is filling up with a low mobility fluid faster than the slow layer. Thus the fast layer resistance to flow is increasing faster than the slow layer resistance, causing the fast layer flow rate to decrease. For $M^0 > 1$, the situation is exactly reversed. Of course, for $M^0 = 1$, there are no changes in mobility, and the ratio of fast layer rate to total rate stays constant. Mobility ratio can have an effect on E_I even if there is no vertical communication. This effect has qualitatively the same trend as the areal and displacement sweep.

N_L Layers

The above results may be readily generalized to an ensemble of N_L layers. First, we generalize the heterogeneity contrast to be between any two layers ℓ and n

$$r_{n\ell} = \left(\frac{k}{\phi \Delta S}\right)_n \left(\frac{\phi \Delta S}{k}\right)_\ell \tag{6.4-9}$$

At a particular time, if n is the number of the layer breaking through, the front position in all the faster layers is given by

$$x_{D\ell} = 1 + (r_{\ell n} - 1)\left(\frac{1 + M^0}{2}\right), \qquad \ell = 1, \ldots, n \tag{6.4-10a}$$

from Eq. (6.4-7c). Similarly, the position in all the slower layers is

$$x_{D\ell} = \frac{\left\{(M^0)^2 + \dfrac{1 - (M^0)^2}{r_{n\ell}}\right\}^{1/2} - M^0}{1 - M^0}, \qquad \ell = n + 1, \ldots, N_L \tag{6.4-10b}$$

from Eq. (6.4-7a). By letting n take on values between 1 and N_L, calculating all N_L front positions, and substituting these into Eqs. (6.4-1) and (6.4-2), we can construct a plot of E_I versus t_D (see Exercise 6D). The E_I-t_D plot follows from this by connecting these points by straight-line segments. That this procedure is not rigorous may be seen from Fig. 6-10, where the curves between breakthrough and sweepout are slightly curved. But if N_L is large, the error introduced by this procedure will be small. The procedure may be easily modified to calculate E_I in an ensemble having the continuous *F-C* distribution discussed in Sec. 6-3. Using the water–oil ratio

$$\text{WOR} = \frac{\sum_{\ell=1}^{n} q_\ell}{\sum_{\ell=n+1}^{N_L} q_\ell} \tag{6.4-11}$$

as the time variable in place of t_D, Johnson (1956) has presented the vertical sweep efficiency as a function of V_{DP} and M^0 in graphical form.

6-5 VERTICAL EQUILIBRIUM

A useful procedure for making general oil recovery calculations is to invoke the assumption of *vertical equilibrium* (VE) across the cross section of the reservoir the displacement is taking place in. When VE applies, it is possible to combine vertical and displacement sweep efficiencies into a pseudodisplacement sweep, which then may be estimated by the one-dimensional theory of Sec. 5-2. This combination means recovery efficiency E_R becomes

$$E_R = E_A E_I E_D = E_A \tilde{E}_D \tag{6.5-1}$$

where $\tilde{E}_D$ is the *pseudo*displacement sweep efficiency. Of course, the areal sweep efficiency E_A must still be estimated and used in Eq. (6.5-1). We discuss how to combine E_A and $\tilde{E}_D$ in Sec. 6-7. Another consequence of the VE assumption is this represents a state of maximum transverse fluid movement, or crossflow. Thus calculations based on VE are useful in estimating the tendency of crossflow to affect displacements when compared to the noncrossflowing calculations of Sec. 6-4.

The VE Assumption

Formally, vertical equilibrium is a condition where the sum of all the fluid flow driving forces in the direction perpendicular to the direction of bulk fluid flow is zero. We see this condition is more nearly met by flow in reservoirs having large aspect ratios (length to thickness) and good vertical communication. Moreover, Sec. 6-6 shows that several classical displacement calculations in the petroleum literature are, in fact, subsets of the more general theory of vertical equilibrium.

To derive a general VE theory, we restrict ourselves to incompressible, immiscible displacements of oil by water and derive the water saturation profile in the transverse direction (z direction) at a fixed cross section (x position). For the assumptions listed above, the conservation Eq. (2D-1) for water becomes in x-z coordinates,

$$\phi \frac{\partial S_1}{\partial t} + \frac{\partial u_{x1}}{\partial x} + \frac{\partial u_{z1}}{\partial z} = 0 \tag{6.5-2}$$

If we introduce Darcy's law (Eq. 2.2-5) into Eq. (6.5-2) and scale the independent variables x and z as

$$x_D = \frac{x}{L}, \qquad z_D = \frac{z}{H_t} \tag{6.5-3}$$

Eq. (6.5-2) becomes

$$\phi\left(\frac{L^2}{k}\right)\frac{\partial S_1}{\partial t} - \frac{\partial}{\partial x_D}\left(\lambda_{r1}\left(\frac{\partial P_1}{\partial x_D} + L\rho_1 g \sin\alpha\right)\right) - \left(\frac{L^2}{H_t^2 k}\right)\frac{\partial}{\partial z_D}\left(k_z\lambda_{r1}\left(\frac{\partial P_1}{\partial z_D} + H_t\rho_1 g\cos\alpha\right)\right) = 0 \tag{6.5-4}$$

The terms in this equation represent water accumulation, x-direction flow, and z-direction flow, respectively (see Fig. 6-11). We assume flow in the z direction is finite; therefore, if the group L^2/kH_t^2 is large, it follows that the term it multiplies must be small. This means the z-direction water flux is a function of x only, or

$$k_z\lambda_{r1}\left(\frac{\partial P_1}{\partial z} + \rho_1 g\cos\alpha\right) = f(x) \tag{6.5-5}$$

Since the water flux in the z direction is finite, if k_z is large, Eq. (6.5-5) implies

$$\frac{\partial P_1}{\partial z} = -\rho_1 g\cos\alpha \tag{6.5-6}$$

Clearly, the above reasoning breaks down at water saturations near the irreducible water saturation where λ_{r1} is zero. But it is true that the saturation range where Eq. (6.5-6) breaks down is precisely the range where the analogous equation for the oil phase is most relevant. Therefore, the arguments leading to Eq. (6.5-6) should be valid in some average sense when applied to both the water and oil phases.

Assuming the group L^2/kH_t^2 is large is reasonable for many practical cases. But assuming k_z is large strains credibility since for most naturally occurring media k_z is less than k. For permeable media having dispersed shale barriers, k_z can be much smaller than k.

The requirements of large L^2/kH_t^2 and k_z may be combined into a single requirement that the effective length-to-thickness ratio

$$R_L = \frac{L}{H_t}\left(\frac{\bar{k}_z}{\bar{k}}\right)^{1/2} \tag{6.5-7a}$$

be large. In Eq. (6.5-7a), the permeabilities are an arithmetic average for

$$\bar{k} = \frac{1}{H_t}\int_0^{H_t} k\,dz \tag{6.5-7b}$$

and a harmonic average for k_z

$$\bar{k}_z = \frac{H_t}{\displaystyle\int_0^{H_t}\frac{dz}{k_z}} \tag{6.5-7c}$$

A displacement actually approaches VE asymptotically as R_L becomes large. Based on numerical simulation (Zapata, 1981) and analytic solutions (Lake and Zapata, 1987), an R_L greater than 10 is sufficient to ensure that the z-direction sweep efficiency is reasonably well described by VE. You may easily verify that R_L can be large for a wide variety of reservoirs. For example, for a 16.2 hm^2 (40-acre) spacing of five-spot patterns, the injector–producer distance is 285 m (933 ft). If we take this to be L, then for H_t = 6.1 m (20 ft) and $\bar{k}_z = 0.1\bar{k}$, we have $R_L = 14.8$, which is large enough for VE to be a good approximation. By taking the $\bar{k}$ to be a harmonic average over the reservoir interval, it is clear the $\bar{k}_z = R_L = 0$ if there are one or more impermeable barriers (for example, continuous shale layers) within the interval H_t. Clearly, the VE assumption will not apply in this case. But the pseudodisplacement sweep efficiency of the intervals between the barriers may be estimated based on VE, and the combined response of all such intervals may be estimated by the noncommunicating methods of Sec. 6-4.

R_L may be regarded as a ratio of a characteristic time for fluid to cross the reservoir in the x direction to that in the z direction. If R_L is large, saturation or pressure fluctuations in the z direction decay much faster than those in the x direction. Therefore, we neglect the z-direction perturbations. Thus when we say that the VE assumption applies or that the subject reservoir is in vertical equilibrium, we are saying, for the bulk of the reservoir, z-direction fluctuations are negligible. Arguments based on the decay time of perturbations were originally advanced by G. I. Taylor for flow in capillary tubes (Lake and Hirasaki, 1981).

For large R_L, the P_1 profile in the z direction is given by Eq. (6.5-6) for most of the cross sections in the reservoir. This procedure applies equally well to the oil phase, giving

$$\frac{\partial P_1}{\partial z} + \rho_1 g \cos\alpha = 0 = \frac{\partial P_2}{\partial z} + \rho_2 g \cos\alpha \tag{6.5-8}$$

When the definition for oil–water capillary pressure $P_c = P_2 - P_1$ is introduced into this equation, we have

$$\frac{\partial P_c}{\partial z} = -(\rho_1 - \rho_2) g \cos\alpha \equiv -\Delta\rho g \cos\alpha \tag{6.5-9}$$

Equation (6.5-9) implicitly describes the water saturation profile in the z direction since P_c is a known function of water saturation. But this saturation distribution is just what would be observed in the transition zone between oil and water under static conditions. Compare Eqs. (6.5-9) and (2A-1), noting the z and P_c increase in opposite directions. Hence the z-direction saturation profile given by Eq. (6.5-9) is identical to that predicted by assuming no flow in the z direction.

We stated that VE is a condition that causes maximum crossflow of fluids, so it is surprising, to say the least, that the same equation describes the saturation profile under conditions of zero and maximum z-direction flow. The situation is analogous to heat conduction in a metal rod where the driving force for heat transfer is a temperature gradient along the axis of the rod (Coats et al., 1971). If the thermal conductivity of the rod is not zero and no heat flows along the rod, the temperature difference between the ends of the rod is zero. But if heat flows at a fixed finite rate along the rod, and the thermal conductivity of the rod is large, the temperature difference is again small. The latter case is analogous to the VE flow in the z direction of Fig. 6-11 where, since the thermal conductivity is large, the heat transfer rate is maximum; the former case is the analogue to hydrostatic equilibrium.

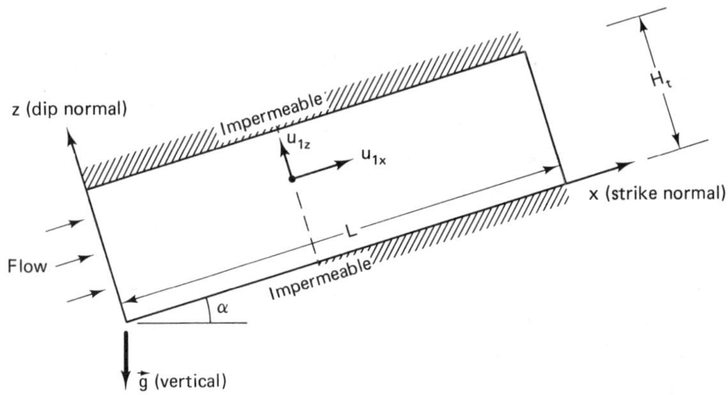

Figure 6-11 Schematic cross-section for vertical equilibrium procedure

Displacement Classification

One of the consequences of VE is a classification of displacements according to degree of segregation. let S_1^+ be some water saturation slightly below $1 - S_{2r}$, and S_1^- slightly above S_{1r}. We can define a capillary transition zone thickness z_{CTZ} as the z-direction distance over which the water saturation changes between these two limits. From Eq. (6.5-9) and Fig. 6-12, this is

$$z_{CTZ} \equiv z\Big|_{S_1^+} - z\Big|_{S_1^-} = \frac{P_c\Big|_{S_1^-} - P_c\Big|_{S_1^+}}{\Delta\rho g \cos\alpha} \tag{6.5-10}$$

We have made the integration of Eq. (6.5-9) assuming the capillary-pressure–water-saturation relation applies throughout z_{CTZ}. In general, the capillary transition zone defined by Eq. (6.5-10) is not the same as that existing at the original water–oil contact, down structure to the left in Fig. 6-11. The idea of the capillary transition zone in a VE reservoir allows the definition of two broad classes of displacements (Dake, 1978). If $z_{CTZ} >> H_t$, the water saturation profiles in the z direction are essentially flat, and the flow is said to be *diffuse*. If $z_{CTZ} << H_t$, the capillary transition zone is small with respect to the reservoir thickness, and the flow is *segregated*. These definitions suggest ideas similar to the definitions of sharpening and spreading waves in Sec. 5-2 except that the latter definitions apply to cross-sectional averaged satura-

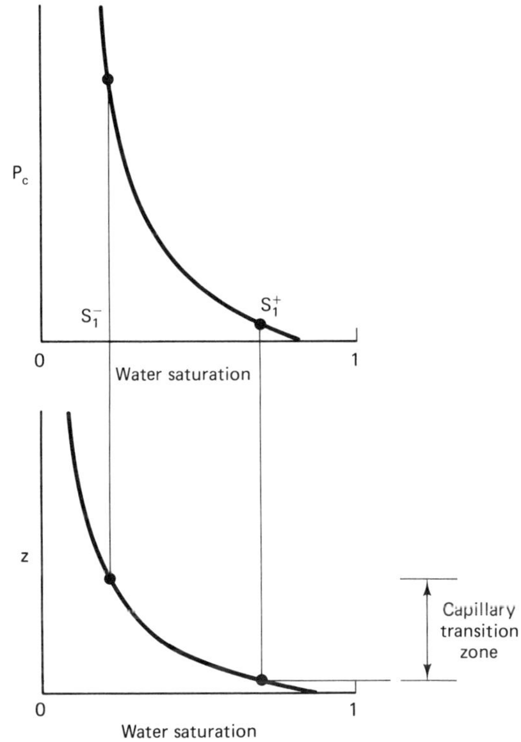

Figure 6-12 Schematic of capillary transition zone

tion waves. The mixing or transition zones in Sec. 5-2 were in the x direction only and were largely caused by chromatographic effects inherent in the permeable medium oil–water fractional flow curves. The capillary transition zone defined by Eq. (6.5-10) is in the z direction and defined by the capillary-pressure–water-saturation relation, the dip angle, and the density difference.

Saturation Profile

Let's now consider the integration of Eq. (6.5-9) at the three different cross sections A, B, and C in Fig. 6-13. In this figure, flow is from right to left for ease of illustration. We take S_{1A}, S_{1B}, and S_{1C} to be the water saturations at the bottom ($z = 0$) of the reservoir at the indicated cross sections $x = x_A$, x_B, and x_C. Because of the direction of flow, and because the initial water saturation is near the irreducible value $S_{1A} > S_{1B} > S_{1C}$. The water saturation profile at each of these cross sections is given implicitly from Eq. (6.5-9)

$$P_c(S_1(x_k, z)) = P_c(S_{1k}) + \Delta\rho\, gz \cos \alpha, \qquad k = A, B, \text{ or } C \qquad (6.5\text{-}11)$$

We do not, at this point, know the x direction position of the $z = 0$ water saturations, which we indirectly determine below. But we can schematically sketch in lines connecting constant values of S_1, as indicated in Fig. 6-13. For positive values of the

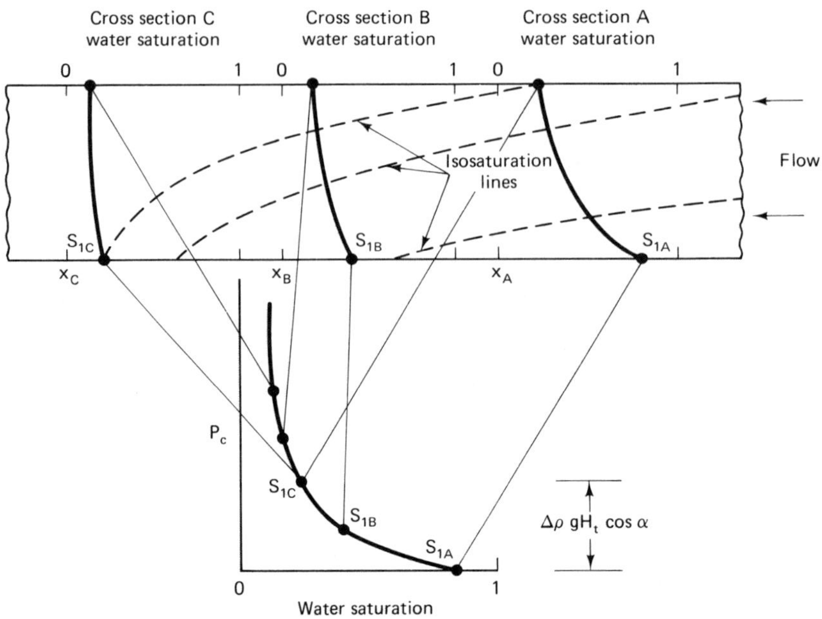

Figure 6-13 Schematic of z-direction water saturation profiles at various cross sections

density difference, the usual case, the isosaturation lines suggest an underrunning of the oil by the injected water. This underrunning, or gravity tongue, is a persistent feature of reservoirs in which gravity forces are strong. Tonguing occurs even in reservoirs that have no dip $\cos \alpha = 1 (D_z = -z)$. The extent of the tonguing is greatly influenced by the shape of the capillary pressure curve. In Sec. 6-6, we discuss a special case of the VE theory in which capillary forces are negligible, and the gravity tonguing occurs as segregated flow.

Pseudoproperties

To use the z-direction S_1 profile, we must convert the original two-dimensional Eq. (6.5-2) to an equivalent one-dimensional equation. Let's integrate Eq. (6.5-2) over the interval thickness H_t and divide the equation by H_t

$$\frac{1}{H_t}\int_0^{Ht} \frac{\partial S_1}{\partial t} dz + \frac{1}{H_t}\int_0^{Ht} \frac{\partial u_{x1}}{\partial x} dz + \frac{1}{H_t}\int_0^{Ht} \frac{\partial u_{z1}}{\partial z} dz = 0 \tag{6.5-12}$$

Since H_t is a constant, the integration and differentiation in the first term commutes, and Eq. (6.5-12) becomes

$$\overline{\phi}\frac{\partial \overline{S}_1}{\partial t} + \frac{\partial \overline{u}_{x1}}{\partial x} = 0 \tag{6.5-13}$$

Terms involving z-direction water flux do not appear in Eq. (6.5-13) since all fluxes vanish at the upper and lower impermeable boundaries of the reservoir. In Eq. (6.5-13), the averages are

$$\bar{S}_1 = \frac{1}{H_t\bar{\phi}} \int_0^{H_t} \phi S_1 \, dz \tag{6.5-14a}$$

$$\bar{\phi} = \frac{1}{H_t} \int_0^{H_t} \phi \, dz, \qquad \bar{u}_{1x} = \frac{1}{H_t} \int_0^{H_t} u_{x1} \, dz \tag{6.5-14b}$$

In these definitions, and in those that follow, all averages are arithmetic averages except the water saturation, which is weighted by the porosity. Introducing the definitions for dimensionless independent variables

$$x_D = \frac{x}{L}, \qquad t_D = \int_0^t \frac{\bar{u}_x \, dt}{\bar{\phi} L} \tag{6.5-15}$$

into Eq. (6.5-13) yields,

$$\frac{\partial \bar{S}_1}{\partial t_D} + \frac{\partial \bar{f}_1}{\partial x_D} = 0 \tag{6.5-16}$$

where $\bar{u}_x = \bar{u}_{x1} + \bar{u}_{x2}$, and $\bar{f}_1 = \bar{u}_{x1}/\bar{u}_x$ is a cross-sectional averaged water fractional flow function. Eq. (6.5-16) is identical to Eq. (5.2-5a) and can be solved in the same manner as the Buckley-Leverett and Welge integration procedures once we define $\bar{f}_1$ in terms of $\bar{S}_1$.

Consider the cross-sectional averaged total flux multiplied by H_t with Darcy's law substituted for the local flux

$$H_t\bar{u}_x = -\int_0^{H_t} k\lambda_{r2}\left(\frac{\partial P_2}{\partial x} + \rho_2 g \sin\alpha\right)dz - \int_0^{H_t} k\lambda_{r1}\left(\frac{\partial P_1}{\partial x} + \rho_1 g \sin\alpha\right)dz \tag{6.5-17}$$

We can express the x-direction oil phase pressure gradient in terms of the water phase pressure gradient and factor to give

$$H_t\bar{u}_x = -\int_0^{H_t} k(\lambda_{r2} + \lambda_{r1})\frac{\partial P_1}{\partial x} \, dz - \int_0^{H_t} k\lambda_{r2}\frac{\partial P_c}{\partial x} \, dz - g \sin\alpha \int_0^{H_t} k(\lambda_{r2}\rho_2 + \lambda_{r1}\rho_1)dz \tag{6.5-18}$$

But from Eq. (6.5-6), it follows that

$$\frac{\partial^2 P_1}{\partial x \, \partial z} = \frac{\partial^2 P_1}{\partial z \, \partial x} = \frac{\partial}{\partial z}\left(\frac{\partial P_1}{\partial x}\right) = 0;$$

hence under VE, the water phase pressure gradient in the x direction is independent of z, as are both $\partial P_2/\partial x$ and $\partial P_c/\partial x$. All gradients may be factored from the integrations and solved for as

$$-\frac{\partial P_1}{\partial x} = \frac{H_t\bar{u}_x + \left(\frac{\partial P_c}{\partial x}\right)\int_0^{H_t} \lambda_{r2}k\, dz + g \sin\alpha \int_0^{H_t} k(\lambda_{r2}\rho_2 + \lambda_{r1}\rho_1)dz}{\int_0^{H_t} k(\lambda_{r2} + \lambda_{r1})dz} \tag{6.5-19}$$

The pressure gradient of Eq. (6.5-19) substituted into the averaged water flux

$$H_t\bar{u}_{x1} = \left(-\frac{\partial P_1}{\partial x}\right)\int_0^{H_t} k\lambda_{r1}\, dz - g\sin\alpha \int_0^{H_t} k\lambda_{r1}\rho_1\, dz \tag{6.5-20}$$

gives

$$\bar{f}_1 = \frac{\bar{u}_{x1}}{\bar{u}_x} = \frac{(\overline{k\lambda_{r1}})}{\overline{k(\lambda_{r1} + \lambda_{r2})}}\left\{1 + \frac{(\overline{k\lambda_{r2}})}{\bar{u}_x}\left(\frac{\partial P_c}{\partial x}\right) - \frac{(\overline{k\lambda_{r2}})\,\Delta\rho g \sin\alpha}{\bar{u}_x}\right\} \tag{6.5-21}$$

Comparing this equation with Eq. (5.3-1) suggests the following definitions for *pseudorelative permeabilities*

$$\tilde{k}_{r1} = \frac{1}{H_t\bar{k}}\int_0^{H_t} kk_{r1}\, dz \tag{6.5-22a}$$

$$\tilde{k}_{r2} = \frac{1}{H_t\bar{k}}\int_0^{H_t} kk_{r2}\, dz \tag{6.5-22b}$$

The capillary pressure in Eq. (6.5-21) is the capillary-pressure–water-saturation relation for any z position in the reservoir. It does not matter which z position since $\partial P_c/\partial x$ is equal at all z positions. From this, it does not follow that the capillary-pressure–water-saturation relation is the same in all z positions since these can vary with permeability. The capillary pressure in Eq. (6.5-21) is often regarded as a pseudofunction, even though it is an actual local curve, since it must be a function of $\bar{S}_1$.

To use the one-dimensional theory of Sec. 5-2 on these equations, we must neglect the x-direction capillary pressure term in Eq. (6.5-21). This omission is not equivalent to neglecting capillary pressure entirely since the capillary pressure in the z direction determines, in part, the z-direction saturation profile. Though it seems inconsistent to maintain capillary pressure in the z direction and neglect it in the x direction, one can show by scaling arguments similar to those used in Sec. 5-3 that when the conditions for VE apply, z-direction effects are far more important than x-direction effects (Yokayama and Lake, 1981).

The procedure for calculating pseudorelative permeability curves ($\tilde{k}_{r1}$ and $\tilde{k}_{r2}$ versus $\bar{S}_1$) is as follows:

1. Select a water saturation at the bottom of the reservoir S_{1k}.
2. Determine the z-direction water saturation profile $S_1(x_k, z)$ at cross section k using Eq. (6.5-11) and the capillary-pressure–water-saturation relation.

3. Calculate the average water saturation at cross section k, $\bar{S}_1(x_k)$, from Eq. (6.5-14a) and from the z-direction porosity profile.
4. Calculate the pseudorelative permeabilities corresponding to $\bar{S}_{1k}$ from Eq. (6.5-22) and the z-direction permeability profile.

Steps 1–4 give a single point on the pseudorelative permeability curve. To construct the entire curve, we repeat the procedure with different values of S_{1k}. The procedure gives all possible water saturation profiles and average water saturations for the reservoir (see Fig. 6-13) though it does not give the x positions of these quantities, which come from solving the one-dimensional Eq. (6.5-13). Though the averaging procedure is fairly straightforward, most of the integrations in it must be evaluated numerically in the absence of analytic functions for the capillary pressure and relative permeability curves (see Exercise 6F).

Once the pseudorelative permeabilities are constructed, the pseudodisplacement sweep efficiency $\tilde{E}_D$ follows from Eqs. (5.1-2) and (5.2-24) with the appropriately averaged quantities appearing in place of the local quantities.

You should appreciate the generality of the VE approach, for we now have a means for calculating and combining displacement E_D and vertical E_I sweep efficiencies with little more trouble than calculating the displacement sweep alone. VE can greatly simplify oil recovery calculations in desktop procedures and numerical simulations (Coats et al., 1971); however, the entire procedure is restricted to reservoirs having a large R_L.

The generalized VE approach for EOR processes has yet to be worked out. (For miscible flow, see Lake and Hirasaki, 1981.)

6-6 SPECIAL CASES OF VERTICAL EQUILIBRIUM

Though the VE procedure in Sec. 6-5 is quite general, being restricted to reservoirs having constant properties in the x direction and a large R_L, several VE flows are special cases. Since these cases are useful in understanding many EOR processes, in this section we review them and show how they follow from the general theory.

Homogeneous with Large Transition Zone

In this case, k and ϕ are both constant in the reservoir, and $z_{CTZ} >> H_t$. From the procedure given above, the saturation profile in the z direction will be essentially flat, and the saturation at the reservoir bottom will not differ much from the average saturation. In this case, the pseudorelative permeabilities $\tilde{k}_{rj}$ become the local (or REV) relative permeabilities k_{rj}. Large z_{CTZ} would be the rule in most of the longer laboratory core floods. In the shorter core experiments, VE is usually not a good assumption, but S_1 may still be uniform in a cross section since the S_1 profile has not had much time to distort.

Homogeneous, Uniform with No Transition Zone

Easily the most celebrated of the VE theories is the theory of gravity tonguing, or underrunning, originally proposed by Dietz (1953). This theory was first proposed as an alternative to the Buckley-Leverett theory, but it is actually a special case of the VE theory because a finite time is required for the conditions underlying the theory to apply. Since the publication of the original Dietz paper, the theory has been applied to gravity overrunning by a miscible gas process (Hawthorne, 1960), and other work has been published describing the approach to VE conditions (Crane et al., 1963). In this section, we restrict ourselves to the water-displacing-oil case though the overrunning case can be similarly developed.

The key assumption in the Dietz theory is the absence of a transition zone, or $z_{CTZ} = 0$. This condition can be accurate only for conditions where the capillary pressure is small (well-sorted or high-permeability media). The sharp transition zone or macroscopic interface resulting from this condition suggests the theory is applicable to any displacement where simultaneous flow of more than one component or phase is absent at any point in the reservoir. If P_c is identically zero, Eq. (6.5-8) cannot be satisfied at any point in the reservoir since the oil and water densities are not, in general, equal. The resolution of this is to let Eq. (6.5-6) apply to zones flowing water and to let the analogous equation for oil apply to zones flowing oil. Figure 6-14 shows the relevant cross section and these zones.

At any cross section containing the tongue, the average water saturation from Eq. (6.5-14a) is

$$\bar{S}_1 = \frac{1}{H_t}[b(1 - S_{2r}) + S_{1r}(H_t - b)] \tag{6.6-1}$$

and the pseudorelative permeability functions from Eq. (6.5-22) are

$$\tilde{k}_{r1} = k_{r1}^{o}\left(\frac{b}{H_t}\right) \tag{6.6-2a}$$

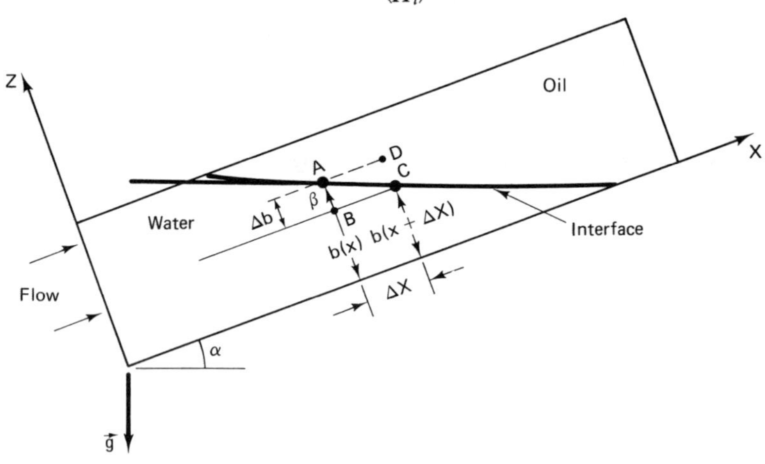

Figure 6-14 Schematic cross section of a water tongue

$$\tilde{k}_{r2} = k^{o}_{r2}\left(\frac{H_t - b}{H_t}\right) \tag{6.6-2b}$$

The interface height b may be eliminated between Eqs. (6.6-1) and (6.6-2) to give

$$\tilde{k}_{r1} = k^{o}_{r1}\left(\frac{\bar{S}_1 - S_{1r}}{1 - S_{1r} - S_{2r}}\right), \qquad \tilde{k}_{r2} = k^{o}_{r2}\left(\frac{1 - S_{2r} - \bar{S}_1}{1 - S_{1r} - S_{2r}}\right) \tag{6.6-3}$$

Thus the pseudorelative permeabilities are straight-line functions of the average water saturation.

We can also derive the tilt angle β of the oil–water interface. Consider the rectangle $ABCD$ of height Δb and width Δx shown in Fig. 6-14. The dimensions Δx and Δb are small (we pass to zero limit below) so that the interface between points A and C is the diagonal of the rectangle. Along the BC side of the rectangle, the x-direction water flux is

$$u_{x1} = -\frac{kk^{o}_{r1}}{\mu_1}\left(\frac{P_C - P_B}{\Delta x} + \rho_1 g \sin\alpha\right) \tag{6.6-4a}$$

and along the AD side, the x-direction oil flux is

$$u_{x2} = -\frac{kk^{o}_{r2}}{\mu_2}\left(\frac{P_D - P_A}{\Delta x} + \rho_2 g \sin\alpha\right) \tag{6.6-4b}$$

In the limit of $\Delta x \to 0$, these two fluxes approach a common value u_x since there can be no accumulation at the interface. Further, the pressures at A and B, and at D and C, are related because of the VE conditions (Eq. 6.5-8)

$$P_B - P_A = \rho_1 g \Delta b \cos\alpha, \qquad P_C - P_D = \rho_2 g \Delta b \cos\alpha \tag{6.6-5}$$

The four equations (Eqs. 6.6-4 and 6.6-5) may be combined to eliminate the four pressures. This procedure gives

$$\tan\beta = \frac{(u_{x1} - u_{x2}M^{o})\mu_1}{(kk^{o}_{r1}\Delta\rho g)\cos\alpha} + \tan\alpha \tag{6.6-6}$$

The tangent of the tilt angle is defined as

$$\tan\beta = +\lim_{\Delta x \to 0} \frac{\Delta b}{\Delta x} \tag{6.6-7}$$

β is defined to be positive and can take on the entire range of values between 0° and 90°. If β is greater than 90°, the tongue is overrunning, and this procedure must be repeated with the displacing fluid above the resident fluid.

For $\beta > 0$—that is, the interface is not parallel to the x axis—the interface reaches a stabilized shape where β is independent of both time and z position. This limit is not an automatic consequence of VE, but the time interval between the onset of the VE conditions and the attainment of the stabilized interface shape appears to be small (Crane et al., 1963). When this steady-state tilt angle β_s is reached, the x-direction fluxes u_{x1} and u_{x2} become independent of z and equal to the cross-sectional average flux $\bar{u}_x$. Equation (6.6-6) then becomes

$$\tan \beta_s = \frac{1 - M^0}{M^0 N_g^0 \cos \alpha} + \tan \alpha \tag{6.6-8}$$

where N_g^0 and M^0 are gravity numbers and mobility ratios defined in Eq. (5.2-3).

Equation (6.6-8) approaches the correct limits of an interface perpendicular to the x direction for $N_g^0 = 0$ (no tonguing) and of a horizontal interface for $M^0 \rightarrow 1$. In the case of a stable gravity tongue, the cross-sectional average water saturation profile approaches a "constant pattern" mixing zone, whereas the directly analogous case of a one-dimensional displacement with straight-line relative premeabilities approaches a shock front. This is a consequence of the finite length of time required for the VE conditions to apply in the tonguing case.

For $\beta < 0$, the interface completely underruns the oil and is said to be unstable. The condition for stability is, from Eq. (6.6-8),

$$M^0 - 1 < M^0 N_g^0 \sin \alpha \tag{6.6-9}$$

The equality form of Eq. (6.6-9) naturally leads to definitions of a critical endpoint mobility ratio $M_c^0 = M^0|_{\beta_s = 0}$

$$M_c^0 = \frac{1}{1 - N_g^0 \sin \alpha} \tag{6.6-10a}$$

and of a critical flux or rate $u_c = u_x|_{\beta_s = 0}$

$$u_c = \frac{\Delta \rho g k k_{r1}^0}{\mu_1 (M^0 - 1)} \sin \alpha \tag{6.6-10b}$$

The conditions to prevent complete underrunning of the oil by the water are $u_x < u_c$ or $M^0 < M_c^0$. Equation (6.6-10a) indicates gravity stabilization is possible even when $M^0 > 1$. Equation (6.6-10b), in particular, is used in estimating the flooding rates in gravity-stabilized miscible displacements.

Layered, Uniform Horizontal Media with $P_c = z_{CTZ} = 0$

For this case of the permeable medium consisting of N_L layers, each of contrasting thickness h_ℓ, permeability k_ℓ, and porosity, ϕ_ℓ, the integrals in the definitions (Eqs. 6.5-14 and 6.5-22) become finite sums

$$\tilde{k}_{rj} = \frac{1}{H_t \bar{k}} \sum_{\ell=1}^{N_L} (khk_{rj})_\ell, \qquad j = 1 \text{ or } 2 \tag{6.6-11a}$$

$$\bar{S}_1 = \frac{1}{H_t \bar{\phi}} \sum_{\ell=1}^{N_L} (\phi h S_1)_\ell \tag{6.6-11b}$$

The definitions (Eqs. 6.6-11a and 6.6-11b) are valid regardless of the ordering of the layers; hence we assume, without loss of generality, they are ordered with decreasing velocity as in Sec. 6-3.

Since neither gravity nor capillary pressure is present, Eq. (6.5-8) is trivially satisfied, and there is no constraint on the saturations in the z direction. To resolve this, Hearn (1971) assumed segregated flow within a layer, as in Fig. 6-15(a). The definitions become

$$\tilde{k}_{r1} = \frac{1}{H_t\bar{k}} \sum_{\ell=n+1}^{N_L} (khk_{r1}^0)_\ell, \qquad \tilde{k}_{r2} = \frac{1}{H_t\bar{k}} \sum_{\ell=1}^{n} (khk_{r2}^0)_\ell \tag{6.6-12a}$$

$$\bar{S}_1 = \frac{1}{H_t\bar{\phi}} \left\{ \sum_{\ell=1}^{n} (\phi h(1 - S_{2r}))_\ell + \sum_{\ell=n+1}^{N_L} (\phi h S_{1r})_\ell \right\} \tag{6.6-12b}$$

where n is the number of the slowest layer (smallest k/ϕ) flowing water at a given cross section. Thus the average water saturation and pseudorelative permeabilities are parametrically functions of n and can be regarded as functions of each other in this way.

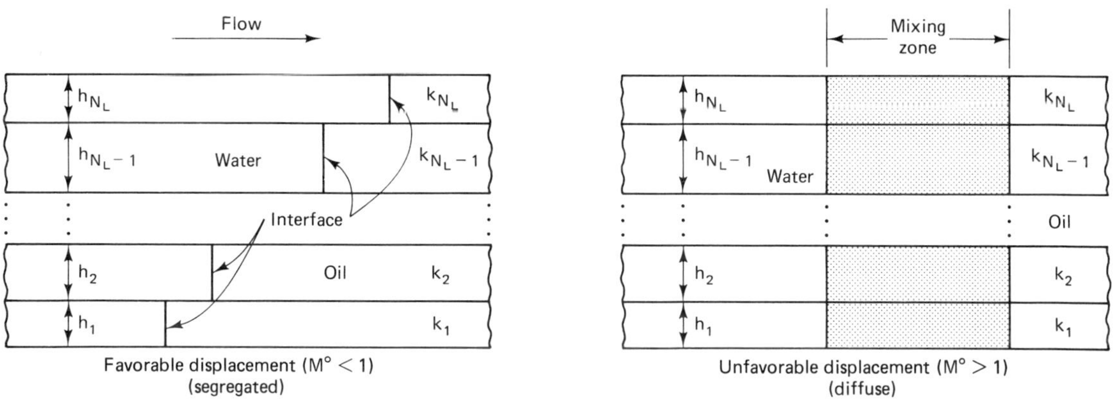

Figure 6-15 Schematic cross section of VE in stratified reservoir with no capillary and gravity effects

Based on arguments related to the direction of flow caused by the viscous pressure driving forces, Zapata and Lake (1981) have shown that assuming segregated flow within a layer is correct when the displacement is favorable $M^0 < 1$. In fact, if VE holds, it is possible for the M^0 to be so low that the effect of the heterogeneities is entirely suppressed (see Exercise 6E). But when the displacement is unfavorable, the viscous forces cause a mixing zone to develop between the front in the fastest layer and that in the slowest layer (Fig. 6-15b). This mixing zone causes the vertical sweep efficiency to be actually greater than the corresponding segregated flow case since the mixing zone attenuates the unfavorable mobility ratio. That diffuse flow can occur in VE displacements in the absence of capillary pressure is a major revelation in the understanding of these processes. The implication is clear that such crossflow might be a source of mixing in all unstable flows.

Stratified, Uniform with $\Delta\rho = 0$ and Constant Mobility

Here, there are no gravity forces to counteract the z-direction imbibition, and the z-direction water saturation profile is uniform within each layer. But because of the variable properties in the z direction, the $P_c - S_1$ function changes. Figure 6-16(a) illustrates this change for the four-layer medium shown. From Eq. (6.5-9), the capillary pressure (not the capillary pressure function) is a constant through any cross section. As indicated in Fig. 6-16, if the constant is known, this specifies the water saturation in each layer at that cross section. Because the mobility is constant, the x-direction viscous pressure gradient is independent of both position and time. For this case, the average water saturation and pseudorelative permeability curves are given by Eq. (6.6-11), but each of the water saturations $S_{1\ell}$ are determined by the relation P_c = constant and the P_c-S_1 relation. Again, the average water saturation and pseudorelative permeabilities are parametrically related through this constant. This procedure yields an immiscible mixing zone between the most advanced and the least advanced front, as shown in Fig. 6-16(b).

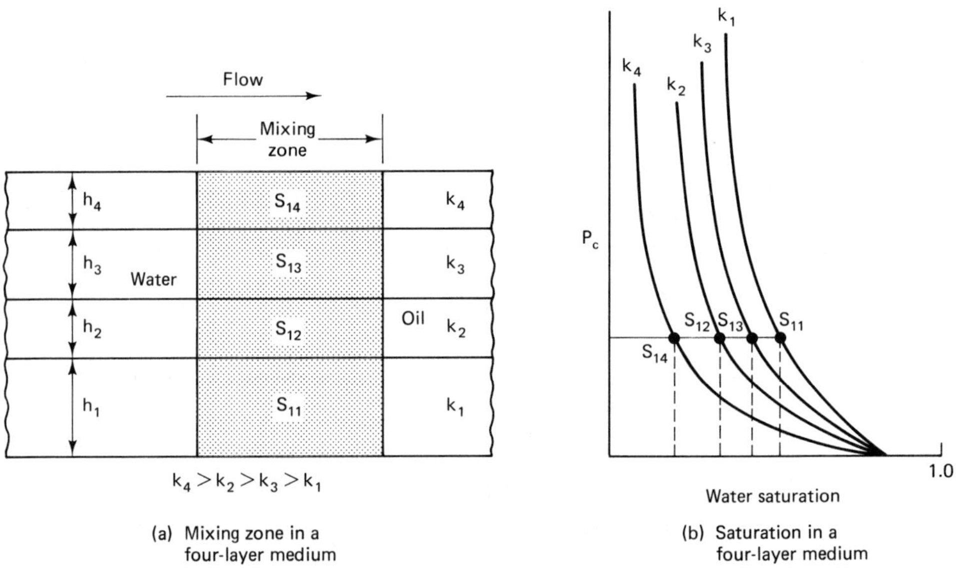

Figure 6-16 Schematic of stratified cross section with no gravity and viscous forces

6-7 COMBINING SWEEP EFFICIENCIES

In this section, we seek to provide an estimate of the recovery efficiency $E_R = E_D E_V$, from Eq. (2.5-5), as a function of dimensionless time by combining vertical, areal, and displacement sweeps. As we mentioned in Sec. 6-1, this procedure is complicated because all three sweep efficiencies depend on one another, and all must be evaluated at times different from that at which the recovery efficiency is desired. If the reservoir is layered and noncommunicating, we could, of course, calculate the areal sweep efficiency of each layer and then average the ϕh of each layer times its areal sweep to obtain the volumetric sweep efficiency. The procedure we describe here includes this method as a special case, but it is valid for combining all types of sweep efficiency curves, not just those for layer-cake models.

Our procedure is based on the idea of apparent pore volumes first presented by Claridge (1972). We assume we have independently determined curves for E_A, E_I, and E_D as functions of t_D. For E_D, it is more convenient to work in the average water saturations $\bar{S}_1$, but there is no loss of generality since the two are related through Eq. (5.1-2). Here, we are restricted to sweep efficiency functions that depend on dimensionless time, heterogeneity, capillary pressure, and so on but that do not depend explicitly on rate or fluid velocity.

Combining Areal and Vertical Sweep

The definition for volumetric sweep efficiency is repeated here as $E_V = E_A E_I$. From Fig. 6-1 (or Fig. 6-17), E_A depends on the z position in the reservoir, and E_I on a particular cross section between the injector and producer. Rather than directly determining these positions, we seek to determine that value of the dimensionless time argument at which the respective values of E_A and E_V will give average values. Therefore, we can rewrite Eq. (6.1-3) as

$$E_V(t_D) = E_A(t_{DA})E_I(t_{DI}) \tag{6.7-1}$$

where t_{DA} and t_{DI} are dimensionless times based on the apparent pore volumes for areal and vertical sweep, respectively.

Figure 6-17(a) schematically shows the positions of a pistonlike displacement at or after breakthrough. Imagine that the shaded volumes have no porosity or permeability, then the volumetric sweep efficiency is equal to the vertical sweep efficiency. The ultimate volume to be swept out, for this rather oddly shaped reservoir then, at infinite throughput is the unshaded or apparent pore volume. But this is just the E_A times the total pore volume; hence the dimensionless time for E_I is

$$t_{DI} = \frac{t_D}{E_A} \tag{6.7-2a}$$

By a similar argument, though it is much more difficult to show parallel cross sections, the dimensionless time for the E_A is

$$t_{DA} = \frac{t_D}{E_I} \tag{6.7-2b}$$

The equation for the volumetric sweep may be written in combined form as

$$E_V = E_A\left(\frac{t_D}{E_I}\right) \cdot E_I\left(\frac{t_D}{E_A}\right) \tag{6.7-3}$$

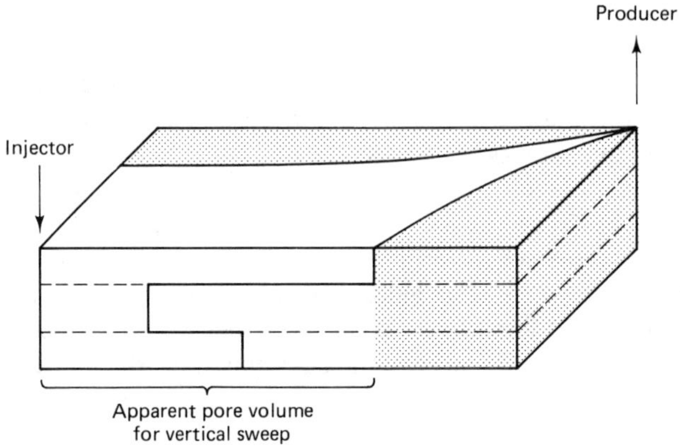

(a) Schematic showing the reference volume for t_{DI}

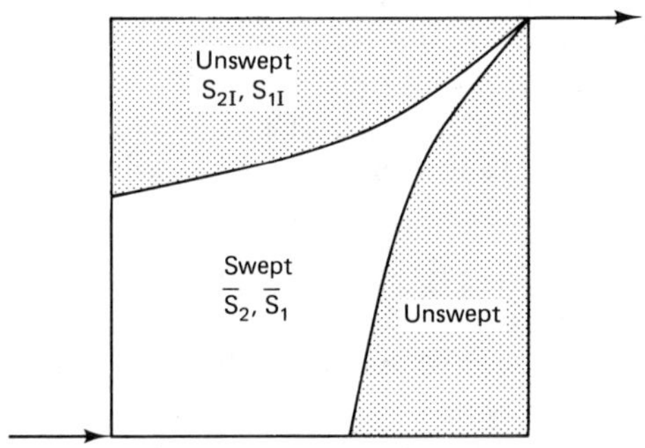

(b) Schematic showing swept and unswept volumes

Figure 6-17 Schematics for combining sweep efficiencies

Since E_A and E_I appear as both multiplications and in the arguments, constructing a volumetric sweep function requires a trial-and-error procedure.

1. Determine the cumulative injection at breakthrough t_D^0. This is just the product of E_A and E_I at their respective breakthrough values, E_A^0 and E_I^0.
2. Pick some t_D after breakthrough t_D^0.
3. Select a trial E_I.
4. Calculate the areal sweep efficiency from $E_A = E_A(t_D/E_I)$.
5. Calculate the vertical sweep from $E_I = E_I(t_D/E_A)$ using E_A.

If E_I agrees with that assumed in step 3, $E_A \cdot E_I$ is the volumetric sweep efficiency at t_D. If E_I is different from step 3, select a new trial E_I, and return to step 4.

Experience has shown that the procedure converges within two to three iterations for typical E_A and E_I functions using direct substitution. By repeating the procedure for several values of t_D, a volumetric sweep efficiency curve may be calculated (see Exercise 6H).

Combining Pseudodisplacement and Areal Sweep

The pseudodisplacement sweep efficiency $\tilde{E}_D$ may be determined from the VE theory of Sec. 6-5. The recovery efficiency follows from this as

$$E_R = \tilde{E}_D(t_{DA})E_A(t_{DD}) \tag{6.7-4}$$

where t_{DA} and t_{DD} are the dimensionless times based on the apparent pore volumes appropriate for the particular sweep efficiency.

Combining a displacement and areal sweep in the manner described here is again a generalization of the procedure proposed by Claridge (1972) and is repeated in Chap. 7 for a miscible flood. The dimensionless time for $\tilde{E}_D$ is the same as Eq. (6.7-2a), but for E_A, we must view the displacement differently. Consider Fig. 6.17(b), which shows an areal view of a displacement divided into a swept and an unswept region. The unswept region contains oil and water saturations at the values present at the initiation $(S_1, S_2)_I$ of the displacement. We identify the saturations $\bar{S}_1$, $\bar{S}_2$ in the swept region with the cross-sectional averaged saturations determined from $\tilde{E}_D$

$$\bar{S}_1 = S_{1I} + \tilde{E}_D(1 - S_{1I}) \tag{6.7-5}$$

At a particular time, the pore volume available to flow for a pistonlike displacement whose front occupies the same position as that shown in Fig. 6-17(b) is the water volume in the swept region. (Another way of viewing this is to suppose the oil saturation $\bar{S}_2$ is part of the immobile phase.) Therefore, the dimensionless time for E_A is now

$$t_{DD} = \frac{t_D}{\bar{S}_1} \tag{6.7-6}$$

The procedure for calculating the recovery efficiency is similar to that given above.

1. Calculate the cumulative injection at breakthrough t_D°. This is equal to the product of the breakthrough values of $\tilde{E}_D$ and E_A.
2. Pick some t_D after breakthrough t_D°.
3. Select a trial $\tilde{E}_D$, and calculate $\bar{S}_1$ from Eq. (6.7-6).
4. Calculate the areal sweep efficiency from $E_A = E_A(t_D/\bar{S}_1)$.
5. Calculate the pseudodisplacement sweep from $\tilde{E}_D = \tilde{E}_D(t_D/E_A)$.

If the $\tilde{E}_D$ agrees with that in step 3, the recovery efficiency is the product of $\tilde{E}_D$ and E_A. If $\tilde{E}_D$ does not agree, return to step 3 with a new trial value.

Combining Vertical, Areal, and Displacement Sweep Efficiencies

If all three efficiencies are independently available, the above procedure may easily be generalized as

$$E_R(t_D) = E_A\left(\frac{t_D}{E_I \bar{S}_1}\right)E_I\left(\frac{t_D}{E_A \bar{S}_1}\right)E_D\left(\frac{t_D}{E_V}\right) \tag{6.7-7}$$

The procedure now requires a two-level trial and error, which is equivalent to first combining E_A and E_I and then combining E_V and E_D. The final result in Eq. (6.7-7) is independent of the order the combinations are carried out in.

As a conclusion to this section, we remind you of the limitations inherent in these procedures. First, we must have independently specified functions of E_A, E_I, and E_D, and these functions must be independent of explicit rate dependence. If a rate dependence is present, the function E_I will depend on the particular pathline it was evaluated on. Perhaps we could evaluate on a pathline having a fluid velocity representative of the entire pattern (this is commonly tried), but there is considerable uncertainty about what this representative value is even in the most well-defined displacements. Recall that, particularly in the VE approaches, the dependence of the sweep efficiencies on rate may not be particularly evident (for example, the Dietz theory is strongly rate dependent, but this is not evident from the general VE approach when capillary pressure becomes small). Further recall that independent specifications of each of the three efficiencies are available through relatively idealized calculations (see Secs. 6-4 through 6-6) for extremes in certain physical properties or through physical models. When any of the above conditions are seriously violated—and their violation significantly affects the results—one must resort to numerical simulation, from which the oil recovery could be directly calculated.

A second more subtle, and perhaps more serious, limitation of the combined sweep efficiency approach deals with scaling. Scaling simply means any of the sweep efficiencies, however determined, must themselves be adjusted for the considerably different scale between the laboratory experiment or analytical calculation and the field application. For example, few of the independent determinations of E_D or E_I account for the nonuniformities surely present in a field displacement.

A classical example of this scale effect involves applying E_D to a viscously unstable field-scale displacement. Much theoretical and experimental work has gone into showing that the size of the instabilities formed, and indeed, whether they propagate or not, is a function of the characteristic lengths of the laboratory experiment or calculation. Thus unless the scaling is such that both effects are the same in the laboratory and in the field (a sometimes impossible task), the lab-derived E_D will be too optimistic. We cover the subject of viscous instabilities in the next section.

6-8 INSTABILITY PHENOMENA

No EOR process is free from some sort of instability. Hence substantial effort has gone into minimizing or preventing instabilities (using polymer to drive surfactants and alkaline agents, or foaming agents to drive CO_2 and steam) and into predicting the oil recovery if fingering is inevitable. We discuss predicting the results of a fingering process in Chap. 7 in connection with solvent flooding where instability phenomena have received the most attention. In this section, we deal with the formation of fingers.

We use the term *fingering* to describe the bypassing of a resident fluid by a displacing agent in a homogeneous, nonuniform medium. The actual bypassing region is a finger. This definition encompasses instabilities caused by both viscous forces (viscous fingers) and gravity forces (gravity fingers) but does *not* include bypassing by permeability heterogeneities. This definition is a little more rigid than that used in the literature, but we believe the inherent distinction is useful because fingering can be prevented from displacements, whereas bypassing caused by heterogeneities cannot (though it can be reduced). In this section, we deal with isothermal flows; in Chap. 11, we discuss the stability of a nonisothermal displacement.

A Necessary Condition for Stability

In keeping with the notion that fingering is a general phenomenon, consider the incompressible, dissipation-free displacement of fluid 2 by fluid 1 in a dipping reservoir, as shown in Fig. 6-18. This figure is a cross section of a displacement, but fingering can occur in either the vertical or areal sense. There is no z-direction communication in this problem. We also consider a perturbation of length ϵ of the displacement front (caused, perhaps, by an isolated nonuniformity in the permeability field), and strive to determine the conditions under which $\epsilon(t)$ will grow or decay as a function of time. The actual fingering phenomenon is, of course, much more random and chaotic than that shown in Fig. 6-18, as evidenced by an areal view of a fingering displacement in a quarter five-spot model shown in Fig. 6-19. Neverthe-

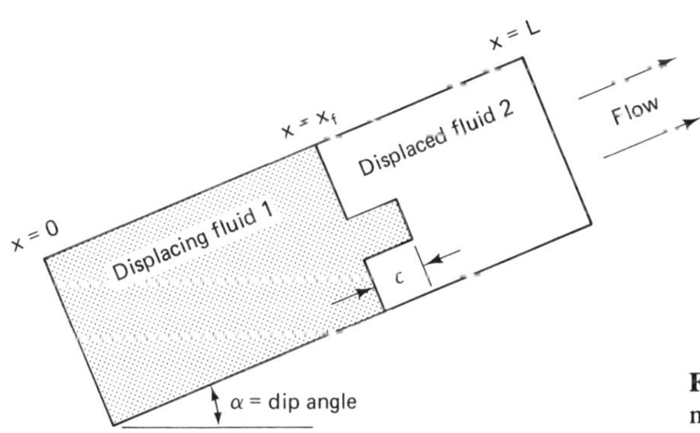

Figure 6-18 Viscous fingering schematic

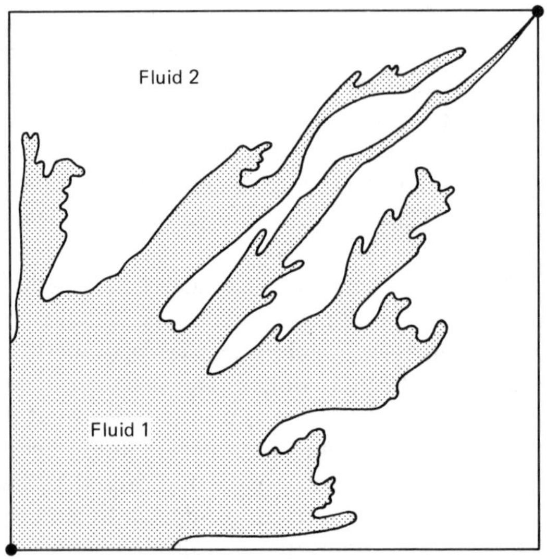

Figure 6-19 Viscous fingering in a quarter five-spot model, $M^\circ = 17$ (from Habermann, 1960)

less, the simple geometry of Fig. 6-18 is tractable to mathematical analysis and yields insights into the more complex situations.

To solve for the conditions ϵ will grow or decay under, we proceed by the moving boundary technique discussed by Collins (1976). In the region behind the displacing fluid front, $x < x_f$, the conservation of fluid 1 gives

$$\frac{\partial u_{xj}}{\partial x} = 0 \tag{6.8-1}$$

where $j = 1$ for $x < x_f$, and $j = 2$ for $x > x_f$. The accumulation terms in both equations are zero since there is no change in concentration in the respective regions. For the same reason, when we substitute Darcy's law into these equations, they become

$$\frac{\partial}{\partial x}\left(\frac{\partial P_j}{\partial x} + \rho_j g \sin \alpha\right) = 0, \qquad j = 1 \text{ or } 2 \tag{6.8-2}$$

The solutions to Eq. (6.8-2) will be of the form

$$P_j = (a_j - \rho_j g \sin \alpha)x + b_j, \qquad j = 1 \text{ or } 2 \tag{6-8.3}$$

where a_j and b_j are integration constants to be determined with appropriate boundary conditions. If P_0 and P_L are the pressures at the reservoir inlet and outlet, respectively, then b_j can be determined as

$$b_1 = P_0 \tag{6.8-4a}$$

$$b_2 = P_L - (a_2 - \rho_2 g \sin \alpha)L \tag{6-8.4b}$$

Using these relations, and requiring continuous x velocities across the front

$$u_{x1}\big|_{x=x_f} = u_{x2}\big|_{x=x_f} = u_x \tag{6.8-5}$$

gives, once again using Darcy's law,

$$M^0 a_1 = a_2 \tag{6.8-6}$$

Equation (6.8-6) determines a_1, for we must have continuity of pressure at x_f in the absence of capillary pressure

$$P_1\big|_{x_f} = P_2\big|_{x_f} \tag{6.8-7}$$

Inserting Eq. (6.8-3) into Eq. (6.8-7) and using Eqs. (6.8-4) and (6.8-6), yields

$$a_1 = \frac{-\Delta P + \rho_2 g \sin\alpha (L - x_f) + \rho_1 g \sin\alpha x_f}{M^0 L + (1 - M^0)x_f} \tag{6.8-8}$$

where $\Delta P = P_0 - P_L$ is the overall pressure drop. The rate of frontal advance is from Darcy's law

$$\frac{dx_f}{dt} = \frac{u_x|_{x_f}}{\phi\Delta S} = \frac{k\lambda_{r1}}{\phi\Delta S}\,\frac{\Delta P + g\sin\alpha[\Delta\rho(L - x_f) - \rho_1 L]}{M^0 L + (1 - M^0)x_f} \tag{6.8-9}$$

Equation (6.8-9) applies to any point on the displacement front. We could have equally well developed an expression for a point on the perturbation front

$$\frac{d(x_f + \epsilon)}{dt} = \frac{k\lambda_{r1}}{\phi\Delta S}\,\frac{\Delta P + g\sin\alpha[\Delta\rho(L - x_f - \epsilon) - \rho_1 L]}{M^0 L + (1 - M^0)(x_f + \epsilon)} \tag{6.8-10}$$

Equation (6.8-10) is identical to Eq. (6.8-9) except $x_f + \epsilon$ has replaced x_f everywhere. The rate of change of the perturbation is

$$\frac{d\epsilon}{dt} = \frac{d(x_f + \epsilon)}{dt} - \frac{dx_f}{dt} = \dot{\epsilon} \tag{6.8-11}$$

which yields, when Eqs. (6.8-9) and (6.8-10) are substituted,

$$\dot{\epsilon} = -\frac{k\lambda_{r1}}{\phi\Delta S}\,\frac{\Delta P(1 - M^0) + Lg\Delta\rho\sin\alpha - Lg\rho_1(1 - M^0)\sin\alpha}{[M^0 L + (1 - M^0)x_f]^2}\,\epsilon \tag{6.8-12}$$

In Eq. (6.8-12), we have assumed $\epsilon << x_f$ with the corresponding simplification. Equation (6.8-12) could be integrated, but for our purpose it is sufficient to investigate only the sign of $\dot{\epsilon}$. The perturbation will grow if $\dot{\epsilon} > 0$, will remain constant if $\dot{\epsilon} = 0$, and will decay if $\dot{\epsilon} < 0$. From the equality of these three choices, we find the condition of neutral stability as

$$-(\Delta P)_c = \frac{L\Delta\rho g\sin\alpha}{1 - M^0} - Lg\rho_1\sin\alpha \tag{6.8-13}$$

where $(\Delta P)_c$ is a critical pressure drop. The superficial velocity corresponding to this is the *critical rate* u_c

$$u_c \equiv -k\lambda_{r1}^{\circ}\left[\frac{-(\Delta P)}{L} + \rho_1 g \sin\alpha\right] = \frac{k\lambda_{r1}^{\circ}\Delta\rho g \sin\alpha}{M^0 - 1} \tag{6.8-14}$$

Using the critical rate, the conditions for finger growth may be restated

$$u_x \begin{cases} > u_c & \text{(unstable)} \\ = u_c & \text{(neutral)} \\ < u_c & \text{(stable)} \end{cases} \tag{6.8-15}$$

where we have also used Darcy's law to express u_x in these inequalities.

Note the similarity between Eq. (6.8-14) and Eq. (6.6-10b), the corresponding critical rate for gravity tonguing. Analogous expressions can be worked out for almost any segregated flow conditions, so this similarity should not be regarded as merely fortuitous. But the differences in the two flows should be kept in mind. The critical rate in Eq. (6.8-14) is based on an unstable displacement in a reservoir having no z-direction communication; that in Eq. (6.6-10b) is the consequence of a VE displacement in a reservoir with very good communication.

To further investigate the stability issue, let's write the condition for stability (finger decay) as

$$(M^0 - 1)u_x < k\lambda_{r1}^{\circ}\,\Delta\rho g \sin\alpha \tag{6.8-16}$$

The superficial velocity u_x in this inequality is always positive, but the density difference can be negative (less dense fluid displacing more dense), as can the dip angle (displacing down dip). Of course, M^0 can take on only positive values though over quite a large range. Table 6-2 shows typical signs of M^0 and $\Delta\rho$ for various EOR processes. Immediately it follows from Eq. (6.8-16) that the condition for stability in a horizontal reservoir is simply $M^0 < 1$. This condition is used universally throughout the EOR literature to describe a stable displacement, particulary in laboratory floods, though the more general Eq. (6.8-16) is actually the most appropriate form (Hill, 1952).

TABLE 6-2 TYPICAL VALUES FOR MOBILITY RATIOS AND DENSITY DIFFERENCES BY PROCESS TYPE

	$M^0 < 1$	$M^0 > 1$
$\Delta\rho > 0$	Waterflood Polymer flood Micellar polymer	Waterflood Polymer flood
$\Delta\rho < 0$	Foam	Steam

Considering the signs possible for α and $\Delta\rho$, we can divide the stability possibilities into four cases, as Table 6-3 shows. Case 1 is unconditionally stable regardless of the values of $\Delta\rho g \sin\alpha$ and M^0 as $\Delta\rho g \sin\alpha$ is positive, and $M^0 < 1$. Similarly, if $\Delta\rho g \sin\alpha < 0$ and $M^0 > 1$, case 4, the displacement is unconditionally

TABLE 6-3 POSSIBLE CASES FOR A STABLE DISPLACEMENT

Case			
1	$M^0 < 1$	$\Delta\rho g \sin\alpha > 0$	Stable
2	$M^0 > 1$	$\Delta\rho g \sin\alpha > 0$	Conditionally stable (type I)
3	$M^0 < 1$	$\Delta\rho g \sin\alpha < 0$	Conditionally stable (type II)
4	$M^0 > 1$	$\Delta\rho g \sin\alpha < 0$	Unstable[a]

[a] Infinite lateral boundaries
Note: Write stability criterion as $(M^0 - 1)u_x < k\lambda_1^\circ \Delta\rho g \sin\alpha$. For $\alpha = 0$ (no dip), the stability criterion becomes $M^0 < 1$.

unstable. The more interesting cases are 2 and 3, which we call type I and type II conditional stability.

For type I stability, if we divide through Eq. (6.8-16) by the positive quantity $(M^0 - 1)$, the stability criterion is written for u_x as in Fig. 6-20. The criterion is an upper bound for u_x and a plot of sweep efficiency (vertical, areal, or volumetric) versus the dimensionless rate u_D

$$u_D = \frac{u_x(M^0 - 1)}{k\lambda_{r1}^\circ \Delta\rho g \sin\alpha} \tag{6.8-17}$$

shows that E_V remains essentially constant until $u_D = 1$ and then decreases thereafter. Since increasing the displacement velocity causes the instability to form, we see that viscous forces destabilize the displacement ($u_D > 1$), whereas gravity forces tend to stabilize the displacement ($u_D < 1$). The resulting instability is a *viscous instability* or *finger*. For type II conditional stability, a similar plot (Fig. 6-21) shows sweep efficiency decreasing for decreasing u_D, beginning a precipitous decline at $u_D = 1$. This is because the stability criterion is now a lower bound since $(M^0 - 1)$ is now negative. For type II conditional stability, viscous forces stabilize the dis-

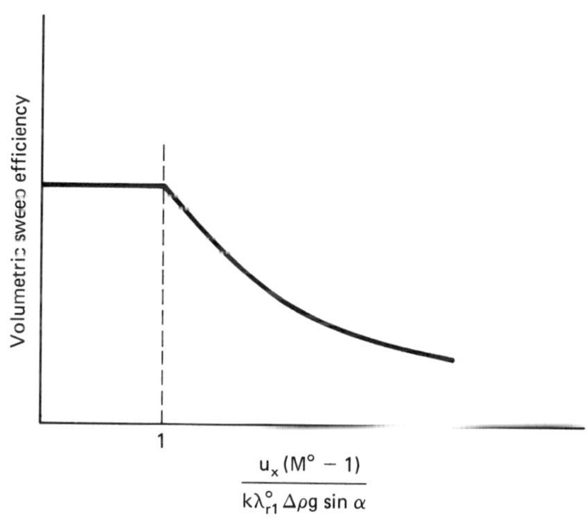

Figure 6-20 Type I conditional stability

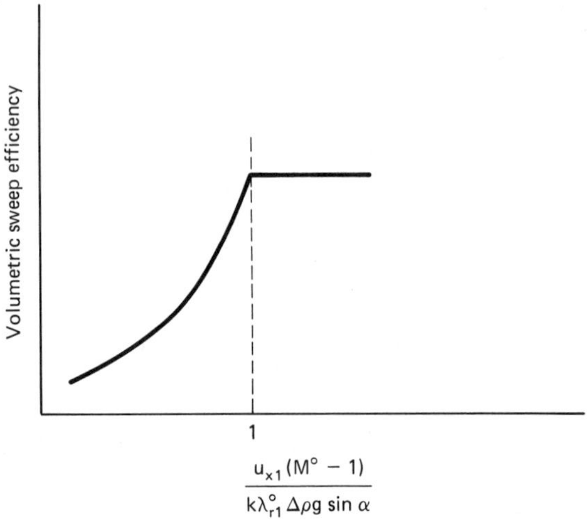

Figure 6-21 Type II conditional stability

placement, and gravity forces destabilize. The resulting instabilty is a gravity instability.

For certain values of the parameters, then, both types of displacements are or can be made stable. The conditional stability is most useful in determining a maximum rate in a dipping displacement where $M^0 > 1$. But usually, this rate is below that required for economic oil production. For type II stability, a larger rate is required, but in practice, this situation is not commonly encountered.

Critical Wavelength

Whereas $u_x < u_c$ is a necessary and sufficient condition for stability, the condition $u_x > u_c$ is, unfortunately, a necessary condition only for instability. This condition is because dissipative effects in flows in media of limited lateral extent tend to suppress fingering. This effect means fingering may be abnormally suppressed in laboratory displacements compared to the same displacement under field conditions. One may legitimately wonder, then, about the purpose of doing laboratory experiments on unstable displacements when this scale effect is not considered.

To investigate this scale effect, we reproduce an argument based on linear stability analysis originally given by Chouke et al. (1959) and then by Gardner and Ypma (1982).

Based on a linear stability analysis of a downward secondary miscible displacement of oil by a less viscous and less dense solvent in a homogeneous, uniform medium, the critical wavelength λ_c of an unstable miscible displacement is

$$\lambda_c = 4\pi \frac{M^0 + 1}{M^0 - 1}\left(\frac{K_\ell}{u_x - u_c}\right) \tag{6.8-18}$$

where the dispersion coefficient K_ℓ is taken to be isotropic. Since the displacement is unstable, we must have $M^0 > 1$ and $u_x > u_c$ so that λ_c is always positive.

The analogous expression for an initially sharp immiscible displacement was also determined by Chouke et al. (1959) and reproduced in greater detail by Peters (1979)

$$\lambda_c = \frac{C}{3}\left[\frac{k\lambda_{r1}^{o}\sigma_{12}}{(M^0 - 1)(u_x - u_c)}\right]^{1/2} \tag{6.8-19}$$

The constant C in Eq. (6.8-19) is called Chouke's constant by Peters, who also determined values $C = 25$ for immiscible displacements with no residual water initially present, and $C = 190$ with irreducible water present. Clearly, the critical wavelength is greater with irreducible water initially present, but the reason for this stabilizing effect is not well understood.

The necessary and sufficient conditions for a type I instability to form based on this analysis are now

$$M^0 > 1 \quad \text{or} \quad u_x > u_c \quad \text{and} \quad \lambda_c < (H_t)_{\max} \tag{6.8-20}$$

where $(H_t)_{max}$ is the maximum lateral extent of the permeable medium. One may readily show (see Exercise 6I) that λ_c is of the order of a few centimeters for typical conditions. Thus, if fingering is desired in a displacement, one must take special precautions that conditions (Eq. 6.8-20) are met. This usually means running displacements at excessively high rates, compared to field rates, or in systems having at least one large transverse dimension. Such a system is the Hele Shaw cell, in which the displacement of Fig. 6-19 is occurring. But if the intent is to suppress fingering, systems having very small transverse dimensions, such as the slim tube experiments we discuss in Chap. 7, are preferable.

Three things are important about both the derivation of critical velocity and wavelength. First, neither says anything about how fingers propagate once they are formed. A finger forms, bifurcates into two branches, one of these dominates (or shields) the other, and the dominant one then bifurcates again to repeat the process (Homsy, 1987). If continued indefinitely, a single finger with numerous appendages representing the bifurcations will result. Figure 6-19 suggests the bifurcation through the various levels of fingers each superimposed on the next larger scale. The smallest scale corresponds to the critical wavelength.

Second, both the critical wavelength and velocity derivations depended on the perturbation being small. It is impossible to say from this what the response to a large perturbation would be, and we can be assured that such large perturbations do exist. Thus Eq. (6.8-20) should also be regarded as only necessary conditions.

Finally, the issues of fingering and heterogeneity cannot be rigorously separated. After all, heterogeneity caused the perturbation in Fig. 6-16 even though we proceeded as though the reservoir was homogeneous. The merging of the fingering and heterogeneity issues is one of the most interesting topics in EOR research, in Chap. 7, we discuss some primitive attempts at this merging.

6-9 SUMMARY

That volumetric sweep efficiency is a complex issue accounts for the scarcity of treatment in this text compared to displacement efficiency. Three factors account for this complexity: a strong dependency on operational issues, nonlinear and irregular geometries, and the difficulty in capturing realistic heterogeneities. Numerical simulators can handle all three of these issues to some extent even though some questions remain about how to represent heterogeneity in simulation models.

There is little in the behavior of the volumetric sweep efficiency of actual reservoir displacements that cannot be at least qualitatively understood through the material we present here. Examples of such behavior are reservoirs with high-permeability thief zones that behave essentially as a two-layer medium, generally high-permeability reservoirs dominated by gravity that conform well to the Dietz theory, low-permeability reservoirs in which crossflow tends to be unimportant, and high-permeability reservoirs with large well spacing that tend to the VE limit rather quickly.

Above all, the recognition of bypassing—through channeling, viscous fingering, gravity segregation, or some combination of these—is important, for this seems to occur in a good many waterfloods and EOR projects.

EXERCISES

6A. *Using Areal Sweep Correlations.* Use the areal sweep efficiency correlations for a confined five-spot in this exercise.

(a) Plot areal sweep efficiency E_A versus dimensionless time t_D for a mobility ratio of 6.5.

(b) If the pattern pore volume is $10^6 m^3$, and the average injection rate is 500 m^3/D, plot cumulative oil recovery (SCM) versus time (months or years). Assume the displacement is pistonlike, vertical sweep is 1, and the pore volume given above is movable. The residual water and oil saturations are 0.2 and 0.3, respectively.

6B. *Heterogeneity Measures of Normal Distributions.* As Table 6-1 shows, permeability often is distributed normally rather than lognormally. When this happens, the cumulative frequency distribution function (Eq. 6.3-5) becomes

$$\Lambda = \frac{1}{2}\left\{1 - \text{erf}\left[\frac{r - \bar{r}}{\sqrt{2\nu_N}}\right]\right\} \tag{6B-1}$$

where $\bar{r}$ is the average permeability–porosity ratio, and ν_N is the variance of the normal distribution. Using Eqs. (6B-1), (6.3-3), and (6.3-4), derive formulas for the Lorenz and Dykstra-Parsons coefficients in terms of ν_N.

6C. *Vertical Sweep Efficiency in a Two-Layer Reservoir*

(a) Derive Eq. (6.4-4) for flow in layer l in a horizontal reservoir.

(b) Calculate and plot the vertical sweep efficiency E_I and the fraction of total flow going into the high velocity layer for a two-layer horizontal reservoir with $k_1 = 2k_2$, $\phi_1 = \phi_2$, $\Delta S_1 = \Delta S_2$, and $h_1 = 3h_2$. Take $M^0 = 0.5$.

6D. *Vertical Sweep Efficiency in a Noncommunicating Reservoir*. For a reservoir having no vertical communication, calculate and plot the vertical sweep efficiency versus dimensionless cumulative water injected for the following five-layer cross section:

h_l(m)	ϕ_l	$k_l(\mu m^2)$
5	0.2	0.100
10	0.22	0.195
2	0.23	0.560
15	0.19	0.055
4	0.15	0.023

The endpoint mobility ratio is 0.5.

6E. *Vertical Equilibrium for Continuous Layers*. For a reservoir for which the VE Hearn model applies with $M^0 < 1$ and $\alpha = 0$,

(a) Show that if the permeability distribution is continuous, the cross-sectional averaged water fractional flow may be written as

$$\bar{f}_1 = \left(1 + \frac{(1 - C)}{H_K M^0 C}\right)^{-1} \tag{6E-1}$$

where H_K is the Koval heterogeneity factor (Fig. 6-8).

(b) Recalculate and plot the vertical sweep efficiency for the two-layer model of part (b) in Exercise 6C. Use $M^0 = 0.5$.

(c) In a two-layer horizontal reservoir, show that the effects of the heterogeneity contrast may be completely suppressed (that is, the fronts travel at equal velocities in both layers) if

$$M^0 < \frac{k_2}{\phi_2}\frac{\phi_1}{k_1} \tag{6E-2}$$

where 1 and 2 represent the high and low velocity layers, respectively.

6F. *Calculating Pseudorelative Permeabilities*. For the discrete permeability–porosity data of Exercise 6D,

(a) Calculate and plot the pseudorelative permeabilites for a waterflood in a horizontal reservoir using the VE Hearn model.

(b) Calculate and plot the vertical sweep efficiency for this flood.

(c) Repeat part (a) for a nonzero capillary pressure function given by

$$P_c = \sigma_{12}\left(\frac{\phi}{k}\right)^{1/2} \cos\theta(1 - S)^4 \tag{6F-1}$$

where σ_{12} is the oil–water interfacial tension, θ is the contact angle, and

$$S = \frac{S_1 - S_{1r}}{1 - S_{1r} - S_{2r}} \tag{6F-2}$$

(d) Calculate and plot the vertical sweep efficiency for part (c).

Additional data for this problem are $\Delta\rho = 0$, $S_{1r} = S_{2r} = 0.2$, $\mu_1 = 1$ mPa-s,

$\mu_2 = 10$ mPa-s, $k^\circ_{r1} = 0.05$, $k^\circ_{r2} = 0.9$, and the relative permeability curves are given as

$$k_{r1} = k^\circ_{r1} S^2, \qquad k_{r2} = k^\circ_{r2}(1 - S) \tag{6F-3}$$

6G. *Deriving Pseudorelative Permeabilities.* The water–oil capillary-pressure–water-saturation function often may be represented as

$$P_c = K\left(\frac{1}{S^2} - 1\right) \tag{6G-1}$$

where K is a constant, and S is the reduced saturation (Eq. 6F-2). If the VE assumptions apply and the reservoir is homogeneous,

(a) Derive the water saturation profile in the dip normal or z direction in terms of a water saturation at the bottom of the reservoir (S_{1B} or S_B).

(b) Derive an expression for the average water saturation as a function of S_{1B} or S_B.

(c) If the local (laboratory-measured) relative permeabilities are approximated by Eq. (6F-3), show that the oil and water pseudorelative permeabilities expressed in terms of the average saturation of part (b) are

$$\tilde{k}_{r1} = \frac{k^\circ_{r1}}{N^\circ_g} \ln\left\{1 + \frac{N^\circ_g \bar{S}^2}{\left(1 - \dfrac{N^\circ_g \bar{S}^2}{4}\right)^2}\right\}, \qquad \tilde{k}_{r2} = k^\circ_{r2}(1 - \bar{S})$$

where

$$N^\circ_g = \frac{\Delta\rho g \cos\alpha\, H_t}{K}$$

(d) For $N^\circ_g = 1$ and $M^0 = 4$, calculate and plot the pseudodisplacement sweep efficiency versus dimensionless time. The dip angle of the reservoir is zero.

6H. *Combining Sweep Efficiencies.* The vertical sweep efficiency curve for a pistonlike displacement is shown in Fig. 6H. Combine this curve with the areal sweep efficiency curve of Exercise 6A to give the volumetric sweep efficiency curve.

6I. *Viscous Fingering Calculations*

(a) Calculate the critical rate for a miscible displacement having the following properties:

$k = 0.12\ \mu\text{m}^2$

$M^0 = 50$

Oil-solvent density difference $= -0.8$ g/cm^3

Solvent mobility $= 10$ (mPa-s)$^{-1}$

Dip angle $= -10°$

(b) If the superficial velocity in the above displacement is 0.8 μm/s, calculate the critical wavelength from stability theory. Take the dispersion coefficient to be 10^{-5} cm^2/s.

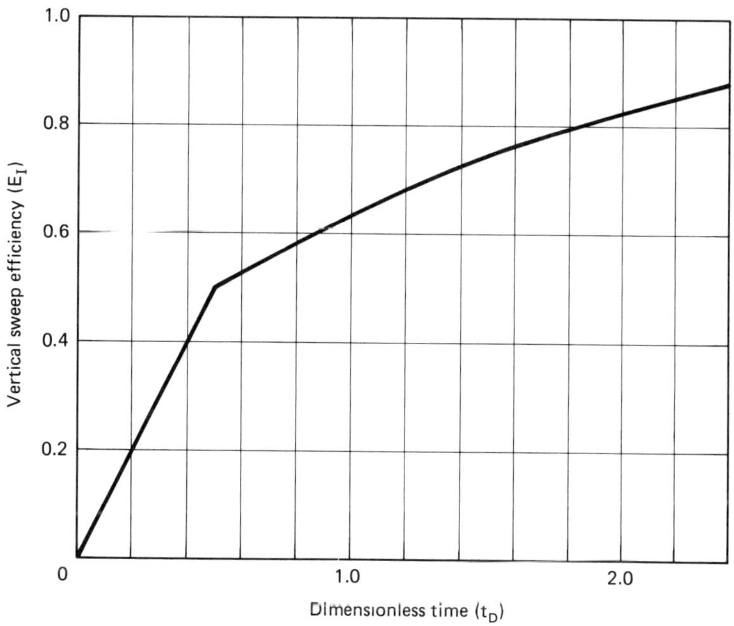

Figure 6H Vertical sweep efficiency function for Exercise 6H

7

Solvent Methods

One of the earliest methods for producing additional oil is through the use of solvents to extract the oil from the permeable media. In the early 1960s, interest centered on injecting liquified petroleum gas (LPG) in small "slugs" and then displacing the LPG by a dry "chase" gas. This process became economically less attractive as the value of the solvent increased. In the late 1970s, interest in solvent methods resurged because of an increased oil price and more confidence in the ability to estimate oil recovery. During this period, the leading solvent became carbon dioxide though several other fluids were used also (Stalkup, 1985).

Two fluids that mix together in all proportions within a single-fluid phase are *miscible*. Therefore, miscible agents would mix in all proportions with the oil to be displaced. But most practical miscible agents exhibit only partial miscibility toward the crude oil itself, so we use the term *solvent* flooding in this text. Many solvents, of course, will become miscible with crude under the right conditions, but all solvents of commercial interest are immiscible to an aqueous phase.

Solvent flooding refers to those EOR techniques whose main oil recovering function is because of extraction, dissolution, vaporization, solubilization, condensation, or some other phase behavior change involving the crude. These methods have other, sometimes very important, oil recovery mechanisms (viscosity reduction, oil swelling, solution gas drive), but the primary mechanism must be extraction.

This oil extraction can be brought about by many fluids: organic alcohols, ketones, refined hydrocarbons, condensed petroleum gas (LPG), natural gas and liquified natural gas (LNG), carbon dioxide, air, nitrogen, exhaust gas, flue gas, and others. In this chapter, we emphasize miscible flooding with gaseous solvents CO_2, CH_4, and N_2, but you should remember there are many potential agents.

7-1 GENERAL DISCUSSION OF SOLVENT FLOODING

Considering the wide variety of solvents, process types, and reservoirs, our discussion must ignore one or more interesting variations. Thus in this section, we discuss CO_2 solvent flooding, and later, we indicate more general aspects of solvent flooding.

Figure 7-1 shows an idealized vertical cross section between an injection and production well. By far the most common application of solvent methods is in a displacement mode as shown, but injection and production through the same wells have been reported (Monger and Coma, 1986). Solvent injection commences into a reservoir in some stage of depletion, most commonly at residual oil or true tertiary conditions. Most solvent floods are in reservoirs containing light crudes (less than 3 mPa-s oil viscosity) though there are exceptions (Goodrich, 1980). The solvent may be introduced continuously in undiluted form, alternated with water in the water-alternating-gas (WAG) process as in Fig. 7-1, or even injected simultaneously with water through paired injection wells. Water is injected with the solvent in this fashion to reduce the usually unfavorable mobility ratio between the solvent and the oil. Carbon dioxide, in particular, can be injected dissolved in water in a distinctly immiscible fashion that recovers oil through swelling and viscosity reduction (Martin, 1951).

If the solvent is completely (first-contact) miscible with the oil, the process has a very high ultimate displacement efficiency since there can be no residual phases (see Sec. 5-4). If the solvent is only partially miscible with the crude, the total composition in the mixing zone (miscible zone in Fig. 7-1) between the solvent and the oil can change to generate or develop miscibility in situ. Regardless of whether the displacement is developed or first-contact miscible, the solvent must immiscibly displace any mobile water present with the resident fluids.

The economics of the process usually dictates that the solvent cannot be injected indefinitely. Therefore, a finite amount or *slug* of solvent is usually followed by a *chase* fluid whose function is to drive the solvent toward the production wells. This chase fluid—N_2, air, water, and dry natural gas seem to be the most common choices—may not itself be a good solvent. But it is selected to be compatible with the solvent and because it is available in large quantities. The similarity between the chase fluid in solvent flooding and the mobility buffer drive in micellar-polymer flooding is evident in Figs. 7-1 and 9-1.

Though the process shown in Fig. 7-1 appears relatively simple, the displacement efficiency and volumetric sweep efficiency are quite complex. In Secs. 7-6 and 7-8, we apply the theory of Chaps. 5 and 6 to solvent flooding, but first we must discuss selected physical properties of solvents and solvent–crude oil systems.

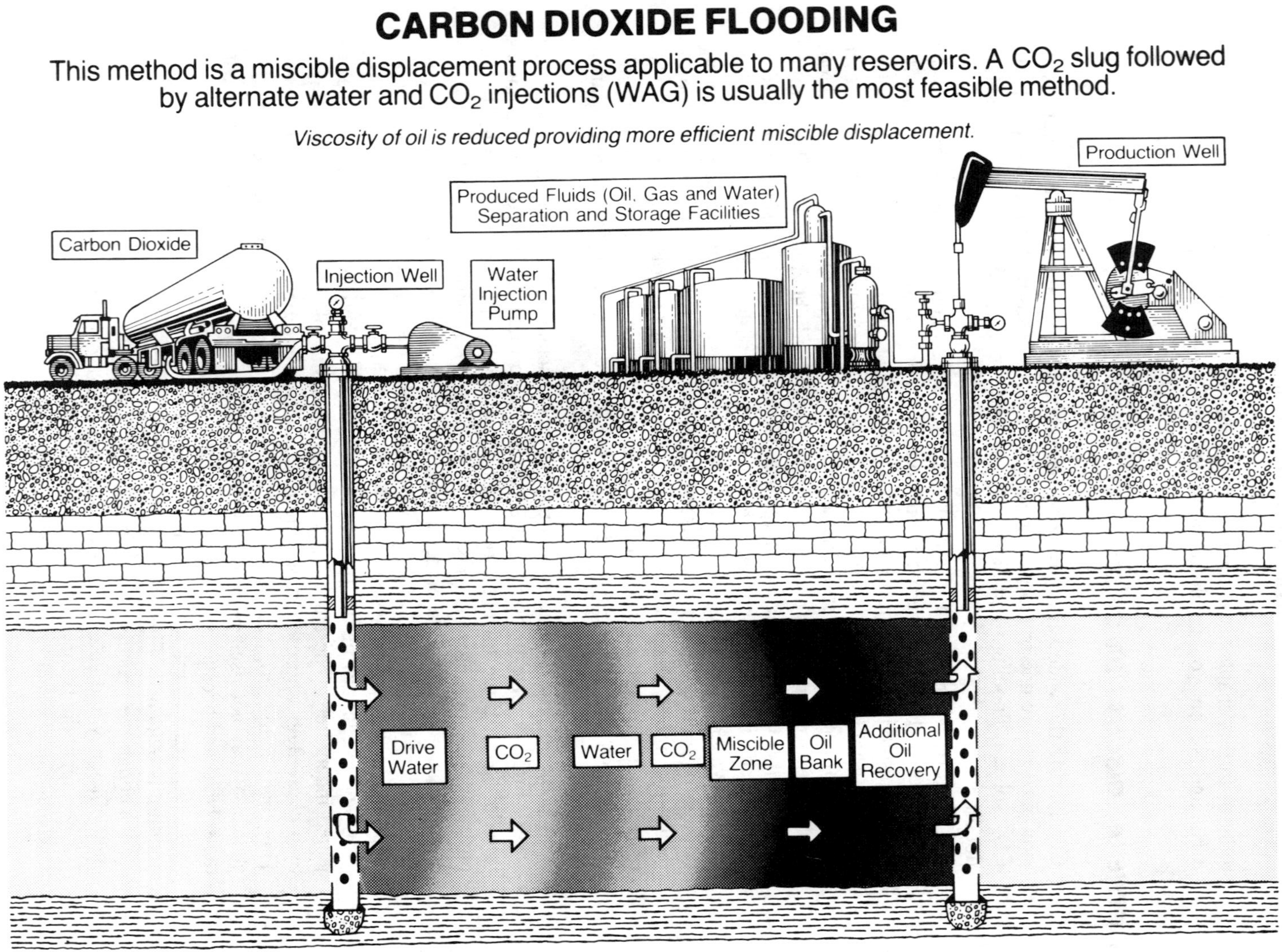

Figure 7-1 Schematic of a solvent flooding process (drawing by Joe Lindley, U.S. Department of Energy, Bartlesville, Okla.)

7-2 SOLVENT PROPERTIES

Figure 7-2 shows phase behavior data (P-T diagram) for various pure components and air. For each curve, the line connecting the triple and critical points is the vapor pressure curve; the extension below the triple point is the sublimation curve (see Sec. 4-1). The fusion curve is not shown. The pressure–temperature plot for air is really an envelope, but its molecular weight distribution is so narrow that it appears as a line in Fig. 7-2. Flue gas is also a mixture of nitrogen, carbon monoxide, and carbon dioxide with a similarly narrow molecular weight distribution; its P-T curve would fall near the nitrogen curve in Fig. 7-2.

The critical pressures for most components fall within a relatively narrow range of 3.4–6.8 MPa (500–1,000 psia), but critical temperatures vary over a much wider range. The critical temperatures of most components increase with increasing molecular weight. Carbon dioxide (molecular weight, $M_W = 44$) is an exception to this trend with a critical temperature of 304 K (87.8°F), which is closer to the critical temperature of ethane ($M_W = 30$) than to propane ($M_W = 44$). (See Vukalovich and Altunin (1968) for a massive compilation of CO_2 properties.) Most reservoir applications would be in the temperature range of 294–394 K (70–250°F) and at pressures greater than 6.8 MPa (1,000 psia); hence air, N_2, and dry natural gas will all be supercritical fluids at reservoir conditions. Solvents such as LPG, in the molecular weight range of butane or heavier, will be liquids. Carbon dioxide will usually be a supercritical fluid since most reservoir temperatures are above the critical temperature. The proximity to its critical temperature gives CO_2 more liquidlike properties than the lighter solvents.

Figures 7-3 and 7-4 give compressibilities factors for air and carbon dioxide, respectively. From these the fluid density ρ_3 can be calculated

$$\rho_3 = \frac{PM_W}{zRT} \tag{7.2-1}$$

The formation volume factor at any temperature and pressure B_3, a specific molar volume, also follows

$$B_3 = z\frac{P_s}{P}\frac{T}{T_s} \tag{7.2-2}$$

In Eq. (7.2-2), T_s and P_s are the standard temperature and pressure, respectively. All fluids become more liquidlike, at a fixed temperature and pressure, as the molecular weight increases. The anomalous behavior of CO_2 is again manifest by comparing its density and formation volume factor to that of air. For CO_2 at 339 K (150°F) and 17 MPa (2,500 psia), $\rho_3 = 0.69$ g/cm^3, and $B_3 = 2.69$ dm^3/SCM. The values for air at the same temperature and pressure are $\rho_3 = 0.16$ g/cm^3, and $B_3 = 7.31$ dm^3/SCM. The CO_2 density is much closer to a typical light oil density than is the air density; hence CO_2 is much less prone to gravity segregation during a displacement than is air. Usually, gravity segregation in a CO_2 flood is more likely where the water saturation is high since CO_2 tends to segregate more from water than oil.

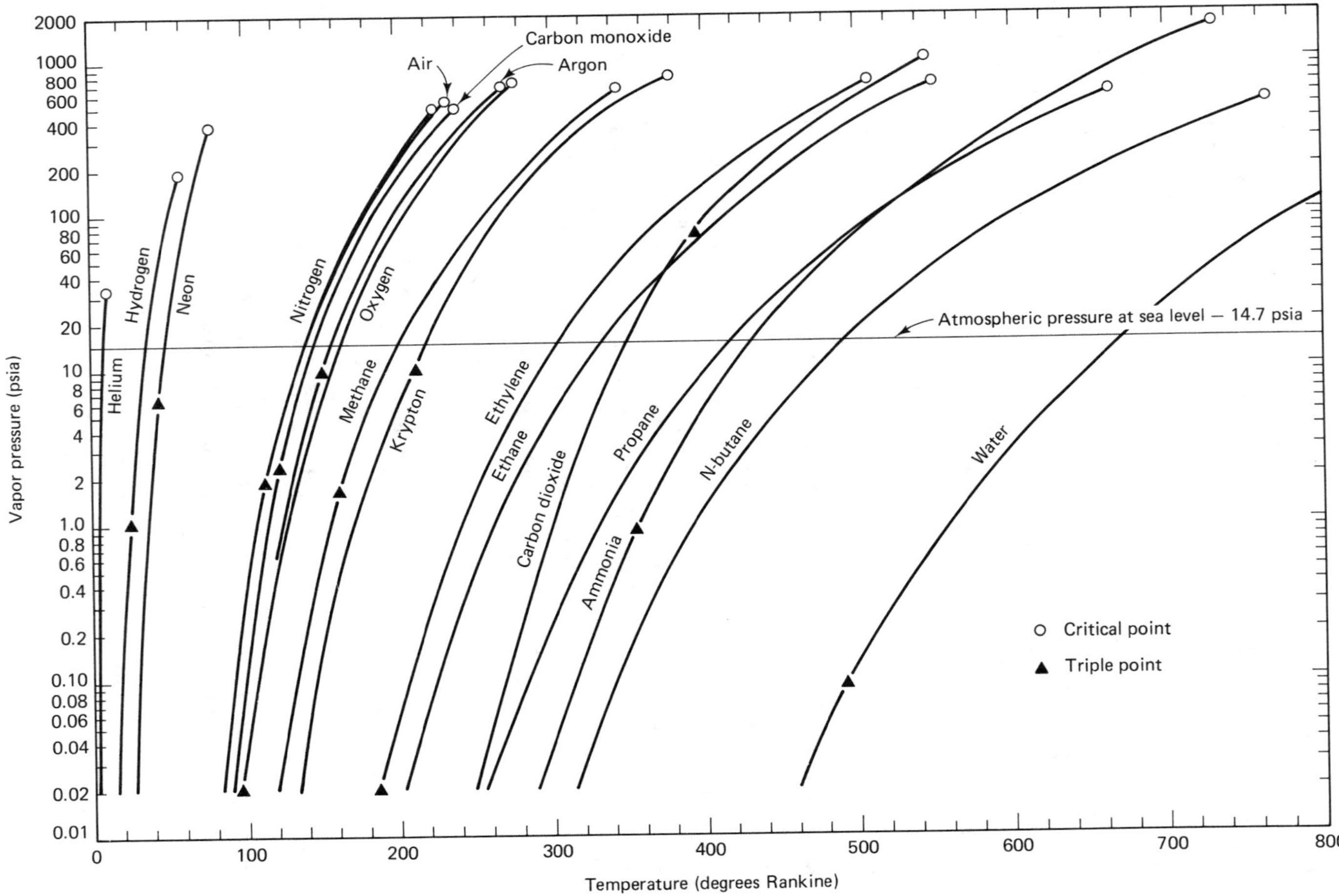

Figure 7-2 Vapor pressure curves for various substances (from Gibbs, 1971)

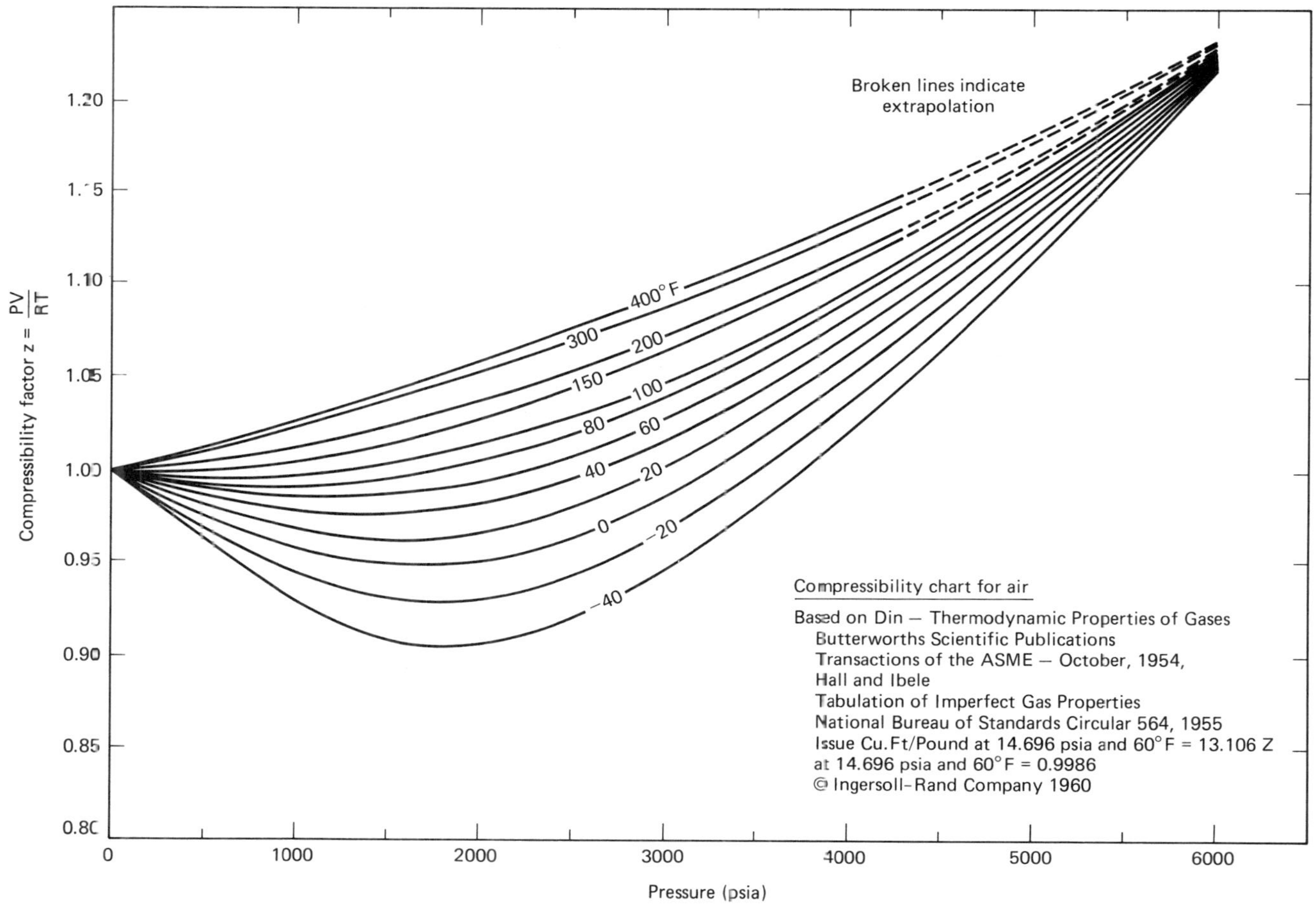

Figure 7-3 Compressibility chart for air (from Gibbs, 1971)

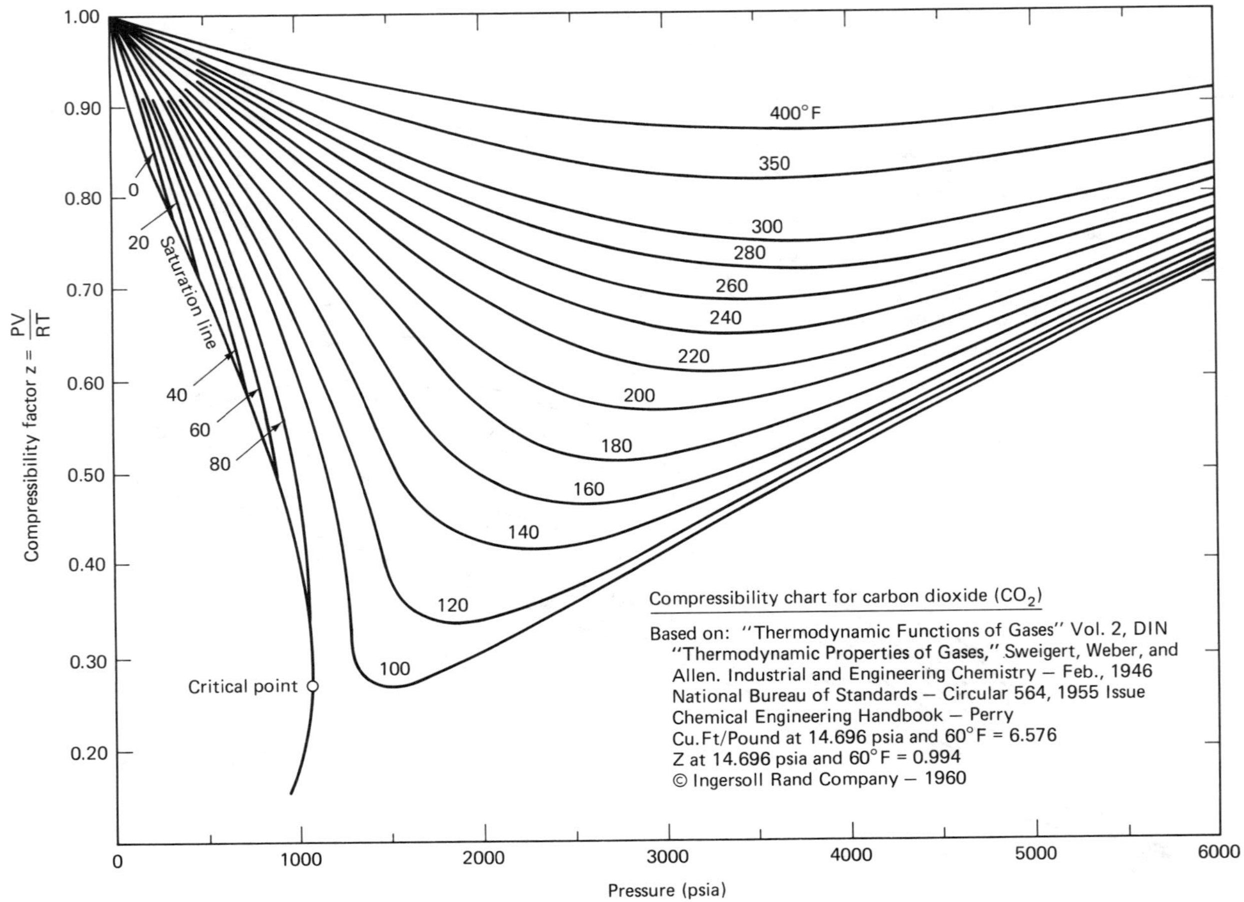

Figure 7-4 Compressibility chart for carbon dioxide (CO_2) (from Gibbs, 1971)

From the formation volume factors, 370 SCM of CO_2 is required to fill one cubic meter of reservoir volume, whereas only 140 SCM air is required at the same temperature and pressure. Thus about three times as many moles (recall that B_3 is a specific molar volume) of CO_2 are required to fill the same reservoir volume as air.

Figures 7-5 and 7-6 give the viscosities of a natural gas mixture and pure CO_2. Over the pressure and temperature range shown, which includes the conditions of interest in EOR, the viscosities of natural gas, and CH_4, air, flue gas, and N_2 are about the same. But the CO_2 viscosity is generally two or three times higher. Relative to a hydrocarbon liquid or water viscosity, the values are still low, so there should be no appreciable difference in the ease of injection of these solvents. However, the CO_2–crude-oil mobility ratio will be two or three times smaller than the other light solvents; hence volumetric sweep efficiency will generally be better for CO_2. (For the correlations for other solvents and solvent mixtures, see McCain, 1973; Reid et al., 1977; and Gas Processors Suppliers Association, 1973.)

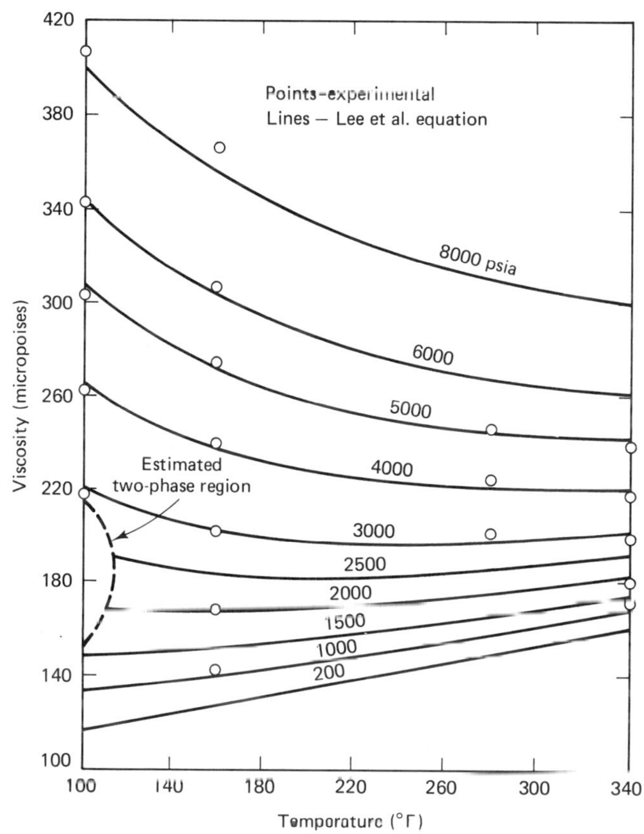

Figure 7-5 Viscosity of a natural gas sample (from Lee et al., 1966)

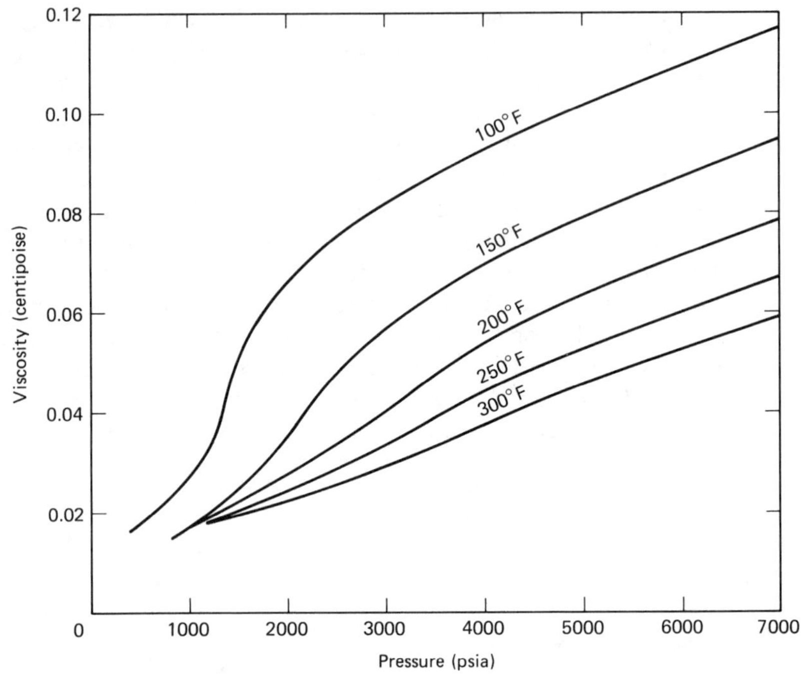

Figure 7-6 Viscosity of carbon dioxide as a function of pressure at various temperatures (from Goodrich, 1980)

7-3 SOLVENT–CRUDE-OIL PROPERTIES

In Secs. 4-1 and 4-2, we discussed general aspects of phase behavior for pure components and mixtures. In this section, we give specific features of solvent–crude phase behavior necessary for the development in later sections.

Figures 7-7 and 7-8 show pressure–composition (P-z) diagrams for two different solvent–crude systems. Recall that these diagrams are plots, at constant temperature, of pressure versus the overall mole percent of solvent in contact with a crude oil. These plots show the number and types of phases and the volume percent liquid. Figure 7-7 is for the recombined Wasson crude oil at 105°F (314 K), and Fig. 7-8 is for the Weeks Island "S" Sand crude at 225°F (381 K). Other diagrams are reported elsewhere (Turek et al., 1980; Orr and Jensen, 1982). The data in Figs. 7-7 and 7-8 represent behavior typical of low- and high-temperature systems. Recall that no water is present during the phase behavior measurements. For mixtures, the mole percent can represent both phase and overall concentrations.

The P-z diagrams have the same general form regardless of the temperature. The left vertical axis gives the phase behavior of the CO_2-free crude; thus the bubble point of the recombined Wasson crude at 314 K (105°F) is 6.81 MPa (1,000 psia) from Fig. 7-7. The right vertical axis similarly gives pure CO_2 properties, which

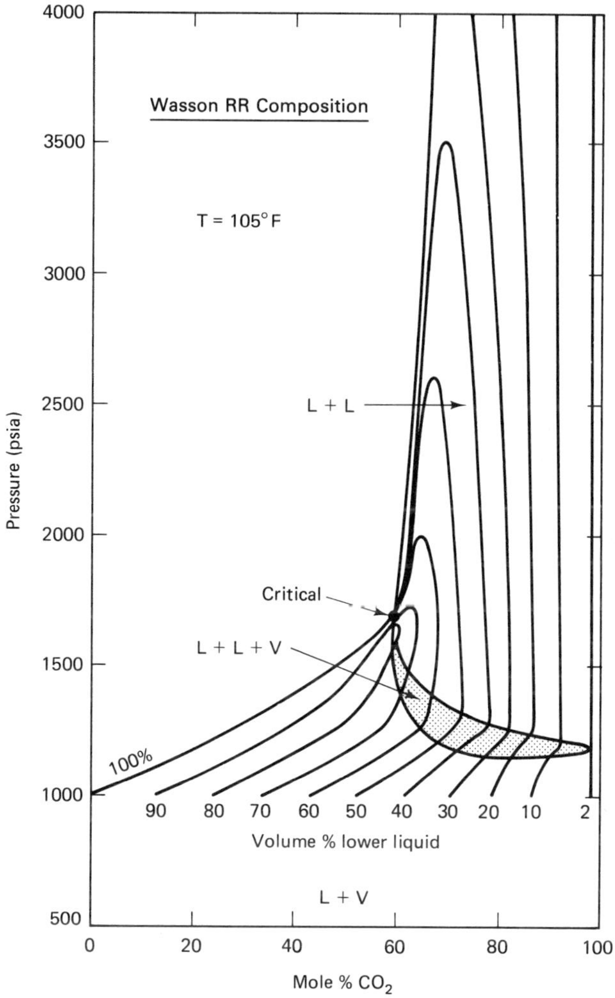

Figure 7-7 P-z diagram for recombined Wasson crude, CO_2 system (from Gardner et al., 1981)

will be a single-phase fluid for Figs. 7-7 and 7-8 since both are above the CO_2 critical temperature. At low pressures and for all CO_2 concentrations, except very near the right axis, the mixture is two-phase liquid and vapor. The liquid volume quality lines are also shown. At high pressures and low CO_2 concentration, the mixture is single phase. At about 60% CO_2, a critical point exists through which pass two single-phase boundaries. The CO_2 composition at this point is the critical composition for the fixed temperature and indicated pressure. The phase boundary line below the critical point is a bubble point curve, and that above is a dew point curve. Thus the upper left corner of the P-z diagram is a supercritical fluid region. The system could form a liquid phase as the light component increases in concentration at a constant pressure greater than the critical. This change is a type of retrograde behavior.

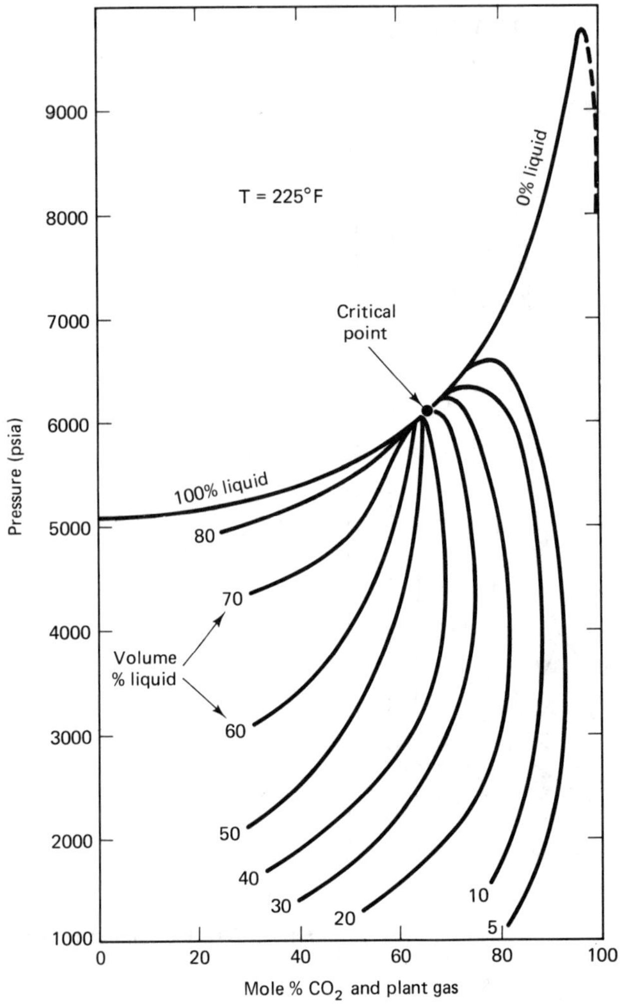

Figure 7-8 Phase envelope for Weeks Island "S" Sand crude and 95% CO_2, 5% plant gas at 225°F (from Perry, 1978)

Though few P-z diagrams are for solvents other than CO_2 in the literature, it appears, based on the N_2–crude-oil data in Fig 7-9, that the above qualitative character applies to other solvents as well. The critical pressure for the N_2 solvent mixture in Fig. 7-9 is much larger (off the scale) than either of the critical pressures of the CO_2 systems in Figs. 7-7 and 7-8.

The main difference between the low- and high-temperature phase behavior is the presence, in Fig. 7-7, of a small three-phase region just below and to the right of the critical point. These phases are two liquids—a light or upper phase and a heavy or lower phase—and a vapor phase. Such behavior has generally not been observed at high temperatures (Fig. 7-8) (Turek et al., 1980). Moreover, at low temperatures, a small amount of solid precipitate can exist over some composition and pressure ranges. The precipitate is composed mainly of asphaltenes, the *n*-heptane insoluble

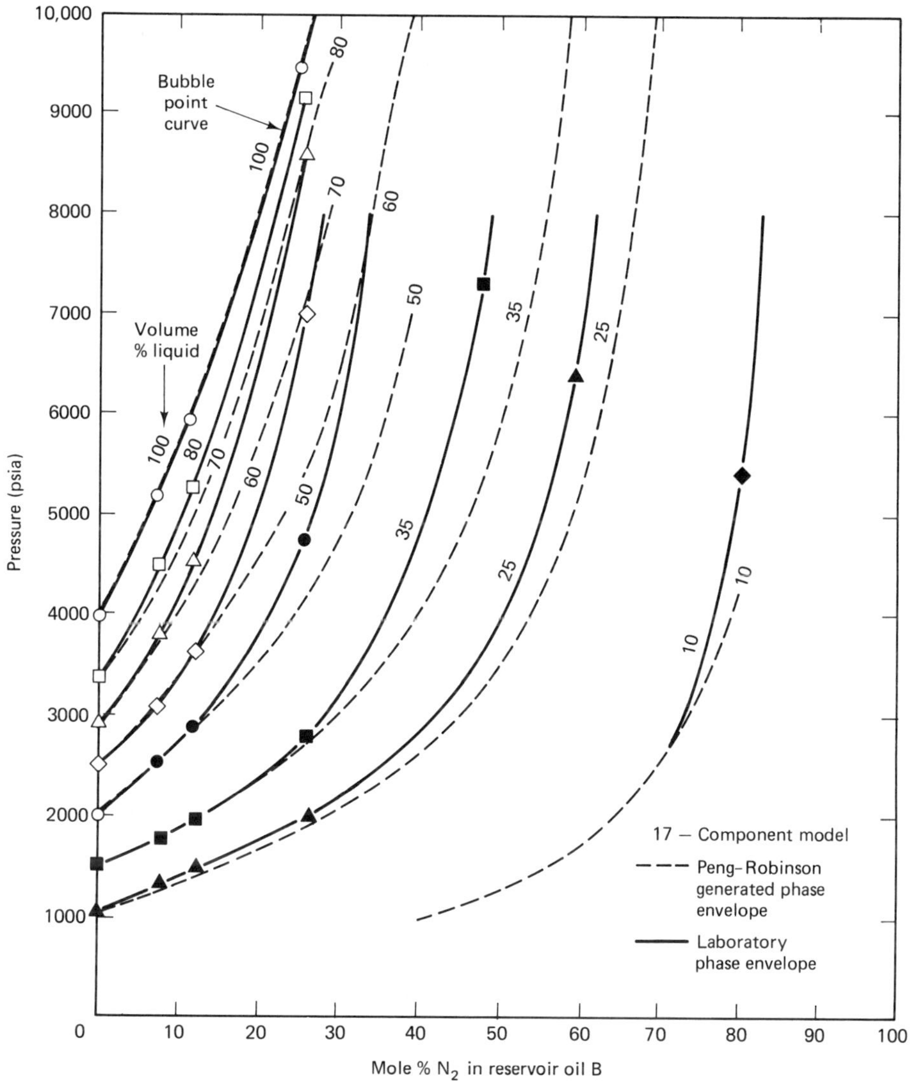

Figure 7-9 Pz diagram for reservoir fluid B-nitrogen system at 164°F (347 K) (from Hong, 1982)

fraction of crude oil (Hirshberg et al., 1982). The region of precipitate formation may overlap the three-phase region. This behavior offers a complication to the displacement process and may even present operational problems since the solid precipitate can cause formation plugging.

Consider now a displacement of a crude by a pure solvent in a permeable medium at some time before solvent breakthrough. The conditions at the injection end of the medium plot on the right vertical axis of the P-z diagram, and those at the production end plot on the left axis at some lower pressure. Conditions in the medium between these extremes are *not* represented on the P-z diagram since the relative amounts of each hydrocarbon component do not remain constant during a displacement, as they do in the PVT measurements of Figs. 7-7 through 7-9. Therefore, the diagrams are not particularly useful for displacement classification, which is based on the ternary diagrams we describe next. Still, one can see qualitatively from these diagrams that completely miscible displacements—those that are a single phase for all solvent concentrations—would require high reservoir pressures, in excess of 66.7 MPa (9,800 psia) for the data in Fig. 7-8.

Ternary diagrams are more useful in classifying solvent floods because they impart more compositional information than do P-z diagrams. Figures 7-10 through 7-12 show representations of these. On these diagrams the solvent–crude mixture is represented by three components; a light component on the top apex, an intermediate crude fraction on the right apex, and a heavy crude fraction on the left apex. The exact split between intermediate and heavy crude components is immaterial to the general features of the phase equilibria or to the miscibility classification. In Figs. 7-10 and 7-11, the split is between the C_6 and C_7 molecular weight fractions. There-

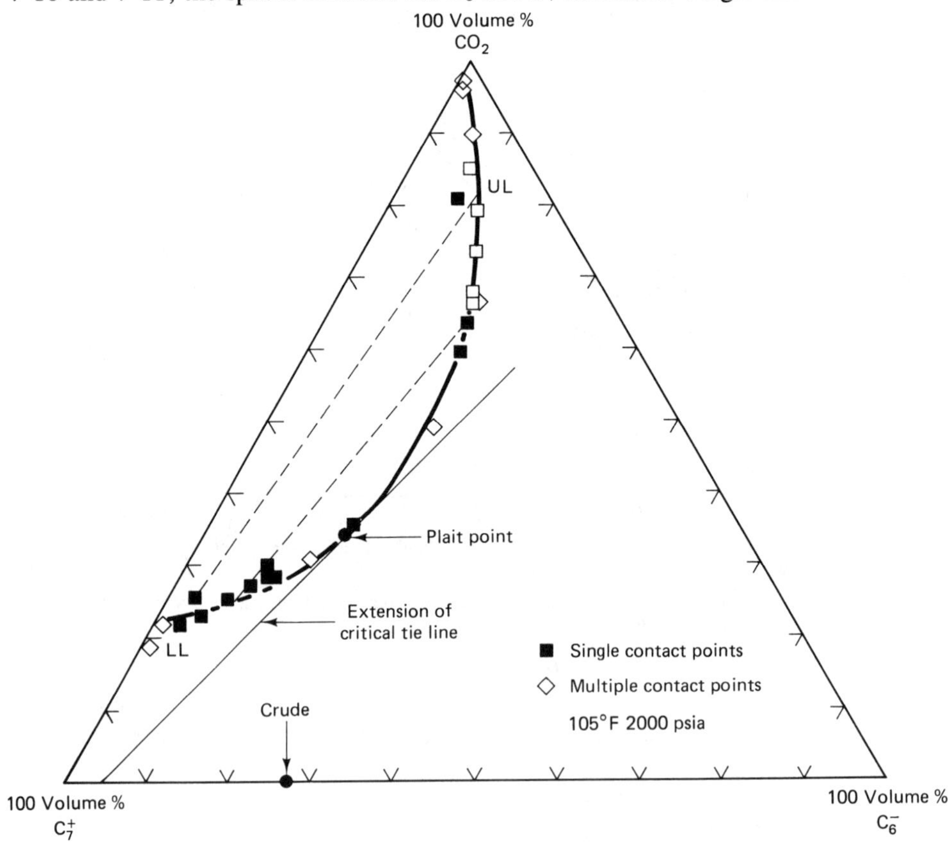

Figure 7-10 Ternary equilibria for CO_2-recombined Wasson crude mixture (Gardner et al., 1981)

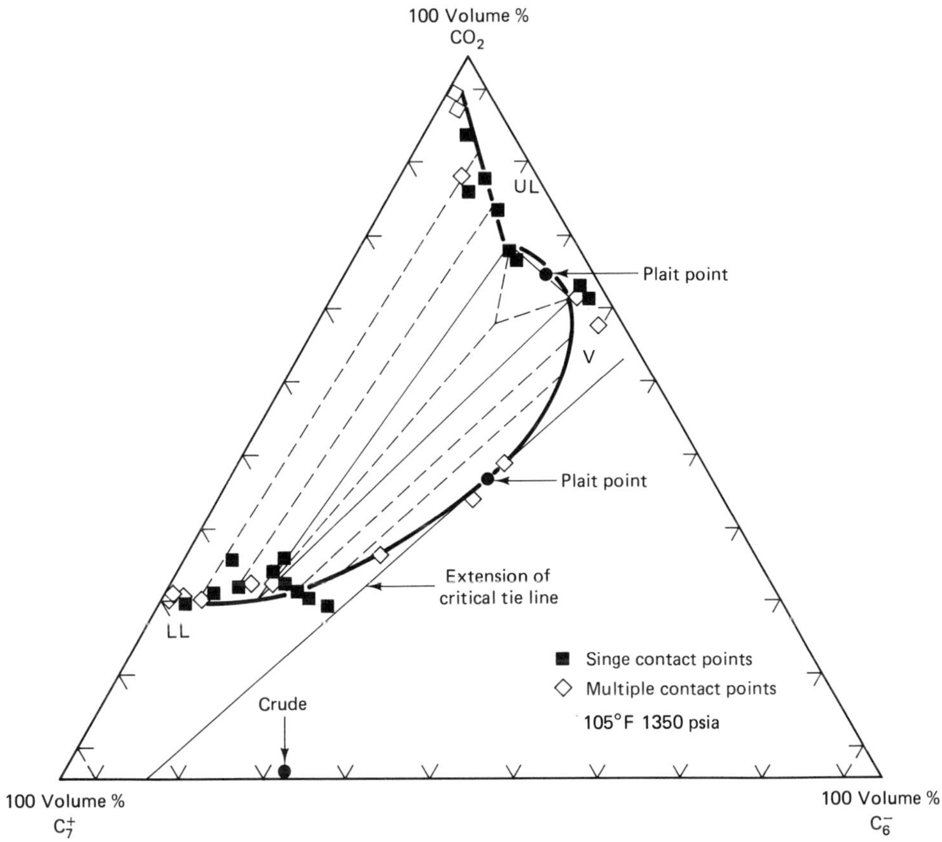

Figure 7-11 Ternary equilibria for CO_2-recombined Wasson crude system (from Gardner et al., 1981)

fore, none of the corners of these ternaries are pure components, hence the designation pseudocomponents. As before, no water is on the diagrams. In addition to those given here, ternary diagrams are in the literature in several other sources: for alcohol solvents (Holm and Csaszar, 1965; Taber and Meyer, 1965), for natural gas solvents (Rowe, 1967), for CO_2 (Metcalfe and Yarborough, 1978; Orr et al., 1981; Orr and Silva, 1982), for N_2 solvents (Ahmed et al., 1981), and for mixtures of CO_2, SO_2, and CH_4 (Sazegh, 1981).

A good example of CO_2–crude-oil equilibria is shown in Fig. 7-10 for the recombined Wasson crude (compare Figs. 7-10 and 7-11 with Fig. 7-8, the P-z diagram for the same mixture). In these solvent–crude systems, the phase equilibria is strongly dependent on reservoir temperature and pressure (recall that the ternary is at constant T and P). Typically, though, the pressure is larger than the cricondenbar of the light intermediate component pseudobinary; hence these two components are miscible in all proportions. The pressure is smaller than that of the light–heavy binary, and there is a region of limited miscibility or two-phase behavior along the

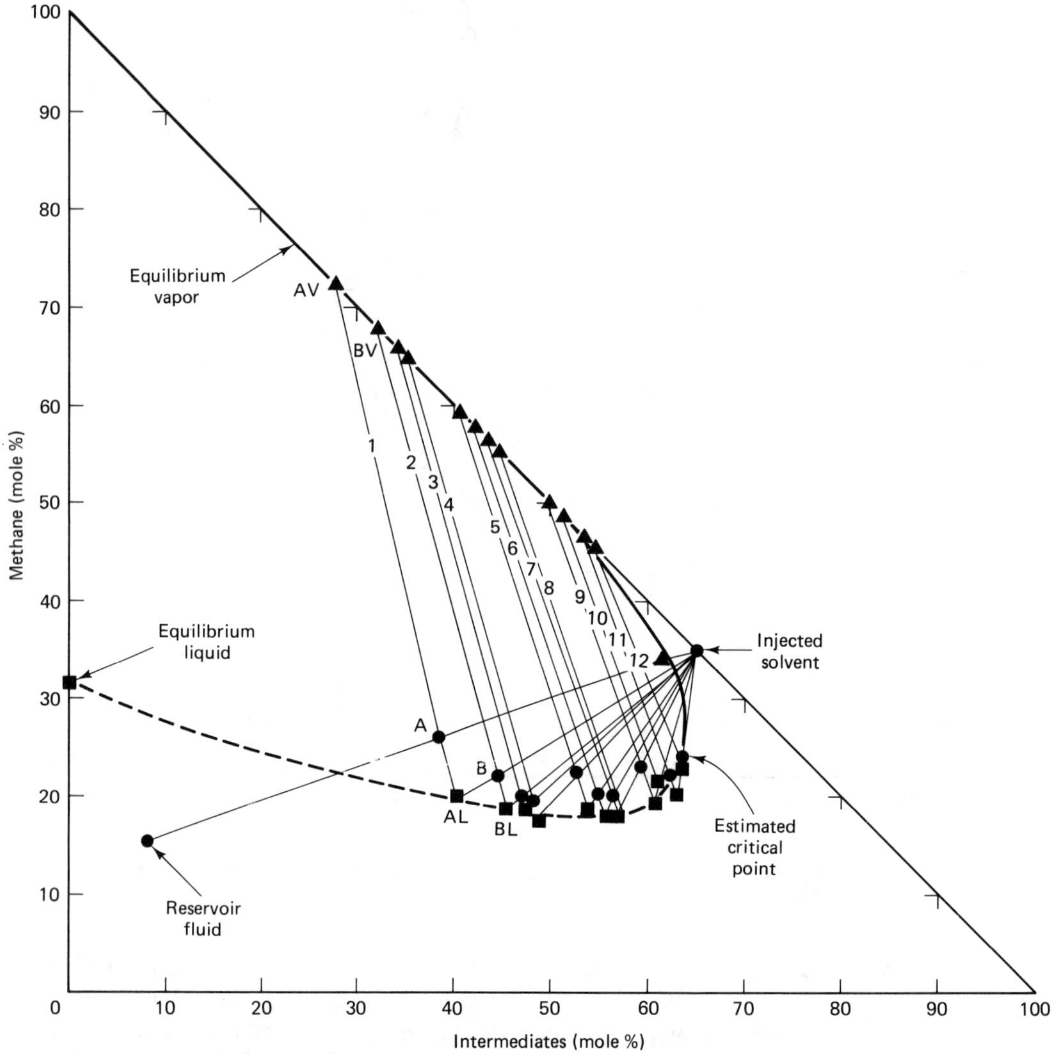

Figure 7-12 Methane–crude oil ternary phase behavior (from Benham et al., 1961)

light–heavy axis. This region of two-phase behavior extends into the interior of the ternary and is bounded by a binodal curve (see Sec. 4-3). Within the binodal curve, there are tie lines whose ends represent the composition of the equilibrium phases. These shrink to a plait point where the properties of the two phases are indistinguishable. The plait point is the critical mixture at this temperature and pressure.

Of great importance in what follows is the critical tie line, the fictitious tie line tangent to the binodal curve at the plait point. The critical tie line is the limiting case of the actual tie lines as the plait point is approached. As pressure increases, the two-phase region shrinks—that is, light–heavy miscibility increases. No general statement is possible about the effect of temperature though the two-phase region generally increases with increasing temperature. For low pressure and low temperature, a three-phase region can intrude into the two-phase region (Fig. 7-11).

These general characteristics apply for solvents other than CO_2 (Fig. 7-12). The composition of the reservoir crude can be placed on the ternary, as can the composition of the solvent. In doing this, we are neglecting the pressure change that is, of course, an essential ingredient in making the fluids flow in the reservoir. Even with this approximation, all compositions in the solvent crude mixing zone do not lie on a straight line connecting the initial and injected. This is because the composition changes are affected by the phase behavior. In fact, these changes are the basis for the classification of solvent displacements that we give in the next few paragraphs (Hutchinson and Braun, 1961).

We represent a one-dimensional displacement of a crude by a solvent on the schematic ternary diagram in Fig. 7-13. The crude is in the interior of the ternary, indicating some of the light component is present initially in the crude. If a straight-line *dilution* path between the solvent and the crude does not intersect the two-phase

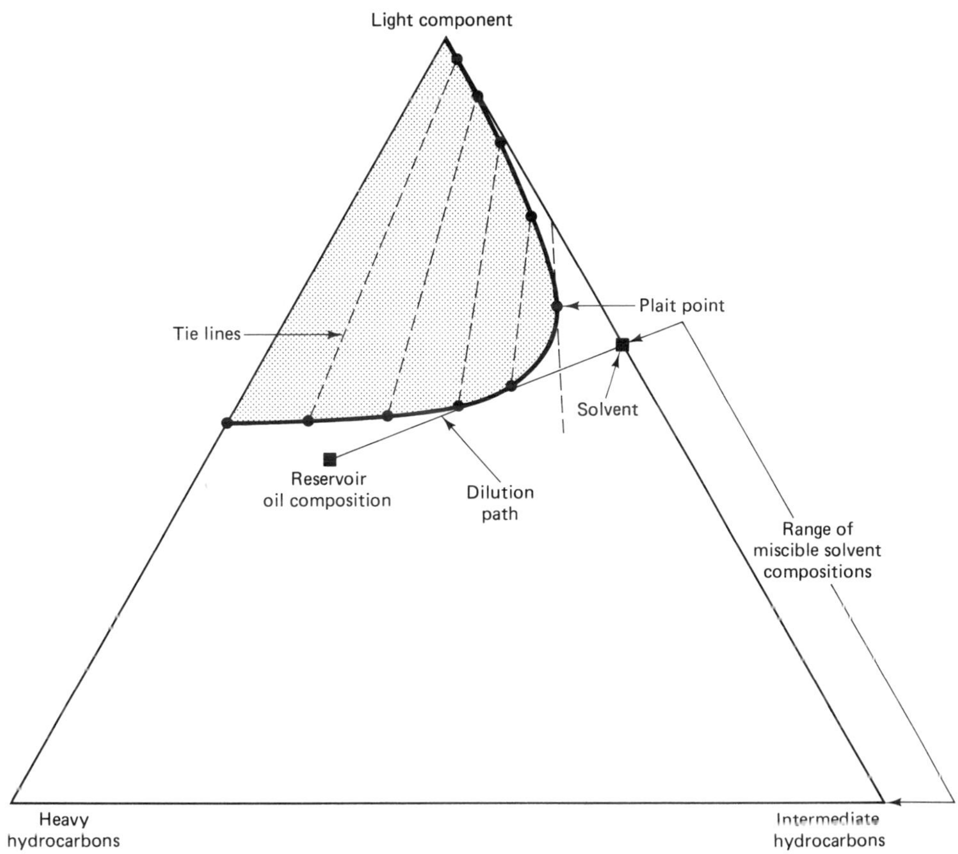

Figure 7-13 Schematic of the first-contact miscible process

region, the displacement will consist of a single hydrocarbon phase that changes in composition from crude to undiluted solvent through the solvent oil mixing zone. The dilution path is linear (see Sec. 7-6) since the only mechanism for mixing is dispersion, there being no water or fractional flow effects associated with the single hydrocarbon phase. A displacement that occurs entirely within one hydrocarbon phase is *first-contact miscible*. There is a range of solvent compositions that will be first-contact miscible with the crude at this temperature and pressure.

Suppose the solvent consists entirely of the light component (Fig. 7-14). The displacement is not first-contact miscible since the dilution path passes through the two-phase region. Imagine a series of well-mixed cells that represent the permeable medium in a one-dimensional displacement. The first cell initially contains crude to which we add an amount of solvent so that the overall composition is given by M_1. The mixture will split into two phases, a gas G_1 and a liquid L_1, determined by the equilibrium tie lines. The gas G_1 will have a much higher mobility than L_1, and this phase moves preferentially into the second mixing cell to form mixture M_2. Liquid L_1 remains behind to mix with more pure solvent. In the second cell mixture, M_2 splits into gas G_2 and liquid L_2, G_2 flows into the third cell to form mixture M_3, and

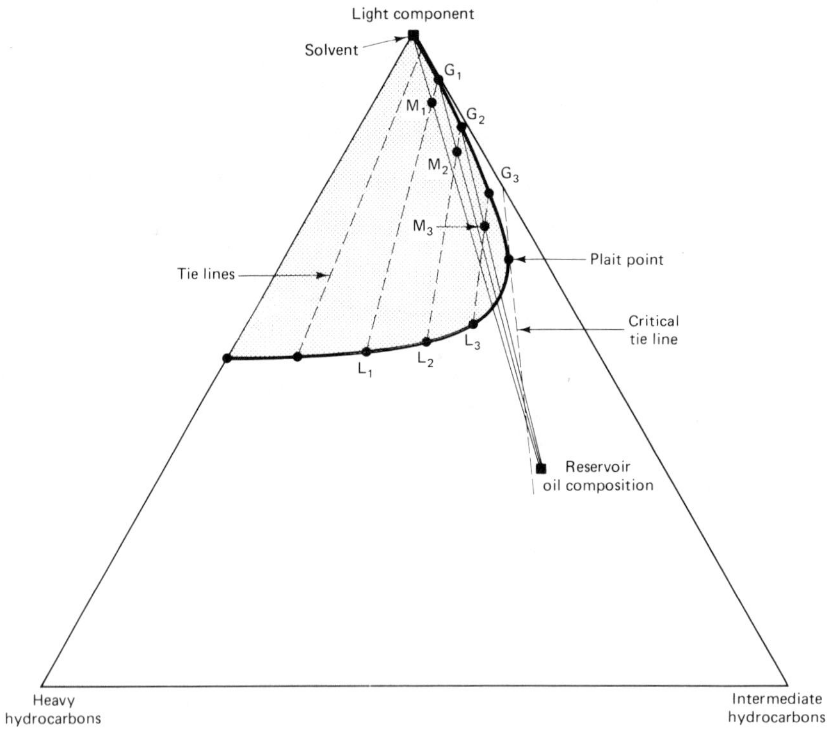

Figure 7-14 Schematic of the vaporizing gas drive process (adapted from Stalkup, 1983)

so forth. At some cell beyond the third (for this diagram), the gas phase will no longer form two phases on mixing with the crude. From this point forward, all compositions in the displacement will be on a straight dilution path between the crude and a point tangent to the binodal curve. The displacement will be first-contact miscible with a solvent composition given by the point of tangency. The process has *developed miscibility* since the solvent has been enriched in intermediate components to be miscible with the crude. Since the intermediate components are vaporized from the crude, the process is a *vaporizing gas drive*. Miscibility will develop in this process as long as the injected solvent and crude are on opposite sides of the critical tie line.

Suppose the crude and solvent compositions are again on opposite sides of the critical tie line but reversed from the vaporizing gas drive (Fig. 7-15). In the first mixing cell, the overall composition M_1 splits into gas G_1 and liquid L_1. Gas G_1 moves on to the next mixing cell as before, and liquid L_1 mixes with fresh solvent to form mixture M_2. Liquid L_2 mixes with fresh solvent, and so forth. Thus in the first mixing cell, this mixing process will ultimately result in a single-phase mixture.

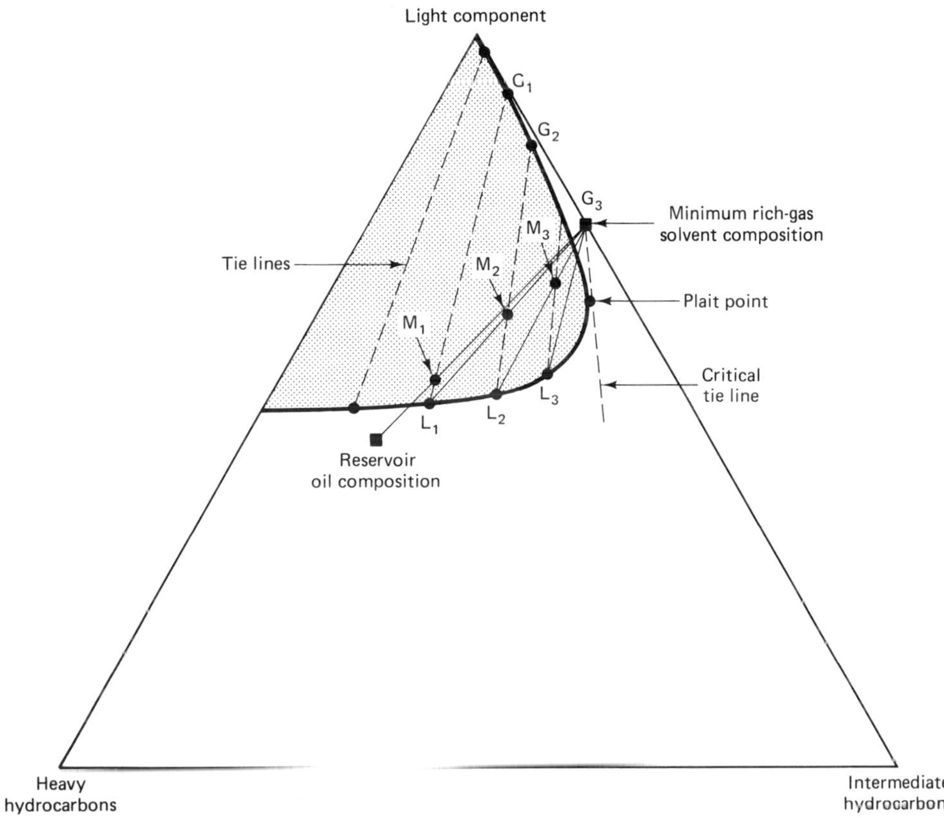

Figure 7-15 Schematic of the rich-gas drive process (adapted from Stalkup, 1983)

Since the gas phase has already passed through the first cell, the miscibility now develops at the rear of the solvent–crude mixing zone as a consequence of the enrichment of the liquid phase in intermediate components there. The front of the mixing zone is a region of immiscible flow owing to the continual contacting of the gas phases G_1, G_2, and so on, with the crude (this is also true at the rear of the mixing zone in the vaporizing gas drive). The process in Fig. 7-15 is the *rich gas drive* process since intermediates were added to enrich the injected solvent. Since these intermediates condense into the liquid phase, the process is sometimes called a *condensing gas drive*. Figure 7-12 shows that more than twelve contacts are necessary to develop miscibility in an actual system.

Figure 7-16 shows a schematic of an immiscible displacement. The crude and solvent are in single-phase regions, but both are on the two-phase side of the critical tie line. Now the initial mixture M_1 in the first mixing cell will form gas G_1, which will flow forward to form mixture M_2, and so forth. This gas is being enriched in intermediate components at the leading edge (forward contacts) of the solvent–crude mixing zone as in a vaporizing gas drive. But the enrichment cannot proceed beyond the gas-phase composition given by the tie line whose extension passes through the

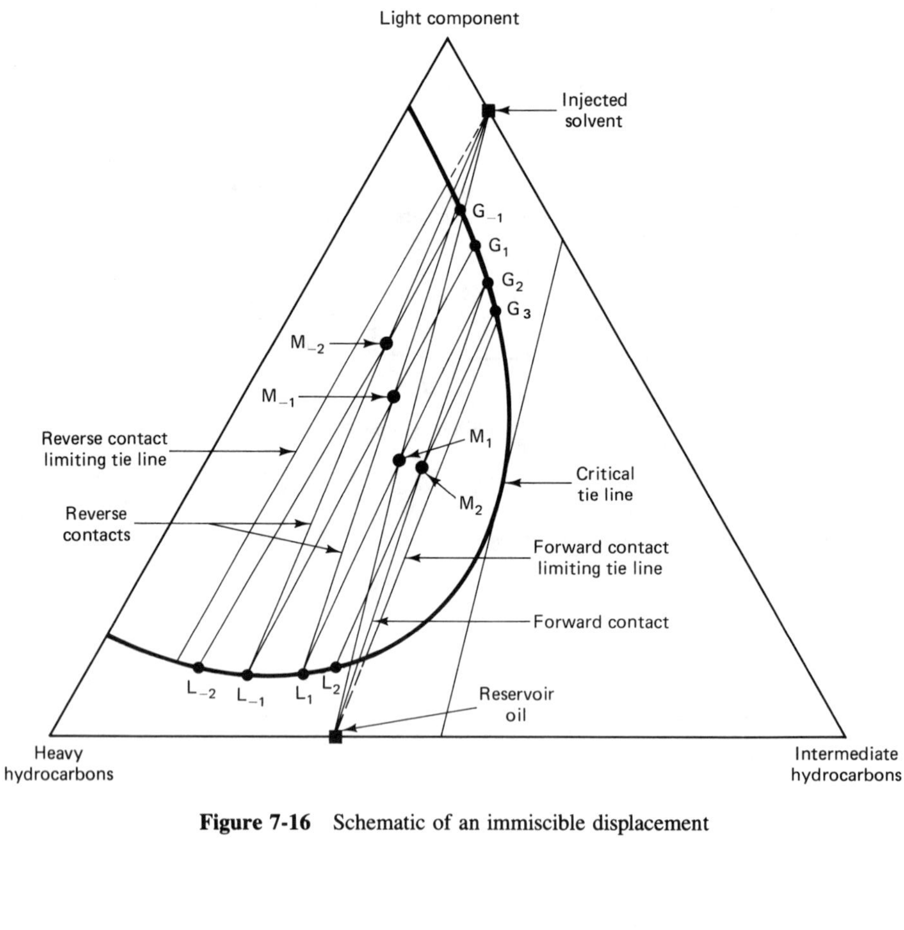

Figure 7-16 Schematic of an immiscible displacement

composition crude. At the forward contacts, there will be an immiscible displacement of the crude by a mixture on the limiting tie line. Back at the first mixing cell, liquid L_1 mixes with solvent to form mixture M_{-1}, just as in the condensing gas drive. The displacement is immiscible here since a single-phase solvent is displacing a two-phase mixture. The liquid phase becomes progressively stripped of intermediates (L_{-1}, L_{-2} and so on) until it reaches another limiting tie line. The displacement is entirely immiscible, then, at both the forward and reverse contacts. The intermediate components are in a gas phase near the production end of the permeable medium, and in a liquid phase at the injection end. An immiscible flood entirely devoid of injected intermediates is a *dry* gas flood.

Figure 7-17 summarizes the classification of solvent displacements. A dilution path (I_2-J_3) that does not pass through the two-phase region is a first-contact miscible displacement. A dilution path entirely on the two-phase side of the critical tie line constitutes immiscible displacement (I_1-J_1). When initial and injected compositions are on opposite sides of the critical tie line, the displacement is either a vaporizing

Dilution path	Type
I_1-J_1	Immiscible
I_1-J_2	Multiple contact (developed) Miscibility (rich gas)
I_2-J_1	Multiple contact (developed) Miscibility (vaporizing gas)
I_2-J_3	First-contact miscible

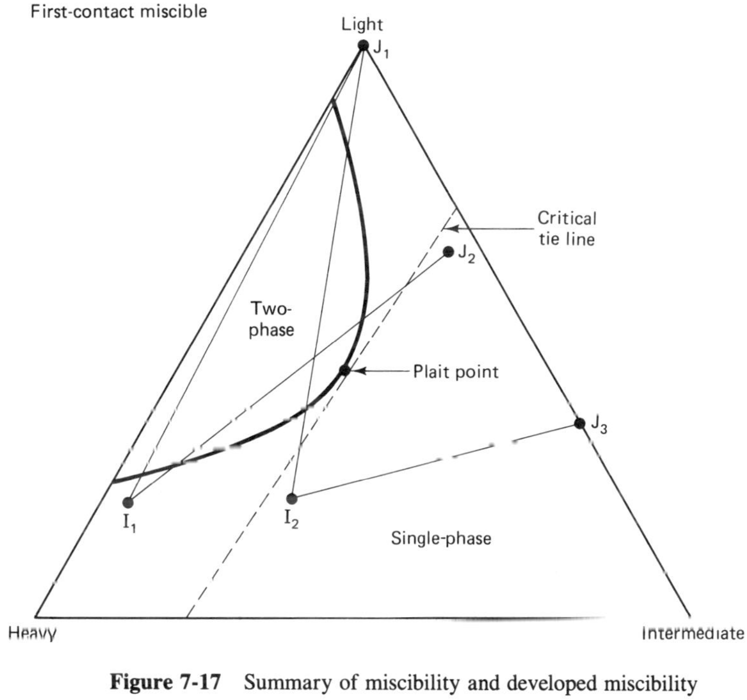

Figure 7-17 Summary of miscibility and developed miscibility

gas drive (I_2-J_1) or a condensing gas drive (I_1-J_2). The last two cases are developed or multiple-contact miscible displacement.

At the conditions shown in Fig. 7-10, CO_2 displaces oil as a vaporizing gas drive. At comparable conditions (Figs. 7-12 and 7-18), N_2 and CH_4 are usually an immiscible solvent. The CH_4 in Fig. 7-12 can be converted into a condensing gas drive by adding about 35 mole % intermediates.

The solvent flooding classifications given here are corroborated by simple wave theory (see Sec. 7-7) and experimental results. Figure 7-19 shows the effluent history of three CO_2 floods in a Berea core. The oil in this displacement was a mixture of 25 mole % C_1, 30 mole % C_4, and 45 mole % C_{10}. The three runs were at 10.2 MPa (1,500 psia) (run 4), an immiscible displacement; 12.9 MPa (1,900 psia) (run 5), a first-contact miscible displacement; and 11.6 MPa (1,700 psia) (run 6), a vaporizing gas drive. The temperature was 344 K (160°F) for all runs.

The effluent histories in Fig. 7-19 are plots of C_1, C_4, and C_{10} concentrations, normalized by their initial values, versus the hydrocarbon pore volumes (HCPV) of CO_2 injected (see Table 5-1). If the dilution path between the oil and the solvent were a straight line, the normalized concentration of all displaced components would be identical. They are identical in the first-contact miscible run 5. The vaporizing gas drive run 6 shows that the normalized concentrations of the heavy component C_{10} declines slightly before the C_4 curve declines (Fig. 7-19b). (The run 6 composi-

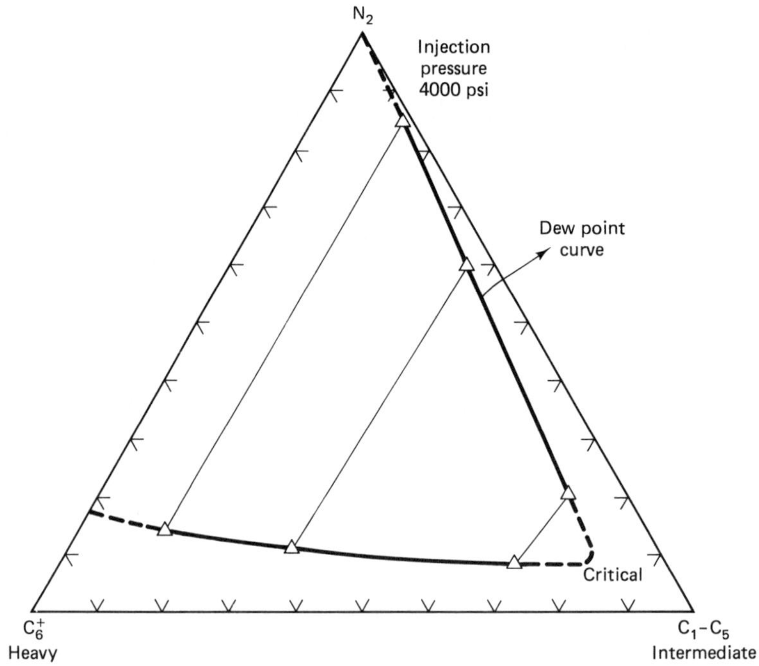

Figure 7-18 Ternary equilibria for N_2–crude-oil mixture (from Ahmed et al., 1981)

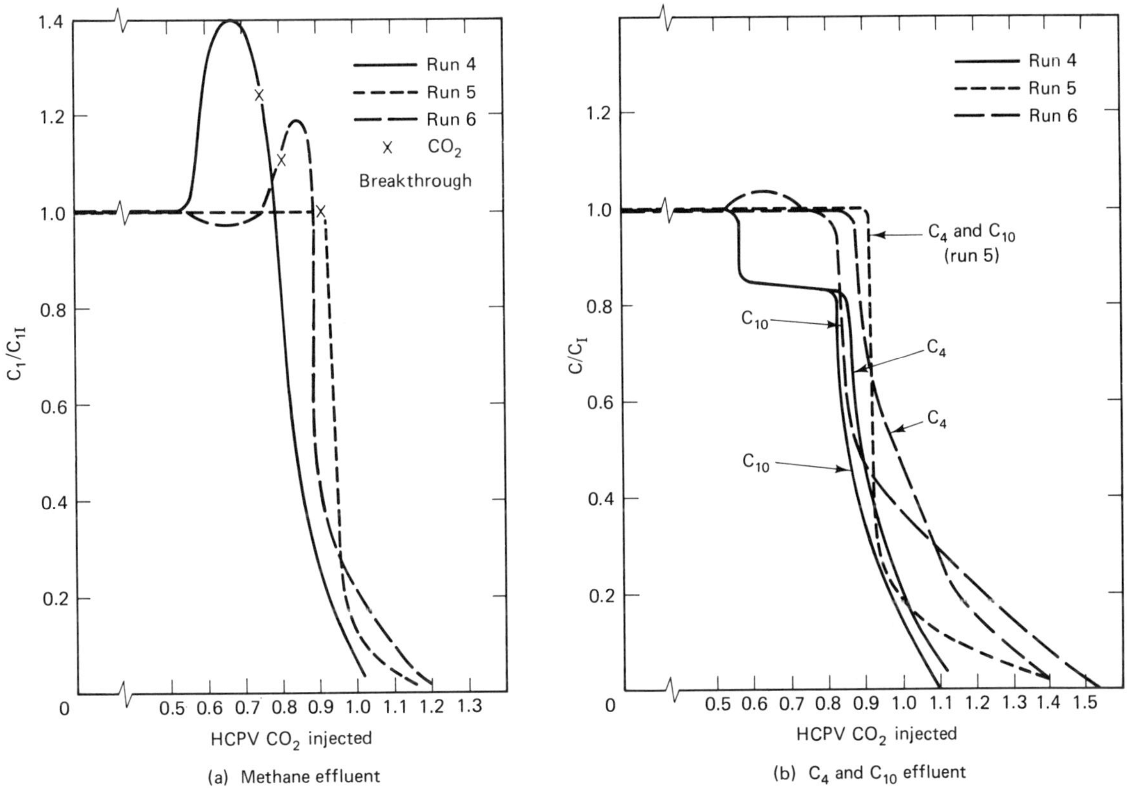

Figure 7-19 Effluent histories from laboratory displacement run 4 immiscible, run 5 first-contact miscible, run 6 multiple-contact miscible (from Metcalfe and Yarborough, 1978)

tion at 0.9 HCPV is relatively rich in C_4.) In addition, the light components C_1 go through a maximum at about the same point (Fig. 7-19a). The C_1 maximum is even more pronounced in the immiscible displacement probably because the fluids can now be saturated with respect to C_1. A similar effect should occur in a condensing gas drive process though the enrichment will now occur at the rear of the mixing zone.

The immiscible displacement and vaporizing gas drive process are similar; however, the oil recovery (displacement efficiency) in the immiscible run (80%) was considerably smaller than that for either the first-contact (97%) or the vaporizing gas drive run (90%). Developed miscibility displacements can give oil recoveries approaching first-contact miscible displacements; immiscible processes are usually much poorer.

Immiscible displacements have merit since pressure requirements are not large, the solvents are usually less expensive, and they can recover some oil. The principal recovery mechanisms for immiscible solvents are (1) a limited amount of vaporization and extraction, (2) oil viscosity reduction, (3) oil swelling, (4) solution gas drive during pressure decline, and (5) interfacial tension lowering. All immiscible dis-

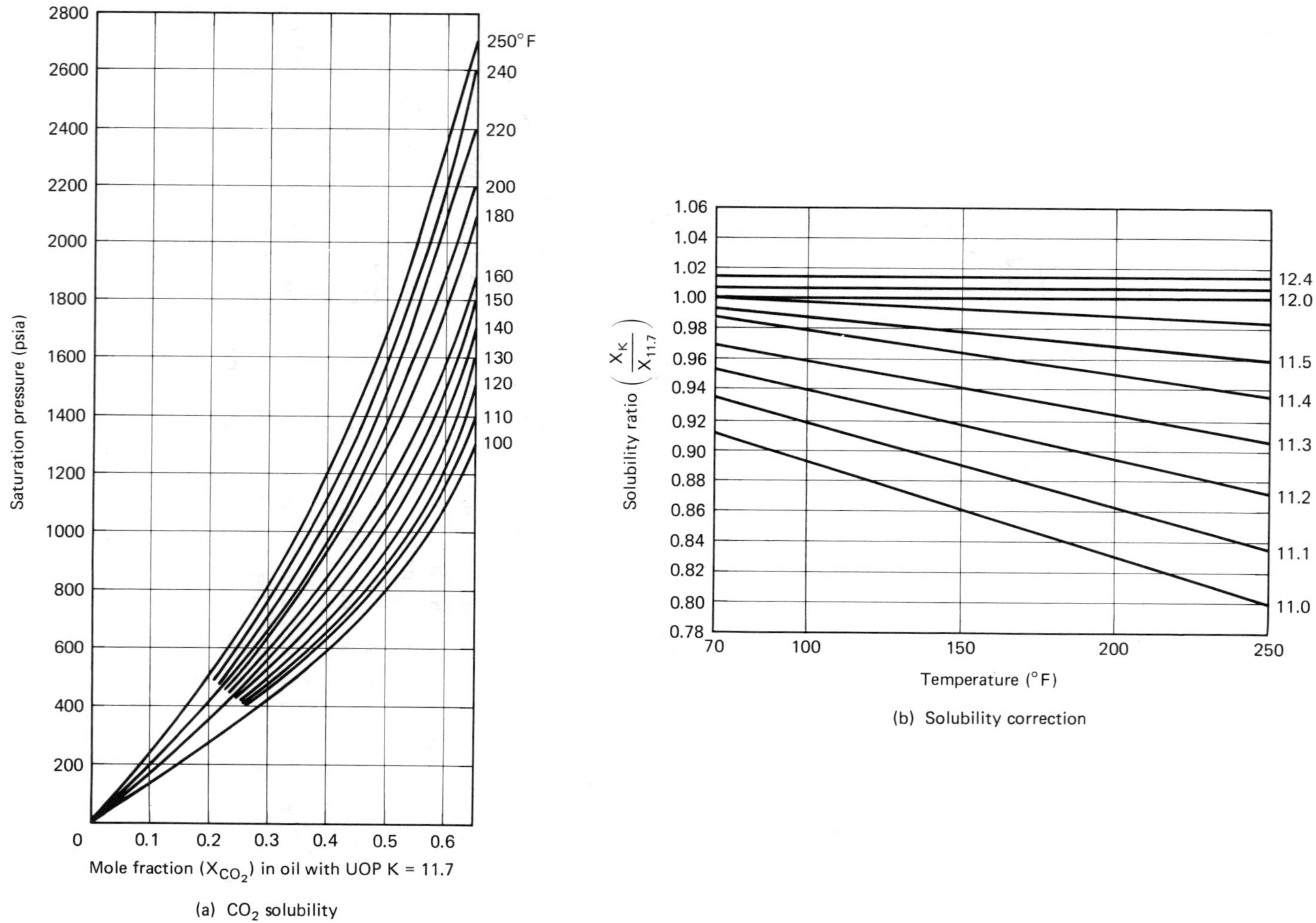

Figure 7-20 Solubility (mole fraction) of carbon dioxide in oils as a function of UOP number (from Simon and Graue, 1965)

placements recover oil in this manner though the data showing these effects are most complete on CO_2-immiscible displacements (Simon and Graue, 1965).

Figures 7-20 through 7-22 show experimental data that emphasize immiscible recovery mechanisms 1–3. Figure 7-20(a) shows the solubility of CO_2 in oil versus temperature and saturation pressure for a crude with a Universal Oil Characteriza-

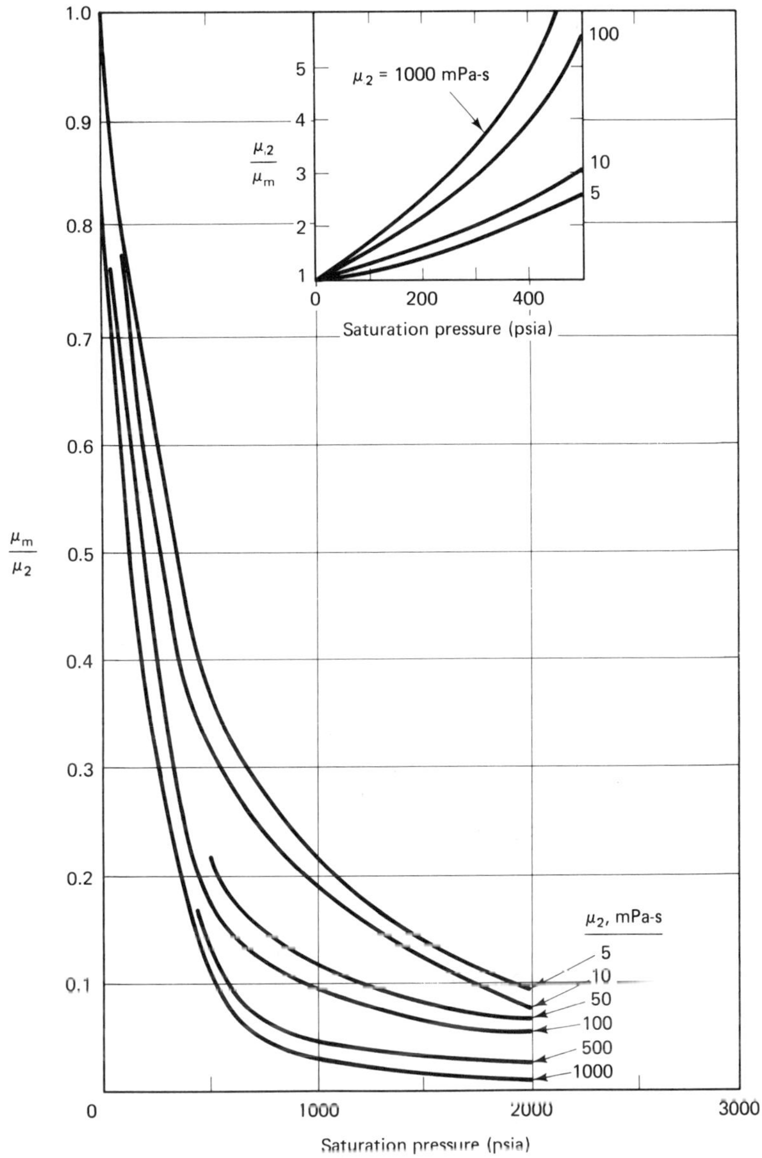

Figure 7-21 Viscosity correlation charts for carbon-dioxide–oil mixtures (from Simon and Graue, 1965)

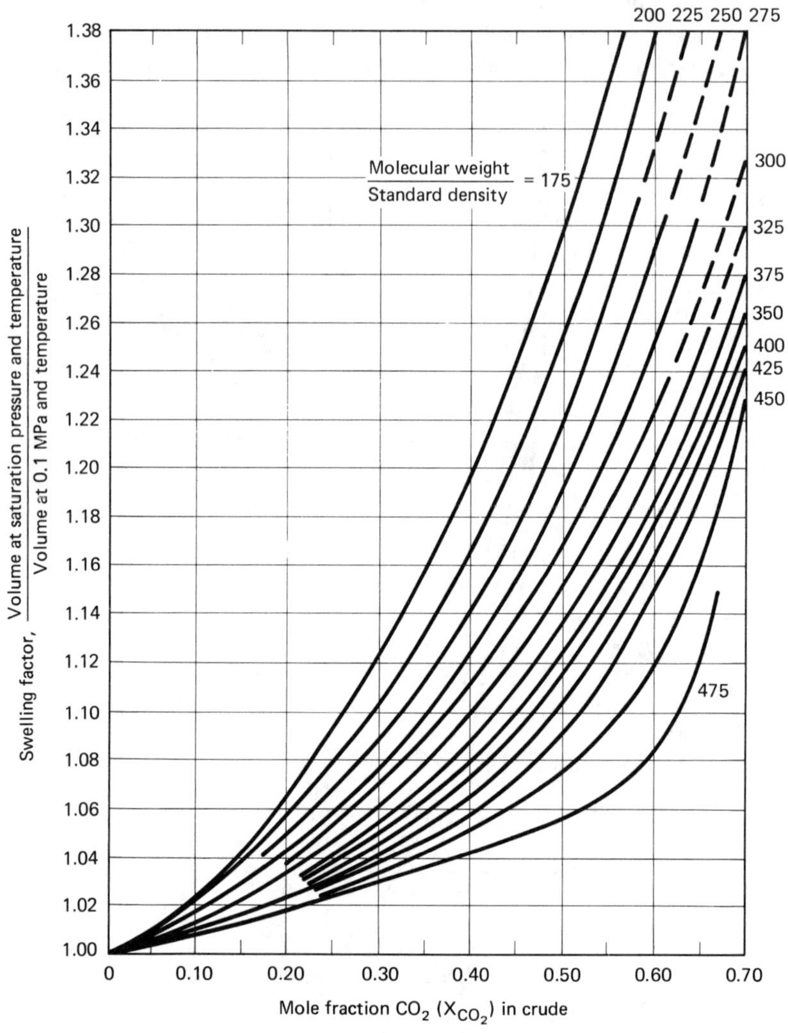

Figure 7-22 Swelling of oil as a function of mole fraction of dissolved carbon dioxide (from Simon and Graue, 1965)

tion factor (K) of 11.7. This factor is the ratio of the cube root of the average boiling point in degrees R to the specific gravity. It can be related to API gravity and viscosity (Watson et al., 1935). The saturation pressure is the bubble point pressure; hence Fig. 7-20(a) is giving the maximum solubility of CO_2 at the indicated temperature and pressure. Figure 7-20(b) corrects the solubility data to other characterization factors. Figure 7-21 gives the viscosity ratio of a CO_2-swollen crude (μ_m in this figure) to the CO_2-free crude (μ_o) as a function of pressure. For moderate saturation pressures, the viscosity reduction is pronounced, particularly for large crude viscosities.

Figure 7-22 illustrates the oil swelling mechanism by giving crude swelling factors correlated with ratios of molecular weight to standard density (g/cm^3). Similar data on the swelling of crude by N_2 are given by Vogel and Yarborough (1980).

Figures 7-20 through 7-22 are complementary. Let's estimate the CO_2 solubility, oil viscosity reduction, and swelling factor for a crude oil at 389 K (150°F) and 8.2 MPa (1,200 psia). Recall that we are calculating the properties of a liquid hydrocarbon phase immiscible with CO_2. Therefore, the overall CO_2 mole fraction must be large enough to be in the two-phase region of the ternary diagram. The relevant physical properties of the crude are as follows: molecular weight = 130, UOP characterization factor K = 11.8, specific gravity = 0.70, normal boiling point = 311 K (100°F), and viscosity = 5 mPa-s. This gives a CO_2 solubility of 55 mole % from Fig. 7-20. This solubility causes the oil viscosity to decrease to 1 mPa-s from Fig. 7-21, and the oil to swell by about 33% from Fig. 7-22. (For additional data on the properties of crude containing immiscible solvents, see Holm, 1961; de Nevers, 1964; Holm and Josendal, 1974; and Tumasyn et al., 1969.)

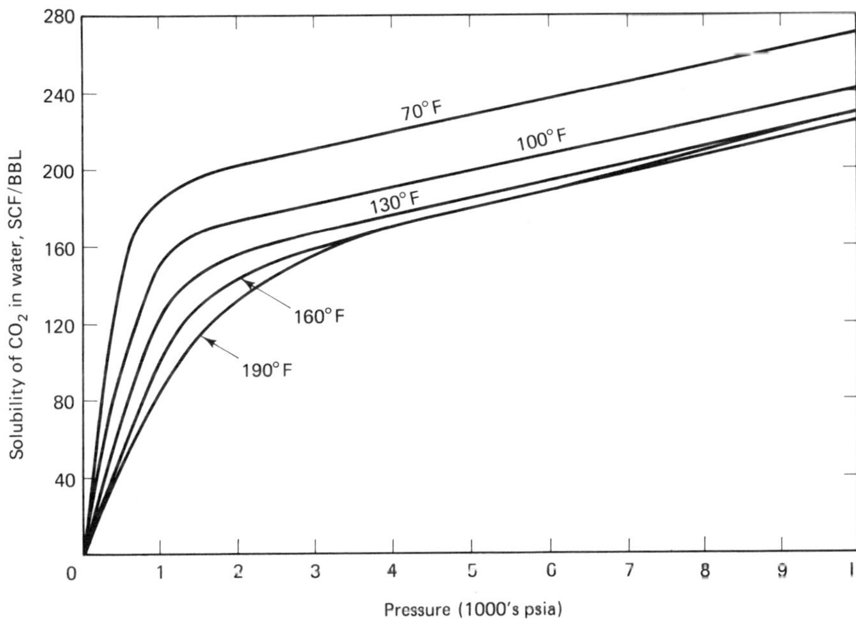

Figure 7-23 Solubility of carbon dioxide in water (from Crawford et al., 1963)

7-4 SOLVENT–WATER PROPERTIES

The solubility of CO_2 in water is a function of temperature, pressure, and water salinity (McRee, 1977). Figure 7-23 shows this solubility as a solution gas–water ratio. The data in Fig. 7-23 give the maximum CO_2 solubility at the indicated temperature and pressure; hence the horizontal axis is actually saturation pressure. The data are entirely equivalent to the data in Fig. 7-21(a) for CO_2–oil mixtures. The solution gas–water ratio may be readily converted into mole fraction.

Carbon dioxide is the only solvent with appreciable solubility in water over EOR temperature and pressure ranges (Culberson and McKetta, 1951). The CO_2 increases the viscosity of water slightly (Tumasyn et al., 1969) and decreases the density (Parkinson and de Nevers, 1969). This density change has been shown (Welch, 1982) to be less than that predicted by ideal solution theory. Neither the change in viscosity nor the change in density is likely to affect oil recovery very much.

7-5 SOLVENT PHASE BEHAVIOR EXPERIMENTS

Solvent phase behavior does not solely determine the character of a solvent flood, but it is of such fundamental importance that we devote a section to some of the common experiments used to measure phase behavior. This discussion leads naturally to the most frequently reported characteristic of solvent phase behavior—minimum miscibility pressure.

Single Contact

In a single-contact experiment, a known amount of solvent is charged into a transparent pressure cell containing a known amount of crude oil. After equilibrium is established at the desired temperature and pressure, a small amount of each phase is withdrawn. The phase compositions represent the ends of an equilibrium tie line. Only the composition of one phase need be measured since the composition of the other phase can be calculated from material balance. Single-contact experiments are useful for measuring P-z diagrams since the pressure can be changed, at fixed overall composition, by changing the cell volume. If the experiment is repeated for various amounts of solvent, the single-contact experiment traces a dilution path on a ternary diagram between the solvent and crude.

Multiple Contact

The multiple-contact experiment duplicates the process described in Sec. 7-3 under miscible process classification. In it (Fig. 7-24), known amounts of solvent and crude are charged to a transparent pressure cell as in the single-contact experiment, but after equilibration, the upper phase is decanted and mixed in a second cell with fresh crude. The lower phase in the cell is similarly mixed with fresh solvent. The upper phase is repeatedly decanted in this manner to simulate, discretely, the mixing that would take place at the forward contacts of the solvent–crude mixing zone. The successive mixings with the lower phase are the reverse contacts. All contacts are at fixed temperature and pressure.

From Fig. 7-24, the multiple-contact experiment for Fig. 7-10, the solvent en-

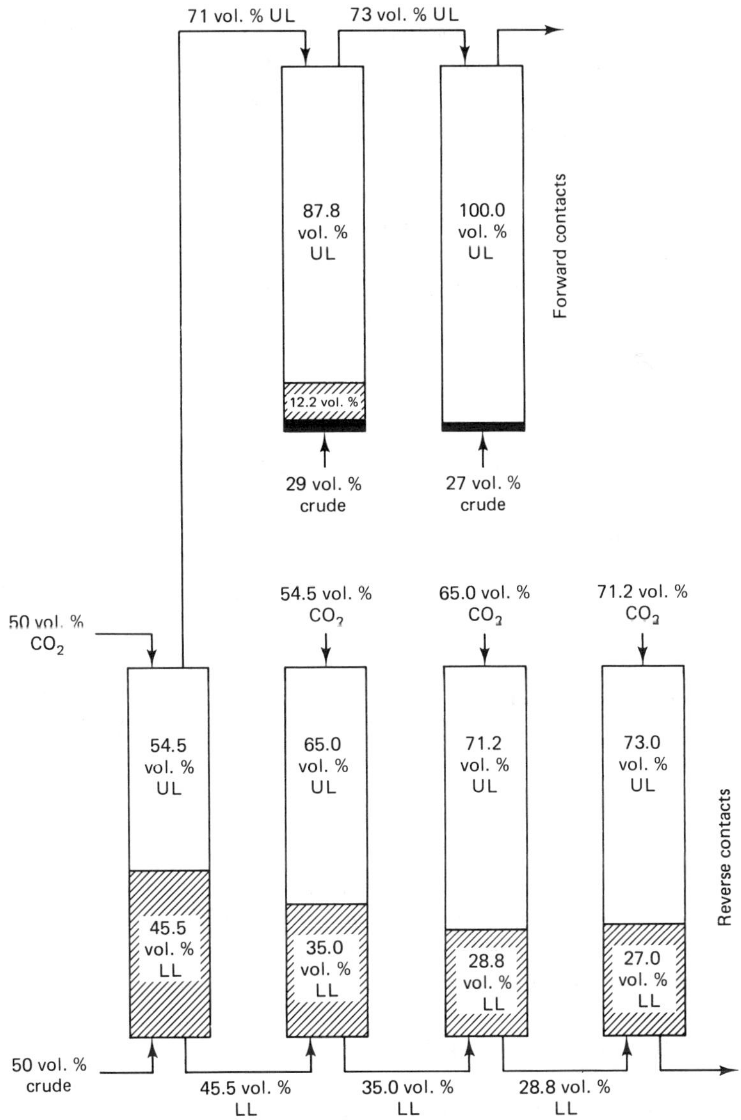

Figure 7-24 Multiple-contact experiment in 105°F (2,000 psia) (from Gardner et al., 1981)

richment in the forward contacts or the crude enrichment in the reverse contacts can cause one of the phases to disappear. This is exactly what is predicted by the arguments used in the process classification section: A single phase cell in the forward contacts indicates a vaporizing gas drive; in the reserve contacts, a condensing gas drive; and two or more phases in all contacts, an immiscible process. If the original cell is single phase for all combinations of solvent and crude oil, the process is first-contact miscible.

The experiment depends somewhat on the initial charges to the first cell, so the results are no more than indications of process classification. If phase compositions are measured at every step, the binodal curve and tie lines on a ternary diagram are established. Agreement between single- and multiple-contact experiments, as in Fig. 7-10, substantiates the pseudocomponent representation of the multicomponent equilibria.

Both single- and multiple-contact experiments place a premium on visual observations, but with careful selection of the initial volumes, these experiments are convenient ways to determine complete ternary equilibria data. Orr and Silva (1982) have proposed a method to measure phase behavior through continuous contacting.

Slim Tube

Filling the gap between the above static measurements and core floods are the *slim tube* experiments. These experiments are crude displacements by solvent, in the absence of water, at fixed temperature. The permeable medium consists of beads or unconsolidated sands packed in tubes of very thin cross section and, frequently, large length. The displacements are run with a fixed pressure at the one end of the system, and because the permeability of the medium is large, pressure gradients are negligible. Table 7-1 shows characteristics of selected slim tube experiments.

The overriding feature of slim tube experiments is the large aspect ratio (length-to-diameter ratio). This is intended to suppress viscous fingering since the long length means there is sufficient time during the displacement for all perturbations to be suppressed by transverse dispersion. Small wavelength perturbations will not form at all since the tube diameter is smaller than the critical wavelength (see Sec. 6-8).

The slim tube experiment, then, is designed to provide an unambigous measure of solvent displacement efficiency. But because of both the highly artificial nature of the permeable medium and the experimental conditions (no water), this is not a realistic displacement efficiency. The results are best regarded as a dynamic measure of phase behavior properties.

Minimum Miscibility Pressure

Although effluent compositions can be monitored during a slim tube displacement, by far the most common information derived from the experiments is the *minimum miscibility pressure* (MMP). Since solvent miscibility increases with pressure, ultimate oil recovery should also increase with pressure. This, in fact, happens, but there is a pressure above which a further pressure increase causes only a minimal increase in oil recovery. The pressure at which oil recovery levels out is the MMP, or minimum *dynamic* miscibility pressure. MMP is variously defined as

- The pressure at which the oil recovery at $t_D = 1.2$ PV of CO_2 injected was equal to or very near the maximum final recovery obtained in a series of tests (Yellig and Metcalfe, 1980)

TABLE 7-1 CHARACTERISTICS OF SLIM TUBE DISPLACEMENT EXPERIMENTS (ADAPTED FROM ORR ET AL., 1982)

Author(s)*	Length (meters)	ID (cm)	Geometry	Packing (mesh)	Permeability (μm^2)	Porosity (%)	Rate (cm/hr)	$\frac{K_l \phi}{uL}$
Rutherford (1952)	1.5	1.98	Vertical tube	50–70 mesh Ottawa sand	24	35	37	0.2
Yarborough and Smith (1970)	6.7	0.46	Flat coil	No. 16 AGS	2.74		66	11.9
Holm and Josendal (1974)	14.6 25.6	0.59		No. 60 Crystal sand			3.81	109.1
Holm and Josendal (1982)	15.8	0.59	Coil	No. 60 Crystal sand	20	39	101–254	12.5 14.9
Huang and Tracht (1974)	6.1	1.65			1.78	43	4.7	4.8
Yellig and Metcalfe (1980)	12.2	0.64 OD	Flat coil	160–200 mesh sand	2.5		5.2–10.2	33.4 58.1
Peterson (1978)	17.1	0.64		60–65 mesh sand	19			
Wang and Locke (1980)	18.0	0.62	Spiral coil	80–100 mesh	13	35	381	12.3
Orr and Taber (1981)	12.2	0.64	Spiral coil	170–200 mesh glass beads	5.8	37	42	13.5
Gardner, Orr, and Patel (1981)	6.1	0.46	Flat coil	230–270 mesh glass beads	1.4	37	32 64	15.0
Sigmund et al. (1979)	17.9	0.78		140 mesh	5	42		

* References in Orr et al.

- The pressure that causes 80% oil recovery at CO_2 breakthrough and 94% recovery at a gas-to-oil ratio of 40,000 SCF/stb (Holm and Josendal, 1974)
- The pressure that causes 90% oil recovery at t_D = 1.2 HCPV of CO_2 injected (Williams et al., 1980)

Others (Perry, 1978; Yellig and Metcalfe, 1980) emphasize the qualitative nature of the miscibility pressure determination. The importance of the exact definition is unknown; all definitions show the same trends in correlations.

The results of slim tube experiments are giving the minimum pressure necessary for the displacement to develop miscibility. Thus the MMP corresponds to the pressure at which the critical tie line passes through the crude composition. This pressure is considerably less than that required for complete or first-contact miscibility (compare the MMP plots with the P-z diagrams). This is the origin of the plateau on the oil-recovery–pressure plot: Any further pressure increase does not increase oil recovery since above the MMP the displacement will tend from developed to first-contact miscibility. These observations are also supported by compositional measurements wherein the properties (viscosity, density, and composition) of phases produced below the MMP become closer to one another as the MMP is approached.

The CO_2 MMP is determined by temperature, pressure, solvent purity, and molecular weight of the heavy fraction of the reservoir crude. Generally, the MMP increases with temperature and heavy fraction molecular weight. Holm and Josendal (1974 and 1982) note that the development of miscibility for CO_2 solvents is the result of extracting hydrocarbon components into a CO_2-rich phase. Therefore, at a given temperature and crude composition, sufficient compression must be applied to the solvent to promote solvency with the crude. This solvency is manifest by the CO_2 density at the temperature of the test. Figure 7-25(a) shows the CO_2 density required to develop miscibility at a given temperature with the C_5–C_{30} percent of the C_5^+ crude fraction. The CO_2 density can be connected to MMP through Fig. 7-4 or Fig. 7-25(b). CO_2 MMP is affected by the type of hydrocarbons (aromatic or paraffinic) in the crude but to a lesser degree than by temperature and CO_2 density (Monger, 1985).

Several works have presented determinations of MMP for impure CO_2. Figure 7-26 shows the results of the effects of N_2, CH_4, H_2S and H_2S-CH_4 mixtures on the CO_2 MMP. Methane and particularly nitrogen increase the CO_2 MMP, whereas H_2S decreases it. Whether an impurity increases or decreases the MMP depends on whether the solvency of the solvent has been enhanced. Solvency is improved (MMP decreases) if CO_2 is diluted with an impurity whose critical temperature is more than that of CO_2. Solvency deteriorates (MMP increases) if CO_2 is diluted with an impurity with a critical temperature less than CO_2. Compare the trends in Fig. 7-26 with the critical temperatures in Fig. 7-2.

The above idea of solvency can be used to estimate the MMP of an impure CO_2 solvent. Sebastian et al. (1984) have correlated the diluted CO_2 MMP by the following:

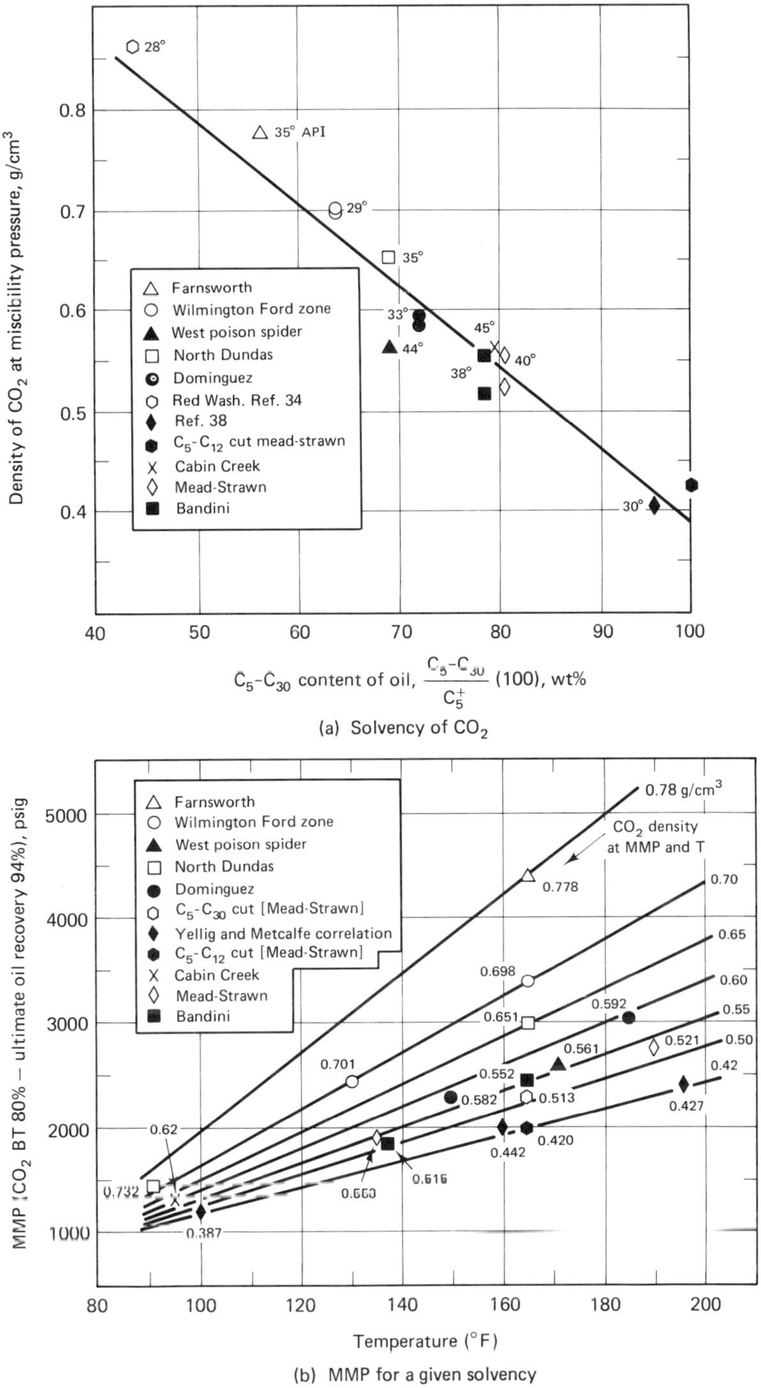

Figure 7-25 Density of CO_2 required for miscible displacement of various oils at 90° to 190°F (from Holm and Josendal, 1982)

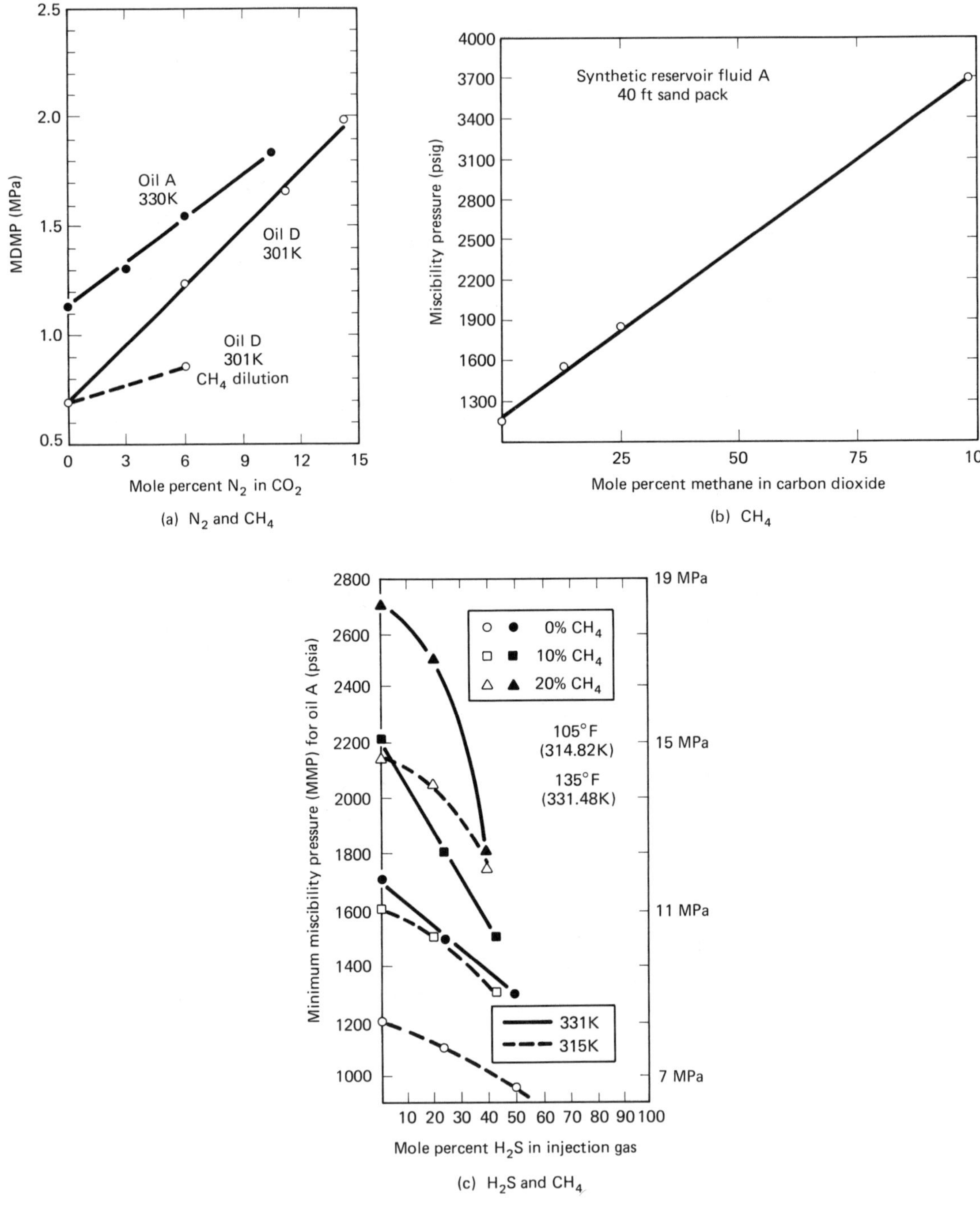

Figure 7-26 Effect of impurities on CO_2 minimum miscibility pressure (from Johnson and Pollin, 1981; Whitehead et al., 1980; and Metcalfe, 1981)

$$\frac{P_{MM}}{(P_{MM})_{CO_2}} = 1.0 - (2.13 \times 10^{-2})(T_{pc} - T_c) + (2.51 \times 10^{-4})(T_{pc} - T_c)^2 - (2.35 \times 10^{-7})(T_{pc} - T_c)^3 \tag{7.5-1}$$

where $T_{pc} = \Sigma_i T_{ci} y_i$ is the pseudocritical temperature of the mixture, and y_i is the mole fraction of species i in the solvent. The denominator of the left side of Eq. (7.5-1) can be estimated from Fig. 7-25. (For other correlations, see Johnson and Pollin, 1981.) No MMP correlation is especially accurate; errors as much as 0.34 MPa (50 psia) are common.

Minimum Enrichment Correlations

For a dry gas process, slim tube results will give an estimate of the amount of intermediates that must be added to develop miscibility in a condensing gas drive. Such experiments were precursors to the MMP experiments (Benham et al., 1961). The oil recovery plot would consist of several experiments each with a successively richer injected solvent but each at constant pressure. When the solvent composition coincided with the tie line extension (through the reverse contacts), oil recovery would cease to increase as the solvent becomes richer in intermediates.

Figure 7-27 is one of 12 plots from Benham et al. (1961) that shows the maximum methane concentration permissible in an LPG solvent that will develop miscibility with the subject crude. These authors correlated the maximum dilution (or

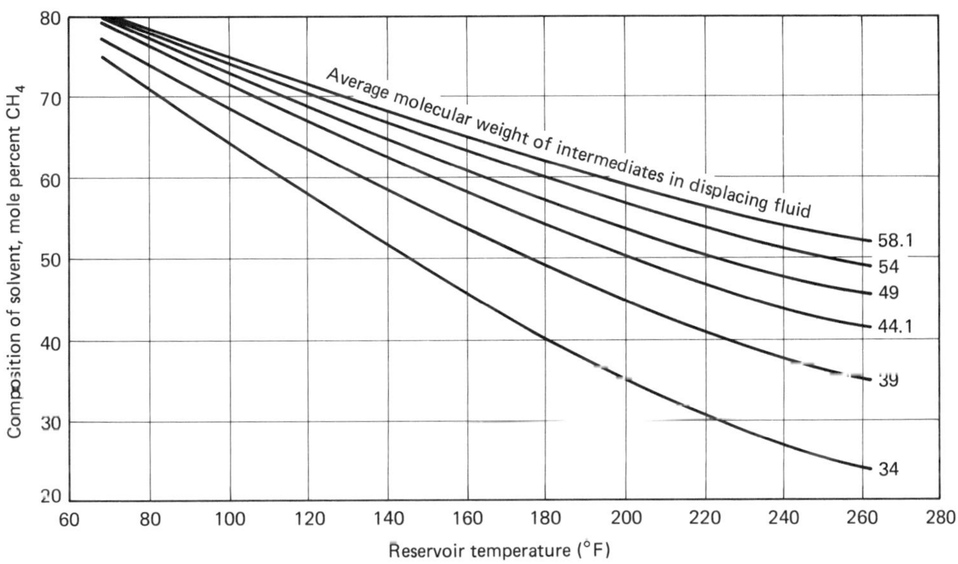

Figure 7-27 Maximum methane dilution in LPG solvent for developed miscibility at 2,500 psia and for a reservoir fluid whose C_5^+ component molecular weight is 240 (from Benham et al., 1961)

minimum enrichment) with temperature, pressure, molecular weight of the intermediate component in the solvent, and molecular weight of the C_5^+ fraction in the crude. The minimum dilution increases with decreasing C_5^+ molecular weight, pressure, and temperature, and it increases with increasing intermediate molecular weight.

Each of these trends follows from the trends in the phase behavior and the position of the crude and solvent on the ternary diagram. And each may be quantitatively established on true ternaries with accurate thermodynamic properties. But the pseudocomponent representation of more than three components on a ternary is not rigorous, and this leads to some difficulty in quantitatively predicting both the minimum dilution and the MMP on actual systems.

7-6 DISPERSION AND SLUG PROCESSES

In the next few sections, we look in detail at how a miscible solvent behaves during oil displacement. You should remember that first-contact and developed miscibility solvent behave very much alike.

Dilution Paths

The concentration of species i in a first-contact miscible displacement is from Eq. (5.5-15)

$$C_i = C_{iI} + \frac{(C_{iJ} - C_{iI})}{2}\left[1 - \operatorname{erf}\left(\frac{x_D - t_D}{2\sqrt{\dfrac{t_D}{N_{\text{Pe}}}}}\right)\right] \tag{7.6-1}$$

For this equation to be valid, we cannot have viscous fingering, layering, or gravity tonguing; hence it is restricted to constant viscosity and density floods in one-dimensional media. In Eq. (7.6-1), x_D is the dimensionless length, t_D the dimensionless time in fractional pore volumes, N_{Pe} the Peclet number, and the subscripts I and J refer to initial and injected conditions, respectively.

If we let the component subscript i refer to the light, intermediate, and heavy pseudocomponents of Sec. 7-3, we can easily show from Eq. (7.6-1) that dilution paths are straight lines on a pseudoternary diagram. Eliminating the term in brackets among the three equations gives

$$\frac{C_1 - C_{1I}}{C_{1J} - C_{1I}} = \frac{C_2 - C_{2I}}{C_{2J} - C_{2I}} = \frac{C_3 - C_{3I}}{C_{3J} - C_{3I}} \tag{7.6-2}$$

The C_i in Eq. (7.6-2) lie on a straight line in composition space; hence the dilution path of Sec. 7-3 is linear.

Superposition

Solvents are usually too expensive to be injected continuously. Thus a typical displacement consists of a finite amount or slug of solvent followed by a less expensive chase fluid. The concentration of a slug follows from Eq. (7.6-1) and the principle of *superposition*. This principle applies to linear partial differential equations, which Eq. (7.6-1) is an approximate solution to. We can, in fact, derive the concentration response of an infinite number of step changes in the influent concentration (see Exercise 7C), but we restrict our discussion here to the case of a single solvent slug displaced by a chase fluid.

Let I, J, and K denote the concentrations of species i in the original fluid, the slug, and the chase fluids, respectively. Superposition states that the sum of individual solutions to a linear differential equation is also a solution to the equation. This seems easy enough to do in practice, but we must take care in selecting the boundary conditions of the individual solutions to give the correct composite solution. Figure 7-28 shows the influent or imposed boundary conditions of the single front problem

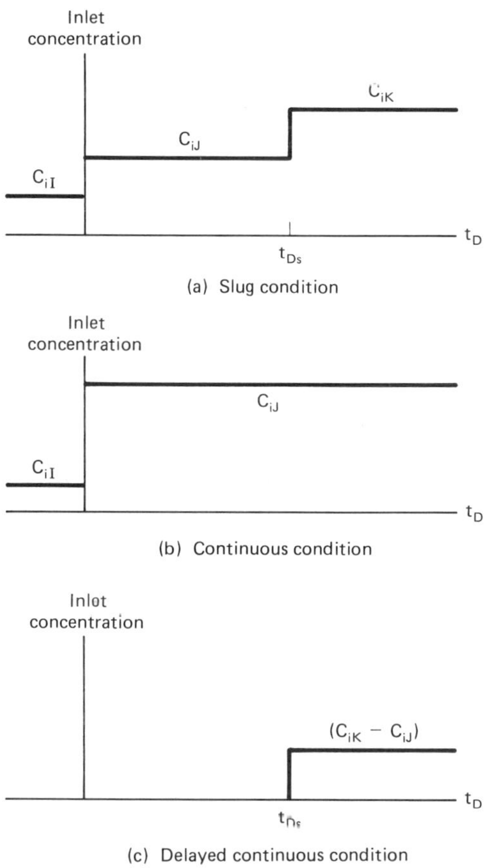

Figure 7-28 Schematic of influent boundary conditions for slugs

(Fig. 7-28b) and that of the composite solution (Fig. 7-28c). The composite solution gives $C_i(x_D, t_D)$ for the imposed conditions in Fig. 7-28(a), simply the sum of the solutions to the conditions in Figs. 7-28(b) and 7-28(c), respectively. The solution to the imposed conditions in Fig. 7-28(b) is Eq. (7.6-1), and that of the imposed condition in Fig. 7-28(c) is

$$C_i = \frac{C_{iK} - C_{iJ}}{2}\left[1 - \operatorname{erf}\left(\frac{x_D - (t_D - t_{Ds})}{2\sqrt{\frac{(t_D - t_{Ds})}{N_{Pe}}}}\right)\right] \tag{7.6-3}$$

By superposition $C_i(x_D, t_D)$ for the influent condition in Fig. 7-28(a) is the sum of Eqs. (7.6-1) and (7.6-3)

$$C_i = \frac{C_{iI} + C_{iK}}{2} + \left(\frac{C_{iI} - C_{iJ}}{2}\right)\operatorname{erf}\left(\frac{x_D - t_D}{2\sqrt{\frac{t_D}{N_{Pe}}}}\right)$$

$$+ \left(\frac{C_{iJ} - C_{iK}}{2}\right)\operatorname{erf}\left(\frac{x_D - (t_D - t_{Ds})}{2\sqrt{\frac{t_D - t_{Ds}}{N_{Pe}}}}\right), \quad t_D > t_{Ds} \tag{7.6-4}$$

Equation (7.6-4) is valid for any value of the injected concentrations.

Frequently, we are interested in the concentration of the solvent at the midpoint between $x_D = t_D$ and $x_D = t_D - t_{Ds}$. Evaluating Eq. (7.6-4) at $x_D = t_D - t_{Ds}/2$ yields this midpoint concentration $\overline{C}_i$

$$\overline{C}_i = \frac{C_{iI} + C_{iK}}{2}\left[1 - \operatorname{erf}\left(\frac{t_{Ds}}{4\sqrt{\frac{t_D}{N_{Pe}}}}\right)\right] + C_{iJ}\operatorname{erf}\left(\frac{t_{Ds}}{4\sqrt{\frac{t_D}{N_{Pe}}}}\right) \tag{7.6-5}$$

This equation is valid only for relatively small t_{Ds} where the difference between the square roots of t_D and $t_D - t_{Ds}$ in the denominator of the error function argument is not large. If $C_{iJ} > C_{iI}$ and $C_{iJ} > C_{iK}$, the midpoint concentration is usually called the *peak* concentration. For $C_{iI} = C_{iK} = 0$, the peak concentration falls with increasing time according to

$$\overline{C}_i = C_{iJ}\operatorname{erf}\left(\frac{t_{Ds}}{4\sqrt{\frac{t_D}{N_{Pe}}}}\right) \tag{7.6-6}$$

The error function may be replaced by its argument for small values of the argument. In this event, the peak concentration falls in inverse proportion to the square root of time. Since $x_D = t_D - t_{Ds}/2$ at the peak concentration, this is equivalent to $\overline{C}_i$ falling in proportion to the inverse square root of the distance traveled.

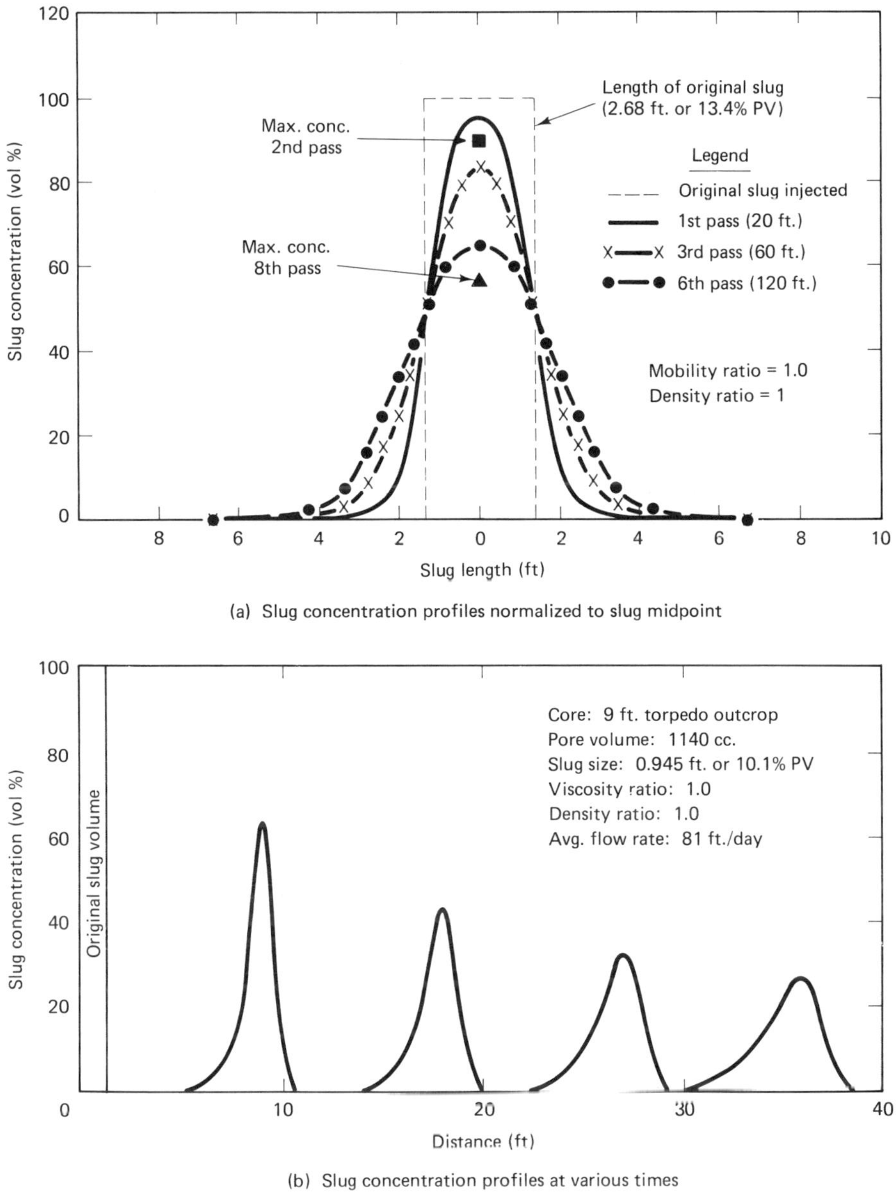

Figure 7-29 Miscible slug concentration profiles for matched viscosity and density displacements (from Koch and Slobod, 1956)

The peak concentration falling below C_{iJ} is the consequence of overlapping front and rear mixing zones. Figure 7-29 shows experimental concentration profiles from a miscible slug displacement at different throughputs. Figure 7-29(a) has the concentration profiles normalized to the midpoint position $x_D = t_D - t_{Ds}/2$ on the horizontal axis. The areas under all curves are the same (material balance is preserved), but the peak concentration falls as the number of passes (travel distance) increases. The unnormalized profiles in Fig. 7-29(b) show that the peak concentration falls approximately as the inverse square root of t_D in experimental floods.

The midpoint concentrations also trace a straight line in the pseudoternary diagram since the error function arguments in Eq. (7.6-5) may be eliminated to give

$$\frac{\overline{C}_1 - \dfrac{C_{1K} + C_{1I}}{2}}{C_{1J} - \dfrac{C_{1K} + C_{1I}}{2}} = \frac{\overline{C}_2 - \dfrac{C_{2K} + C_{2I}}{2}}{C_{2J} - \dfrac{C_{2K} + C_{2I}}{2}} = \frac{\overline{C}_3 - \dfrac{C_{3K} + C_{3I}}{2}}{C_{3J} - \dfrac{C_{3K} + C_{3I}}{2}} \tag{7.6-7}$$

This equation says that as time increases, the midpoint concentration traces a straight line between the injected slug concentration C_{iJ} and the average concentration of the

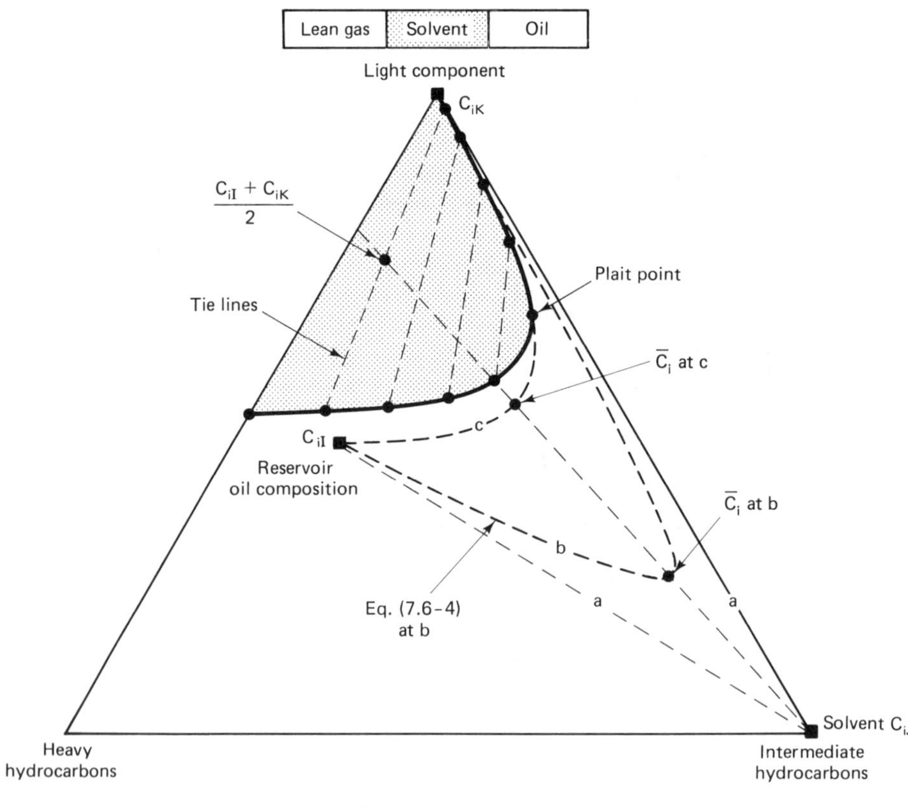

Figure 7-30 Dilution of solvent slug by mixing (from Stalkup, 1983)

fluids ahead of and behind the slug. The midpoint concentrations at successive times a, b, and c are shown in Fig. 7-30, as are the dilution paths given by Eq. (7.6-4). The dilution paths become straight line segments from C_{iJ} to $\overline{C}_i$ and then from $\overline{C}_i$ to C_{iK} for t_{Ds} small. These considerations are valid only so long as the entire dilution path stays in the single-phase region of the diagram. It is not necessary for $\overline{C}_i$ to fall into the two-phase region for the displacement to lose first-contact miscibility (see Exercise 7E).

7-7 TWO-PHASE FLOW IN SOLVENT FLOODS

Two or more phases are all too common in solvent floods. When this happens, the dispersion theory of Sec. 7-6 does not apply. But general conclusions about such displacements are still possible based on the coherent or simple wave theory first introduced in Sec. 5-6. This theory neglects dissipative effects of any kind; hence we omit dispersion in the following discussion and restrict our treatment to centered simple waves (see Sec. 5-4 for definitions).

We treat two cases of two-phase flow in miscible displacements: (1) solvent floods in the absence of an aqueous phase and (2) first-contact miscible displacements in the presence of an aqueous phase. In both cases, fluid displacement takes place in a one-dimensional permeable medium at constant temperature and with incompressible fluids and solid.

Solvent Floods in the Absence of an Aqueous Phase

In this section, we give a theoretical base for the classifications of Sec. 7-3. Consider a three-component system consisting of an intermediate hydrocarbon ($i = 2$), a light hydrocarbon ($i = 3$), and a heavy hydrocarbon that can form no more than two phases at constant temperature and pressure. As we discussed in Sec. 4-3, the overall concentrations C_i, the phase concentrations C_{ij}, and the saturations S_j ($j = 2$ or 3) can be conveniently represented on ternary diagrams.

The topology within the two-phase region is important for this problem. Figure 7-31 shows a ternary diagram with a two-phase region exaggerated to point out certain landmarks. Within the two-phase region is a family of quality lines that do not intersect and converge at the plait point. The binodal curve itself is a quality line. There are also lines denoting the residual saturations of the two phases. These lines do not, in general, coincide with quality lines since residual saturations must decrease as the plait point is approached (see Sec. 3-4). This decrease is because the interfacial tension between the two phases must vanish at the plait point. Along each tie line, there exists a curve relating the fractional flow of one of the phases to its saturation. Three of these curves, along tie lines A-A', B-B' and C-C', are in the upper left insert to Fig. 7-31. The shape of the fractional flow curves is not determined by the phase behavior alone, but the curves become straighter (more miscible-like) with smaller residual phases along tie lines near the plait point. Because phase

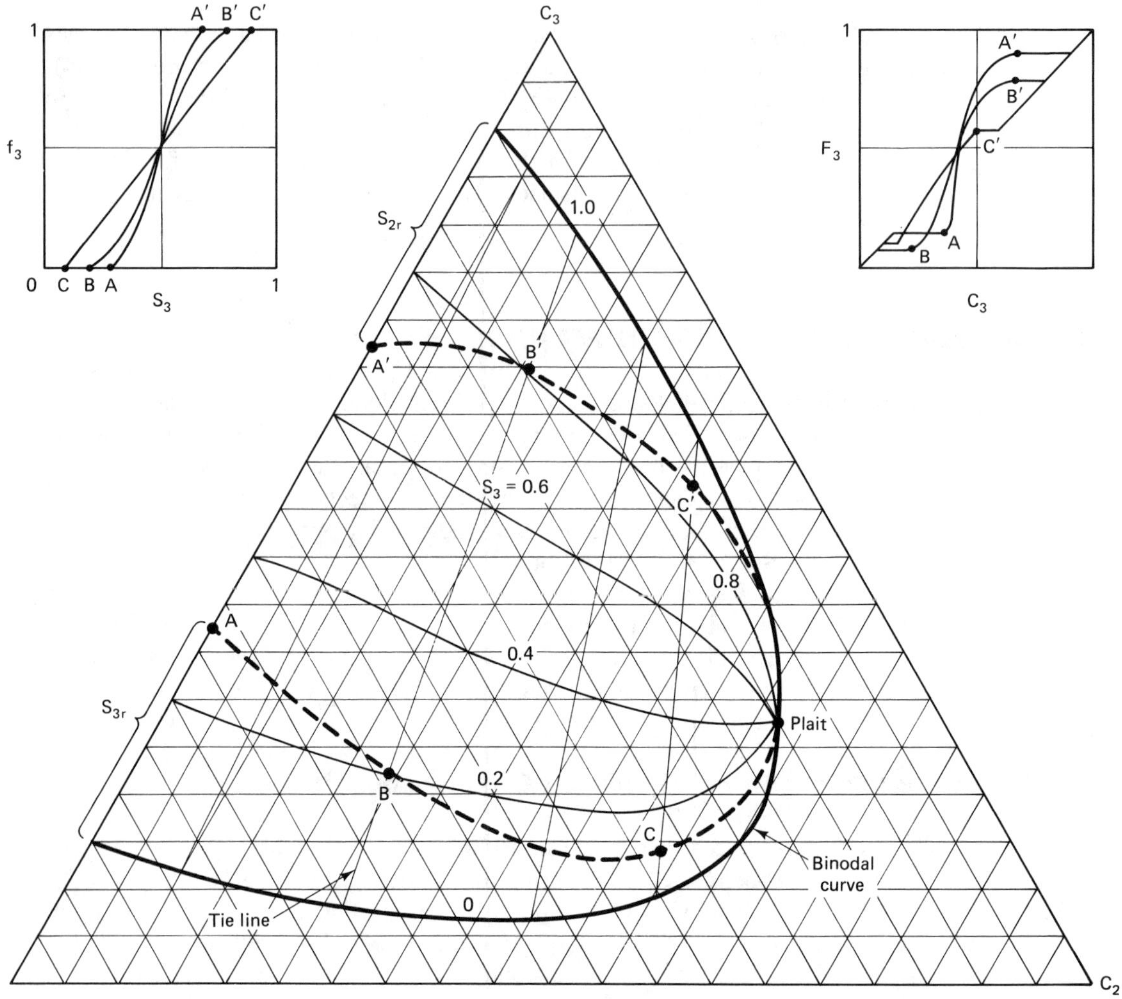

Figure 7-31 Landmarks on a two-phase ternary

compositions are constant along tie lines, the C_{ij}, S_j and f_j can be converted to fractional flux and overall concentration through Eqs. (5.4-3). The upper right insert of Fig. 7-31 shows an F_3-C_3 along the three tie lines.

This ternary system has only two independent components, which we arbitrarily take to be C_2 and C_3. The coherence condition (Eqs. 5.4-5 and 5.6-14) for this case becomes

$$v_{C_2} = \frac{dF_2}{dC_2} = \frac{dF_3}{dC_3} = v_{C_3} \tag{7.7-1}$$

Using the condition $f_2 + f_3 = 1$, and the definitions for overall flux and concentration (Eqs. 5.4-3a and 5.4-3c), Eq. (7.7-1) can be rewritten as (Helfferich, 1982)

$$v_{C_2} = \frac{f_2(dC_{22} - dC_{23}) + dC_{23} + (C_{22} - C_{23})df_2}{S_2(dC_{22} - dC_{23}) + dC_{23} + (C_{22} - C_{23})dS_2}$$
$$= \frac{f_2(dC_{32} - dC_{33}) + dC_{33} + (C_{32} - C_{33})df_2}{S_2(dC_{32} - dC_{33}) + dC_{33} + (C_{32} - C_{33})dS_2} = v_{C_3} \tag{7.7-2}$$

The curve in the ternary composition space that a displacement follows (the composition route) is quite complex, but certain segments (composition paths) are readily apparent from Eq. (7.7-2).

1. Unit velocity paths. These occur along any direction in the single-phase region (all directions are coherent) or along the binodal curve. In both cases, $f_2 = S_2 = 1$ or $f_3 = S_3 = 1$, depending on the side of the plait point, and the composition velocity is

$$v_{C_2} = v_{C_3} = 1 \tag{7.7-3}$$

Equation (7.7-3) is the same result as Eq. (5.4-7).

Within the two-phase region is an *equivelocity* path where $f_2 = S_2$. This path is the intersection of a straight line through $f_2 = S_2 = 0$ and $f_2 = S_2 = 1$ and the family of fractional curves (Fig. 7-31). It converges to the plait point, but it does not, in general, coincide with a quality line.

2. Tie line paths. On tie lines in the two-phase region, $dC_{ij} = 0$. This also satisfies Eq. (7.7-2). On these paths, the concentration velocities are

$$v_{C_i} = \frac{df_2}{dS_2}, \qquad i = 2, 3 \tag{7.7-4}$$

Equation (7.7-4) is the same as the saturation velocity in the Buckley-Leverett theory (Eq. 5.2-10).

Two other types of paths are not so easily derived. Both follow from integrating the composition path curve

$$\frac{dC_2}{dC_3} = \frac{v_{\tilde{C}}^{\pm} - F_{22}}{F_{23}} \tag{7.7-5a}$$

where the composition velocity is

$$v_{\tilde{C}}^{\pm} = \frac{1}{2}\{(F_{22} + F_{33}) \pm [(F_{33} - F_{22})^2 + 4F_{23}F_{32}]^{1/2}\} \tag{7.7-5b}$$

and

$$F_{22} = \left(\frac{\partial F_2}{\partial C_2}\right)_{C_3}$$

and so on

3. Singular curves. Along these curves, the velocity of the fast and slow paths is equal. The curves follow from setting the discriminant of Eq. (7.7-5b) equal to

zero. Singular curves are composition paths since they can be generated from Eq. (7.7-5a) as long as $F_{23} \neq 0$.

4. Non-tie line paths. Within the two-phase region, there are also composition paths whose trajectories are not readily apparent from the above equations. Figure 7-32 shows each of these paths.

If the tie lines extend to a common point, the concentration velocity along the nontie line paths is constant (Cere and Zanotti, 1985). The phase equilibria is now represented by Eq. (4.4-27b), which we repeat for this special case

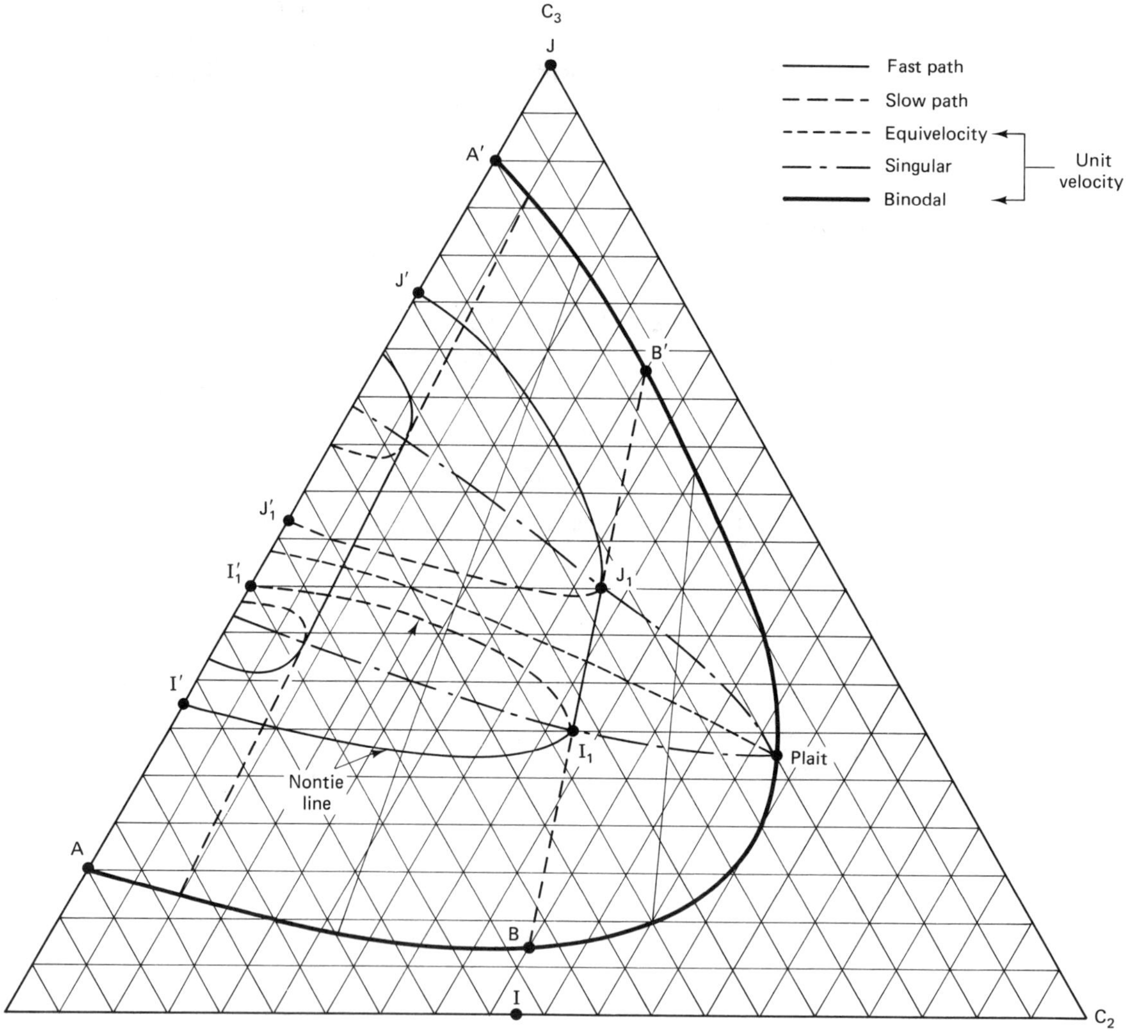

Figure 7-32 Composition path in two-phase ternary equilibria

$$C_{3j} - C_3^0 = \eta(C_{2j} - C_2^0), \qquad j = 2 \text{ or } 3 \tag{7.7-6}$$

A particular tie line is represented by a particular value of η. If we introduce the definitions for overall flux and concentration into the coherence condition Eq. (7.7-1), we have

$$\frac{dF_2}{dC_2} = \frac{d(C_{22}f_2 + C_{23}f_3)}{d(C_{22}S_2 + C_{23}S_3)} = \frac{d(C_{32}f_2 + C_{33}f_3)}{d(C_{32}S_2 + C_{33}S_3)} = \frac{dF_3}{dC_3} \tag{7.7-7}$$

We can substitute Eq. (7.7-6) into the third term to give, after some rearrangement and identification,

$$\frac{dF_2}{dC_2} = \frac{d\eta(F_2 - C_2^0) + \eta dF_2}{d\eta(C_2 - C_2^0) + \eta dC_2} \tag{7.7-8}$$

Our task is to find the combination of variables that makes this equation an identity.

Immediately we see that the equivelocity and tie line paths are returned from Eq. (7.7-8), for the conditions $F_2 = C_2^0 = C_2$ and $d\eta = 0$ clearly satisfy the equation. But the existence of both paths is more general than this since it follows from Eq. (7.7-2).

The nontie line paths are defined by the following equations:

$$\begin{aligned} dF_2 &= d\eta(F_2 - C_2^0) + \eta dF_2 \\ dC_2 &= d\eta(C_2 - C_2^0) + \eta dC_2 \end{aligned} \tag{7.7-9}$$

Eliminating η between these two equations gives an ordinary differential equation relating F_2 and C_2 along the nontie line path

$$\frac{dF_2}{dC_2} = \frac{F_2 - C_2^0}{C_2 - C_2^0} \tag{7.7-10}$$

Integrating this equation yields a linear relation between F_2 and C_2

$$F_2 - C_2^0 = I_c(C - C_2^0) \tag{7.7-11a}$$

where I_c is an integration constant independent of either F_2 or C_2. Immediately it follows from Eqs. (7.7-11a) and (7.7-1) that the velocity along the nontie line path is constant and that the constant velocity is, in fact, the integration constant. The path itself is given by the linear relation

$$F_i - C_i^0 = v_C(C_i - C_i^0), \qquad i = 2, 3 \tag{7.7-11b}$$

Since the above development applies for either independent species, we drop the subscript on v_c.

As Fig. 7-32 shows, the entire two-phase region is covered with a net of nontie line paths along each of which the velocity is constant. Some of these paths cross the tie line paths, but others merge continuously with it at a point where the velocity along both paths is equal. The curve defining the locus of these intersection points is given by

$$\left(\frac{dF_2}{dC_2}\right)_{\text{tie line}} = \left(\frac{F_2 - C_2^0}{C_2 - C_2^0}\right)_{\text{nontie line}} \tag{7.7-12}$$

from Eqs. (7.7-4) and (7.7-11b). This curve is the singular curve discussed above wherein the discriminant of Eq. (7.7-5b) vanishes.

Figure 7-33 shows two fractional flux curves going from points J to I along the lines A-A' and B-B'. These were selected because they are on tie lines that extend to the points J and I, respectively. The curves consist of three segments: a portion of unit slope corresponding to the single-phase regions in Fig. 7-32, horizontal portions corresponding to single-phase flow in the presence of another residual phase, and a curved portion corresponding to two-phase flow. The curve with the more compressed curved portion corresponds to the tie line nearest the plait point.

Figure 7-33 also shows the construction for the singular point as suggested by Eq. (7.7-12). Since the slope of a tie line path is the coherent velocity, the tie line

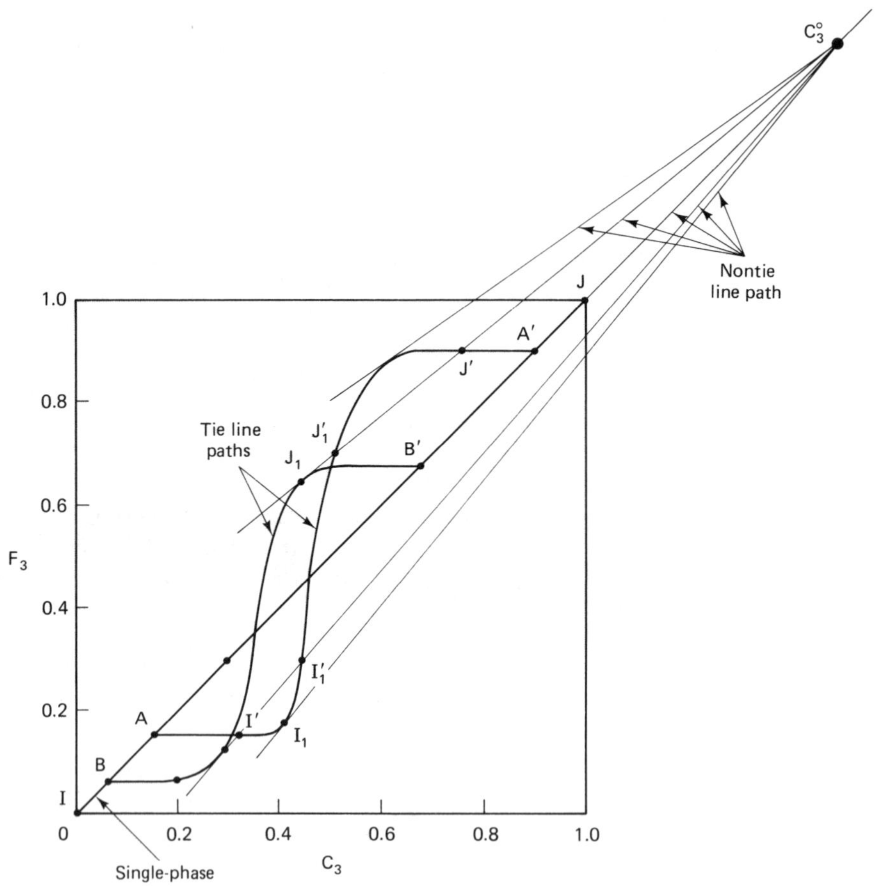

Figure 7-33 Fractional flux curves for Fig. 7-32

paths are slow outside the singular point intersections, and fast within the intersections. Understanding how a given tie line path can be both fast and slow is important for what follows.

You should compare both the above constructions with those in Fig. 5-12(d). Just as there are a variety of possibilities in Fig. 5-12, because of the variety of fractional flow shapes, so there are several possible behaviors for the singular curves. For example, a fractional curve without an inflection (Fig. 5-12a) will have one singular curve that coincides with a residual phase saturation curve.

We use the curves in Fig. 7-33 to select the only physically possible composition routes in Fig. 7-32. Before doing this, we remind you of the principles established in Sec. 5-6 for centered simple waves.

1. The composition route must stay on the composition path segments in the ternary diagram.
2. The composition velocity must decrease monotonically in the upstream direction. (This rule is actually a special case of the more general statement that all concentrations must be single valued.)
3. The correct composition route must be insensitive to infinitesimal perturbations in concentration. (This rule was not needed in Chap. 5, but it is here.)

We build the composition path from I to J in three segments, each of which must satisfy these rules.

Consider first the displacement of J_1 by J. There are an infinite number of paths between J and J_1, but we consider only the two extreme routes $J \rightarrow J' \rightarrow J_1$ and $J \rightarrow B' \rightarrow J_1$. The second path is nonphysical because it involves a fast segment (in the single-phase region) upstream of the slow segment between B' and J_1. We could resolve this by putting a shock directly from J to J_1 but this would no longer follow the composition paths. Route $J \rightarrow J' \rightarrow J_1$ also contains fast paths upstream of slow paths, but the resulting resolution into shocks (Fig. 7-34a) remains on the composition paths. In fact, this is the only route between the two extremes we discuss here that remains so because the switch from one tie line to the other, J' to J_1, takes place along a nontie line path.

We see that the general use of non-tie line paths is to switch between tie lines. Figure 7-34(b) shows a composition profile for this displacement. The displacement shocks across the residual phase 2 saturation, causing complete recovery of this phase. This recovery, which takes place in the absence of lowered interfacial tension and developed miscibility, occurs because phase 2 dissolves into the injected phase J. Dissolution waves are normally inefficient since their propagation velocity is slow (Fig. 7-34).

These comments also apply to the path from I_1' to I. Of the two extreme routes, $I_1' \rightarrow I_1 \rightarrow I$ and $I_1' \rightarrow A \rightarrow I$, only the latter yields a route along which the shock resolutions will remain on the composition path segments. Figure 7-35 shows the construction and corresponding profiles.

By comparing the routes $J \rightarrow J_1$ and $I_1' \rightarrow I$, we see both entry and exit from a

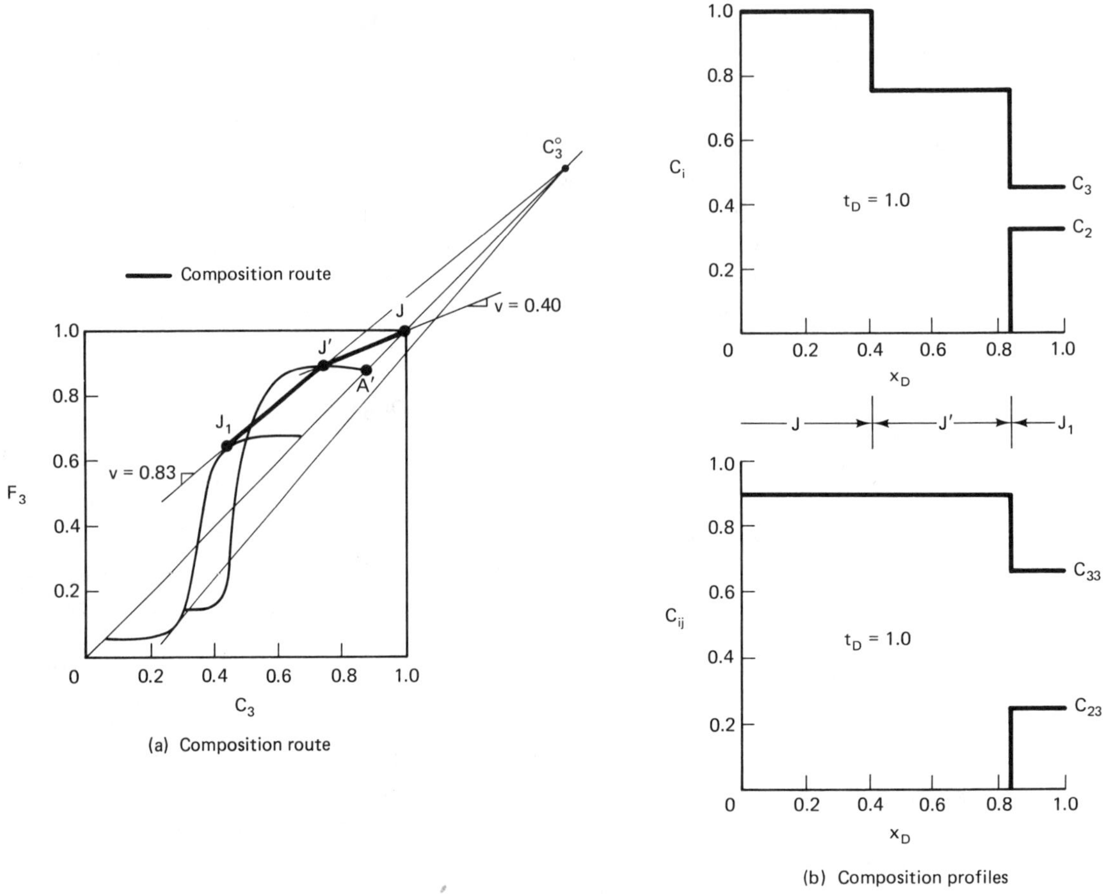

Figure 7-34 Composition route and profiles for displacement $J \rightarrow I_1$

two-phase region take place along tie line extensions. Further, both entry and exit contain slow shocks, which are the direct result of the slow segments of the fractional flux curves. If I_1' were above the equivelocity curve in Fig. 7-35(a), the route would follow a fast shock along the tie line nearest the plait point.

The third segment, J_1 to I_1, follows a single tie line path whose velocity is given by the Buckley-Leverett construction (Fig. 7-36) from Eq. (7.7-4).

In a sense, the above constructions, particularly I_1' to I, are misleading when applied to the entire displacement from I to J because the rules for centered simple waves must apply globally rather than individually to segments. To see this, consider the four possible composition routes from J to I: $J \rightarrow J' \rightarrow J_1 \rightarrow I$, $J \rightarrow J_1' \rightarrow J_1 \rightarrow I$, $J \rightarrow I_1' \rightarrow I_1 \rightarrow I$, and $J \rightarrow I' \rightarrow I_1 \rightarrow I$. After you carefully consider each case, with shock segments interspersed, the requirement of a monotonically increasing concentration velocity forces you to see the only correct choice is $J \rightarrow J' \rightarrow J_1 \rightarrow I$. The immiscible displacement $J \rightarrow I$ will consist of two shock segments between which is a small spreading wave; these are sketched schematically in Fig. 7-37a.

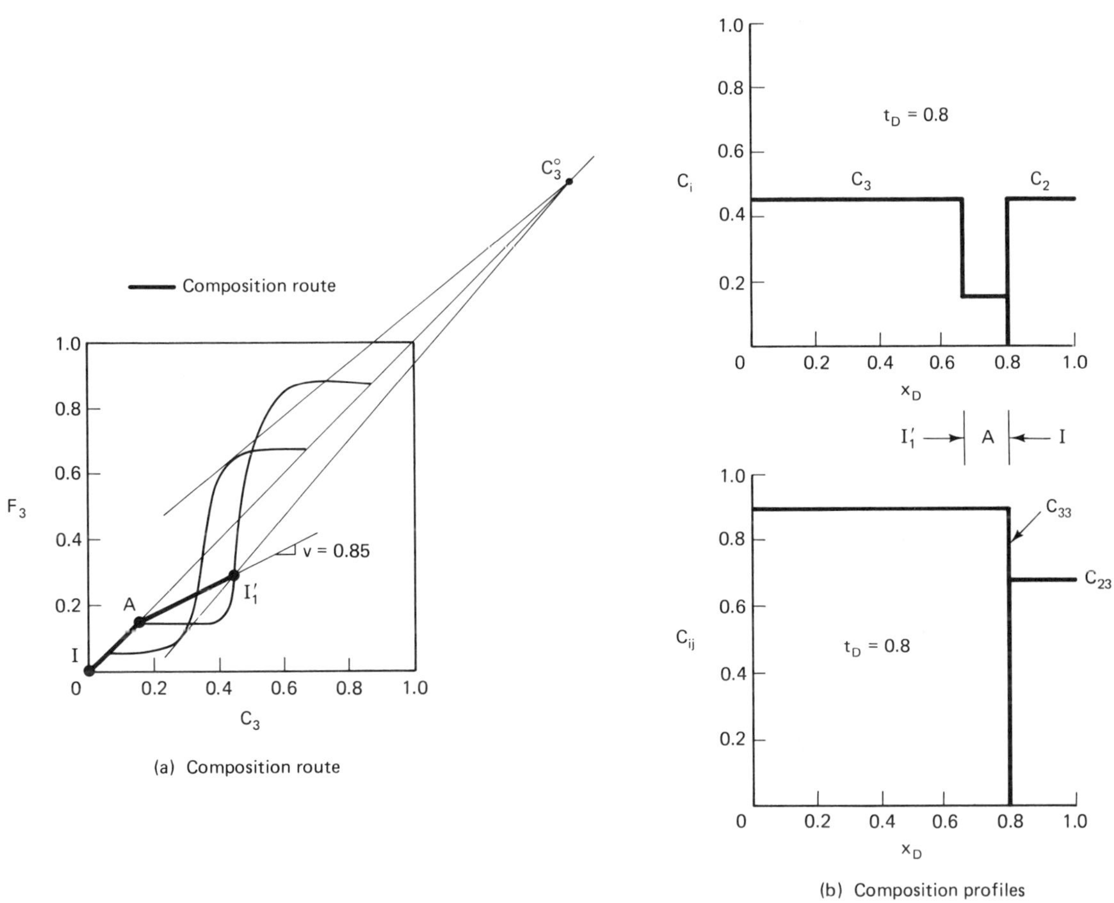

Figure 7-35 Composition route and profiles for displacement $I_1' \rightarrow I$

As in all MOC problems, an infinite number of mathematical solutions exist, but by assumption, only one physical solution exists. Finding the physical solution involves trial and error according to the following procedure:

1. Locate a tentative composition route between the injected and initial conditions on the fractional flux diagram. This route consists of segments that conform to the paths we discussed above.
2. Resolve all physical inconsistencies along the tentative route with shocks. In doing this, assume the differential and integral composition routes are the same.
3. Discard any tentative solution in which the shock resolution leads to a route that does not follow the composition path sequence. When this happens, return to step 1 with another tentative route. The correct solution is usually clear after a few trials.

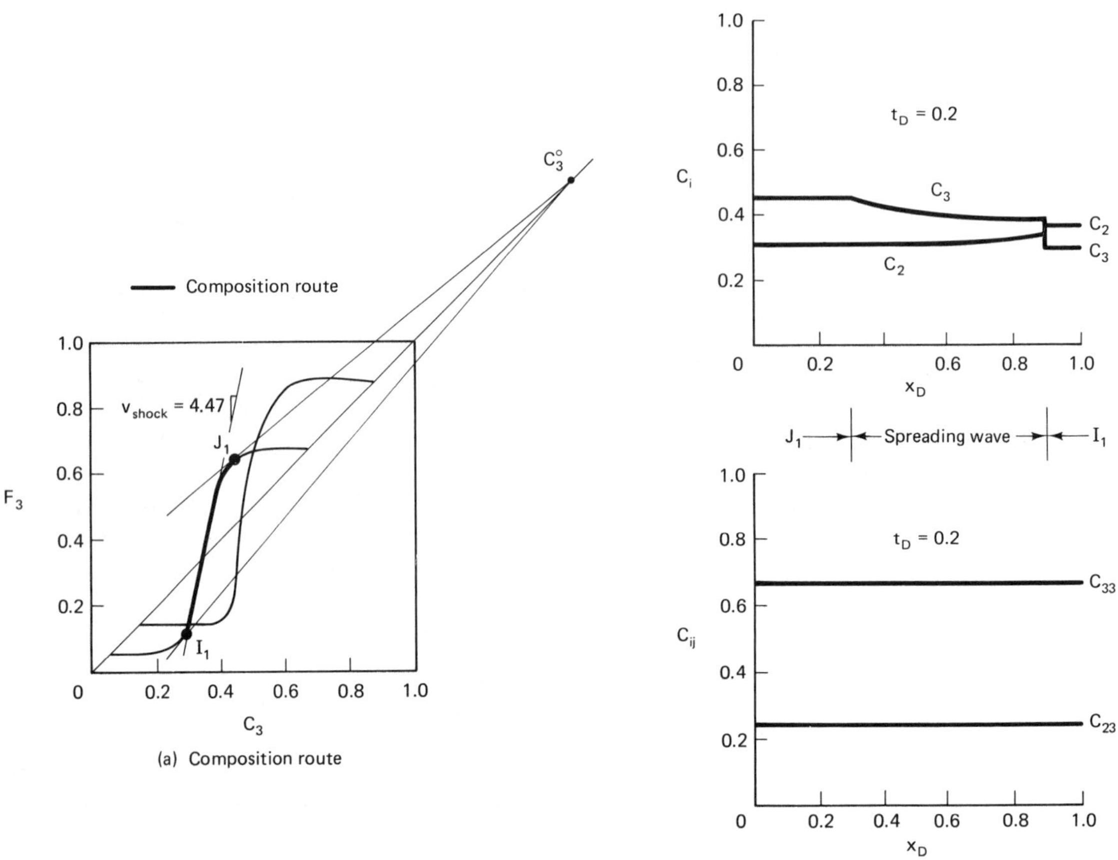

Figure 7-36 Composition route and profiles for displacement $I_1 \rightarrow I'$

These rules enable us to sketch the composition routes for the three types of displacements (Fig. 7-37). The composition route of the immiscible displacements (Fig. 7-37a) both enters and exits the two-phase region on tie line extensions. The entering segment is an extremely slow shock (a solubilization wave), which is the consequence of the residual phase saturations. If the system includes more than three true components—that is, at least one apex was a pseudocomponent—the displacement would not revert to single-phase behavior as suggested by Fig. 7-37(a) (Gardner and Ypma, 1982). Compare Fig. 7-37(a) to Fig. 7-16.

The vaporizing gas drive process (Fig. 7-37b) shows a composition route that approaches the binodal curve on a tie line extension and then follows the binodal curve until it reaches a point on a straight line tangent to the initial composition. Compare Fig. 7-37(b) to Fig. 7-14.

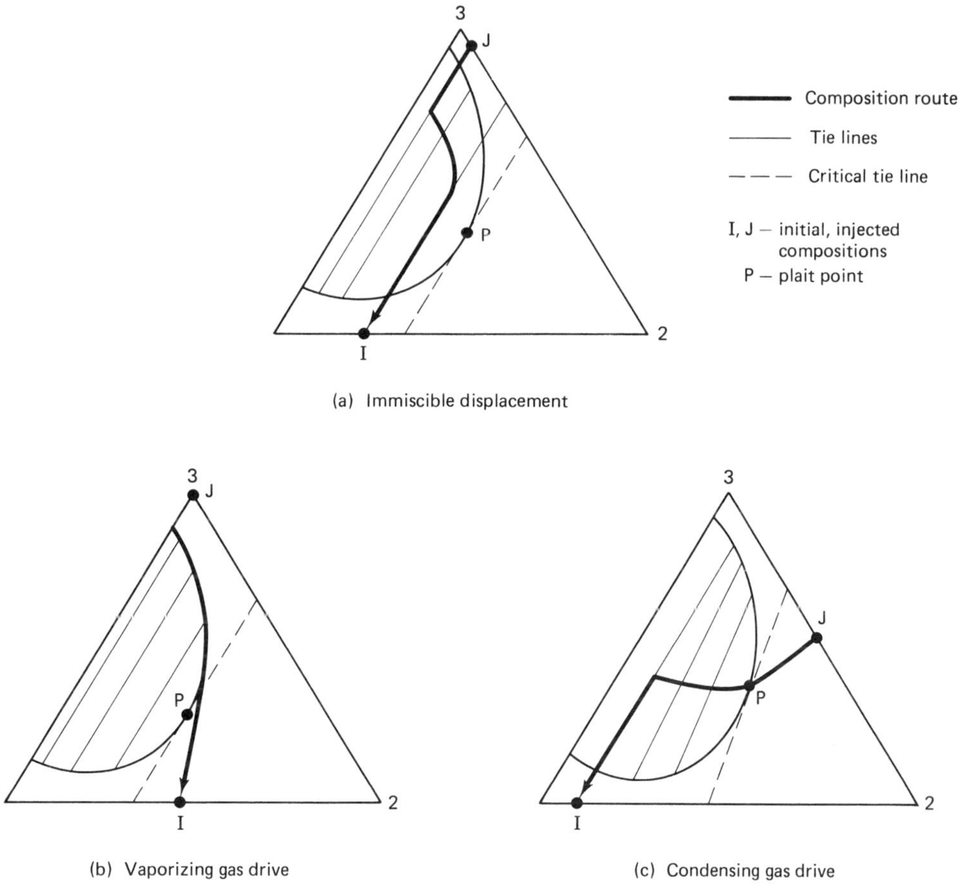

Figure 7-37 Composition routes for immiscible and developed miscibility processes

In the condensing gas drive process (Fig. 7-37c), the composition route enters the two-phase region through the plait point (Hutchinson and Braun, 1961), follows the equivelocity path, and then exits from the two-phase region on a tie line extension. The route definitely passes through the two-phase region, but it does so as a shock since the concentration velocity on the equivelocity curve is unity, and the tie line extension segment is a shock. Compare Fig. 7-37(c) to Fig. 7-15. Auxiette and Chaperon (1981) give an experimental investigation of these processes.

Both developed miscibility cases will appear as a first-contact displacement in the absence of dissipation. The similarity between developed and first-contact displacements justifies using first-contact approximations on *all* the developed miscibility displacements we discuss below.

First-Contact Miscible Displacements in the Presence of an Aqueous Phase

Water does not affect hydrocarbon phase behavior, and the water solubility of most solvents is small. But the inevitable presence of an aqueous phase can affect displacement behavior through fractional flow effects. In this section, we investigate the effects of an aqueous phase on a first-contact miscible displacement. Although the treatment here can be given formally, as was that discussed above, we present instead an entirely equivalent, but more direct, approach based on fractional flow curves.

To do this, we assume incompressible fluids and rock, no dissipative effects, and solvent–water relative permeabilities are the same as oil–water relative permeabilities. Thus a water–solvent fractional flow f_1^s differs from a water–oil fractional flow f_1 only by the difference between the use of solvent and oil viscosities and densities. Figure 7-38 shows both the f_1 and f_1^s curves based on relative permeabilities from Dicharry et al. (1972). Because the relative permeabilities do not change, residual phase saturations of both the aqueous and oleic phases are invarient. The initial condition I in the one-dimensional displacement is uniform with water cut f_{1I}.

We take an arbitrary injection condition J to be comprised of some prespecified proportion of solvent and water f_{1J}^s given on the solvent–water curve. Injecting water and solvent together in the so-called water-alternating-gas (WAG) process is commonly used in solvent floods. The solvent–water mixture has better volumetric sweep efficiency and is less prone to viscous fingering than solvent alone (Caudle and Dyes, 1958). The volumetric flow rate ratio of water to solvent in the injected fluid is the *WAG ratio* W_R, given by

$$f_{1J}^s = \frac{W_R}{1 + W_R} \tag{7.7-13}$$

In Eq. (7.7-13) and hereafter, we assume no solubility of solvent ($i = 3$) or oil ($i = 2$) in the aqueous ($i = j = 1$) phase, and we assume no solubility of water in the hydrocarbon ($j = 2$) phase.

In an actual WAG process, the water and solvent are usually injected in alternate slugs so that the cumulative volumes of solvent and water really define the WAG ratio rather than Eq. (7.7-13). The differences in displacement behavior caused by simultaneous injection rather than alternating injection have been investigated by Welch (1982).

Since the displacement is first-contact miscible, the wave between the injected solvent and the oil is indifferent. Hence the oil–solvent wave velocity is

$$v_3 = \frac{1 - f_{1J}^s}{1 - S_{1J}} \tag{7.7-14}$$

from Eq. (5.4-5b). Equation (7.7-14) neglects solvent adsorption. v_3 can also be written in terms of the change in water saturation change across the solvent–oil front

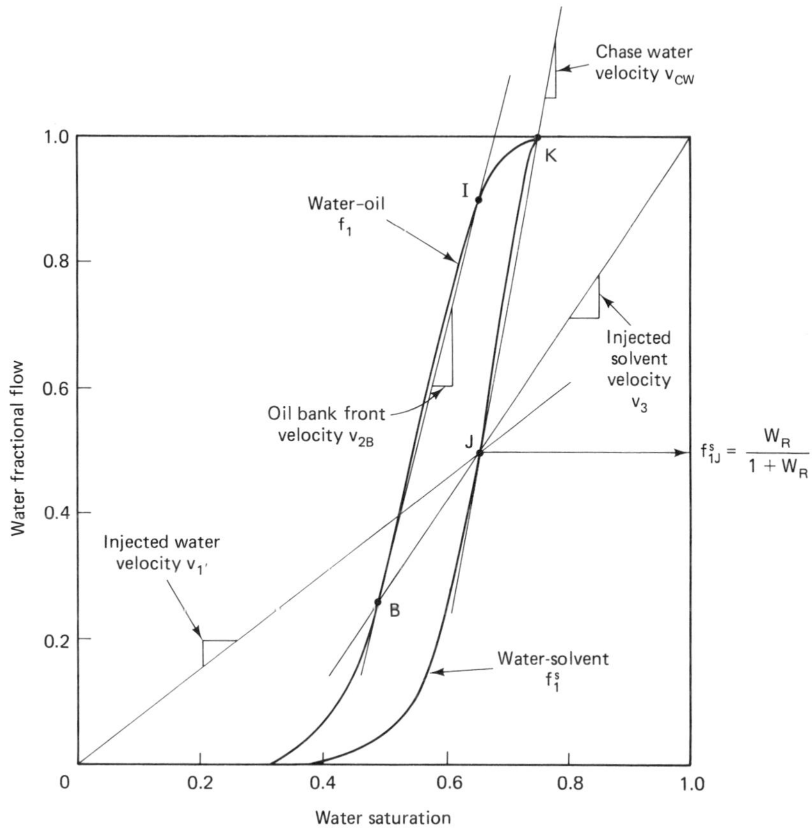

Figure 7-38 Schematic fractional flow construction for first-contact miscible displacements in the presence of an aqueous phase

$$v_3 = \frac{f^s_{1J} - f_{1B}}{S_{1J} - S_{1B}} \tag{7.7-15}$$

Equation (7.7-14) is the equation of a straight line from the upper right-hand corner of the fractional flow plot through the injected conditions (Fig. 7-38). Equating Eq. (7.7-14) to Eq. (7.7-15) says if this line is continued, its intersection with the water–oil fractional flow curve will give the water and oil saturation and fractional flow in the region ahead of the solvent–oil wave. Since $1 - f_{1B}$ is larger than $1 - f_{1I}$, the displaced oil forms a region of high oil saturation or *oil bank* ahead of the solvent–water front.

The leading edge of this oil bank flows with specific velocity v_{2B} given by

$$v_{2B} = \frac{f_{1I} - f_{1B}}{S_{1I} - S_{1B}} \tag{7.7-16}$$

also from Eq. (5.4-5b). This is the equation of a straight line from the initial conditions I to the oil bank, point B in Fig. 7-38. Since the injected water miscibly displaces the resident water, the specific velocity of the displaced water wave $v_{1'}$ is the straight line from the lower left corner of the fractional flow plot to the injected conditions (compare the lines for $v_{1'}$ and v_3 with case B in Fig. 5-12).

The velocity of the connate water banked up by the injected water is

$$v_{1'} = \frac{f^s_{1J}}{S_{1J}} \tag{7.7-17}$$

which is also shown in Figs. 7.38 and 7.39. The water ahead of this wave is banked-up connate water which, for a secondary solvent flood ($f_{1I} = 0$), constitutes a waterflood ahead of the solvent front. Caudle and Dyes (1958) verified experimentally that injecting at a WAG ratio so that the banked-up connate water does not propagate faster than the solvent resulted in optimal oil recovery.

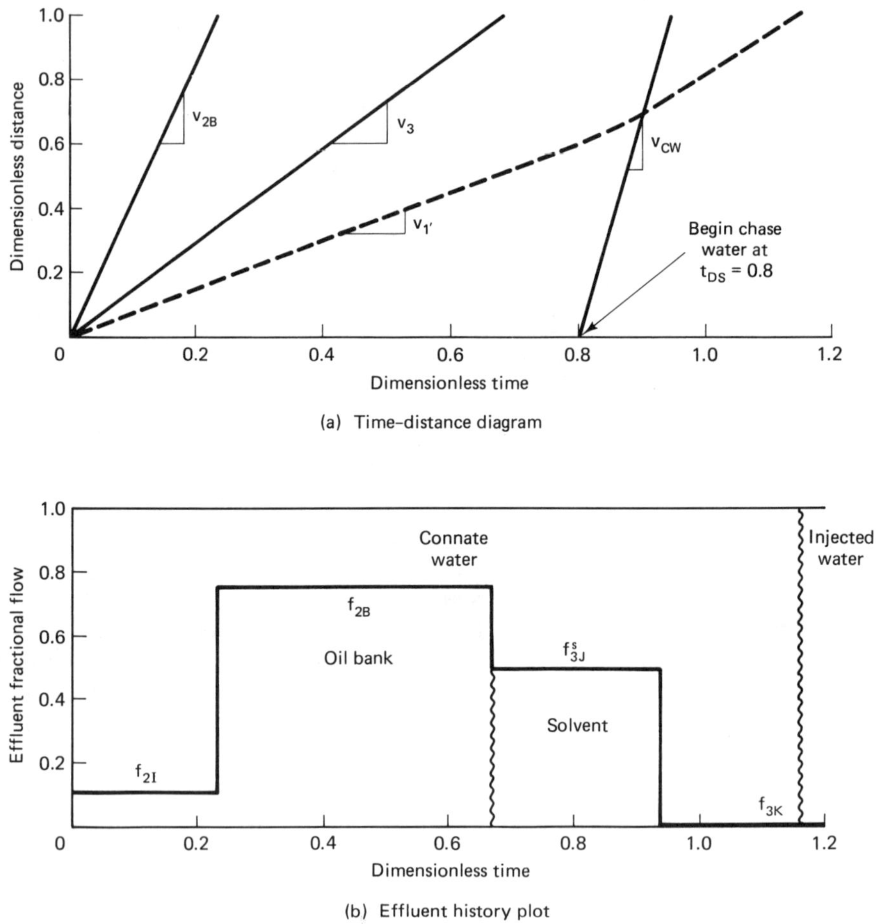

Figure 7-39 Time–distance diagram and effluent history plot for the displacement in Fig. 7-38

It is also possible to treat slug behavior with fractional flow theory. Suppose after injecting at condition J for some dimensionless time t_{Ds}, we follow with chase water at condition K on the diagram. Since the slope of the solvent–water fractional flow curve is monotonically decreasing from K to J, this displacement front is a shock with specific velocity

$$v_{cw} = \frac{1 - f_{1J}^s}{1 - S_{2r} - S_{1J}} \tag{7.7-18}$$

whose straight-line construction is also shown in Fig. 7-38. If the chase fluid were to be a second gas having the same properties as and first-contact miscible with the solvent, the velocity of the chase-fluid–solvent front would be given by the slope of the f_1^s-S_1 curve since the slope of this curve is monotonically increasing from $S_1 = S_{1r}$ to the injected conditions.

Figure 7-39 shows the time–distance diagram and effluent history for the displacement in Fig. 7-38. We have taken $t_{Ds} = 0.8$ to avoid interference between the oil-bank–solvent and the solvent chase water waves. In the effluent history plot, the miscible displacement fronts are designated by a wavy line.

Several general observations follow from these plots. First, the ultimate oil recovery is complete—that is, the final condition in the system is zero oil saturation. Of course, this is the natural consequence of first-contact miscible displacements where no residual phases are allowed. Second, oil production ceases when solvent breaks through. The moderately early solvent breakthrough is the consequence of the pore space inaccessible to the solvent caused by the presence of irreducible water: With no water present, the solvent slug always breaks through near $t_D = 1$. Based on hydrocarbon pore volumes $(1 - S_{1r})$, the solvent in Fig. 7-39 breaks through at $t_D = 0.96$ HCPV, which is much more in line with dispersion theory. Finally, the amount of solvent produced (0.14 PV) is considerably less than the amount injected (0.4 PV). This reduction is the consequence of trapping of the hydrocarbon-miscible solvent by the chase water. If the solvent slug size t_{Ds} were less than about 0.6, the chase-water–solvent front would have overtaken the solvent–oil-bank front and trapped some oil. Such an observation suggests a procedure whereby we could select the minimum solvent slug size ($t_{Ds} = 0.53$ in Fig. 7-39) that effects complete oil recovery.

7-8 SOLVENT FLOODS WITH VISCOUS FINGERING

Unfortunately, first-contact miscible displacements actually behave considerably differently than that shown in Fig. 7-38. Figure 7-40 shows the experimental results of a developed miscible displacement in a Berea core in which oil initially at residual conditions is displaced by a CO_2 solvent in a $W_R = 0$ displacement. The deviation of this displacement from a straight-line composition route in the ternary diagram was small. In the experimental displacement, the water cut was initially 1.0 and decreased to about 0.15 at $t_D = 0.15$. The water cut remained essentially constant until

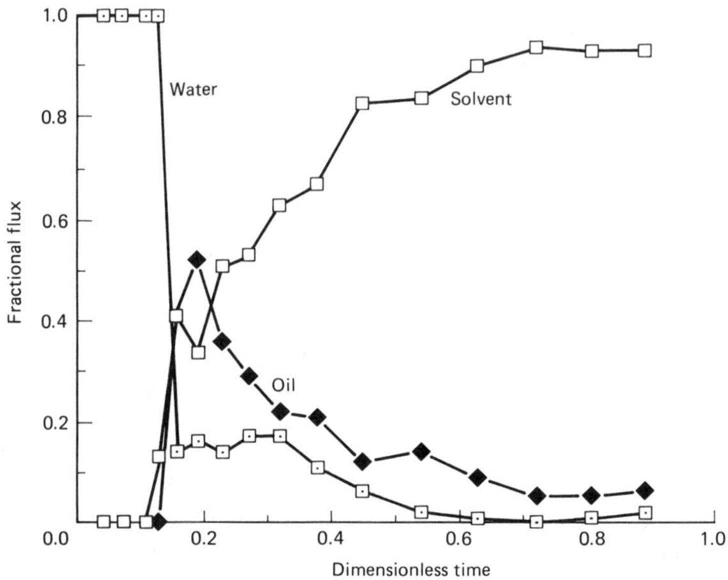

Figure 7-40 Effluent history of a carbon dioxide flood (from Whitehead et al., 1981)

about $t_D = 0.33$, at which point, it decreased gradually to 0. But when the water cut originally fell to $t_D = 0.14$, both oil and solvent broke through. This leaves a remaining oil saturation of about 0.25 at termination. It is unclear if 100% oil recovery would have been obtained had the experiment been continued. Several pore volumes of solvent injection would have been required, however.

The primary cause of the simultaneous oil and solvent breakthrough and prolonged oil recovery in experimental displacements is viscous fingering. In Sec. 6-8, we concluded that miscible displacements with typical solvents were always unstable, barring a gravity stabilization or a boundary effect, because the solvent–oil mobility ratio is greater than 1. Here we give descriptions of the character of simultaneous oil and solvent flow after the onset of fingering.

Heuristic Models

Because of the chaotic nature of viscous fingering, a rigorous mathematical theory is not possible. The behavior of a fingering displacement may be estimated by various heuristic theories, including (1) a modification of fractional flow theory (Koval, 1963), (2) rate-controlled mass transfer between solvent and oil fingers (Dougherty, 1963), (3) defining a suitably weighted mixture viscosity (Todd and Longstaff, 1972), (4) accounting for mixing in fingers directly (Fayers, 1984), and (5) defining a composition-dependent dispersion coefficient (Young, 1986).

In this section, we deal exclusively with the Koval theory; we leave the others as an exercise. By excluding the others, we do not imply the Koval approach is superior since all involve empirical parameters that must be determined by history matching. However, the Koval theory is in common use, and it fits naturally into our fractional flow theme.

The mixing zone length (the dimensionless distance between prespecified values of a cross-sectionally averaged concentration profile) of a fingering displacement, in the absence of boundary effects, grows in proportion to time. This observation prompted Koval to instigate a fractional flow theory for viscous fingering. If viscous fingers initiate and propagate, their growth in horizontal plane flow would look something like the cross section in Fig. 7-41, where the oil and solvent are in segregated flow. The displacement is first-contact miscible, with no dissipation, and without water present. If dissipation can vertically smear the fingers, the mixing zone will grow in proportion to the square root of time, as in dispersion theory. This growth can be quite small if longitudinal dispersion is small or the system length is large (Hall and Geffen, 1965).

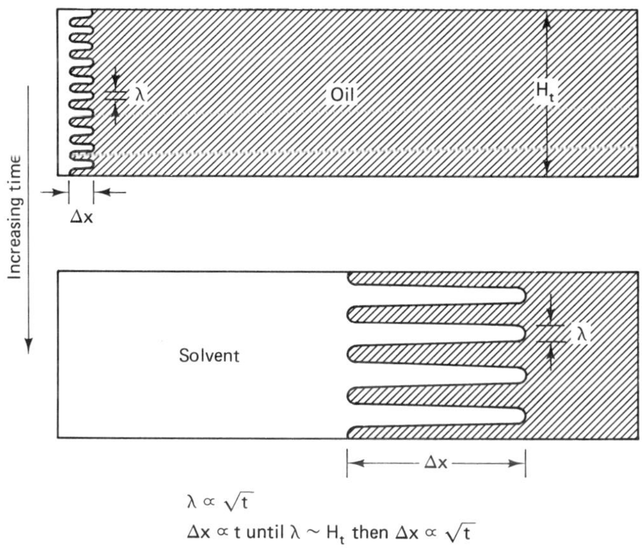

Figure 7-41 Idealization of viscous finger propagation (from Gardner and Ypma, 1982)

With these qualifications, the volumetric flow rate of solvent across a vertical plane within the mixing zone is

$$q_3 = -\frac{A_3 k}{\mu_3}\left(\frac{\partial P}{\partial x}\right) \tag{7.8-1a}$$

and that of oil is

$$q_2 = -\frac{A_2 k}{\mu_2}\left(\frac{\partial P}{\partial x}\right) \tag{7.8-1b}$$

where A_3 and A_2 are cross-sectional areas of oil and solvent. There are no relative permeabilities or capillary pressures in these equations since the displacement is first-contact miscible. These equations assume a horizontal displacement. The fractional flow of solvent in the oleic phase across the same vertical plane is

$$f_{32} = \frac{q_3}{q_3 + q_2}$$

by definition, which, when Eq. (7.8-1) is substituted, yields

$$f_{32} = \frac{A_3/\mu_3}{A_3/\mu_3 + A_2/\mu_2} \tag{7.8-2}$$

Equation (7.8-2) assumes the x-direction pressure gradients are equal in the oil and solvent fingers. Because the displacement is in plane flow, the oil and solvent cross-sectional areas are proportional to average concentrations, or

$$f_{32} = \left(1 + \frac{1}{\nu}\left(\frac{1 - \overline{C}_{32}}{\overline{C}_{32}}\right)\right)^{-1} \tag{7.8-3}$$

where ν is the oil–solvent viscosity ratio, and $\overline{C}_{32}$ is the average solvent concentration in the oleic phase across the cross section.

Equation (7.8-3) is a description of the segregated flow fingering in Fig. 7-41. Koval had to modify the definition of ν to match experimental displacements. The final form of the solvent fractional flow is

$$f_{32} = \left(1 + \frac{1}{K_{val}}\left(\frac{1 - \overline{C}_{32}}{\overline{C}_{32}}\right)\right)^{-1} \tag{7.8-4}$$

where K_{val} is the Koval factor.

Koval Corrections

The Koval factor modifies the viscosity ratio to account for local heterogeneity and transverse mixing in the following fashion:

$$K_{val} = H_K \cdot E \tag{7.8-5}$$

The parameter E changes the viscosity ratio to account for local mixing

$$E = (0.78 + 0.22\ \nu^{1/4})^4 \tag{7.8-6}$$

The consequence of Eq. (7.8-6) is that the numerical value of E is usually smaller than that of ν. That is, the effect of fingering is not as severe as it appears from the original viscosity ratio. The 0.22 and 0.78 factors in Eq. (7.8-8) seem to imply the solvent fingers contain, on the average, 22% oil, which causes the viscosity ratio attenuation through the quarter-power mixing rule. In fact, Koval eschewed this interpretation by remarking that the numerical factors were simply to improve the agreement with experimental results. This would seem to restrict Eq. (7.8-6) to the exact class of experiments reported by Koval. Remarkably, Claridge (1980) has shown that the 0.22–0.78 factors accurately describe fingering displacements over large ranges of transverse dispersion. Very likely the finger dilution is being caused by viscous crossflow since the mechanism is consistent with linear mixing zone growth (Waggoner and Lake, 1987).

The heterogeneity factor H_K corrects the reduced viscosity ratio for the local heterogeneity of the medium. Selecting the correct value for H_K is the most subjective feature of the Koval theory. In Fig. 6-8, the heterogeneity factor was calculated from the Dykstra-Parsons coefficient. It has also been correlated with the longitudinal Peclet number (Gardner and Ypma, 1982).

The fractional flow expression (Eq. 7.8-4) is the same as the water fractional flow in a waterflood where the oil and water have straight-line relative permeabilities. For such a case (see Exercise 5E), the Buckley-Leverett equation (Eq. 5.2-10) may be integrated analytically to give the following expression for effluent fractional flow:

$$f_{32}\big|_{x_D=1} = \begin{cases} 0, & t_D < \dfrac{1}{K_{val}} \\ \dfrac{K_{val} - \left(\dfrac{K_{val}}{t_D}\right)^{1/2}}{K_{val} - 1}, & \dfrac{1}{K_{val}} < t_D < K_{val} \\ 1, & K_{val} < t_D \end{cases} \tag{7.8-7}$$

The oil fractional flow is $1 - f_{32}\big|_{x_D=1}$. This equation has been compared to experimental data in the original Koval paper and elsewhere (Claridge, 1980; Gardner and Ypma, 1982).

Koval with Mobile Water

The Koval theory applies to first-contact miscible displacements in the absence of flowing water. The theory may be readily generalized to fingering first-contact miscible displacements with water present by modifying the overall flux and concentration definitions (see Sec. 5-4). The overall flux for oil and solvent becomes

$$F_2 = (1 - f_{32})f_2 \tag{7.8-8a}$$

$$F_3 = f_{32}f_2 \tag{7.8-8b}$$

where f_1 and f_2 are the actual water and hydrocarbon fractional flow functions, and f_{32} is given by Eq. (7.8-4). To be consistent with Eq. (7.8-6), the hydrocarbon phase viscosity in both f_1 and f_2 is given by the quarter-power mixing rule.

The overall concentrations of the oil and solvent are

$$C_2 = (1 - \overline{C}_{32})S_2 \tag{7.8-9a}$$

$$C_3 = \overline{C}_{32}S_2 \tag{7.8-9b}$$

The water concentration is simply S_1 because there is no solvent solubility in the water phase. Equations (7.8-8) and (7.8-9), substituted into the conservation equations for oil and solvent, may then be solved by the simple wave procedure discussed in Sec. 5-7 for the oil–gas–water problem.

Figure 7-42 shows the effluent fluxes for four displacements using this procedure. Figure 7-42(a) is for a non-WAG secondary flood, which is simply the results of the original theory (Eq. 7.8-7). Figure 7-42(b) is for a tertiary non-WAG dis-

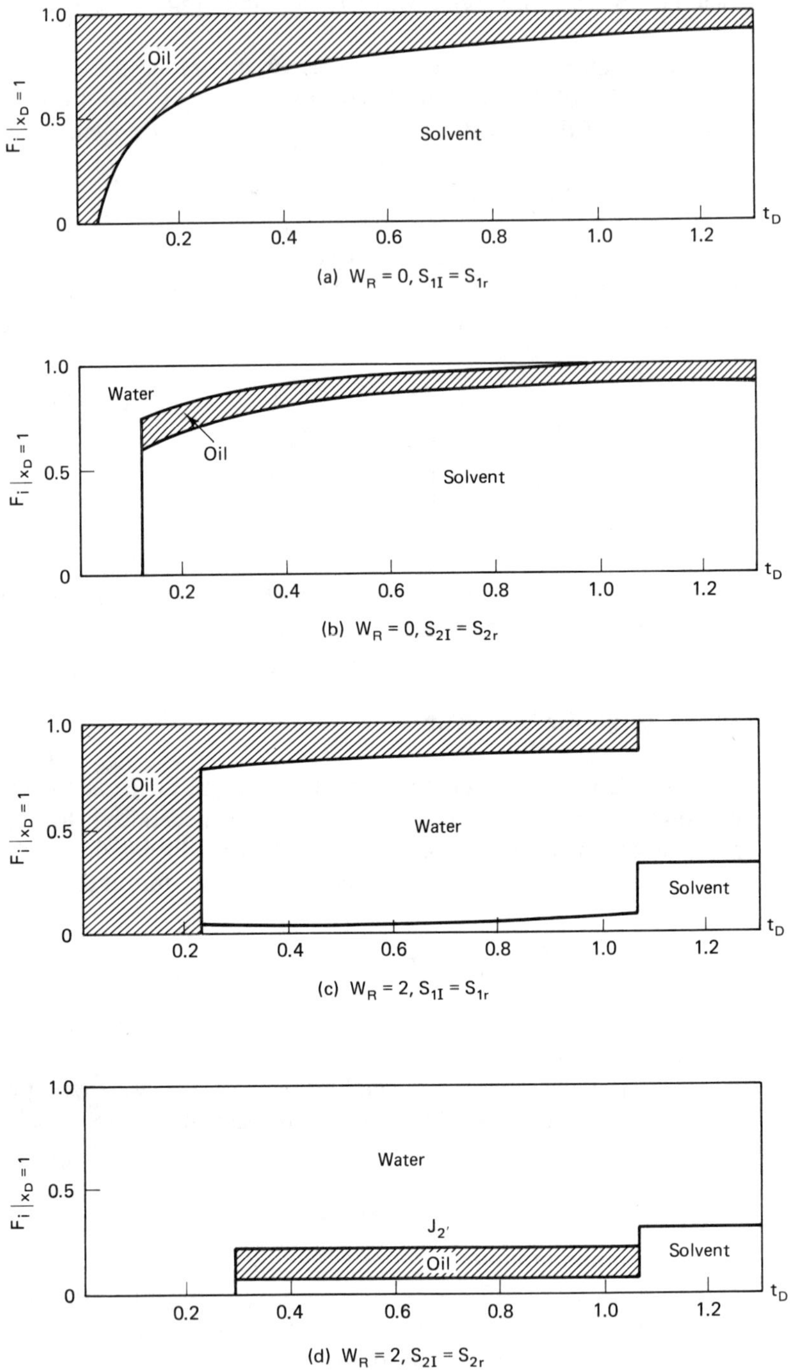

Figure 7-42 Effluent histories for four fingering cases

placement. Figures 7-42(c) and 7-42(d) are for secondary and tertiary WAG displacements ($W_R = 2$). The oil and water relative permeabilities in Figs. 7-39 and 7-42 are the same so that comparing Fig. 7-39 and Fig. 7-42(d) should reveal the effect of fingering on a first-contact miscible displacement with water present.

For both cases, the oil is produced as a bank of constant cut. But the bank oil cut is smaller for the fingering displacement, and the oil breakthrough and complete sweepout times are later. In the fingering case, oil and solvent break through together though the solvent is at low cut. By comparing Figs. 7-42(a) and 7-42(c) and Figs. 7-42(b) and 7-42(d), we see that, regardless of the initial conditions, the WAG procedure delays solvent breakthrough and hastens complete oil recovery.

Based on the comparisons in Fig. 7-42, it appears that WAG is universally better than injecting solvent alone, particularly when the solvent efficiency is considered. However, the presence of an initial mobile water saturation causes a residual oil saturation to even a first-contact displacement (see Sec. 7-9), and it is possible that the WAG process will cause this also.

Other methods besides WAG to improve mobility control in miscible flooding include the use of polymers (Heller et al., 1984) and foams. To date only foams have been extensively investigated, and since foams are envisioned to drive a variety of EOR processes, we delay their discussion until Chap. 10, where they more naturally fit after micellar-polymer flooding.

7-9 SOLVENT FLOODING RESIDUAL OIL SATURATION

A residual oil saturation in solvent flooding can come about by two broad phenomena: (1) a local heterogeneity (dead-end pores) in the permeable medium and (2) an interaction of dispersion or viscous fingering with the phase behavior. The former phenomenon occurs in first-contact miscible displacements, and the latter in developed miscible flood.

The definition of residual oil in a miscible flood (a paradoxical quantity) is slightly different from that in a waterflood. In a waterflood, residual oil is left behind as capillary-trapped globs, and no amount of throughput will displace this oil without some imposed change in the local capillary number. In a first-contact or developed miscible flood, all the oil, even that "trapped" by whatever mechanism, will eventually be recovered through extraction if enough solvent is injected. By residual oil in a miscible flood, then, we mean that quantity of oil left behind a solvent flood at some practical extreme of oil cut, oil rate, water–oil ratio, or gas–oil ratio (the data in Fig. 7-40 are up to a gas–oil ratio of about 550 SCM/SCM). Admittedly, this lacks the precision of the waterflood definition, but from the practical view of recovering oil economically, this distinction is not serious. By this definition, oil severely bypassed by a viscous finger is residual oil. Since we discussed capillary-trapped residual oil earlier in Sec. 3.4, we discuss other causes here.

Local Heterogeneity

To investigate the effects of local heterogeneity on trapped oil saturation, researchers have conducted experiments in laboratory cores on first-contact miscible displacements (Raimondi and Torcaso, 1964; Stalkup, 1970; Shelton and Schneider, 1975; Spence and Watkins, 1980). In these experiments, viscous fingering was suppressed by gravity stabilization or by matching the viscosity and density of the displacing and displaced fluids.

The miscible flood residual oil was found to depend on several things, the most important of which is the presence of a high mobile water saturation. A sample of this experimental data (Fig. 7-43) plots trapped oil saturation S'_{2r}, normalized by the waterflood residual oil saturation, versus the steady-state flowing water saturation. The normalized trapped oil saturations approach unity at high water saturations in these data and are close to zero for water saturations less than 50%. The steepness of the curves and the magnitudes of the residual saturations at high water saturation are of concern in displacements where a high water saturation is present (tertiary floods or WAG floods). The data in Fig. 7-43 were from displacements in strongly water-wet media. In oil-wet or intermediate-wet media, the trapping is not nearly as pronounced. Thus the trapped oil saturation has been correlated with capillary pressure curve hysteresis (Shelton and Schneider, 1975), a fractional relative permeability ratio (Raimondi and Torcaso, 1964), and dimensionless oil bank saturation (Stalkup,

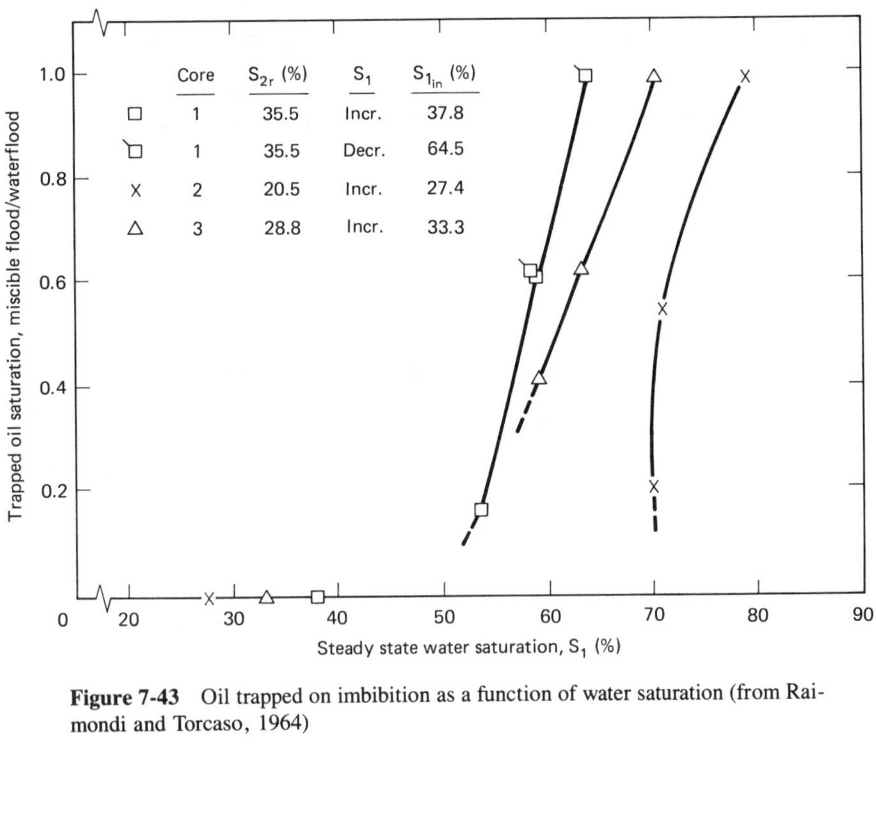

Figure 7-43 Oil trapped on imbibition as a function of water saturation (from Raimondi and Torcaso, 1964)

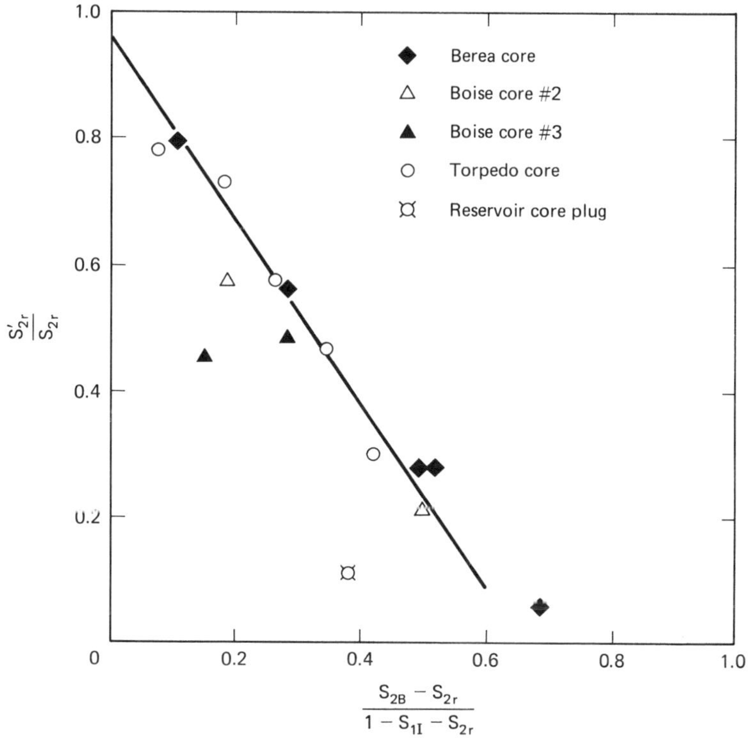

Figure 7-44 Influence of oil bank and residual oil saturation on the total stagnant hydrocarbon saturation (from Stalkup, 1970)

1970). Figure 7-44 shows the correlation of trapped or stagnant oil saturation with dimensionless oil bank saturation. S_{2B} is the oil bank saturation determined from the graphical construction in Fig. 7-39 and should contain corrections for the wettability of the medium since wettability is contained in the fractional flow curves. The oil bank saturation should contain corrections for injected water since the WAG ratio also affects the construction.

The most common interpretation for the effect of mobile water in miscible flood trapped oil saturation is that on a microscopic basis the water shields, or blocks, the solvent from contacting the oil. This explanation also qualitatively accounts for the effect of wettability since the oil and water phases are, depending on the wettability, differently distributed in the medium. In water-wet media, oil is contained in the large pores mostly away from the rock surfaces. The water phase is far more connected compared to the oil phase and thus could serve as a shield to oil originally present in pores not in the main flow channels. For oil-wet media, the phase distribution is reversed—the oil phase is the more continuous, and water is a less effective shield.

The interpretation of water blocking stagnant pores is somewhat like the dead-end pore model used to explain the behavior of water-free, first-contact displacements. The capacitance or dead-end pore model was originally proposed to explain the concentration "tail" observed in the breakthrough curves of first-contact, stable miscible displacements. This tail is more pronounced in carbonates than in sandstones (Fig. 7-45) because the pore structure of a typical carbonate is more heterogeneous (Spence and Watkins, 1980). Mathematical solutions fit the breakthrough curves well (Fig. 7-45) even though the physical interpretation of the parameters in the dead-end pore model has been questioned (Coats and Smith, 1964). The dead-

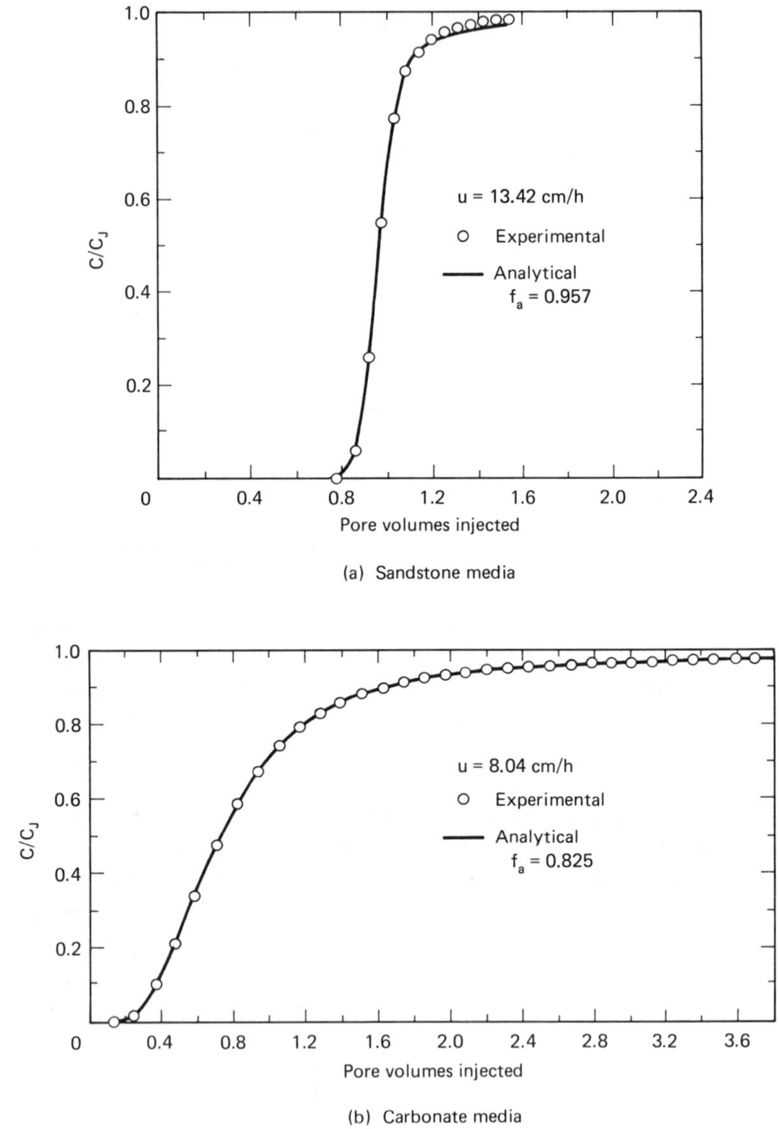

Figure 7-45 Typical breakthrough curves (from Spence and Watkins, 1980)

end pore model also qualitatively explains other features of first-contact miscible flood trapping, so we summarize the mathematical theory here.

Consider a stable, first-contact miscible displacement, in the absence of water, flowing in a permeable medium where a fraction f_a of the pore space is available to flow and a fraction $(1 - f_a)$ is stagnant. Solvent can flow from or into the stagnant or dead-end pores only by diffusion, represented by a mass-transfer coefficient k_m. The conservation equation for solvent becomes in the absence of dispersion

$$\phi f_a \frac{\partial C_{32}}{\partial t} + u \frac{\partial C_{32}}{\partial x} = -k_m(C_{32} - C_{3s})$$
$$(1 - f_a)\frac{\partial C_{3s}}{\partial t} = k_m(C_{32} - C_{3s}) \tag{7.9-1}$$

where C_{32} and C_{3s} are the solvent concentrations in the flowing and dead-end pores. With dimensionless distance and time, these equations become

$$f_a \frac{\partial C_{32}}{\partial t_D} + \frac{\partial C_{32}}{\partial x_D} = -N_{\mathrm{Da}}(C_{32} - C_{3s})$$
$$(1 - f_a)\frac{\partial C_{3s}}{\partial t_D} = N_{\mathrm{Da}}(C_{32} - C_{3s}) \tag{7.9-2}$$

where $N_{\mathrm{Da}} = k_m L\phi/u$ is the *Damkohler number,* a dimensionless quantity that is a ratio of the rates of diffusion from the dead-end pores to the bulk fluid flow. Equation (7.9-2) is a two-parameter (f_a and N_{Da}) representation of flow without dispersion. Deans (1963) gives the analytic solution to Eq. (7.9-2) subject to a step change in influent solvent concentration

$$\frac{C_{32}}{C_{3J}} = \begin{cases} 0, & Z \leqq 0 \\ 1 - e^{-Z} \int_0^Y e^{-\xi} - I_0(2\sqrt{\xi Z})\, d\xi, & Z \geqq 0 \end{cases} \tag{7.9-3}$$

where $Z = N_{\mathrm{Da}}(t_D - x_D f_a)/(1 - f_a)$, $Y = N_{\mathrm{Da}} x_D$, and I_0 is the modified Bessel function of the first kind, zero order. Equation (7.9-3) says the solvent concentration changes abruptly from zero to $C_{32}/C_{3J} = e^{-Y}$ at $Z = 0$.

The solvent effluent history (at $x_D = 1$) is from Eq. (7.9-3)

$$\frac{C_{32}}{C_{3J}} = \begin{cases} 0, & t_D \leqq f_a \\ 1 - e^{-Z} \int_0^{N_{\mathrm{Da}}} e^{-\xi} - I_0(2\sqrt{\xi Z})\, d\xi & t_D \geqq f_a \end{cases} \tag{7.9-4}$$

Figure 7-46 plots Eq. (7.9-4) for fixed f_a and various N_{Da}. For very small N_{Da}, the breakthrough curve behaves normally with the pore space contracted by $(1 - f_a)$. For this case, the miscible flood trapped oil saturation would simply be $(1 - f_a)$ times the oil saturation in the dead-end pores since the solvent cannot enter the stagnant pores. But for very large N_{Da}, the effect of the stagnant pore space vanishes since mass transfer to and from the flowing fraction is rapid. In this extreme, the trapped oil saturation should vanish.

These observations partly explain the dependence of miscible flood trapped oil

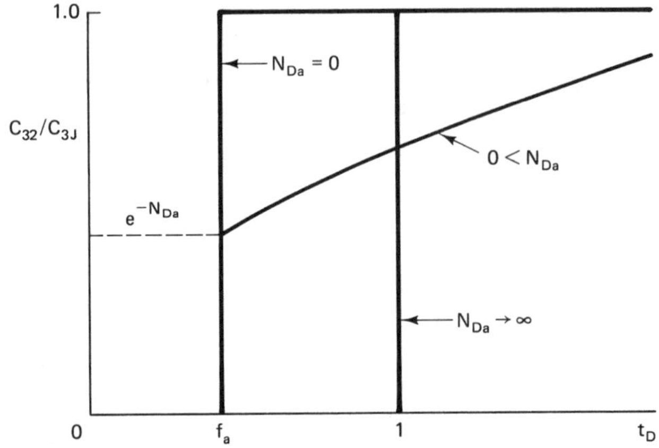

Figure 7-46 Effluent solvent concentration for fixed flowing fraction f_a and various N_{Da}; no dispersion

saturation on velocity and system length. As suggested by the definition of the Damkohler number, the trapped oil saturation should decrease with decreasing velocity and increasing system length. At field-scale conditions, large length and small velocity, the Damkohler number is usually much larger than in a laboratory experiment. Thus laboratory experiments may be overestimating miscible flood trapped oil saturation.

Including dispersion in Eq. (7.9-1) requires a numerical solution (Coats and Smith, 1964). Of course, the solutions so obtained fit experimental data better than Eq. (7.9-4) but do not alter the general conclusions.

The effect of water blocking is difficult to see from the preceding mathematics. For conceptual clarity, it is best to separate the water-blocking and dead-end pore effects by dividing the permeable medium pore space into flowing, isolated, and dendritic fractions (Salter and Mohanty, 1982). The flowing pore space is the fraction through which a phase flows into and from at least one pore throat. The dendritic fraction is connected to the flowing fraction through a mass transfer coefficient as above but does not exhibit flow itself. The isolated fraction of a phase is completely surrounded by the other phase through which no diffusion can occur. The amounts and properties of all fractions are functions of the phase saturations, the wettability of the medium, and the saturation history. Generally, the isolated and dendritic fractions vanish as the nonwetting phase saturation increases. But these two nonflowing fractions can occupy most of the total pore space at low nonwetting phase saturations.

Phase Behavior Interference

When the miscibility of a displacement is developed, the analysis is considerably complicated because, besides the water-blocking effect, a solvent flood can now trap oil by interactions with the phase behavior. Fig. 7-47 gives results from a combined experimental and theoretical study of Gardner et al. (1981) that shows the results of CO_2 displacements at two different pressure and dispersion levels. At both pressures,

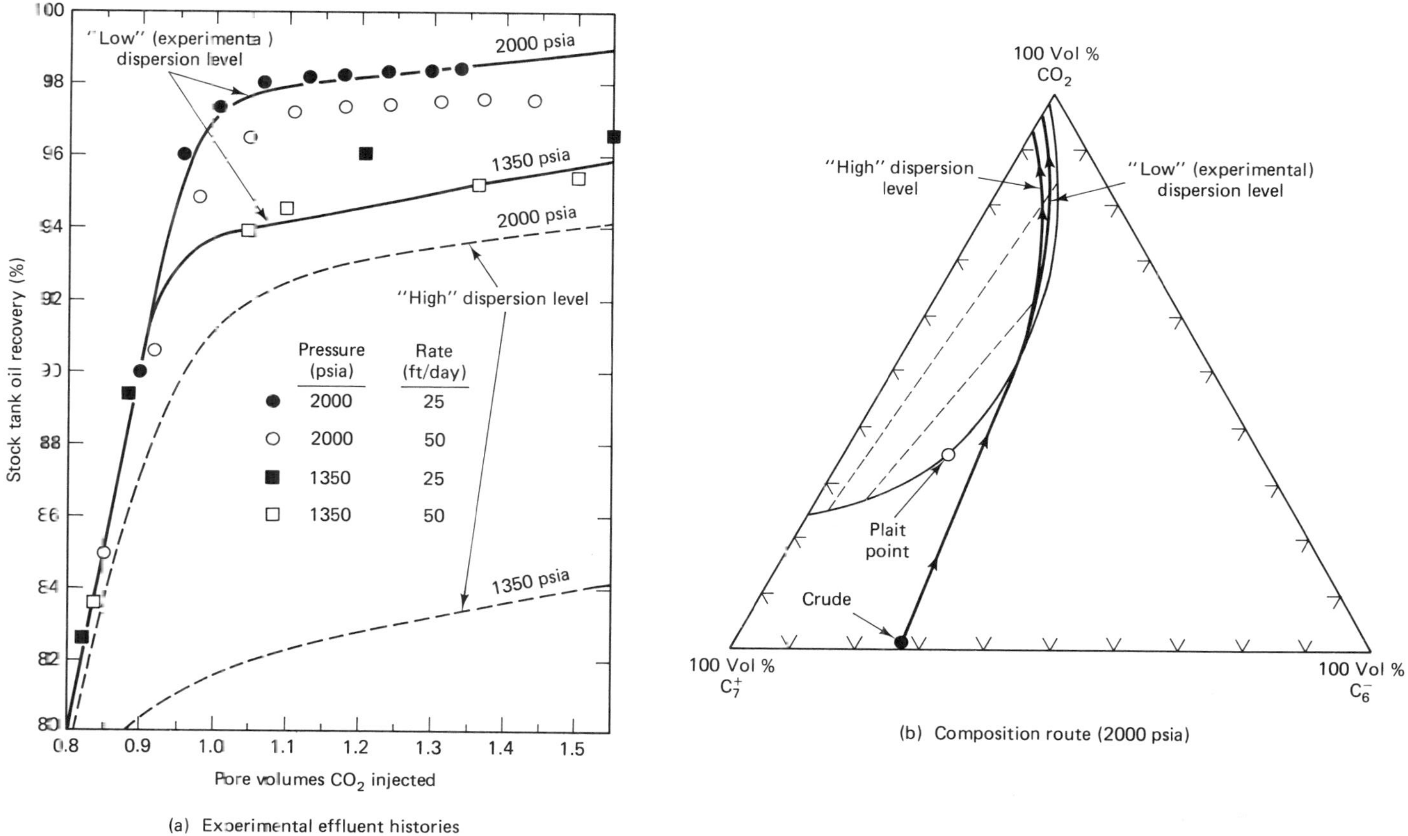

Figure 7-47 Results of CO_2 displacements at two different pressure and dispersion levels (from Gardner et al., 1981)

the displacements are vaporizing gas drives. Still, the lower pressure gives a measurably lower oil recovery than the higher pressure. The effect is relatively insensitive to rate, and there was no mobile water, indicating the lower recovery is caused by something more than the dead-end pore effect.

Figure 7-47(b) shows the composition route for the 13.6 MPa (2,000 psia) displacement in Fig. 7-47(a). Dispersion causes the composition route for this developed miscibility displacement to enter the two-phase region (compare this to the no-dispersion extreme in Fig. 7-37b). This intrusion will lower oil recovery because the trapped phase saturations within the two-phase region are large, the interfacial tension between the two hydrocarbon phases being large. Though the effect of dispersion on the experimental data ("low" dispersion level) is relatively minor, the simulated effect at the high dispersion level is pronounced.

The displacements in Fig. 7-47 were gravity stabilized so that it would be proper to ignore viscous fingering. That this phenomenon also contributes to the trapped oil saturation in an unstable displacement is demonstrated by the work of Gardner and Ypma (1982). Figure 7-48 shows literature data on trapped miscible oil saturation plotted versus residence ($L\phi/u$) time for several secondary CO_2 floods. The decrease in trapped oil saturation with residence time is very much like the decrease associated with increasing N_{Da} in the first-contact miscible floods discussed earlier. But the displacements in Fig. 7-48 were generally not stable, and there was no mobile water present.

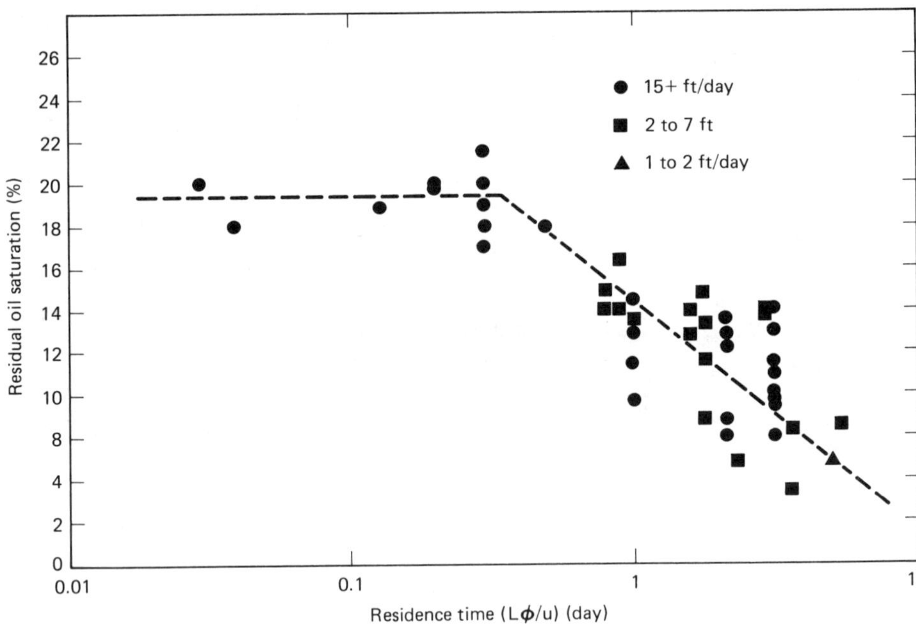

Figure 7-48 Literature data on trapped miscible flood oil saturation versus residence time (from Gardner and Ypma, 1982)

Gardner and Ypma interpret the large residual oil saturations at small residence times to be the consequence of a synergistic effect between the phase behavior and viscous fingering. They argue that in the longitudinal direction at the tip of the viscous finger, miscibility between the solvent and crude oil develops much like that shown in Fig. 7-37b. In the transverse direction, mixing takes place because of transverse dispersion and, perhaps, viscous crossflow. As we have seen, mixing due to dispersion causes straight-line dilution paths on pseudoternary diagrams (see Fig. 7-13). Such mixing does not cause developed miscibility unless very long residence times or very high transverse dispersion is allowed. Thus oil is first swept out by the longitudinal movement of a finger, the tip of which contains the light-enriched CO_2 solvent, and then reflows back into the finger from the transverse direction into a region of pure CO_2. Since CO_2 and crude are not first-contact miscible, multiple phases form in the finger, and trapping occurs. In fact, in simulations, it was observed that the trapped oil was actually present in highest amounts in the regions where the solvent fingers had passed because of this resaturation and phase behavior effect. Though this seems paradoxical—that the largest remaining oil saturation is where the solvent has swept—the contention is supported by correlating the data in Fig. 7-48 against a transverse dispersion group, reproducing this correlation with simulation, and finally, matching the effluent history of laboratory floods with the simulation results. Interestingly, the composition routes of zones both inside and outside the fingers passed well into the interior of the two-phase region of the ternary. When transverse dispersion is large, the transverse mixing takes place before the sol-

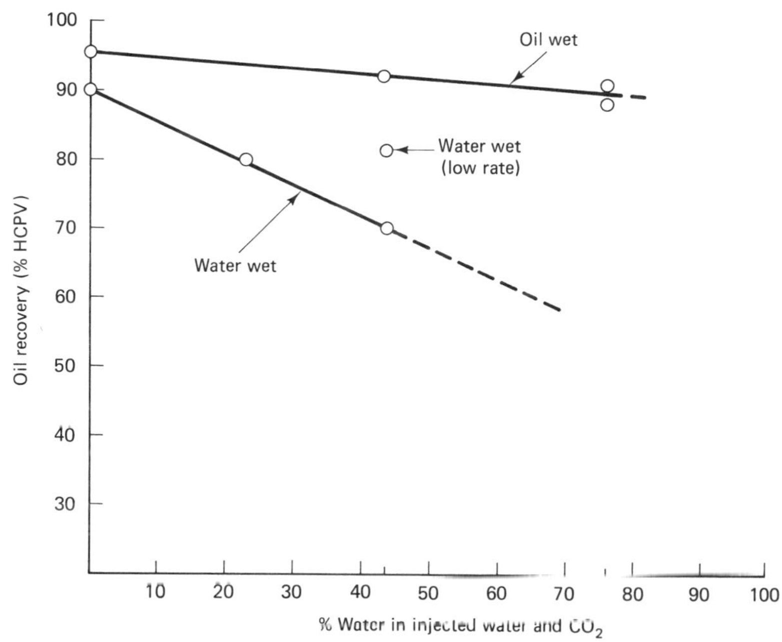

Figure 7-49 Oil recovery versus injected water fraction for tertiary CO_2 displacements in water-wet and oil-wet media (Tiffin and Yellig, 1982)

vent fingers have emptied of the displacing mixture, and trapped oil saturation goes down.

Undoubtedly, the interaction with phase behavior, dispersion, and viscous fingering all play a part in understanding these complex phenomena. Still, it seems persuasive that the wettability of the medium plays a central role, particularly since there seems to be a wettability effect in even the most complicated developed miscible, unstable, displacements (Fig. 7-49).

7-10 ESTIMATING FIELD RECOVERY

In this section, we combine the effects of areal sweep efficiency and displacement efficiency.

Assume we have a plot of average solvent and oil concentration versus dimensionless time in a one-dimensional displacement. This can be from an overall material balance of a laboratory experiment or from the fractional flow calculation in Secs. 7-7 and 7-8. Figure 7-52 shows the average concentrations from the experimental data in Fig. 7-40. The solid lines are the fractional flow solution.

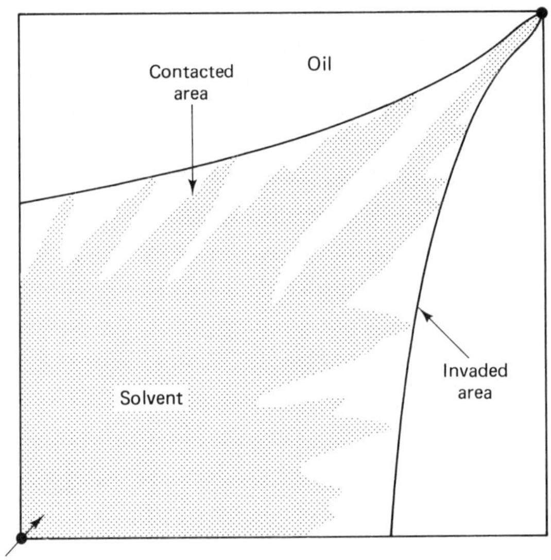

Figure 7-50 Schematic illustration of contacted and invaded area in quarter 5-spot pattern

In this section, we illustrate the correction of this data for areal sweep efficiency only. The procedure for correcting for vertical sweep is similar except we must now use a volumetric sweep efficiency function rather than an areal sweep efficiency function. The correction based on areal sweep would also be correct if the average concentration curves are corrected for vertical sweep, that is, were they averaged over a cross section using pseudofunctions.

Since we are explicitly including viscous fingering in the average concentration function, it is important *not* to include it in the areal sweep correlation also. Claridge anticipated this event by defining an "invaded area" sweep efficiency as shown in Fig. 7-51. He determined that the areal sweep correlation of Caudle and Witte (1959) most nearly approximated the invaded area sweep and derived equations to describe it for flow in a confined five-spot. He also gave a procedure for combining areal and displacement sweep for secondary, non-WAG displacements. Our procedure is a generalization of Claridge's to first-contact floods of arbitrary WAG ratios and arbitrary initial conditions.

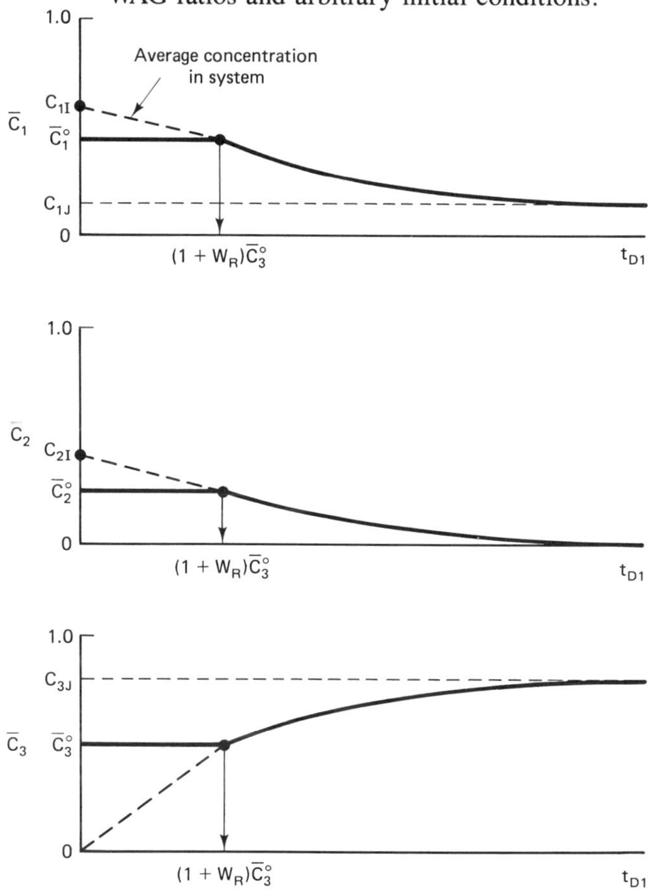

Figure 7-51 Schematic of the behavior of average concentrations

The invaded area is defined by a curve connecting the extreme tips of the viscous fingers (Fig. 7-50) and given by the product of E_A, the invaded areal sweep efficiency, and the pattern area. E_A as a function of dimensionless time and mobility ratio is given in Claridge's paper; it is not repeated here though it could be given graphically for a particular case. In Fig. 7-50, the contacted area is that actually occupied by the solvent fingers.

Central to the procedure is the idea of average concentrations behind the front. We define these to be the average concentrations in the invaded zone

$$\bar{C}_i = \frac{\text{Volume of component } i \text{ in invaded zone}}{\text{Volume of invaded zone}} \tag{7.10-1}$$

The average concentrations in the contacted area are $(C_1, C_2, C_3)_J$, the injected concentrations. The $\overline{C}_i$ are equal to the average concentration functions after breakthrough; before this, they are constant and equal to their breakthrough value $\overline{C}_i^0$ (see Fig. 7-52).

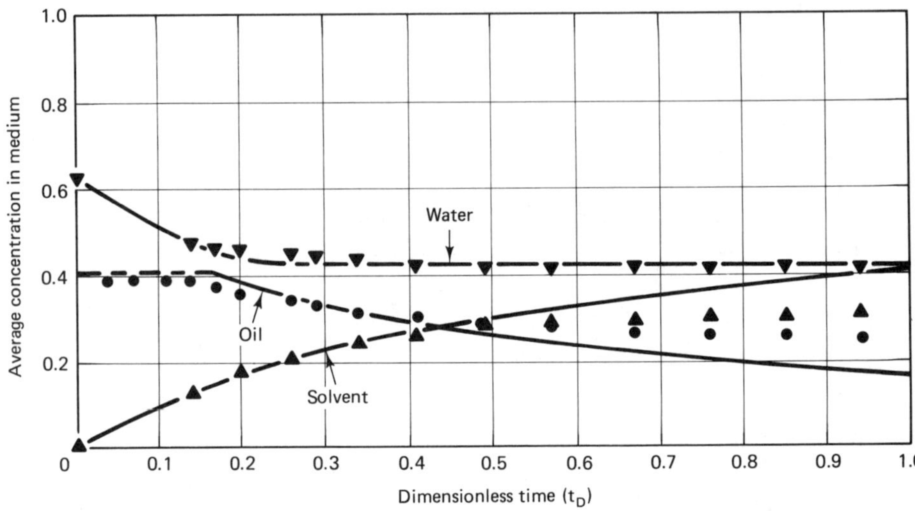

Figure 7-52 Average concentration from the experimental displacement in Fig.7-40

Since E_A and $\overline{C}_i$ are both known functions of time, the cumulative production of component i is

$$N_{pi} = \int_0^t F_{iJ}\, q dt - V_p E_A(\overline{C}_i - \overline{C}_{ij}), \qquad i = 1, 2, 3 \qquad (7.10\text{-}2)$$

from an overall material balance (Eq. 2.5–2). In Eq. (7.10-2), q is the injection–production rate, V_p is the total pore volume, and t is time. All volumes in this equation are in reservoir volumes. For oil, in particular, we can write

$$N_{p2D} = E_A\left(1 - \frac{\overline{C}_2}{C_{2I}}\right) \qquad (7.10\text{-}3)$$

where N_{p2D} is the cumulative oil produced expressed as a fraction of oil in place at the start of solvent injection $(N_{p2}/V_p C_{2I})$. To express oil recovery as a fraction of original (at discovery) oil in place, Eq. (7.10-3) should be multiplied by the ratio of C_{2I} to $(1 - S_{1r})$, the original oil saturation.

We begin here to use t_{D1} as the time variable for the C_i, and t_{D2} for E_A since, in general, neither function depends explicitly on the actual dimensionless time t_D in Eq. (7.10-3). We relate t_{D1}, t_{D2}, and t_D to one another below.

Breakthrough occurs at $t_{D1} = (1 + W_R)\overline{C}_3^0$ where $\overline{C}_3^0$ is the average solvent concentration behind the front at or before breakthrough. In Fig. 7-52, plotted curves show the average concentration in the one-dimensional system versus t_{D1} (dotted lines). In the following development, we do not use system average concentrations; we show them in Fig. 7-52 for completeness. Average system concentrations and average concentrations behind the front coincide after breakthrough.

Imagine a continuous one-dimensional permeable medium with $\overline{C}_2 = \overline{C}_2(t_{D1})$ and $\overline{C}_3 = \overline{C}_3(t_{D1})$ known. An appropriate definition for t_{D1} is

$$t_{D1} = \frac{\text{Volume solvent + Water injected}}{\text{Volume invaded}} \tag{7.10-4}$$

If we identify the flow-excluded regions with the uninvaded regions in Fig. 7-51, the dimensionless time t_{D1} becomes

$$t_{D1} = \frac{t_D}{E_A} \tag{7.10-5}$$

On the other hand, consider a homogeneous five-spot pattern with $E_A = E_A(t_{D2})$ known, into which solvent and water are being simultaneously injected. If the oil and water in the invaded region are regarded as part of the rock matrix, the appropriate dimensionless time t_{D2} becomes

$$t_{D2} = \frac{\text{Volume solvent injected}}{\text{Volume solvent in invaded region}} \tag{7.10-6}$$

which may be decomposed into

$$t_{D2} = \frac{\text{Volume solvent injected}}{\text{Volume solvent + Water injected}} \times \frac{\text{Volume solvent + Water injected}}{\text{Volume of pattern}} \times \frac{\text{Volume of pattern}}{\text{Volume solvent in invaded region}} \tag{7.10-7}$$

After breakthrough, t_{D2} may be written in our terminology as

$$t_{D2} = \frac{t_D}{(1 + W_R)\overline{C}_3} \tag{7.10-8}$$

Equations (7.10-5) and (7.10-8) are the relations among the various dimensionless times. Claridge calls t_{D1} and t_{D2} the *apparent* pore volumes injected for the appropriate variable. t_D may be eliminated between Eqs. (7.10-5) and (7.10-8) to give

$$t_{D1} = \frac{t_{D2}}{E_A}(1 + W_R)C_3 \tag{7.10-9}$$

The definitions in Eqs. (7.10-5) and (7.10-8) may be verified by observing that when breakthrough happens in an areal sense, it also happens in a one-dimensional sense. Thus at breakthrough, we have $t_{D2} = E_A$ and $\overline{C}_3 = \overline{C}_3^0$ from which it follows that $t_{D1} = (1 + W_R)\overline{C}_3^0$ from Eq. (7.10-9). Figure 7-52 shows this is indeed the correct dimensionless breakthrough time for the one-dimensional system.

The procedure to calculate the correct t_{D1}, t_{D2}, and t_D is iterative.

1. Estimate the mobility ratio $\overline{M}$ to be used in the areal sweep correlation. We take this to be the mobility ratio based on the average concentrations behind the front at breakthrough

$$\overline{M} = \frac{(\lambda_{rt})_{\overline{C}_i^0}}{(\lambda_{rt})_{\overline{C}_{iI}}} \qquad (7.10\text{-}10a)$$

This requires a knowledge of the relative permeability curves. If these are not available, $\overline{M}$ may be estimated from the one-dimensional data as

$$\overline{M} \cong \frac{(q/\Delta P)_{\text{final}}}{(q/\Delta P)_I} \qquad (7.10\text{-}10b)$$

where q = total volumetric rate, and ΔP = overall pressure drop. The numerical value of $\overline{M}$ does not change during the calculation.
2. For this value of $\overline{M}$, find the breakthrough areal sweep efficiency $t_{D2}^0 = E_A^0$. The dimensionless breakthrough time for the combined areal and displacement sweep is $t_D^0 = E_A^0(1 + W_R)\overline{C}_3^0$ from Eq. (7.10-8). The iterative calculations begin at t_D^0.
3. Fix $t_D > t_D^0$.
4. Pick $t_{D1} > \overline{C}_3^0(1 + W_R)$.
5. Calculate $\overline{C}_3(t_{D1})$ from the one-dimensional results.
6. Calculate t_{D2} from Eq. (7.10-8).
7. Estimate $E_A(t_{D2})$ from the areal sweep correlation.
8. Calculate t_{D1} from Eq. (7.10-5).
9. Test for convergence. If the t_{D1} estimated in steps 4 and 8 differ by less than some small preset tolerance, the procedure has converged; if not, reestimate t_{D1}, and return to step 4.
10. Calculate cumulative oil produced from Eq. (7.10-3), and calculate the combined fraction flow of each component from

$$F_i^T = \left(1 - \frac{dE_A}{dt_{D2}}\right)F_i(t_{D1}) + \frac{dE_A}{dt_{D2}}F_{iJ} \qquad (7.10\text{-}11)$$

11. Increment t_D, and return to step 3 for a later time. The entire procedure continues until t_D is larger than some preset maximum. The procedure converges in two to four iterations per step by simple direct substitution. The combined fractional flow in Eq. (7.10-11) represents contributions from the invaded zone (first term) and the uninvaded zone (second term), with all expressions being evaluated at consistent values of t_{D1}, t_{D2}, and t_D. The F_i terms in Eq. (7.10-11) are from the one-dimensional curves, and the derivatives are numerically evaluated. Once $F_i^T(t_D)$ is known, we calculate component rates in standard volumes as

$$q_i = \frac{qF_i^T}{B_i}, \qquad i = 1, 2, 3 \qquad (7.10\text{-}12a)$$

corresponding to a real time t by inverting

$$t_D = \int_0^t \frac{q\,dt}{V_p} \tag{7.10-12b}$$

Figure 7-53 shows the results of the corrections for areal sweep applied to the data in Fig. 7-52. The y-axis plots cumulative oil produced as a fraction of initial oil in place at the start of solvent injection, and the x-axis plots each of the three dimensionless times. The combined areal and displacement sweep case breaks through earlier than the other two and, except for early time, is everywhere smaller. The combined fractional oil recovery at a particular dimensionless time is not simply the product of displacement and areal sweep at that time. The correct dimensionless time for consistently evaluating the latter two is given by Eqs. (7.10-5) and (7.10-8). For this particular case, Fig. 7-53 indicates the combined oil recovery is roughly equally dependent on areal and displacement sweep efficiencies.

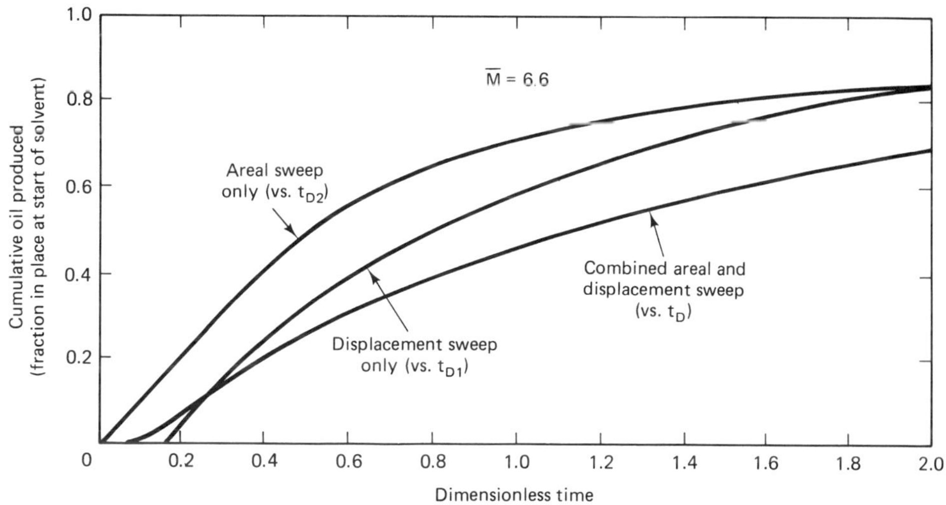

Figure 7-53 Calculated cumulative oil produced

7-11 CONCLUDING REMARKS

Solvent methods currently occupy a large fraction of implemented EOR methods. For certain classes of reservoirs—low permeability, fairly deep, and with light oil—they are clearly the method of choice. Future technology, particularly related to gravity stabilization and mobility control methods, could expand this range somewhat, but the target oil is nevertheless immense.

The topics of special importance in this chapter are the solvent flooding classifications, the usefulness of the minimum miscibility pressure correlations, and viscous fingering. The importance of viscous fingering remains largely unappreciated in large-scale displacement because of the obscuring effects of heterogeneity;

however, it is undoubtedly true that this phenomenon, perhaps in conjunction with others, accounts for the large discrepancy between lab-scale and field-scale oil recoveries. The material on dispersion and slugs and on solvent-water-oil fractional flow can form the basis for many design procedures. Of course, both topics easily lend themselves to the graphical presentation which is an essential part of this text.

EXERCISES

7A. *Immiscible Solvent*. A particular crude oil has a specific gravity of 0.76, normal boiling point of 324 K (124°F), molecular weight of 210 kg/kg-mole, and viscosity of 15 mPa-s. At 8.16 MPa (1,200 psia) and 322 K (120°F), estimate

(a) The CO_2 solubility in the oil

(b) The viscosity of the saturated CO_2–crude-oil mixture

(c) The swelling factor of the mixture

(d) The CO_2 water solubility, and express this as a mole fraction

Use the Simon and Graue correlations (Figs. 7-20 through 7-22) and the water solubility correlations (Fig. 7-23).

7B. *Calculating Minimum Miscibility Pressure*. An analysis of a particular separator oil is given below, including analyses at two different solution gas levels. Using the 1982 Holm and Josendal correlation (Fig. 7-25), estimate the minimum miscibility pressure (MMP) for the separator oil and the oil with 53.4 and 106.9 SCM dissolved gas/SCM dissolved oil. The reservoir temperature is 344 K (160°F).

	Weight percent		
Component	Separator oil	Oil + 53.4 SCM gas/SCM oil	Oil + 106.9 SCM gas/SCM oil
C_1		21.3	53.0
C_2		7.4	18.4
C_3		6.1	15.1
C_4		2.4	6.0
C_5–C_{30}	86	54.0	6.5
C_{31^+}	14	8.8	1.1

What can you conclude about the effect of solution gas on the MMP? How would you explain this with a ternary diagram?

7C. *Superposition and Multiple Slugs*. Using the principle of superposition applied to M influent step changes to a one-dimensional medium, show the composite solution to the convective–diffusion equation is

$$C_i = \frac{C_{i0} + C_{iM}}{2} - \frac{1}{2}\sum_{j=1}^{M} (C_{ij} - C_{ij-1})\phi\left(t_D - \sum_{k=1}^{j} t_{Dk}\right) \tag{7C-1}$$

where C_{ij} = injected concentration of component i during time interval j (C_{i0} is the same as C_{iI}), and t_{Dj} = duration of interval j, and where

$$\phi(t_D) = \text{erf}\left\{\frac{x_D - t_D}{2\sqrt{\frac{t_D}{N_{Pe}}}}\right\}$$

Eq. (7C-1) is valid only for $t_D > \sum_{k=1}^{j} t_{Dk}$.

7D. *Dilution Paths on Ternary Diagrams*. Plot for the following:

(a) Concentration profiles at $t_D = 0.5$ for the displacement of an oil of composition $(C_2, C_3)_I = (0.1, 0)$ by a small slug ($t_{Ds} = 0.1$) of composition $C_{2J} = 1.0$, which is then followed by a chase gas of composition $C_{3K} = 1.0$. Take the Peclet number to be 100.

(b) The dilution path of the concentration profile in part (a) on a ternary diagram as in Fig. 7-30.

7E. *Rich Gas Dilution*. Based on the ternary diagram in Fig. 7E with initial oil composition $(C_2, C_3,)_I = (0.1, 0)$,

(a) Determine the minimum intermediate component concentration (C_{2J}) that may be used in a continuous mixture of dry gas and intermediate displacing fluid that will ensure developed miscibility.

(b) Using the C_{2J} of part (a) as a lower bound, estimate the solvent slug size necessary to ensure first-contact miscibility at $t_D = 1$ for a series of C_{2J} values. Plot the total amount of intermediate injected ($C_{2J}t_{Ds}$) versus the slug size to determine an optimum. Take the Peclet number to be 1,000.

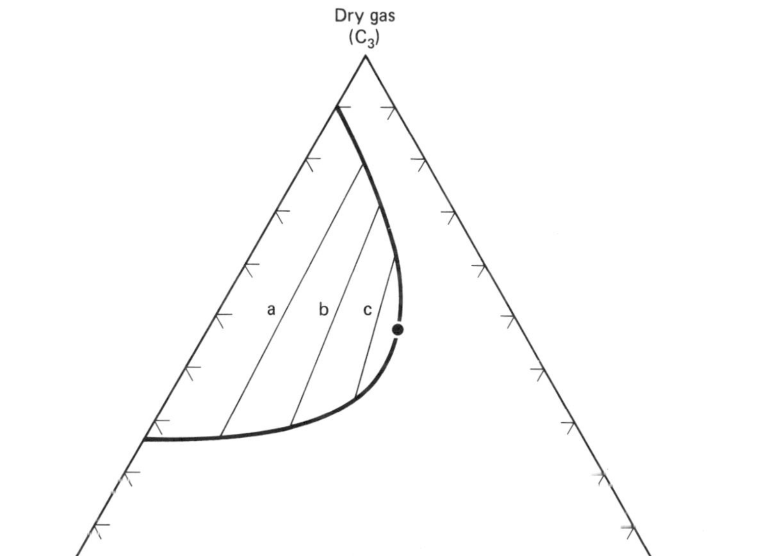

Figure 7E Ternary diagram for rich gas design problem

7F. *Fractional Flow Solution of Immiscible Displacement* The fractional flow curves along the three tie lines in Fig. 7E are shown in Fig. 7F. The straighter curves (with the smaller residual phase saturations) are nearer to the plait point. Phase 3 is that richest in component 3.

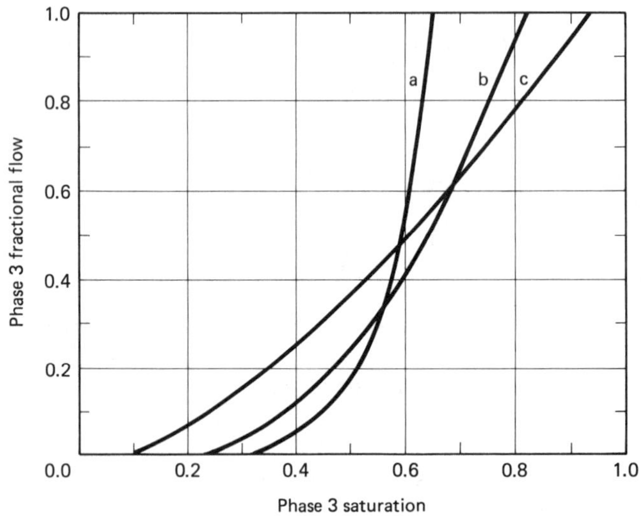

Figure 7F Fractional flow curve for Exercise 7F

(a) On the ternary diagram, sketch residual saturation lines, the singular curve(s), the equivelocity path, and as many nontie line paths as possible.

(b) Plot all the possible fractional flux curves you can. The initial oil composition is $(C_2, C_3)_I = (0.28, 0)$, and the injected solvent composition is $(C_2, C_3)_J = (0.09, 0.91)$. These compositions are on extensions of the lines farthest and nearest the plait point, respectively.

(c) Pick the physically possible solutions from the curves of part (b), and plot saturation and concentration profiles at $t_D = 0.8$.

7G. *WAG Calculations*. Figure 7G gives representative relative permeability curves for the Slaughter Estate Unit (SEU). The water, oil, and solvent viscosities are 0.5, 0.38, and 0.037 mPa-s, respectively.

(a) Plot the water–oil and water–solvent fractional flow curves. Assume the relative permeability curves for these pairs are the same and take $\alpha = 0$.

(b) Determine the optimal WAG ratio for a first-contact miscible secondary displacement in the absence of viscous fingering and dispersion.

(c) If the optimal WAG ratio is used, calculate the minimum solvent–water slug size (t_{Ds}) for complete displacement. The chase fluid is water.

(d) If the solvent–water slug size is 50% greater than that calculated in part (c), plot the time–distance diagram and effluent history for this displacement.

(e) Estimate the miscible flood trapped oil saturation S'_{2r} from Fig. 7-44.

7H. *Solvent Velocity with Water–Oil Solubility*

(a) Show that by including the solvent water solubility and the solubility of the solvent in a trapped oil saturation, the solvent specific velocity (Eq. 7.7-14) becomes

$$v_3 = \frac{1 - f_{1J}^s(1 - C_{31})}{1 - S_{1J}(1 - C_{31}) - S'_{2r}(1 - C_{32})} \tag{7H-1}$$

where C_{31} = solvent solubility in water $= R_{31}B_3/B_1$, and C_{32} = solvent solubility in oil $= R_{32}B_3/B_2$. R_{ij} is the solubility of component i in phase j in standard volumes of i per standard volumes of j. See Fig. 7-20 and 7-23.

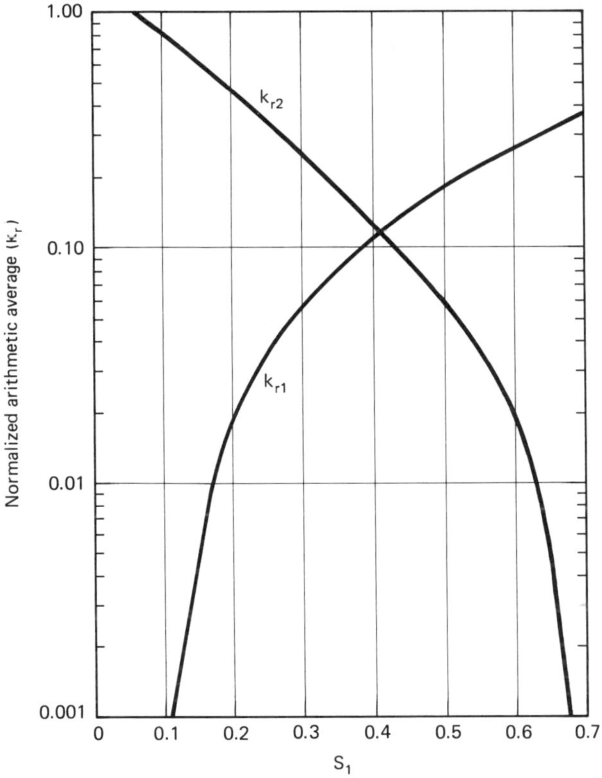

Figure 7G Slaughter Estate Unit relative permeability curves (from Ader and Stein, 1982)

(b) Using the S'_{2r} from part (e) of Exercise 7G, and taking $R_{31} = 17.8$ SCM/SCM, $R_{32} = 214$ SCM/SCM, $B_3 = 10^{-3}$ m³/SCM, $B_1 = 1$ m³/SCM, and $B_2 = 1.2$ m³/SCM, repeat parts (b–d) of Exercise 7G.

(c) Repeat parts (c) and (d) of Exercise 7G if the chase fluid is a gas having the identical properties of the solvent instead of water.

7I. *Carbonated Waterflooding Fractional Flow.* One of the earlier EOR techniques is displacement by CO_2-saturated water. This technique is amenable to fractional flow analysis (de Nevers, 1964).

(a) Show that the specific velocity of a pistonlike carbonated water front is given by

$$v_{\Delta C_3} = \frac{1 - \dfrac{K_{21}^3}{K_{21}^3 - 1}}{1 - S_{2r} - \dfrac{K_{21}^3}{K_{21}^3 - 1}} \tag{7I-1}$$

Equation (7I-1) assumes flow behind the front is at a CO_2-saturated residual oil phase.

(b) By matching the specific velocity of the oil bank rear to Eq. (7I-1) show the oil bank saturation and fractional flow must satisfy

$$v_{\Delta C_3} = \frac{1 - \dfrac{f_1 - C_{32}}{1 - C_{32}}}{1 - \dfrac{S_1 - C_{32}}{1 - C_{32}}} \tag{7I-2}$$

In these equations, K_{21}^3 is the volumetric partition coefficient of CO_2 ($i = 3$) between the water ($j = 1$) and oil ($j = 2$) phases, and C_{32} is the volume fraction of CO_2 in the oil. $f_1(S_1)$ is the water fractional flow curve.

(c) Estimate C_{32} and K_{21}^3 from Fig. 7-20 at 15 MPa and 340 K. You may assume ideal mixing in both phases.

(d) Calculate and plot the effluent oil cut of a carbonated waterflood in a one-dimensional permeable medium with initial (uniform) oil cut of 0.1.

(e) On the same graph, plot the effluent oil fractional flow of a noncarbonated waterflood. Finally, plot the incremental oil recovery (IOR) versus t_D.

For this problem, use the following parameters in the exponential relative permeability curves: $n_1 = n_2 = 2$, $k_{r1}^0 = 0.1$, $\phi = 0.2$, $k_{r2}^0 = 0.8$, $\mu_1 = 0.8$ mPa-s, $\mu_2 = 5$ mPa-s, $S_{1r} = S_{2r} = 0.2$, and $\alpha = 0$. The oil molecular weight is 200 kg/kg-mole, its density is 0.78 g/cm³, and the UOP factor is 11.2.

7J. *Viscous Fingering and Displacement Efficiency*. Using the Koval theory (Eq. 7.8-7), plot the effluent history of a first-contact miscible displacement where the oil–solvent viscosity ratio is 50, and the heterogeneity factor is 5.

7K. *Viscous Fingering by Mixing Parameter*. In the Todd-Longstaff (1972) representation of viscous fingering, the Koval factor K_{val} in Eq. (7.8-4) is replaced by K_{TL} where

$$K_{TL} = \frac{M_{2e}}{M_{3e}} = \nu^{1-\omega} \tag{7K-1}$$

where M_{2e} and M_{3e} = effective solvent and oil viscosities in the mixing zone, ν = viscosity ratio, and ω = mixing parameter ($0 < \omega < 1$).

(a) Repeat Exercise 7J with $\omega = 1/3$.

(b) Determine the correspondence between K_{val} and K_{TL} by setting $K_{TL} = K_{val}$ in Eq. (7.8-5) and plotting ω versus ν for various H_k.

7L. *Dispersion as a Normal Distribution*. One view of dispersion is that it is the result of the mixing of a large number of fluid particles along independent paths. If so, the distribution of particles should follow a normal distribution. In this exercise, we show that the equations in Sec. 7-6 reduce to such a form.

(a) Show that Eq. (7.6-4) applied to a unit slug $C_{iI} = C_{iK} = 0$ and

$$t_{Ds} C_{iJ} = 1 \tag{7L-1}$$

reduces to

$$C_i = \frac{1}{2t_{Ds}} \left\{ \operatorname{erf}\left[\frac{x_D - (t_D - t_{Ds})}{2\sqrt{\dfrac{t_D}{N_{Pe}}}}\right] - \operatorname{erf}\left[\frac{x_D - t_D}{2\sqrt{\dfrac{t_D}{N_{Pe}}}}\right]\right\} \tag{7L-2}$$

for $t_D >> t_{Ds}$.

(b) Using the definition for the error function (Eq. 5.5-14), show that Eq. (7L-2) becomes

$$C_i = \left(\frac{N_{\text{Pe}}}{4\pi t_D}\right)^{1/2} e^{-[(x_D - t_D)^2/(4t_D/N_{\text{Pe}})]} \qquad \text{(7L-3)}$$

as $t_{Ds} \rightarrow 0$. Equation (7L-3) says the distribution of a large number of particles at $x_D = 0$ initially approaches a normal distribution with mean position $x_D = t_D$ and a standard deviation of $2\sqrt{t_D/N_{\text{Pe}}}$.

7M. *Calculating Solvent Oil Recovery*. Figure 7M shows the volumetric sweep efficiency of a tertiary solvent displacement.

(a) Using the procedure in Sec. 7-10, estimate and plot cumulative oil recovery (fraction of oil in place at start of solvent injection) and oil cut versus dimensionless time. Use the average concentrations of Fig. 7-50.

(b) If the oil formation volume factor is 1.2 m^3/SCM, the reservoir pore volume is 160 hm^3, and the average injection rate is 80 m^3/day, calculate and plot the cumulative oil produced and oil rate versus time.

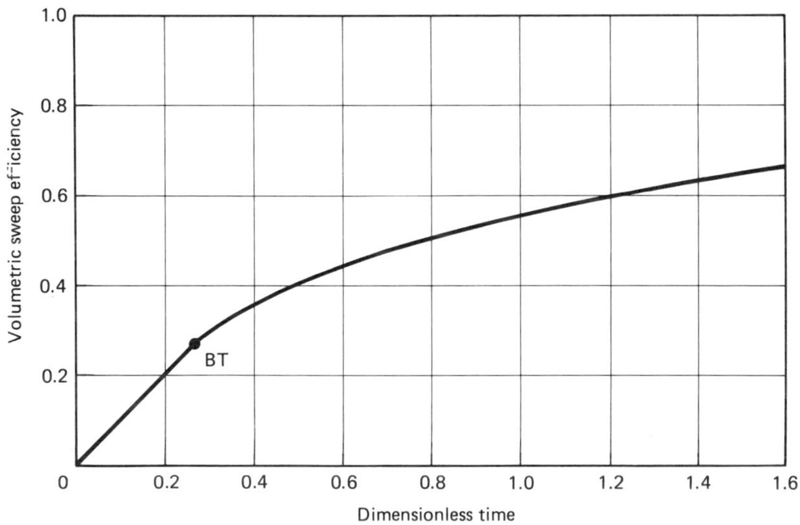

Figure 7M Volumetric sweep efficiency for miscible displacement

8

Polymer Methods

Polymer flooding consists of adding polymer to the water of a waterflood to decrease its mobility. The resulting increase in viscosity, as well as a decrease in aqueous phase permeability that occurs with some polymers, causes a lower mobility ratio. This lowering increases the efficiency of the waterflood through greater volumetric sweep efficiency and a lower swept zone oil saturation. Irreducible oil saturation does not decrease although the remaining oil saturation does, approaching S_{2r} for both waterflooding and polymer flooding. The greater recovery efficiency constitutes the economic incentive for polymer flooding when applicable. Generally, a polymer flood will be economic only when the waterflood mobility ratio is high, the reservoir heterogeneity is high, or a combination of these two occurs.

Polymers have been used in oil production in three modes.

1. As near-well treatments to improve the performance of water injectors or watered-out producers by blocking off high-conductivity zones.
2. As agents that may be cross-linked in situ to plug high-conductivity zones at depth in the reservoir (Needham et al., 1974).

   These processes require that polymer be injected with an inorganic metal cation that will cross-link subsequently injected polymer molecules with ones already bound to solid surfaces.
3. As agents to lower water mobility or water–oil mobility ratio.

The first mode is not truly polymer flooding since the actual oil-displacing agent is not the polymer. Certainly most polymer EOR projects have been in the third mode, the one we emphasize here. We discussed how lowering the mobility ratio affects displacement and volumetric sweep efficiency in Chaps. 5 and 6.

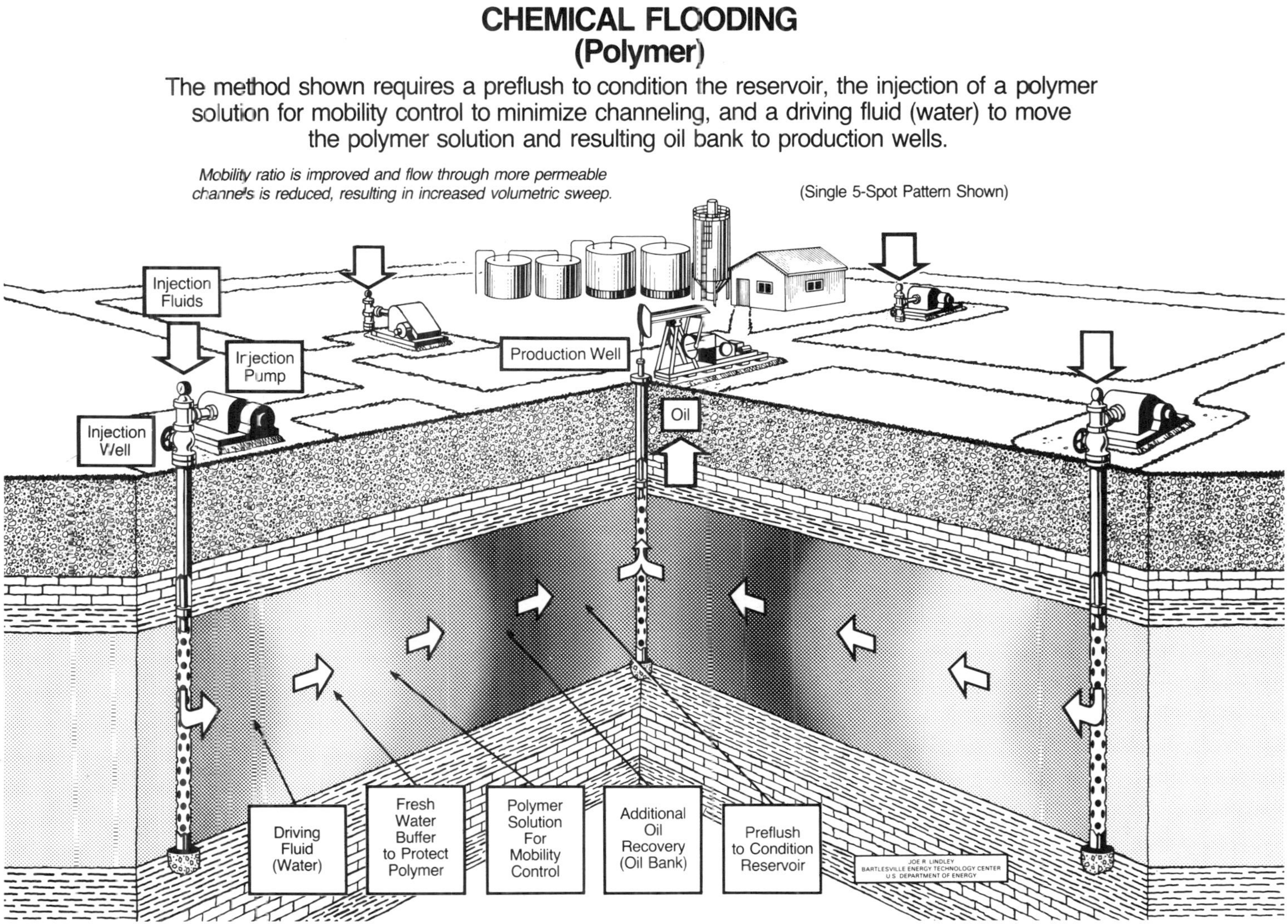

Figure 8-1 Schematic illustration of polymer flooding sequence (drawing by Joe Lindley, U.S. Department of Energy, Bartlesville, Okla.)

Figure 8-1 shows a schematic of a typical polymer flood injection sequence: a preflush usually consisting of a low-salinity brine; an oil bank; the polymer solution itself; a freshwater buffer to protect the polymer solution from backside dilution; and finally, chase or drive water. Many times the buffer contains polymer in decreasing amounts (a grading or taper) to lessen the unfavorable mobility ratio between the chase water and the polymer solution. Because of the driving nature of the process, polymer floods are always done through separate sets of injection and production wells.

Mobility is lowered in a polymer flood by injecting water that contains a high molecular weight, water-soluble polymer. Since the water is usually a dilution of an oil-field brine, interactions with salinity are important, particularly for certain classes of polymers.

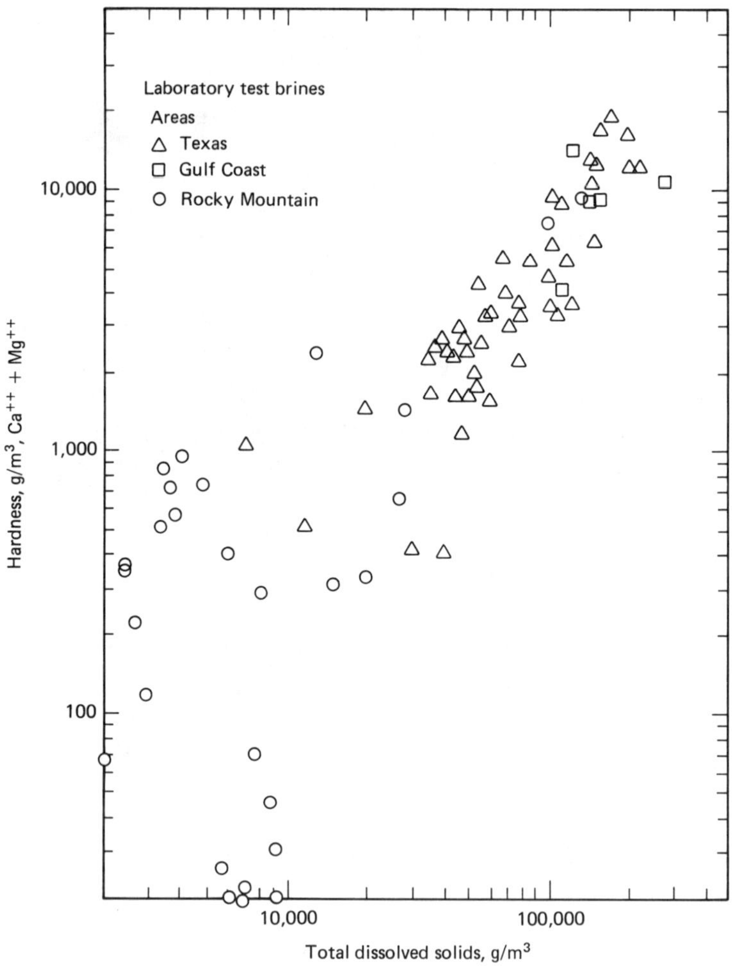

Figure 8-2 Salinities from representative oil-field brines (from Gash et al., 1981)

Salinity is the total dissolved solids (TDS) content of the aqueous phase. Figure 8-2 shows typical values. Virtually all chemical flooding properties depend on the concentrations of specific ions rather than salinity only. The aqueous phase's total divalent cation content (hardness) is usually more critical to chemical flood properties than the same TDS concentration. Figure 8-2 also shows typical brine hardnesses.

Because of the high molecular weight (1 to 3 million), only a small amount (about 500 g/m^3) of polymer will bring about a substantial increase in water viscosity. Further, several types of polymers lower mobility by reducing water relative permeability in addition to increasing the water viscosity. How polymer lowers mobility, and the interactions with salinity, can be qualitatively illustrated with some discussion of polymer chemistry.

8-1 THE POLYMERS

Several polymers have been considered for polymer flooding: Xanthan gum, hydrolyzed polyacrylamide (HPAM), copolymers (a polymer consisting of two or more different types of monomers) of acrylic acid and acrylamide, copolymers of acryl amide and 2-acrylamide 2-methyl propane sulfonate (AM/AMPS), hydroxyethylcellulose (HEC), carboxymethylhydroxyethylcellulose (CMHEC), polyacrylamide (PAM), polyacrylic acid, glucan, dextran polyethylene oxide (PEO), and polyvinyl alcohol. Although only the first three have actually been used in the field, there are many potentially suitable chemicals, and some may prove to be more effective than those now used.

Nevertheless, virtually all the commercially attractive polymers fall into two generic classes: polyacrylamides and polysaccharides (biopolymers). In the remainder of this discussion, we deal with these exclusively. Figure 8-3 shows representative molecular structures.

Polyacrylamides

These are polymers whose monomeric unit is the acrylamide molecule. As used in polymer flooding, polyacrylamides have undergone partial hydrolysis, which causes anionic (negatively charged) carboxyl groups (—COO^-) to be scattered along the backbone chain. The polymers are called partially hydrolyzed polyacrylamides (HPAM) for this reason. Typical degrees of hydrolysis are 30%–35% of the acrylamide monomers; hence the HPAM molecule is negatively charged, which accounts for many of its physical properties.

This degree of hydrolysis has been selected to optimize certain properties such as water solubility, viscosity, and retention. If hydrolysis is too small, the polymer will not be water soluble. If it is too large, its properties will be too sensitive to salinity and hardness (Shupe, 1981).

The viscosity increasing feature of HPAM lies in its large molecular weight.

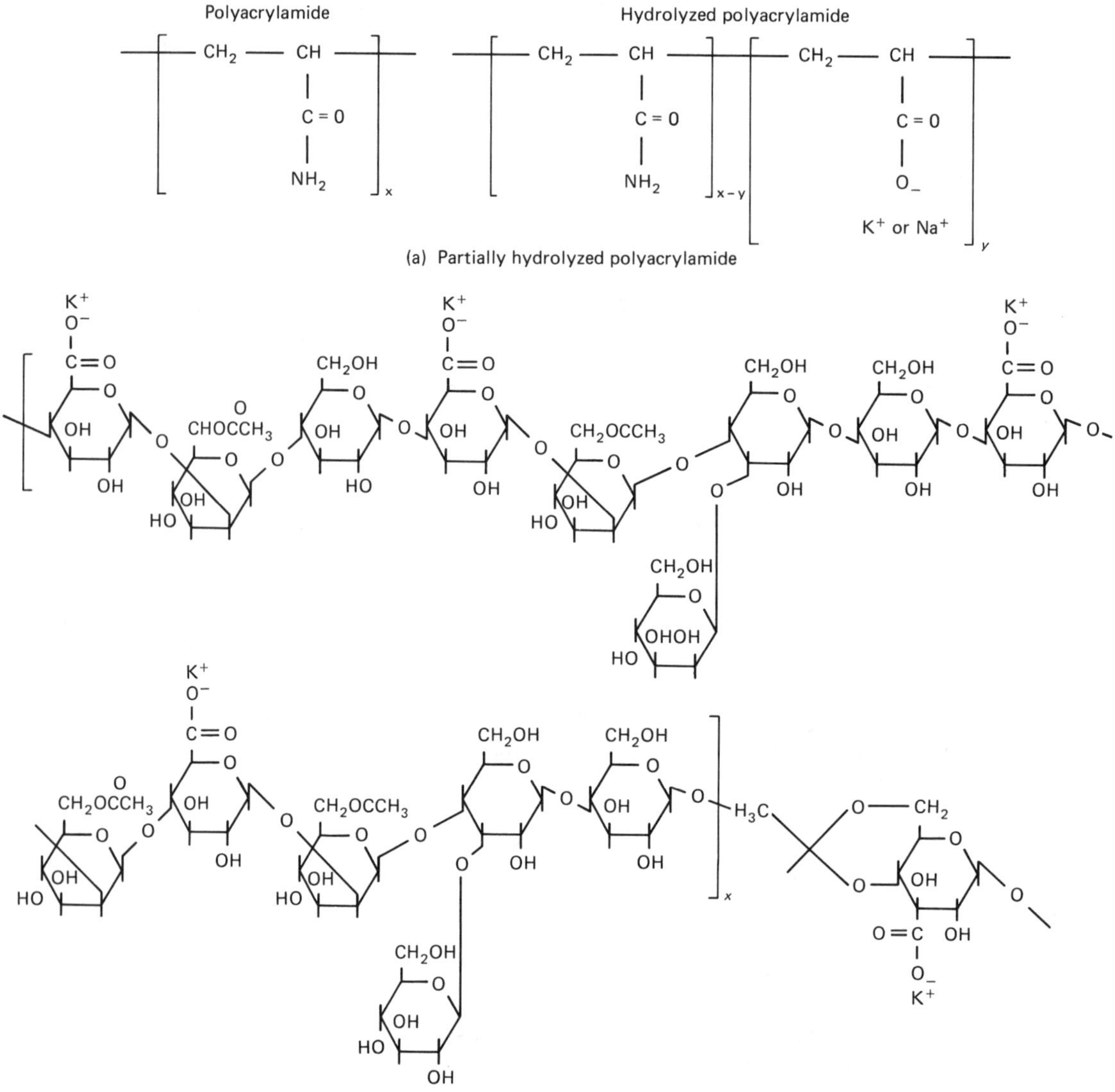

(a) Partially hydrolyzed polyacrylamide

(b) Polysaccharide (biopolymer)

Figure 8-3 Molecular structures (from Willhite and Dominguez, 1977)

This feature is accentuated by the anionic repulsion between polymer molecules and between segments on the same molecule. The repulsion causes the molecule in solution to elongate and snag on others similarly elongated, an effect that accentuates the mobility reduction at higher concentrations.

If the brine salinity or hardness is high, this repulsion is greatly decreased through ionic shielding since the freely rotating carbon–carbon bonds (Fig. 8-3a) allow the molecule to coil up. The shielding causes a corresponding decrease in the effectiveness of the polymer since snagging is greatly reduced. Virtually all HPAM properties show a large sensitivity to salinity and hardness, an obstacle to using HPAM in many reservoirs. On the other hand, HPAM is inexpensive and relatively resistant to bacterial attack, and it exhibits permanent permeability reduction.

Polysaccharides

These polymers are formed from the polymerization of saccharide molecules (Fig. 8-3b), a bacterial fermentation process. This process leaves substantial debris in the polymer product that must be removed before the polymer is injected (Wellington, 1980). The polymer is also susceptible to bacterial attack after it has been introduced into the reservoir. These disadvantages are offset by the insensitivity of polysaccharide properties to brine salinity and hardness.

Figure 8-3(b) shows the origin of this insensitivity. The polysaccharide molecule is relatively nonionic and, therefore, free of the ionic shielding effects of HPAM. Polysaccharides are more branched than HPAM, and the oxygen-ringed carbon bond does not rotate fully; hence the molecule increases brine viscosity by snagging and adding a more rigid structure to the solution. Polysaccharides do not exhibit permeability reduction. Molecular weights of polysaccharides are generally around 2 million.

Today, HPAM is less expensive per unit amount than polysaccharides, but when compared on a unit amount of mobility reduction, particularly at high salinities, the costs are close enough so that the preferred polymer for a given application is site specific. Historically, HPAM has been used in about 95% of the reported field polymer floods (Manning et al., 1983). Both classes of polymers tend to chemically degrade at elevated temperatures.

Polymer Forms

The above polymers take on three distinctly different physical forms: powders, broths and emulsions. Powders, the oldest of the three, can be readily transported and stored with small cost. They are difficult to mix because the first water contacting the polymer tends to form very viscous layers of hydration around the particles, which greatly slow subsequent dissolution. Broths are aqueous suspensions of about 10 wt.% polymer in water which are much easier to mix than powders. They have the disadvantage of being rather costly because of the need to transport and store large volumes of water. Broths are quite viscous so they can require special mixing facilities. In fact, it is this difficulty which limits the concentration of polymer in the broth. Emulsion polymers, the newest polymer form, contain up to 35 wt. % polymer solution, suspended through the use of a surfactant, in an oil-carrier phase. Once this water-in-oil emulsion is inverted (see Fig. 9-5), the polymer concentrate can be mixed with make-up water to the desired concentration for injection. The emulsion flows with roughly the same viscosity as the oil carrier, which can be recycled.

8-2 POLYMER PROPERTIES

In this section, we present qualitative trends, quantitative relations, and representative data on the following properties: viscosity relations, non-Newtonian effects, polymer transport, inaccessible pore volume, permeability reduction, chemical and biological degradation, and mechanical degradation.

Viscosity Relations

Figure 8-4 shows a plot of Xanflood viscosity versus polymer concentration. This type of curve has traditionally been modeled by the Flory-Huggins equation (Flory, 1953)

$$\mu_1' = \mu_1[1 + a_1 C_{41} + a_2 C_{41}^2 + a_3 C_{41}^3 + \cdots +] \tag{8.2-1}$$

where C_{41} is the polymer concentration in the aqueous phase, μ_1 is the brine (solvent) viscosity, and a_1, a_2, and so on are constants. In the remainder of this chapter we drop the second subscript 1 on the polymer concentration since polymer is always in an aqueous phase. The usual polymer concentration unit is g/m^3 of solution, which is approximately the same as ppm. The linear term in Eq. (8.2-1) accounts for the dilute range where the polymer molecules act independently (without entanglements). For most purposes, Eq. (8.2-1) can usually be truncated at the cubic term.

For a 1,000 g/m^3 Xanflood solution at 0.1 s^{-1} in 1 wt % NaCl brine at 24°C,

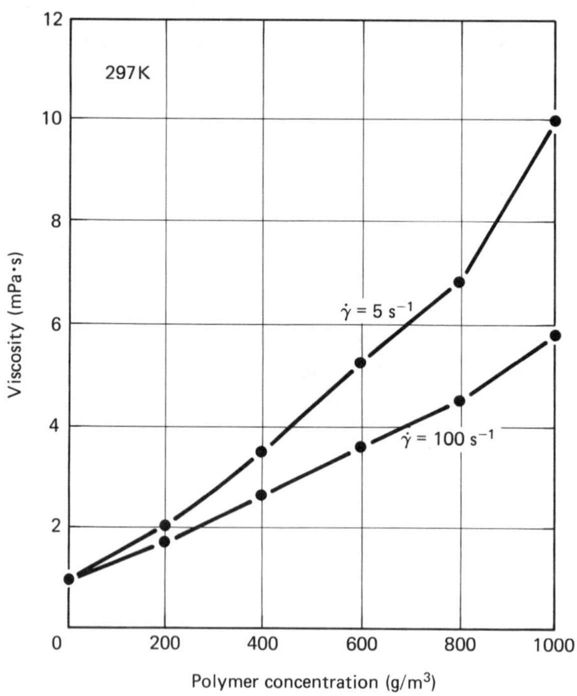

Figure 8-4 Xanflood viscosity versus concentration in 1% NaCl brine (from Tsaur, 1978)

the viscosity is 70 mPa-s (70 cp) from Fig. 8-4. Compared to the brine at the same conditions, this is a substantial increase in viscosity brought about by a relatively dilute concentration (recall that 1,000 g/m^3 = 0.1 wt %). Xanflood at these conditions is an excellent thickener.

A more fundamental way of measuring the thickening power of a polymer is through its intrinsic viscosity, defined as

$$[\mu] = \underset{C_4 \to 0}{\text{limit}} \left[\frac{\mu_1' - \mu_1}{\mu_1 C_4} \right] \tag{8.2-2}$$

From its definition, $[\mu]$ is a measure of the polymer's intrinsic thickening power. It is insensitive to the polymer concentration. The intrinsic viscosity for the Xanflood polymer under the conditions given above is 70 dl/g, the units being equivalent to reciprocal weight percent. Intrinsic viscosity is the same as the a_1 term in Eq. (8.2-1).

For any given polymer–solvent pair, the intrinsic viscosity increases as the molecular weight of the polymer increases according to the following equation (Flory, 1953):

$$[\mu] = K' M_w^a \tag{8.2-3}$$

The exponent varies between about 0.5 and 1.5 and is higher for good solvents such as freshwater. K' is a polymer-specific constant.

The above relationships are useful for characterizing the polymer solutions. For example, the size of the polymer molecules in solution can be estimated from Flory's (1953) equation for the mean end-to-end distance

$$d_p = 8(M_w[\mu])^{1/3} \tag{8.2-4}$$

This equation, being empirical, presumes certain units; $[\mu]$ must be in dl/g, and d_p is returned in Angstroms (10^{-10} m). This measure of polymer size is useful in understanding how these very large molecules propagate through the small pore openings of rocks. The molecular weight of Xanthan gum is about 2 million. From Eq. (8.2-4), d_p is about 0.4 μm. This is the same size as many of the pore throats in a low-to-moderate permeability sandstone. As a result, we would expect to, and in fact do, observe many polymer–rock interactions.

Non-Newtonian Effects

Figure 8-5 shows polymer solution viscosity μ_1' versus shear rate $\dot{\gamma}$ measured in a laboratory viscometer at fixed salinity. At low shear rates, μ_1' is independent of $\dot{\gamma}$ ($\mu_1' = \mu_1^0$), and the solution is a Newtonian fluid. At higher $\dot{\gamma}$, μ_1' decreases, approaching a limiting ($\mu_1' = \mu_1^\infty$) value not much greater than the water viscosity μ_1 at some critical high shear rate. This critical shear rate is off-scale to the right in Fig. 8-5. A fluid whose viscosity decreases with increasing $\dot{\gamma}$ is *shear thinning*. The shear thinning behavior of the polymer solution is caused by the uncoiling and unsnagging

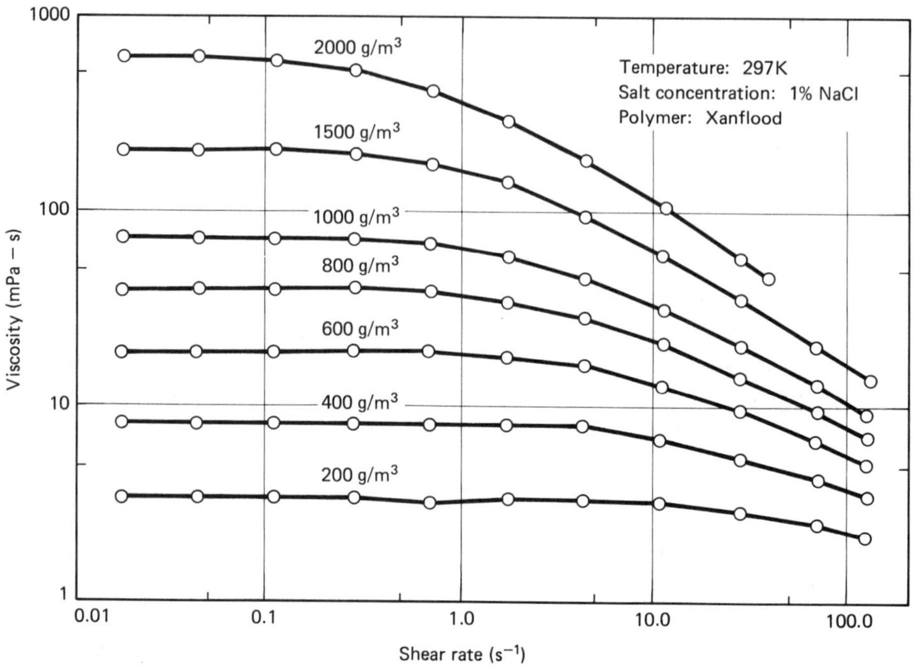

Figure 8-5 Polymer solution viscosity versus shear rate and polymer concentration (from Tsaur, 1978)

of the polymer chains when they are elongated in shear flow. Below the critical shear rate, the behavior is part reversible.

Figure 8-6 shows a viscosity–shear-rate plot at fixed polymer concentration with variable NaCl concentration for an AMPS polymer. The sensitivity of the viscosity to salinity is profound. As a rule of thumb, the polymer solution viscosity decreases a factor of 10 for every factor of 10 increase in NaCl concentration. The viscosity of HPAM polymers and HPAM derivatives are even more sensitive to hardness, but viscosities of polysaccharide solutions are relatively insensitive to both.

The behavior in Figs. 8-5 and 8-6 is favorable because, for the bulk of a reservoir's volume, $\dot{\gamma}$ is usually low (about 1–5 s^{-1}), making it possible to attain a design mobility ratio with a minimal amount of polymer. But near the injection wells, $\dot{\gamma}$ can be quite high, which causes the polymer injectivity to be greater than that expected based on μ_1^0. The relative magnitude of this enhanced injectivity effect can be estimated (Sec. 8-3) once quantitative definitions of shear rate in permeable media, and shear-rate–viscosity relations are given.

The relationship between polymer-solution viscosity and shear rate may be described by a *power-law* model

$$\mu_1' = K_{pl}(\dot{\gamma})^{n_{pl}-1} \tag{8.2-5}$$

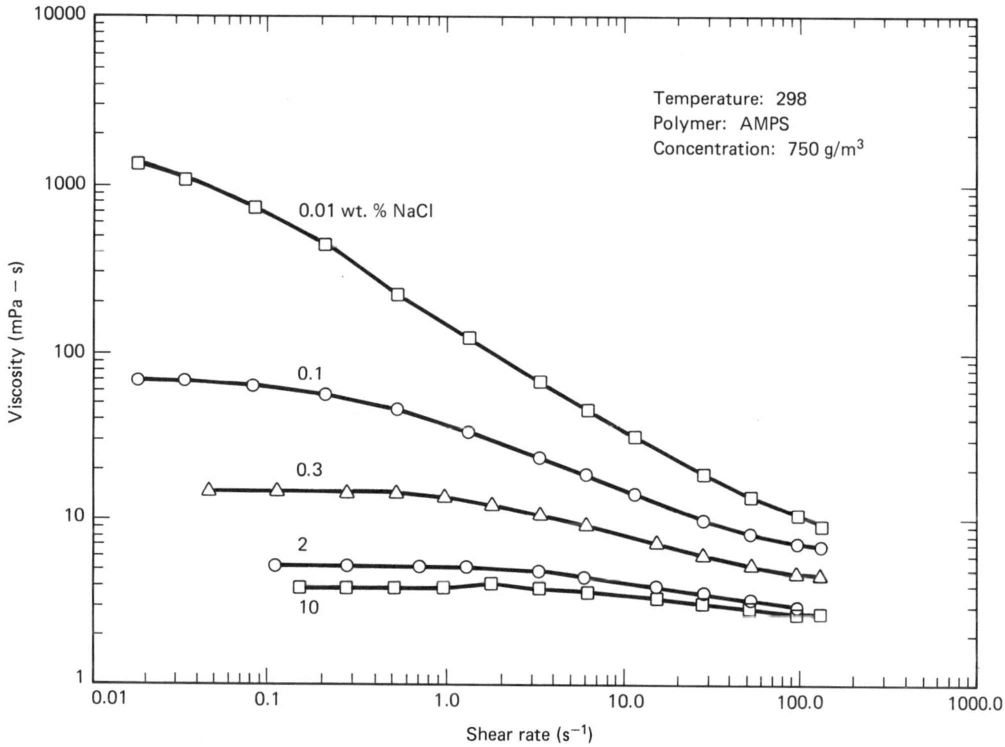

Figure 8-6 Polymer solution viscosity versus shear rate at various brine salinities (from Martin et al., 1981)

where K_{pl} and n_{pl} are the power-law coefficient and exponent, respectively. For shear thinning fluids, $0 < n_{pl} < 1$; for Newtonian fluids, $n_{pl} = 1$, and K_{pl} becomes the viscosity. $\dot{\gamma}$ is always positive. Equation (8.2-5) applies only over a limited range of shear rates: Below some low shear rate, the viscosity is constant at μ_1^0, and above the critical shear rate, the viscosity is also constant μ_1^∞.

The truncated nature of the power law is awkward in some calculations; hence another useful relationship is the *Meter model* (Meter and Bird, 1964)

$$\mu_1' = \mu_1^\infty + \frac{\mu_1^0 - \mu_1^\infty}{1 + \left(\dfrac{\dot{\gamma}}{\dot{\gamma}_{1/2}}\right)^{n_M - 1}} \tag{8.2-6}$$

where n_M is an empirical coefficient, and $\dot{\gamma}_{1/2}$ is the shear rate at which μ_1' is the average of μ_1^0 and μ_1^∞. As with all polymer properties, all empirical parameters are functions of salinity, hardness, and temperature.

When applied to permeable media flow, the above general trends and equations continue to apply. μ_1' is usually called the apparent viscosity μ_{app} and the effective shear rate $\dot{\gamma}_{eq}$ is based on capillary tube concepts, as we derived in Sec. 3-1 for Newtonian fluids. For power-law fluids, the procedure is identical (see Exercise 8B) except the beginning equation is Eq. (8.2-5). We give only the results here.

The apparent viscosity of a flowing polymer solution is

$$\mu_{app} = H_{pl} u^{n_{pl}-1} \tag{8.2-7}$$

where (Hirasaki and Pope, 1974)

$$H_{pl} = K_{pl} \left(\frac{1 + 3n_{pl}}{n_{pl}} \right)^{n_{pl}-1} (8k_1 \phi_1)^{(1-n_{pl}/2)} \tag{8.2-8}$$

The right side of Eq. (8.2-7) is $K_{pl} \dot{\gamma}_{eq}^{n_{pl}-1}$, which yields the equivalent shear rate for flow of a power-law fluid

$$\dot{\gamma}_{eq} = \left(\frac{1 + 3n_{pl}}{n_{pl}} \right) \frac{u}{\sqrt{8k_1 \phi_1}} \tag{8.2-9}$$

In both Eqs. (8.2-8) and (8.2-9), k_1, the aqueous phase permeability, is the product of the phase's relative permeability and the absolute permeability. ϕ_1, the aqueous phase porosity, is ϕS_1.

The only difference between the equivalent shear rate and that for the Newtonian fluid (Eq. 3.1-11) is the first term on the right-hand side. This factor is a slowly varying function of n_{pl}; hence the Newtonian shear rate affords an excellent approximation of the shear rate in non-Newtonian flow.

Even though $\dot{\gamma}_{eq}$ has units of reciprocal time, shear rate is essentially a steady-state representation since it can be realized in steady laminar flow in a tube. Thus the constitutive Eqs. (8.2-5) and (8.2-6) are representing purely viscous effects since an instantaneous change in $\dot{\gamma}_{eq}$ causes a similar change in μ_1'. In reality, fluctuations in $\dot{\gamma}_{eq}$, or elastic effects, do affect polymer properties; these we discuss separately below.

Polymer Transport

All polymers experience retention in permeable media because of adsorption onto solid surfaces or trapping within small pores. Polymer retention varies with polymer type, molecular weight, rock composition, brine salinity, brine hardness, flow rate, and temperature. Field-measured values of retention range from 7 to 150 μg polymer/cm^3 of bulk volume, with a desirable retention level being less than about 20 μg/cm^3. Retention causes the loss of polymer from solution, which can also cause the mobility control effect to be lost—a particularly pronounced effect at low polymer concentrations. Polymer retention also causes a delay in the rate of the polymer and generated oil bank propagation (see Sec. 8-4).

For more quantitative work, we represent polymer adsorption by a *Langmuir-type* isotherm

$$C_{4s} = \frac{a_4 C_4}{1 + b_4 C_4} \tag{8.2-10}$$

where C_4 and C_{4s} are the species concentrations in the aqueous and on the rock phases. The units of adsorption can take on a variety of forms, but mass of polymer per mass of rock is most common. In our notation, this is $\omega_{4s}/(1 - \omega_{4s})$, strictly speaking, but ω_{4s} is very much smaller than 1. The units conversion between C_{4s}(g/m^3) and ω_{4s}(μg/g-rock) are embedded in the constants a_4 and b_4. The b_4 in Eq. (8.2-10) controls the curvature of the isotherm, and the ratio a_4/b_4 determines the plateau value for adsorption (Fig. 8-7).

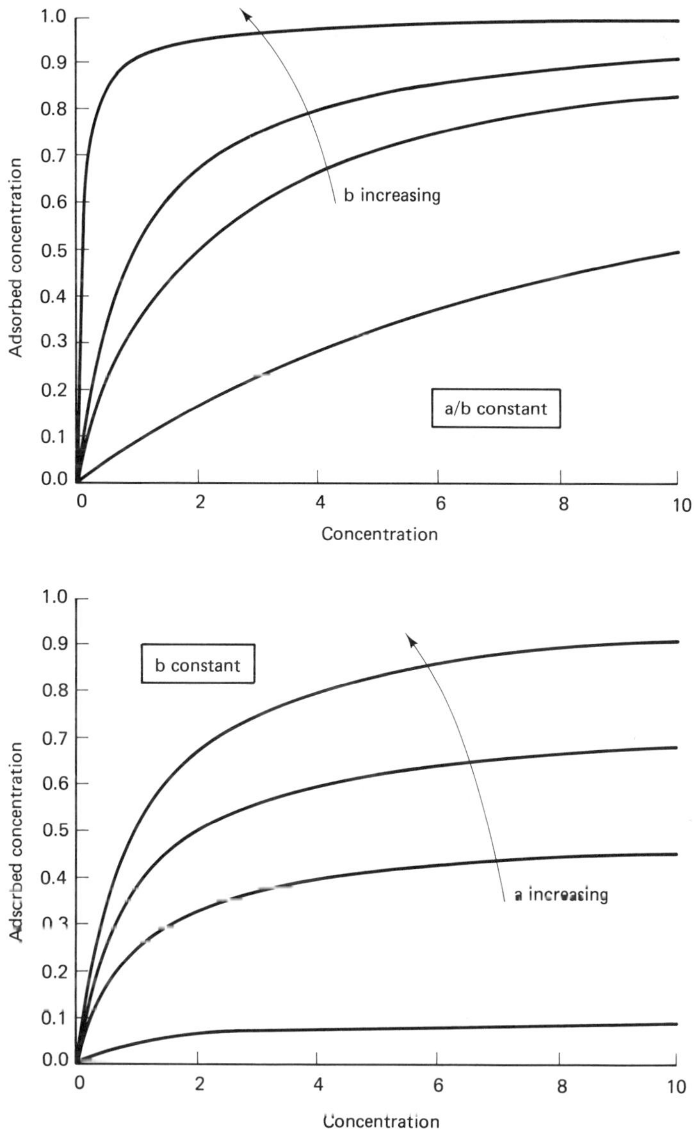

Figure 8-7 Typical Langmuir isotherm shapes

In the original Langmuir theory, the plateau adsorption corresponded to monolayer coverage of the surface by physical adsorption (see Exercise 8C). Considering the anionic character of water-soluble polymers, the adsorption is more likely to be chemical adsorption described by an exchange isotherm like Eq. (3.5-4). In fact, polymer adsorption does increase with increasing salinity and hardness but measured surface coverages are much smaller than monolayer coverage. Moreover, it is unknown if adsorption is reversible. Hence Eq. (8.2-10) and Fig. 8.7 are simply empirical representations of physical observations. (This is the origin of the term *Langmuir-type*.) Typical polymer adsorption isotherms are quite steep; that is, they attain their plateau value at very low C_4. The values given above for polymer adsorption are referring to the plateau adsorption.

Equation (8.2-10) is a general isotherm function. The specific form depends on the units of the retention; unfortunately, no standard form exists for this. Common ways to report retention are

$$\frac{\text{Mass polymer}}{\text{Mass solid}} = \frac{\omega_{4s}}{(1 - \omega_{4s})}$$

$$\frac{\text{Mass polymer}}{\text{Surface area}} = \frac{\omega_{4s}}{a_v}$$

$$\frac{\text{Mass polymer}}{\text{Bulk volume}} = \omega_{4s}\rho_s(1 - \phi)$$

$$\frac{\text{Mass polymer}}{\text{Pore volume}} = \frac{\omega_{4s}\rho_s(1 - \phi)}{\phi} = C_{4s}$$

$$\frac{\text{Volume polymer solution}}{\text{Pore volume}} = \frac{\omega_{4s}\rho_s(1 - \phi)}{(\phi C_4)} = D_4$$

The last of these is often called the *frontal advance loss*.

Inaccessible Pore Volume

Offsetting the delay caused by retention is an acceleration of the polymer solution through the permeable medium caused by *inaccessible pore volume* (IPV). The most common explanation for IPV is that the smaller portions of the pore space will not allow polymer molecules to enter because of their size. Thus a portion of the total pore space is uninvaded or inaccessible to polymer, and accelerated polymer flow results. A second explanation of IPV is based on a wall exclusion effect whereby the polymer molecules aggregate in the center of a narrow channel (Duda et al., 1981). The polymer fluid layer near the pore wall has a lower viscosity than the fluid in the center, which causes an apparent fluid slip.

IPV depends on polymer molecular weight, medium permeability, porosity, and pore size distribution and becomes more pronounced as molecular weight increases and the ratio of permeability to porosity (characteristic pore size) decreases. In extreme cases, IPV can be 30% of the total pore space.

Permeability Reduction

For many polymers, viscosity–shear-rate data derived from a viscometer (μ_1' versus $\dot{\gamma}$) and those derived from a flow experiment (μ_{app} versus $\dot{\gamma}_{eq}$) will yield essentially the same curve. But for HPAM, the viscometer curve will be offset from the permeable medium curve by a significant and constant amount. The polymer evidently causes a degree of permeability reduction that reduces mobility in addition to the viscosity increase.

Actually, permeability reduction is only one of three measures in permeable media flow (Jennings et al., 1971). The *resistance factor* R_F is the ratio of the injectivity of brine to that of a single-phase polymer solution flowing under the same conditions

$$R_F = \frac{\lambda_1}{\lambda_1'} = \lambda_1 \mu_{app} = \left(\frac{k_1}{\mu_1}\right) \cdot \left(\frac{\mu_1'}{k_1'}\right) \tag{8.2-11}$$

For constant flow rate experiments, R_F is the inverse ratio of pressure drops; for constant pressure drop experiments, R_F is the ratio of flow rates. R_F is an indication of the total mobility lowering contribution of a polymer. To describe the permeability reduction effect alone, a *permeability reduction factor* R_k is defined as

$$R_k = \frac{k_1}{k_1'} = \frac{\mu_1}{\mu_1'} R_F \tag{8.2-12}$$

A final definition is the *residual resistance factor* R_{RF}, which is the mobility of a brine solution before and after (λ_{1a}) polymer injection

$$R_{RF} = \frac{\lambda_1}{\lambda_{1a}} \tag{8.2-13}$$

R_{RF} indicates the permanence of the permeability reduction effect caused by the polymer solution. It is the primary measure of the performance of a channel-blocking application of polymer solutions. For many cases, R_k and R_{RF} are nearly equal, but R_F is usually much larger than R_k because it contains both the viscosity-enhancing and the permeability-reducing effects.

The most common measure of permeability reduction is R_k, which is sensitive to polymer type, molecular weight, degree of hydrolysis, shear rate, and permeable media pore structure. Polymers that have undergone even a small amount of mechanical degradation seem to lose most of their permeability reduction effect. For this reason, qualitative tests based on screen factor devices are common to estimate polymer quality.

The screen factor device is simply two glass bulbs mounted into a glass pipette as shown in Fig. 8-8. Into the tube on the bottom of the device are inserted several fairly coarse wire screens through which the polymer solution is to drain. To use the device, a solution is sucked through the screens until the solution level is above the upper timing mark. When the solution is allowed to flow freely, the time required to pass from the upper to the lower timing mark t_d is recorded. The screen factor for

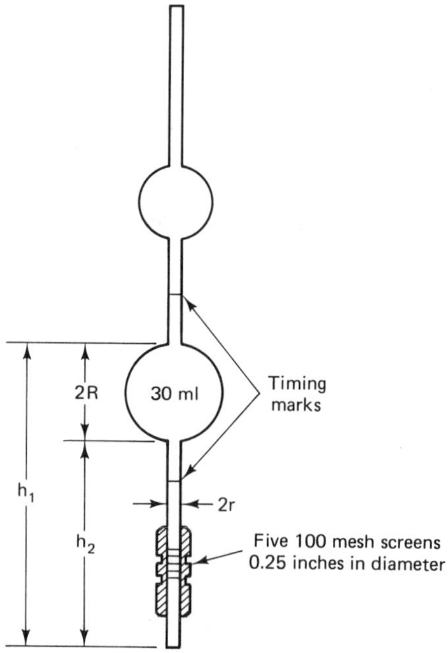

Figure 8-8 Screen factor device (adapted from Foshee et al., 1976)

the polymer solution is then defined as

$$S_F = \frac{t_d}{t_{ds}} \tag{8.2-14}$$

where t_{ds} is the similar time for the polymer-free brine.

Because of the normalization, screen factors are independent of temperature, device dimensions, and screen coarseness, and they are fairly independent of screen spacing. The screen factor is not independent of polymer concentration, but its primary intent is to measure the time-dependent portion of the polymer's solution configuration; that is, it measures the rate at which a polymer molecule returns to its steady-state flow configuration after it has been perturbed. This relaxation time is evidently very fast for the polysaccharides because they do not have a measurable screen factor even at high concentrations. HPAMs have much slower relaxation times because their screen factors can be large even at the same viscosity as a polysaccharide solution. The above explanations are consistent with the chemical properties of the two polymer groups given in Sec. 8-1 and can be used to deduce the sensitivity of screen factors to brine salinity and hardness.

Screen factors are particularly sensitive to changes in the polymer molecule itself. One definition of polymer quality is the ratio of the degraded to the undegraded screen factors. This use is important for screen factor devices, particularly in locations that prohibit more sophisticated equipment.

Another use for screen factors is as a correlator for R_F and R_{RF} (Fig. 8-9). The explanation for such a correlation is consistent with that given above on polymer re-

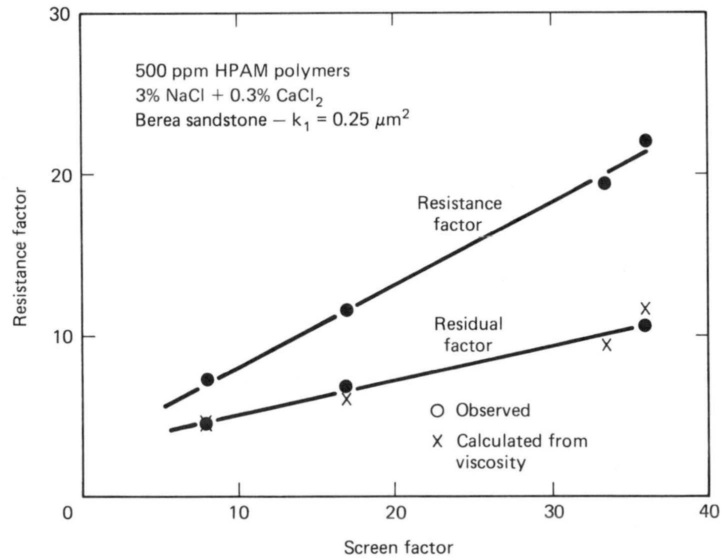

Figure 8-9 Correlation of resistance factors with screen factors (from Jennings et al., 1971)

laxation. On a pore scale, steady flow in permeable media is actually a succession of contracting and diverging channels. The frequency with which the solution experiences these contractions, compared with the polymer relaxation time, determines the degree of permeability reduction. Such an effect also qualitatively explains the increase in viscometer viscosity at very high shear rates (Hirasaki and Pope, 1974).

The relaxation time argument cannot completely account for permeability reduction because such effects have been observed in glass capillaries. For this case, permeability reduction seems to be caused by polymer adsorption, which decreases the effective pore size (see Exercise 8E).

A reasonable question is whether permeability reduction is a desirable effect. R_k is difficult to control, being sensitive to even small deteriorations in the polymer quality. Moreover, an extremely large R_k will cause injectivity impairment. But it is possible to achieve a predesignated degree of mobility control with less polymer if $R_k > 1$. If M_T^0 is a design or target endpoint mobility ratio,

$$M_T^0 = \left(\frac{k_1'}{\mu_1^0}\right)\left(\frac{\mu_2}{k_2^0}\right) = \frac{M^0|_{R_k=1}}{R_k} = \frac{M^0}{R_{RF}} \tag{8.2-15}$$

In this equation, $M^0|_{R_k=1}$ is the mobility ratio of a polymer having no permeability reduction, and M^0 is the endpoint water–oil mobility ratio. Clearly, if $R_k > 1$, the polymer viscosity μ_1^0 can be smaller than if $R_k = 1$, which indicates a given concentration of HPAM will have a lower mobility ratio than polysaccharide under conditions where both polymers have the same flowing viscosity. Note that the limiting viscosity μ_1^0 is used to estimate M^0 from Eq. (8.2-15).

Chemical and Biological Degradation

The average polymer molecular weight can be decreased, to the detriment of the overall process, by chemical, biological, or mechanical degradation. We use the term *chemical degradation* to denote any of several possible processes such as thermal oxidation, free radical substitution, hydrolysis, and biological degradation.

For a given polymer solution, there will be some temperature above which the polymer will actually thermally crack. Although not well established for most EOR polymers, this temperature is fairly high, on the order of 400 K. Since the original temperature of oil reservoirs is almost always below this limit, of more practical concern for polymer flooding is the temperature other degradation reactions occur at.

The average residence time in a reservoir is typically very long, on the order of a few years, so even slow reactions are potentially serious. Reaction rates also depend strongly on other variables such as pH or hardness. At neutral pH, degradation often will not be significant, whereas at very low or very high pH, and especially at high temperatures, it may be. In the case of HPAM, the hydrolysis will destroy the carefully selected extent of hydrolysis present in the initial product. The sensitivity to hardness will increase, and viscosity will plummet. For Xanthan gum, hydrolysis is even more serious since the polymer backbone is severed, resulting in a large decrease in viscosity.

TABLE 8-1 SELECTED BACTERICIDES AND OXYGEN SCAVENGERS (ADAPTED FROM *ENHANCED OIL RECOVERY*, NATIONAL PETROLEUM COUNCIL, 1984)

Bactericide	Oxygen scavengers
Commonly used	
Acrolein	Hydrazine
Formaldehyde	Sodium bisulfite
Sodium dichlorophenol	Sodium hydrosulfite
Sodium pintachlorophenol	Sulfur dioxide
Proposed or infrequent use	
Acetate salts of coco amines	
Acetate salts of coco diamines	
Acetate salts of tallow diamines	
Alkyl amino	
Alkyl dimethyl ammonium chloride	
Alkyl phosphates	
Calcium sulfate	
Coco dimethyl ammonium chloride	
Gluteraldehyde	
Paraformaldehyde	
Sodium hydroxide	
Sodium salts of phenols	
Substituted phenols	

Oxidation or free radical chemical reactions are usually considered the most serious source of degradation. Therefore, oxygen scavengers and antioxidants are often added to prevent or retard these reactions. These chemicals are strong reducing agents and have the additional advantage of reducing iron cations from the +3 to the +2 state. They, in turn, help prevent gelation, agglomeration, and other undesirable effects that can cause wellbore plugging and reduced injectivity. Wellington (1980) has found that alcohols such as isopropanol and sulfur compounds such as thiourea make good antioxidants and free radical inhibitors.

Laboratory results indicate Xanthan can be stabilized up to about 367 K, and HPAM to about 394 K. In the case of Xanthan, the results depend strongly on the precise conditions such as salinity and pH, with high salinity and pH between 7 and 9 being preferred. Obviously, one should test the particular polymer solution under the particular reservoir conditions of interest to establish the expected behavior.

Biological degradation can occur with both HPAM and polysaccharides, but is more likely with the latter. Variables affecting biological degradation include the type of bacteria in the brine, pressure, temperature, salinity, and the other chemicals present. As in waterflooding, the preventive use of biocide is highly recommended. Often too little biocide is used or it is started too late, and the ensuing problems become almost impossible to correct. Table 8-1 lists typical polymer flooding additions.

Mechanical Degradation

Mechanical degradation is potentially present under all applications. It occurs when polymer solutions are exposed to high velocity flows, which can be present in surface equipment (valves, orifices, pumps, or tubing), downhole conditions (perforations or screens), or the sand face itself. Perforated completions, particularly, are a cause for concern as large quantities of polymer solution are being forced through several small holes. For this reason, most polymer injections are done through open-hole or gravel-pack completions. Partial preshearing of the polymer solution can lessen the tendency of polymers to mechanically degrade. Because flow velocity falls off quickly with distance from an injector, little mechanical degradation occurs within the reservoir itself.

All polymers mechanically degrade under high enough flow rates. But HPAMs are most susceptible under normal operating conditions, particularly if the salinity or hardness of the brine is high. Evidently, the ionic coupling of these anionic molecules is relatively fragile. Moreover, elongational stress is as destructive to polymer solutions as is shear stress though the two generally accompany each other. Maerker (1976) and Seright (1983) have correlated permanent viscosity loss of a polymer solution to an elongational stretch rate–length product. On a viscosity-shear-rate plot (purely shear flow), mechanical degradation usually begins at shear rates equal to or somewhat less than the minimum viscosity shear rate.

8-3 CALCULATING POLYMER FLOOD INJECTIVITY

The economic success of all EOR processes is strongly tied to project life or injection rate, but polymer flooding is particularly susceptible. In many cases, the cost of the polymer itself is secondary compared to the present value of the incremental oil. Because of its importance, many field floods are preceded by single-well injectivity tests. Here we give a simple technique for analyzing injectivity tests based on the physical properties given in the previous section.

The injectivity of a well is defined as

$$I \equiv \frac{i}{\Delta P} \tag{8.3-1}$$

where i is the volumetric injection rate into the well, and ΔP is the pressure drop between the bottom-hole flowing pressure and some reference pressure. Another useful measure is the relative injectivity

$$I_r = \frac{I}{I_1} \tag{8.3-2}$$

where I_1 is the water injectivity. I_r is an indicator of the injectivity decline to be anticipated when injecting polymer. Both I and I_r are functions of time, but the long-time limit of I_r for a Newtonian polymer solution is simply the viscosity ratio if skin effects are small. However, the ultimate I_r for an actual polymer solution can be higher than this because of shear-thinning.

We make several simplifying assumptions, many of which can be relaxed (Bondor et al., 1972). The well, of radius R_w, whose injectivity we seek, is in a horizontal, homogeneous, circular drainage area of radius R_e. The pressures at R_e and R_w are P_e and P_{wf}, respectively. P_e is constant (steady-state flow), but P_{wf} can vary with time. The fluid flowing in the reservoir is a single aqueous phase, at residual oil saturation, which is incompressible with pressure-independent rheological properties. Dispersion and polymer adsorption are negligible although the polymer can exhibit permeability reduction. The flow is one-dimensional and radial. Finally, the entire shear rate range in the reservoir lies in the power-law regime; hence Eq. (8.2-7) describes the apparent viscosity.

Subject to these assumptions, the continuity equation (Eq. 2.4-11) reduces to

$$\frac{d}{dr}(ru_r) = 0 \tag{8.3-3}$$

where u_r is the radial volumetric flux. This equation implies the volumetric rate is independent of r and equal to i since

$$i = 2\pi r H_t u_r \tag{8.3-4a}$$

Equation (8.3-4a) is a consequence of the incompressible flow assumption; however, i is not independent of time. Let us substitute Darcy's law for u_r in Eq. (8.3-4a)

$$i = -\frac{2\pi r H_t k_1'}{\mu_{app}}\frac{dP}{dr} = -\frac{2\pi r H_t k_1}{H_{pl} u_r^{n_{pl}-1} R_k}\frac{dP}{dr} \tag{8.3-4b}$$

from Eq. (8.2-7). This equation has been defined so that i is positive. The permeability reduction factor is introduced through Eq. (8.2-12). Eliminating u_r with Eq. (8.3-4a) yields an ordinary differential equation, which may be integrated between the arbitrary limits of P_1 at r_1 and P_2 at r_2.

$$P_2 - P_1 = \left(\frac{i}{2\pi H_t}\right)^{n_{pl}} \frac{H_{pl} R_k}{k_1(1 - n_{pl})}\left(r_1^{1-n_{pl}} - r_2^{1-n_{pl}}\right) \tag{8.3-5a}$$

The Newtonian flow limit, $n_{pl} = 1 = R_k$ and $H_{pl} = \mu_1$, of this equation is the familiar steady-state radial flow equation,

$$P_2 - P_1 = \frac{i\mu_1}{2\pi k_1 H_t}\ln\left(\frac{r_1}{r_2}\right) \tag{8.3-5b}$$

We now apply these equations to the polymer flood injectivity.

At some time t during the injection, the polymer front (assumed sharp) is at radial position R_p where

$$\textstyle\int_0^t i\,dt = \pi(R_p^2 - R_w^2)H_t\phi(1 - S_{2r}) \tag{8.3-6}$$

The left side of this equation is the cumulative volume of polymer solution injected. Therefore, Eq. (8.3-5a) applies in the region $R_w < r < R_p$, and Eq. (8.3-5b) applies in the annular region $R_p < r < R_e$. With the appropriate identification of variables, we have for the second region

$$P|_{R_p} - P_e = \frac{i\mu_1}{2\pi k_1 H_t}\ln\left(\frac{R_e}{R_p}\right) \tag{8.3-7a}$$

and for the first

$$P_{wf} - P|_{R_p} = \left(\frac{i}{2\pi H_t}\right)^{n_{pl}} \frac{H_{pl}}{k_1(1 - n_{pl})R_k}\left(R_p^{1-n_{pl}} - R_w^{1-n_{pl}}\right) \tag{8.3-7b}$$

where $P|_{R_p}$ is the pressure at the polymer–water front. Adding these two equations gives the total pressure drop from R_w to R_e.

$$P_{wf} - P_e = \left(\frac{i}{2\pi H_t}\right)^{n_{pl}} \frac{H_{pl} R_k}{k_1(1 - n_{pl})}\left(R_p^{1-n_{pl}} - R_w^{1-n_{pl}}\right) + \frac{i\mu_1}{2\pi k_1 H_t}\left(\ln\left(\frac{R_e}{R_p}\right) + s_w\right) \tag{8.3-8}$$

where s_w, the intrinsic skin factor of the well, has been introduced to account for well damage.

Equation (8.3-8) substituted into the injectivity definition (Eq. 8.3-1) gives

$$I^{-1} = \left(\frac{i}{2\pi H_t}\right)^{n_{pl}} \frac{H_{pl} R_k}{i(1 - n_{pl})\, k_1} (R_p^{1-n_{pl}} - R_w^{1-n_{pl}}) + \frac{\mu_1}{2\pi k_1 H_t}\left(\ln\left(\frac{R_e}{R_p}\right) + s_w\right) \tag{8.3-9}$$

The water injectivity I_1 is given by Eqs. (8.3-1) and (8.3-5b), with $r_1 = R_w$ and $r_2 = R_e$. This and I, calculated from Eq. (8.3-9), yield an expression for I_r through Eq. (8.3-2). Both I and I_r relate to the cumulative polymer solution injection (or to time) through Eq. (8.3-6).

8-4 FRACTIONAL FLOW IN POLYMER FLOODS

The fractional flow treatment of polymer floods resembles the water–solvent treatment in Sec. 7-7. The only major complications are the addition of terms for polymer retention and inaccessible pore volume (IPV). In this section, we apply the usual fractional flow assumptions: one-dimensional flow, incompressible fluid and rock, and nondissipative mixing.

Single-Phase Flow

First, consider the case of a water-soluble species that is being adsorbed from solution via a Langmuir-type isotherm. The isotherm is given by Eq. (8.2-10).

Let the flow be such that species concentration C_{4I} is being displaced by concentration C_{4J} in single-phase flow where $C_{4J} > C_{4I}$. From Eq. (5.4-5a), the specific velocity of concentration C_4 is

$$v_{C_4} = \left(1 + \frac{(1-\phi)\rho_s}{\phi}\frac{d\omega_{4s}}{dC_4}\right)^{-1} = \left(1 + \frac{dC_{4s}}{dC_4}\right)^{-1}$$

From Eq. (8.2-10), the specific velocity becomes

$$v_{C_4} = \left(1 + \frac{C_4}{(1 + b_4 C_4)^2}\right)^{-1} \tag{8.4-1}$$

But since $C_{4J} > C_{4I}$, we have $v_{C_4}|_J > v_{C_4}|_I$, and the displacement is a shock; if $C_{4J} < C_{4I}$ (see Exercise 8J), it would be a spreading wave. But for C_{4J} displacing C_{4I}, the front between C_{4J} and C_{4I} moves with specific velocity

$$v\Delta_{C_4} = \left(1 + \frac{(1-\phi)\rho_s}{\phi}\frac{\Delta\omega_{4s}}{\Delta C_4}\right)^{-1} = \left(1 + \frac{\Delta C_{4s}}{\Delta C_4}\right)^{-1} \tag{8.4-2}$$

from Eq. (5.4-5b). In this equation, $\Delta(\ ) = (\ )_J - (\ )_I$. If, as is usually the case for polymer floods, $C_{4I} = 0$, Eq. (8.4-2) reduces to

$$v_{\Delta C_4} = \frac{1}{1 + \dfrac{(1-\phi)\rho_s}{\phi}\left(\dfrac{\omega_{4s}}{C_4}\right)_J} \equiv \frac{1}{1 + D_4} \quad (8.4\text{-}3)$$

where D_4 is the frontal advance loss for the polymer. It is also called the retardation factor because adsorption causes the front velocity to be lower than that of the ideal miscible displacement (see Sec. 5-4). D_4 is one of the most useful concepts in both polymer and micellar-polymer flooding because it expresses retention in pore volume units which are consistent with slug size.

Two-Phase Flow

The fractional flow treatment will consist of two phases (aqueous $j = 1$ and oleic $j = 2$) and three components (brine $i = 1$, oil $i = 2$, and polymer $i = 4$). Let the permeable medium have a uniform original water saturation of S_{1I}. We inject an oil-free polymer solution ($S_{1J} = 1 - S_{2r}$). The initial overall polymer concentration is 0, and the polymer concentration in the aqueous phase is C_{4J}. Polymer and water do not dissolve in the oil ($C_{12} = C_{42} = 0$); the oil has no solubility in the aqueous phase ($C_{21} = 0$).

Effect of IPV The aqueous phase porosity is ϕS_1. Only a portion of this pore volume fraction, ($\phi S_1 - \phi_{\text{IPV}}$), is accessible to the polymer; hence the overall polymer concentration per unit bulk volume is

$$W_4 = (\phi S_1 - \phi_{\text{IPV}})\rho_1 \omega_{41} + (1 - \phi)\rho_s \omega_{4s} \quad (8.4\text{-}4a)$$

Similarly, the overall water concentration is

$$W_1 = (\phi S_1 - \phi_{\text{IPV}})\rho_1(1 - \omega_{41}) + \phi_{\text{IPV}}\rho_1 \quad (8.4\text{-}4b)$$

since only water is present in the excluded pore volume ϕ_{IPV}. But the IPV can be easily neglected in Eq. (8.4-4b) because the polymer concentration is very small ($\omega_{11} \cong 1$). The overall oil concentration and Eqs. (8.4-4a) and (8.4-4b) sum to the porosity as required by the assumption of incompressible flow.

Oil Displacement The polymer itself alters neither the water nor the oil relative permeabilities because, as we have seen in Sec. 3-4, the apparent viscosity cannot be increased enough to change residual phase saturations. Moreover, when permeability reduction is significant, it applies over the entire saturation range but only to the wetting phase (Schneider and Owens, 1982). We may, therefore, construct a polymer-solution-oil (polymer–oil) water fractional flow curve simply by using the apparent viscosity in place of the water viscosity and dividing k_{r1} by R_k. Figure 8-10 shows both the water–oil ($f_1 - S_1$) and polymer–oil ($f_1^p - S_1$) fractional flow curves.

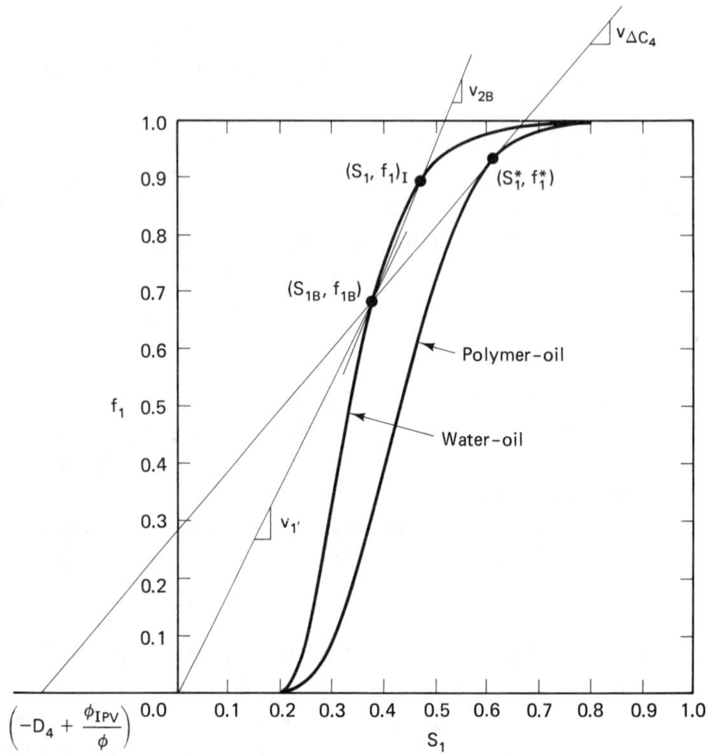

Figure 8-10 Graphical construction of polymer flooding fractional flow

Because the polymer adsorption is Langmuir-like, and because the polymer displaces the connate water miscibly, the polymer front is pistonlike and has specific velocity

$$v_{\Delta C_4} = \frac{f_1^p(S_1^*)}{S_1^* + D_4 - \phi_e} \tag{8.4-5a}$$

where D_4 is the polymer retardation factor defined in Eq. (8.4-3), and

$$\phi_e = \frac{\phi_{\text{IPV}}}{\phi} \tag{8.4-5b}$$

S_1^* and $f_1^p(S_1^*)$ are the water saturations and fractional flows at the polymer shock front. S_1^* may also be regarded as a point in the spreading portion of the mixed polymer–oil wave given by the Buckley-Leverett equation, whence from Eq. (8.4-5) we can define S_1^*

$$v_{\Delta C_4} = \frac{f_1^p(S_1^*)}{S_1^* + D_4 - \phi_e} = \left(\frac{df_1^p}{dS_1}\right)\bigg|_{S_1^*} = v_{C_1} \tag{8.4-6}$$

since S_1^* is also in the shock portion of the polymer–oil wave. The Buckley-Leverett treatment in Sec. 5-2 used a similar argument.

Equation (8.4-6) will also determine the oil bank saturation since S_2 will change discontinuously with velocity given by

$$v_{\Delta C_2} = \frac{f_1^p(S_1^*) - f_1(S_{1B})}{S_1^* - S_{1B}} = v_{C_1} \tag{8.4-7}$$

Equations (8.4-6) and (8.4-7) are particular statements of the coherence condition (Eq. 5.6-14).

As in the solvent–water treatment in Sec. 7-7, the velocity of the front of the oil (or water) bank is given by

$$v_{\Delta C_2} = \frac{f_{1B} - f_{1I}}{S_{1B} - S_{1I}} = v_{\Delta C_1} \tag{8.4-8}$$

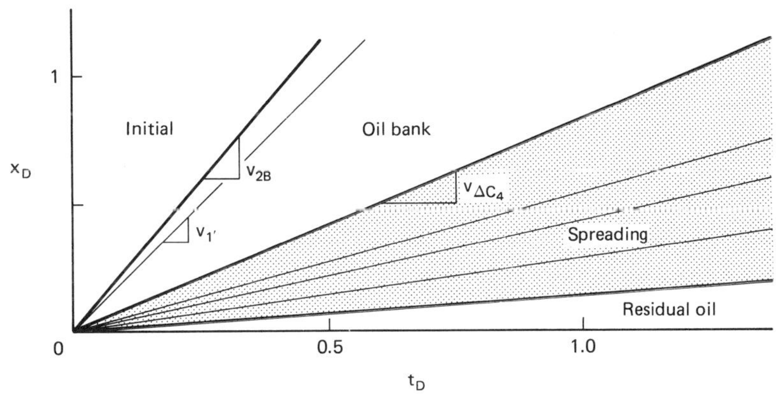

(a) Time-distance diagram

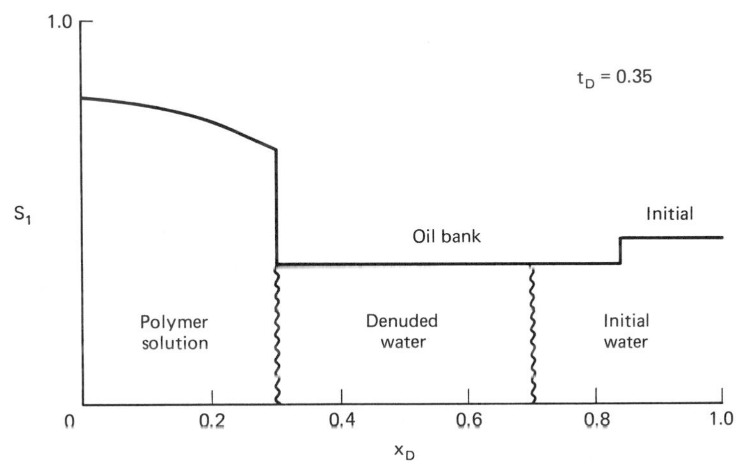

(b) Saturation and concentration profile

Figure 8-11 Figures for the fractional flow curves in Fig. 8-10 for a pistonlike oil bank front. The construction proceeds in the same manner as in Sec. 7-7. Figure 8-11 shows the time–distance diagram and a composition profile at $t_D = 0.35$ for the construction in Fig. 8-10.

Though relatively direct, the construction in Figs. 8-10 and 8-11 has several important insights into polymer floods.

1. The oil bank breakthrough time (reciprocal of the oil bank specific velocity $v_{\Delta C_2}$) increases as S_{1I} increases, suggesting polymer floods will be more economic if they are begun at low initial water saturation. Of course, the lower S_{1I}, the higher the mobile oil saturation, also a favorable indicator for polymer floods.
2. Adsorption (large D_4) causes a delay of all fronts. D_4 can be large if the porosity is low, the retention is high, or the injected polymer concentration C_{4J} is low. Usually, C_{4J} is so low that D_4 can be high even if retention is moderate.
3. Inaccessible pore volume causes an acceleration of all fronts, exactly opposite to retention. In fact, retention and IPV can exactly cancel so that the polymer front and the denuded water front $v_{1'}$ (Fig. 8-10) travel at the same velocity.
4. Both D_4 and IPV influence the oil bank saturation, which in turn, influences the oil bank mobility and the desired injected polymer concentration.

8-5 ELEMENTS OF POLYMER FLOOD DESIGN

Polymer flood design is a complex subject. But most of the complexity arises from reservoir-specific aspects of a particular design. In this section, we deal in generalities that apply to all types of polymer flooding.

A polymer flood design procedure will follow these six steps.

1. Screen the candidate reservoirs. The distinction between technical and economic feasibility is important. Technical feasibility means a given reservoir can be polymer flooded regardless of the funds available. Economic feasibility means the project has a good chance of being profitable. Technical feasibility is measured by a series of binary screening parameters (see National Petroleum Council, 1984). But for polymer flooding, there are only two: the reservoir temperature should be less than about 350 K to avoid degradation, and the reservoir permeability should be greater than about 0.02 μm^2 to avoid plugging. Economic feasibility can be estimated by simple hand calculations (as in the fractional flow method) or through using predictive models (Jones et al., 1984), which requires deciding how the polymer is to be used.
2. Decide on the correct mode. The choices are (a) mobility control (decrease M), (b) profile control (improve the permeability profile at the injectors or producers), or (c) some combination of both. We have not discussed profile control, but the concepts and goals are similar to polymer flooding. We want to inject an agent that will alter the permeability so that more fluid will go into the tight rock than into the high-permeability rock. We can do this by using gels, polymers, and solids and by using selective perforation. When selective

perforation is ineffective or incompletely effective, we use chemical agents or solids.

3. Select the polymer type. The requirements for EOR polymers are severe. An outline of the principal ones is as follows:
   (a) Good thickening. This means high mobility reduction per unit cost.
   (b) High water solubility. The polymers must have good water solubility under a wide range of conditions of temperature, electrolyte composition, and in the presence of stabilizers.
   (c) Low retention. All polymers adsorb on reservoir rocks to various degrees. Retention may also be caused by plugging, trapping, phase separation, and other mechanisms. Low here means less than 20 μg/g.
   (d) Shear stability. During flow through permeable media, stress is applied to the polymer molecules. As we discussed, if this is excessive, they may mechanically break apart or permanently degrade, resulting in less viscosity. HPAM is especially subject to shear degradation.
   (e) Chemical stability. Polymers, like other molecules, can chemically react, especially at high temperature and in the presence of oxygen. Antioxidants are used to prevent this.
   (f) Biological stability. Both HPAM and polysaccharides can be degraded by bacteria, but the latter are more susceptible. Biocides are required to prevent this.
   (g) Good transport in permeable media. This catchall includes essentially the ability to propagate the polymer through the rock intact and without excessive pressure drop or plugging. Good transport also means good injectivity and no problems with microgels, precipitates, and other debris.

   Obviously, no one polymer can universally meet these requirements for all reservoir rocks. Thus we must tailor the polymer to the rock to some extent. Some general guidelines are possible for minimum standards, but the ultimate criterion must be economics.

4. Estimate the amount of polymer required. The amount, the total mass in kilograms to be injected, is the product of the slug size, the pore volume, and the average polymer concentration. Ideally, the amount would be the result of an optimization study that weights the present value of the incremental oil against the present value of the injected polymer. Each iteration of the optimization procedure requires estimating the polymer concentration in initial portion (spike) of the slug and estimating the volume of the polymer slug (spike plus rate of taper).
   (a) Estimating the spike concentration. Suppose we have decided on a target mobility ratio that might come from simulation studies (see Chap. 6) or simply injectivity limitations. If the target mobility ratio is M_T

$$M_T = \frac{(\lambda_{rt})_{\text{polymer}}}{(\lambda_{rt})_{\text{oil bank}}} = \frac{(\lambda'_{r1} + \lambda_{r2})_{S_1^*}}{(\lambda_{r1} + \lambda_{r2})_{S_{1B}}} \tag{8.5-1}$$

   Estimating the spike concentration simply means picking the value of injected polymer concentration that gives the correct M_T in this equation.

The translation between apparent viscosity follows from permeability reduction factor correlations and shear rate data as in Fig. 8-4. The latter must be evaluated at a sheer rate corresponding to the median velocity in the flood—usually the low shear rate plateau. Estimating the denominator of Eq. (8.5-1), the oil bank relative mobility, is a little more difficult.

One procedure is to estimate the oil bank saturation through the graphical procedure of Sec. 8-4, and then estimate the oil bank mobility from the relative permeability curves evaluated at this saturation

$$(\lambda_{rt})_{OB} = \left(\frac{k_{r1}}{\mu_1} + \frac{k_{r2}}{\mu_2}\right)\Bigg|_{S_{1B}} \tag{8.5-2}$$

This procedure is iterative inasmuch as S_{1B} depends on the polymer–oil fractional flow curve. This, in turn, depends on the polymer apparent viscosity whose value we are estimating in Eq. (8.5-1). Fortunately, the dependence between S_{1B} and apparent viscosity is weak, and a trial-and-error procedure should converge rapidly.

A second procedure is to base the total mobility of the oil bank on the minimum in the total relative mobility curve (Gogarty et al., 1970). The minima in such curves do not, in general, correspond to the oil bank saturation from fractional flow theory. However, taking M_T based on the minimum will yield a conservative design since the mobility ratio with the actual oil bank saturation will always be less than or equal to M_T. The method has the advantage of simplicity since it is noniterative.

Both methods require care in measuring relative permeability curves since hysteresis can render the drainage and imbibition k_r's different (Chang et al., 1978). Such hysteresis effects are particularly difficult to reproduce when the initial water saturation begins at an intermediate value. The second method is also commonly used in micellar-polymer design (see Fig. 9-34).

(b) Estimate the polymer slug volume. One way to do this is to simply let the slug volume be somewhat larger than the retention. Although this is the basic premise in designing a micellar slug, retention is not the dominant factor in polymer slug sizing. The major control affecting slug size is viscous fingering between the chase water and the polymer spike.

In predicting the extent of fingering, all the problems in estimating the rate of finger propagation that we discussed in Sec. 7-8 apply. Once again, we apply the Koval model, but here the effective mobility ratio must be modified to account for the polynomial mixing expressed in Eq. (8.2-1).

$$E = (1 + a_1 C_4 + a_2 C_4^2 + a_3 C_4^3 + \cdots +)R_{RF} = \frac{K_{val}}{H_K} \tag{8.5-3}$$

where $C_4 = 0.22\ C_{4J}$. The use of this equation, particularly the constant mixing factor 0.22, is relatively untested in polymer flooding. We have assumed complete analogy between the first-contact miscible flooding case and the unstable chase water displacement in this regard.

We use the time–distance diagram to sketch slug sizing alternatives. An obvious sizing technique is to begin chase water injection just as the polymer breaks through (Fig. 8-12a). This is excessively conservative since much full-strength polymer is produced. A second possibility is to adjust the polymer slug size so that the polymer and chase water break through simultaneously (Fig. 8-12b), leading to the following equation for slug size

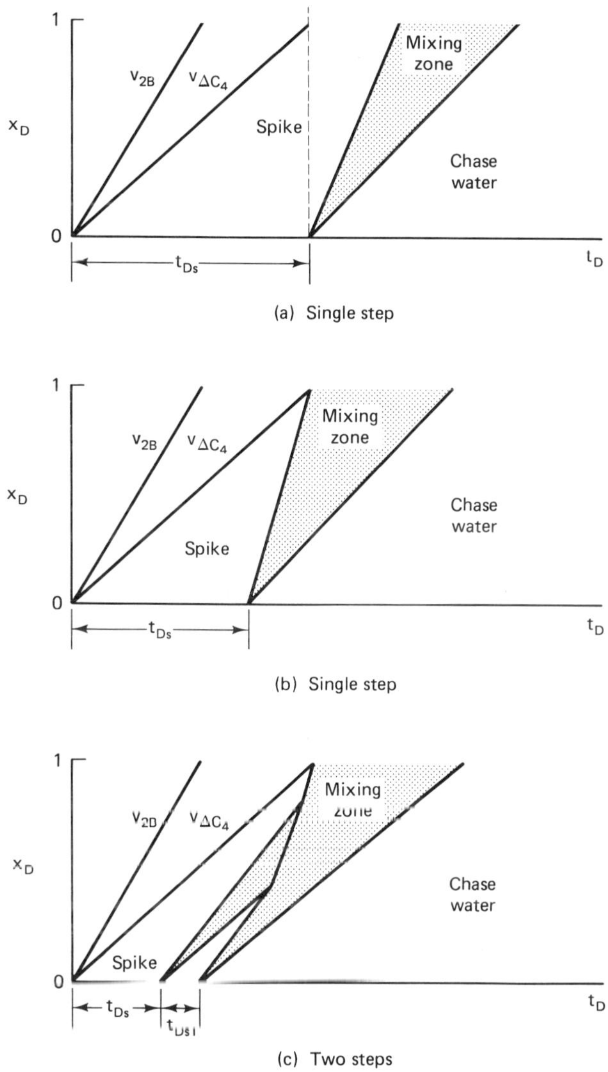

Figure 8-12 Time–distance diagrams for polymer grading

$$t_{Ds} = 1 - S_{2r} - D_4 - \frac{1 - S_{2r}}{K_{val}} \tag{8.5-4}$$

where we have taken $S_1^* = 1 - S_{2r}$ and $\phi_{IPV} = 0$. An equally viable alternative is to grade the polymer back to chase water in steps. Figure 8-12(c) shows two such steps with the size of the spike and the intermediate step adjusted so that the chase water again breaks through with the polymer. The isoconcentration lines become curved after there is wave interference.

The case in Fig. 8-12(c) uses less polymer than that in Fig. 8-12(b). In fact, Claridge (1978) has shown that a continuously graded polymer drive uses the least amount of polymer. Such grading is impractical except as a limiting case to compare it to the no-grading case (Fig. 8-12b). But a succession of finite grading steps is extremely difficult to deal with theoretically because of the numerous degrees of freedom present. That is, the engineer must decide on the N number of steps, the N slug volumes, and the $N - 1$ intermediate concentrations. In practice, single-step (no-grading) polymer floods, and logarithmic grading (Mungan, 1968) are the most common procedures.

5. Design polymer injection facilities. Getting a good quality solution is, of course, important, but the cost of the injection facilities is usually small compared to well and chemical costs.

   The three essential ingredients are mixing facilities, filtration, and injection equipment. The type of mixing apparatus depends on the polymer. For solid polymers, a skid-mounted solid mixer is required. Concentrates or emulsion polymers require somewhat less sophistication although the latter may require some emulsion breaking. Filtration largely depends on the success of the mixing, but ordinarily it is no more stringent than what is required by waterflooding. But if exotic and difficult filtration is required, the complexity and cost can become significant. Injection equipment is the same as that for waterflooding. All surface and downhole equipment should be modified to avoid all forms of degradation.

6. Consider the reservoir. Little is required here beyond the usual waterflood considerations such as the optimal well pattern and spacing, completion strategy, pattern allocation (balance), reservoir characterization, and allowable injection rates.

   Optimal values of these quantities imply precise values that will result in the maximum rate of return on investment. Since several quantities are involved, it is usually not possible to perform optimizations on everything. Hence most of the parameters must be fixed by other considerations (such as striving for a target mobility ratio). But for the most sensitive quantities, optimization is required.

   Figure 8-13 shows a schematic optimization for the amount of polymer injected. The vertical axes plot both an economic measure, such as the cumulative incremental discounted cash flow (DCF), and incremental oil recovery

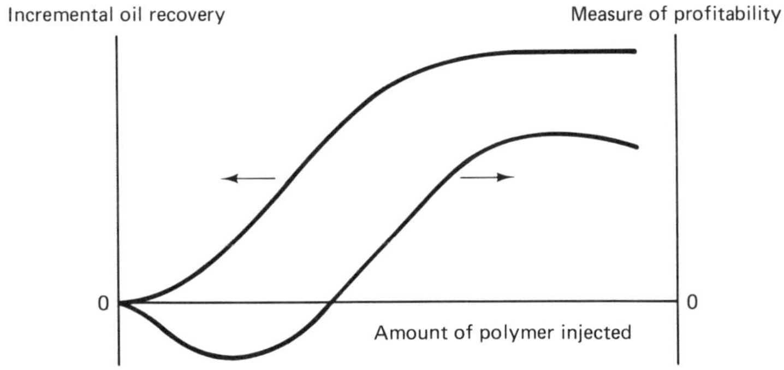

Figure 8-13 Schematic incremental oil recovered and economic trends for a mobility control flood.

(IOR) versus the amount of polymer injected. The IOR curve is monotonically increasing from zero. The DCF curve begins at zero, decreases for small polymer amounts, and then rises to a maximum at substantially larger amounts. After this point, the DCF decreases monotonically. The DCF decreases initially because the entire expense of the polymer is assessed in the initial stages of a project when little incremental oil has yet been produced. This front-end loading effect is present in all EOR processes, particularly chemical floods. Such a curve is highly instructive because it counters a tendency to short-cut the amount of polymer injected if the initial economics are unfavorable. Unfortunately, many actual polymer flood applications have used less than the optimum amount of polymer.

8-6 FIELD RESULTS

The incremental oil recovery (IOR) from a polymer flood is the difference between the cumulative oil actually produced and that which would have been produced by a continuing waterflood (see Exercise 8L). Thus for a technical analysis of the project, it is important to establish a polymer flood oil rate decline and an accurate waterflood decline rate. Figure 8-14 shows the IOR for the North Burbank polymer flood.

Table 8-2 summarizes other field results on more than 250 polymer floods based on the comprehensive survey of Manning et al. (1983). The table emphasizes oil recovery data and screening parameters used for polymer flooding. Approximately one third of the reported projects are commercial or field-scale floods. The oil recovery statistics in Table 8-2 show average polymer flood recoveries of 3.56% remaining (after waterflood) oil in place and about 1 m^3 of IOR for each kilogram of polymer injected with wide variations in both numbers. The large variability reflects the emerging nature of polymer flooding in the previous decades. Considering the average polymer requirement and the average costs of crude and polymer, it appears

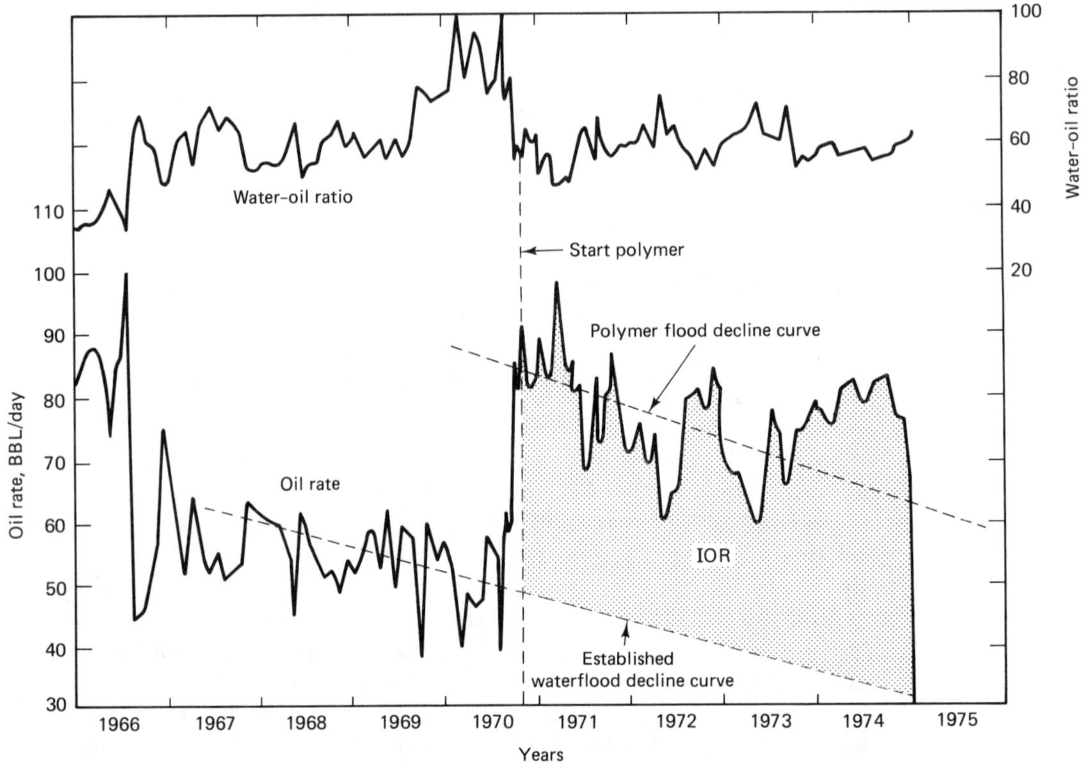

Figure 8-14 Tertiary polymer flood response from North Burbank Unit, Osage County, Okla. (from Clampitt and Reid, 1975)

that polymer flooding should be a highly attractive EOR process. However, such costs should always be compared on a discounted basis, reflecting the time value of money. Such a comparison will decrease the apparent attractiveness of polymer flooding because of the decreased injectivity of the polymer solutions.

8-7 CONCLUDING REMARKS

In terms of the number of field projects, polymer flooding is the most common enhanced oil recovery technique in existence. The reasons for this are that, short of waterflooding, polymer flooding is the simplest technique to apply in the field and requires a relatively small capital investment. Most of the field projects have been small, however, as has the amount of oil recovered, a fact that should be expected from the treatment given in this chapter. Nevertheless, there can exist significant potential for an acceptable rate of return even when recovery is low.

The most important property covered in this chapter is the non-Newtonian behavior of polymer solutions, because such behavior impacts on the polymer require-

TABLE 8-2 POLYMER FLOOD STATISTICS (ADAPTED FROM MANNING, 1982)

Value standard (units)	Number projects*	Mean	Minimum	Maximum	Standard deviation
Oil recovery (% remaining OIP)	50	3.56	0	25.3	5.63
Polymer utilization (m^3/kg polymer)	80	0.94	0	12.81	1.71
Oil recovery (m^3/hm^3 bulk volume)	88	3.1	0	24.3	4.72
Permeability variation (fraction)	118	0.70	0.06	0.96	0.19
Mobile oil saturation (fraction)	62	0.27	0.03	0.51	0.12
Oil viscosity (mPa-s)	153	36	0.072	1,494	110.2
Resident brine salinity (kg/m^3 TDS)	10	40.4	5.0	133.0	33.4
Water-to-oil mobility ratio (dimensionless)	87	5.86	0.1	51.8	11.05
Average polymer concentration (g/m^3)	93	339	51	3,700	343
Temperature (K)	172	319	281	386	302
Average permeability (μm^2)	187	0.349	0.0015	7.400	0.720
Average porosity (fraction)	193	0.20	0.07	0.38	0.20

*Partial data available on most projects; includes both commercial and pilot projects

ments through the design mobility ratio, and on the ability to accurately forecast the rate of polymer injection. Polymer injection rate determines project life which, in turn, determines the economic rate of return. Injectivity estimates along with estimates of mobile oil saturation and the likelihood that polymer will remain stable in a given application are the most important determinants in polymer flooding success.

EXERCISES

8A. *Calculating Shear Rates*. Calculate the equivalent shear rate under the following conditions:

(a) In an open-hole completion (entire well cylinder open to flow) where q = 16 m^3/day, R_w = 7.6 cm, and net pay H_t = 15.25 m.

(b) In the field where the interstitial velocity is 1.77 μm/s.

(c) Using the data for Xanflood at 297 K and 1% NaC1 (Fig. 8-5), estimate the effective permeable medium viscosity at the above conditions for a 600 g/m^3 polymer solution.

(d) Suppose the well in part (a) is perforated with 1 cm (ID) holes over its entire net pay at a density of 4 holes/m. Assuming a uniform fluid distribution, estimate the shear rate in the perforations.

(e) Comparing the results of parts (a) and (d), what do you conclude about the preferred completion technique in polymer flooding? Use $k_1 = 0.1\ \mu m^2$, $\phi = 0.2$, and $S_1 = 1.0$ in all parts.

8B. *Derivation of Power Law in Permeable Media*. Equation (8.2-9) may be derived in the same manner as Eq. (3.1-11). The procedure is as follows:

(a) Show that a force balance on an annular element of a single-phase fluid flowing through a tube (as in Fig. 3.1) in laminar steady-state flow is

$$\frac{1}{r}\frac{d(r\tau_{rz})}{dr} = \frac{\Delta P}{L} \tag{8B-1}$$

where τ_{rz} is the shear stress on the cylindrical face at r, and $\Delta P/L$ is the pressure gradient. This equation, when integrated, yields

$$\tau_{rz} = \frac{\Delta P}{2L} r \tag{8B-2}$$

The shear stress must be finite at $r = 0$.

(b) The power-law expression relating shear stress to shear rate is

$$\tau_{rz} = K_{pl}\,\dot{\gamma}^{n_{pl}-1} \tag{8B-3}$$

where

$$\dot{\gamma} = -\frac{dv}{dr} \tag{8B-4}$$

is the shear rate. Show that combining Eqs. (8B-2) through (8B-4) leads to a differential equation whose solution is

$$v(r) = \left(\frac{\Delta P}{2LK_{pl}}\right)^{1/n_{pl}}\left(\frac{n_{pl}}{1 + n_{pl}}\right)\left(R^{\frac{1+n_{pl}}{n_{pl}}} - r^{\frac{1+n_{pl}}{n_{pl}}}\right) \tag{8B-5}$$

This equation has used the no-slip condition $v(R) = 0$.

(c) Using Eq. (8B-5), show that the shear rate at the wall of the tube depends on the average velocity as

$$\dot{\gamma}_{\text{wall}} = \frac{1 + 3n_{pl}}{n_{pl}}\left(\frac{\bar{v}}{R}\right) \tag{8B-6}$$

(d) When the equivalent radius from Eq. (3.1-4) is substituted, this gives

$$\dot{\gamma}_{eq} = \left(\frac{1 + 3n_{pl}}{n_{pl}}\right)\frac{u}{(8k_1\phi_1)^{1/2}} \tag{8B-7}$$

With appropriate variable identifications, this equation yields Eqs. (8.2-7) and (8.2-8) when substituted into

$$\mu_{app} = \frac{\tau_{rz}}{\dot{\gamma}_{eq}} \tag{8B-8}$$

8C. *Langmuir Calculations*. The Langmuir isotherm and various other insights may be derived fairly simply. Suppose a permeable medium in contact with a solution containing an adsorbing species consists of a fixed number of surface sites. A fraction θ of these sites is covered when the solution concentration of an adsorbing species is C.

(a) Let the rate of adsorption be $k_f C(1 - \theta)$ and the rate of desorption be $k_r \theta$. k_f and k_r are the forward and reverse rate constants. At equilibrium, the forward and reverse reaction rates are equal. Show that the fractional surface coverage is

$$\theta = \frac{\dfrac{k_f}{k_r} C}{1 + \dfrac{k_f}{k_r} C} \tag{8C-1}$$

(b) Show that θ may be related to ω_s, adsorption in mass per unit of rock mass by

$$\theta = \frac{d_p^2 N_A \rho_r}{a_v M_w} \omega_s \tag{8C-2}$$

where ρ_r is the adsorbed species density, a_v is the specific surface area of the medium, M_w is the molecular weight of the adsorbed species, and N_A is Avogadro's number. Assume the adsorbed species exists on the surface as a monolayer of cubes of diameter d_p.

(c) If the observed polymer adsorption is 18 μg/g-rock, calculate θ. Take the medium to have the Berea properties tabulated in Table 3-5. You must derive the effective polymer sphere diameter from the intrinsic viscosity (Eq. 8.2-4) and the data in Fig. 8-4. The polymer molecular weight is 2 million.

(d) What can you conclude about the nature of the adsorption of polymers from this?

8D. *Complications to Langmuir Isotherm*

(a) Suppose there is Langmuir adsorption of a single adsorbing species with a finite mass transfer rate r_{mt} between the bulk solution and the solid–fluid interface given by

$$r_{mt} = h(C - \tilde{C}) \tag{8D-1}$$

In this expression, C and $\tilde{C}$ are the bulk and interface concentrations, and h is the specific mass transfer coefficient. Show that an isotherm relating θ to C has the same form as Eq. (8C-1) but with k_f' replacing k_f where

$$\frac{1}{k'} = \frac{1}{k_f} + \frac{1}{h} \tag{8D-2}$$

You must assume the rate of adsorption is equal to r_{mt}.

(b) Show that if $h \to \infty$, the isotherm approaches the expression derived in Exercise 8C.

(c) Consider now a case where $h \to \infty$ and there are $i = 1, \ldots, N_C$ adsorbing species, each competing for a fixed number of sites. Derive the Langmuir isotherm relating the adsorbed concentration of species i, θ_i, to its bulk concentration C_i.

(d) Use this expression to justify the fractional coverages calculated in part (d) of Exercise 8C.

8E. *Simplified Permeability Reduction.* One of the explanations for permeability reduction is that the effective pore size is decreased (or the effective grain diameter increased) because of the adsorption of a layer of polymer on the rock surface. In the following, take the medium to be comprised as spheres of diameter D_p:

(a) Derive an expression for the permeability reduction factor R_k based on the polymer adsorbing as a uniform layer of thickness δ on the rock surface. You must use the hydraulic radius concept developed in Sec. 3-1.

(b) Make two plots, at $\phi = 0.1$ and 0.2, of polymer adsorption (in mg polymer/g rock) versus R_k. Take the density of the adsorbed polymer to be 1.5 g/cm^3 and the density of the rock to be 2.5 g/cm^3.

8F. *Representation of Linear Viscoelasticity.* A powerful conceptual model of a liquid that has some elastic effects is the Maxwell model, which is the series combination of a spring and a dashpot

where F is the force sustained by the model, and ϵ_1 and ϵ_2 are the strains (dimensionless deformations). Let the spring be a linear elastic element so that

$$F = k\epsilon_1 \tag{8F-1}$$

likewise, the dashpot is a Newtonian viscous element

$$F = \mu\dot{\epsilon}_2 \tag{8F-2}$$

where k and μ are the spring constant and viscosity of the element. Because of the series arrangement, the force supported by both elements is the same; however, the total strain ϵ is

$$\epsilon = \epsilon_1 + \epsilon_2 \tag{8F-3}$$

(a) Show that the relationship between the time behavior of the force and the strain is

$$\mu\dot{\epsilon} = \theta\dot{F} + F \tag{8F-4}$$

In this equation, $\theta = \mu/k$ is the relaxation time of the model, and $\dot{\epsilon}$ is the time derivative of ϵ.

(b) To integrate this, we treat $\dot{\epsilon}$ as a known function of time. Show that the general solution is

$$F(t) = e^{-t/\theta}F(0) + ke^{-t/\theta}\int_0^t e^{\xi/\theta}\frac{d\epsilon}{d\xi}\,d\xi \tag{8F-5}$$

The next three steps complete the analogy between the Maxwell model and viscoelastic flow.

(c) If the rate of strain is constant and the initial force of the model is zero, show that

$$F(t) = \mu\dot{\epsilon}(1 - e^{-t/\theta}) \tag{8F-6}$$

(d) The apparent viscosity of the model is defined as $F/\dot{\epsilon}$. Show from Eq. (8F-4) that this becomes

$$\mu_{app} = \frac{\mu}{1 + \theta\dfrac{\dot{F}}{F}} \tag{8F-7}$$

(e) Use this equation and Eq. (8F-6) to show that

$$\mu_{app} = \frac{\mu}{1 + N_{\text{Deb}}} \tag{8F-8}$$

The quantity in the denominator of Eq. (8F-8) is the Deborah number

$$N_{\text{Deb}} = \frac{\theta}{t} \tag{8F-9}$$

This number, the ratio of relaxation time to undisturbed flow time around a rock grain, is a measure of viscoelastic effects in permeable media flow when the characteristic flow time t has been replaced by $\phi D_p/u$.

8G. *Analysis of Screen Factor Device.* The screen factor device in Fig. 8-8 may be analyzed as a permeable medium experiencing gravity drainage. The volume V of fluid in the bulb at any height $h (h_1 > h > h_2)$ is

$$V = \frac{\pi}{3}(h - h_2)^2(3R - h + h_2) \tag{8G-1}$$

from the bulb geometry. If we treat the screen pack as a permeable resistive element, the flux through the screens is

$$u = -\left(\frac{k\rho g h}{L\mu_{app}}\right) \tag{8G-2}$$

(a) Since $u = -1/\pi r^2\,(dV/dt)$, show from these equations that the height h is the solution to

$$\frac{dh}{dt}[(h - h_2)(2R - (h - h_2))] = \frac{r^2 \rho g k h}{\mu_{app} L} \tag{8G-3}$$

L in these equations is the height of the screen pack.

(b) Neglecting the drainage times in the tubes above and below the lower bulb, derive an expression for the drainage time for a Newtonian fluid. The drainage time t_d is defined as

$$t_d = t\,|_{h=h_2} - t\,|_{h=h_1} \tag{8G-4}$$

(c) Repeat part (b) with a viscoelastic fluid whose apparent viscosity is

$$\mu_{app} = \frac{H_{VE}}{1 + bu} \tag{8G-5}$$

In view of Eqs. (8G-4) and (8.2-14), show that the screen factor SF is given by

$$SF = \frac{H_{VE}}{\mu_1} + \frac{k\rho g}{\mu_1 L}\,bI \tag{8G-6}$$

where I is a geometric factor. From Exercise 8F, the screen factor is directly proportional to the fluid's relaxation time.

8H. *Injectivity Calculation*
Use the following data for the Coalinga HX sand (Tinker et al., 1976):

$\phi = 0.28$	$k_1 = 0.036\ \mu\text{m}^2$
$K_{pl} = 7.5$ mPa-s(s)$^{n_{pl}-1}$	$\mu_1 = 0.64$ mPa-s
$n_{pl} = 0.8$	$H_t = 2.44$ m
$R_k = 3$	$R_w = 10$ cm
$R_e = 284$ m	$i = 30\ \text{m}^3/\text{D}$
	$S_{2r} = 0.2$

(a) Calculate the relative injectivity I_r versus cumulative polymer injected. Plot I_r versus t_D (up to $t_D = 0.5$) on linear graph paper.

(b) Show that when $R_p = R_e$, the Newtonian polymer case ($n_{pl} = 1$) reduces to

$$I_r = \frac{\mu_1}{K_{pl} R_k} \tag{8H-1}$$

(c) Plot the Newtonian polymer case for the HX sand on the same plot as in part (a).

8I. *Improvements to Injectivity Calculations*. If the shear rate range in a cylindrical reservoir is outside the power-law range, the following truncated form of Eq. (8.2-7) must be used:

$$\mu_{app} = \begin{cases} \mu_1^0, & u < u_0 \\ H_{pl} u^{n_{pl}-1}, & u_0 < u < u_\infty \\ \mu_1^\infty, & u > u_\infty \end{cases} \tag{8I-1}$$

where u_0 and u_∞ are superficial velocities which define the limits of the power law range.

(a) Repeat the derivation in Sec. 8-3 for I and I_r using Eq. (8I-1), assuming both the maximum and minimum velocities fall outside the power-law range.

(b) For numerical simulation, it may be more convenient to define injectivity in terms of the average reservoir pressure $\bar{P}$ rather than P_e (Bondor et al., 1972). Rederive the expression for I defined in this manner.

(c) For large numerical simulations, the entire non-Newtonian range of polymer behavior is confined within one grid block of the well. This being the case, the non-Newtonian effect can be effectively expressed as a time-varying skin factor in terms of an average polymer "saturation." Derive an expression for this skin factor.

8J. *Transport of Adsorbing Slugs*. The leading edge of a polymer slug adsorbing as a Langmuir isotherm is self-sharpening.

(a) Show that the rear of the slug ($C_K < C_J$) is a spreading wave.

(b) If the Langmuir parameters in Eq. (8.2-10) are $a = 2$ and $b = 20$, plot the time–distance diagram and effluent history of $t_{Ds} = 0.4$ slug displacement. Take $C_I = C_K = 0$ and $C_J = 1$.

(c) The propagation of slugs satisfies an overall material balance

$$t_{Ds} = \frac{\int_0^\infty (C + C_s)\, dx_D}{C_J} \tag{8J-1}$$

Use the analogy to the Welge integration in Sec. 5-2 to show that Eqs. (8.4-1) and (8.4-2) satisfy this identically. In all these calculations, take the flow to be single phase with the usual fractional flow assumptions.

8K. *Asymptotic Mixing Zone Length* (Lake and Helfferich, 1978). Stabilized mixing zones occur in miscible displacements if the transported species adsorbs according to a Langmuir isotherm. The spreading caused by dispersion is balanced by the sharpening caused by adsorption. In the following, take the dimensionless material balance of an adsorbing species to be

$$\frac{\partial(C + C_s)}{\partial t_D} + \frac{\partial C}{\partial x_D} - \frac{1}{N_{\text{Pe}}}\frac{\partial^2 C}{\partial x_D^2} = 0 \tag{8K-1}$$

where C and C_s are the solution and adsorbed concentrations for an adsorbing species. C is normalized so that the injected concentration is unity, $C_J = 1$, and $C_I = 0$.

(a) Show that Eq. (8K-1) may be transformed to a moving coordinate system (x_D', t_D), where $x_D' = x_D - v_{\Delta C} t_D$, and $v_{\Delta C}$ is the shock velocity of C. This gives

$$\frac{\partial(C + C_s)}{\partial t_D} - v_{\Delta C}\frac{\partial C_s}{\partial x_D'} + (1 - v_{\Delta C})\frac{\partial C}{\partial x_D'} - \frac{1}{N_{\text{Pe}}}\frac{\partial^2 C}{\partial (x_D')^2} = 0 \tag{8K-2}$$

(b) The displacement will asymptotically approach stabilized flow where the time derivatives in Eq. (8K-2) are zero. Show that in this limit the resulting ordinary differential equation may be integrated to give

$$\Delta x_D = \frac{1}{N_{\text{Pe}}}\int_{0.9}^{0.1} \frac{dC}{(1 - v_{\Delta C})C - v_{\Delta C} C_s} \tag{8K-3}$$

Equation (8K-3) uses the boundary conditions $C(+\infty) = dC(+\infty)/dx_D = 0$ and the definition of dimensionless mixing zone given in Eq. (5.2-15a).

(c) When $C_J = 1$, it is convenient to write the Langmuir isotherm (Eq. 8.2-10) so that the plateau adsorption appears in the equation in place of the parameter a

$$C_s = \frac{(1 + b)C_{sJ} C}{1 + bC} \tag{8K-4}$$

where C_{sJ} is the maximum adsorbed concentration. Substitute Eq. (8K-4) into Eq. (8K-3), and perform the indicated integration to show that

$$\Delta x_D = \frac{1}{N_{\text{Pe}}}\left\{\frac{1 + C_{sJ}}{C_{sJ}}\right\}\left(1 + \frac{2}{b}\right)\ln(9) \tag{8K-5}$$

where this equation has used a form of $v_{\Delta C}$ consistent with Eq. (8K-4).

(d) Take Eq. (8K-5) in the limits of $b \to \infty$, $b \to 0$, and $N_{\text{Pe}} \to \infty$, and justify each answer on physical grounds.

8L. *Fractional Flow and Incremental Oil*

(a) Calculate the polymer frontal advance lag D_4 when the maximum polymer adsorption is 38 g/m^3 (bulk volume), the injected polymer concentration is 1200 g/m^3, and the porosity is 0.2.

(b) Using the D_4 of part (a) and the water–oil relative permeabilities in Fig. 8L, calculate the effluent history of polymer and oil for a polymer flood with $\mu_1^0 = 30$ mPa-s. Take the oil and water viscosities to be 20 and 1 mPa-s, respectively, the dip angle to be 0, the permeability reduction factor to be 1, and the initial water saturation to be 0.4.

(c) The technically correct way to evaluate a polymer flood is by the incremental oil recovery (IOR)

$$\text{IOR} = \begin{pmatrix}\text{Polymer flood}\\ \text{oil produced}\end{pmatrix} - \begin{pmatrix}\text{Waterflood}\\ \text{oil produced}\end{pmatrix} \qquad (8L\text{-}1)$$

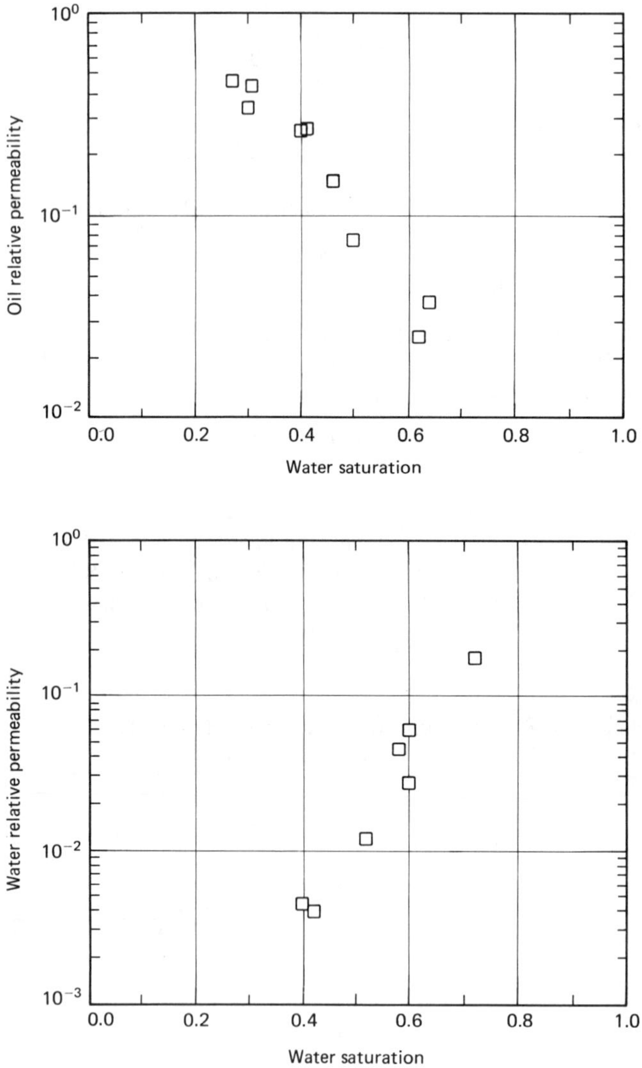

Figure 8L Relative permeabilities for Exercise 8L (from El Dorado, 1977).

Calculate and plot IOR (in SCM) versus time (years). Take the pore volume to be 1.6×10^6 m^3, the injection rate constant at 480 SCM/day, and all formation volume factors to be 1.0 m^3/SCM.

8M. *Fractional Flow and Slugs*. Fractional flow theory can be used to gain insight into the behavior of polymer slugs, under idealized conditions, and into the polymer utilization factor.

(a) Assume the polymer is to be injected as a slug. If the chase water displaces the polymer as an ideal miscible displacement at residual oil saturation, show that the polymer chase water front travels with specific velocity

$$v_{CW} = \frac{1}{1 - S_{2r}} \tag{8M-1}$$

if the polymer adsorption is irreversible and excluded pore volume negligible.

(b) Show that the polymer slug size just needed to satisfy adsorption is equal to D_4.

(c) The data to use in the remainder of this exercise are

$a = 1$ cm^3/g-rock	$C_{4J} = 800$ g/m^3
$b = 100$ cm^3/mg	$\rho_s = 2.65$ g/cm^3
$\phi = 0.2$	

Plot the time–distance and effluent histories (oil and polymer) if the slug size used is one half that demanded by adsorption. Use the fractional flow curves and initial condition of Exercise 8L.

8N. *Polymer Flood Design*. You want to design a polymer flood in a reservoir containing an oil and brine whose viscosities are 25 mPa-s and 0.38 mPa-s, respectively, at reservoir temperature of 73°C. The relative permeability curves of Fig. 8L apply, and conditions indicate the Xanflood data in Figs. 8-4 and 8-5 are satisfactory for this reservoir.

(a) Plot the total relative mobility curves. If the desired mobility ratio is 0.7, estimate the polymer concentration required to bring this about. Use the data in Fig. 8-5, and recall that μ_1'/μ_1 is essentially independent of temperature.

(b) Estimate the power-law parameters K_{pl}, n_{pl}, and H_{pl} for the polymer solution in part (a).

(c) The flood is to be done at a constant volumetric injection rate of 20 m^3/D. Estimate and plot as a function of volume injected the bottom-hole injection pressure in MPa. Justify the shape of this curve on physical grounds.

(d) For an open-hole completion, estimate the shear rate the polymer solution will be exposed to. Does this portend mechanical degradation of the polymer?

Take the reservoir to be circular with $R_e = 950$ m and $P_e = 18$ MPa. Additional properties are $k = 0.05$ μm^2, $s_w = 0$, $R_w = 5$ cm, $H_t = 42$ m, $\phi = 0.2$, and $S_{2r} = 0.3$.

9

Micellar-Polymer Flooding

From the earliest days, it was recognized that capillary forces caused large quantities of oil to be left behind in well-swept zones of waterflooded oil reservoirs. Capillary forces are the consequence of the interfacial tension (IFT) between the oil and water phases that resists externally applied viscous forces and causes the injected and banked-up connate waters to locally bypass oil. Similarly, early efforts of enhanced oil recovery strove to displace this oil by decreasing the oil-water IFT. Though many techniques have been proposed and field tested, the predominant EOR technique for achieving low IFT is micellar-polymer (MP) flooding.

Lowering interfacial tension recovers additional oil by reducing the capillary forces that leave oil behind any immiscible displacement. This trapping is best expressed as a competition between viscous forces, which mobilize the oil, and capillary forces, which trap the oil. The local capillary number N_{vc}, the dimensionless ratio of viscous to capillary forces, determines the residual oil and water saturations through a capillary desaturation curve (CDC). Section 3-4 gives general features about the CDC and N_{vc}. In this chapter, we specialize those results to MP flooding. Recall that ultralow IFTs are required—of the order of 1 μN/m—and that these values can be attained only through highly surface-active chemicals.

9-1 THE MP PROCESS

MP flooding is any process that *injects* a surface-active agent (a surfactant) to bring about improved oil recovery. This definition eliminates alkaline flooding (see Chap. 10) where the surfactant is generated in situ and other EOR processes where lowering the capillary forces is not the primary means of oil recovery.

MP flooding has appeared in the technical literature under many names: detergent, surfactant, low-tension, soluble oil, microemulsion, and chemical flooding. We use the term *micellar-polymer* flooding because it is the least ambiguous (chemical flooding, for example, could describe all nonthermal EOR processes) and most comprehensive (no other name implies the polymer component). Moreover, several names imply a specific sequence and type of injected fluids as well as the specific nature of the oil-recovering MP slug itself. Though there are differences among processes, in this chapter we emphasize the similarities since they are more numerous and important.

Figure 9-1 shows an idealized version of an MP flooding sequence. The process is usually applied to tertiary floods and is always implemented in the drive mode (not cyclic or huff 'n puff). The complete process consists of the following:

Preflush. A volume of brine whose purpose is to change (usually lower) the salinity of the resident brine so that mixing with the surfactant will not cause loss of interfacial activity. Preflushes have ranged in size from 0% to 100% of the floodable pore volume (V_{pf}) of a reservoir. In some processes, a sacrificial agent is added to lessen the subsequent surfactant retention (Holm, 1982).

MP slug. This volume, ranging from 5% to 20% V_{pf} in field applications, contains the main oil-recovering agent, the primary surfactant. Several other chemicals (Fig. 9-1) are usually needed to attain the design objectives. We discuss the purpose of these chemicals in more detail later.

Mobility buffer. This fluid is a dilute solution of a water-soluble polymer whose purpose is to drive the MP slug and banked-up fluids to the production wells. All the polymer flooding technology discussed in Chap. 8 carries over to designing and implementing the mobility buffer. Thus in this chapter, we deal relatively little with the mobility buffer though there is good evidence (see Fig. 9-33) that this volume is very important to the oil recovering ability of the en-

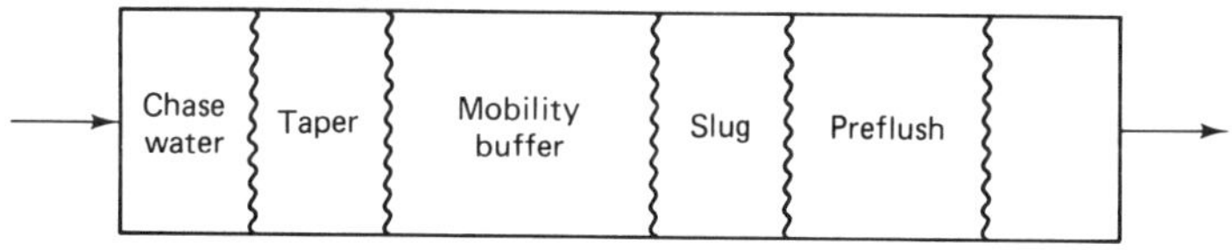

Mobility buffer	Slug	Preflush
250-2500 g/m^3 polymer	1-20% Surfactant	Electrolyte (Na^+, Ca^{++}, etc.)
0-1% Alcohol	0-5% Alcohol	Sacrificial chemicals
Stabilizers	0-5% Cosurfactant	0-100% V_{pf}
Biocide	0-90% Oil	
0-100% V_{pf}	Polymer	
	5-20% V_{pf}	

Figure 9-1 Idealized cross section of a typical micellar-polymer flood (from Lake, 1984)

tire sequence. The target oil for an MP flood—the residual oil—is different from that of a polymer flood—the movable oil.

Mobility buffer taper. This is a volume of brine that contains polymer, grading from that of the mobility buffer at the front end (the spike) to zero at the back. The gradual decrease in concentration mitigates the effect of the adverse mobility ratio between the mobility buffer and the chase water.

Chase water. The purpose of the chase water is simply to reduce the expense of continually injecting polymer. If the taper and mobility buffer have been designed properly, the MP slug will be produced before it is penetrated by this fluid.

9-2 THE SURFACTANTS

Since much is required of the MP surfactant, we discuss surfactant solutions here. This discussion can be no more than a précis of the voluminous literature on surfactant properties. (For more on oil-recovering surfactants, see Akstinat, 1981.)

A typical surfactant monomer is composed of a nonpolar (lypophile) portion, or *moiety*, and a polar (hydrophile) moiety; the entire monomer is sometimes called an *amphiphile* because of this dual nature. Figures 9-2(a) and 9-2(b) show the

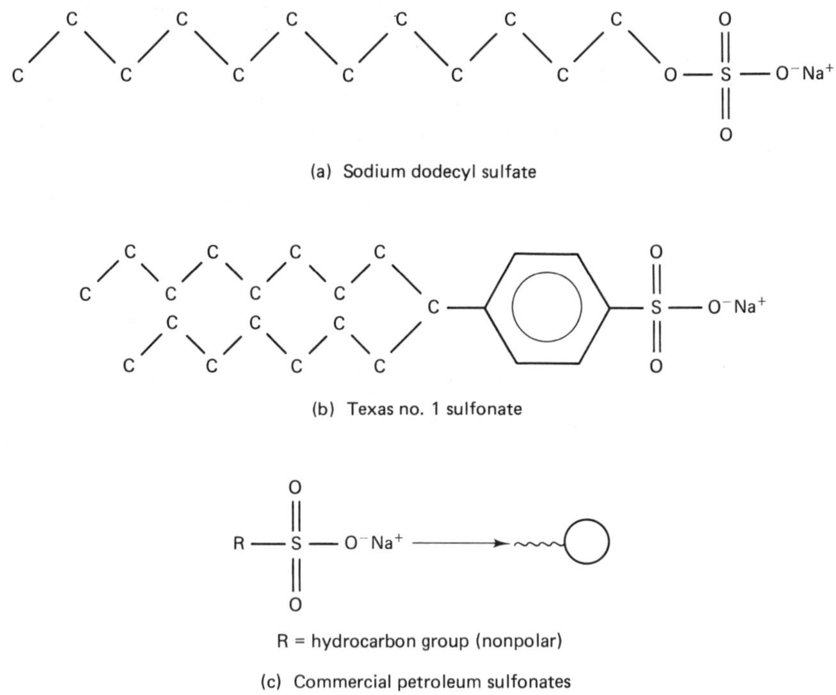

Figure 9-2 Representative surfactant molecular structures (from Lake, 1984)

molecular structure of two common surfactants and illustrate a shorthand notation for surfactant monomers: The monomer is represented by a "tadpole" symbol, with the nonpolar moiety being the tail and the polar being the head.

Surfactants are classified into four groups depending on their polar moieties, (Table 9-1).

Anionics. As required by electroneutrality, the anionic (negatively charged) surfactant molecule (distinct from monomer) is uncharged with an inorganic metal cation (usually sodium) associated with the monomer. In an aqueous solution, the molecule ionizes to free cations and the anionic monomer. Anionic surfactants are the most common in MP flooding because they are good surfactants, relatively resistant to retention, stable, and can be made relatively cheaply.

Cationics. If the polar moiety is positively charged, the surfactants are cationic. In this case, the surfactant molecule contains an inorganic anion to balance the charge. Cationic surfactants are used little in MP flooding because they are highly adsorbed by the anionic surfaces of interstitial clays.

Nonionics. A class of surfactants that have seen extensive MP use, mainly as cosurfactants but increasingly as primary surfactants, is the nonionics. These surfactants do not form ionic bonds but, when dissolved in aqueous solutions, exhibit surfactant properties by electronegativity contrasts between their constituents. Nonionics are much more tolerant of high salinities than anionics and historically have been poorer surfactants.

Amphoterics. This class of surfactants contains aspects of two or more of the other classes. For example, an amphoteric may contain both an anionic group and a nonpolar group. These surfactants have not been used in oil recovery.

TABLE 9-1 CLASSIFICATION OF SURFACTANTS AND EXAMPLES (ADAPTED FROM AKSTINAT, 1981)

Anionics	Cationics	Nonionics	Amphoterics
Sulfonates Sulfates Carboxylates Phosphates	Quaternary ammonium organics, pyridinum, imidazolinium, piperidinium, and sulfononium compounds	Alkyl-, Alkyl- aryl-, acyl-, acylamindo-, acylaminepolyglycol, and polyol ethers Alkanolamides	Aminocarboxylic acids

Within any one class, there is a huge variety of possible surfactants. Figure 9-2 shows some of this variety by illustrating differences in nonpolar molecular weight (C_{12} for the sodium dodecyl sulfate (SDS) versus C_{16} for Texas No. 1), polar moiety identity (sulfate versus sulfonate), and tail branching (straight chain for SDS versus two tails for Texas No. 1) all within the same class of anionic surfactants. Besides these, there are variations in both the position of the polar moiety attachment and the number of polar moieties (monosulfonates versus disulfonates, for example). Even small variations can drastically change surfactant properties. For example, sulfates tend to be less thermally stable than sulfonates. (For more details on the effect of structure on surfactant properties, see Graciaa et al., 1981; Barakat et al., 1983.)

The most common primary surfactant used in MP flooding is petroleum sulfonates. These are anionic surfactants produced by sulfonating a pure organic chemical (sometimes called synthetic sulfonates), an intermediate molecular weight refinery stream, or when appropriate, even a crude oil itself. If R—C=C—H represents the molecular formula of the feedstock, the sulfonation reaction proceeds as

$$\text{R—C=C—H} + \text{SO}_3 \longrightarrow \text{R—C=C—SO}_3^- + \text{H}^+ \qquad (9.2\text{-}1)$$

The reaction can also proceed to saturate the carbon–carbon double bond

$$\text{R—C=C—SO}_3^- + \text{H}_2 \longrightarrow \text{R—CH—CH—SO}_3^- \qquad (9.2\text{-}2)$$

Here we adopt a shorthand notation that shows only the atoms participating in the chemical reaction. The surfactant produced in Eq. (9.2-1) is an α-olefin sulfonate, and that produced in Eq. (9.2-2) is an alkyl sulfonate. If the feedstock is aromatic, the sulfonation produces an alkyl benzene sulfonate

$$\text{R} - \langle\bigcirc\rangle + \text{SO}_3 \longrightarrow \text{R} - \langle\bigcirc\rangle - \text{SO}_3^- + \text{H}^+ \qquad (9.2\text{-}3)$$

The sulfate in these reactions comes from bubbling SO_3 gas through the feedstock or through contact with a solvent the SO_3 is dissolved into. The sulfonation reactions (Eqs. 9.2-1 through 9.2-3) yield a highly acidic aqueous solution through the parallel reactions

$$\text{H}_2\text{O} + \text{SO}_3 \longrightarrow \text{H}_2\text{SO}_4 \qquad (9.2\text{-}4)$$

$$\text{H}_2\text{SO}_4 \longrightarrow \text{H}^+ + \text{HSO}_4^- \longrightarrow 2\text{H}^+ + \text{SO}_4^{2-} \qquad (9.2\text{-}5)$$

The solution is subsequently restored to a neutral pH by adding a strong base, such as NaOH or NH_3, dissolved in water. This neutralization step also provides the counterion for the sulfonate; for the α-olefin sulfonate this is

$$\text{Na}^+ + \text{R—C=C—SO}_3^- \longrightarrow \text{R—C=C—SO}_3\text{—Na} \qquad (9.2\text{-}6)$$

If the feedstock is unrefined, a mixture of surfactant types will result. The mixture can contain a distribution of isomeric forms, molecular weights, and degrees of sulfonation (mono- versus disulfonation). The mixture is extremely difficult to characterize except through several gross properties. Table 9-2 shows some typical properties of commercial sulfonates. Typical molecular weights range from 350 to 450 kg/kg-mole, with the lower values indicating greater water solubility. In some calculations, it is better to use the surfactant equivalent weight (molecular weight divided by charge) instead of the molecular weight. Thus equivalent weight (mass per equivalent) and molecular weight are the same for monosulfonates. Some products contain impurities: unreacted oil from the sulfonation step and water from the neutralization. Part of the surfactant, as purchased, is inactive. Inasmuch as it is the

TABLE 9-2 SELECTED PROPERTIES OF A FEW COMMERCIAL ANIONIC SURFACTANTS

Company	Surfactant name	Molecular weight	Activity (wt %)	Oil (wt %)	Water (wt %)	Salt (wt %)	Type
ALCOLAC	SIPONATE DS-10	350	98.0				Sodium dodecyl benzene sulfonate
ALCOLAC	SIPONATE A168	350	70.0				
CONOCO	AES 14125		58.3	1.7			Alfonic ether sulfates
CONOCO	AES 1412A		60.0	3.0			Alfonic ether sulfates
EXXON	RL 3070	334	60.0	14.0	25.2		Alkyl aryl sodium sulfonates
EXXON	RL 3011	375	64.4	25.2	10.0		Alkyl aryl sodium sulfonates
EXXON	RL 3330	390	66.0	24.0	9.5		Alkyl aryl sodium sulfonates
EXXON	RL 3331	391	65.0	26.5	8.1		Alkyl aryl sodium sulfonates
EXXON	RL 3332	460	60.0	31.4	8.1		Alkyl aryl sodium sulfonates
EXXON	RL 2917	515	65.7	25.7	8.5		Alkyl aryl sodium sulfonates
KAO	LS 3203	330	65.0		53.0		Linear alkyl sulfonate
KAO	LS 3202	480	44.1	54.6	0.06		Linear alkyl sulfonate
LION	LEONOX E		94.0	2.0		2.0	
LION	LEONOX D	350	94.0	2.0		3.0	Alpha olefin sulfonate
LION	LION AJS-2	375	35.0				
LION	LEONOX K	570	30.0				Alfonic ether sulfates
SHELL	ENORDET AOS 310–40	317	38.0		61.0	< 1	Alcohol ethoxy sulfonate
SHELL	ENORDET LXS 370–60	375	60.1		38.0	1.9	Linear alkyl xylene sulfonates
SHELL	ENORDET LXS 395–60	395	60.0		37.4	2.6	Linear alkyl xylene sulfonates
SHELL	ENORDET LXS 420–60	417	60.6		36.6	2.8	Linear alkyl xylene sulfonates
SHELL	ENORDET 3ES–441–60	441	59.3		29.5	1.2	Linear alkyl xylene sulfonates
STEPAN	PS HMW		50.7	24.4	22.1		
STEPAN	PS MMW		53.2	18.4	26.6		
STEPAN	PS 360	360	65.8	18.9	12.4		
STEPAN	PS 420	420	56.1	13.0	28.8		
STEPAN	PS 465	464	58.7	14.9	24.2		
WITCO	TRS 40	330–350	40–43	18.0	40.0		
WITCO	TRS 10–410	415–430	61–63	33.0	4–5		
WITCO	TRS 16	440–470	61–63	32.5	4–5		
WITCO	TRS 18	490–500	61–63	32.5	4–5		

surfactant itself we are interested in, all slug concentrations should report the surfactant concentration only (100% active basis).

In the following discussion, we ignore distinctions between surfactant types by simply treating the surfactant as the tadpole structure of Fig. 9-2.

If an anionic surfactant is dissolved in an aqueous solution, the surfactant dissociates into a cation and a monomer. If the surfactant concentration is then increased, the lypophilic moieties of the surfactant begin to associate among themselves to form aggregates or micelles containing several monomers each. A plot of surfactant monomer concentration versus total surfactant concentration (Fig. 9-3) is a curve that begins at the origin, increases monotonically with unit slope, and then levels off at the *critical micelle concentration* (CMC). Above the CMC, all further increases in surfactant concentration cause increases only in the micelle concentration. Since CMCs are typically quite small (about 10^{-5} to 10^{-4} kg-moles/m^3), at nearly all concentrations practical for MP flooding, the surfactant is predominantly in the micelle form. This is the origin of the name micellar-polymer flooding. The representations of the micelles in Fig. 9-3 and elsewhere are schematic. The actual structures of the micelles are not static and can take on various forms.

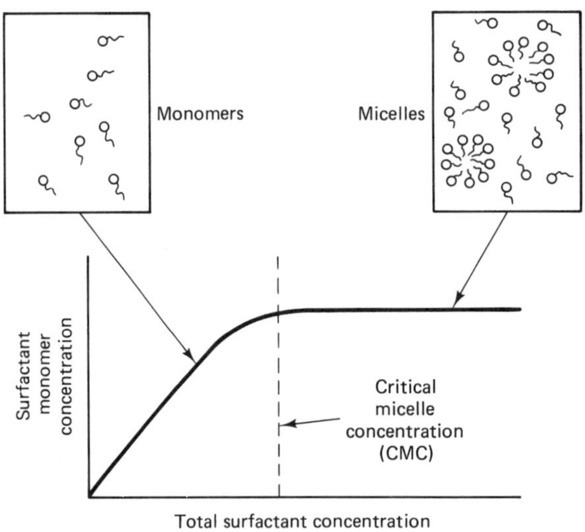

Figure 9-3 Schematic definition of the critical micelle concentration (from Lake, 1984)

When this solution contacts an oleic phase (the term *oleic phase* indicates this phase can contain more than oil), the surfactant tends to accumulate at the intervening interface. The lipophilic moiety "dissolves" in the oleic phase, and the hydrophilic in the aqueous phase. The surfactant prefers the interface to the micelle; however, only small surfactant concentrations are needed to saturate the interface. The dual nature of the surfactant is important since the accumulation at the interface causes the IFT between the two phases to lower. The IFT between the two phases is a function of the excess surfactant concentration at the interface. The excess is the difference between the interface and bulk concentration. The interface blurs in much the same manner as do vapor–liquid interfaces near a critical point.

The surfactant itself and the attending conditions should be adjusted to maximize this effect, but this affects the solubility of the surfactant in the bulk oleic and aqueous phases. Since this solubility also impinges on the mutual solubility of brine and oil, which also affects IFTs, this discussion leads naturally to the topic of surfactant–oil–brine phase behavior. Curiously, the surfactant concentration itself plays a rather minor role in what follows compared to the temperature, brine salinity, and hardness. This is true of many micellar properties.

9-3 SURFACTANT–BRINE–OIL PHASE BEHAVIOR

Surfactant–brine–oil phase behavior is conventionally illustrated on a ternary diagram (see Sec. 4-3). By convention, the top apex of the ternary diagram represents the surfactant pseudocomponent ($i = 3$), the lower left represents brine ($i = 1$), and the lower right represents oil ($i = 2$). Table 9-3 summarizes these and other notational conventions.

TABLE 9-3 NOTATION AND COMMON UNITS FOR MP FLOODING

i	Species	Concentration unit
1	Water	Volume fraction
2	Oil	Volume fraction
3	Surfactant	Volume fraction
4	Polymer	Weight percent or g/m^3
5	Anion	meq/cm^3-pore volume
6	Divalents	meq/cm^3-pore volume
7	Cosurfactant	Volume fraction
8	Monovalents	meq/cm^3-pore volume

j	Phase
1	Aqueous
2	Oleic
3	Microemulsion

MP phase behavior is strongly affected by the salinity of the brine. Consider the sequence of phase diagrams, Figs. 9-4 through 9-6, as the brine salinity is increased. The phase behavior we now describe was originally given by Winsor (1954) and adapted to MP flooding later (Healy et al., 1976; Nelson and Pope, 1978).

At low brine salinity, a typical MP surfactant will exhibit good aqueous-phase solubility and poor oil-phase solubility. Thus an overall composition near the brine–oil boundary of the ternary will split into two phases: an excess oil phase that is essentially pure oil and a (*water-external*) microemulsion phase that contains brine, surfactant, and some solubilized oil. The solubilized oil occurs when globules of oil occupy the central core of the swollen micelles. The tie lines within the two-phase region have a negative slope. This type of phase environment is variously called a Winsor type I system, a lower-phase microemulsion (because it is more dense than the excess oil phase), or a type II(−) system. We adopt the type II(−) terminology here. II means no more than two phases can (not necessarily will) form, and (−) means the tie lines have negative slope (Fig. 9-4). The plait point in such a system P_R is usually located quite close to the oil apex. Any overall composition above the binodal curve is single phase.

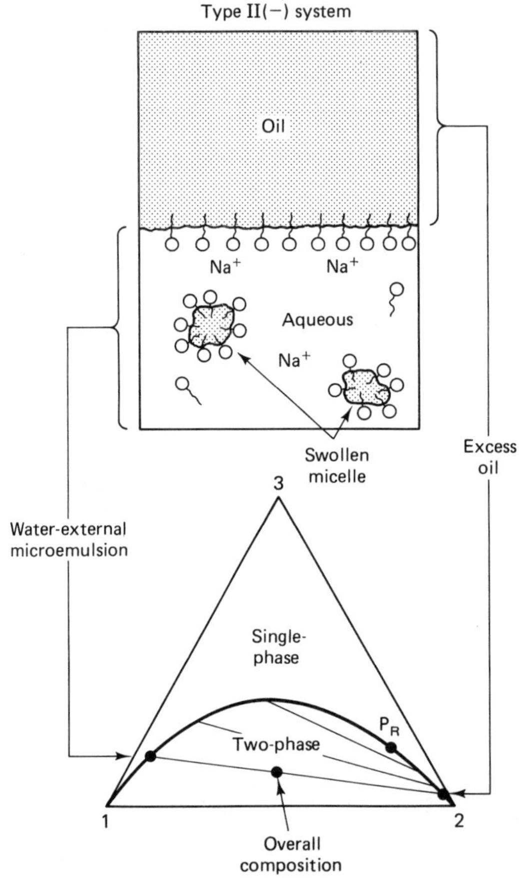

Figure 9-4 Schematic representations of the type II(-) system (from Lake, 1984)

For high brine salinities (Fig. 9-5), electrostatic forces drastically decrease the surfactant's solubility in the aqueous phase. An overall composition within the two-phase region will now split into an excess brine phase and an (*oil-external*) micro-emulsion phase that contains most of the surfactant and some solubilized brine. The brine is solubilized through the formation of inverted swollen micelles, with brine at their cores. The phase environment is a Winsor type II system, an upper-phase microemulsion, or a type II(+) system. The plait point P_L is now close to the brine apex.

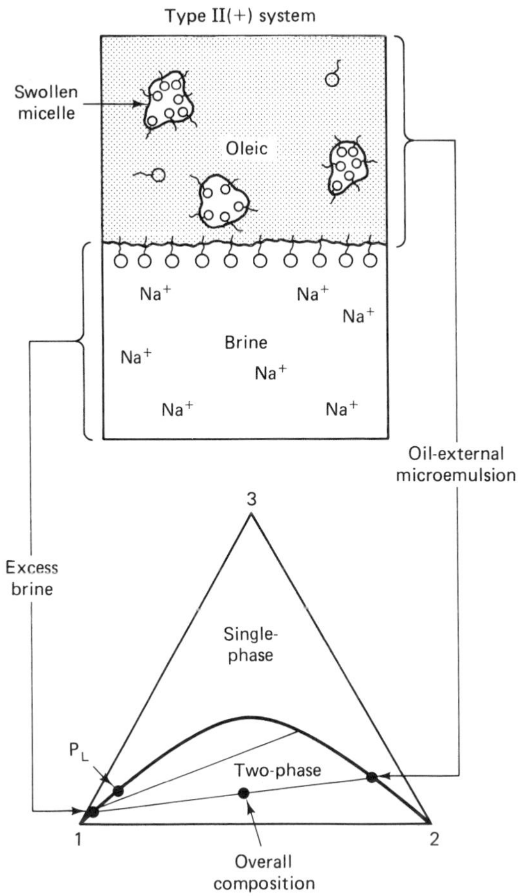

Figure 9-5 Schematic representation of high-salinity type II(+) system (from Lake, 1984)

The two extremes presented thus far are roughly mirror images: The microemulsion phase is water-continuous in the type II(−) systems and oil-continuous in type II(+) systems. The induced solubility of oil in a brine-rich phase, a type II(−) system, suggests an extraction mechanism in oil recovery. Though extraction does play some role, it is dwarfed by the IFT effect discussed below, particularly when phase behavior at intermediate salinities is considered.

At salinities between those of Figs. 9-4 and 9-5, there must be a continuous change between type II(−) and II(+) systems. The obvious change of a counterclockwise tie line rotation and corresponding plait point migration is incorrect; there is no salinity where the solubility of the surfactant in the brine- and oil-rich phases are exactly equal. But there is a range of salinities where a third surfactant-rich phase is formed (Fig. 9-6). An overall composition within the three-phase region separates into excess oil and brine phases, as in the type II(−) and II(+) environments, and into a microemulsion phase whose composition is represented by an invariant point. This environment is called a Winsor type III, a middle-phase microemulsion, or a type III system. To the upper right and left of the three-phase region

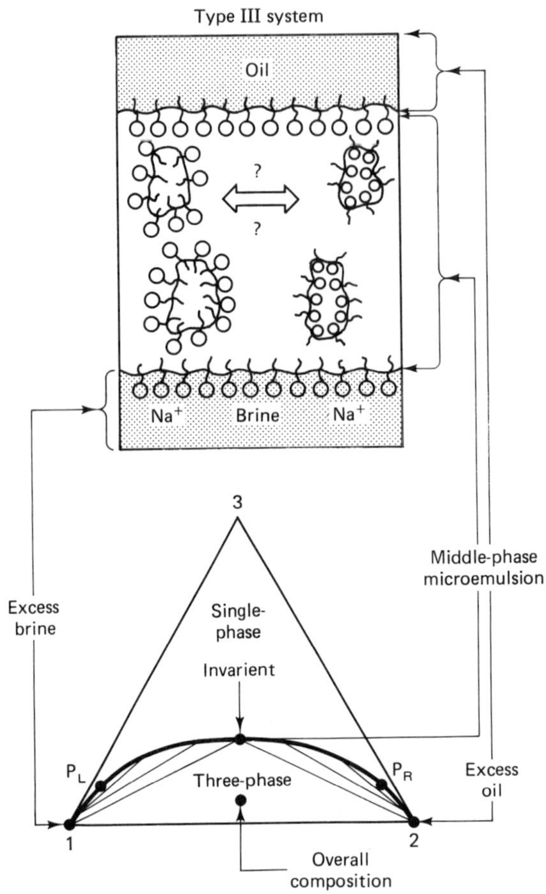

Figure 9-6 Schematic representation of optimal-salinity type III system (from Lake, 1984)

are type II(−) and II(+) lobes wherein two phases will form as before. Below the three-phase region, there is a third two-phase region (as required by thermodynamics) whose extent is usually so small that it is neglected (Anderson et al., 1976). In the three-phase region, there are now two IFTs between the microemulsion and oil σ_{32} and the microemulsion and water σ_{31}.

Figure 9-7, a prism diagram, shows the entire progression of phase environments from type II(−) to II(+). The type III region forms through the splitting of a critical tie line that lies close to the brine–oil boundary as the salinity increases to C_{Sel} (Bennett et al., 1981). A second critical tie line also splits at C_{Seu} as salinity is decreased from a type II(+) environment. Over the type III salinity range, the invariant point M migrates from near the oil apex to near the brine apex before disappearing at the respective critical tie lines. Equally important, as the migration takes place, the surfactant concentration in the microemulsion phase goes through a minimum near where brine–oil ratio at the invariant point becomes 1.

The migration of the invariant point implies essentially unlimited solubility of

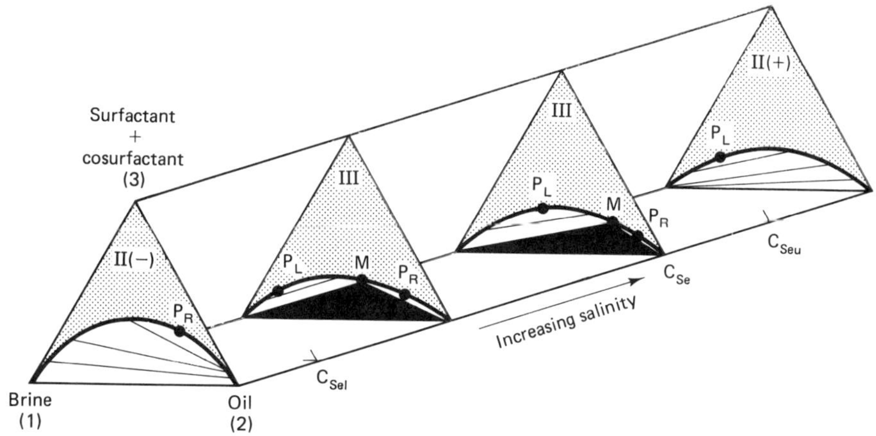

Figure 9-7 Pseudoternary or "tent" diagram representation of micellar-polymer phase behavior (from Lake, 1984)

brine and oil in a single phase. This has generated intense research into the nature of the type III microemulsion (Scriven, 1976). The middle-phase microemulsion cannot be simultaneously oil- and water-external. Somewhere between C_{Sel} and C_{Seu}, the micelles contained therein undergo an inversion, and many phase properties (for example, electrical conductivity) abruptly change from being characteristic of the water to being characteristic of the oil. Moreover, several other properties (see Fig. 9-13) take on extreme values. Though logically appealing, the phase inversion salinity does not necessarily indicate optimal salinity.

Several variables other than brine electrolyte content can bring about the Fig. 9-7 phase environment shifts. In general, changing any condition that enhances the surfactant's oil solubility will cause the shift from type II(−) to II(+). We discuss some of the more important below.

Surfactant Structure

In general, increasing the importance of the nonpolar end of the surfactant will increase oil solubility. Such changes include increasing the nonpolar molecular weight, decreasing the tail branching (Graciaa et al., 1981), decreasing the number of polar moieties (from disulfonates to monosulfonates), and decreasing the strength of the polar moiety.

Two common measures of the competition between the hydrophile and lipophile indicate oil solubility. The surfactant's charge density is the number of dissociated ions per molecule divided by the molecular size. Surfactant brine solubility goes up as charge density increases. A second measure is the hydrophile–lipophile balance (HLB) number. For certain types of surfactant (for example, nonionics), the HLB number is simply related to molecular structure. But for others, the HLB number cannot be uniquely defined apart from the oil and brine it is competing for

(Shinoda and Kunieda, 1979). Though both measures have enjoyed a degree of success, they are difficult to apply to petroleum sulfonate because of the many chemical species contained therein.

Cosurfactants

One of the first uses for cosurfactants was to adjust the surfactant pseudocomponent so that the II(−) → II(+) transition occurs at different salinities. A water soluble cosurfactant (for example, tertiary amyl alcohol, a second petroleum sulfate, or *n*-butanol) also causes the surfactant to be more water soluble. Higher molecular weight alcohols cause increased oil solubility (Salter, 1977). Bourrel et al. (1978) have derived mixing rules for the properties of surfactant–cosurfactant mixtures.

Oil Properties

If the oil can be made more polar, it will act as a better solvent for the surfactant, hastening the II(−) → II(+) transition. There are several measures for this tendency. High specific gravity crudes tend to be rich in organic acids; thus surfactant oil solubility is lower in high gravity oils (Puerto and Reed, 1982). Similarly, low specific volume crudes behave in the same fashion (Nelson, 1982). Cash et al. (1976) devised a measure of oil effects on surfactant–brine–oil phase behavior by comparing the transitions observed with a crude to a refined hydrocarbon. The surfactant in all cases is Texas No. 1 (Fig. 9-2) in a NaCl brine. If the transition from II(−) → II(+) for a crude occurs at the same salinity as the linear alkane, the alkane carbon number (ACN) of the refined oil and the equivalent alkane carbon number (EACN) of the crude are equal. Therefore, EACN is relatively easy to measure and gives an indication of the model oil to be used in formulation. The same idea can be used to categorize surfactants (Graciaa et al., 1981).

Decreasing Temperature

There is little generality in the tendency for the surfactant to dissolve in oil as temperature increases. For most anionics, higher temperatures mean more brine solubilities (Nelson and Pope, 1978). This trend is reversed for most nonionics.

Decreasing Pressure

MP phase behavior, being an all-liquid system, is relatively insensitive to pressure. But Nelson (1982) has noted a substantial pressure effect in gassy crudes. Interestingly, the trend here parallels that of the oil properties given above: As the specific volume of the oil increases (through decreased pressure), the surfactant becomes more water soluble.

Decreasing the surfactant's oil solubility will cause the reverse of these changes. Thus Fig. 9-7 could be redrawn with any of the above variables on the base of the prism with the variable C_{Se} increasing in the direction of increased oil solubility. These observations have occupied a very great share of the MP literature. Their utility will become apparent under our discussion of IFTs in the next section.

9-4 NONIDEAL EFFECTS

In much the same manner as the ideal gas law approximates the behavior of real gases, Figs. 9-4 through 9-6 are approximations to actual MP phase behavior. Though nonidealities are significant in many instances, in this section, we mention only the most important.

1. At high surfactant concentrations or low temperatures (Scriven, 1976; Healy and Reed, 1974) or even in the presence of pure surfactants (Salter, 1983), phases other than those in Fig. 9-7 have been observed. These phases tend to be high-viscosity liquid crystals or other condensed phases. The large viscosities are detrimental to oil recovery since they can cause local viscous instabilities during a displacement or decreased injectivity. Frequently, low-to-medium molecular weight alcohols (*cosolvents*) are added to MP formulations to "melt" these undesirable viscosities. Because most alcohols are weak surfactants, the term *cosurfactant* has enjoyed popular usage for these additions, as it has for the addition of other surfactants. When the brine contains polymer, a condensed phase occurs at low surfactant concentration because of exclusion of the polymer from the microemulsion phases. Cosurfactants can be used to eliminate this polymer–surfactant incompatibility (Trushenski, 1977).
2. When cosurfactants are present, it is often inappropriate to lump all the chemicals into the surfactant apex of the prism in Fig. 9-7. If the cosurfactants do not partition with the primary surfactant during a displacement, much of the benefit from adding the chemical is lost; hence surfactant–cosurfactant separation effects are an important concern. Efforts to account for the preferential partitioning of the cosurfactant include a quaternary phase behavior representation (Salter, 1978) and a *pseudophase* theory (Hirasaki, 1982).
3. The type III salinity limits (C_{Sel} and C_{Seu}) are functions of surfactant concentration. This dependency may be visualized by tilting the vertical triangular planes in Fig. 9-7 about their bases. This is sometimes called the *dilution effect*.

   One way to graphically represent the dilution effect is through the *salinity requirement* diagram (Fig. 9-8). This diagram is a plot of overall surfactant concentration C_3 (horizontal axis) versus the salinity (vertical axis). All other variables are held constant. Figure 9-8 represents salinity as percent dilution of a particular high salinity brine. The upper curve shows the boundary between the types II(+) and III environments or a curve of C_{Seu} versus C_3. Simi-

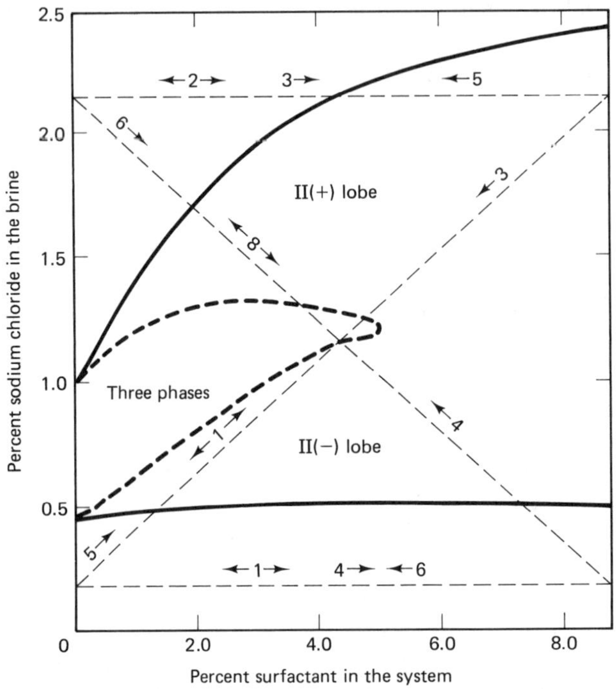

Figure 9-8 Salinity-requirement diagram (from Nelson, 1982)

larly, the lower curve shows C_{Sel} versus C_3; hence the region between the two curves gives the extent of the type III region as a function of C_3. Other plots (Glover et al., 1979; Bourrel et al., 1978) plot the extent of observed three-phase behavior in a similar fashion. Figure 9-8 also shows a three-phase region within the type III region.

The MP system in Fig. 9-8 shows a type III region that decreases with salinity. For other surfactants and brines, the trend can be entirely opposite (Bourrel et al., 1978). For ideal MP phase behavior, neither C_{Seu} nor C_{Sel} should depend on C_3—that is, the salinity requirement diagram should consist of two horizontal lines. Frequently, the behavior of soft brines will approximate this since the dilution effect is particularly pronounced when the brine contains significant quantities of divalent ions.

4. The phase behavior shifts are specific to the exact ionic composition of the brine, not simply to the total salinity. Hence just as in polymer flooding, it is insufficient to characterize the brine as merely "fresh" or in terms of its total dissolved solids content. For anionic surfactants, other anions in solution have little effect on the MP phase behavior, but cations readily cause phase environment changes. Divalent cations (calcium and magnesium are the most common) are usually 5–20 times as potent as monovalent cations (usually sodium). Divalents are usually present in oil-field brines in smaller quantities than

monovalents (Fig. 8-1), but their effect is so pronounced that it is necessary, as a minimum, to separately account for salinity—total dissolved solids—and hardness—total divalent cation concentration. Nonconstant monovalent–divalent ratios will also cause electrolyte interactions with clay minerals through cation exchange. The disproportionate effects of the salinity and hardness are accounted for by defining a weighted sum of the monovalent and divalent concentrations as an "effective" salinity C_{Se}.

The salinity effects discussed here are much less significant with nonionic surfactants where there are no ionic associations. Even for anionics, they can be greatly attenuated by adding nonionic cosurfactants. The cosurfactant monomers add into the micelle between the larger primary surfactant monomers, thus lessening the charge density of the micelle surface and making the "mixed" micelle more like a nonionic.

9-5 PHASE BEHAVIOR AND INTERFACIAL TENSION

You may be wondering what this discussion of MP phase behavior has to do with the goal of recovering oil through lowered IFT. Early MP flooding literature contains much information about the techniques of measuring IFTs and what causes them to be low (Cayias et al., 1975). IFTs depend on the types and concentration of surfactant, cosurfactant, electrolyte, oil, polymer, and temperature. However, in surely one of the most significant advances in all MP technology, all IFTs have been shown to directly correlate with the MP phase behavior. Healy and Reed (1974) originally proposed the correlation, which has been theoretically substantiated by Huh (1979) and experimentally verified by several others (for example, Glinsmann, 1979; Graciaa et al., 1981).

A practical benefit of this correlation is immediately realized: Relatively difficult measurements of IFTs can be largely supplanted by relatively easy phase behavior measurements. Indeed, in the recent literature, the behavior of IFTs has been inferred by a narrower subset of phase behavior studies based on the solubilization parameter (Bourrel et al., 1978). As important as this benefit is, a more important benefit is that the correlation logically provides a basis for MP design. We discuss design in Sec. 9-13.

To investigate further the relation between IFTs and phase behavior, let C_{23}, C_{13}, and C_{33} be the volume fractions of oil, brine, and surfactant in the microemulsion phase. According to Figs. 9-4 through 9-6, the microemulsion phase is present at all salinities; hence all three quantities are well defined and continuous. For systems containing alcohol, C_{33} is the surfactant coordinate less the cosurfactant content. Solubilization parameters between the microemulsion-oleic phases S_{32}, for type II(−) and III phase behavior, and between the microemulsion-aqueous phases S_{31} for type II(+) and III are defined as

$$S_{32} = \frac{C_{23}}{C_{33}} \tag{9.5-1a}$$

$$S_{31} = \frac{C_{13}}{C_{33}} \tag{9.5-1b}$$

The interfacial tensions between the corresponding phases, σ_{32} and σ_{31}, are empirical functions only of S_{32} and S_{31}. Figure 9-9 shows a typical correlation.

Figure 9-10 shows the corresponding behavior of the solubilization parameters and IFTs in a different manner. Consider a locus at constant oil, brine, and surfactant overall concentrations in Fig. 9-7, but with a variable salinity. If nonideal effects are unimportant and the locus is at low surfactant concentration and intermediate brine–oil ratios, σ_{32} will be defined from low salinity up to C_{Seu}, and σ_{31} from C_{Sel} to high salinities. Both IFTs are the lowest in the three-phase type III region between C_{Sel} and C_{Seu} where both solubilization parameters are also large. Further, there is a precise salinity where both IFTs are equal at values low enough (about 1 μN/m) for good oil recovery. This salinity is the *optimal salinity* C_{Sopt} for this partic-

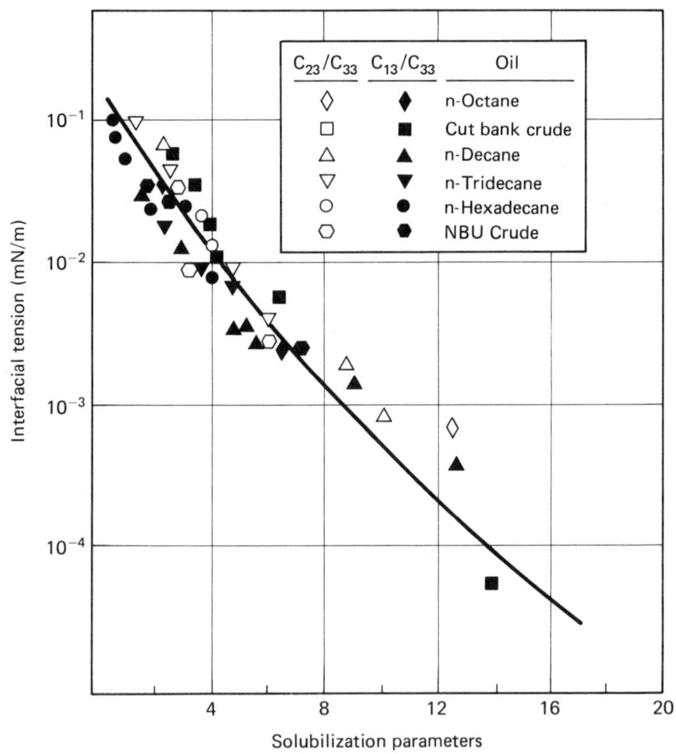

Figure 9-9 Correlation of solubilization parameters with interfacial tensions (from Glinsmann, 1979)

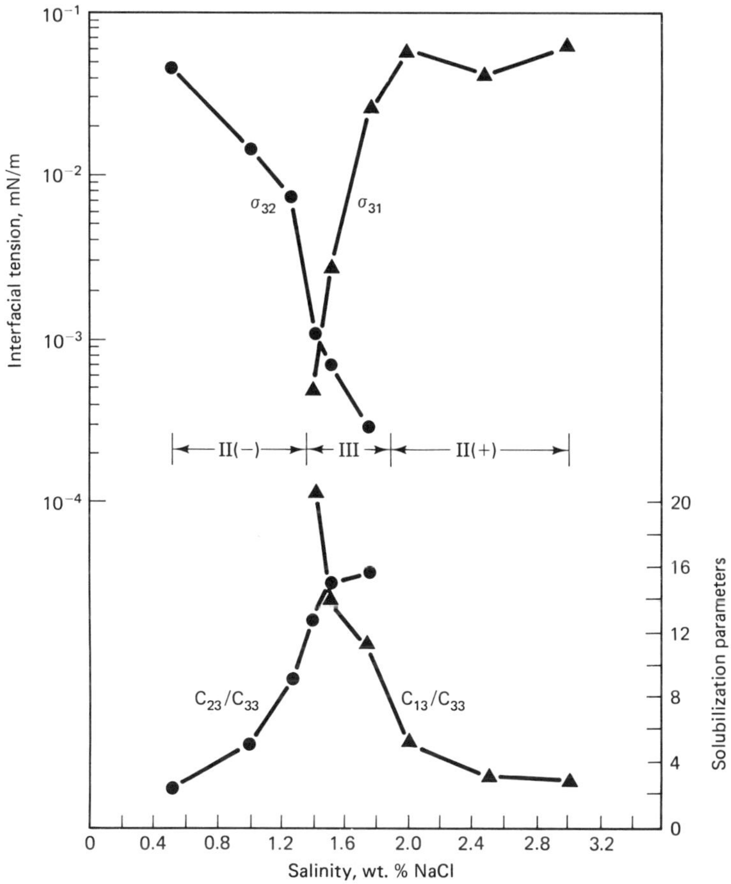

Figure 9-10 Interfacial tensions and solubilization parameters (from Reed and Healy, 1977)

ular surfactant–brine–oil combination, and the common IFT is the *optimal* IFT. Optimal salinities have been defined on the basis of equal IFTs, as in Fig. 9-10, equal solubilization ratios (Healy et al., 1976), equal contact angles (Reed and Healy, 1979), and the midpoint between C_{Seu} and C_{Sel}. Fortunately, all definitions of optimal salinity give roughly the same value.

The optimal salinity based on solubilization parameters also corresponds to the salinity where oil recovery in a core flood is a maximum. Figure 9-11 illustrates this oil-recovery optimal salinity. The middle panel, Fig. 9-11b, shows a plot similar to the upper panel in Fig. 9-10 for a different surfactant system; the lower panel shows the oil recovery for a series of constant-salinity core floods. The optimal salinity based on solubilization parameters, IFTs, and oil recovery agree well. Since optimal phase behavior salinity is the same as maximum oil recovery salinity, clearly one of the goals of an MP design is to generate this optimal salinity in the presence of the

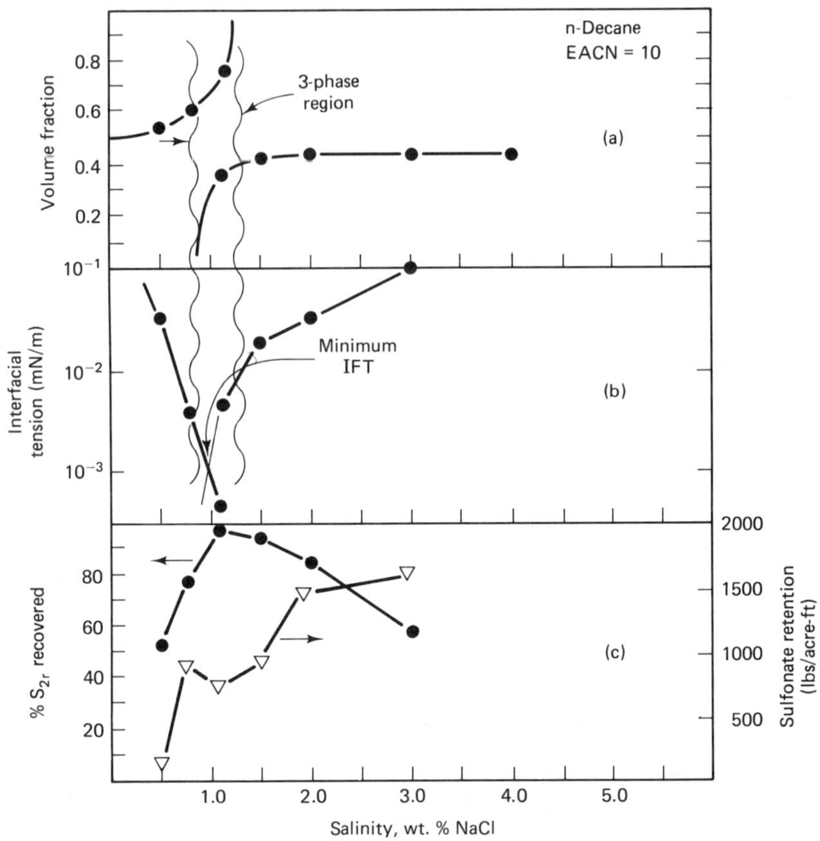

Figure 9-11 Correlation of phase volume and IFT behavior with retention and oil recovery (from Glinsmann, 1979)

surfactant. The optimal salinity does not correspond to minimum surfactant retention (Fig. 9-11c), but this is because of competing effects as we mention below.

Because of the dilution effect mentioned above, maximum oil recovery is really where the combination of electrolyte, surfactant, and cosurfactant concentrations bring about maximum solubilization parameters. Hence one should speak of optimal *conditions* rather than optimal salinity. The optimal salinity terminology is deeply embedded within the MP literature, but it is precise only for the ideal Fig. 9-7 phase behavior. Do not confuse the optimal salinity C_{Sopt}, an intrinsic property of the oil–brine–surfactant combination, with the prevailing salinity C_{Se}, an independent variable in the MP design.

Optimal salinities can vary greatly depending on the nature of the surfactant and brine pseudocomponents. But it is dismaying that for many commercially attractive surfactants in most MP candidate reservoirs, the optimal salinity is smaller than the resident brine salinity. Optimal salinities can be raised by adding to the slug any chemical that increases the primary surfactant's brine solubility. Adding cosurfactants to the MP slug normally increases the optimal IFT.

The notion of optimal conditions is directly connected to the phase behavior of MP systems. Even properties of MP systems apparently unrelated to phase behavior (retention, for example) are functions of salinity, cosurfactant concentration, and temperature. This observation leads to the interesting speculation that *all* MP properties (retention, phase behavior, IFT, mobilities) correlate to optimal salinity and, perhaps, to solubilization parameters.

9-6 OTHER PHASE PROPERTIES

Our understanding of MP phase behavior follows from the ternary representation of the Winsor phase behavior progression. Other representations are common, particularly to show phase properties.

A very useful phase-behavior representation is the *volume fraction diagram* (VFD) (Fig. 9-12). Imagine a point of fixed overall composition (parallel to the salinity axis) in the ternary planes in Fig 9-6. The volumes of each phase are observed and plotted as the brine salinity changes. Starting with a low salinity, the VFD shows a succession of decreasing oleic-phase volume and increasing aqueous-phase volume with some three-phase overlap in the middle. If the overall surfactant concentration is low and the brine–oil ratio (WOR) is about 1, the appearance of the lower brine phase corresponds approximately to the onset of the type III region (C_{Sel}), and the disappearance of the upper oleic phase corresponds approximately to the termination of the type III region (C_{Seu}). The salinity at which the brine and oleic phases have equal volumes is a good approximation of the optimal salinity if the surfactant and cosurfactant concentration is low enough. Compare the VFD in Fig. 9-11a to Fig. 9-12b.

Varying salinity while holding other variables constant is sometimes called a salinity scan. Varying the salinity is the most common presentation of the VFD; however, a derivative of the VFD, in which the cosurfactant concentration is varied in place of the salinity, is sometimes useful. To minimize the number of measurements, each scan can be relatively coarse (about ten measurements) and then supplemented with fill-in measurements to refine the estimate of the important events.

Of course, any phase property can be plotted in place of the phase volumes. Figure 9-13 shows the microemulsion-phase viscosity as a function of salinity. Over this range, the microemulsion phase, as defined above, is continuous and shows a viscosity maximum at a salinity near the optimal. The maximum indicates molecular ordering in the phase that seems to be the strongest at the phase inversion salinity. Such maxima can be either beneficial, if it can be used to provide mobility control in the slug, or detrimental, if it leads to excessively viscous fluids. It was to counteract the latter tendency that cosurfactants were first added to MP slugs. Over the same salinity range in Fig. 9-13, the excess phase viscosities do not change appreciably.

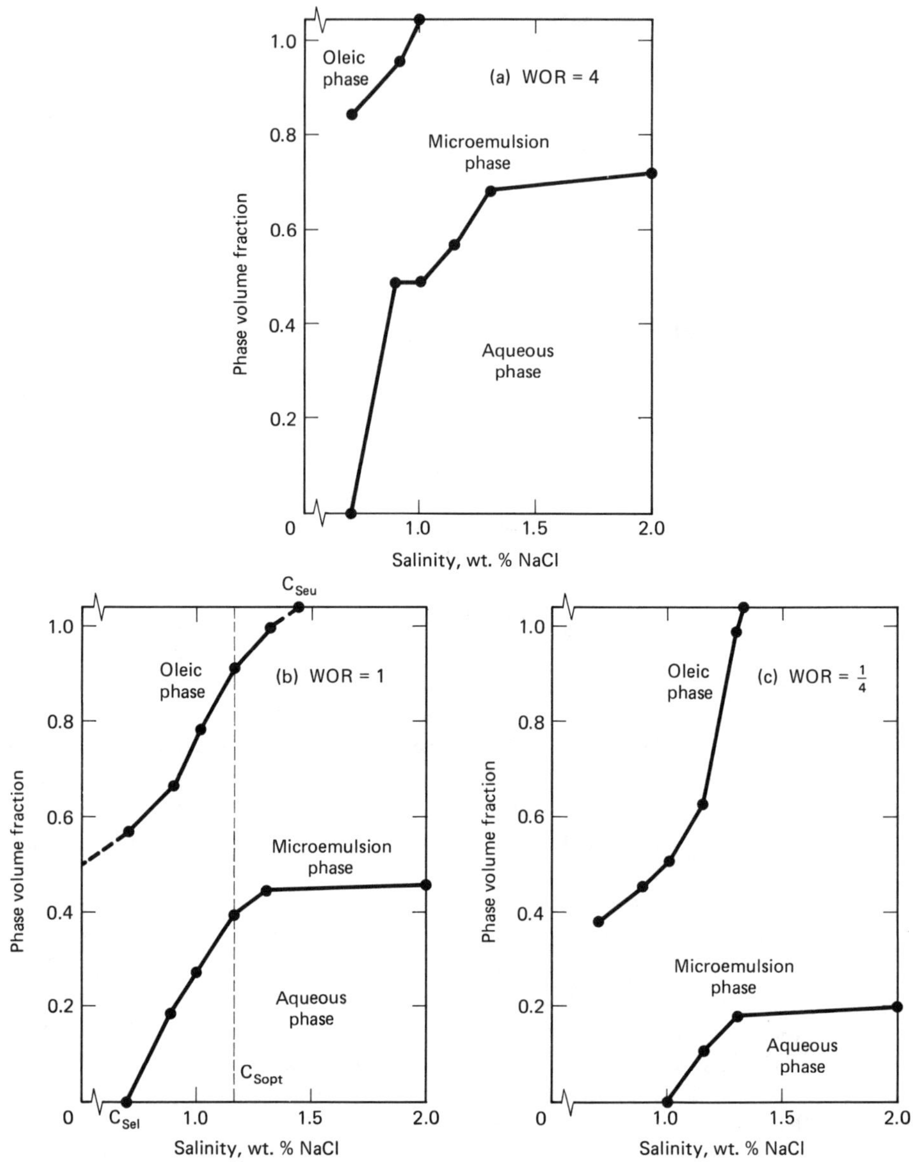

Figure 9-12 Phase volume diagrams (salinity scans) at three water–oil ratios (from Englesen, 1981)

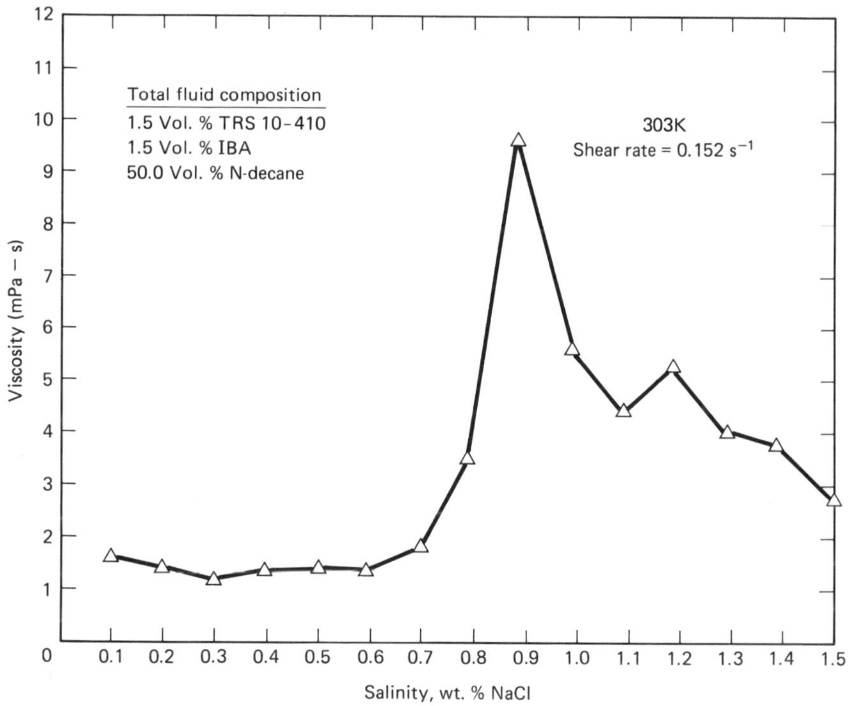

Figure 9-13 Microemulsion phase viscosity as a function of salinity (from Jones, 1981)

9-7 QUANTITATIVE REPRESENTATION OF MICELLAR PROPERTIES

Prediction of MP response rests on being able to quantitatively represent the foregoing behavior in equations. Given the complexities, many of which we have only alluded to, it is not possible to derive comprehensive representations. Here we seek to capture the major features by following these assumptions.

1. All fluids are incompressible and mix ideally.
2. Temperature does not change, and the phase behavior is insensitive to pressure. These restrictions mean phase behavior is driven only by changes in the effective salinity C_{Se}.
3. The ternary equilibria in Fig. 9-11 apply. For the moment, we neglect the nonideal effects.
4. The height of the binodal curve passes through a minimum near the optimal salinity C_{Sopt}. The minimum forces the optimal salinity based on IFT and phase

behavior to be equal since the solubilization parameters will be a maximum at C_{Sopt}.

5. The splitting of the critical tie line as the type III system forms or disappears is so close to the ternary base that we can take the incipient invariant points to coincide with the left and right apexes at these events.

We strive for a representation that captures the basics of the MP behavior without becoming burdened with an excessive number of parameters. Such simplicity means equations are used to describe as much of the behavior as posssible, and we choose these equations to have a small number of adjustable parameters. Though many of the equations are empirical, we strive to make limiting cases theoretically rigorous.

Salinity Events

We form the equations so that the adjustable parameters have physical significance on the previously described diagrams. The effective salinity corresponding to the type II(−) to III transition C_{Sel} is approximately the salinity on a VFD where the third microemulsion phase appears. C_{Seu} is where the microemulsion phase disappears. For best approximation for both quantities, use a VFD with a water–oil ratio of 1. At this water–oil ratio, the optimal salinity C_{Sopt} is the C_{Se} where the excess phases have equal volumes.

In what follows, the effective salinities are normalized by C_{Sopt}. The resulting dimensionless effective salinities

$$C_{SeD} = \frac{C_{Se}}{C_{Sopt}} \tag{9.7-1}$$

are those that control the phase behavior. Clearly, C_{SeD} can take on any positive value and is equal to 1 at optimal conditions.

Other events coming from the VFD relate to the surfactant maximum coordinate on the binodal curve at low, optimal, and high salinities. At $C_{SeD} = 1$, the surfactant coordinate of the invariant point is

$$C_{3M} = \frac{C_3}{S_3} = C_{3\max 1} \tag{9.7-2}$$

where S_3 is the volume fraction (saturation) of the microemulsion phase. The oil and brine coordinates at optimal conditions are

$$C_{1M} = C_{2M} = \frac{1 - C_{3M}}{2} \tag{9.7-3}$$

These equations assume the excess phases are free of surfactant.

At low salinity, the height of the binodal curve is

$$C_{3\max0} = \left.\frac{C_3}{S_3}\right|_{C_{Sel}} \tag{9.7-4a}$$

and the similar quantity at high salinity is

$$C_{3\max2} = \left.\frac{C_3}{S_3}\right|_{C_{Seu}} \tag{9.7-4b}$$

Equations (9.7-4a) and (9.7-4b) generally provide underestimations of the binodal curve heights. Figure 9-14 shows the quantities in Eqs. (9.7-2) and (9.7-4).

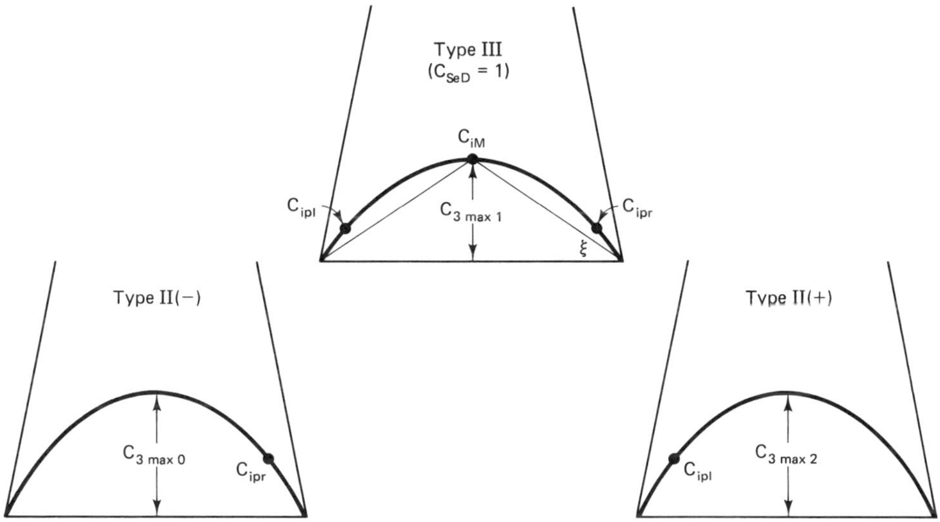

Figure 9-14 Definition of quantities for phase-behavior representation

Events not observable from the VFDs are the plait point locations. We assume the oil coordinates of the plait point vary linearly between the limits of C_{SeuD} and C_{SelD} as shown in Fig. 9-15. C_{2PL} and C_{2PR}, the left and right plait point oil coordinates, apply to the type II(−) and II(+) systems, respectively. Figure 9-15 also shows the assumed linear variation of the invarient point oil coordinate C_{2M}. The superscript * refers to low and high salinity limiting cases. Typical values of C_{2PL} and C_{2PR} are 0.05 and 0.95, respectively.

These seven parameters (C_{SelD}, C_{SeuD}, $C_{3\max0}$, $C_{3\max1}$, $C_{3\max2}$, C_{2PL}, and C_{2PR}) are sufficient to define the phase behavior with a few additional assumptions.

Binodal Curve

We use the same formalism to represent the binodal curve in all phase environments. For type III, this means (Fig. 9-14) the two two-phase lobes are defined by a

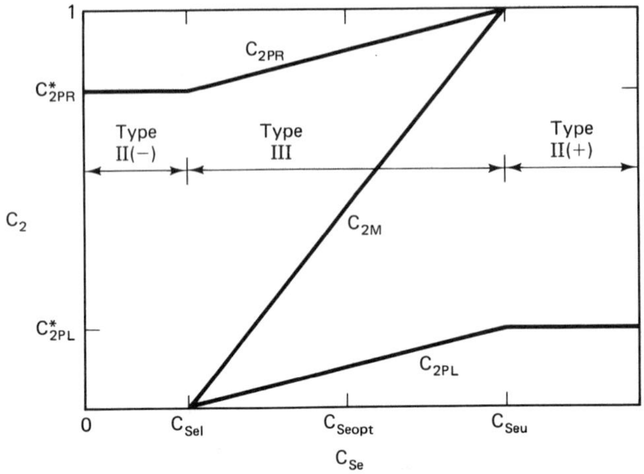

Figure 9-15 Migration of plait and invariant points with effective salinity

continuous curve. For simplicity, let the Hand representation from Eq. (4.4-23) define the binodal curve with $B_H = -1$.

$$\frac{C_{3j}}{C_{2j}} = A_H\left(\frac{C_{1j}}{C_{3j}}\right), \qquad j = 1, 2, \text{ or } 3 \tag{9.7-5}$$

Equation (9.7-5) forces the binodal curve to be symmetric. Solving Eq. (9.7-5) for C_{3j} in terms of C_{2j} gives (recalling that $C_{1j} + C_{2j} + C_{3j} = 1$)

$$C_{3j} = \frac{1}{2}([(C_{2j}A_H)^2 + 4C_{2j}A_H(1 - C_{2j})]^{1/2} - C_{2j}A_H) \tag{9.7-6}$$

Since the curve is symmetric, the maximum C_{3j} occurs when $C_{2j} = C_{1j}$ or, alternatively, when $C_{2j} = (1 - C_{3j})/2$, as in Eq. (9.7-3). This substituted into Eq. (9.7-6) gives

$$A_{Hm} = \left(\frac{2\,C_{3\max m}}{1 - C_{3\max m}}\right)^2 \tag{9.7-7}$$

where $m = 0$, 1, or 2 corresponding to the salinity extremes. The A_{Hm}'s are linearly interpolated as

$$\begin{aligned} A_H &= A_{H0} + (A_{H1} - A_{H0})C_{SeD}, & C_{SeD} &\leqq 1 \\ A_H &= A_{H1} + (A_{H2} - A_{H1})(C_{SeD} - 1), & C_{SeD} &\geqq 1 \end{aligned} \tag{9.7-8}$$

Tie Lines in Two-Phase Systems

Since the treatment for the II(+) system is identical, let us deal with the tie lines in a II(−) system only, but we use C_{2PL} instead of C_{2PR}. Again, using the Hand representation, but with $F_H = 1$, the phase distribution (Eq. 4.4-24) now becomes

$$\frac{C_{32}}{C_{22}} = E_H\left(\frac{C_{33}}{C_{13}}\right) \tag{9.7-9}$$

This equation applies at the plait point from which we have $E_H = C_{1PL}/C_{2PL} = (1 - C_{2PL} - C_{3PL})/C_{2PL}$. Since the plait point is also on the binodal curve, Eq. (9.7-6) applies to give

$$E_H = \frac{1 - C_{2PR} - \frac{1}{2}[(A_H C_{2PR})^2 + 4A_H C_{2PR}(1 - C_{2PR})^{1/2} - A_H C_{2PR}]}{C_{2PR}} \tag{9.7-10}$$

C_{2PR} being defined as a function of salinity, this equation and Eq. (9.7-8) give the salinity dependence of E_H.

Type III

The three-phase portion of this environment poses no difficulties since the excess phases are pure by assumption, and the composition of the microemulsion phase is given by the coordinates of the invariant point. For a given C_{SeD}, C_{2M} is fixed, and C_{3M} follows from Eq. (9.7-6).

The two-phase lobes are somewhat more trouble. Once again, we consider only the II(−) lobe since the II(+) lobe is analogous. Let's suppose the Hand representations (Eqs. 9.7-5 and 9.7-9) apply to transformed concentrations (denoted by superscript prime) where

$$C'_{2j} = C_{2j} \sec \xi \tag{9.7-11a}$$

$$C'_{3j} = C_{3j} - C_{2j} \tan \xi \tag{9.7-11b}$$

$$C'_{1j} = 1 - C'_{2j} - C'_{3j}, \qquad j = 2 \text{ or } 3 \tag{9.7-11c}$$

The angle ξ in these equations is from Fig. 9-14

$$\tan \xi = \frac{C_{3M}}{C_{1M}} \tag{9.7-12a}$$

or, alternatively,

$$\sec \xi = \frac{(C_{1M}^2 + C_{3M}^2)^{1/2}}{C_{1M}} \tag{9.7-12b}$$

These relations allow the parameter E_H to be expressed in terms of the untransformed coordinates of the plait point as

$$E_H = \frac{C'_{1PR}}{C'_{2PR}} = \frac{1 - (\sec \xi - \tan \xi)C_{2PR} - C_{3PR}}{C_{2PR} \sec \xi} \tag{9.7-13}$$

When the relation between C_3 and C_2 (Eq. 9.7-6) is used, this gives E_H as a function of salinity in the type II(−) lobe ($C_{Sel} < C_{Se} < C_{Seu}$). You can verify that these manipulations for the type III lobes merge continuously with the two-phase environments.

9-8 ADVANCED MP PHASE BEHAVIOR

For ideal surfactant–brine–oil systems, phase boundaries and optimal salinity would be independent of brine salinity. This observation means plots of phase boundaries and optimal salinity versus overall surfactant concentration—the salinity requirement diagram (SRD)—would consist of horizontal lines. Petroleum sulfonate systems generally do not manifest this type of behavior (Fig. 9-8 or Fig. 9-16); however, this nonideality can be explained by the pseudophase theory (Hirasaki, 1982; Camilleri et al., 1987). The theory also illustrates the correct measure of optimal conditions, accounts for preferential partitioning of a cosurfactant among the various phases, and fits nicely into the formalism of the previous section.

Figure 9-16(b) shows a three-phase type III system with a water-external mid-

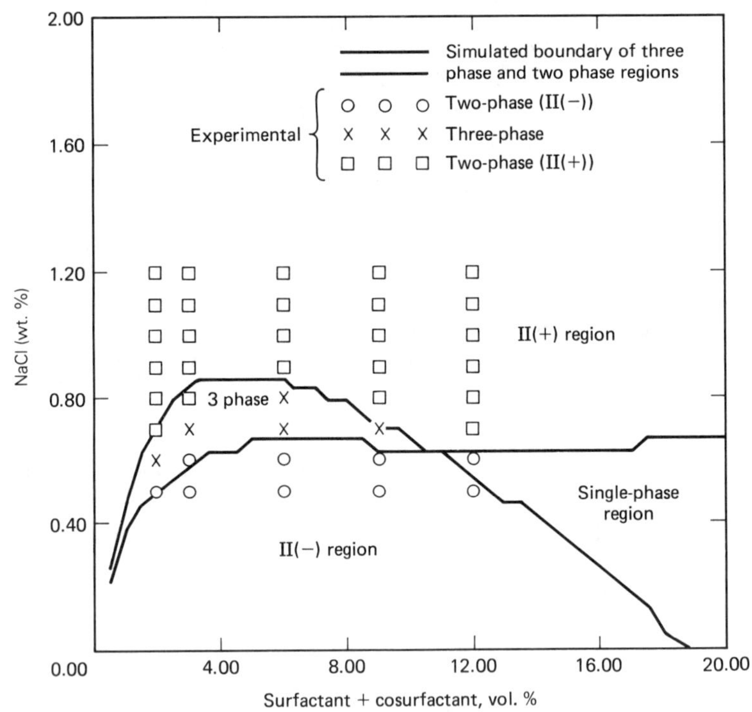

Figure 9-16 Salinity requirement diagram for brine, decane, isobutanol, TRS 10-410. Surfactant/alcohol = 1, Na/Ca = 10 (equivalents), Water–oil ratio = 1 (from Prouvost, 1984)

dle phase, and Fig. 9-16(a) shows the pseudophase representation. Since the only volume occupying components in the system are water, oil, surfactant, and cosurfactant, the system is naturally represented on a quaternary diagram (Fig. 9-16a). All charged species, except those we mention below, exist in unassociated form. The system consists of three pseudocomponents.

1. An oleic pseudocomponent consisting of the excess oil phase and the oil at the center of the swollen micelles.
2. An aqueous pseudocomponent consisting of the excess brine phase and the brine in the microemulsion. This phase contains all charged species not associated with the micelles. Both the oleic and aqueous pseudocomponents can contain cosurfactant but neither contains surfactant.
3. An interfacial pseudocomponent consisting of the surfactant, cosurfactant, and counterions associated with the micelles. Micelles containing two or more surfactant types are mixed micelles.

The theory has three separate facets: definition of effective salinity, cosurfactant partitioning, and cation association with the mixed micelles.

Definition of Effective Salinity

The phase rule (Eq. 4.1-2) states there are two ($N_C = 7$, $N_P = 3$, $N_R = 2$) degrees of freedom for an optimal surfactant system at fixed temperature and pressure. Thus there must be two variables specified to fix optimal conditions. The phase rule gives no indication of what the two degrees of freedom should be except that they should be intensive thermodynamic variables, an observation that rules out overall concentrations.

Glover et al. (1979) present experimental data that suggests the divalent cations bound to the micelles are the most direct indicator of optimality. They suggest optimal salinity decreases linearly with f_6^3, the fraction of the total divalent cations bound to the micelles.

$$\text{Optimal salinity} \sim -\beta_6 f_6^3 \tag{9.8-1}$$

where β_6 is a positive constant. Moreover, ample experimental evidence (Baviere et al., 1981) suggests optimal salinity varies linearly with cosurfactant concentration.

$$\text{Optimal salinity} \sim \beta_7 f_7^3 \tag{9.8-2}$$

where f_7^3 is the fraction of the cosurfactant associated with the micelles. The constant β_7 can be positive if the cosurfactant is more water-soluble than the surfactant, and it can be negative otherwise. Equations (9.8-1) and (9.8-2) suggest the following combination for the optimal salinity expressed as the anion concentration in the aqueous phase:

$$C_{51} = C_{51}^*(1 - \beta_6 f_6^3)(1 + \beta_7 f_7^3) \tag{9.8-3}$$

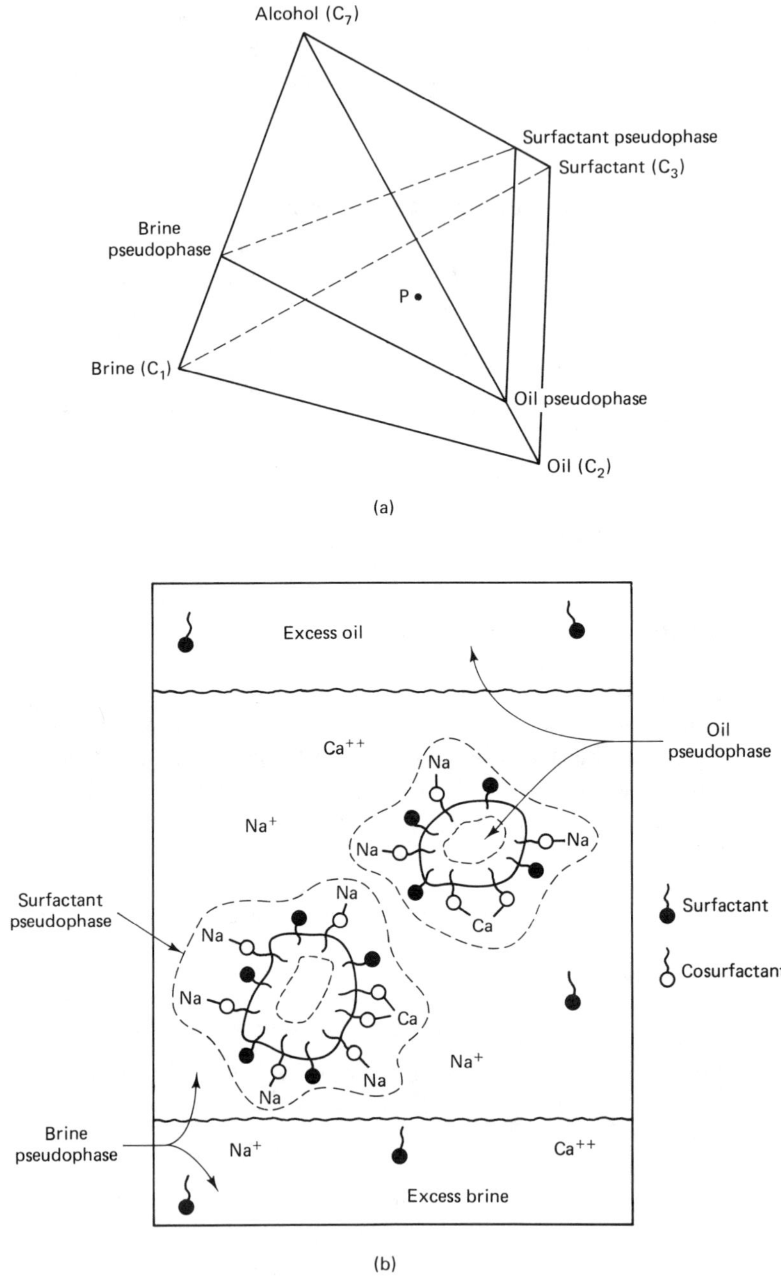

Figure 9-17 Schematic representation of pseudophase theory for surfactant–brine–oil–cosurfactant systems

In this equation, C_{51}^* is the optimal anion concentration in the absence of divalents and cosurfactants. Equation (9.8-3) suggests a definition of effective salinity

$$C_{Se} = \frac{C_{51}}{(1 - \beta_6 f_6^3)(1 + \beta_7 f_7^3)} \qquad (9.8\text{-}4)$$

C_{Se} is the effective salinity used as a normalizing factor in Sec. 9-7. The remaining tasks are to define f_6^3 and f_7^3.

Cosurfactant Partitioning

In the following, we use C_i^j to designate pseudophase compositions. Let's estimate the pseudocomponent compositions at point $P(C_1, C_2, C_3, C_7)$ in Fig. 9-16(a). If the pseudocomponents are the apexes of the indicated triangle,

$$C_7 = C_1 C_7^1 + C_2 C_7^2 + C_3 C_7^3 \qquad (9.8\text{-}5)$$

That the pseudocomponent concentrations occupy the role of phase saturations in the equation accounts for why the theory is called the pseudophase theory. Now let's use partition coefficients to eliminate two of the pseudocompositions

$$K_{21}^7 = \frac{C_7^2}{C_7^1}, \qquad K_{31}^7 = \frac{C_7^3}{C_7^1} \qquad (9.8\text{-}6)$$

Ideally, the partition coefficients should be equal to the cosurfactant partition coefficients in the absence of surfactant (Prouvost, 1984). These substituted into Eq. (9.8-6) give the cosurfactant concentration in the aqueous pseudophase

$$C_7^1 = C_7 \left(\sum_{j=1}^{3} C_j K_{j1}^7 \right)^{-1} \qquad (9.8\text{-}7a)$$

Because this pseudophase contains only cosurfactant and water

$$C_1^1 = 1 - C_7^1 \qquad (9.8\text{-}7b)$$

Equations (9.8-7b) and (9.8-6) can be used to calculate all pseudophase compositions from overall compositions and the partition coefficients.

To use the equations of the previous section, we must express the overall and phase compositions in terms of the pseudocomponents. We define C_{Pi} as

$$C_{Pi} = \frac{\text{Volume of } i + \text{Volume of 7 associated with } i}{\text{Total volume}} \qquad (9.8\text{-}8)$$

which gives

$$C_{Pi} = C_i(1 + C_7^i), \qquad i = 1, 2, 3 \qquad (9.8\text{-}9)$$

The C_{Pi} are overall concentrations and are to be used directly in the strict ternary representation. The equations collapse to the Sec. (9-6) equations when cosurfactant and divalent concentrations are zero. Further, Eq. (9.8-9), summed over three phases, equals unity from Eq. (9.8-7a).

The fractional cosurfactant associated with the micelle follows directly from this also. By definition

$$f_7^3 = \frac{\text{Volume of 7 in pseudophase 3}}{\text{Volume of pseudophase 3}} \tag{9.8-10}$$

which is simply

$$f_7^3 = \frac{1}{1 + \dfrac{1}{C_7^3}} \tag{9.8-11}$$

also by definition.

Divalent Cation Association

Competition for anionic sites on the micelle surface is through electrical forces. Hence a cation exchange law of the following form applies:

$$\left(\frac{(C_9^3)^2}{C_6^3} = \beta^3 C_3^3 \frac{(C_9)^2}{C_6}\right)_1 \tag{9.8-12}$$

This equation is a form of Eq. (3.5-4) in which the constant (β^3) is multiplied by a factor that will convert the volume fraction C_3^3 into units of meq/L^3 of pore volume. Equation (9.8-12) assumes the cosurfactant is nonionic and all the surfactant is available to exchange.

Two types of electroneutrality now apply: on the micelle surface

$$C_{61}^3 + C_{81}^3 = C_{31} \tag{9.8-13a}$$

and in the bulk aqueous phase

$$C_{61} + C_{81} = C_{51} \tag{9.8-13b}$$

These form three equations in four unknowns, from which it becomes possible to solve for the bound divalents in terms of the unassociated species concentrations

$$C_{61}^3 = C_{31}^3 + \frac{1}{2}[r_{86}^3 - ((r_{86}^3)^2 + 4r_{86}^3 C_{31}^3)^{1/2}] \tag{9.8-14a}$$

where

$$r_{86}^3 = \beta^3 C_3^3 \frac{(C_{81}^3)^2}{C_{61}^3} \tag{9.8-14b}$$

Compare these equations to Eq. (3.5-7). Once the left side of Eq. (9.8-14a) is known, the fraction of the total divalents bound to the micelle follows from

$$f_6^3 = \frac{C_6^3}{C_3} \tag{9.8-15}$$

and C_{Se} can be estimated from Eq. (9.7-3).

The above theory will fit experimental data very well. Figure 9-17 shows the agreement between estimated and calculated phase boundaries for a system of petroleum sulfonate, decane, isobutanol, Na, and Ca. To construct this match, Prouvost (1984) assumed the above theory applies to phase boundaries as well as to optimal salinities. The theory and experiment agree well even though the SRD is far from ideal.

9-9 HIGH CAPILLARY NUMBER RELATIVE PERMEABILITIES

A transport property that deserves treatment in a separate section is the high capillary number relative permeability. In this section, we discuss two- and three-phase experimental results based on the work of Delshad et al. (1987) (see Sec. 3-3 for a discussion of low capillary number relative permeabilities).

Few theoretical relations exist for relative permeabilities in general, much less for those at high capillary number. We do know the extreme values of relative permeability functions occur at residual phase saturations. The latter are functions of capillary number N_{vc} through the capillary desaturation curve (CDC) (see Sec. 3-4). Further, for very high values of N_{vc}, we expect the relative permeabilities to approach straight-line functions between zero and unit endpoints with no residual phase saturations. For low N_{vc}, the relative permeabilities should return to the two- or three-phase high IFT functions. The variation between these extremes is not well established.

High N_{vc} relative permeabilities are difficult to measure. In one type of experiment, the large N_{vc} may be attained by increasing the flow rate. This technique causes experiments to proceed rapidly since, as we saw in Sec. 3-4, N_{vc} must increase by several factors of 10 before a significant effect occurs. Such high rates are clearly unrepresentative of typical reservoir fluid velocities. If the high N_{vc} is established by lowering the IFT, the experiments tend to be dominated by transient composition changes. In principle, these transients could be analyzed by the methods given in Sec. 9.10, but this requires knowing the relative permeabilities, whose measurement is the point of the experiment.

The most reliable measurement is of steady-state relative permeabilities using preequilibrated fluids. For micellar fluids in two-phase flow, this consists of displacing a composition on one end of a tie line with another on the same tie line at constant salinity. When the effluent and injected fractional flows are equal, and transients caused by nonideal phase behavior are gone, the relative permeability to the flowing phases may be calculated from the measured effluent cuts and pressure drop. A similar provision exists in the three-phase ideal systems where all compositions are in equilibrium at constant salinity. Of course, such transients may take some time to die out; thus steady-state experiments can be time consuming. The uniform saturations established by such a procedure follow from material balance or, preferably, tracer data interpreted by a suitable numerical model (Delshad et al., 1987).

Despite these difficulties, high N_{vc} relative permeabilities for two-phase flow have been rather intensively measured, but three-phase data are rare. Figure 9-18 shows steady-state relative permeabilities to brine, oil, and microemulsion phases for both two- and three-phase flow. The permeable medium was strongly water wet in both cores A and B at high N_{vc} conditions. $N_{vc} = 0.01$ at the optimal salinity used in the experiments. The micellar system under test closely followed ideal phase behavior. From these high N_{vc} data, several observations can be made.

1. The residual phase saturations are nonzero. Of course, these values are points on the CDC. Except for the oleic phase, whose endpoint was already high in the water-wet medium, the endpoint relative permeabilities are substantially different from their low N_{vc} values.
2. The high N_{vc} relative permeabilities are not straight lines. The curves in these figures are the matches of the exponential forms Eq. (3.3-4) to the data. But the exponents n_1 and n_2 in these equations are not substantially different from their low N_{vc} values.
3. The two- and three-phase data follow essentially the same curves.
4. The relative permeability for all three phases are functions of their own saturations. This observation is at odds with the high N_{vc} behavior of three-phase gas, oil, and water flows (Stone, 1970).
5. Probably the most surprising conclusion is that the excess brine phase was not the most strongly wetting phase as it was under low N_{vc} conditions. This observation is supported by a variety of observations not present in Fig. 9-17. However, the microemulsion and excess brine residual phase saturations have about the same value at $N_{vc} = 0.01$.
6. The shape of the microemulsion curve is concave downward. This observation is highly atypical of relative permeabilities and can be explained only as wall or interfacial slippage.

For approximate calculation, let the exponential relative permeabilities of Eq. (3.3-4) approximate two-phase high N_{vc} behavior. Suppose the CDC of a type II(−) system is represented by Fig. 3-19, with $(N_{vc})_c$ and $(N_{vc})_t$ corresponding to the wetting state of phase j. We can define linear interpolants for the endpoints and the curvatures. For example, the endpoints vary according to

$$(k_{r2}^0)' = (k_{r2}^0) + \left(1 - \frac{S_{3r}'}{S_{3r}}\right)(1 - k_{r2}^0) \tag{9.9-1a}$$

$$(k_{r3}^0)' = (k_{r3}^0) + \frac{S_{2r}'}{S_{2r}}(1 - k_{r3}^0) \tag{9.9-1b}$$

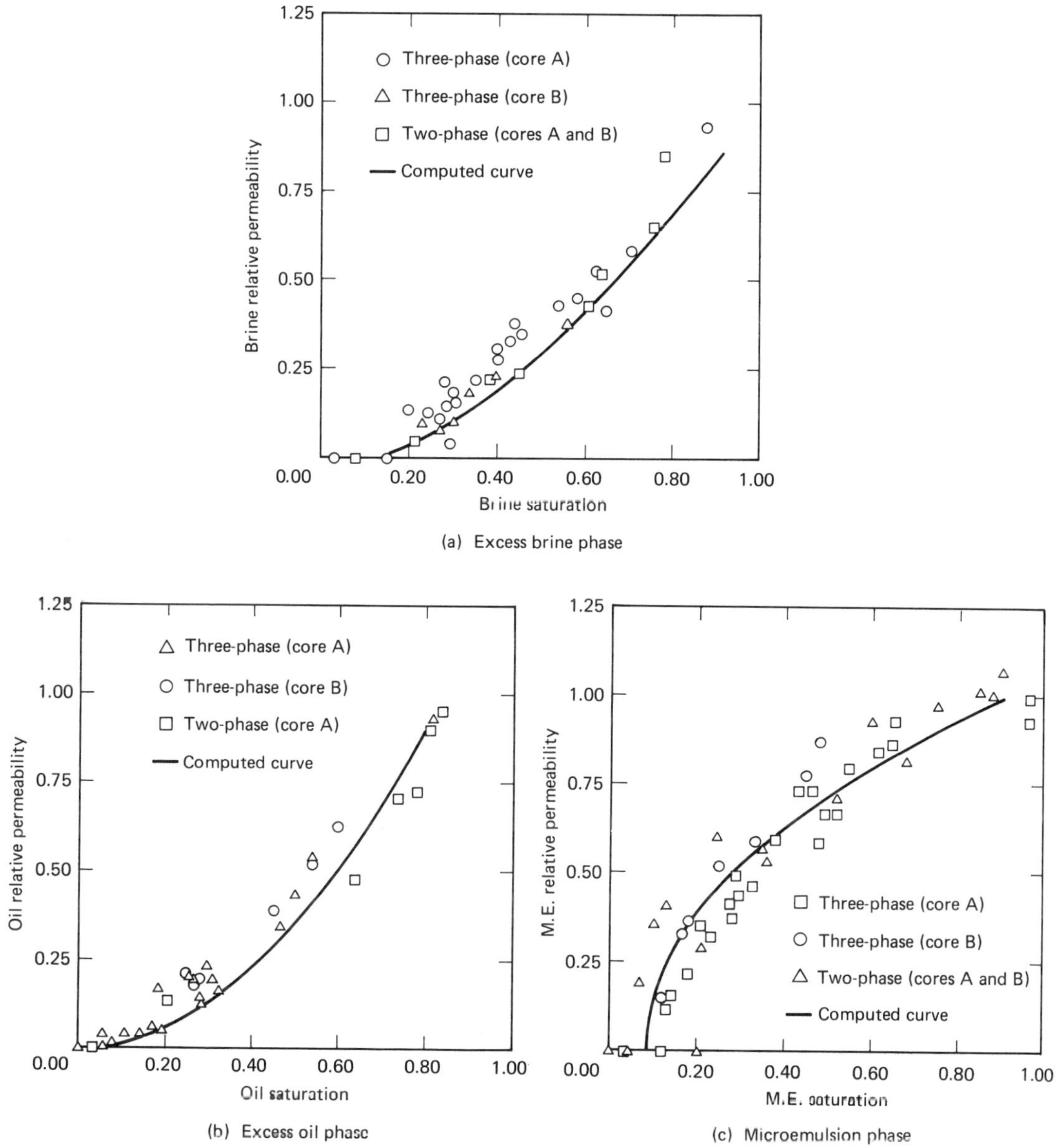

Figure 9-18 Two- and three-phase relative permeabilities (from Delshad et al., 1987)

where S'_{2r} and S'_{3r} are the high N_{vc} residual phase saturations. This approximate linear relation has been substantiated by Stegemeier (1976). The nonunit curvatures of the relative permeabilities seem to persist beyond the point of zero residual phase saturations; hence it seems reasonable that the logarithm of N_{vc} itself be used as an interpolating function

$$n_j' = n_j + (1 - n_j) \log\left[\frac{N_{vc}}{(N_{vc})_c}\right], \qquad j = 2 \text{ or } 3 \qquad (9.9\text{-}2)$$

Relations for type II(+) systems, where $j = 1$ or 3, follow analogous arguments.

We can now estimate two-phase relative permeabilities from phase behavior, a solubilization parameter correlation, the CDC curves, and low N_{vc} relative permeability curves. Suppose we know the overall composition of a type II(−) system that splits into two equilibrium phases. The phase compositions follow the ternary diagram. These can be converted to solubilization parameters using Eq. (9.4-1) and then into IFTs from the appropriate correlations. We use this to calculate N_{vc}, and the CDCs to estimate residual phase saturations. The high N_{vc} curves follow from Eqs. (9.9-1) and (9.9-2). If additional data are available about viscosities, dip angles, and densities, we can easily calculate phase fractional flows.

For three-phase flow, even such rough estimates are not warranted. Theoretical models by Hirasaki et al. (1983) and Delshad et al. (1987), though plausible in limiting senses, account neither for the intermediate wetting of the excess brine phase nor for the observation that the phase relative permeabilities are functions only of their own saturations. Clearly, we are hindered by a lack of understanding about the pore-level nature of high capillary number flows.

9-10 FRACTIONAL FLOW THEORY IN MICELLAR-POLYMER FLOODS

Fractional flow theory can be just as insightful for MP floods as for the solvent and polymer floods we covered in Secs. 7-7 and 8-4. In fact, there are so many similarities to those processes that we draw heavily on the material in those sections.

To make the analysis, we invoke the usual fractional flow assumptions: incompressible fluid and rock, one-dimensional flow, and no dissipative effects. In addition, we neglect the presence of the polymer drive (the polymer treatment can be added as an exercise), assume three-component MP floods with a step change in concentration at the origin of a time–distance diagram, and treat only those floods with constant phase behavior environment. Further, we neglect surfactant retention until later in this section where we invoke more restrictive assumptions about the phase behavior. To shorten the development, we cover only the high-salinity type II(+) floods. Fractional flow treatment for three-phase MP floods has not been extensively investigated (Giordano and Salter, 1984), but it could be so treated with the numerical technique of Sec. 5-7.

Ternary Landmarks

Figure 9-19(a) shows the basic phase and saturation behavior. This is very much like the behavior in Fig. 7-31 except that the miscibility gap extends entirely across the bottom edge of the ternary, and of course, water is explicitly included on the dia-

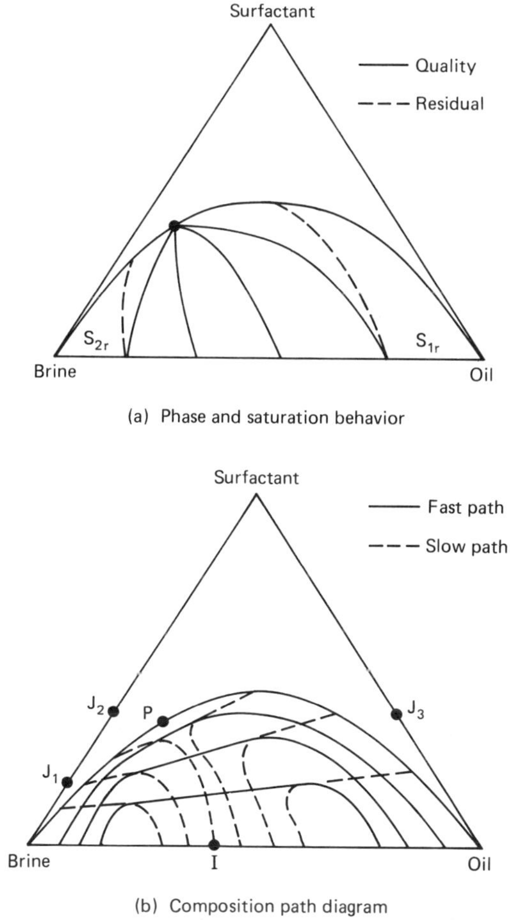

Figure 9-19 Ternary diagram and composition paths for micellar–polymer system

gram. All phase diagrams in this section have exaggerated two-phase regions. One significant difference with solvent flooding is lines of residual oleic and aqueous phase saturations merge with the binodal curve at some distance from the plait point. This happens because the oleic-aqueous capillary number increases (IFT decreases) rapidly as the plait point is approached, which causes S_{2r} and S_{1r} to approach zero (see Sec. 3-4). For continuous surfactant injection, as we are treating here, this issue is entirely secondary. But for finite slugs in highly dissipative displacements—that is, the realistic cases—the rate of approach to zero S_{2r} is very important. The aqueous-oleic fractional flow curve follows from the large N_{vc} relative permeabilities we discussed in Sec. 9-9.

MP Flooding without Retention

The relative permeability behavior does not affect the qualitative features of the composition path diagram (Fig. 9-19b). The development in Sec. 7-7 applies directly: We see the presence of "hair-pin" fast paths along tie lines, slow paths on either side, and a succession on nontie line paths. Because of the graphical possibilities, we assume the component distribution between phases is given by a family of straight lines intersecting at $C_3 = 0$ and $C_2 = C_2^0$ (the tie line envelope is a point on the C_3 axis)

$$C_{3j} = \eta(C_{2j} - C_2^0), \qquad j = 1 \text{ or } 2 \tag{9.10-1}$$

from which we have $C_1^0 = 1 - C_2^0$. The parameter η is the slope of the phase distribution line.

To review briefly, the component velocities along a tie line are

$$v_{C_i} = \frac{dF_i}{dC_i}, \qquad i = 1, 2, 3 \tag{9.10-2}$$

The nontie line paths carry the constant specific velocities given by

$$v_{\Delta C_i} = \frac{F_i - C_i^0}{C_i - C_i^0}, \qquad i = 1, 2, 3 \tag{9.10-3}$$

At the tangent intersection of the tie line and nontie line paths, we must have

$$\frac{F_i - C_i^0}{C_i - C_i^0} = \frac{dF_i}{dC_i} \tag{9.10-4}$$

which defines the two singular curves and allows the location of the appropriate constructions on a fractional-flux–overall-concentration plot. Other paths include the binodal curve itself and the equivelocity curve where $f_1 = S_1$.

In Fig. 9-19(b), we illustrate behavior for fractional flux curves whose S-shape persists even to low IFT. Our task is to string together the paths so that the composition route leads to monotonically decreasing composition velocities.

We focus on the three different injection conditions. Condition J_1 is an aqueous (oil-free) surfactant solution below the critical tie line extension, J_2 is an aqueous surfactant solution above the extension, and J_3 is an oleic (brine-free) slug below the extension. Conditions J_1 and J_2 represent low- and high-concentration aqueous surfactant solutions, and condition J_3 is an oil-soluble solution. In each case, the initial condition will be at I, a uniform tertiary condition.

Figure 9-20 shows the composition route and the S_2 and C_3 profiles at fixed t_D for the low-concentration surfactant displacement. Starting at the injection condition, the composition route enters the two-phase region along a tie line extension, switches to the nontie line path at the second singular point on the tie line, switches again to the fast path along the ternary base, and then to the initial condition I. For typical fractional flux curves, this causes a shock to an oil bank saturation S_{2B}, and a mixed wave from S_{2B} to $S_2 = 0$. Following the tie line causes the curious effect that the flowing surfactant concentration can be greater than the injected concentration. The displacement can also be relatively inefficient if the spreading portion of the oil bank rear is large.

For the high-concentration surfactant displacement (Fig. 9-21), the composition route passes through the plait point, follows the equivelocity path to the oil

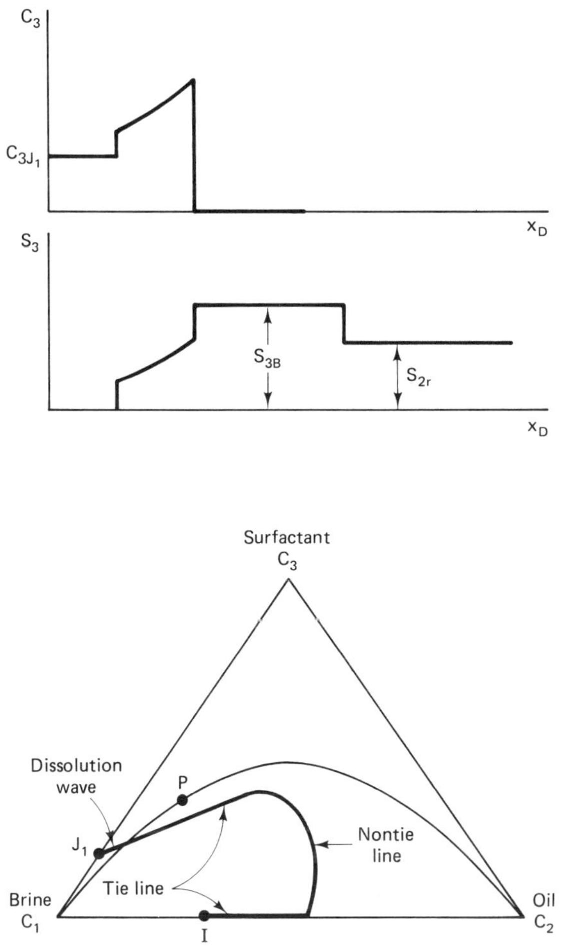

Figure 9-20 Composition route and profiles for low-concentration surfactant flood

bank saturation, and then on to the initial condition. This displacement is directly analogous to a condensing gas drive miscible displacement since the surfactant wave is indifferent and moves with unit specific velocity (compare this displacement with the lower panel in Fig. 7-35b). As such, it is highly efficient; however, the greater efficiency is bought with a higher surfactant concentration. The oil bank saturation is also somewhat lower than in Fig. 9-20.

The oleic surfactant behavior is shown in Fig. 9-22. Here the composition route also enters along a tie line extension, branches to a non-tie line path at the first singular point, and then on to the oil bank and the initial condition. In many respects, this displacement is the mirror image of that in Fig. 9-20. However, the ultimate microemulsion phase saturation is unity, meaning the oil bank saturation S_{2B} is between the initial and 1.0. The surfactant concentration decreases monotonically in this displacement, which as in Fig. 9-20, can also be inefficient.

There is great variety of behavior in the displacement character even under the

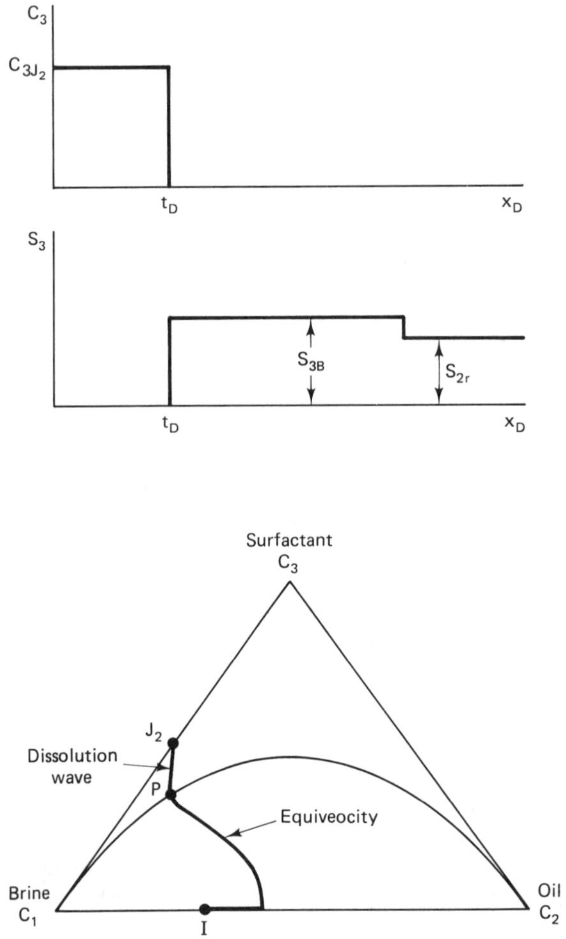

Figure 9-21 Composition route and profiles for high-concentration surfactant flood

restrictive assumptions invoked here. Some of this variety is present in the constructions used to infer Figs. 9-20 through 9-23. The cases for type II(−) are analogous. The nature of the composition route does not change with the shape of the fractional flux curve even though the latter greatly affects the efficiency of the displacement.

MP Flooding with Retention

Adding retention complicates the analysis because the composition route no longer follow tie lines. But by making a few additional assumptions, we can develop a fractional flow solution that uses fractional flow curves instead of fractional fluxes.

Let us now analyze the type II(−) system where the right plait point is in the oil corner of the ternary, and the amount of solubilized oil in the microemulsion phase is negligible. The aqueous and microemulsion phase are now equivalent ($S_1 = S_3$). If the injected slug composition is below a tangent from the binodal curve

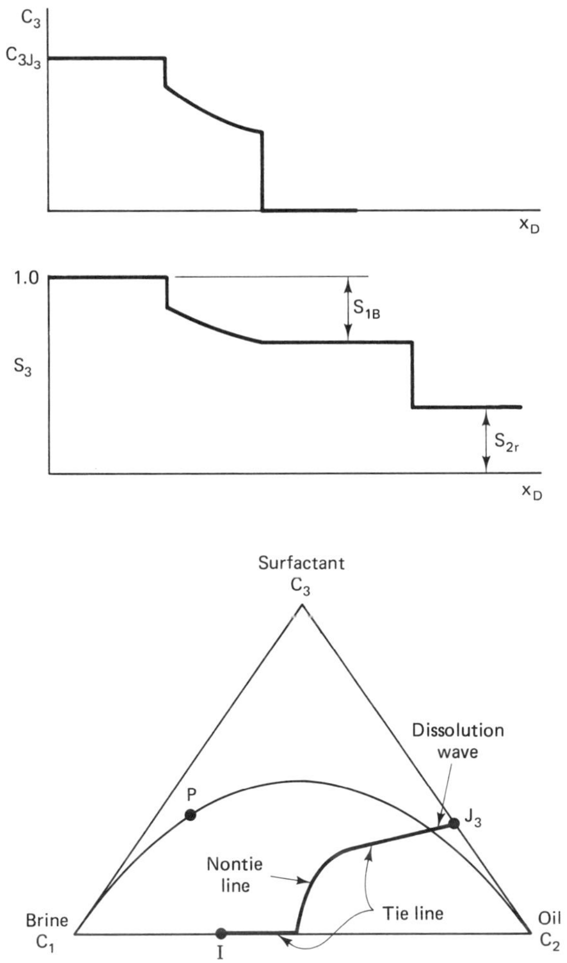

Figure 9-22 Composition route and profiles for high-concentration oleic surfactant flood

at the plait point, it must necessarily be on a tie line even if it contains no oil. Let the residual oil saturation on this tie line be S'_{2r}, the ultimate value of a low IFT (high N_{vc}) aqueous-phase fractional flow curve as shown in Fig. 9-24. This figure also shows the water-oil fractional flow curve f_1 along the tie line on the base of the ternary. Since this aqueous slug miscibly displaces the irreducible water, the velocity of the corresponding indifferent wave is

$$v_{C_3} = \frac{f_1^s}{S_1 + D_3} \tag{9.10-5}$$

from Eq. (5.4-8a). Note that f_1^s is the microemulsion (aqueous) phase high N_{vc} fractional flow. In this equation, D_3 is the surfactant's frontal advance loss given by

$$D_3 = \frac{1 - \phi}{\phi} \frac{C_{3s}}{C_{3J}\rho_s} \tag{9.10-6}$$

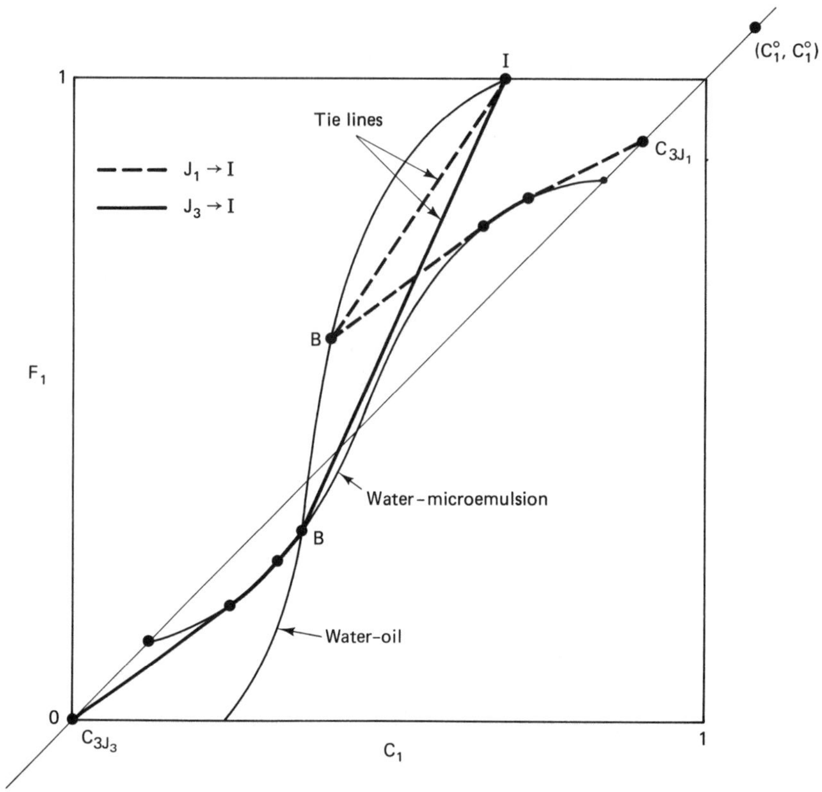

Figure 9-23 Fractional flux and composition routes for aqueous and oleic surfactant displacements

The most general case occurs when the rear of the oil bank travels as a mixed wave. At the leading edge of the spreading portion of this wave, the specific velocity in Eq. (9.10-5) must be equal to the specific oil velocity at some saturation S_1^* given implicitly by

$$\left(\frac{df_1^s}{dS_1} = \frac{f_1^s}{S_1 + D_3}\right)_{S_1^*} \tag{9.10-7}$$

The specific velocity of the shock portion of the oil bank rear is

$$v_{\Delta C_2} = \frac{f_{2B} - f_2^s(S_2^*)}{S_{2B} - S_2^*} \tag{9.10-8}$$

This must be equal to v_{C_3} evaluated at $S_2^* = 1 - S_1^*$. If the oil bank front is a shock, it travels with velocity given by

$$v_{\Delta C_2} = \frac{f_{2B} - f_{2I}}{S_{2B} - S_{2I}} \tag{9.10-9}$$

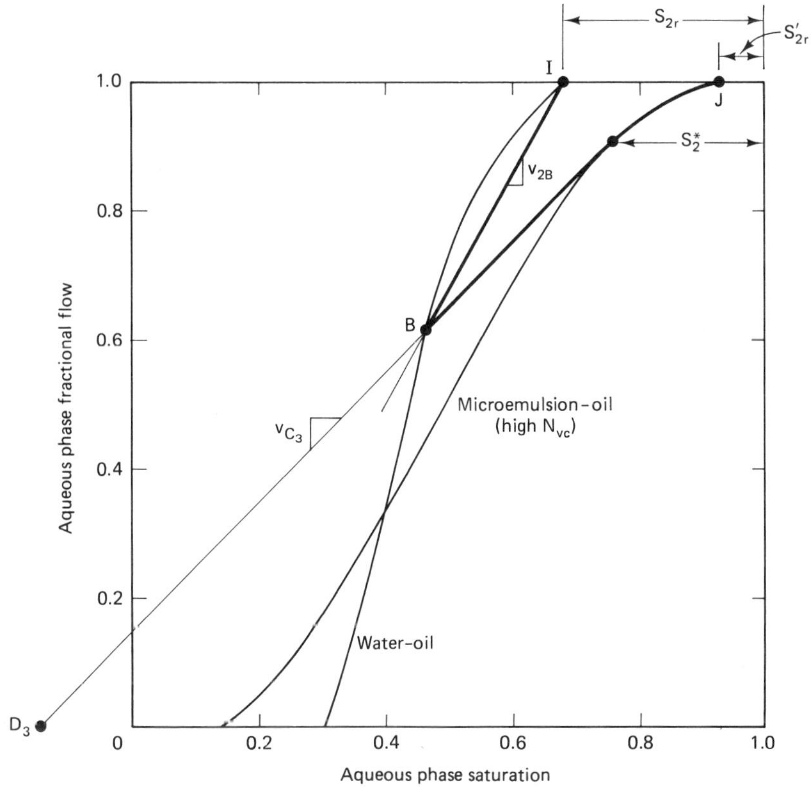

Figure 9-24 Graphical construction for simplified II(−) surfactant displacements

These equations are entirely suggestive of the polymer flooding construction in Sec. 8-4. This parallel is also apparent from comparing the construction given in Fig. 9-24 with the one in Fig. 8-10. The construction of the time–distance and profile diagrams corresponding to Fig. 9-24 is left as an exercise.

An issue not dealt with in Sec. 8-8 is the minimum slug size needed to satisfy retention. Let's suppose the surfactant displacement is pistonlike, that is, $S_{2I} = S'_{2r} = S_2^*$. The minimum surfactant slug size is attained when the similarly pistonlike surfactant–polymer front overtakes the surfactant front at the injection end of the medium. This gives a minimum surfactant slug size of $t_{Ds} = D_3$, meaning the frontal advance loss is an expression of the retention capacity of the medium expressed in units consistent with the slug size. Therefore, knowing D_3 is the beginning point in estimating surfactant requirement in MP flooding. The above result does not depend on the existence of a pistonlike surfactant front.

9-11 ROCK-FLUID INTERACTIONS

Brine salinity and hardness would have far less importance to MP flooding if the host permeable medium were unreactive. Unfortunately, in all but the most artificial cases, reservoir minerals provide an almost limitless source of monovalent and divalent cations as well as ample sites for surfactant retention. Two sources of cations are mineral dissolution and cation exchange. Dissolution usually occurs at such a low level that it can be neglected in MP floods (but not in high-pH floods). Cation exchange is rarely negligible. For this reason, we discuss it at some length in this section. In the second half of the section, we deal with surfactant retention.

Cation Exchange

We treat the simplest case of monovalent–divalent exchange in single-phase flow in the absence of surfactant or oil. The displacement satisfies the fractional flow assumptions. (For more complicated treatment, see Pope et al., 1978.)

Each point in the permeable medium must satisfy solution electroneutrality

$$C_5 = C_6 + C_8 \tag{9.11-1a}$$

and electroneutrality on the clays

$$Z_v = C_{6s} + C_{8s} \tag{9.11-1b}$$

The units on all concentrations are in equivalents per unit pore volume, and the superfluous phase subscript has been dropped. These equations imply the monovalent, divalent, and anion concentrations are not independent; hence for convenience, we choose to proceed with the divalent and anion concentrations as the dependent variables. At local equilibrium, each point in the medium must also satisfy

$$\frac{C_{8s}^2}{C_{6s}} = K_N \frac{C_8^2}{C_6} \tag{9.11-2}$$

which is simply the cation exchange isotherm of Eq. (3.5-4). Using Eq. (9.11-1), we can express the adsorbed divalent concentration as

$$C_{6s} = Z_v\left(1 + \frac{K_N r}{Z_v}\left[\frac{1}{2} - \left(\frac{Z_v}{K_N r} + \frac{1}{4}\right)^{1/2}\right]\right) \tag{9.11-3a}$$

where

$$r = \frac{C_8^2}{C_6} = \frac{(C_5 - C_6)^2}{C_6} \tag{9.11-3b}$$

Equations (9.11-3) are the basic equilibrium representations.

Let's now consider the displacement of solution I by solution J under the above conditions. The coherence conditions (Eq. 5.6-14)

$$\left(1 + \frac{dC_{5s}}{dC_5}\right)^{-1} = v_{C_5} = v_{C_6} = \left(1 + \frac{dC_{6s}}{dC_6}\right)^{-1} \tag{9.11-4}$$

are satisfied at all points in the medium. Equation (9.11-4) implies

$$\frac{dC_{5s}}{dC_5} = \frac{dC_{6s}}{dC_6} = \lambda \tag{9.11-5}$$

where λ is the eigenvalue for this problem. In matrix form, Eq. (9.11-5) becomes

$$\begin{pmatrix} C_{55} & C_{56} \\ C_{65} & C_{66} \end{pmatrix} \begin{pmatrix} dC_5 \\ dC_6 \end{pmatrix} = \lambda \begin{pmatrix} dC_5 \\ dC_6 \end{pmatrix} \tag{9.11-6}$$

where $C_{65} = (\partial C_{6s}/\partial C_5)_{C_6}$, and so on. The matrix on the left side has a row of zeros because the anion does not adsorb. Solving Eq. (9.11-6) for the eigenvalues gives

$$\lambda^- = 0, \qquad \lambda^+ = C_{56} \tag{9.11-7}$$

From this, it is obvious that $\lambda^+ > \lambda^-$ and that the wave corresponding to λ^- is faster than that corresponding to λ^+ from Eq. (9.11-4).

The eigenvector corresponding to each eigenvalue gives the concentration change across each wave. For the fast wave, inserting λ^- gives

$$C_{65}\, dC_5 + C_{66}\, dC_6 = 0 \tag{9.11-8a}$$

and from the slow wave, inserting λ^+ gives

$$dC_5 = 0 \tag{9.11-8b}$$

Immediately we see that the anion concentration is constant across the slow wave since $dC_5 = 0$. The fast wave interpretation is only a little less obvious. Equation (9.11-8a) is the change in C_5 and C_6 that would occur at constant C_{6s}. We can see this by setting the total differential dC_{6s}

$$dC_{6s} = C_{66}\, dC_6 + C_{65}\, dC_5 \tag{9.11-9}$$

equal to zero. The result (Eq. 9.11-9) is independent of the form of the exchange isotherm.

The coherent solution, therefore, predicts two waves: an indifferent salinity wave that moves at unit velocity across which the clay concentration remains constant and an exchange wave where the clay changes to be in equilibrium with the injected solution at constant anion concentration. The character of the latter wave depends on the direction in which the concentration velocity is increasing through Eqs. (9.11-3) and (9.11-4).

For example, consider Fig. 9-25(a), which shows the composition space for the two cation system plotted on a ternary representation. Lines of constant anion concentration are parallel to the right edge, and lines of constant clay composition are curves converging to the lower left corner. These curves are described by Eq. (9.11-3). Both sets of lines form the composition path diagram. The bold line segments I-J' (salinity wave) and J'-J (exchange wave) form the composition route. This particular sequence is selected because it is the only one for which the concentration velocities monotonically decrease from I to J. The exchange wave is spreading if the concentration velocity decreases from J' to J; otherwise, it is a shock. Figure 9-25(b) shows a time–distance diagram for the former case.

To illustrate the accuracy of the above predictions, Fig. 9-26 shows the effluent histories of two laboratory core floods through which are flowing solutions containing only calcium (divalents), sodium (monovalents), and chloride (anion). In both

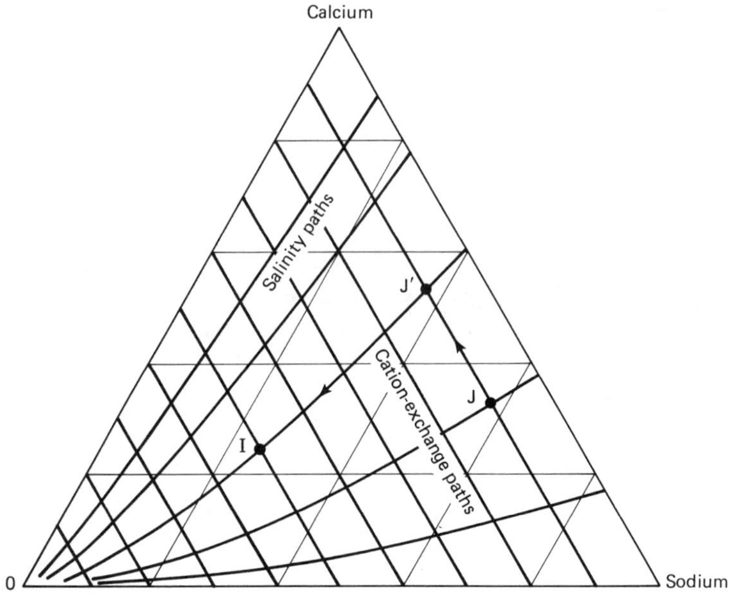

(a) Composition path diagram

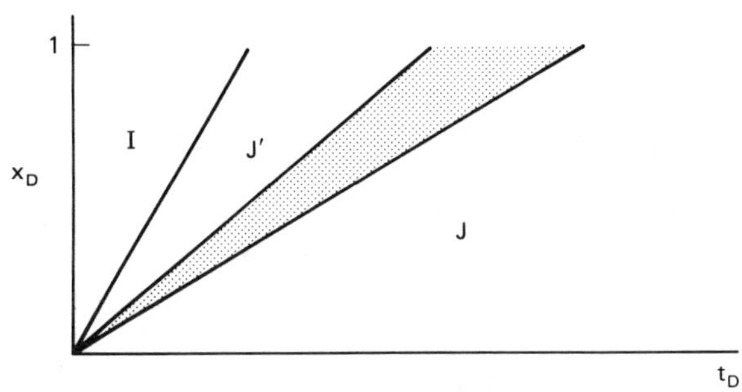

(b) Time-distance diagram

Figure 9-25 Diagrams for two exchanging cation case (from Pope et al., 1978)

cases, the injected calcium was the same; hence in the absence of exchange, the effluent calcium concentration should not change. But because the anion concentration changes, cation exchange occurs, and the effluent calcium does change. In Fig. 9-26(a), the increased anion concentration causes calcium to be expelled from the clays. In Fig. 9-26(b), the reverse occurs. In both cases, the prediction based on the coherent treatment agrees well with the observed results. Calculated results including dispersion match even better (Lake and Helfferich, 1978).

These results hold immense practical significance for MP flooding in general and the use of low-salinity preflushes in particular. One of the intentions of a preflush is to remove divalent cations so that the slug can work more effectively. However, the above theory suggests the following hindrances:

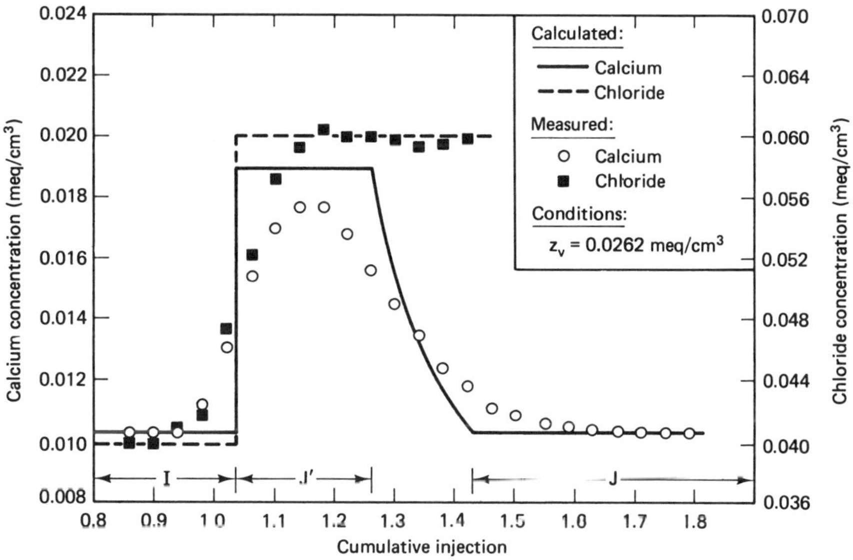

(a) Spreading exchange wave

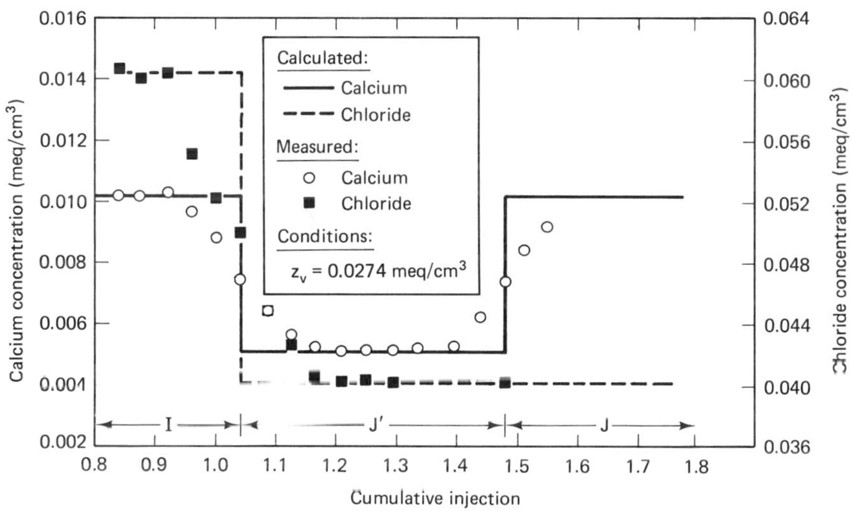

(b) Sharpening exchange wave

Figure 9-26 Comparison between theory and experiment for two exchanging cation displacement (from Pope et al., 1978)

1. It is entirely possible to inject a low-salinity preflush that actually loads the clays with divalent cations because changes in the ratio r determine clay loading. If r decreases, the clays will take up divalents, which regardless of the salinity, are available for subsequent release into solution.
2. Even if r decreases so that the clays unload divalents, this normally takes a large preflush because the exchange wave velocity is very slow at typical cation exchange capacities and brine concentrations.
3. If the injected solution is entirely devoid of divalents, the clays will still only partially unload because the dissolution of a small amount of divalent-containing minerals acts as a persistent source of hardness.

One philosophy for preflushes is to avoid upsetting the clays at all costs. Doing this is simple in principle: One just injects the preflush, slug, and polymer drive at the same r ratio as exists in the formation brine. But in practice, this procedure is complicated by dispersion-induced mixing (Lake and Helfferich, 1978) and by exchange of divalents with micelles (see Sec. 9-8).

Surfactant Retention

Surfactant retention is probably the most significant barrier to the commercial application of MP flooding. The problem here is one of selectivity. The surfactants should have good selectivity for oil–water interfaces, but they should also have poor selectivity for fluid–solid interfaces.

Surfactants are retained through four mechanisms.

1. On metal oxide surfaces (Fig. 9-27), the surfactant monomer will physically adsorb through hydrogen bonding and ionically bond with cationic surface sites. At higher surfactant concentrations, this association includes tail-to-tail interactions with the solution monomers, resulting in proportionally greater adsorption. At and above the CMC, the supply of monomers becomes constant, as does the retention. The Langmuir-type isotherm of adsorption versus overall surfactant concentration resembles the CMC plot in Fig. 9-4, which can be expressed as

$$C_{3s} = \frac{a_3 C_3}{1 + b_3 C_3} \tag{9.11-10}$$

where a_3/b_3 represents the plateau adsorption value. C_3 here is the surfactant concentration in the liquid phase wetting the substrate. The parameter b_3 is large, being related to the CMC, which is very small compared to practical surfactant concentrations (see Fig. 9-28). The surfactant isotherm therefore attains its plateau at such a low C_3 that it may be usefully represented as a step function. This form of retention should be reversible with surfactant concentration. The parameters a_3 and b_3 are functions of salinity since they depend on the number of surface sites available for adsorption.

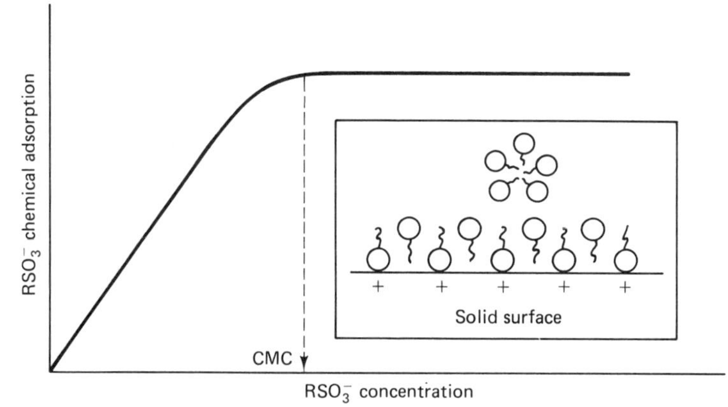

Figure 9-27 Surfactant adsorption on metal oxide surfaces (adapted from Harwell, 1983)

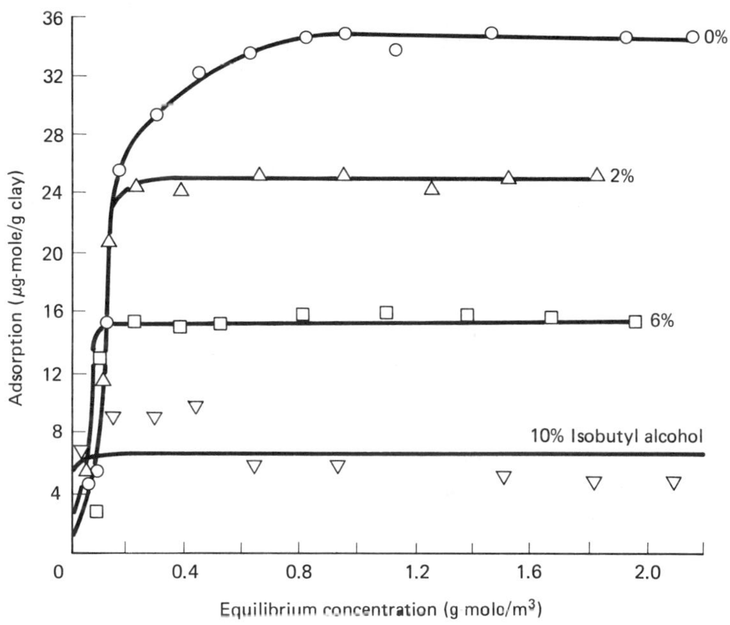

Figure 9-28 Effect of cosurfactant on surfactant retention. Surfactant is 4-phenyl dodecyl benzene sulfonate. (adapted from Fernandez, 1978)

2. In hard brines, the prevalence of divalent cations causes the formation of surfactant–divalent complexes

$$2\text{R-SO}_3^- + \text{M}^{2+} \longrightarrow \text{M(R-SO}_3)_2\downarrow \tag{9.11-11}$$

which have a low solubility in brine. Precipitation of this complex will lead to retention. When oil is present, it can compete for the surfactant. Of course, the precipitate must also compete with the micelles for the surfactant (Somasundrun et al., 1979).

3. At hardness levels somewhat lower than those required for precipitation, the preferred multivalent surfactant will be a monovalent cation that can chemically exchange with cations originally bound to the reservoir clays (Hill and Lake, 1978).

$$\begin{aligned} \text{R-SO}_3^- + \text{M}^{2+} &\longrightarrow \text{MR-SO}_3^+ \\ \text{Na-Clay} + \text{MR-SO}_3^+ &\longrightarrow \text{MR-SO}_3\text{-Clay} + \text{Na}^+ \end{aligned} \tag{9.11-12}$$

This effect is not unlike the divalent-micelle effect we discussed in Sec. 9-8. The surfactant bound to the clays will exhibit tail–tail interaction as in Fig. 9-27.

As a consequence of the ionic bonding and tail–tail interactions, adding a cosurfactant will reduce both types of retention (Fig. 9-28). Cosurfactants perform this service in two ways: (1) by filling surface sites that might otherwise be occupied by surfactant and (2) by mitigating the tail-to-tail associations. The retention expressed by Eq. (9.11-12) can also be lessened by filling the clay sites with a more preferred metal cation. This form of retention is reversible with both M^{2+} and surfactant concentration.

4. In the presence of oil in a II(+) phase environment, the surfactant will reside in the oil-external microemulsion phase. Because this region is above the optimal salinity, the IFT is relatively large, and this phase and its dissolved surfactant can be trapped. Figure 9-29 illustrates this phenomenon. The filled squares represent the surfactant injected, and the open squares the surfactant retained in a series of constant-salinity core floods. Retention increases smoothly with salinity (both a_3 and b_3 are functions of salinity) until 3% NaCl, at which point it increases so substantially that all the injected surfactant is retained. 3% NaCl is just above C_{Seu} for this system; hence the deviation can be nicely explained by phase trapping. A similar phase trapping effect does not occur in the II(−) environment because the aqueous mobility buffer miscibly displaces the trapped aqueous-external microemulsion phase. Using less than optimal salinities can, therefore, eliminate phase trapping. This form of retention is strongly affected by the MP phase behavior.

Most studies of surfactant retention have not made these mechanistic distinctions. Therefore, which mechanism predominates in a given application is not obvious. *All* mechanisms retain more surfactant at high salinity and hardness, which in turn, can be attenuated by adding cosurfactants. Precipitation and phase trapping can be eliminated by lowering the mobility buffer salinity at which conditions the chemical adsorption mechanism on the reservoir clays is predominant. In this event, there should be some correlation of surfactant retention with reservoir clay content. Fig. 9-30 attempts to make this correlation by plotting laboratory and field surfactant retention data against clay fraction. The correlation is by no means perfect since it ig-

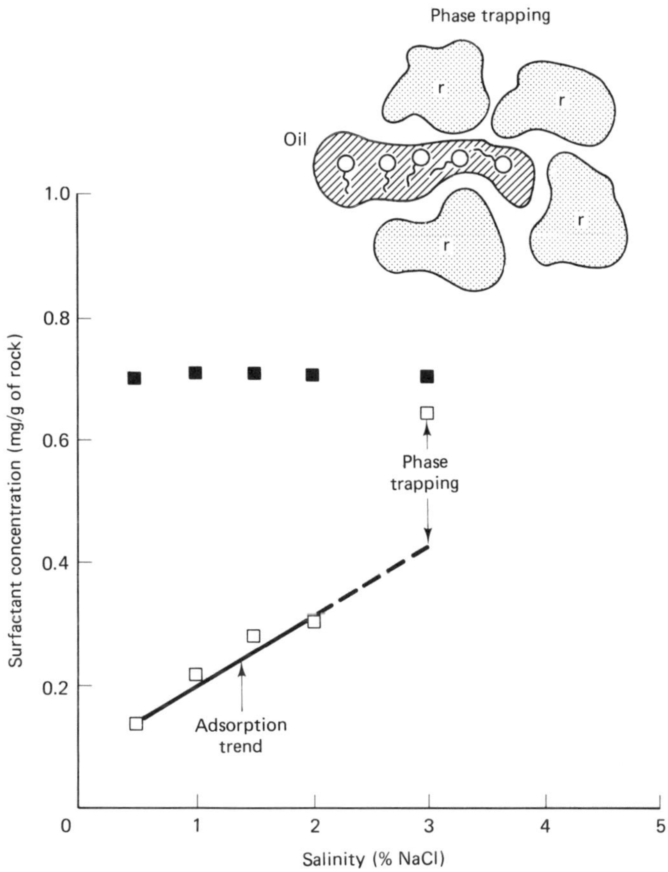

Figure 9-29 Surfactant retention caused by phase trapping; 3% NaC1 is a type II(+) microemulsion system (from Glover, et al., 1979)

nores variations in MP formulation and clay distribution as well as salinity effects. However, the figure does capture a general trend useful for first-order estimates of retention. In addition, note that the difference between lab- and field-measured retention is not significant. This observation implies that surfactant retention can be effectively measured in the laboratory.

A useful way to estimate the volume of surfactant required for an MP slug is through the dimensionless frontal advance lag D_3 defined in Eq. (9.10-6). C_{3s} is the surfactant retention from Fig. 9-30 (the plateau value a_3/b_3 on the appropriate isotherm), ϕ is the porosity, C_{3J} is the surfactant concentration in the MP slug, and ρ_s is the surfactant density. D_3 is a fraction that expresses the volume of surfactant retained at its injected concentration as a fraction of the floodable pore V_{pf}. For optimal surfactant usage, the volume of surfactant injected should be large enough to contact all V_{pf} but small enough to prevent excessive production of the surfactant.

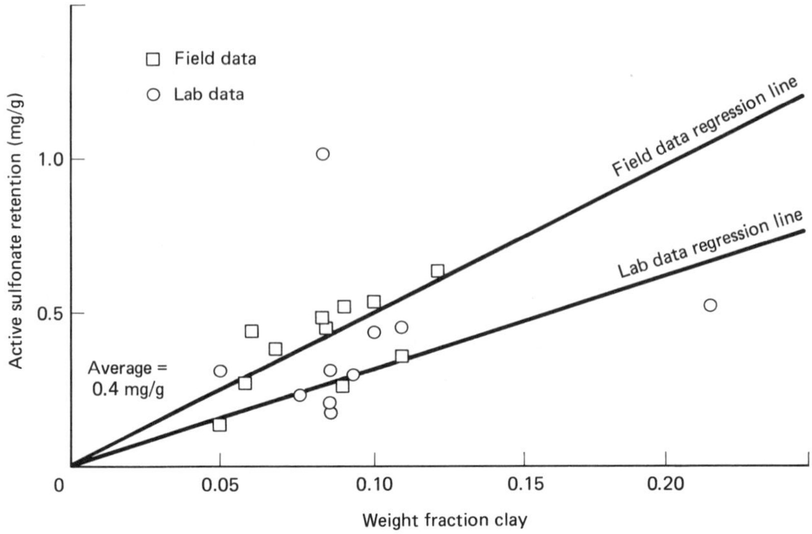

Figure 9-30 Overall surfactant retention correlated with clay content (adapted from Goldburg et al., 1983)

Besides wasting an expensive chemical, the produced surfactant can cause severe produced emulsions. Thus the MP slug size should be no less but not much greater than D_3. The total amount of surfactant injected is independent of injected surfactant since, from Eq. (9.10-6), $D_3 C_{3J}$ is independent of C_{3J}.

9-12 TYPICAL PRODUCTION RESPONSES

In this section, we review responses of typical laboratory core and field flood showing the important features and expectations of MP flooding.

Laboratory Flood

Fig. 9-31 plots an effluent response of a typical MP flood in a Berea core showing oil cut, produced surfactant (Mahogany AA), cosurfactant, (isopropyl alcohol), polymer, and chloride anion concentrations. All concentrations have been normalized by their respective injected values. The chloride indicates the salinity in this flood. At the top of the figure is the phase environment of the produced fluids. The slug size is $t_{Ds} = 0.1$, and the horizontal axis is t_D, the volume of fluid injected since the start of the slug expressed as a fraction of the core's pore volume. There was no preflush. (For further details of this and similar core floods, see Gupta, 1980.)

Figure 9-31 shows a typical, though by no means optimal, oil recovery experiment. Before surfactant injection, the core was waterflooded so that it produces no oil initially. Oil breaks through at about $t_D = 0.2$, with relatively sustained cuts of

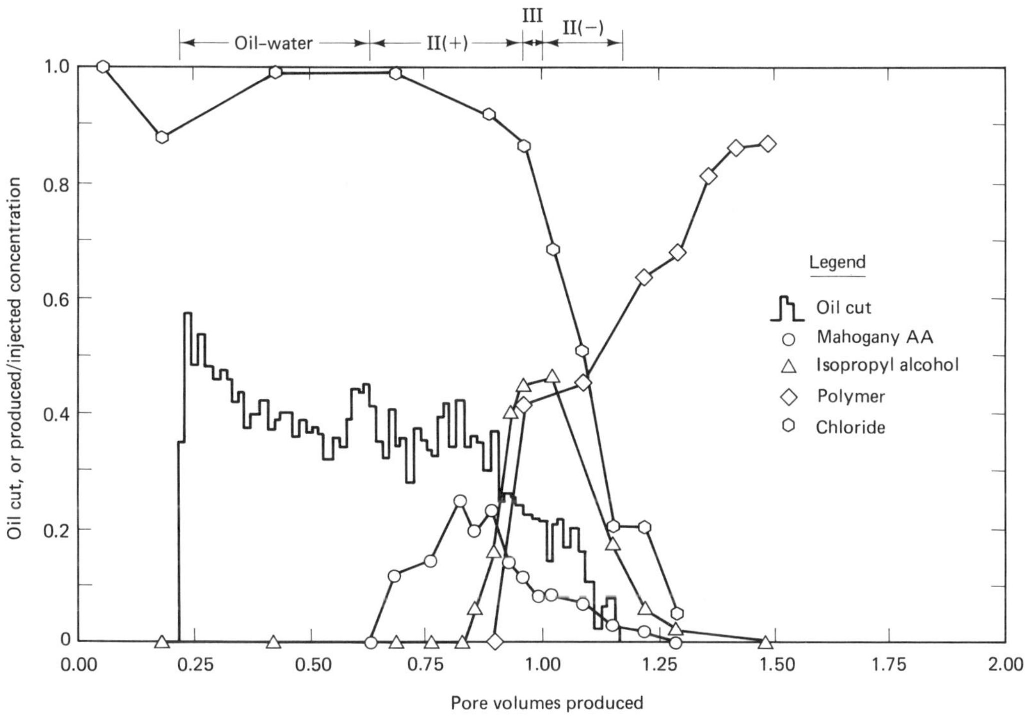

Figure 9-31 Typical core-flood production response (from Gupta, 1980)

about 40% until about $t_D = 0.6$, at which point the surfactant appears. The behavior in this portion of the flood is consistent with the fractional flow theory in Sec. 9-10. About 60% of the produced oil is free of the injected chemicals. That 40% of the oil is produced with the surfactant indicates a viscous instability apparently caused by nonideal phase behavior. A well-designed flood will produce 80% to 90% of the oil ahead of the surfactant. Even here, though, the oil is invariably produced early and at fairly low cuts in laboratory experiments.

Surfactant breaks through at $t_D = 0.6$, reaches its maximum produced concentration of 30% of the injected concentration at $t_D = 0.8$, and ceases at $t_D = 1.5$. The total amount of surfactant produced is about one half that injected, which indicates substantial, though not excessive, retention.

The surfactant is preceding both the chloride and polymer by about 0.3 V_p. This separation indicates preferential partitioning of the cosurfactant between the aqueous and microemulsion phases (see Sec. 9-8 on phase behavior nonidealities). Though this did not drastically affect oil recovery, which is in excess of 90% of the residual oil, the separation is not a favorable indication for this design. A good MP design should show simultaneous production of all MP slug constituents as well as good oil recovery.

Field Response

As a field example, consider Fig. 9-32, which shows the produced fluid analyses of well 12-1 in the Bell Creek (Carter and Powder River counties, Montana) MP flood. This flood used a high oil content MP slug preceded by a preflush that contained sodium silicate to lessen surfactant retention and reduce divalent cation concentration. Well 12-1 was a producer in the center of an unconfined single five-spot pattern. (For further details on the flood, see Holm, 1982; Aho and Bush, 1982.)

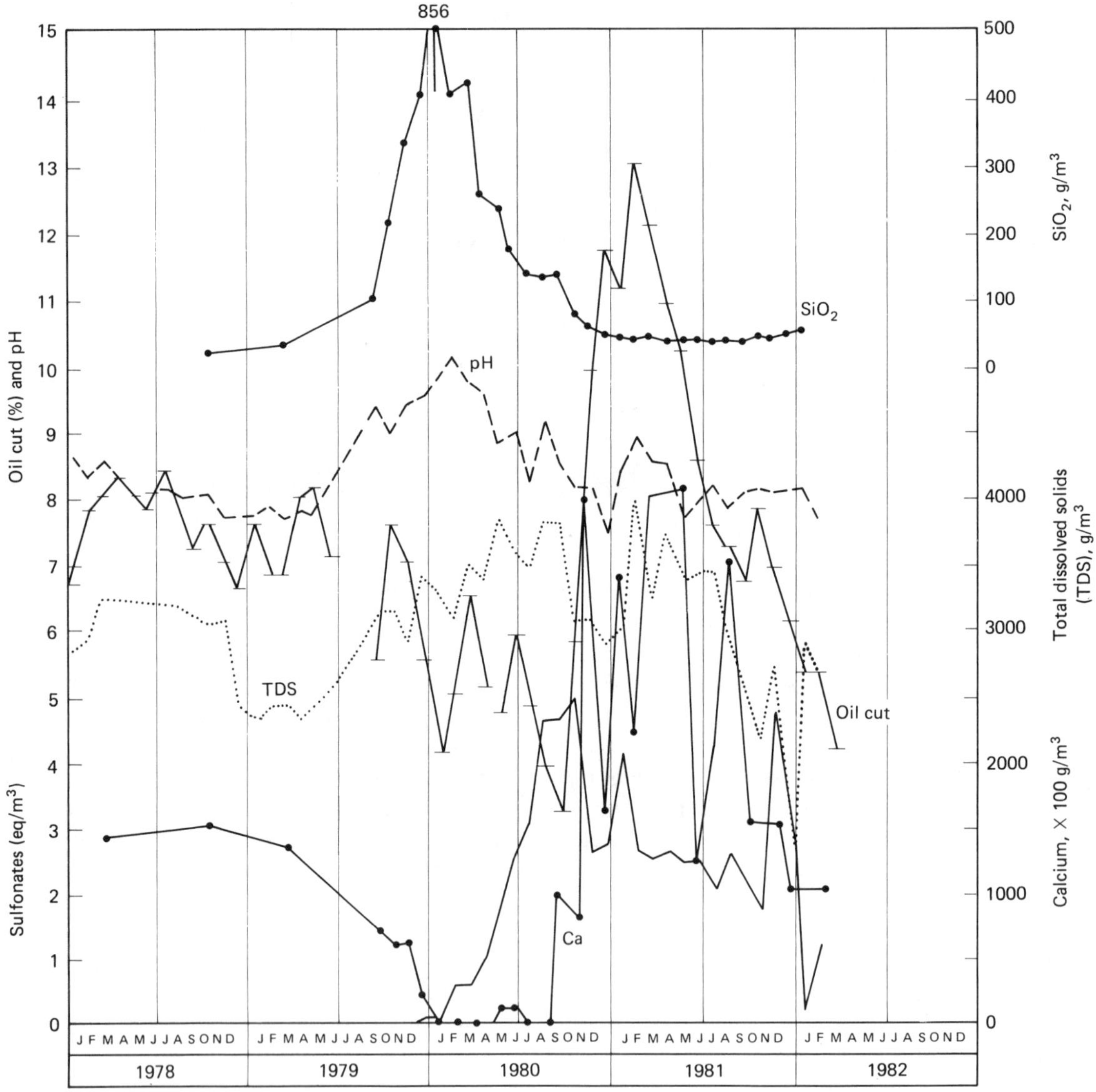

Figure 9-32 Production response from Bell Creek Pilot (from Holm, 1982)

Before MP slug injection in February, 1979, well 12-1 was experiencing low and declining oil cuts. Beginning in late 1980, MP oil response reversed the decline and reached peak cuts of about 13% about six months later. The pre-MP decline must be clearly established to accurately evaluate the MP oil recovery, an unnecessary step in evaluating the core flood. Moreover, compared to the core flood, there is no evident clean oil production; surfactant production actually preceded the oil response. Simultaneous oil and surfactant production is a persistent feature of field MP floods probably because of heterogeneities and dispersive mixing. The surfactant is preceding the oil in Fig. 9-32 because of preferentially water-soluble disulfonate components in the MP slug. The peak oil cut is invariably lower in field floods (13% in Fig. 9-32 versus nearly 60% in Fig. 9-31).

Other significant features in Fig. 9-32 are the evident presence of the preflush preceding the MP slug, inferred from the maxima in the pH and silicate concentrations, and the very efficient removal of the calcium cations ahead of the surfactant. But when oil production commenced, calcium rose roughly to its premicellar level.

Figure 9-33 shows ultimate oil recovery efficiency E_R (ultimate oil produced divided by oil in place at start of MP process) from a survey of more than 40 MP field tests correlated as a function of mobility buffer slug size t_{DMB}. Similar analyses on other process variables showed no or weak correlation (Lake and Pope, 1979). The strong correlation in Fig. 9-33 indicates the importance of mobility control in MP design. Though we have largely ignored mobility control in this chapter, it is

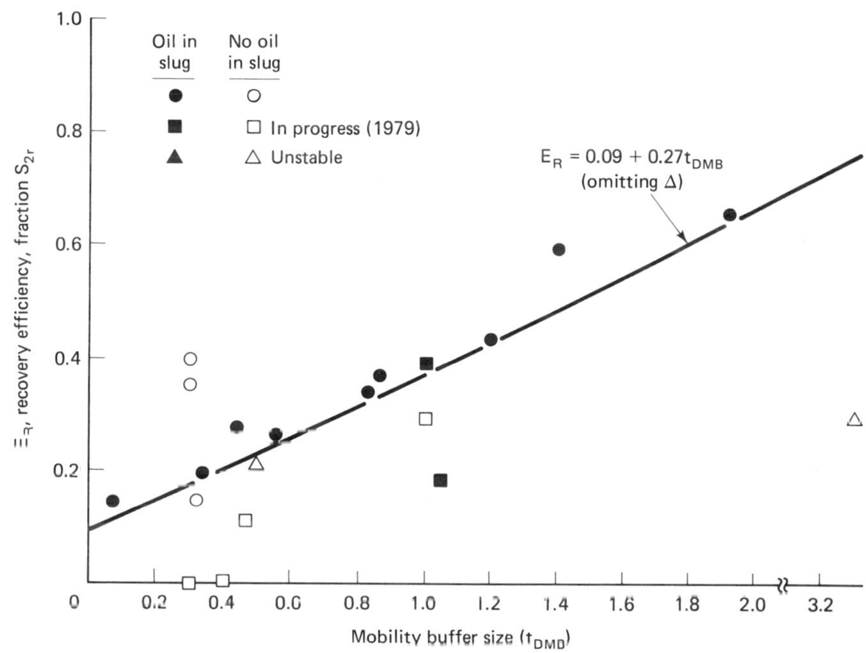

Figure 9-33 Recovery efficiencies from 21 MP field tests (adapted from Lake and Pope, 1979)

clearly an important variable. Note from Fig. 9-33 that the high oil content slugs have generally been driven by polymer drives larger than the high water content slugs.

Ultimate oil recovery efficiency averages about 30% of residual oil saturation in field tests (Fig. 9-33). Since oil recovery efficiency can be quite high in core floods, it seems that the peak oil cut and ultimate oil recovery efficiency in a technically successful MP field flood will average about one third of their respective values in core floods.

9-13 DESIGNING AN MP FLOOD

A successful MP flood must achieve three things for efficient oil recovery (Gilliland and Conley, 1975).

1. The MP surfactant slug must propagate in an interfacially active mode (i.e., at optimal conditions).
2. Enough surfactant must be injected so that some of it is unretained by the permeable media surfaces.
3. The active surfactant must sweep a large portion of the reservoir without excessive dissipation because of dispersion or channeling.

The first of these objectives is met through the formulation step of the MP design procedure; the second two objectives are met through scale up. Though there is considerable overlap, the formulation step consists mainly of test tube experiments and core floods; the scale-up steps consist mainly of core floods and numerical simulations.

Generating Optimal Conditions

There are three techniques for generating optimal conditions in MP floods.

1. Raise the MP slug optimal salinity to that of the resident brine salinity in the candidate reservoir. Philosophically, this procedure is the most satisfying of the three possibilities, and it is usually the most difficult. Though the subject of intensive research, surfactants having high optimal salinities that are not, at the same time, unstable at reservoir conditions, excessively retained by the solid surfaces, or expensive are yet to be discovered. Field successes with synthetic surfactants have demonstrated the technical feasibility of this approach (Bragg et al., 1982). A second way to make the optimal salinity of the MP slug closer to the resident brine salinity is to add cosurfactant. This approach is the most common implementation to date; however, as we mentioned, there are

penalties in surfactant–cosurfactant separation, loss of interfacial activity and expense.

2. Lower the resident salinity of a candidate reservoir to match the MP slug's optimal salinity. This common approach is the main purpose of the preflush step illustrated in Fig. 9-1. A successful preflush is appealing because, with the resident salinity lowered, the MP slug would displace oil wherever it goes in the reservoir, and retention would also be low. Preflushes generally require large volumes to significantly lower the resident salinity owing to mixing effects and cation exchange (see Sec. 9-11). With some planning, the function of preflush could be accomplished during the waterflood preceding the MP flood.
3. Use the most recent salinity gradient design technique for generating active MP slugs (Paul and Froning, 1973; Nelson and Pope, 1978; Hirasaki et al., 1983). This technique tries to dynamically lower the resident salinity to optimal during the course of the displacement by sandwiching the MP slug between the overoptimal resident brine and an underoptimal mobility buffer salinity. Table 9-4 illustrates the results of experimental core floods for different sequences of salinities. The experiment numbers on this table match the uncircled numbers in Fig. 9-8. Three core floods—numbers 3, 6, and 7—stand out both with respect to their low ultimate saturation and surfactant retention. The common feature of all these experiments is that the salinity of the polymer drive is underoptimal. In fact, no other variable, including, paradoxically, surfactant slug concentration, has such a similarly strong effect (Pope et al., 1982). The salinity gradient design has several other advantages: it is resilient to design and process uncertainties, provides a favorable environment for the polymer in the mobility buffer, minimizes retention, and is indifferent to the surfactant dilution effect.

TABLE 9-4 PHASE-ENVIRONMENT TYPE AND MP FLOOD PERFORMANCE FOR THE SALINITY-REQUIREMENT DIAGRAM IN FIG. 9-8 (FROM NELSON, 1982)

Chemical flood number	Phase type promoted by the			Residual oil saturation after chemical flood (% PV)	Injected surfactant retained by the core (%)
	Waterflood brine	Chemical slug	Polymer drive		
1	II(−)	II(−)	II(−)	29.1*	52
2	II(+)/III	II(+)/III	II(+)/III	25.2*	100*
3	II(+)/III	II(+)/III	II(−)	2.0†	61*
4	II(−)	II(−)	II(+)/III	17.6*	100*
5	II(−)	II(+)/III	II(+)/III	25.0	100
6	II(+)/III	II(−)	II(−)	5.6†	59†
7	II(−)	II(+)/III	II(−)	7.9*	73*
8	II(+)/III	II(−)	II(+)/III	13.7†	100*

*Average of duplicates
†Average of triplicates

Injecting Enough Surfactant

The first aspect of overcoming retention is to design the flood so that retention is as low as possible. This includes minimizing the chemical and physical adsorption effects discussed above and eliminating phase trapping by propagating the slug in a low-salinity environment. Cosurfactants and sacrificial agents in a preflush may also be appropriate. Once a low surfactant retention value is in hand, enough surfactant must be injected so that some of it transports to the production wells. As in polymer flooding, there are two aspects to this issue: the slug's surfactant concentration and the slug size.

Strong theoretical or practical reasons for selecting the slug surfactant concentration do not exist. The concentration must be large enough so that a type III region can form when the salinity is optimal but small enough so that the slug can be easily handled and transported. The latter requirement usually means the slug is single-phase and not excessively viscous and the surfactant does not precipitate.

Perhaps a more stringent lower bound on surfactant concentration is in its relative rate of propagation. The frontal advance loss of D_3 contains surfactant concentration in the denominator. This means the rate of slug propagation, as well as the maximum oil cut calculated from fractional flow theory (Fig. 9-24), decreases as concentration decreases. Because of the worth of the oil, the resulting delay in oil production is a liability to the process even if the ultimate oil recovery were unaffected. This argument suggests the concentration should be as large as possible, and the slug size should be correspondingly small. But extremely small slugs would seem to be sensitive to dispersive mixing in the reservoir.

Once the slug concentration is set, the slug size follows from the value of D_3, as in Sec. 9-11. To satisfy retention, the slug size, based on floodable pore volume, must be somewhat larger than retention. Of course, how much larger is a strong function of the prevailing economics and reservoir characteristics. (For a graphical procedure, see Jones, 1972.)

Maintaining Good Volumetric Sweep

Figure 9-33 attests that the importance of this issue, particularly with respect to the mobility buffer, cannot be overstated.

The mobility control agent in the slug can be polymer or oil as in Fig. 9-13. Whatever the agent, it is of paramount importance that the slug–oil bank front be made viscously stable since small slugs cannot tolerate even a small amount of fingering. Thus we seek a slug less mobile than the oil bank it is to displace. To provide a margin of safety in estimating the oil bank mobility, use the minimum in the total relative mobility curves (see Sec. 3.3) to base the mobility control on. Such curves (Fig. 9-34) show that the minimum can be substantially less than the total relative mobility of either endpoint. Since these curves are subject to hysteresis, it is important that the relative permeability curves be measured in the direction of increasing oil saturation for tertiary floods.

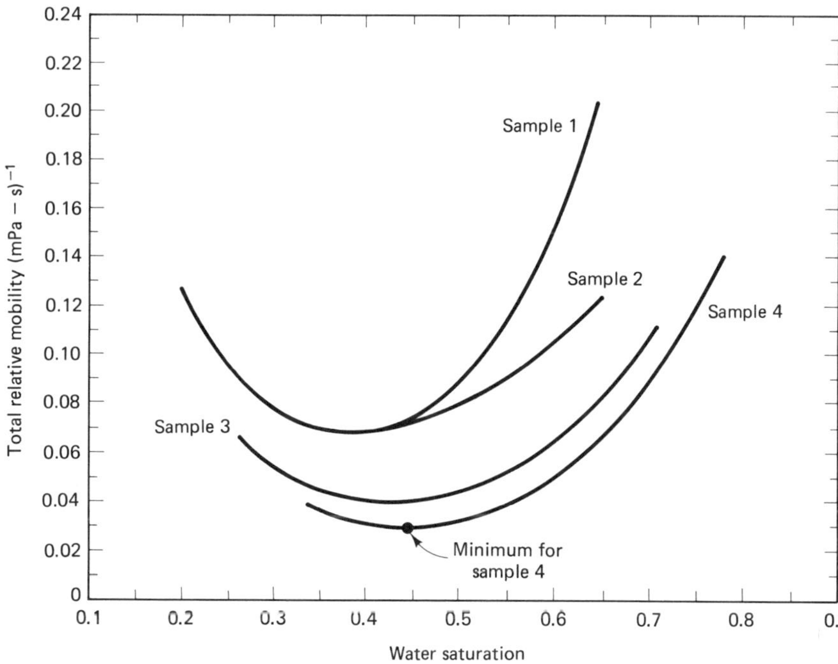

Figure 9-34 Total relative mobilities for samples of the same reservoir (from Gogarty et al., 1970)

Sizing the mobility buffer proceeds like the polymer drive we discussed in Sec. 8-5. Here the spike portion of the buffer must have mobility equal to or less than the slug mobility. Since the latter depends on the degree of oil desaturation, the buffer mobility cannot be designed independently of the slug.

9-14 MAKING A SIMPLIFIED RECOVERY PREDICTION

In the next few paragraphs, we describe a simple procedure to estimate oil recovery and oil rate–time curves for an interfacially active MP process. Since interfacial activity may be lost in innumerable ways, the procedure will be most accurate for processes that clearly satisfy the first design goal in Sec. 9-13. The procedure has two steps: estimating the recovery efficiency of an MP flood and then proportioning this recovery according to injectivity and fractional flow to give an oil rate–time curve. (For further details of the procedure, see Paul et al., 1982.)

Recovery Efficiency

The recovery efficiency E_R of a tertiary $(S_{2I} - S_{2R})$ MP flood is the product of a volumetric sweep efficiency E_V, a displacement efficiency E_D, and a mobility buffer efficiency E_{MB}

$$E_R = E_D E_V E_{MB} \tag{9.14-1}$$

Each quantity must be independently calculated.

Displacement efficiency. The displacement efficiency of an MP flood is the ultimate (time-independent) volume of oil displaced divided by the volume of oil contacted

$$E_D = 1 - \frac{S'_{2r}}{S_{2r}} \tag{9.14-2}$$

where S'_{2r} and S_{2r} are the residual oil saturation to an MP and a waterflood, respectively. S_{2r} must be known, but S'_{2r} can be obtained from a large slug (free from the effects of surfactant retention) laboratory core flood. Low values of S'_{2r} indicate successful attainment of good interfacial activity in the MP slug. If core flood results are not available, S'_{2r} may be estimated from a CDC using a "field" capillary number (Lake and Pope, 1978) based on the median velocity in a confined five-spot pattern.

$$N_{vc} = \frac{0.565 q\sigma}{H_t \sqrt{A_p}} \qquad \text{(dimensionless)} \tag{9.14-3}$$

Here, q is the volumetric injection rate and A_p is the pattern area. For approximate calculation, assume $\sigma = 1\ \mu\text{N/m}$ in Eq. (9.14-3). The CDC chosen to estimate S'_{2r} should be consistent, as much as possible, with conditions of the candidate reservoir.

Volumetric sweep efficiency. Volumetric sweep efficiency E_V is the volume of oil contacted divided by the volume of target oil. E_V is a function of MP slug size t_{Ds}, retention D_3, and heterogeneity based on the Dykstra-Parsons coefficient V_{DP}.

Consider the layered medium in Fig. 9-35 into which is injected an MP slug of size t_{Ds}. If the flow is apportioned by kh, and there is no crossflow, the slug size in layer ℓ is

$$t_{Ds_\ell} = t_{Ds}\left(\frac{k}{\phi}\right)_\ell \left(\frac{\overline{\phi}}{\overline{k}}\right) = t_{Ds} F'_\ell \tag{9.14-4}$$

where Eq. (9.14-4) has introduced the derivative of the flow-capacity–storage-capacity curve (F-C curve) first discussed in Sec. 6-3. Besides invoking a continuous permeability–porosity distribution, using F-C curves implies the layers are arranged as decreasing (k/ϕ). If $t_{Ds_m} > D_3$ in a particular layer m, it will be completely swept; otherwise, the layer's volumetric sweep will be in proportion to t_{Ds_m}/D_3

$$E_{Vm} = \begin{cases} 1, & t_{Ds_m} > D_3 \\ \dfrac{t_{Ds_m}}{D_3}, & t_{Ds_m} \leqq D_3 \end{cases} \tag{9.14-5}$$

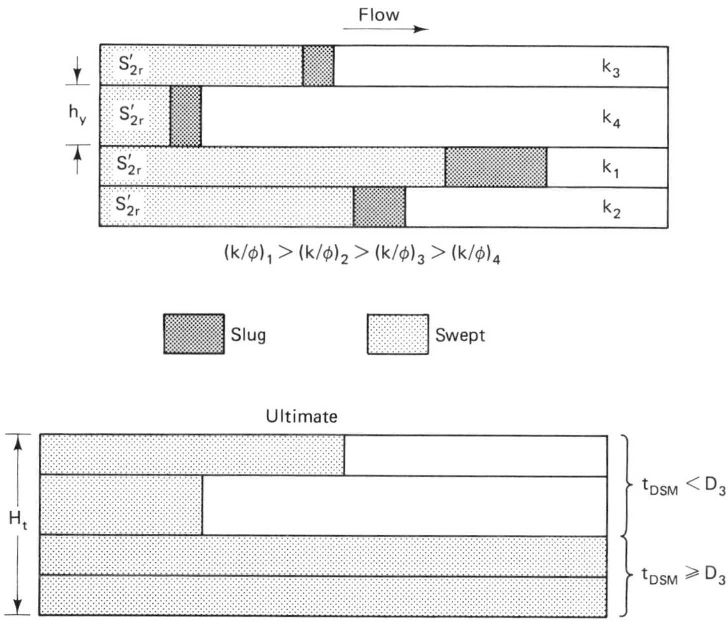

Figure 9-35 Schematic representation of MP slug sweep in a layered medium

Figure 9-35 illustrates this division. E_{Vm} summed over $n = 1, \ldots, m, \ldots, N_L$ layer after being weighted by $(\phi h)_m$ gives

$$E_V = C_m + \frac{t_{Ds}}{D_3}(1 - F_m) \tag{9.14-6}$$

To calculate E_V with t_{Ds}, D_3, and F-C curve known, begin by finding the layer number m where $t_{Ds_m} = D_3$. This determines the coordinates $(F, C)_m$ in Eq. (9.14-6) to determine E_V. Equations (6.3-11) and (6.3-12) establish a relation between the F-C curves, the heterogeneity factor H_K, and V_{DP} for a lognormal continuous permeability distribution; thus E_V in Eq. (9.14-6) can be related directly to V_{DP}. Figure 9-36 shows this relationship. V_{DP} may be estimated from geologic study, matching the prior waterflood, or core data (see Table 6-1). The D_3 is from Eq. (9.10-6).

Mobility buffer efficiency. The mobility buffer efficiency E_{MB} is a function of E_V and V_{DP}

$$E_{MB} = (1 - E_{MBe})\left[1 - \exp\left(\frac{-0.4 t_{DMB}}{E_V^{1.2}}\right)\right] + E_{MBe} \tag{9.14-7a}$$

and

$$E_{MBe} = 0.71 - 0.6 V_{DP} \tag{9.14-7b}$$

where E_{MBe} is the mobility buffer efficiency extrapolated to $t_{DMB} = 0$, and t_{DMB} is the mobility buffer volume, fraction V_{pf}. Equation (9.12-7) was obtained by numerical simulation.

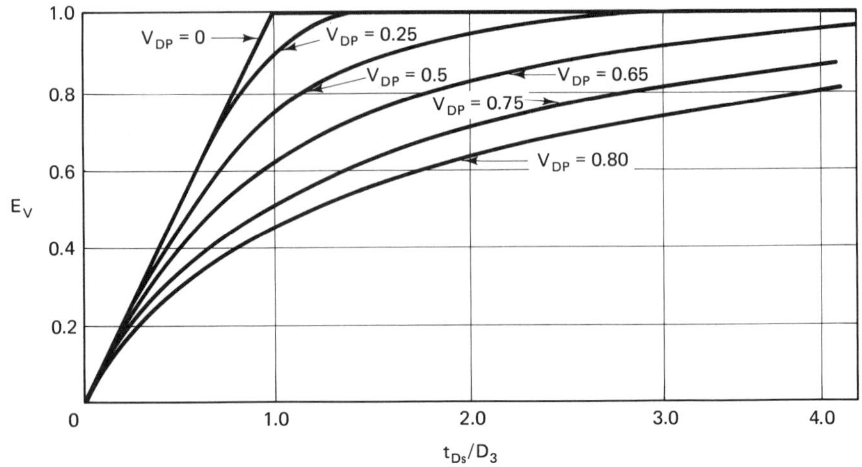

Figure 9-36 Effect of slug size–retention ratio on vertical sweep efficiency (from Paul et al., 1982)

The recovery efficiency E_R now follows from Eq. (9.14-1), which may be checked for reasonableness against Fig. 9-33.

Calculation of an Oil-Rate–Time Plot

The production function (oil rate q_2 versus time) is based on E_R and the following procedure. We assume the dimensionless production function is triangular with oil production beginning when the oil bank arrives. From here, q_2 increases linearly to a peak (maximum) oil cut when the surfactant breaks through and then decreases linearly to the sweep-out time. The triangular shape is imposed by the reservoir heterogeneity.

The first step is to calculate the dimensionless oil bank and surfactant breakthrough times for a homogeneous flood

$$t_{D_B} = \left(\frac{S_{2_B} - S_{2I}}{f_{2_B} - f_{2I}}\right) t_{Ds} \tag{9.14-8a}$$

$$t_{D_s} = 1 + D_3 - S'_{2r} \tag{9.14-8b}$$

where t_{D_B} is the dimensionless oil bank arrival time, and t_{D_s} is the surfactant arrival time. S_{2B} and f_{2B} may be estimated from the simplified fractional flow theory (see Sec. 9-10) or directly from laboratory experiments.

The second step is to correct these values for the heterogeneity of the candidate reservoir using the heterogeneity factor H_K defined in Eq. (6.3-11).

The corrected breakthrough times are now

$$\hat{t}_{D_B} = \frac{t_{D_B}}{H_K} \tag{9.14-9a}$$

$$\hat{t}_{Ds} = \frac{t_{Ds}}{H_K} \tag{9.14-9b}$$

and the peak oil cut f_{2pk} is

$$\hat{f}_{2pk} = \frac{\left(H_K - H_K\left(\frac{t_{D_B}}{t_{D_s}}\right)^{1/2}\right)}{(H_K - 1)} f_{2B} \tag{9.14-10}$$

The symbol ^ represents a quantity in a layered medium.

The final step is to convert the dimensionless production function to oil rate q_2 versus time t. This follows from

$$q_2 = q\hat{f}_2 \tag{9.14-11a}$$

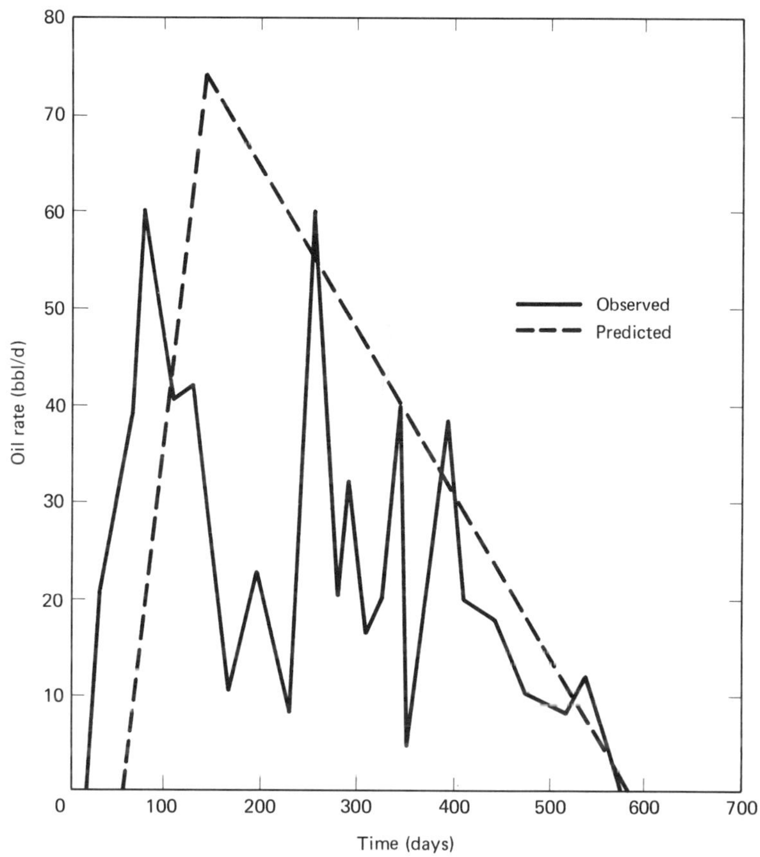

Figure 9-37 Comparison between predicted and observed oil-rate–time responses for the Sloss micellar–polymer pilot (from Paul et al., 1982)

$$t = \frac{V_{pf}\hat{t}_D}{q} \tag{9.14-11b}$$

Here $\hat{f}_2$ and $\hat{t}_D$ are any points on the triangular oil recovery curve that begins at $(\hat{t}_{DB}, 0)$, peaks at $(\hat{t}_{Ds}, \hat{f}_{2pk})$, and ends at $(\hat{t}_{Dsw}, 0)$. $\hat{t}_{Dsw}$, the dimensionless time at complete oil sweepout, is selected to make the area under the $\hat{f}_2$-$\hat{t}_D$ curve equal to E_R,

$$\hat{t}_{Dsw} = \hat{t}_{D_B} + \frac{2E_R S_2}{\hat{f}_{2pk}} \tag{9.14-12}$$

Figure 9-37 compares the results of this procedure with the Sloss MP pilot.

9-15 CONCLUDING REMARKS

In terms of the number of design decisions required, micellar-polymer flooding is the most complicated enhanced oil recovery process. This complexity, along with reservoir heterogeneity and the need for a rather large capital investment, make micellar-polymer flooding a high-risk process. Consequently, recent years have seen a decline in interest in the process. The potential for the process is immense, however, even slightly exceeding that of thermal methods, at least in the United States. Moreover, both polymer and *MP* flooding seem uniquely suited for light-oil reservoirs in isolated areas of the world.

Reservoirs amenable to micellar-polymer flooding contain light- to medium-weight oils with moderate to high permeability. Since injectivity is essential in this process as in polymer flooding, we seek reservoirs with depth sufficient to tolerate high injection pressures but not so deep as to promote thermal degradation. Finally, the process is sensitive to high brine salinities, although this can be dealt with somewhat by suitable surfactant/polymer selection and design.

The important topics in this chapter deal with the association of interfacial activity with brine salinity and hardness through phase behavior, the importance of surfactant retention, and the need for good mobility control. In a sense, the design criteria given in Sec. 9-13 apply to all EOR processes, but it is only in micellar-polymer flooding that all criteria seem to apply with equal severity. Finally, the screening estimation of recovery in Sec. 9-14 is a useful yet simple tool for assessing the suitability of a reservoir and for estimating the risk associated with the process.

EXERCISES

9A. *The Units of MP Flooding*. A particular petroleum sulfonate surfactant has an average molecular weight of 400 kg/kg-mole, a density of 1.1 g/cm^3, and a monosulfonate-to-disulfonate mole ratio of 4. Express the overall surfactant concentration of a 5 volume percent aqueous solution in g/cm^3, kg-moles/cm^3, meq/cm^3, mole fraction, and mass fraction.

9B. *Surfactant Equilibria and Aggregation*. Relatively simple models can reveal much about surfactant equilibria. The surfactant is a monosulfonate in this problem.

(a) The aggregation of surfactant monomers into micelles in a NaCl brine may be represented by the following reaction:

$$N_A(\text{Na}^+ + \text{RSO}_3^-) \rightleftarrows (\text{RSO}_3\text{Na})_{N_A} \qquad (9B\text{-}1)$$

where N_A is the aggregation number. Using the definition for total surfactant (monomer + micelles), derive an expression between total and monomer sulfonate concentrations. If the equilibrium constant for Eq. (9B-1) is 10^{15} and $N_A = 10$, estimate the critical micelle concentration. The total sodium concentration is 10,000 g/m^3.

(b) Consider a more complicated situation where 0.3175 kg-moles/m^3 monosulfonate surfactant solution is added to a NaCl brine. In a NaCl brine solution, five species can form: surfactant monomer (RSO_3^-), surfactant micelles [$(\text{RSO}_3\text{Na})_{N_A}$], free sodium-surfactant (RSO_3Na), precipitated sodium-surfactant ($\text{RSO}_3\text{Na}\downarrow$), and free sodium ($\text{Na}^+$). Calculate the concentration of each species when the overall sodium concentration is 100 g/m^3. Use the data in part (a) for the monomer–micelle reaction, and take the equilibrium constant for the sodium-sulfonate formation to 3×10^6 and the solubility product for the precipitate to be 10^{-8}.

(c) Repeat the calculation of part (b) if the overall sodium concentration is 100,000 g/m^3. What can you conclude about the effect of high salinities on surfactant precipitation?

9C. *Phase Ratios for Hand's Rule*. In Sec. 4.4, we saw that flash calculations for vapor–liquid equilibria required using flash vaporization ratios or *K*-values. The analogous quantities for Hand's rule are *phase ratios* defined as

$$R^j_{ik} = \frac{C_{ij}}{C_{kj}} \qquad (9C\text{-}1)$$

for components i and k. Nearly all the flash calculation can be formulated in terms of the phase ratios. Assume a type II(−) phase behavior (j = 2 or 3) in the following:

(a) Show that the Hand equations for the binodal curve (Eq. 4.4-23) and the component distribution (Eq. 4.4-24) can be written as

$$R^j_{32} = A_H(R^j_{31})^B, \qquad j = 2 \text{ or } 3 \qquad (9C\text{-}2)$$

$$R^2_{32} = E_H(R^3_{31})^F \qquad (9C\text{-}3)$$

(b) We can interchange the roles of phase concentrations and phase ratios. Show that the consistency relation $\Sigma^3_{i=1} C_{ij} = 1$ reduces to

$$C_{ij} = \left(\sum_{k=1}^{3} R^j_{ki}\right)^{-1}, \qquad j = 2 \text{ or } 3 \qquad (9C\text{-}4)$$

where $R^j_{ii} = 1$.

There are 18 phase ratios in two-phase systems. But only 4 of these are independent since

$$R^j_{ik} = (R^j_{ki})^{-1} \quad \text{and} \quad R^j_{ik} = \frac{R^j_{im}}{R^j_{km}} \qquad (9C\text{-}5)$$

As is consistent with the phase rule, specifying any one of these will determine the others and all phase concentrations through Eq. (9C-4). Solve for the phase concentrations when $R_{31}^{2} = 5$. Take $A_H = 0.5$, $B_H = -1.5$, $E_H = 0.137$, and $F_H = 0.65$. Note that the phase ratios for the microemulsion phase are the same as solubilization parameters.

9D. *Using the Hand Equations*

(a) For the Hand parameters $A_H = 2$, $B_H = -0.5$, $E_H = 600$, and $F_H = 2.3$, plot the binodal curve and at least two tie lines on triangular coordinates. Flash calculations in two-phase regions require an additional constraint over those in Exercise 9C. The constraint here is the equation of a tie line

$$S_1 = \frac{C_i - C_{i3}}{C_{i1} - C_{i3}}, \qquad i = 1, 2, \text{ or } 3 \tag{9D-1}$$

for type II(+) systems (where $j = 1$ replaced $j = 2$ in Eqs. [9C-2] and [9C-3]). Any two of these may be used for a flash calculation, for example,

$$\frac{C_1 - C_{13}}{C_{11} - C_{23}} - \frac{C_2 - C_{23}}{C_{21} - C_{23}} = f \tag{9D-2}$$

The flash consists of picking the correct phase ratio (see Exercise 9C) so that $f = 0$ with the C_i known.

(b) Calculate the compositions and amounts of each phase present if the overall composition C_i is (0.45, 0.45, 0.1).

9E. *Two-Phase Flash Calculation (Plait Point in Corner).* In a type II(+) system with the plait point in the brine corner, we have $C_{11} = 1$, and $C_{21} = C_{31} = 0$. The phase distribution Eq. (9.7-9) now becomes superfluous, as does the binodal Eq. (9.7-5) for the aqueous phase. The entire Hand representation collapses to

$$\frac{C_{33}}{C_{23}} = A_H\left(\frac{C_{33}}{C_{13}}\right)^{BH} \tag{9E-1}$$

(a) Show that the tie line equation for this special case reduces to

$$C_{23} = C_{33}\left(\frac{C_2}{C_3}\right) \tag{9E-2}$$

$$C_{13} = 1 - C_{33}\left[\frac{(1 - C_1)}{C_3}\right] \tag{9E-3}$$

which express the microemulsion phase concentrations as ratios of each other.

(b) Show that Eqs. (9E-1) through (9E-3) may be used to solve explicitly for the surfactant concentration in the microemulsion phase as

$$\frac{1}{C_{33}} = \frac{1 - C_1}{C_3} + \left(\frac{C_3}{A_H C_2}\right)^{-1/B_H} \tag{9E-4}$$

(c) For an overall composition of $C_i = (0.45, 0.45, 0.1)$, solve Eq. (9E-4) for the phase composition and the saturation of the aqueous phase. Take $A_H = 2$ and $B_H = -0.5$.

(d) Compare the results of part (c) with the results of part (b) of Exercise 9D. What do you conclude about approximating this phase behavior with the plait point in one of the corners of the ternary?

9F. *Equilibrium Calculations with Simplified Phase Behavior*. Use the simplified Hand representations with $B_H = -1$ and $F_H = 1$ in the following. Further, take the left and right oil coordinates of the plait point to be 0.05 and 0.95, respectively; the low-, optimal-, and high-salinity binodal curve heights to be 0.2, 0.1, and 0.2, respectively; and the lower- and upper-effective salinity limits to be 0.06% and 1.4% NaCl. The optimal salinity is at the midpoint between these two. Make all the calculations at a salinity of 0.08% NaCl where the phase environment is type III.

(a) Calculate the Hand parameter A_H and the coordinates of the two plait points and of the invariant point.

(b) Plot the binodal curve and the three-phase region on a ternary diagram.

(c) Calculate the phase concentrations and saturations at an overall concentration of $C_i = (0.65, 0.3, 0.05)$.

(d) Repeat part (c) at an overall concentration of $C_i = (0.44, 0.44, 0.12)$.

Plot both points on the diagram of part (b).

9G. *Phase Behavior and IFT*. Fig. 9G shows the bottom half of six surfactant–brine–oil mixtures. These diagrams are on rectangular coordinates having a greatly expanded vertical scale. C_{Se} is the salinity in wt. % NaCl. In the following, the surfactant concentration is 0.05 volume fraction:

(a) Calculate and plot volume fraction diagrams at brine–oil ratios of 0.2, 1.0, and 5.

(b) At a brine–oil ratio of 1, calculate and tabulate the solubilization parameters.

(c) Use the correlation in Fig. 9-9 to convert the solubilization parameters to interfacial tensions. Plot these solubilization parameters against salinity, and estimate the optimal salinity.

(d) Plot the IFTs in part (c) against salinity on semilog paper. Estimate the optimal salinity based on IFT and the optimal IFT.

(e) Compare the optimal salinities in parts (c) and (d) to the midpoint salinity. The latter is the salinity halfway between C_{Seu} and C_{Sel}.

9H. *Fractional Flow Construction for Type II(−) Systems*. Fig. 9H shows water fractional flux curves for a type II(−) MP system for which all tie lines extend to the common point $C_i^o = (0.1, 1.1, 0)$.

(a) Calculate and plot an overall water concentration (C_1) profile at oil bank breakthrough and an effluent water flux (F_1) for the following cases:

Case	Injected composition (J)		Initial composition (I)	
	C_2	C_3	C_1	C_2
1	0	0.10	0.66	0.34
2	0.97	0.03	0.66	0.34
3	0	0.10	0.20	0.80

For all cases, the displacement satisfies the fractional flow assumptions, the surfactant is not retained by the permeable medium, and the surfactant injection is con-

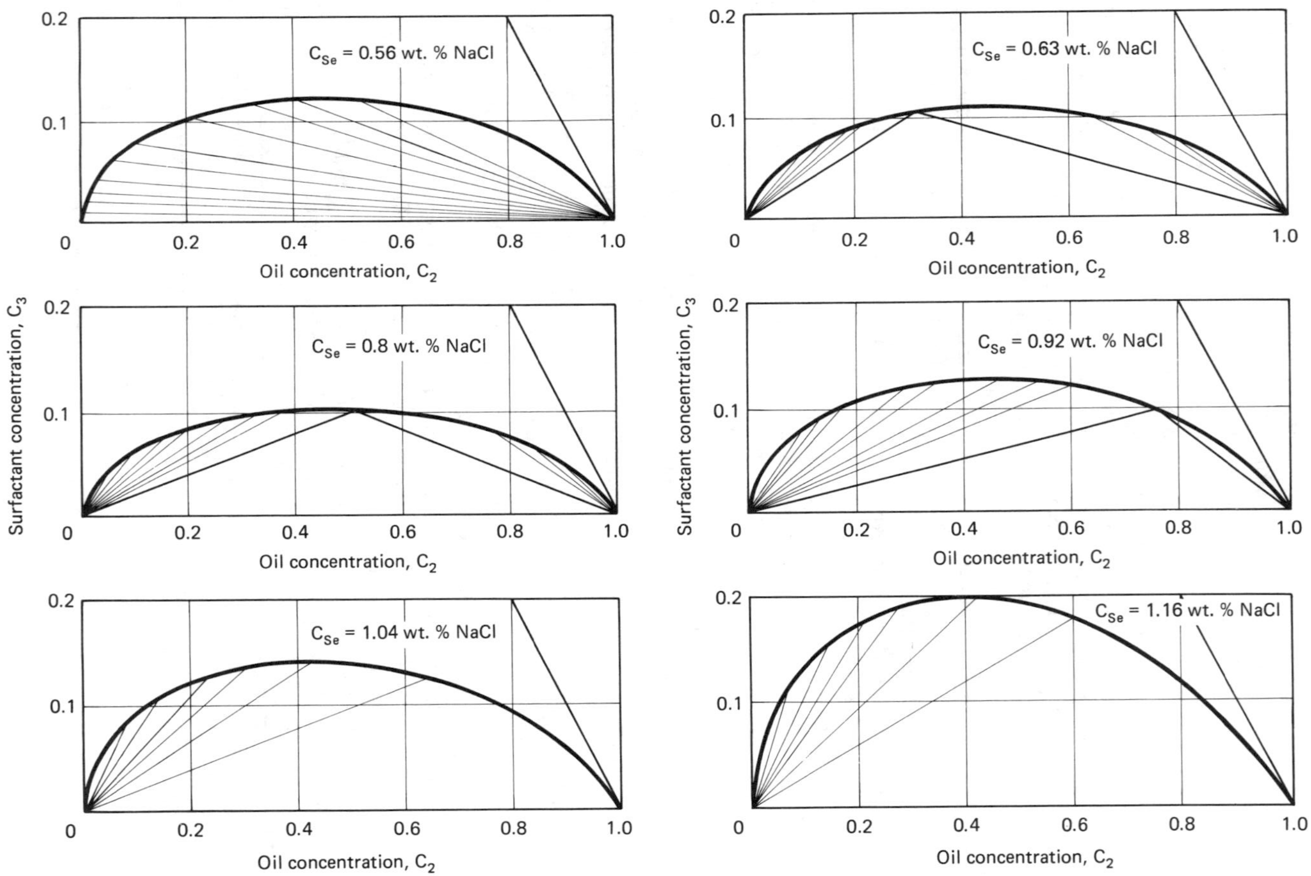

Figure 9G Ternary diagrams at various salinities (from Engleson, 1981)

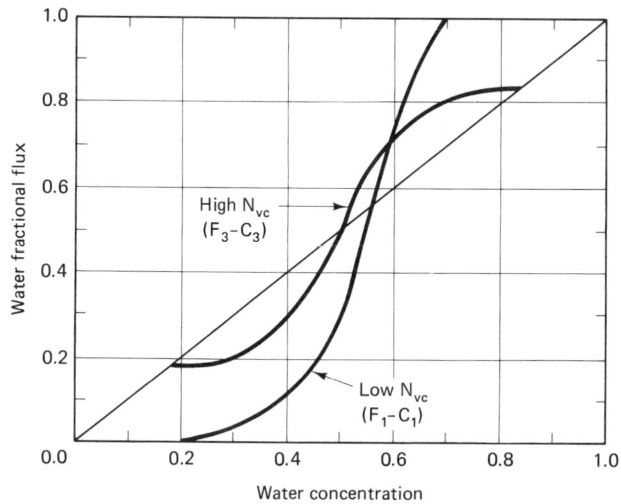

Figure 9H Water fractional flux for Exercise 9H

tinuous. All injected compositions lie on extensions of the tie lines whose fractional flux curves are shown in the figure.

(b) On the water concentration profiles of part (a), sketch (no calculation necessary) the microemulsion phase saturation S_3 profile.

(c) On the water effluent histories of part (a), sketch the overall surfactant C_3 effluent history.

9I. *Two-Phase II(−) Fractional Flow*. Use the data in Figs. 9G and 9-9 in the following. Take the oil-free injected slug concentration to be 0.05 volume fraction surfactant, and the salinity to be constant at 0.56% NaCl. The surfactant is in an ideal mixture. The low N_{vc} relative permeability curves are given by

$$S_{2r} = 0.3, \qquad k_{r2}^0 = 0.8, \qquad n_2 = 1.5$$

$$S_{3r} = 0.2, \qquad k_{r3}^0 = 0.1, \qquad n_3 = 5$$

Phase 3 is water when N_{vc}. The displacement occurs at a superficial velocity of 10 μm/s. The microemulsion, oil, and water viscosities are 2, 5, and 1 mPa-s. The medium is horizontal. Use Fig. 3-19 as the capillary desaturation curve.

(a) Estimate and plot the relative permeability curves corresponding to the tie lines the initial and injection conditions are on. Use the high N_{vc} relative permeabilities of Eqs. (9.9-1) and (9.9-2).

(b) Estimate and plot the microemulsion fractional flow curves along the two tie lines in part (a).

(c) Plot the time–distance diagram and a composition profile at oil bank breakthrough for this displacement if the injection is continuous surfactant. Use the simplified fractional flow analysis of Eqs. (9.10-5) through (9.10-9). Take $D_3 = 0.1$.

9J. *Slugs and Simplified Fractional Flow*. Use the simplified fractional flow of Eqs. (9.10-5) through (9.10-9) in the following. The displacement is a constant II(−) phase environment consisting of an oil-free surfactant slug followed by a polymer drive. The water-, oil-, and microemulsion-phase viscosities are 1, 5, and 10 mPa s, respectively, and the relative permeability data at low and high N_{vc} are

	Oleic phase			Microemulsion phase		
	S_{2r}	k_{r2}^0	n_2	S_{3r}	k_{r3}^0	n_3
Low N_{vc}	0.3	0.8	1.5	0.2	0.1	5.0
High N_{vc}	0.05	0.9	1.2	0.1	0.6	2.5

(a) Estimate the polymer solution viscosity in the mobility buffer if the mobility ratio between the slug and drive is to be 0.8. The polymer has no permeability reduction effect.
(b) Calculate and plot the three aqueous-phase fractional flow curves (water–oil, microemulsion–oil, polymer-solution–oil) based on the data in part (a) and the polymer solution viscosity.
(c) Estimate the minimum slug size required to entirely sweep the one-dimensional medium with slug. Take $D_3 = 0.2$ and $D_4 = 0.1$. There is no polymer in the slug.
(d) Calculate and plot the time–distance diagram if the slug size is one half that estimated in part (c).
(e) Calculate and plot saturation profiles at $t_D = 0.3$ and 0.8 for the conditions of part (d).

9K. *Fractional Flow with Oil-Soluble Slug.* For this exercise, take the displacement to be constant type II(+) phase environment ($j = 1$ or 3) with the plait point in the brine corner. The surfactant is now dissolved in a predominantly oleic phase.
(a) Show that the surfactant-specific velocity is analogous to Eq. (9.10-5)

$$v_{\Delta C_3} = \frac{1 - f_1^s}{1 - S_1 + D_3} \tag{9K-1}$$

and the oil bank saturation is given by the solution to

$$v_{\Delta C_2} = \frac{C_{2J} f_1^s - f_{1B}^s - C_{3J}}{C_{2J} S_1 - S_{1B} - C_{3J}} = \frac{1 - f_1^s}{1 - S_1 + D_3} \tag{9K-2}$$

(b) Illustrate the graphical solution of Eq. (9K-2) on an aqueous-phase fractional flow plot. What is the effect of injected oil concentration on the oil bank saturation? Justify this observation on physical grounds.
(c) Figure 9K shows high- and low-N_{vc} fractional flow curves for a particular displacement. Based on these curves, calculate and plot an oleic-phase saturation profile at $t_D = 0.5$. Take $D_3 = 0.1$ and the surfactant injection to be continuous.

9L. *Preflush Size Estimation.* The composition of an initial reservoir brine and a possible preflush solution are as follows:

Species	Reservoir brine (I), meq/cm^3	Preflush (J), meq/cm^3
Na^+	0.02	0.01
Ca^{2+}	0.06	0.005
Cl^-	0.08	0.015

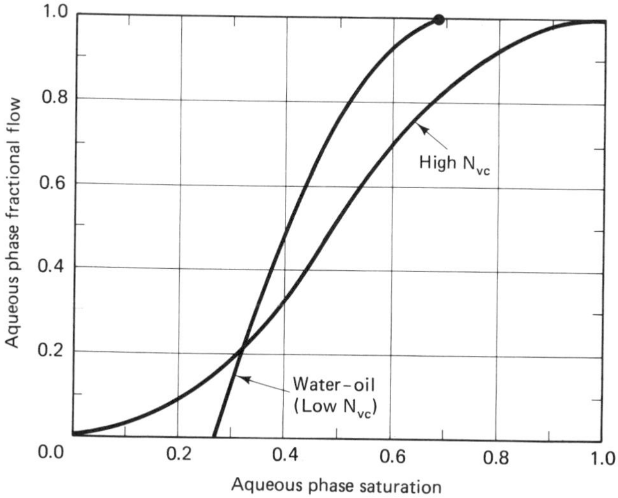

Figure 9K Aqueous-phase fractional flow curves for Exercise 9K

The cation exchange capacity of the reservoir is $Z_v = 0.05$ meq/cm^3 of pore volume. The cation exchange satisfies Eq. (9.11-3) with $K_N = 0.1$. Assume single-phase flow of an ideal solution that contains only the species explicitly stated above.

(a) Sketch this displacement in composition space as in Fig. 9-25a.

(b) Estimate the pore volumes of fluid J required to reduce the effluent calcium concentration to the injected value. What percentage of the clays are in the calcium form at this point?

(c) Calculate and plot the time–distance diagram for this displacement.

(d) State whether you think this would be an effective preflush for an MP flood.

9M. *Importance of Mobility Control in MP Floods*. In the absence of other data, high N_{vc} relative permeabilities for a type II(−) system may be approximated by straight lines through the points $(S'_{3r}, 0)$ and $(1 - S_{2r}, k^0_{r3})'$ for the aqueous phase and through $(S_{3r}, k^0_{r2})'$ and $(1 - S'_{2r}, 0)$ for the oleic phase.

(a) Plot two high-N_{vc} fractional flow curves for aqueous-phase ($j = 3$) viscosities of 5 and 50 mPa-s. Take $\mu_2 = 5$, $\mu_3 = 0.8$ mPa-s, $S'_{3r} = 0.15$, $S'_{2r} = 0.05$, $(k^0_{r3})' = 0.8$, and $(k^0_{r2})' = 0.6$. The medium has no dip.

(b) Using the El Dorado relative permeabilities of Fig. 8L, illustrate the effects of good mobility control on an MP flood by calculating oil saturation profiles for the two cases in part (a) at $t_D = 0.3$. The frontal advance lag $D_3 = 0.16$. The injected aqueous surfactant is continuous.

9N. *Performance Prediction*. Use the following information to perform a screening estimation of oil recovery on an MP project. The water–oil relative permeability data in Fig. 9K is appropriate.

(a) Estimate the swept zone oil displacement efficiency E_D if the injection rate per pattern is 65 m^3/day. The pattern area is 8.1 hm^2, and the formation thickness is 2 m. Take the IFT to be 1 μN/m, and use the CDC in Fig. 3-19 for the nonwetting phase.

(b) Calculate the volumetric sweep efficiency E_V. Take the Dykstra-Parsons coefficient to be 0.5, the slug size to be 0.16, and $D_3 = 0.12$.

(c) Estimate the recovery efficiency based on the above if the mobility buffer size is 0.8 PV.

(d) Calculate and plot the oil production rate versus time.

10

Other Chemical Methods

Surfactants have a larger role in enhanced oil recovery than lowering interfacial tension. They can be used to change wettability, promote emulsification and entrainment, lower bulk-phase viscosity, and stabilize dispersions. In this chapter, we discuss two chemical methods that, like micellar-polymer flooding, rest on effective use of surfactants.

The first method, foam flooding, uses surfactants to reduce gas-phase mobility through formation of stable gas–liquid foams. Interfacial tension lowering is not a significant mechanism. In alkaline flooding, the second method, IFT lowering is important, but the surfactant is generated in situ unlike MP flooding. Alkaline flooding closely parallels MP flooding in the importance of phase behavior. Both foam and alkaline flooding require an understanding of the nature and performance of surfactants (see Sec. 9-2).

10-1 FOAM FLOODING

Gas–liquid foams offer an alternative to polymers for providing mobility control in micellar floods. In addition, and perhaps more importantly, foams can be used as mobility control agents in miscible floods and well treatment and have been both proposed and field tested as mobility control agents in thermal floods. Because most of the properties of foam stem from adding surfactants to an aqueous phase, and the background of these agents was given in Sec. 9-2, we devote a section in this chapter to discussing foams for all EOR applications.

Foams are dispersions of gas bubbles in liquids. Such dispersions are normally quite unstable and will quickly break in less than a second. But if surfactants are added to the liquid, stability is greatly improved so that some foams can persist indefinitely. Surfactants used as foaming agents have many of the attributes described for micellar-polymer flooding. For aqueous foams, it is usually desirable for the surfactant to have a somewhat smaller molecular weight to enhance water solubility. Fried (1960) and Patton et al. (1981) give extensive lists of surfactants that have mobility control potential. Of course, we should expect that such surfactants also possess many of the undesirable features of the micellar-polymer surfactants, particularly with respect to sensitivity to highly saline brine, temperature, oil type, and retention.

10-2 FOAM STABILITY

The stability of a foam may be understood by viewing the liquid film separating two gas bubbles in cross section, as in Fig. 10-1(b). The polar head groups of the surfactant are oriented into the interior of the film, and the nonpolar tails toward the bulk gas phase. Except, perhaps, in the case of dense, hydrocarbonlike gases, the tails are not actually in the exterior fluid as shown, but most surfactant molecules are ori-

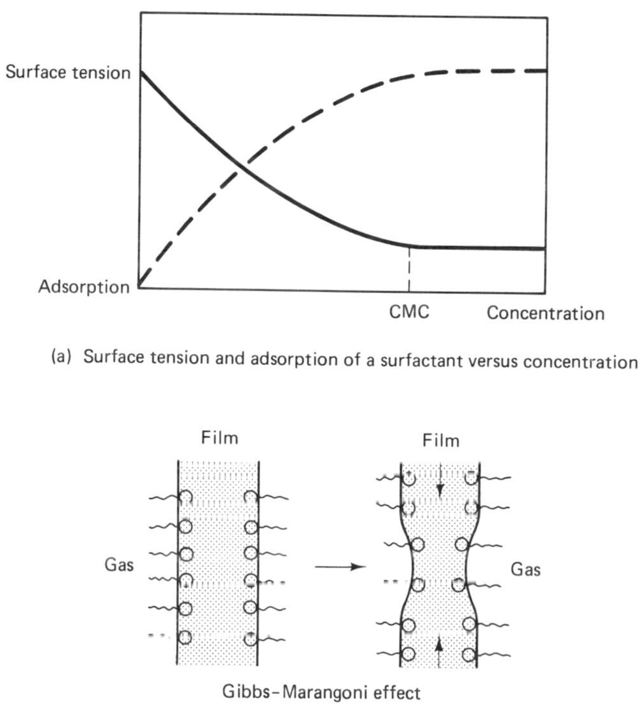

Figure 10-1 The mechanism of film stability (adapted from Overbeek, 1972)

ented as shown just inside the film boundaries. Figure 10-1(a) shows that the gas–liquid surface tension is a decreasing function of surface adsorption, which is defined as the difference between surface and bulk concentration. Suppose some external force causes the film to thin as shown in Fig. 10-1(b). The film surface area increases locally, causing the surface surfactant concentration to decrease and the surface tension to increase. The increase promotes a surface tension difference along the film boundaries and causes the film to regain its original configuration. This restoration is the Gibbs-Marangoni effect (Overbeek, 1972). Clearly, the surface tension at the gas–liquid interfaces plays an important role in film stability. Very low surface tensions would not be favorable; fortunately, gas–liquid surface tensions are rarely lower than 20 mN/m even with the best foaming surfactants. In the absence of external forces, the film is in a state of equilibrium caused by a balance between the repulsion forces of the electrical layer just inside the film boundary and the attractive van der Wall forces between the molecules in the film (Fig. 10-2).

The above can be stated in terms of the pressure in the film. In this and all following similar arguments, we take the pressure in the gas phase to be constant owing to its relatively low density (if the foam is static) or low viscosity (if the foam is in motion). Since capillary pressure is inversely proportional to interfacial curvature, the pressure in the thinned portion of the film is lower than in the adjacent flat portion. This causes a pressure difference between the thinned and unthinned portions within the film, flow, and healing.

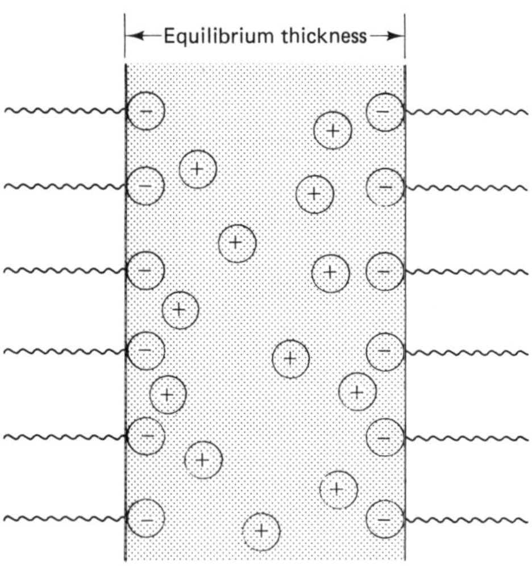

Figure 10-2 Electrical double layer in film leads to a repulsion between surfaces (from Overbeek, 1972)

Spontaneous Foam Collapse

If the film becomes substantially smaller than the equilibrium thickness the film will collapse. Thinning can spontaneously come about by several mechanisms.

Usually, gas bubbles are present in the foam with a nonuniform size distribution. Since capillary forces are inversely proportional to interface curvature, small bubbles will have a higher pressure than large bubbles. This pressure difference causes a chemical potential difference, which in turn, causes gas to diffuse through the liquid from the small to the large bubbles. The small bubbles will eventually disappear, and the films surrounding the large bubbles will thin to the point that they eventually collapse. This process should be enhanced by a wide range of bubble sizes and by large solubility and diffusivity of the gas in the liquid phase. Therefore, foams made up of gases that have a low water solubility, such as nitrogen, would be more stable than foams made from a gas that has a larger solubility, such as carbon dioxide.

A second factor that promotes film thinning is drainage of the film liquid because of external forces. Early in the drainage, the drainage is caused mainly by gravity, where films are relatively thick and film boundaries uniformly curved. In later stages, the formerly spherical gas bubbles take on a polyhedral shape, and large pressure differences occur between the film boundary curvatures at the sides of the bubbles and at the junctions of the films. These junctions are called *Plateau borders,* and the large curvatures at the border means the junctions possess a locally low pressure. The pressure difference between the film side and the junctions further accelerates film collapse.

These factors will ultimately cause the foam to collapse spontaneously. Patton et al. (1981) report on the rate of spontaneous collapse of a large number of foams as a function of surfactant type, temperature, and pH. The half-life of the foam heights reported in their static tests range from 1 to about 45 minutes. They report that anionic surfactants have greater stability than nonionics, and that the stability of sulfonate foams is greatly affected by water hardness. Foams were generally more unstable at high temperatures, and many could be stabilized only by adding a cosurfactant.

Induced Collapse

Besides these spontaneous mechanisms, several external effects will also cause the gas–liquid film to collapse.

Local heating will cause a local decrease in the surface adsorption, which will cause the film to break in a manner that is the reverse of the stabilizing influence of the Gibbs-Marangoni effect in Fig. 10-1.

Adding certain surface-active agents (foam breakers) to the liquid will replace the surface molecules in Fig. 10-2 with molecules that do not contribute enough to the repulsive forces to keep the film stable. Under certain conditions, oil may perform this function and a changing electrolyte balance may be a factor.

If an aqueous foam film is touched with a hydrophobic surface (Fig. 10-3a), the resulting contact angle and curvature causes a pressure gradient away from the contact point. When the surface is withdrawn, this gradient causes the film to rapidly retract, collapsing the film. The same experiment, performed with a hydrophilic surface, causes the film rupture to heal when the surface is withdrawn (Fig. 10-3b). Thus aqueous foams would likely be difficult to form in oil-wet permeable media. For these applications, using oleic foams would be more desirable.

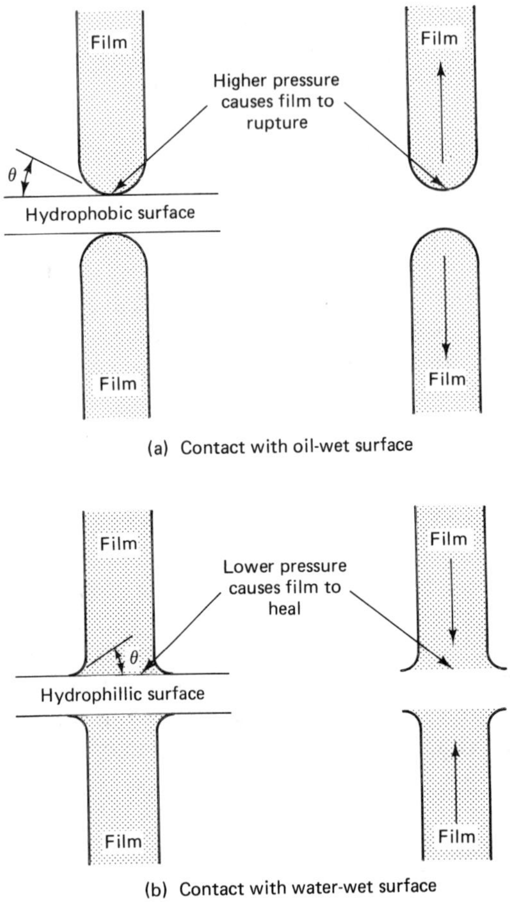

Figure 10-3 Influence of solid surface on film stability

10-3 FOAM MEASURES

Physically, foams are characterized by three measures.

1. *Quality,* the volume of gas in a foam expressed as a percent or fraction of the total foam volume. The quality can vary with both temperature and pressure because the gas volume can change, and gas dissolved in the liquid phase can come out of solution. Foam qualities can be quite high, approaching 97% in many cases. A foam with quality greater than 90% is a dry foam. This definition of quality is consistent with that given in Sec. 3.2 for two-phase equilibria.
2. The average bubble size, called the foam *texture*. There is a large range possible for the texture ranging from the collodial size (0.01–0.1 μm) up to that of a macroemulsion. Figure 10-4 shows typical bubble size distributions. The tex

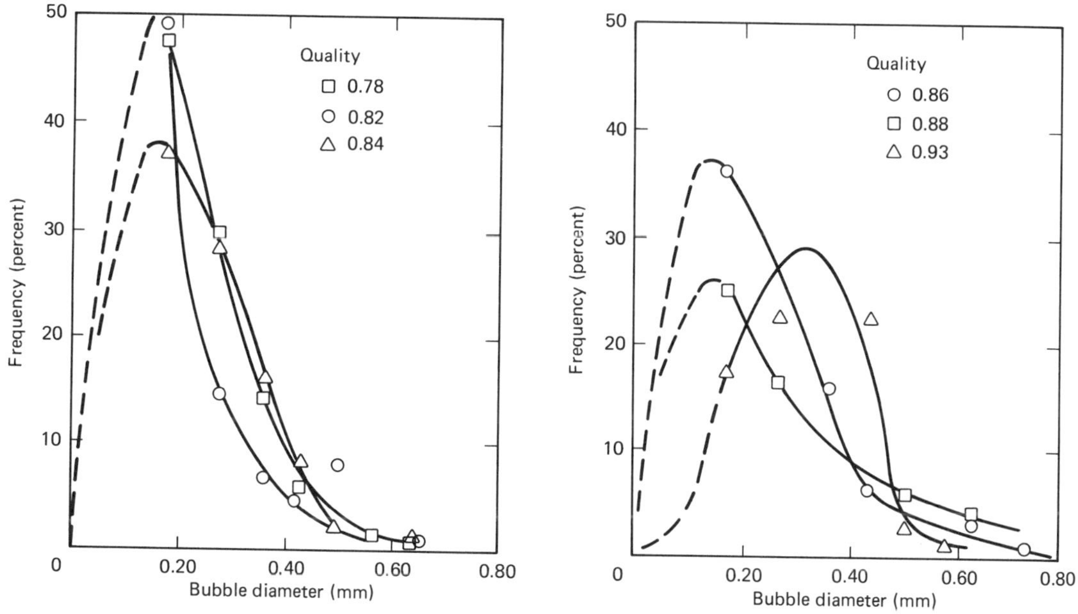

Figure 10-4 Bubble size frequency distributions (from David and Marsden, 1969)

ture determines how the foam will flow through a permeable medium. If the average bubble size is much smaller than the pore diameter, the foam flows as dispersed bubbles in the pore channels. If the average bubble size is larger than the pore diameter, the foam flows as progression of films that separate individual gas bubbles. Considering typical foam textures and pore sizes, the latter condition is more nearly realized, particularly for high-quality foams. Compare Figs. 3-7 and 10-4.

3. The range of bubble sizes. As we mentioned, foams with a large distribution range are more likely to be unstable.

No doubt you have noticed the parallels between these foam measures and local permeable media properties. Foam quality is the analogue to porosity, texture to the mean pore size, and the bubble size distribution to the pore size distribution. Many of the properties of foam flow in permeable media can be explained by comparing the relative magnitudes of the analogous quantities.

10-4 MOBILITY REDUCTION

Foams flowing in permeable media can drastically reduce the mobility of a gas phase. The sample data in Fig. 10-5 show this effect. This figure shows the steady-state mobility of foams of differing quality in Berea cores at three different perme-

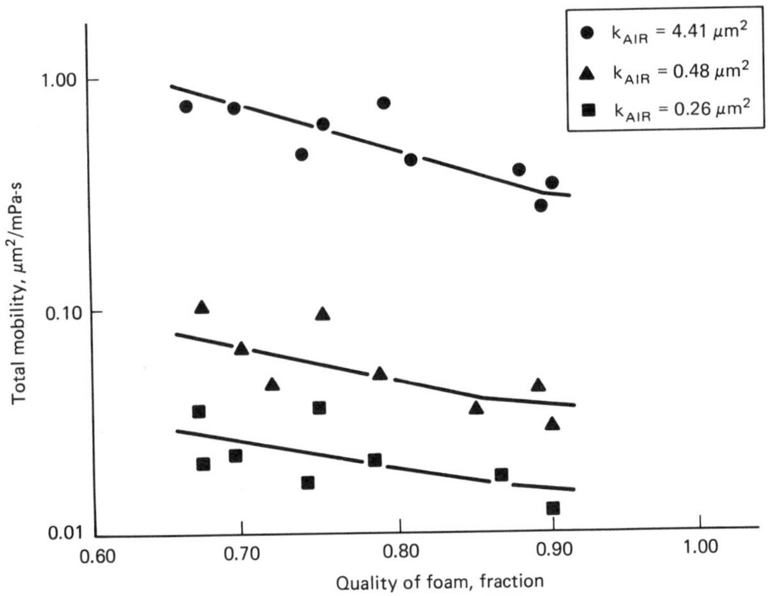

Figure 10-5 Effective permeability–viscosity ratio versus foam quality for consolidated permeable media and 0.1% aerosol foam (from Khan, 1965)

abilities. The observed mobilities are less than the single-phase mobility of water (this would be slightly less than the air permeabilities shown in the figure) and are substantially less than the single-phase mobility of the gas phase. Thus the mobility of the foam is lower than that of either of its constituents. The mobility of the foam decreases with increasing quality until the films between the gas bubbles begin to break. At this point (not shown in Fig. 10-5), the foam collapses, and the mobility increases to the gas mobility. Foams are effective in reducing the mobility at all three permeability levels in Fig. 10-5, but the effect of foam quality is relatively larger at the highest permeability. This is a consequence of the disparity between the texture of the foam and the mean pore size of the medium (Fried, 1960).

The mobility reduction caused by the foam can be viewed as an increase in the effective viscosity of a single-phase flow or as a decrease in the gas phase permeability. In the first case, the effective viscosity comes from dividing the measured mobility into the single-phase gas permeability. In the second case, the gas phase permeability follows from multiplying the measured mobility by the gas viscosity. Figure 10-6 shows representative data of the second type; it shows the gas phase permeability, both with and without foam, and gas saturation plotted against the liquid injection rate. The foam causes a great decrease in gas permeability at the same rate and even at the same gas saturation (middle curve) compared to the surfactant-free displacement. On the other hand, Fig. 10-7 shows that neither the gas saturation nor the presence of the foaming agent affects the aqueous phase relative permeability.

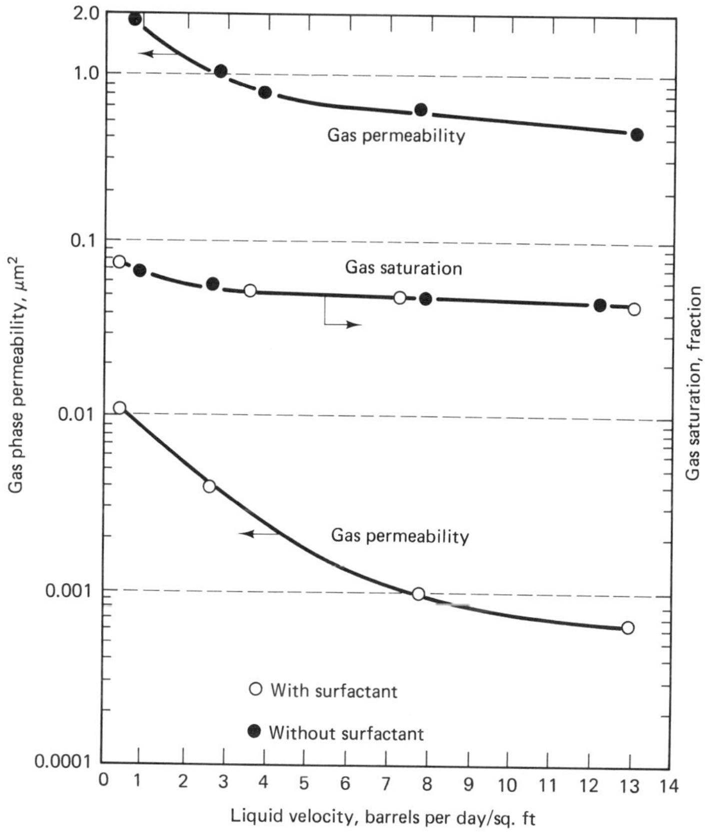

Figure 10-6 Effect of liquid flow rate and gas saturation of gas permeability with and without surfactant (Bernard and Holm, 1964)

The low mobilities in foam flow are postulated to be caused by at least two different mechanisms: (1) the formation of or the increase in a trapped residual gas phase saturation and (2) a blocking of pore throats due to the gas films. From Fig. 10-6, it appears that the effect of a trapped gas saturation, which would lower the mobility through a relative permeability lowering, is much smaller than the pore throat blocking effect. But the trapped gas phase saturation effect may become important during the later stages of a displacement where the lower pressures could cause more gas to come out of solution.

The mobility reduction of foams, viewed as a viscosity enhancement, has been studied by several researchers in capillary tubes (Holbrook et al., 1981). Figure 10-8 shows a sample of this type of data. On a log–log plot of shear stress versus shear rate, a slope of 1 indicates a Newtonian fluid (see Eq. 8B-3 with $n_{pl} = 1$). The foams in Fig. 10-8 all have slopes of less than 1, indicating a pseudoplastic or shear thinning type of flow. The level of the curves on such a plot indicates the overall re-

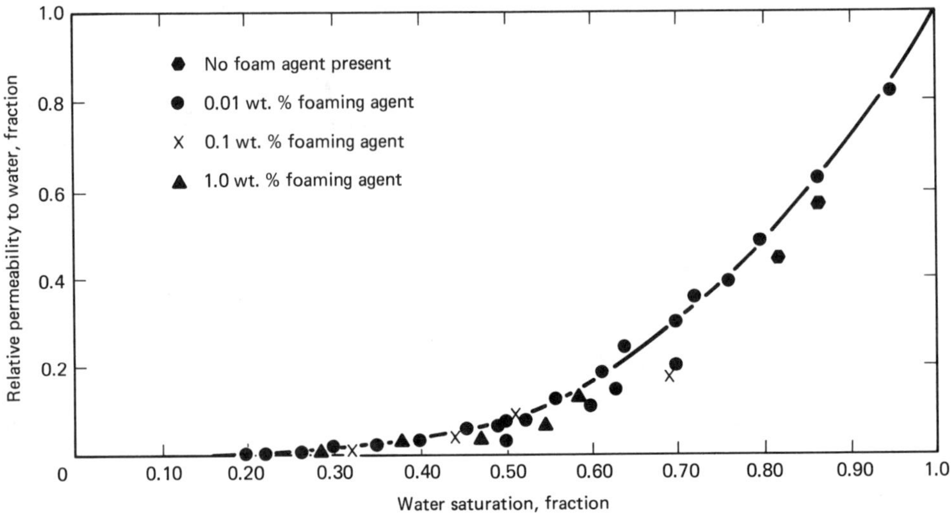

Figure 10-7 Effect of foaming agent on water relative permeability (from Bernard et al., 1965)

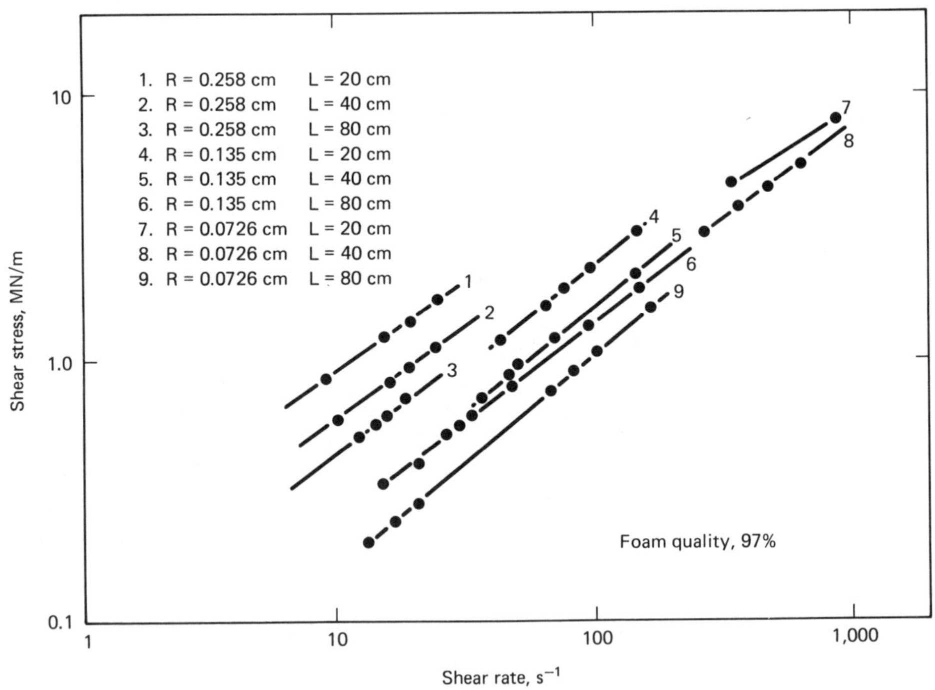

Figure 10-8 Shear stress versus shear rate for foam flow in capillary tubes (from Patton et al., 1981)

sistance to flow in the form of the power-law coefficient. The overall coefficient is largest in the large radius tubes, just as the mobility reduction effect was most pronounced in the highest permeability cores in Fig. 10-5. Figure 10-8 also shows a dependence on tube length that is probably the result of a partially collapsing foam.

The rheological behavior of foams in capillary tubes can best be understood through the analysis of the flow of a single bubble (Fig. 10-9). Under static conditions, the film would be unstable as required by the interfacial tension gradient discussed above. Under flowing conditions, viscous forces push fluid into the slip layer between the tube wall and inner film boundary so that the bubble is elongated somewhat, and the film thickness is generally larger than when the bubble was static. Using theoretical arguments based on a Newtonian fluid and an inviscid gas, Hirasaki and Lawson (1983) have shown that the film thickness increases as the bubble velocity to the 2/3 power. Considering no-slip flow in the capillary with a stationary bubble in a tube moving with velocity v, the relation between the force F required to move the bubble and v is

$$\frac{F}{A} = \frac{\mu v}{\Delta r} \tag{10.4-1}$$

where Δr is the film thickness, and A is the contact area between the film and the tube wall. Using the results of the Hirasaki and Lawson theory and solving for the viscosity in Eq. (10.4-1) gives the proportionality

$$\frac{F}{A} \sim \mu v^{1/3} \tag{10.4-2}$$

This relation predicts the apparent viscosity F/Av of a foam flowing in a capillary tube decreases with increasing velocity. The implication of this is that the shear thinning effect observed in capillary tubes is actually a consequence of the film thickening as velocity increases. Thus one must be extremely careful when interpreting capillary tube data for permeable media properties. A second implication of Eq. (10.4-2)

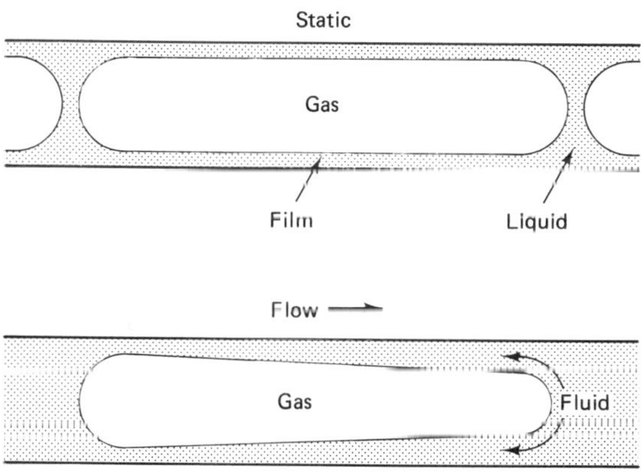

Figure 10-9 Behavior of a foam bubble under static and flowing conditions in a capillary tube (from Hirasaki and Lawson, 1983)

is that foam texture, which determines the spacing between films and A, occupies an even greater importance in determining rheological behavior than does foam quality.

10-5 ALKALINE FLOODING

The second chemical EOR process is high-pH or alkaline flooding, a schematic of which is shown in Fig. 10-10. As in polymer and MP flooding, there is usually a brine preflush to precondition the reservoir, a finite volume of the oil-displacing chemical, and a graded mobility buffer driving agent; moreover, the entire process is usually driven by chase water. For both high-pH and MP flooding, the surfactant is injected, whereas in high-pH flooding, it is generated in situ.

High pHs indicate large concentrations of the hydroxide anions (OH^-). The pH of an ideal aqueous solution is defined as

$$pH = -\log[H^+] \tag{10.5-1}$$

where the concentration of hydrogen ions $[H^+]$ is in kg-moles/m^3 water. As the concentration of OH^- is increased, the concentration of H^+ decreases since the two concentrations are related through the dissociation of water

$$K_1 = \frac{[OH^-][H^+]}{[H_2O]} \tag{10.5-2}$$

and the water concentration in an aqueous phase is nearly constant. These considerations suggest two means for introducing high pHs into a reservoir: dissociation of a hydroxyl-containing species or adding chemicals that preferentially bind hydrogen ions.

Many chemicals could be used to generate high pHs, but the most commonly used are sodium hydroxide ("caustic" or NaOH), sodium orthosilicate, and sodium carbonate (Na_2CO_3). NaOH generates OH^- by dissociation,

$$NaOH \longrightarrow Na^+ + OH^- \tag{10.5-3a}$$

and the latter two through the formation of weakly dissociating acids (silicic and carbonic acid, respectively), which remove free H^+ ions from solution.

$$Na_2CO_3 \longrightarrow 2Na^+ + CO_3^{2-} \tag{10.5-3b}$$

$$2H_2O + CO_3^{2-} \longrightarrow \underset{\text{(carbonic acid)}}{H_2CO_3} + 2OH^- \tag{10.5-3c}$$

for sodium carbonate.

High-pH chemicals have generally been used in field applications in concentrations up to 5 wt % (injected pHs of 11 to 13) and with slug sizes up to 0.2 PV. These result in amounts of chemical quite comparable to the surfactant usage in MP flooding; however, high-pH chemicals are substantially less costly.

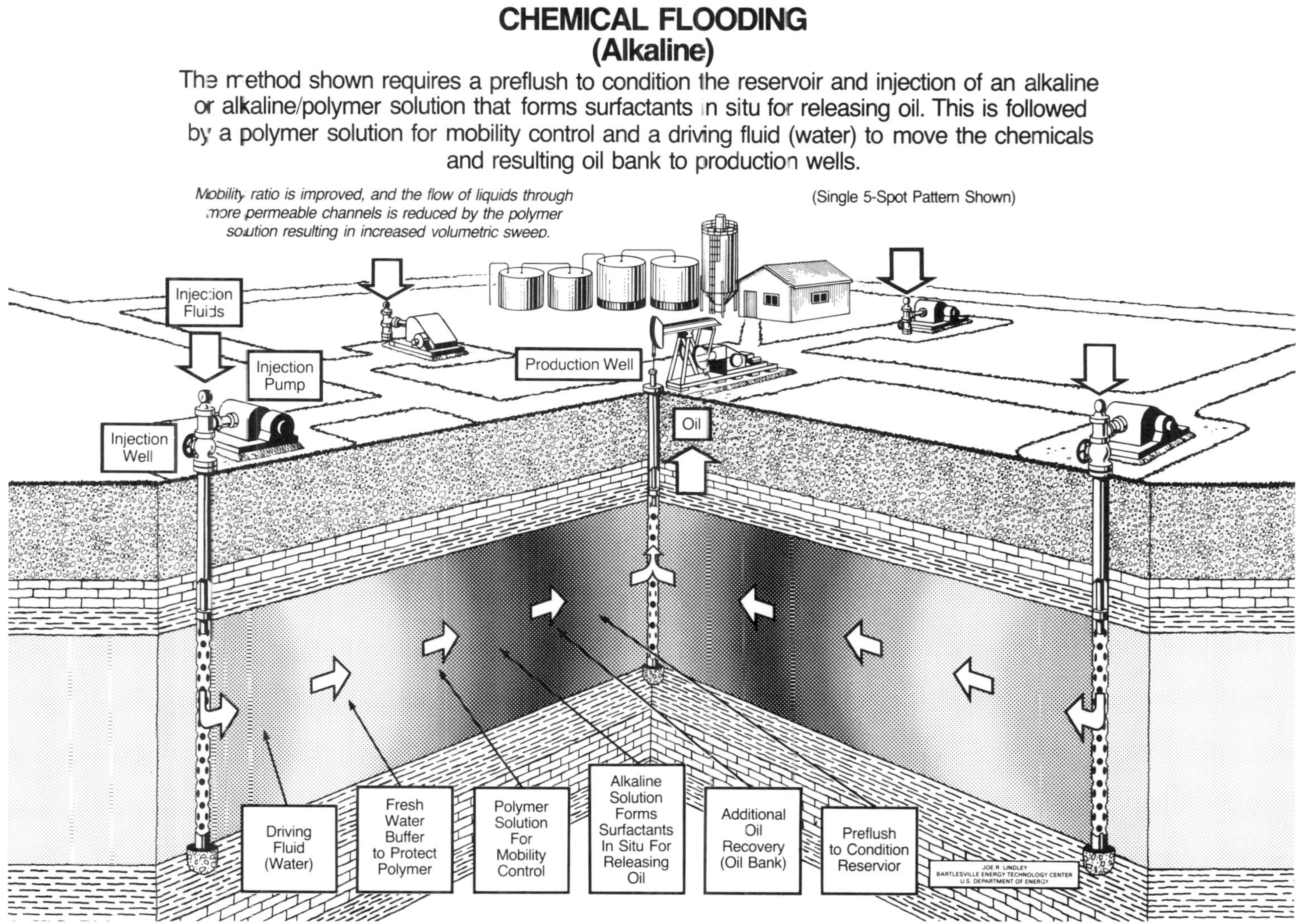

Figure 10-10 Schematic illustration of alkaline flooding (drawing by Joe Lindley, U.S. Dept. of Energy Bartlesville, Oklahoma)

10-6 SURFACTANT FORMATION

OH^- by itself is not a surfactant since the absence of a lipophillic tail makes it exclusively water soluble. If the crude oil contains an acidic hydrocarbon component HA_2, some of this, HA_1, can partition to the aqueous phase where it can react according to (Ramakrishnan and Wassan, 1982)

$$\begin{aligned} HA_2 &= HA_1 \quad \text{(partitioning)} \\ HA_1 &= A_1^- + H^+ \quad \text{(reaction)} \end{aligned} \tag{10.6-1}$$

Though the exact nature of HA_2 is unknown, it is probably highly dependent on crude oil type. The deficiency of hydrogen ions in the aqueous phase will cause the extent of this reaction to be to the right. The anionic species A_1^- is a surfactant that can have many of the properties and enter into most of the phenomena we described in Chap. 9 for MP flooding.

If no HA_2 is originally present in the crude, little surfactant can be generated. A useful procedure for characterizing crudes for their attractiveness to alkaline flooding is through the *acid number*. The acid number is the milligrams of potassium hydroxide (KOH) required to neutralize one gram of crude oil. To make this measurement, the crude is extracted with water until the acidic species HA_2 is removed. The aqueous phase containing HA_1, A_1^-, and H^+ is then brought to $pH = 7$ by adding KOH

$$\begin{aligned} KOH &\longrightarrow OH^- + K^+ \\ HA_1 + OH^- &\longrightarrow A_1^- + H_2O \end{aligned} \tag{10.6-2}$$

For a meaningful value, the crude must be free of acidic additives (scale inhibitor, for example) and acidic gases (CO_2 or H_2S). A good alkaline flooding crude candidate will have an acid number of 0.5 mg/g or greater, but acid numbers as low as 0.2 mg/g may be candidates since only a small amount of surfactant is required to saturate oil–brine interfaces. Figure 10-11 shows a histogram of acid numbers based on the work of Jennings et al. (1974).

10-7 DISPLACEMENT MECHANISMS

Oil recovery mechanisms in high-pH flooding have been attributed to eight separate mechanisms (de Zabala et al., 1982). In this section, we concentrate on only three: IFT lowering, wettability reversal, and emulsion formation. The last two mechanisms are also present in MP flooding but are dwarfed by the low IFT effects. With smaller ultimate oil recoveries, the distinction among effects becomes important in alkaline flooding.

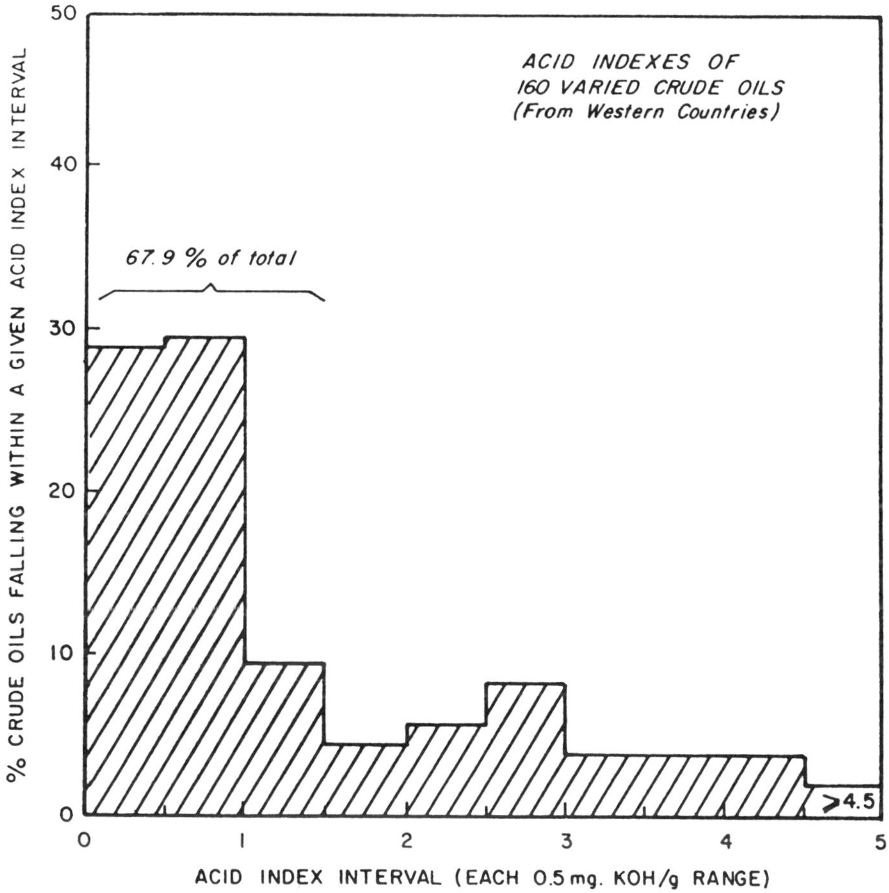

Figure 10-11 Histogram of acid numbers (from Minssieaux, 1976)

IFT Lowering

The generated surfactant A_I^- aggregates at oil–water interfaces that can lower IFT (Ramakrishnan and Wassan, 1982). In general, this lowering is not as pronounced as in MP flooding, but under certain conditions, it can be large enough to produce good oil recovery. Figure 10-12 shows IFT measurements of caustic solutions against a crude oil at various brine salinities. The IFTs are sensitive to both NaOH concentration and salinity, showing minima in the NaOH concentration range of 0.01–0.1 wt %. The decrease in IFT in these experiments is limited by the spontaneous emulsification of the oil–water mixture when the IFT reaches a minimum.

There are many similarities in the low IFT effects in MP and high-pH flooding. The data in Fig. 10-12 suggest an optimal salinity of about 1.0 wt % NaCl

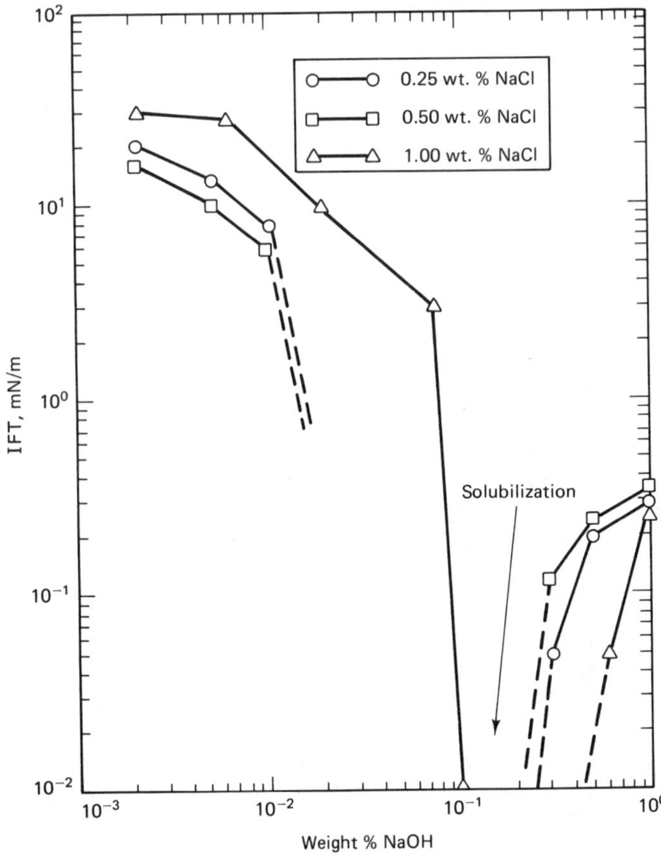

Figure 10-12 Interfacial tensions for caustic–crude–brine systems (from Ramakrishnan and Wasson, 1982)

for a 0.03 wt % NaOH solution. (Compare Fig. 10-12 to Fig. 9-10.) Indeed, Jennings et al. (1974) have shown that there is an optimal NaOH concentration for a given salinity in oil recovery experiments. Moreover, the presence of the emulsification effect when IFTs are low is exactly what one would expect from an MP system at a surfactant concentration above the invariant point in surfactant concentration. This suggests the data in Fig. 10-12 are showing that a type II(−) phase environment exists at low NaOH concentrations, and type II(+) at high NaOH concentrations. Further work is necessary to definitively establish the connection to MP phase behavior since the actual surfactant concentration A_1^- is likely to be much lower in an alkaline system. Still, Nelson et al. (1984) show that a cosurfactant can increase the optimal salinity in a system much as in MP systems.

Wettability Reversal

Owens and Archer (1971) show that increasing the water wetness increases ultimate oil recovery. The wettability was reported as decreasing the water-oil contact angle measured on polished synthetic surfaces. This has also been shown by others using high-pH chemicals (Wagner and Leach, 1959; Ehrlich et al., 1974). The increased oil recovery is the result of two mechanisms: (1) a relative permeability effect that causes the mobility ratio of a displacement to decrease and (2) a shifting of the capillary desaturation curve.

Cooke et al. (1974) have reported improved oil recovery with increased oil wetness. Other data show that oil recovery is a maximum when the wettability of a permeable medium is neither strongly water wet nor strongly oil wet (Lorenz et al., 1974). Considering the latter information, the important factor may be the change in the wettability rather than the final wettability of the medium. In the original wetting state of the medium, the nonwetting phase occupies large pores, and the wetting phase occupies small pores. If the wettability of a medium is reversed, nonwetting fluid will exist in small pores, and wetting fluid in large pores. The resulting fluid redistribution, as the phases attempt to attain their natural state, would make both phases vulnerable to recovery through viscous forces.

Emulsion Formation

Alkaline chemicals can cause improved oil recovery through the formation of emulsions. The emulsification produces additional oil in at least two ways: through a mobility ratio lowering since many of these emulsions have a substantially increased viscosity and through solubilization and entrainment of oil in a flowing aqueous stream. The first mechanism improves displacement and volumetric sweep, as does any other mobility control agent. Local formation of highly viscous emulsions is not desirable since these would promote viscous instability. For an oil-free alkaline solution, the solubilization and entrainment mechanism would be more important when the IFT between the swollen water phase and the remaining crude is low. Figure 10-12 shows that for certain conditions, emulsification and low IFTs occur simultaneously. McAuliffe (1973) showed that emulsions injected in a core and those formed in situ give comparable oil recoveries.

10-8 ROCK–FLUID INTERACTIONS

Interactions of the alkaline chemicals and the permeable media minerals can cause excessive retardation in the propagation of these chemicals through the medium. In this section, we discuss three aspects of rock–fluid interactions: formation of divalent-hydroxide compounds, mineral dissolution, and cation exchange.

Divalent-Hydroxide Compounds

OH^- ions themselves are not appreciably bound to the solid surfaces, but in the presence of multivalent cations they can form hydroxyl compounds

$$M^{x+} + x(OH^-) = M(OH)_x \qquad (10.8\text{-}1)$$

which, being relatively insoluble, can precipitate from solution. (See Table 3-7 for solubilities.) This reaction in turn lowers the pH of the solution and can cause formation damage through pore blockage and fines migration. The anionic surfactant species A_1^- can interact with the inorganic cations in solution just as in MP flooding. However, interactions with the divalent cations usually take precedence, particularly in hard brines or where there are substantial quantities of soluble multivalent minerals. Because of these interactions and those involving surfactants A_1^-, alkaline processes are as sensitive to brine salinity and hardness as are MP processes.

Mineral Dissolution

Unlike MP flooding, high-pH chemicals can react directly with clay minerals and the silica substrate to cause consumption of OH^- ions. The reactions with clays are manifest by the elution of soluble aluminum and silica species from core displacements (Bunge and Radke, 1982). The resulting soluble species can subsequently cause precipitates through hydroxyl reactions as in Eq. (10.8-1) (Sydansk, 1983). The rate of hydroxyl consumption from this "slow" reaction (cation exchange is generally fast enough so that local equilibrium applies) is determined by the following treatment.

Let's consider the rate of migration of hydroxide ions that are reacting with the silicate in the permeable medium. Subject to the usual fractional flow assumptions, the hydroxide concentration C_3 must satisfy

$$\phi \frac{\partial C_3}{\partial t} + u \frac{\partial C_3}{\partial x} = r_3 \qquad (10.8\text{-}2a)$$

in single-phase, one-dimensional flow. This equation is subject to the boundary and initial conditions

$$C_3(x, 0) = C_{3I}, \qquad C_3(0, t) = C_{3J} \qquad (10.8\text{-}2b)$$

We consider only the continuous injection inasmuch as this is sufficient to illustrate the important features of hydroxide transport. (For the treatment of an alkaline slug, see Bunge and Radke, 1982.)

The term r_3 is the rate of consumption of the hydroxide ions because of the following reaction:

$$SiO_2 + 2OH^- \longrightarrow H_2SiO_4^{2-} \qquad (10.8\text{-}3)$$

This reaction should be regarded as generic because (1) the molecular formula for naturally occurring silica is not SiO_2, and because of this, (2) the end product is actually a distribution of several possible soluble silica products. However, Eq.

(10.8-3) captures the essence of hydroxide consumption for reaction with both silicon dioxide and clays.

If, as is assumed here, the solution is undersaturated with respect to the least soluble silica mineral, the reverse reaction of Eq. (10.8-3a) is negligible, and the rate of reaction is

$$r_3 = -k_3' C_3^m \tag{10.8-4}$$

where k_3' is the kinetic reaction rate, and m is the order of the reaction. The SiO_2 concentration should appear in Eq. (10.8-4), but this is present is such a large excess that its concentration remains constant. (For treatment where the silica concentration varies with the progress of the reaction, see Hekim et al., 1982; Yortsos, 1982.) This excess, plus the absence of hydroxide bound to the surface, accounts for the absence of the solid-phase conservation equation. Solid–fluid reactions of the type in Eq. (10.8-4) normally have reaction orders different from the ideal $m = 2$.

Solid–fluid reactions can also be limited by mass transfer from the bulk phase to the surface. Let h_3 be a mass transfer coefficient expressing this rate of transfer. Thus the flux of hydroxide ions to the surface is

$$N_3 = h_3(C_3 - \tilde{C}_3) \tag{10.8-5}$$

where $\tilde{C}_3$ indicates a concentration evaluated at the surface (not a solid concentration). The surface concentration is what determines r_3. If we let the local mass transfer equal the reaction rate at the surface, then

$$h_3(C_3 - \tilde{C}_3) = k_3' \tilde{C}_3^m \tag{10.8-6}$$

In principle, this equation can be solved for $\tilde{C}_3$ and then substituted into Eq. (10.8-4) to give a rate expression entirely in terms of the bulk concentration C_3. But in practice, this process is convenient only for $m = 1$. Following this eventuality, we have

$$r_3 = -k_3 C_3 \tag{10.8-7a}$$

where

$$k_3 = \left(\frac{1}{h_3} + \frac{1}{k_3'}\right)^{-1} \tag{10.8-7b}$$

is the overall kinetic constant. The latter expression says k_3 is the harmonic average of the resistances to mass transfer. If either resistance becomes negligible, k_3 becomes equal to the resistance of the remaining process.

Introducing Eq. (10.8-7a) into Eq. (10.8-2) and reverting to nondimensional variables gives

$$\frac{\partial C_3}{\partial t_D} + \frac{\partial C_3}{\partial x_D} = -N_{Da} C_3 \tag{10.8-8}$$

where

$$N_{\text{Da}} = \frac{k_3 L \phi}{u} \tag{10.8-9}$$

is the *Damkohler* number, the relative rates of mass transfer (to the solid surface), and bulk flow (through the medium). N_{Da} is the fundamental scaling variable for nonequilibrium flows. If N_{Da} is greater than about 10, the flow is essentially in local equilibrium (Jennings and Kirkner, 1984); if it is very small, the reaction may be neglected. All other things being equal, N_{Da} is usually much larger in the field than in the laboratory because of the much larger lengths.

In the limit of kinetically dominated consumption, N_{Da} becomes a constant in one-dimensional flow. In the other extreme of mass transfer limited reactions, N_{Da} is now based on the mass transfer coefficient. A great deal of data indicates mass transfer coefficients are proportional to velocity in low Reynolds number flows (Fedkiw, 1979). In this case, N_{Da} (in fact, the entire dimensionless problem statement) becomes independent of velocity.

Equation (10.8-8) is a nonreducible, hyperbolic differential equation we first discussed in Sec. 5-6. Comparing the notation in Eq. (5.6-3) with Eq. (10.8-8), we find

$$\frac{dt_D}{ds} = 1, \qquad \frac{dx_D}{ds} = 1, \qquad \frac{dC_3}{ds} = -N_{\text{Da}} C_3 \tag{10.8-10}$$

where s is a parameter along a characteristic curve. This equation implies the characteristics are straight lines with unit slope in $x_D - t_D$ space, along which C_3 varies as

$$C_3 = C_{3J} e^{-N_{\text{Da}} s} \tag{10.8-11}$$

Since the characteristics do not cross, the hydroxide cannot propagate faster than a tracer. In terms of the original variables, the solution becomes

$$C_3 = \begin{cases} C_{3I}, & x_D \geqq t_D \\ C_{3J} e^{-N_{\text{Da}} x_D}, & x_D \leqq t_D \end{cases} \tag{10.8-12a}$$

written as a concentration profile and

$$C_3 = \begin{cases} C_{3I}, & t_D \leqq 1 \\ C_{3J} e^{-N_{\text{Da}}}, & t_D \geqq 1 \end{cases} \tag{10.8-12b}$$

written as an effluent history. You may show that Eq. (10.8-12a) satisfies Eq. (10.8-5) and the conditions in Eq. (10.8-2b).

As simple as this treatment is, it illustrates two rather general features of transport in nonequilibrium reacting systems. First, even though shocks do form, the reaction does not delay the propagation of the hydroxide front. This is in contrast to equilibrium processes that do cause front delay. Second, the primary manifestation of the reaction is in the depressed concentration behind the wave. For the simple case considered here, the depression is a direct consequence of the magnitude of N_{Da}. For a large source, the concentration slowly approaches the injected value. But for large N_{Da}, this can happen very slowly. Bunge and Radke (1982) caution that laboratory measured values of hydroxide consumption can be extremely misleading unless the N_{Da} value is the same between the laboratory and the field. Finally, the depressed concentration behind the front is observed even for flows in local equilibrium if the hydroxide reactions are reversible.

Cation Exchange

Another high-pH process intimately associated with clays is cation exchange. As we introduced in Sec. 3-5, cation exchange is well approximated as an equilibrium process—that is, N_{Da} for cation exchange is large. In high-pH flooding, the phenomenon is slightly different since one of the important exchanging cations is also undergoing a reaction with the solvent. The following treatment was originally proposed by Bunge and Radke (1985).

H^+ cations can exchange with other cations bound to anionic clay sites according to

$$\text{Clay—H} + M^+ \rightleftharpoons \text{Clay—M} + H^+ \qquad (10.8\text{-}13a)$$

for monovalent-hydrogen exchange. The corresponding expression for divalent-hydrogen exchange follows from the same procedure.

The isotherm expressing the exchange in Eq. (10.8-13a) is

$$K_N = \frac{C_{Ms} C_3}{C_M C_{3s}} \qquad (10.8\text{-}13b)$$

where we have taken the solution and solid-phase activities to be 1. We can write an equation for the direct uptake of hydroxide anions from solution by eliminating the hydrogen concentration between Eq. (10.5-2) and (10.8-13b)

$$C_3 = \frac{K_N C_{Ms}}{K_1 C_M C_{3s}} \qquad (10.8\text{-}14)$$

where $C_3 = [H^+]$ in consistent units. Similarly, the bound hydrogen concentration C_{3s} can be eliminated by the solid-phase electroneutrality constraint, $Z_v = C_{3s} + C_{Ms}$, to give on solution for the bound monovalents

$$C_{Ms} = \frac{\left(Z_v \dfrac{K_N}{K_1} C_3\right) C_M}{1 + \left(\dfrac{K_N}{K_1} C_3\right) C_M} \qquad (10.8\text{-}15)$$

The equilibrium expressed by Eq. (10.8-15) reflects the following exchange reaction:

$$\text{Clay—M} + H_2O \rightleftharpoons \text{Clay—H} + OH^- + M^+ \qquad (10.8\text{-}16)$$

from which we see that the sorption of an equivalent of monovalent cation corresponds to the disappearance of an equivalent of hydroxide from solution. Therefore, we can identify the left side of Eq. (10.8-15) with a fractional hydroxide uptake.

Equation (10.8-15) is of the form of the Langmuir isotherm we first discussed in Sec. 8-2. It also indicates the exchange is reversible, and the plateau retention is independent of monovalent concentration. The rate of approach to this plateau strongly depends on the monovalent concentration, and in fact, the curvature of the isotherm can be strongly repressed when the salinity is high. Figure 10-13 shows experimentally measured isotherms of hydroxide uptake. All curves except the one indicated are in the absence of added sodium.

Unlike the nonequilibrium reactions discussed above, equilibrium processes do cause retardation. Figure 10-14 shows the combined effect of exchange and reaction in the results of a laboratory flood. Note the delay in the front and the slight depression in pH once the front arrives. The lower-pH floods require more than 3 V_p of fluid injection to attain the injected pH.

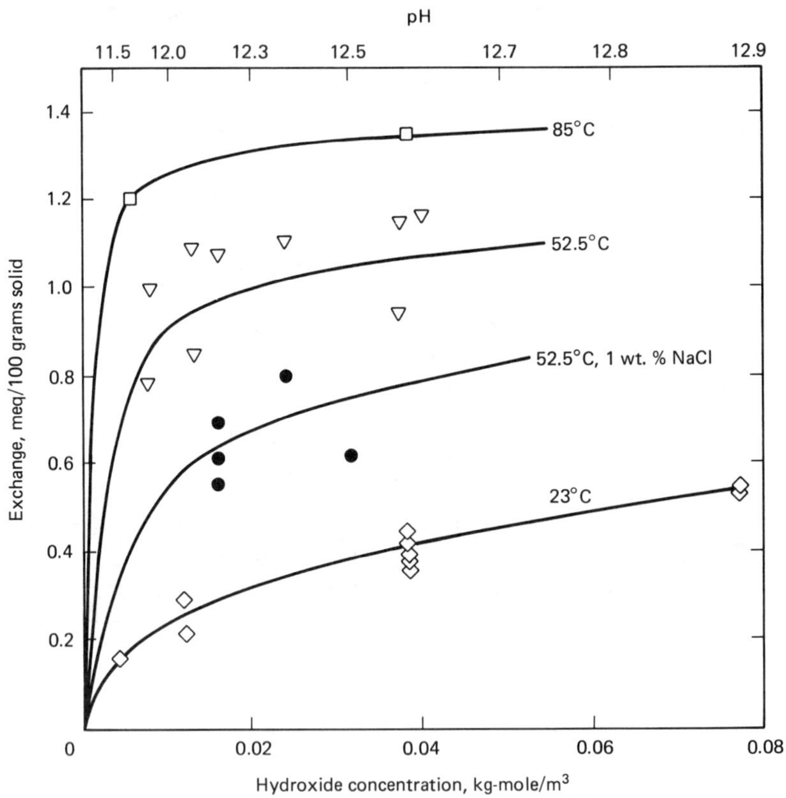

Figure 10-13 Reversible hydroxide uptake for Wilmington, Ranger-zone sand (from Bunge and Radke, 1985)

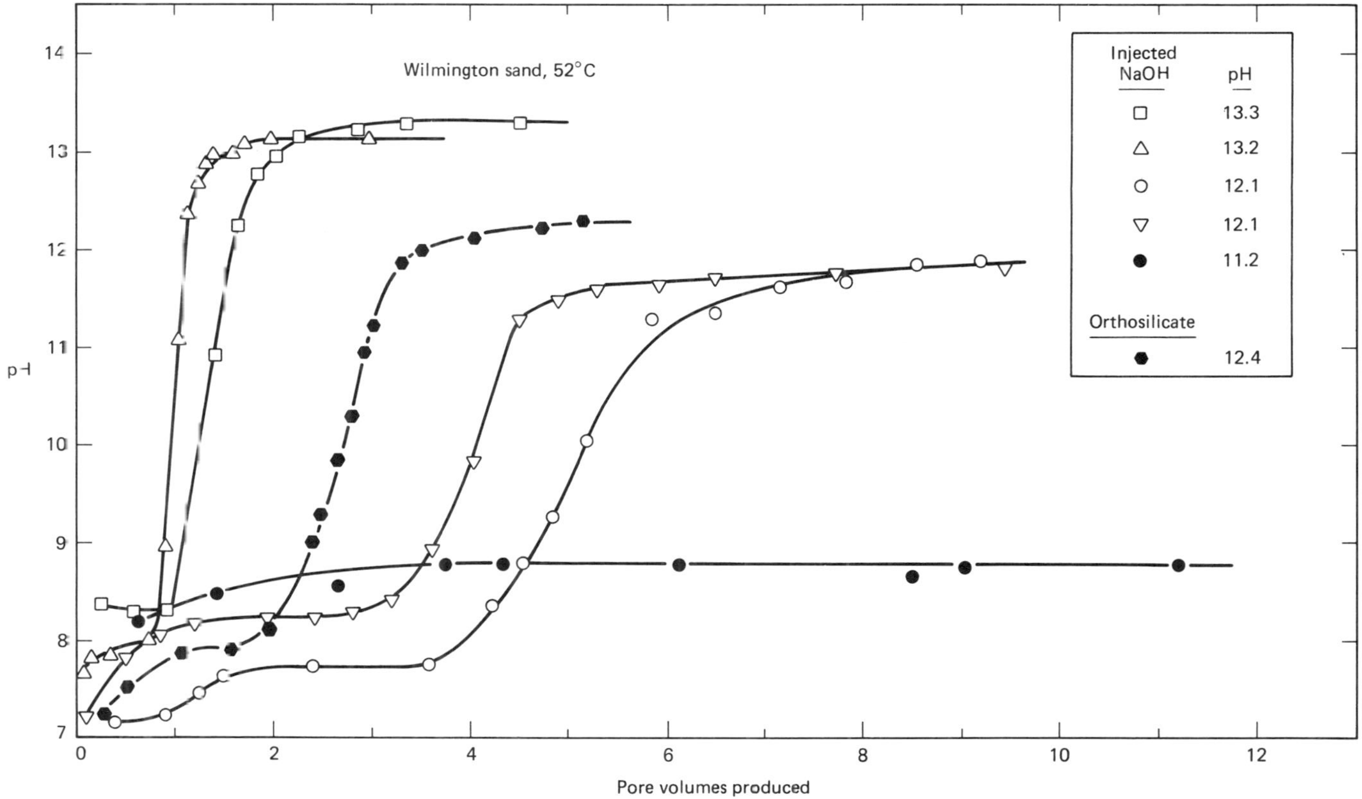

Figure 10-14 Experimental and theoretical effluent histories of pH (from Bunge and Radke, 1982)

Simplified Procedure

The hydroxide delay illustrated in Fig. 10-14 may be readily estimated through a simplified procedure given by Bryant et al. (1986).

First, we give some definitions in Fig. 10-15 for ideal flow. pH *lag* is the difference between the arrival times of an ideal tracer and that of the hydroxide wave at the effluent end of the permeable medium. The *useful* pH is the pH behind the pH wave, and pH *loss* is the difference between the injected pH and the useful pH. As we suggest in the above discussion, pH lag is caused mainly by cation exchange, and pH loss mainly by silicate mineral dissolution.

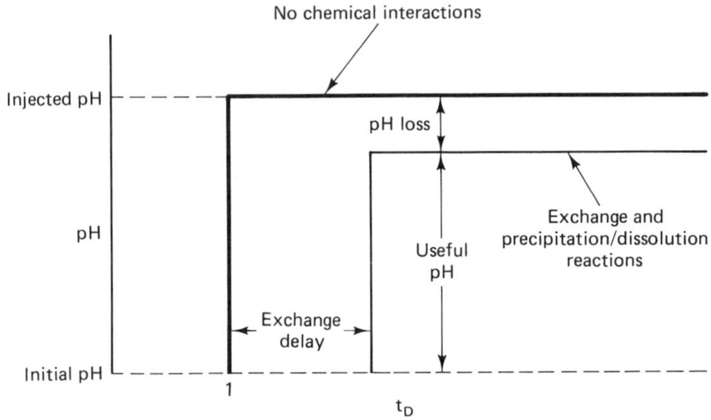

Figure 10-15 Definitions for ideal hydroxide transport (from Bryant et al., 1986)

The simplified procedure is based on local equilibrium. This means the predictions will be conservative—that is, it will predict a somewhat larger lag and loss than what is actually observed. For the purposes of evaluation, this is a desirable feature. Bryant et al. suggest reprecipitation of hydroxides is negligible, and dissolution of silica through Eq. (10.8-3) and the formation of soluble $H_3Si_2O_4^-$ and $Si_2O_4^{2-}$ accounts for most of the consumption of the injected OH^- behind the pH wave. These species are formed through the following two reactions:

$$OH^- + H_4SiO_4 \rightleftharpoons H_3SiO_4^- + H_2O \tag{10.8-17a}$$

$$OH^- + H_3SiO_4^- \rightleftharpoons H_2SiO_4^{2-} + H_2O \tag{10.8-17b}$$

Equations (10.8-17a) and (10.8-17b) are simply itemizations of Eq. (10.8-3). The equilibria for reactions in Eqs. (10.8-3), (10.8-17a), and (10.8-17b) are

$$K^{sp} = [C_{H_4SiO_4}] \tag{10.8-18a}$$

$$K_1 = \frac{[C_{H_3SiO_4^-}]}{[C_3][C_{H_4SiO_4}]} \tag{10.8-18b}$$

$$K_2 = \frac{[C_{H_2SiO_4^{2-}}]}{[C_3][C_{H_3SiO_4^-}]} \tag{10.8-18c}$$

where the concentrations are in equivalents per volume of pore space, the behavior is assumed to be ideal, the activity of the solid SiO_2 is unity, and the concentration of water is a constant lumped into the equilibrium constant. Table 3-7 tabulates the numerical values for the K_1, K_2, and K^{sp}. These constants can be corrected to the desired temperature through Eq. (3.5-13).

The useful pH follows from carrying out a flash calculation involving the above species. Let C_{tJ} denote the sum of the concentrations of OH^-, $H_3SiO_4^-$, and $H_2SiO_4^{2-}$ in the injected aqueous fluid

$$C_{tJ} = [C_3] + [C_{H_3SiO_4^-}] + [C_{H_2SiO_4^{2-}}] \tag{10.8-19}$$

Based on this, the variable C_{tJ} is analogous to the overall molar concentrations we discussed in Sec. 4-4, and Eq. (10.8-19) is equivalent to Eq. (4.4-3). Of course, there are several other anionic species in the plateau; these are either present in minor amounts or do not participate in reactions. In the latter case, they would subtract from the mass balance equations since their equilibrated and preequilibrated concentrations would be the same. A similar argument must apply to the cations because the clays in the plateau region have been equilibrated with the injected solution across the pH wave.

The equilibria expressions (Eqs. 10.8-18) can eliminate all but the hydroxide concentration from Eq. (10.8-19). On solving for C_3, this gives

$$[C_3] = \frac{-(1 + K^{sp}K_1) + [(1 + K^{sp}K_1)^2 + 8K^{sp}K_2C_{tJ}]^{1/2}}{4K^{sp}K_2} \tag{10.8-20}$$

The useful pH follows from Eqs. (10.5-1) and (10.5-2) after Eq. (10.8-20) has been solved. The concentrations in Eq. (10.8-20) must be converted to molality before this transformation. Equation (10.8-20) also shows the advantage of injecting buffered silicate solutions because C_{tJ}, and hence the useful pH, will be higher than in the absence of the silicate. Physically, this means the injected silicate species concentrations are more nearly in equilibrium with the medium, and pH loss is minimal. The laboratory flood of sodium orthosilicate in Fig. 10-14 confirms this effect.

The pH lag is caused by the loading of the clays with hydroxide. The specific velocity of the pH lag follows from the chromatography of a species undergoing transport with Langmuirian sorption (Eq. 8.2-10)

$$v_{\Delta C_3} = \left(1 + \frac{C_{3s} - C_{3sI}}{C_3 - C_{3I}}\right)^{-1} \tag{10.8-21}$$

where C_3 follows from Eq. (10.8-20), and C_{3s} (now the sorbed hydroxide concentration) is from the isotherm value (Fig. 10-13) corresponding to C_3. Unlike the pH loss, pH lag is strongly affected by the other cations present because these will affect the hydroxide isotherm. Breakthrough of the pH wave occurs at the reciprocal of the specific velocity, and the lag is the difference of this number from 1.

Bryant et al. (1986) compare the difference between the calculated and experimental pH loss and lag for six different laboratory floods. As expected, the calculations give pH lags somewhat greater than those observed and useful pHs somewhat higher. Nevertheless, the agreement is satisfactory for most engineering purposes, particularly given the simplicity of the procedure.

10-9 FIELD RESULTS

Figure 10-16 shows the production data from a high-pH flood conducted in the Whittier field. The crude oil was 20° API with a 40 mPa-s viscosity, and the 0.2 wt % NaOH chemical was injected as a 0.23 V_p slug.

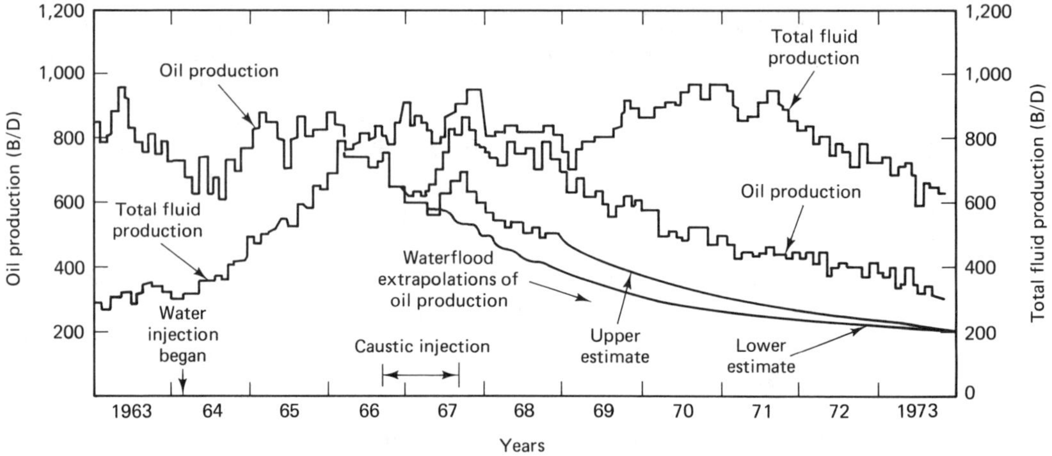

Figure 10-16 Production response from the Whittier field alkaline flood (from Graue and Johnson, 1974)

Many features in this performance are common to the responses of the other chemical flooding processes. The oil production rate declines as the total fluid production increases, indicating a declining oil cut. The oil rate response to the caustic injection is again superimposed on the waterflood decline, which is extrapolated to estimate the incremental oil recovery (IOR). (Figure 10-16 shows two waterflood decline curves, one based on the actual decline and another based on computer simulation.) The 55,750 to 74,860 SCM of oil produced by the caustic injection was considered a success by the operators.

Table 10-1 shows a summary of data from completed high-pH field floods. Note the wide range in reservoir and oil characteristics and in oil saturation at the start of the flood. IOR recovery expressed as a percent of V_p ranges from 0.0006 to 8.0, which translates into recoveries (expressed as a percent of the OIP at the project start) comparable to but slightly smaller than those reported from polymer floods. Of equal importance is the stb of IOR produced per pound of chemical injected (0.015–0.43 in Table 10-1). This is substantially lower than the polymer flood values; however, the cost of the high-pH chemical is also substantially lower.

TABLE 10-1 SUMMARY OF HIGH-PH FIELD TESTS (ADAPTED FROM MAYER ET AL., 1980)

Field location	Type of high-pH material	Acid number (mg KOH/g)	Chemical concentration injected (wt %)	Slug size ($\%V_p$)	IOR (percent pore volume)	Chemical requirement (stb/1b chemical)
Southeast Texas	Na_2CO_3	2.4	3.2			
Harrisburg, Nebraska	NaOH	low	2.0	0.013	0.003	0.03
Northward-Estes, Texas	NaOH	0.22	5.0	15	8.0	0.03
Singleton, Nebraska	NaOH	low	2.0	8	0.023	0.042
Whittier, California	NaOH		0.2	20	0.05–0.07	0.32–0.43
Brea-Olinda, California	Orthoscilicate	0.12				
Orcutt Hill, California	Orthoscilicate	0.6	0.42	0.017	0.0006	0.015–0.030

11

Thermal Methods

Thermal methods, particularly steam drive and steam soak, are easily the most successful enhanced oil recovery processes. In Chap. 1, we saw that steam methods currently account for nearly 80% of the EOR oil from less than one third of the EOR projects. Thermal flooding is commercially successful and has been for almost 30 years. In this chapter, we explore the reasons for this success.

Unfortunately, we can give no more than an overview of this scientifically complex and interesting subject. Several texts (White and Moss, 1983) and monographs (Prats, 1982) are available on thermal flooding alone. Our intent is to apply the twin bases of this book—phase behavior and fractional flow theory—to thermal methods in some detail. In addition, we deal with the important ancillary topic of heat loss.

Thermal methods rely on several displacement mechanisms to recover oil, but the most important is the reduction of crude viscosity with increasing temperature. We can draw several important conclusions from Fig. 11-1, a plot of crude kinematic viscosity (μ_2/ρ_2) versus temperature.

Crude kinematic viscosity decreases dramatically with a rise in temperature. The effect reflects the dynamic viscosity μ_2 change since crude density changes relatively little with temperature. For example, a heavy crude (10°–20° API) that undergoes a temperature increase from 300 to 400 K, easily obtainable in thermal methods, will produce a viscosity well within the flowing range (less than 10 mPa-s). Figure 11-1 greatly compresses the vertical axis simply to plot the observed changes on one scale.

For lighter crudes, the viscosity reduction is less. Therefore, thermal methods

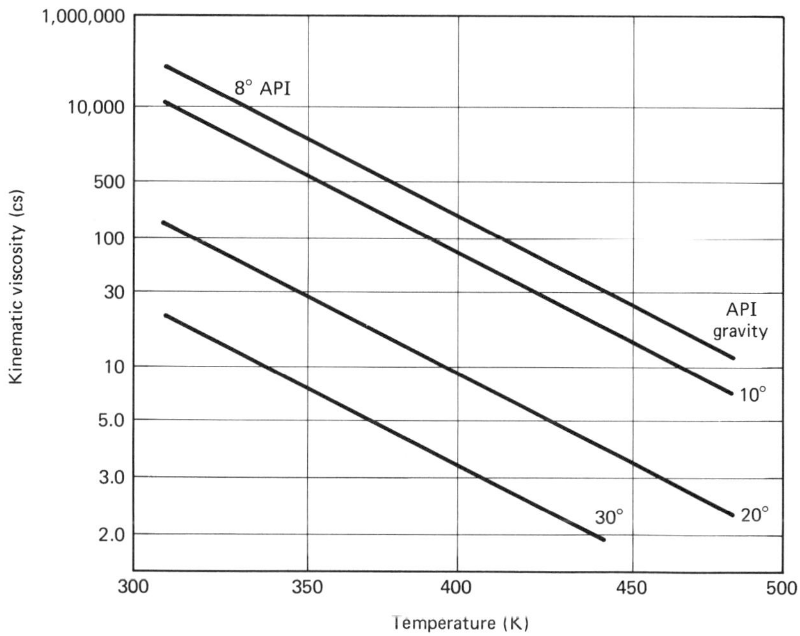

Figure 11-1 Effect of temperature on crude oil viscosity (adapted from Farouq Ali, 1974)

are not nearly so advantageous for these crudes, particularly since waterflooding would probably be an attractive alternative. The viscosity reduction for very heavy crudes (less than 10° API) is strong but still not enough to make them flow economically. Thus there are practical limits on both viscosity extremes.

11-1 PROCESS VARIATIONS

The four basic ways to apply thermal methods are hot water flooding, steam soak, steam drive, and in situ combustion. We deal with only the last three individually since hot water floods occur at some point in all the others.

Steam soak. In a steam soak (stimulation or huff 'n puff) steam is introduced into a well, and then the well is returned to production after a brief shut in period (Fig. 11-2a). The steam heats up a zone near the well and also provides some pressure support for the subsequent production. The shut-in or soak period allows thermal gradients to equalize, but should not be long enough for the pressure to escape. During shut-in, all the injected steam condenses, and the well produces a mixture of hot water and oil. A great advantage of a steam soak is that all the wells can be producing nearly all the time, the injection and soak periods usually being small.

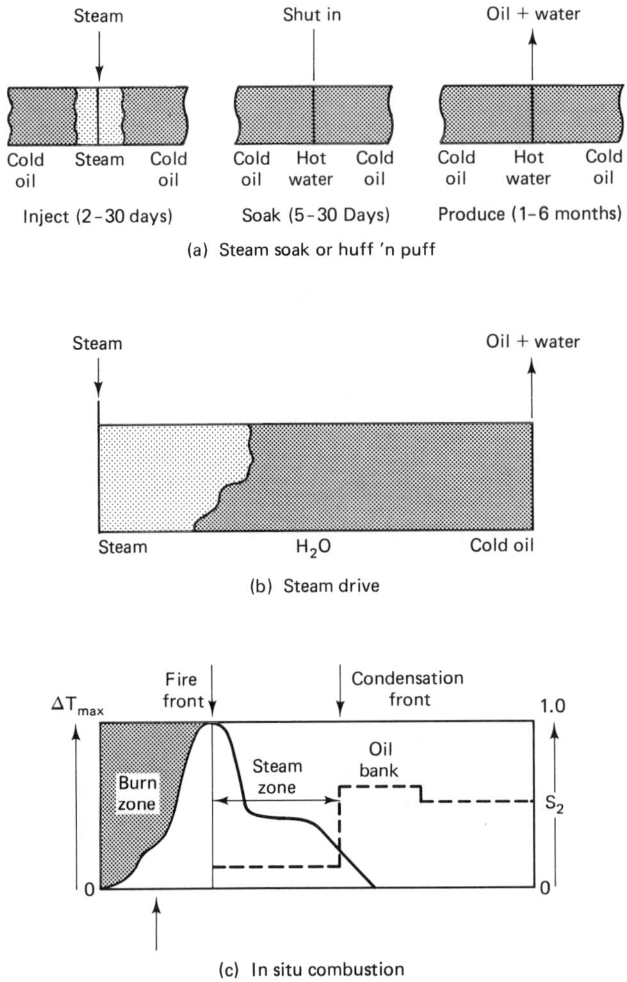

Figure 11-2 Process variations for thermal methods

(a) Steam soak or huff 'n puff

(b) Steam drive

(c) In situ combustion (adapted from Prats, 1982)

Steam drive. A steam drive uses at least two sets of wells, those into which steam is injected and those from which oil is produced (Fig. 11-2b). A steam drive usually results in higher ultimate recoveries than steam soak because it penetrates deeper into the reservoir. For the same reason, well spacing need not be as small in drives as in soaks for equivalent oil recovery. The small spacing partially offsets the disadvantage of sacrificing some of the wells to injection. Since steam drive is present to some extent in all thermal processes, we focus on it in later analyses.

In situ combustion. Figure 11-2(c) is a schematic of a forward in situ combustion process. Usually, some form of oxidant (air or pure oxygen) is introduced into the formation, the mixture then spontaneously or externally ignites, and subsequent injection propagates a fire zone through the reservoir. The fire zone is only a meter or so wide, but it generates very high temperatures. These temperatures vaporize connate water and a portion of the crude, both of which are responsible for some oil displacement. The vaporized connate water forms a steam zone ahead of the burn front, which operates very much like a steam drive. The vaporized oil consists mainly of light components that form a miscible displacement. The reaction products of a high-temperature combustion can also form an in situ CO_2 flood.

For most cases, viscosity reduction is by far the most important cause of additional oil recovered by thermal methods, but other mechanisms can be important; for example, distillation, miscible displacement, thermal expansion, wettability changes, cracking, and lowered oil–water interfacial tension. The relative importance of each mechanism depends on the oil being displaced and the process. Cracking is relatively unimportant in steam processes, with their relatively low temperatures, but it is quite important during in situ combustion, and thermal expansion and distillation become more important as the cold viscosity of the crude decreases.

11-2 PHYSICAL PROPERTIES

Understanding the thermodynamic and transport properties of water and crude oil is necessary to elucidate the mechanisms of thermal methods. We review these properties and their temperature dependence in this section. The most important water properties for our treatment are the steam–water phase envelope, steam quality, and latent heat of vaporization. For crudes, the most important property is the temperature dependence of viscosity.

Water Properties

The temperature rises in a thermal flood because additional energy is introduced or generated in the reservoir. This energy content is well approximated by the water enthalpy. Figure 11-3 shows a pressure–enthalpy diagram for water. This diagram is analogous to the pressure–composition diagrams we discussed in Sec. 4-1, with enthalpy being the composition variable. Figure 11-3 has several important landmarks.

1. Two-phase envelope. The envelope defines the region of two-phase behavior, as does the envelope on the pressure–molar volume diagram in Fig. 4-2. The left boundary is the bubble point curve, and the right is the dew point curve. To the left and right of the envelope are the supercooled liquid and superheated vapor (steam) regions, respectively. Within the two-phase region, temperature and pressure are interdependent.

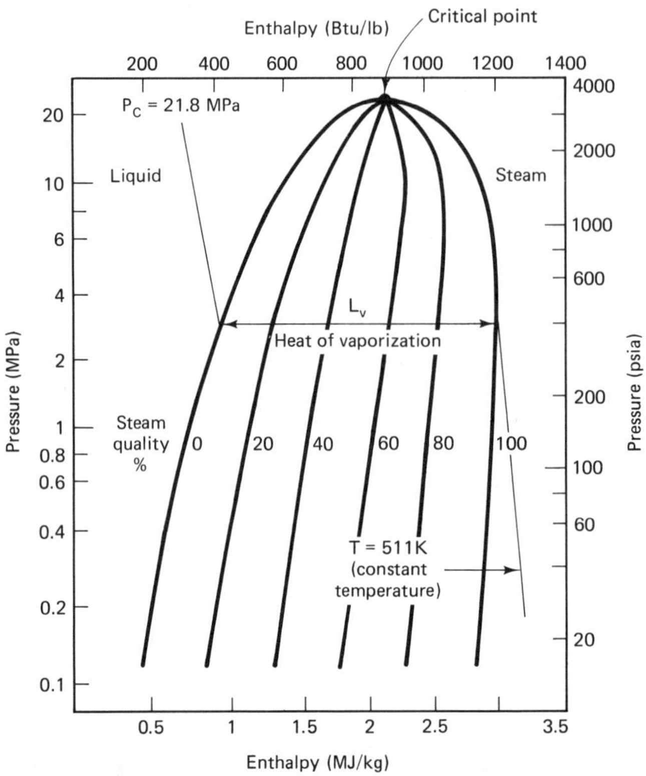

Figure 11-3 Enthalpy–pressure diagram for water (adapted from Bleakley, 1965)

2. Steam quality. Steam quality y is the amount of the total vapor, by weight, expressed as a fraction (or percent) of the mass of liquid plus vapor

$$y = \frac{\rho_3 S_3}{\rho_1 S_1 + \rho_3 S_3} \tag{11.2-1}$$

Quality is normally reported as a percent but, like saturation, is always used in calculations as a fraction. The quality lines within the two-phase envelope represent the relative amount of the total mass that is steam. Lines of constant temperature (only one is illustrated) in Fig. 11-3 fall steeply in the liquid region, are constant across the two-phase envelope, and then fall steeply again in the steam region.

3. Saturated liquid. A liquid is *saturated* if it exists at the temperature and pressure at which steam can be generated. This curve represents 0% steam quality.
4. Saturated vapor. Saturated vapor is water at the temperature and pressure where exactly 100% of the water present has been converted to a vapor.
5. Latent heat. Latent heat of vaporization L_v is the quantity of heat added to a given mass of saturated water (0% quality steam) to convert it to saturated vapor (100% quality steam) at constant temperature. The heat is *latent* because the temperature of the system does not change as the liquid is converted to va-

por. On an enthalpy–pressure diagram, latent heat is the difference in the *x*-coordinates between the dew and bubble point curves in Fig. 11-3 at a particular pressure. The latent heat vanishes at the critical point of water, 21.8 MPa and 647 K.

6. Sensible heat. Sensible heat is the quantity of heat that must be added to a given mass of water to raise its temperature without changing phase. This quantity is *sensible* because a thermometer in the water will sense a temperature increase as heat is added (at a constant pressure) until steam generation begins.

Figure 11-4 shows the pressure-specific volume diagram for water. The saturated vapor curve on the right of the envelope shows that steam density is much smaller than saturated liquid density except very near the critical point. In fact,

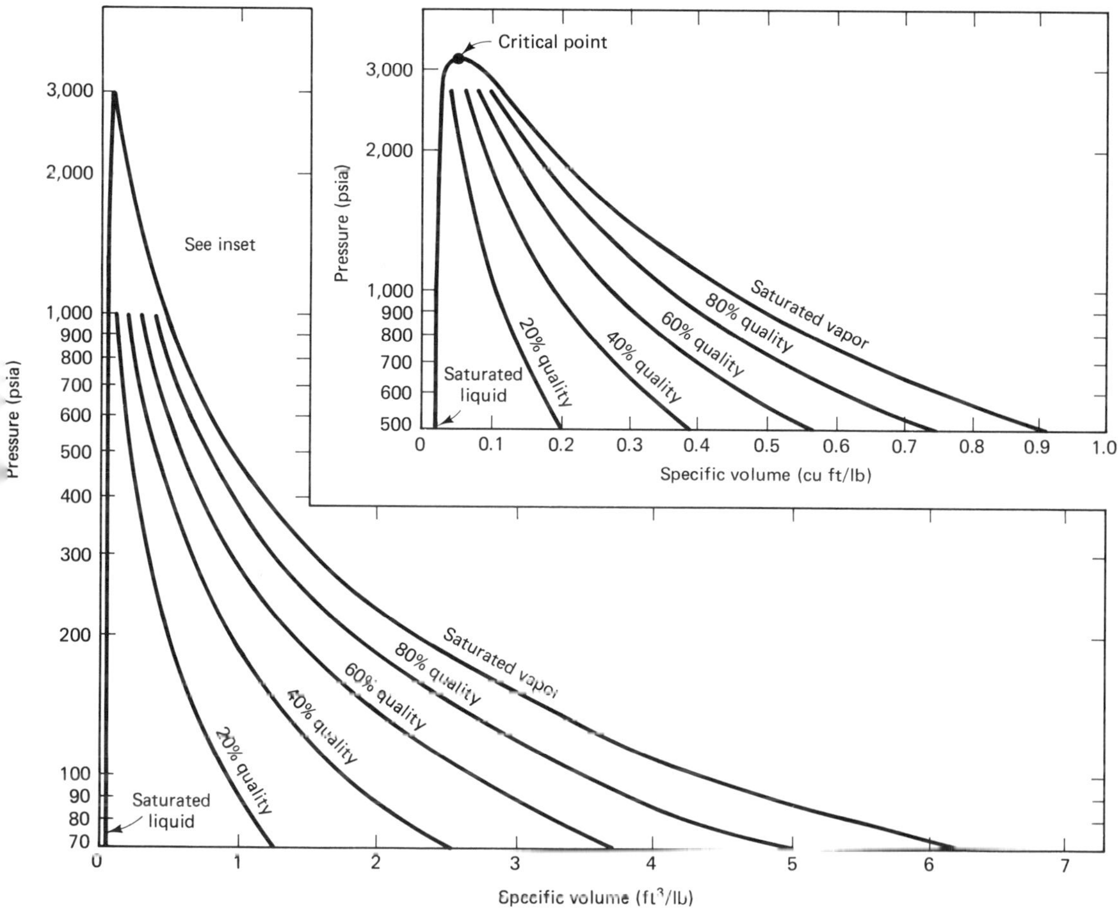

Figure 11-4 Pressure–specific-volume diagram for water (from Bleakley, 1965)

simultaneously considering Figs. 11-3 and 11-4, we see the volumetric enthalpy (ρH) of saturated liquid is actually greater than that for steam.

The physical properties in Figs. 11-3 and 11-4 appear in steam tables; Table 11-1 is an excerpt. In addition, Farouq Ali (1974) has fit analytical expressions to water properties (Table 11-2).

TABLE 11-1 THERMODYNAMIC PROPERTIES OF SATURATED WATER (ABRIDGED FROM KEENAN ET AL., 1969)

Temp. (°F)	Abs. pressure (psia) P	Specific[1] volume (ft³/lb)		Specific[2] enthalpy (Btu/lb)	
		Liquid	Vapor	Liquid	Vapor
32	0.08854	0.01602	3306	0.00	1075.8
35	.09995	.01602	2947	3.02	1077.1
40	.12170	.01602	2444	8.05	1079.3
45	.14752	.01602	2036.4	13.06	1081.5
50	.17811	.01603	1703.2	18.07	1083.7
60	.2563	.01604	1206.7	28.06	1088.0
70	.3631	.01606	867.9	38.04	1092.3
80	.5069	.01608	633.1	48.02	1096.6
90	.6982	.01610	468.0	57.99	1100.9
100	.9492	.01613	350.4	67.97	1105.2
110	1.2748	.01617	265.4	77.94	1109.5
120	1.6924	.01620	203.27	87.92	1113.7
130	2.2225	.01625	157.34	97.90	1117.9
140	2.8886	.01629	123.01	107.89	1122.0
150	3.718	.01634	97.07	117.89	1126.1
160	4.741	.01639	77.29	127.89	1130.2
170	5.992	.01645	62.06	137.90	1134.2
180	7.510	.01651	50.23	147.92	1138.1
190	9.339	.01657	40.96	157.95	1142.0
200	11.526	.01663	33.64	167.99	1145.9
210	14.123	.01670	27.82	178.05	1149.7
212	14.696	.01672	26.80	180.07	1150.4
220	17.186	.01677	23.15	188.13	1153.4
230	20.780	.01684	19.382	198.23	1157.0
240	24.969	.01692	16.323	208.34	1160.5
250	29.825	.01700	13.821	216.48	1164.0
260	35.429	.01709	11.763	228.64	1167.3
270	41.858	.01717	10.061	238.84	1170.6
280	49.203	.01726	8.645	249.06	1173.8
290	57.556	.01735	7.461	259.31	1176.8

TABLE 11-1 CONTINUED

Temp. (°F)	Abs. pressure (psia) P	Specific[1] volume (ft³/lb)		Specific[2] enthalpy (Btu/lb)	
		Liquid	Vapor	Liquid	Vapor
300	67.013	.01745	6.466	269.59	1179.7
310	77.68	.01755	5.626	279.92	1182.5
320	89.66	.01765	4.914	290.28	1185.2
330	103.06	.01776	4.307	300.68	1187.7
340	118.01	.01787	3.788	311.13	1190.1
350	134.63	.01799	3.342	321.63	1192.3
360	153.04	.01811	2.957	332.18	1194.4
370	173.37	.01823	2.625	342.79	1196.3
380	195.77	.01836	2.335	353.45	1198.1
390	220.37	.01850	2.0836	364.17	1199.6
400	247.31	.01864	1.8633	374.97	1201.0
410	276.75	.01878	1.6700	385.83	1202.1
420	308.83	.01894	1.5000	396.77	1203.1
430	343.72	.01910	1.3499	407.79	1203.8
440	381.59	.01926	1.2171	418.90	1204.3
450	422.6	.0194	1.0993	430.1	1204.6
460	466.9	.0196	0.9944	441.4	1204.6
470	514.7	.0198	.9009	452.8	1204.3
480	566.1	.0200	.8172	464.4	1203.7
490	621.4	.0202	.7423	476.0	1202.8
500	680.8	.0204	.6749	487.8	1201.7
520	812.4	.0209	.5594	511.9	1198.2
540	962.5	.0215	.4649	536.6	1193.2
560	1133.1	.0221	.3868	562.2	1186.4
580	1325.8	.0228	.3217	588.9	1177.3
600	1542.9	.0236	.2668	617.0	1165.5
620	1786.6	.0247	.2201	646.7	1150.3
640	2059.7	.0260	.1798	678.6	1130.5
660	2365.4	.0278	.1442	714.2	1104.4
680	2708.1	.0305	.1115	757.3	1067.2
700	3093.7	.0369	.0761	823.3	995.4
705.4	3206.2	.0503	.0503	902.7	902.7

[1] 1 ft³/lb = 0.062 m³/kg

[2] 1 Btu/lb = 2.2 kJ/kg

TABLE 11-2 THERMAL PROPERTIES OF WATER (ADAPTED FROM FAROUQ ALI, 1974)

English, P [=] psia				SI, P [=] MPa				
Quantity x	a	b	Limit psia	Quantity x	a	b	Limit MPa	Percent error
Saturation temperature (°F)	115.1	0.225	300	Saturation temperature −256 (K)	197	0.225	2.04	1
Sensible heat (Btu/lbm)	100	0.257	1,000	Sensible heat (MJ/kg)	0.796	0.257	6.80	0.3
Latent heat (Btu/lbm)	1,318	−0.0877	1,000	Latent heat (MJ/kg)	1.874	−0.0877	6.80	1.9
Saturated steam enthalpy (Btu/lbm)	1,119	0.0127	100	Saturated steam enthalpy (MJ/kg)	2.626	0.0127	2.04	0.3
Saturated steam specific volume (ft^3/lbm)	363.9	−0.959	1,000	Saturated steam specific volume (m^3/kg)	0.19	−0.959	6.80	1.2

Note: $x = aP^b$

Crude Oil Properties

Easily the most important crude oil property for thermal flooding is the viscosity dependence on temperature. As for most liquids, the Andrade (1930) equation captures this dependence

$$\mu_2 = Ae^{B/T} \tag{11.2-2}$$

where T is in absolute degrees. A and B are empirical parameters whose values are determined from two viscosity measurements at different temperatures. For extrapolation or interpolation, Eq. (11.2-2) indicates a semilog plot of viscosity versus T^{-1} should be a straight line.

If only one measurement is available, a coarse estimate of viscosity follows from Fig. 11-5. This single parameter correlation assumes viscosity is a universal function of the temperature change. To use the plot, enter the vertical axis with the known viscosity (4.38 mPa-s in this case), find the x-axis coordinate, move to the right by the temperature increase (101.6°C), and then return to the curve. The y-axis reading is the desired viscosity.

Several other crude oil properties, such as specific heat, volumetric heat capacity, and thermal conductivity, are functions of temperature. Empirical equations to predict these properties include the Gambill (1957) equation for specific heat

$$C_{p2} = \frac{0.7 + 0.0032T}{\rho_2^{0.5}} \tag{11.2-3}$$

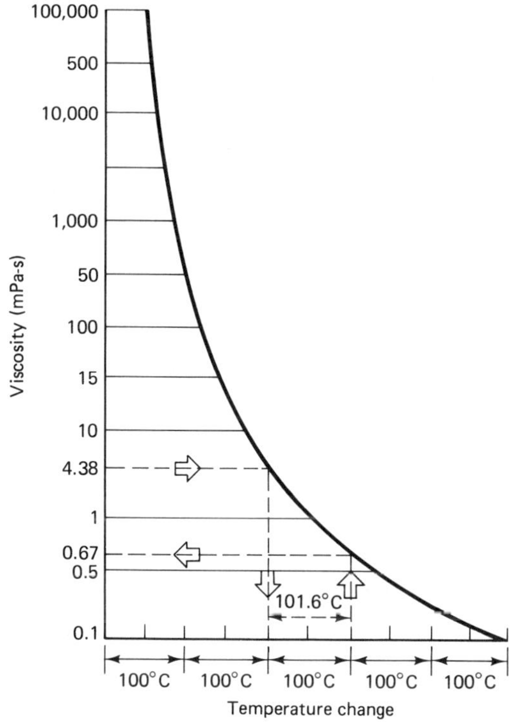

Figure 11-5 Single-parameter viscosity correlation (from Lewis and Squires, 1934)

where C_{p2} is in kJ/kg-K, T in degrees K, and ρ_2 in g/cm^3, and the thermal conductivity (Maxwell, 1950)

$$k_{T2} = 0.135 - 2.5 \times 10^{-5}T \tag{11.2-4}$$

k_{T2} in this equation has units of kJ/m-hr-K. Equation (11.2-4) is based on correlations for heavy fractions. These correlations are generally accurate to within 5%. (For more details on these correlations, see the original references.)

Equations (11.2-3) and (11.2-4) allow estimation of the thermal diffusion coefficient

$$K_{T2} = \frac{k_{T2}}{\rho_2 C_{p2}} \tag{11.2-5}$$

for crude oils. This quantity has units of m^2/s, as do the dispersion coefficients in Eq. (2.2-14).

Rock Properties

The total thermal conductivity of an unconsolidated sand filled with a single phase j is

$$k_{Tt} = 0.0149 - 0.0216\phi + 8.33 \times 10^{-7}k - 10^{-4}\left(\frac{D_{90}}{D_{10}}\right) + 7.77D_{50} + 4.188k_{Tj} + 0.0507k_{Ts} \tag{11.2-6}$$

The parameters in this equation have their usual meanings except that D_{90} and D_{10} are particle diameters smaller than 10% and 90% of the total sample by weight. The units on the total, fluid j, and solid (k_{Tt}, k_{Tj}, k_{Ts}) thermal conductivities are J/m-s-K, the permeability k is in μm^2, and the median grain size D_{50} is in mm.

For fluid-filled consolidated sandstones, the analogous relation is

$$\frac{k_{Tt}}{k_{Td}} = 1 + 0.299\left[\left(\frac{k_{Tj}}{k_{Ta}}\right)^{0.33} - 1\right] + 4.57\left[\frac{\phi}{1-\phi}\frac{k_{Tj}}{k_{Td}}\right]^{0.482}\left[\frac{\rho}{\rho_s}\right]^{-4.30} \tag{11.2-7}$$

where the subscripts a and d refer to air and dry rock. ρ is the density of the liquid-saturated rock. The thermal conductivities in Eqs. (11.2-6) and (11.2-7) are at a reference temperature of 293 K; they are rather weak functions of temperature, but corrections are given in Somerton (1973).

The volumetric heat capacity appears in many thermal flooding calculations. It is defined for all phases, including the solid, as

$$M_{Tj} = \rho_j C_{pj}, \qquad j = 1, \ldots, N_P, s \tag{11.2-8}$$

We have encountered this quantity before in Eq. (2.4-15), which defined a total volumetric heat capacity.

Solid Properties

Table 11-3 gives representative values of density, specific heat, thermal conductivity, and thermal diffusion coefficient for selected media. These values are appropriate for rough estimates of the rock–fluid thermal properties or for comparison to more refined estimates from Eqs. (11.2-6) through (11.2-8). The heat capacity of the solid phase varies relatively little with lithology, but the thermal conductivity can vary by a factor of 2 (compare the values for limestone and siltstone).

The fractional heat transferred into the crude is roughly the ratio of the water volumetric heat capacity M_{T1} to the overall heat capacity M_{Tt} (Eq. 2.4-15). This ratio is mainly a function of the oil saturation S_2 and the porosity ϕ. Using typical values for the heat capacities ($M_{T1} = 3.97$ MJ/m^3-K, $M_{T2} = 1.78$ MJ/m^3-K, and $M_{Ts} = 2.17$ MJ/m^3-K), we see that for a high-porosity, high-S_2 flood, about 20% (M_{T2}/M_{Tt}) of the heat resides in the oil. The value falls to 5% or less for tertiary floods in a low-porosity reservoir. These percentages also suggest guidelines about the best use of thermal methods: they are most efficient in high-porosity reservoirs undergoing secondary flooding.

Considering the success of thermal methods, such low percentages are remarkable. The success of thermal methods, where so little of the heat actually resides in the oil (recall we have not yet corrected the injected heat for losses to the wellbore and adjacent strata), must be due to the effectiveness of reducing oil viscosity (Fig. 11-1).

TABLE 11-3 DENSITY, SPECIFIC HEAT, THERMAL CONDUCTIVITY, AND THERMAL DIFFUSION COEFFICIENT OF SELECTED ROCKS (ADAPTED FROM FAROUQ ALI, 1974)

Rock	Bulk density (g/cm^3)	Specific heat (kJ/kg-K)	Thermal conductivity (J/s-m-K)	Thermal diffusion coefficient (mm^2/s)
Dry				
Sandstone	2.08	0.726	0.831	0.55
Silty sand	1.90	0.801	(0.66)	(0.43)
Siltstone	1.92	0.809	0.649	0.42
Shale	2.32	0.761	0.989	0.56
Limestone	2.19	0.801	1.611	0.92
Fine sand	1.63	0.726	0.593	0.50
Coarse sand	1.74	0.726	0.528	0.42
Water Saturated				
Sandstone	2.27	0.999	2.610	1.15
Silty sand	2.11	1.142	(2.50)	(1.04)
Siltstone	2.11	1.094	(2.50)	(1.08)
Shale	2.38	0.844	1.600	0.79
Limestone	2.38	1.055	3.360	1.34
Fine sand	2.02	1.344	2.607	0.96
Coarse sand	2.08	1.249	2.910	1.12

Note: Values in parentheses are estimated.

11-3 FRACTIONAL FLOW IN THERMAL DISPLACEMENTS

Propagation of Thermal Fronts

Noncombustion heat fronts can propagate in three ways: as hot water, as saturated steam, or as a noncondensable gas. Each has a characteristic velocity of propagation.

Let fluid 3 displace fluid 1 in a one-dimensional medium that has infinite lateral extent. As always, fluid 1 is cold water, but fluid 3 can be hot water, noncondensable gas, or a saturated steam. Fluid 3 has a higher temperature (T^+) than fluid 1 (T^-), and in all cases, the displacement takes place without mixing. This means neither the miscibility (or lack thereof) of the displacement nor its stability is at issue. We further assume conduction is negligible (this eliminates heat losses to adjacent media), displacement takes place at constant pressure, the reference temperature for all enthalpies is T^- (that is, $H_1 = 0$), and finally, all thermal properties are independent of temperature. These assumptions constitute the extension of the fractional flow assumptions to thermal floods.

The equations describing this displacement are the one-dimensional versions of the mass and energy balances, Eqs. (2.1-9) and (2.3-13). These equations are hyperbolic and reducible, with the above assumptions, so we expect coherence to apply for a centered simple wave here also (see Sec. 5-6). Coherence implies the energy and mass waves move with the same velocity. Unfortunately, we must deal with a dimensional velocity since the absence of incompressible fluids does not lead to a clear meaning for the normalizing or tracer velocity. But we can express the front velocity as a multiple of the cold water velocity u_1/ϕ.

Based on Eq. (5.4-5b), the front velocity is

$$v = \frac{1}{\phi}\left\{\frac{\rho_3 u_3 - \rho_1 u_1}{\rho_3 - \rho_1}\right\} \tag{11.3-1}$$

and based on a shock velocity derived from the conservation of energy, the same velocity is

$$v = \frac{1}{\phi}\,\frac{\rho_3 u_3 H_3}{\rho_3 H_3 + \left(\dfrac{1-\phi}{\phi}\right)\rho_s H_s} \tag{11.3-2}$$

Equation (11.3-2) neglects all forms of energy other than thermal and takes enthalpy to be equal to internal energy. $H_s = C_{ps}(T^+ - T^-) = C_{ps}\,\Delta T$ is the specific enthalpy of the solid. Three special cases follow from Eqs. (11.3-1) and (11.3-2).

Fluid 3 is hot water. In this case $\rho_3 = \rho_1$, $H_3 = C_{p1}\,\Delta T$, and Eq. (11.3-2) becomes

$$v_{HW} = \frac{u_3}{\phi}\,\frac{1}{1 + \dfrac{1-\phi}{\phi}\dfrac{M_{Ts}}{M_{T3}}} \tag{11.3-3}$$

Equation (11.3-3) has used the definition (Eq. 11.2-8) of volumetric heat capacity. To convert this to a cold water flux basis, use Eq. (11.3-1) to eliminate u_3/ϕ

$$v_{HW} = \frac{u_1}{\phi}\,\frac{1}{1 + D_{HW}} \tag{11.3-4}$$

where

$$D_{HW} = \left(\frac{1-\phi}{\phi}\right)\frac{M_{Ts}}{M_{T1}} \tag{11.3-5}$$

is the retardation factor for the thermal front. The velocity in Eq. (11.3-4) is independent of the temperature difference. For this case of incompressible flow the heat fronts propagate slower than tracer fronts that would have velocity u_1/ϕ. This slower propagation occurs because the thermal mass of the solid forces D_{HW} to be nonzero and positive. Note the analogy between Eq. (11.3-5) and the retardation factors for polymer Eq. (8.4-3) and micellar-polymer flooding Eq. (9.10-6).

Fluid 3 is steam of quality *y*. Here we have $H_3 = C_{p1}\Delta T + yL_v$, where L_v is the latent heat of vaporization. Substituted into Eq. (11.3-2), this gives

$$v_{SF} = \frac{u_3}{\phi} \frac{C_{p1}\Delta T + yL_v}{C_{p1}\Delta T + yL_v + \frac{1-\phi}{\phi}\frac{\rho_s C_{ps}}{\rho_3}} \tag{11.3-6}$$

Again, eliminating u_3/ϕ with Eq. (11.3-1) gives

$$v_{SF} = \frac{u_1}{\phi} \frac{1}{1 + D_{SF}} \tag{11.3-7}$$

where D_{SF} is the retardation factor for the steam front

$$D_{SF} = \frac{D_{HW}}{1 + h_D} \tag{11.3-8}$$

and h_D is a dimensionless latent heat

$$h_D = \frac{yL_v}{C_{p1}\Delta T} \tag{11.3-9}$$

h_D is the ratio of latent to sensible heat. Since $h_D \geqq 0$, steam fronts ($\Delta T > 0$) move faster than hot water fronts for equivalent u_1/ϕ. That is, L_v causes the front to propagate faster because it stores heat better. D_{SF} is not independent of the temperature difference (through the $C_{p1}\Delta T$ term), nor is it independent of the pressure (through L_v). High pressure steam floods approach the hot water limit since $L_v \rightarrow 0$ as pressure approaches the critical point of water.

We can make one further insight into the flux of steam. From Eq. (11.3-7), we must have $v_{SF} < u_1/\phi$ since D_{SF} must be greater than zero. But when we eliminate v_{SF} from this inequality using Eq. (11.3-1), we find that $u_3 > u_1$. Thus the steam front propagates slower than the cold water interstitial velocity, which in turn, is slower than the steam velocity. Modest front propagation rates require quite high steam fluxes.

Fluid 3 is a noncondensable gas. This case is similar to the hot water case except $H_3 = C_{p3}\Delta T$. A similar procedure to the above yields

$$v_G = \frac{u_1}{\phi} \frac{1}{1 + D_G} \tag{11.3-10}$$

where the retardation factor is now

$$D_G = \frac{1-\phi}{\phi} \frac{M_{Ts}}{M_{T3}} \tag{11.3-11}$$

Because $\rho_3 C_{p3} << \rho_1 C_{p1}$, D_G is much greater than D_{HW}. Hence heated gas floods propagate the slowest of all three cases.

Flow with Oil

In the next few sections, we analyze some simple thermal displacements using fractional flow theory. The basic governing equations are conservation of water

$$\frac{\partial(\rho_1 S_1 + \rho_3 S_3)}{\partial t_D} + \frac{\partial(\rho_1 f_1 + \rho_3 f_3)}{\partial x_D} = 0 \tag{11.3-12a}$$

conservation of oil

$$\frac{\partial(\rho_2 S_2)}{\partial t_D} + \frac{\partial(\rho_2 f_2)}{\partial x_D} = 0 \tag{11.3-12b}$$

and conservation of energy

$$\frac{\partial\left(\rho_1 S_1 H_1 + \rho_2 S_2 H_2 + \rho_3 S_3 H_3 + \dfrac{(1-\phi)}{\phi}\rho_s S_s H_s\right)}{\partial t_D} + \frac{\partial(\rho_1 f_1 H_1 + \rho_2 f_2 H_2 + \rho_3 f_3 H_3)}{\partial x_D} = 0 \tag{11.3-12c}$$

The mass balances follow from the basic one-dimensional fractional flow Eqs. (2.4-3)—with water in a second phase—and the energy balance Eq. (2.4-14b). Of course, to write these, we have had to make some rather restrictive assumptions that deserve special discussion when applied to thermal flooding.

Writing the equations in terms of the fractional flow rather than flux (with the f_j instead of u_j) means we have invoked the fractional flow assumptions. The portion of the fractional flow assumptions least likely to apply in general is that of incompressible fluids and rock. Oil and water (hot or cold) generally can be taken to be incompressible without great error, but steam is highly compressible. Aside from consistency with the other parts of this book and the attendant simplifications, there is no compelling reason for assuming steam to be incompressible. (For a treatment that does not assume incompressible fluids and solid, see Shutler and Boberg, 1972; Aydelotte and Pope, 1983.) However, pressure gradients in steam zones are usually small so that densities therein can be approximately constant. Of course, assuming an incompressible solid means there can be no oil production because of pore compression. The no-dissipation part of the fractional flow assumptions now includes thermal conductivity, which is absent from Eq. (11.3-12c).

We further assume there is no solubility of oil in water and no oil vaporization. We neglect all forms of energy except thermal energy, and we take internal energies equal to enthalpies. Finally, we solve Eq. (11.3-12c) by assuming no lateral heat loss. These assumptions mean we can use the conventional definitions of dimensionless time and position (Eqs. 5.2-6) and the method of coherence to solve for $S_1(x_D, t_D)$ and $T(x_D, t_D)$.

Hot Water Floods

For this case, $S_3 = 0$ and the assumption of incompressible fluids and solid is good. With this, Eq. (11.3-12a) is the only independent material balance; it becomes

$$\frac{\partial(\rho_1 S_1)}{\partial t_D} + \frac{\partial(\rho_1 S_1)}{\partial x_D} = 0 \tag{11.3-13a}$$

which has the saturation velocity

$$v_{S_1} = \left(\frac{\partial f_1}{\partial S_1}\right)_T \tag{11.3-13b}$$

Similarly, writing the energy Eq. (2.4-14c) in a fractional flow form yields

$$\left(M_{T1} S_1 + M_{T2} S_2 + \frac{1-\phi}{\phi} M_{Ts}\right)\frac{\partial T}{\partial t_D} + (M_{T1} f_1 + M_{T2} f_2)\frac{\partial T}{\partial x_D} = 0 \tag{11.3-14a}$$

We have also used the mass balance equations for water and oil to eliminate saturation derivatives. Equation (11.3-13b) implies a velocity for the temperature change

$$v_T = \frac{M_{T1} f_1 + M_{T2} f_2}{M_{T1} S_1 + M_{T2} S_2 + \dfrac{(1-\phi)}{\phi} M_{Ts}} \tag{11.3 14b}$$

v_T is a function of T only through the temperature dependence of the f_j. This temperature dependence is much weaker than the dependence of μ_2 on T. Thus the temperature wave is a shock in hot water displacements.

Nevertheless, coherence says the velocities in Eqs. (11.3-13b) and (11.3-14b) are equal, which yields

$$\frac{df_1}{dS_1} = \frac{f_1 + \dfrac{M_{T2}}{M_{T1} - M_{T2}}}{S_1 + \dfrac{M_{T2} + \dfrac{(1-\phi)}{\phi} M_{Ts}}{M_{T1} - M_{T2}}} \tag{11.3-15}$$

on eliminating $S_2 = 1 - S_1$ and $f_2 = 1 - f_1$. Equation (11.3-15) determines the water saturation S_1^* just behind the hot water front with the construction suggested in Fig. 11-6. The construction is quite analogous to that in Fig. 8-10 for polymer flooding except the straight material balance line does not begin from a point on the x-axis. This feature, which is caused by the difference between water and oil volumetric heat capacities in Eq. (11.3-15), arises because of the different directions of convection of the two components. Water is convecting heat to the thermal front while oil is convecting heat away from it. By our assumptions, convection is the only form of heat transfer occurring in this displacement.

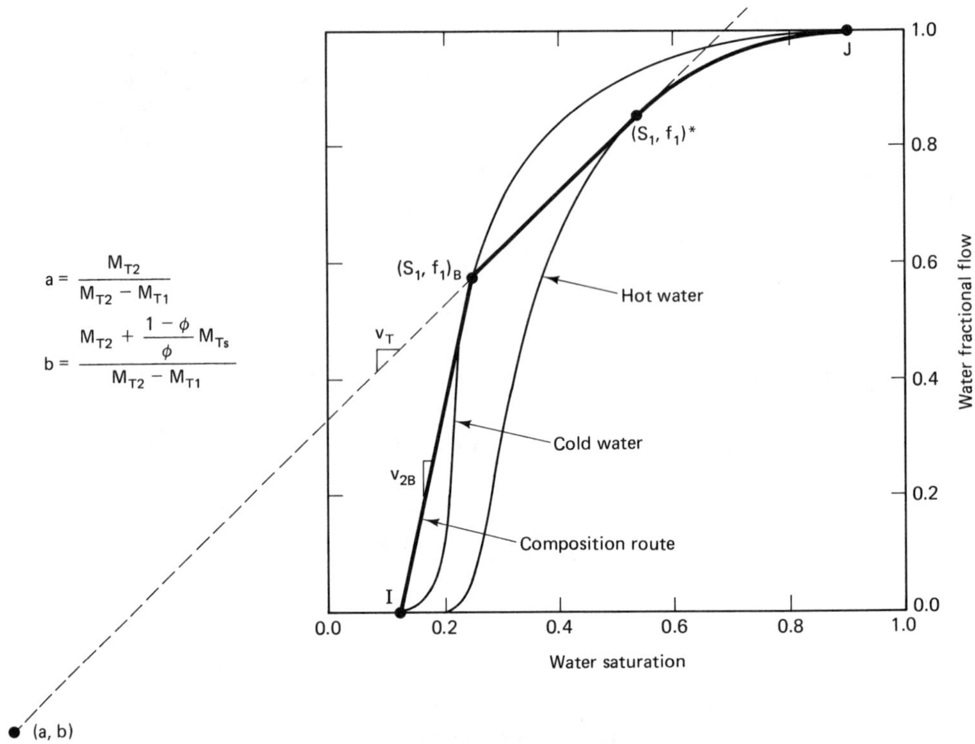

Figure 11-6 Graphical construction of hot water flood

The rear of the oil bank must propagate with the same velocity; hence the extension of the material balance line with the cold oil fractional flow curve gives the oil bank saturation. The leading edge of the cold oil bank follows from the usual secant construction as shown in Fig. 11-6.

Steam Displacements

We anticipate that in the absence of lateral heat loss, a steam front will propagate faster than a hot water front, and there will be no condensation. With heat losses, some condensation can occur, but we save this discussion for Sec. 11-6. Behind the steam front, temperature must be constant because, by assumption, pressure is constant (pressure gradients being negligible). Hence the entire left side of the energy balance becomes

$$\frac{\partial(\rho_3 S_3)}{\partial t_D} + \frac{\partial(\rho_3 f_3)}{\partial x_D} = 0 \tag{11.3-16}$$

in one dimension. From Eqs. (11.3-12a), (11.3-12b), and (11.3-16), we see that the mass of each phase is conserved in the steam zone. But this is exactly the same problem solved in Sec. 5-7 where we considered the flow of water, gas, and oil as immiscible phases. There we constructed the composition path diagram (Fig. 5-21), which illustrated the passing from an initial condition I to an injected condition J in two waves.

Though the solution presented in Fig. 5-21 is the same as the solution to a propagating steam front, they differ in one important respect: The initial condition I in the current problem is no longer given because this is the condition immediately behind the steam front. To find the condition I, we resort to applying the coherence condition across the steam front.

The integral coherence condition for the steam front, written in terms of the oil and water amounts, is

$$\underbrace{\frac{\rho_1(f_1H_1)^+ + \rho_2(f_2H_2)^+ + \rho_3(f_3H_3)^+}{\rho_1(S_1H_1)^+ + \rho_2(S_2H_2)^+ + \rho_3(S_3H_3)^+ + \dfrac{(1-\phi)}{\phi}\rho_s H_s^+}}_{\text{(energy)}} = \underbrace{\frac{\rho_1 f_1^+ + \rho_3 f_3^+ - \rho_1 f_1^-}{\rho_1 S_1^+ + \rho_3 S_3^+ - \rho_1 S_1^-}}_{\text{(water)}} = \underbrace{\frac{f_2^+ - f_2^-}{S_2^+ - S_2^-}}_{\text{(oil)}} \tag{11.3-17}$$

where the + and − represent conditions immediately upstream (the "injected" condition) and downstream of the front. No negative term appears in the energy equation because the reference temperature for enthalpy is T^- by assumption. We can simplify Eq. (11.3-17) by letting $H_3 = H_1 + L_v$, and $H_j = C_{pj}T$ (for $j = 1$ and 2). This yields

$$\frac{(M_{T1}f_1^+ + M_{T2}f_2^+ + M_{T3}f_3^+)T^+ + \rho_3 L_v f_3^+}{(M_{T1}S_1^+ + M_{T2}S_2^+ + M_{T3}S_3^+ + M_{Ts})T^+ + \rho_3 L_v S_3^+} = \frac{\rho_3 + f_1^+(\rho_1 - \rho_3) - \rho_3 f_2^+ - \rho_1 f_1^-}{\rho_3 + S_1^+(\rho_1 - \rho_3) - \rho_3 S_2^+ - \rho_1 S_1^-} = \frac{f_2^+ - f_2^-}{S_2^+ - S_2^-} \tag{11.3-18}$$

Normally, the steam zone temperature T^+ is known, leaving ten unknowns (f_j^+ and S_j^+ for $j = 1, 2, 3$; and f_j^- and S_j^- for $j = 1, 2$) in the two Eqs. (11.3-18). There are five independent relations between fractional flow and saturation—three for the upstream side and two for the downstream side—and, of course, the S_j and f_j on both sides must sum to unity, adding an additional two equations. We are left with an indeterminant system because there are nine total equations in ten unknowns.

One way around this indeterminancy is to invoke additional assumptions regarding the upstream conditions (Shutler and Boberg, 1972). An example of this would be to let $f_2^+ = 0$. Probably the most rigorous way would be to derive additional jump conditions by restoring the dissipative terms and solving the profile in a moving coordinate system (Bryant et al., 1986). Once the upstream conditions (I) are determined, the solution proceeds as in Sec. 5-7.

11-4 HEAT LOSSES FROM EQUIPMENT AND WELLBORES

Heat losses to rock and water easily represent the most significant source of heat loss in thermal methods. Although preventing this is beyond our reach, minimizing heat losses from equipment and wellbores and to adjacent strata *is* within our power.

Equipment losses. Heat is lost from surface equipment such as pipes, fittings, valves, and boilers. Such equipment is routinely insulated so that losses are small except under extreme circumstances. Most heat transfer books give procedures for detailed calculation from surface lines. Table 11-4 gives approximate heat losses that are adequate for most designs.

TABLE 11-4 TYPICAL VALUES OF HEAT LOSSES FROM SURFACE PIPING

		Heat loss, Btu/hr-ft² surface area for inside temperatures of			
Insulation	Conditions	200 °F	400 °F	600 °F	800 °F
Bare metal pipe	Still air, 0 °F	540[1]	1560	3120	
	Still air, 100 °F	210	990	2250	
	10-mph wind, 0 °F	1010	2540	4680	
	10-mph wind, 100 °F	440	1710	3500	
	40-mph wind, 0 °F	1620	4120	7440	
	40-mph wind, 100 °F	700	2760	5650	
		Heat loss, Btu/hr-ft of linear length of pipe at inside temperatures of			
Magnesia pipe insulation, air temperature 80 °F	Standard on 3-in. pipe	50[2]	150	270	440
	Standard on 6-in. pipe	77	232	417	620
	1 ½ in. on 3-in. pipe	40	115	207	330
	1 ½ in. on 6-in. pipe	64	186	335	497
	3 in. on 3-in. pipe	24	75	135	200
	3 in. on 6-in. pipe	40	116	207	322

[1] 1 Btu/hr-ft² ≅ 3.0 J/m² − s

[2] 1 Btu/hr-ft = 0.91 J/m − s

Wellbore losses. Heat losses from the wellbore, on the other hand, can cause a sizable energy debit if the reservoir is deep. We devote the remainder of this section to estimating the wellbore fluid temperature or quality and the rate of heat lost at a given depth.

Estimating heat losses from a wellbore provides an excellent extended example of the application of heat transfer theory and approximate solutions. It consists of three segments: steady-state heat transfer through the drill-hole region, transient heat conduction in the earth adjacent to the well, and an overall heat balance on the fluid in the wellbore itself. With appropriate assumptions, each problem can be solved separately and then merged for the final result. Combining steady-state, transient, and overall balances, as we are about to do, is a *quasi-steady* approximation. The basic equations for steady-state, transient, and macroscopic heat transfer are Eqs. (2.3-5), (2.3-10), and (2.5-8), respectively.

Estimating Overall Heat Transfer Coefficient

Estimates of heat transfer rate through the drill hole region come from the following equation:

$$\Delta \dot{Q} = 2\pi R_{to} U_T (T_f - T_d) \Delta z \tag{11.4-1}$$

where $\Delta \dot{Q}$ is the heat transfer rate (energy units divided by time) through a section of a vertical wellbore Δz in height having an outer tubing radius of R_{to}. U_T is the overall heat transfer coefficient based on the outer surface of the tubing. Using Eq. (11.4-1) requires an estimate of U_T.

Heat transfer through the drill-hole region involves several different resistances between the fluid flowing in the tubing and the formation. Starting with the formation and moving inward, these are a cement zone, casing, annulus, tubing insulation, the tubing itself, and the flowing fluid. Figure 11-7 shows a schematic of this

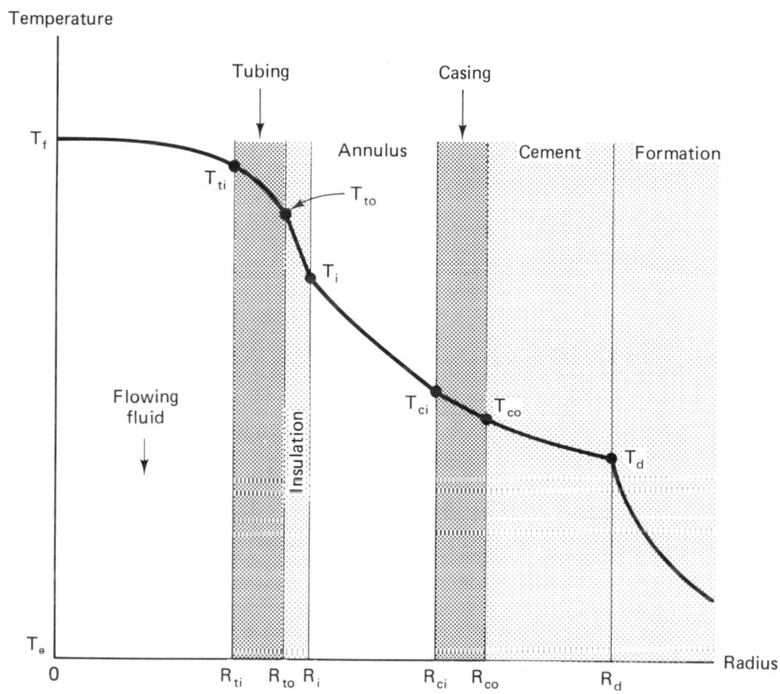

Figure 11-7 Schematic temperature profile in drill hole (adapted from Willhite, 1967)

and definitions of symbols. Equation (11.4-1) is in terms of the temperature difference between the fluid T_f and the temperature at the drill-hole radius T_d.

Following Willhite (1967), we assume radial symmetry in the drill hole, no heat transfer in the z direction, and temperature-independent thermal conductivities. Since the drill-hole region occupies a much smaller volume than the formation, it is reasonable to assume temperature transients here die out much faster than in the formation. Thus we take a steady-state energy balance to apply in the tubing, insulation, casing, and cement

$$\frac{d}{dr}(rq_c) = 0 \tag{11.4-2}$$

where q_c is the radial component of the conductive heat flux $\vec{q}_c$ in Eq. (2.3-8), heat transfer here being solely by conduction. Since the radius-heat flux product is a constant, the heat transfer rate over height z is also a constant

$$\Delta\dot{Q} = 2\pi r\Delta z q_c = -2\pi r k_T \frac{dT}{dr}\Delta z \tag{11.4-3}$$

Equation (11.4-3) may be integrated for the temperature differences between the inside and outside of each region

$$T_{ti} - T_{to} = \frac{\Delta\dot{Q}\ln\left(\dfrac{R_{to}}{R_{ti}}\right)}{2\pi k_{Tt}\Delta z} \quad \text{(tubing)} \tag{11.4-4a}$$

$$T_{to} - T_i = \frac{\Delta\dot{Q}\ln\left(\dfrac{R_i}{R_{to}}\right)}{2\pi k_{Ti}\Delta z} \quad \text{(insulation)} \tag{11.4-4b}$$

$$T_{ci} - T_{co} = \frac{\Delta\dot{Q}\ln\left(\dfrac{R_{co}}{R_{ci}}\right)}{2\pi k_{Tc}\Delta z} \quad \text{(casing)} \tag{11.4-4c}$$

$$T_{co} - T_d = \frac{\Delta\dot{Q}\ln\left(\dfrac{R_d}{R_{co}}\right)}{2\pi k_{Tcem}\Delta z} \quad \text{(cement)} \tag{11.4-4d}$$

k_{Tt} in Eq. (11.4-4a) is the thermal conductivity of the tubing.

Neither the fluid in the tubing nor the fluid in the annulus transfers heat strictly by conduction; hence they must be treated separately. Let the heat transfer rate in these regions be expressed as

$$T_f - T_{ti} = \frac{\Delta\dot{Q}}{2\pi R_{ti}\Delta z h_{Tf}} \quad \text{(flowing fluid)} \tag{11.4-5a}$$

$$T_i - T_{ci} = \frac{\Delta\dot{Q}}{2\pi R_i \Delta z h_{Ta}} \quad \text{(annulus fluid)} \tag{11.4-5b}$$

by analogy to Eq. (11.4-1). h_{Tf} and h_{Ta} are the heat transfer coefficients of the fluids in the tubing and annulus, respectively. They may be estimated mainly through correlations as we discuss below.

We can sum Eqs. (11.4-4) and (11.4-5) to give the overall temperature drop

$$T_f - T_d = \frac{\Delta\dot{Q}}{2\pi\Delta z}\left\{\frac{1}{R_{ti}h_{Tf}} + \frac{\ln\left(\frac{R_{to}}{R_{ti}}\right)}{k_{Tt}} + \frac{\ln\left(\frac{R_i}{R_{to}}\right)}{k_{Ti}} + \frac{1}{R_i h_{Ta}} + \frac{\ln\left(\frac{R_{co}}{R_{ci}}\right)}{k_{Tc}} + \frac{\ln\left(\frac{R_d}{R_{co}}\right)}{k_{Tcem}}\right\} \tag{11.4-6}$$

which inserted into Eq. (11.4-1), gives the overall heat transfer coefficient

$$U_T^{-1} = R_{to}\left\{\frac{1}{R_{ti}h_{Tf}} + \frac{\ln\left(\frac{R_{to}}{R_{ti}}\right)}{k_{Tt}} + \frac{\ln\left(\frac{R_i}{R_{to}}\right)}{k_{Ti}} + \frac{1}{R_i h_{Ta}} + \frac{\ln\left(\frac{R_{co}}{R_{ci}}\right)}{k_{Tc}} + \frac{\ln\left(\frac{R_d}{R_{co}}\right)}{k_{Tcem}}\right\} \tag{11.4-7}$$

This equation expresses the total conductance between the fluid and the formation as a sum of series resistances each weighed by geometrical factors. If any of the zones in Fig. 11-7 are absent (inner and outer radii equal), that term will be absent in Eq. (11.4-7). Moreover, if the thermal conductivity of a component is large, as is usually true with the tubing and casing, the corresponding term in Eq. (11.4-7) will be small. Many times, in fact, a single term will dominate the overall heat transfer coefficient (as might occur in insulated tubing where k_{Ti} is small), but general use of Eq. (11.4-7) requires us to estimate h_{Tf} and h_{Ta}.

Heat Transfer Coefficient in Tubing and Annulus

The major difficulty in using Eq. (11.4-7) is estimating h_{Tf} and h_{Ta} since the other terms are constant. Heat transfer from a flowing fluid is by conduction and convection, and if the flow rate is large, heat is dissipated by viscous heating. Figure 11-8(a) shows schematic velocity and temperature profiles. Theoretical arguments (Bird et al., 1960) suggest h_{Tf} correlates as the following dimensionless equation:

$$N_{\text{Nu}} = f(N_{\text{Pr}}, N_{\text{Re}}, N_{\text{Br}}) \tag{11.4-8}$$

for tubes with large length-to-diameter ratios. The dimensionless groups in Eq. (11.4-8) are

$$\text{Nusselt number} = \frac{R_{ti}h_{Tf}}{k_{Tf}} = N_{\text{Nu}} \tag{11.4-9a}$$

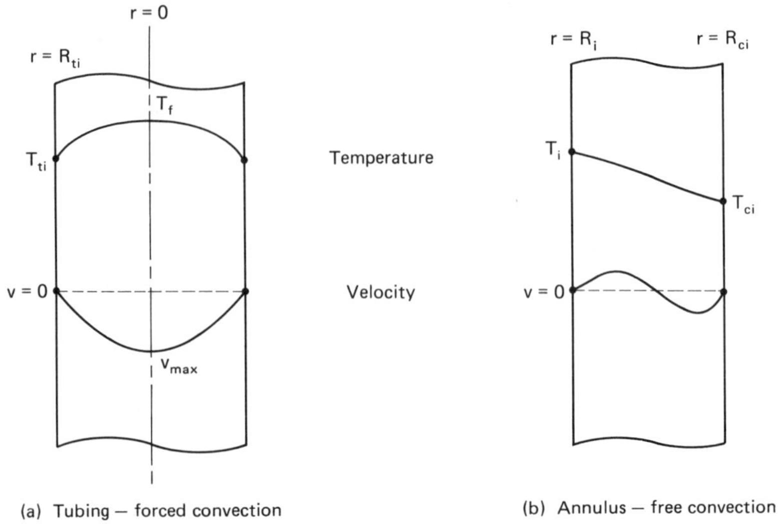

Figure 11-8 Schematic velocity and temperature profiles in tubing and annulus (adapted from Willhite, 1967)

$$\text{Prandtl number} = \frac{C_{pf}\mu_f}{k_{Tf}} = N_{\text{Pr}} \tag{11.4-9b}$$

$$\text{Reynolds number} = \frac{2\rho_f\bar{v}R_{ti}}{\mu_f} = N_{\text{Re}} \tag{11.4-9c}$$

$$\text{Brinkman number} = \frac{\mu_f v_{\max}^2}{k_{Tf}(T_f - T_{ti})} = N_{\text{Br}} \tag{11.4-9d}$$

where the overbar in Eq. (11.4-9c) indicates a volume average. As the naming after persons suggests, these are familiar groups in the heat transfer literature. Each group has a physical interpretation: N_{Nu} is the ratio of total to conductive heat transfer; N_{Pr} is the ratio of convection to conductive heat transfer; N_{Re} is the ratio of inertial to viscous forces in the fluid flow; and N_{Br} is the ratio of viscous heat dissipation to conduction. Of these four, N_{Br} is the only one containing a temperature difference; however, if it is small, as it often is for liquid flows, this dependence is weak. For simple geometry, the specific form of Eq. (11.4-8) can be derived theoretically; in practical cases, the relationship is empirical (Bird et al., 1960). See exercise 11F.

Heat transfer through an annulus is even more complicated. If the annulus is sealed at both ends, there can be no bulk flow; however, the temperature difference between T_i and T_{ci} causes local density differences in the annulus fluid, which causes flow. We call such flow *free* convection to distinguish it from the *forced* convection in the tubing. Figure 11-8(b) shows schematic velocity and temperature profiles for the annulus. Again, a dimensional argument suggests a relation among dimensionless groups, a particular form of which is (Willhite, 1967)

$$N_{\text{Nu}} = 0.049(N_{\text{Gr}}N_{\text{Pr}})^{1/3}N_{\text{Pr}}^{0.074} \tag{11.4-10}$$

for flat plates with large length-to-diameter ratios. The additional group in Eq. (11.4-10) is the Grashof number

$$N_{Gr} = \frac{(R_{ci} - R_i)^3 g \rho_a \beta_T (T_i - T_{ci})}{\mu_a} \tag{11.4-9e}$$

which is the ratio of free convection transport to viscous forces. The parameter β_T is thermal expansion coefficient defined as $-1/\rho_a(\partial\rho_a/\partial T)_P$, and the subscript a refers to the annulus fluid. The fluid properties in N_{Nu}, N_{Pr}, and N_{Gr} are now based on the annulus fluid. The Grashof number contains a temperature difference that is usually unknown a priori; thus in applications, it may be necessary to solve for heat loss by trial and error.

Usually an annulus is air filled, but on occasion, it has been evacuated. When this occurs, heat transfer is almost exclusively through radiation. Radiation is a form of heat flux independent of convection or conduction. Under some circumstances, radiative heat transfer can account for a substantial fraction of the heat transfer.

Heat Conduction in the Formation

The immense thermal mass of the earth surrounding the wellbore, only a small fraction of which is in contact with the reservoir, suggests heat transfer here is transient. In this segment, we repeat a procedure first given by Ramey (1962) for calculating temperatures beyond the drill hole $r > R_d$.

Let heat transfer in the formation be strictly by radial conduction. In the absence of any velocities, Eq. (2.3-10) becomes

$$\frac{\partial T}{\partial t} = \left(\frac{k_T}{\rho C_p}\right)_s \cdot \frac{1}{r}\frac{\partial}{\partial r}\left(r\frac{\partial T}{\partial r}\right) = \frac{K_{Ts}}{r}\frac{\partial}{\partial r}\left(r\frac{\partial T}{\partial r}\right) \tag{11.4-11}$$

where Eq. (2.3-9) has been inserted for the conductive heat flux and Eq. (11.2-5) used for the thermal diffusion coefficient. Equation (11.4-11) also assumes an incompressible, single-phase formation so that a change in internal energy is manifest only as a change in temperature. Once this equation is solved for $T(t, r)$ for $r > R_d$, the heat transfer rate follows from the spatial gradient at $r = R_d$. The following boundary and initial conditions apply to Eq. (11.4-11):

$$T(0, r) = T(t, \infty) = T_e \tag{11.4-12a}$$

$$-k_{Ts}\left(\frac{\partial T}{\partial r}\right)_{r=R_d} = \frac{\Delta\dot{Q}}{2\pi R_d \Delta z} \tag{11.4-12b}$$

The undisturbed external temperature T_e is actually a function of z because of the geothermal gradient

$$T_e = a_T z + T_0 \tag{11.4-13}$$

where a_T is usually about 0.18 K/km, and T_0 is the mean surface temperature. The existence of this gradient implies a constant rate of heat transfer from the earth's core; it also suggests a z-dependency in the problem that is not explicit in the equations. The solution will therefore be for a particular z, but the variation with z arises only when solving the energy balance for the flowing fluid. Equation (11.4-12b) expresses the continuity of heat flux at $r = R_d$. Combining it with Eq. (11.4-1) gives the "conduction" condition.

$$-k_{Ts}\left(\frac{\partial T}{\partial r}\right)_{r=R_d} = \frac{U_T R_{to}(T_f - T_d)}{R_d} \tag{11.4-14}$$

As we discussed, all temperatures are functions of z.

For nonzero R_d, the solution to Eqs. (11.4-11), (11.4-12), and (11.4-14) must be numerical, but once it is known, the heat transfer rate follows from Eq. (11.4-12b) to give

$$\Delta\dot{Q} = \frac{2\pi k_{Ts}(T_d - T_e)\Delta z}{f_T(t_D)} \tag{11.4-15}$$

where f_T is a function of dimensionless time t_D and formation Nusselt number

$$t_D = \frac{K_{Ts}t}{R_d^2} \tag{11.4-16a}$$

$$N_{\text{Nu}} = \frac{R_{to}U_T}{k_{Ts}} \tag{11.4-16b}$$

Figure 11-9 shows the logarithm of f_T plotted versus the logarithm of t_D with N_{Nu} as a parameter.

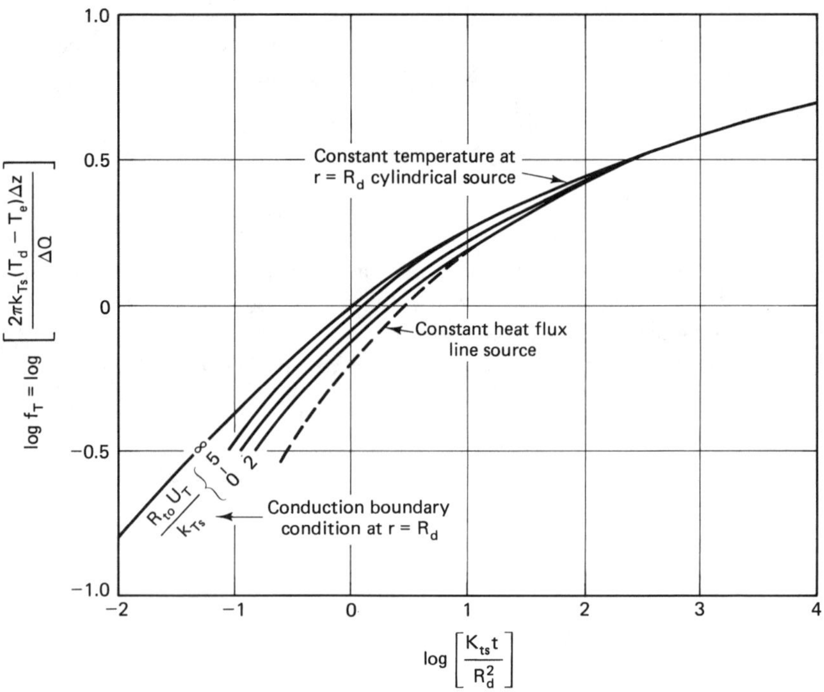

Figure 11-9 Transient heat transfer function (from Ramey, 1962)

Ramey (1964) gives the procedure for using these equations. Let's solve for the inner casing temperature T_{ci} and heat loss rate $\Delta\dot{Q}$ at a given depth and time. We know the radii in Fig. 11-7; the thermal conductivities of the tubing, insulation, casing, and cement zone; the thermal properties of the flowing fluid, the annulus fluid, and the formation; the viscosity and average velocity of the flowing fluid; and the depth z and the bulk fluid temperature T_f. The procedure is as follows:

1. Calculate T_e from Eq. (11.4-13), and calculate N_{Pr} and N_{Re} for the flowing fluid and N_{Pr} for the annulus fluid from Eqs. (11.4-9). Calculate t_D from Eq. (11.4-16a).
2. Assume a value for h_{Ta}, and calculate U_T from Eq. (11.4-7), all other quantities being independent of temperature. If N_{Br} is not small, a value for h_{Tf} must be assumed also.
3. Calculate the formation Nusselt number from Eq. (11.4-16b) and, using this and t_D, estimate f_T from Fig. 11-9. Calculate T_d from

$$T_d = \frac{T_f f_T(t_D) + \left(\dfrac{k_{Ts}}{R_{to} U_T}\right) T_e}{f_T(t_D) + \left(\dfrac{k_{Ts}}{R_{to} U_T}\right)} \tag{11.4-17}$$

   T_f in this equation follows from eliminating $\Delta\dot{Q}$ between Eqs. (11.4-1) and (11.4-15). We can now calculate $\Delta\dot{Q}$ from either equation.
4. With $\Delta\dot{Q}$ and T_d known, the casing temperature T_{ci} and all the others follow successive application of Eqs. (11.4-4) and (11.4-5).

   The solution would now be complete but for the assumed value of h_{Ta} in step 2.
5. Calculate N_{Gr} from Eq. (11.4-9e) and use Eq. (11.4-10) to estimate h_{Ta}. If radiation is important, we would correct for it here.
6. Recalculate U_T from its definition (Eq. 11.4-7). Compare this value to that used in step 2; repeat steps 2–6 with the new value of U_T if agreement is not satisfactory. The convergence test is on U_T, a much weaker function of temperature than h_{Ta}. Convergence should be obtained in fewer than three steps.

Heat Loss from Wellbore

We now focus attention on the element Δz through which heat is passing at rate $\Delta\dot{Q}$. First, we eliminate T_d between Eqs. (11.4-1) and (11.4-15) to give

$$\Delta\dot{Q} = \frac{2\pi k_{Ts} R_{to} U_T}{k_{Ts} + R_{to} k_{Ts} f_T(t_D)} (T_f - T_e)\Delta z \tag{11.4-18}$$

In what follows; we take U_T to be constant for ease of illustration.

If we apply the overall energy balance (Eq. 2.5-6a) to the element Δz, we have

$$A\,\Delta z\,\frac{d}{dt}(\overline{\rho_f U}) + \Delta\dot{H} - \overline{v}\rho_f Ag\Delta z = -\Delta\dot{Q} \tag{11.4-19a}$$

where we have neglected kinetic energy and mechanical work terms. Further, by writing the enthalpy rate entering and leaving Δz as the product of a specific enthalpy and a constant mass flow rate $\dot{m} = \overline{v}\rho_f A$, we have

$$\dot{m}(\Delta\overline{H} - g\Delta z) = -\Delta\dot{Q} \tag{11.4-19b}$$

Equation (11.4-19b) has also dropped the time derivatives by the same quasi-steady-state argument used above for the drill hole.

The simplest heat loss model follows from Eq. (11.4-15) by taking T_f constant at the surface inlet temperature (this makes $\Delta\overline{H} = 0$) and integrating the resulting ordinary differential equation for $d\dot{Q}/dz$ (in the limit $\Delta z \rightarrow 0$) (Ramey, 1964)

$$\dot{Q}(z) = \frac{2\pi k_{Ts} R_{to} U_T}{k_{Ts} + R_{to} U_T f_T(t_D)}\left[(T_f - T_0)z - \frac{a_T z^2}{2}\right] \tag{11.4-20}$$

where we have replaced T_e with Eq. (11.4-13) and T_d with Eq. (11.4-1) before integrating. This equation yields the maximum heat loss rate up to depth z because the temperature difference between T_f and T_e is the maximum possible value. $(T_f - T_0)$ is the difference between inlet and surface temperatures.

For more general cases, let's eliminate $\Delta\dot{Q}$ between Eqs. (11.4-18) and (11.4-19b), which yields, after again taking the limit as z approaches zero,

$$\frac{d\overline{H}}{dz} - g = -\frac{2\pi k_{Ts} R_{to} U_T (T_f - T_e)}{\dot{m}[k_{Ts} + R_{to} U_T f_T(t_D)]} \tag{11.4-21}$$

Equation (11.4-21) is a working equation. The sign convention is that z increases downward, and $\dot{Q}$ is positive when heat is lost from the wellbore. We can invoke Eq. (11.4-21) on several special cases by taking different forms for the specific enthalpy.

If the fluid flowing in the tubing is an ideal gas, as steam would be at low pressure, the enthalpy is independent of pressure

$$d\overline{H} = C_{p3} dT_f \tag{11.4-22a}$$

Substituted into Eq. (11.4-21), this gives

$$\frac{dT_f}{dz} = \frac{g}{C_{p3}} - \frac{2\pi k_{Ts} R_{to} U_T (T_f - T_e)}{C_{p3}\dot{m}[k_{Ts} + R_{to} U_T f_T(t_D)]} \tag{11.4-22b}$$

Equation (11.4-22b) will integrate to

$$T_f = a_T z + T_0 - A_T\left(a_T + \frac{g}{C_{p3}}\right) + \left[(T_f - T)_0 + A_T\left(a_T + \frac{g}{C_{p3}}\right)\right]e^{-z/A_T} \tag{11.4-22c}$$

where

$$A_T(t_D) = \frac{\dot{m}C_{p3}[k_{Ts} + R_{to}U_T f_T(t_D)]}{2\pi k_{Ts} R_{to} U_T} \tag{11.4-22d}$$

and T_{fo} in Eq. (11.4-22c) is the inlet surface temperature at $z = 0$. With T_f now determined as a function of depth, we can integrate Eq. (11.4-19b) for the heat loss down to z. These two equations say the fluid temperature and heat loss vary as an exponential plus a linear term with depth, the rate of change being determined by A_T, which is proportional to the mass flow rate.

If the flowing fluid is a superheated vapor at the inlet surface temperature, Eq. (11.4-22c) will describe its temperature down to the saturation temperature. Below this point, the fluid will be a saturated two-phase mixture some distance down the tubing where the fluid will condense gradually to saturated water as more heat is lost. In this case, the specific enthalpy relates to the steam quality as

$$\overline{H} = H_1 + yL_v \tag{11.4-23a}$$

If pressure is constant, this leads to a relatively simple differential equation in quality (Satter, 1965)

$$\frac{dy}{dz} = \frac{g}{L_v} - \frac{(T_f - T_e)}{A_T} \tag{11.4-23b}$$

where

$$A_T = \frac{\dot{m}L_v[k_{Ts} + R_{to}U_T f(t_D)]}{2\pi R_{to} k_{Ts}} \tag{11.4-23c}$$

Because a change in steam quality at constant pressure must take place at constant temperature, we can integrate Eq. (11.4-23b), with T_f constant at the saturation temperature for the fluid quality in the tubing

$$y = 1 + \left\{ \frac{\left(\frac{gA_T}{L_v}\right) + T_0 - T_f}{A_T} \right\} z + \frac{a_T z^2}{2A_T} \tag{11.4-23d}$$

where $y = 1$ at $z = 0$. The heat loss follows from Eq. (11.4-23a); note that a constant flowing temperature does not imply no heat loss if the fluid is condensing.

Equation (11.4-23d) is deceiving in its simplicity. It has neglected the hydrodynamics of two-phase flow in a vertical pipe and the significant effect that U_T (through h_{Tf}) can change with condensation. Still, the equation is quite instructive, particularly when merged with a heat loss calculation for flowing gas.

Heat is lost from the wellbore because a temperature difference exists between the heated wellbore and the geothermal temperature in the surrounding formation. Figure 11-10 shows the state of a wellbore, into which superheated steam is being injected, as a function of depth and injection time. This calculated result assumes pressure is constant in the wellbore.

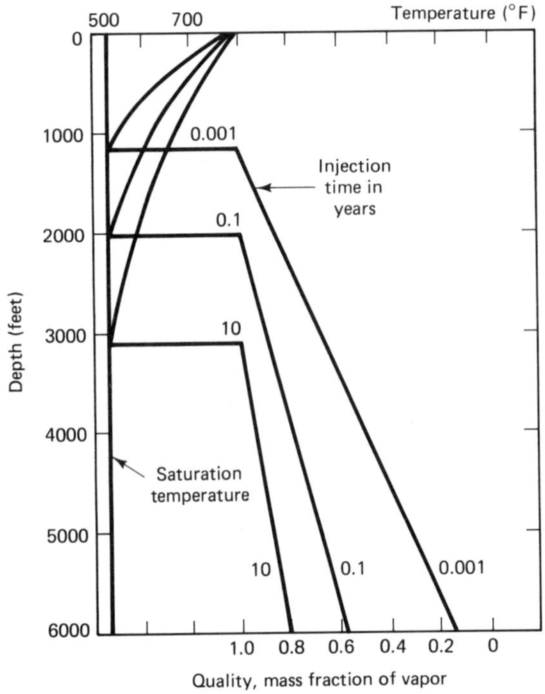

Figure 11-10 Change in temperature or steam quality with depth (from Satter, 1965)

The progression in Figure 11-10 can easily be shown on the enthalpy–pressure diagram in Fig. 11-3. Suppose the wellbore pressure is 3.1 MPa and the surface temperature is 800 K. The temperature falls to the dew point temperature at the wellbore pressure. Equation (11.4-22c) approximates the temperature change, a horizontal line in the superheated steam region in Fig. 11-3. From this point further down, the wellbore temperature becomes constant, and steam quality declines as predicted by Eq. (11.4-23d). The change becomes apparent in Fig. 11-3 where the extension of the horizontal line from the superheated steam region coincides with lines of constant temperature but decreasing quality. At a fixed time, the heat lost per unit mass of steam is given by the difference in the x-coordinates on the enthalpy–pressure diagram. The entire progression moves down the wellbore with increasing time.

Controlling Heat Loss

Heat losses from the wellbore to the surrounding formation can be controlled in three ways.

1. Restrict application. Figures 11-11 and 11-12 indicate deep wells and large producing lives are to be avoided. Steam processes, in particular, are generally not practical at depths more than 1,000 m. If the reservoir depth is not too

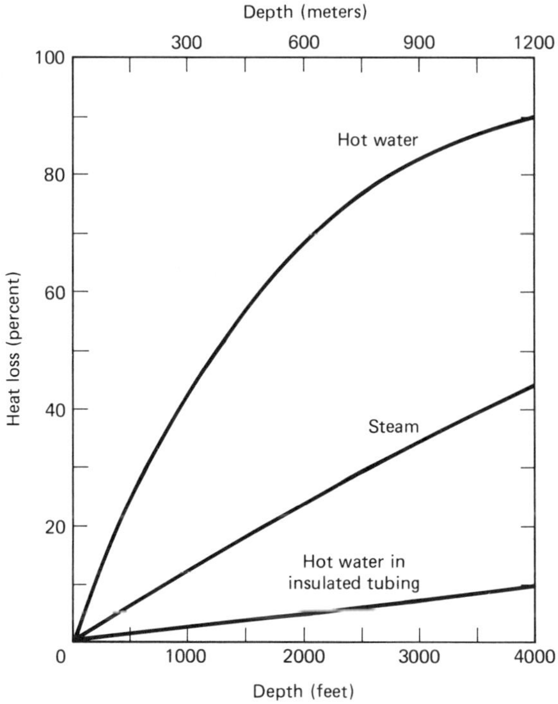

Figure 11-11 Effect of insulation on heat loss (from Ramey, 1962)

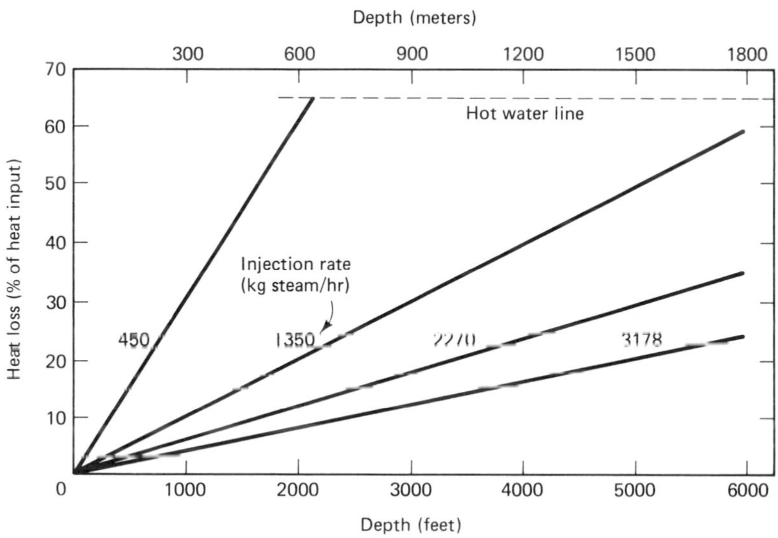

Figure 11-12 Effect of injection rate on heat loss (from Satter, 1965)

large, pattern spacing can be relatively small, which will cause short producing lives. Small spacing will also reduce the amount of heat lost to the adjacent strata.

2. Insulate the casing. The actual mechanisms for wellbore heat loss are conduction from the casing, radiation between the tubing and the casing, and free convection in the annulus. We can suppress all these mechanisms by insulating the casing or tubing from the formation.

   Figure 11-11 shows the dramatic effect that insulation can have. The insulation causes a tenfold reduction in heat losses in hot water injection. The reduction would not be quite as large in steam injection since the vapor heat transfer coefficient is already about one half that of hot water.

   Whether or not insulation is appropriate depends on the benefit in heat saved weighed against the cost of insulation. This, in turn, depends on the type of insulation and the depth of the well. By now it has become common practice to leave an airspace between the annulus and tubing to provide partial insulation.

3. Inject at high rate or high surface pressure. As the injection rate increases, so does the heat transfer from the hot fluid. Several of the equations given above attest to this; examine the fluid temperature change in Eq. (11.4-22b) and the quality change in Eq. (11.4-23d). Heat loss rate increases with $\dot{m}$, but the heat loss rate does not increase nearly as fast as the rate heat is delivered to the formation, so the *relative* loss rate goes down. Figure 11-12 shows the advantages of this strategy; tripling the injection rate reduces the relative heat losses by about a factor of 3. A secondary benefit to be gained from a high injection rate is a short project life.

Two cautions are in order here. The injection pressure must not be so high as to exceed the formation parting pressure. Just as in waterflooding, such parting will introduce high permeability channels into the formation with resulting loss of volumetric sweep efficiency (although this problem is not as severe in thermal as in other EOR methods). The second concern is that a high injection rate in a steam drive will lead to excessive heat losses through the producers if continued after steam breakthrough. The rate of loss from the wellbore must be balanced with the rate of loss through the producers when this happens, but usually the injection rate is reduced once steam breaks through.

An obvious way to avoid wellbore heat losses is to generate the heat in situ or at the bottom of the wellbore. The first technique is the basis for in situ combustion (see Sec. 11-8), which extends the practical depth of thermal floods to about 2,000 m. Below this depth, compression costs tend to be prohibitive. The second technique implies downhole steam generation.

Two types of downhole steam generators produce steam at the sand face. In the direct-fired generator, water and fuel are mixed in a combustion chamber, burned, and the entire mixture (steam, unburned fuel, and combustion products) is then injected. The CO_2 in the combustion products is an EOR agent in its own right, but the device is difficult to operate and maintain. The indirect-fired generator returns the combustion mixture to the surface and, though a little easier to maintain, is clearly more complex. The combustion products for both can represent an environmental hazard.

11-5 HEAT LOSSES TO OVERBURDEN AND UNDERBURDEN

The fourth source of heat loss in thermal methods is loss to the adjacent strata or the overburden and underburden. As in Sec. 11-4, the analysis of this loss is a combination of local and overall heat transfer techniques that leads to a highly practical result. We give here an exposition of the Marx-Langenheim (1959) theory as expounded by Farouq Ali (1966).

The objective of the Marx-Langenheim (ML) theory is to calculate the heated area as a function of time and reservoir properties. The heated area leads then to expressions for oil rate, oil–steam ratio, and energy efficiency (see Exercise 11G). The procedure is most appropriate for steam drives, but the expression for heated area applies for all thermal processes.

Figure 11-13 shows a schematic illustration of the heating. We assume the heated zone in an areally infinite reservoir contains a single phase with negligible horizontal heat conduction. These assumptions result in an ideally sharp temperature profile. We further assume the over- and underburden extend to infinity along both the positive and negative z-axis.

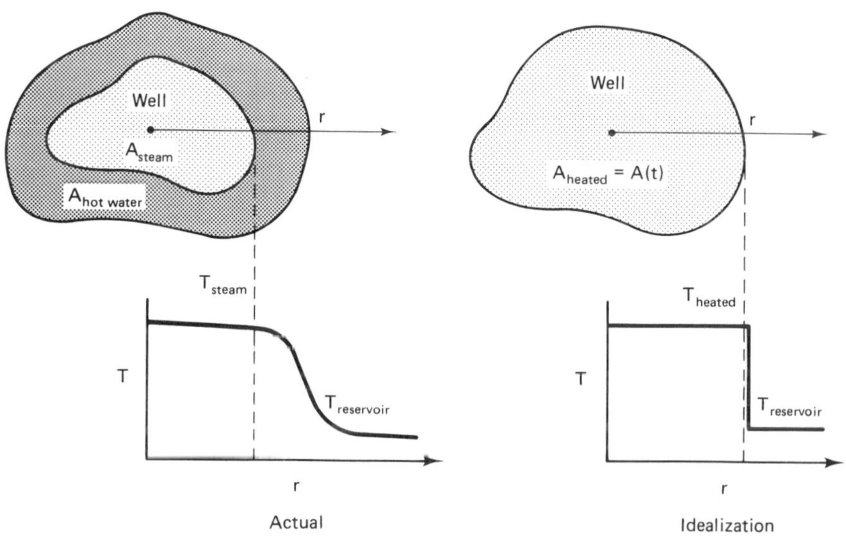

Figure 11-13 Idealization of heated area for Marx-Langenheim theory

The Local Problem

The objective here is to derive an expression for the rate of heat loss $\dot{Q}$ to the over- and underburden and the area $A(t)$ of the heated zone as a function of time. If the over- and underburden are impermeable, heat transfer is entirely through conduction. All fluid velocities and convective fluxes now being zero, the energy balance (Eq. 2.4-14) reduces to the one-dimensional form

$$\frac{M_{Tt}}{k_{Ts}}\frac{\partial T}{\partial t} = \frac{1}{K_{Ts}}\frac{\partial T}{\partial t} = \frac{\partial^2 T}{\partial z^2} \tag{11.5-1}$$

K_{Ts} is the thermal diffusion coefficient of the over- and underburden, and M_{Tt} is the total volumetric heat capacity of the same. Equation (11.5-1) assumes all thermal properties are independent of temperature.

Equation (11.5-1) applies to a vertical segment of the over- and underburden of cross-sectional area ΔA_k whose face at $z = 0$ (the top of the reservoir) is at the original temperature T_I until some time t_k when it is raised to T_J. The boundary conditions for Eq. (11.5-1) are now

$$T(z, 0) = T_I = T(\infty, t); \qquad T(0, t_k) = T_J \tag{11.5-2a}$$

Equation (11.5-2a) neglects the geothermal gradient outside the reservoir, an assumption that implies, since the problem is symmetric, we need not separately treat the over- and underburden. The time scale of the problem may be offset by t_k so that the last boundary condition in Eq. (11.5-2a) becomes

$$T(0, \tau) = T_J \tag{11.5-2b}$$

where $T = T(z, \tau)$, $\tau = t - t_k$, and $\tau > 0$.

Equations (11.5-1) and (11.5-2) are now precisely of the form and boundary condition as Eqs. (5.5-7) and (5.5-12) whose solution we abstract directly as

$$T(z, t) = T_I - (T_I - T_J)\,\text{erf}\left(\frac{z}{2\sqrt{K_{Ts}\tau}}\right) \tag{11.5-3}$$

where τ has replaced the time variable. The rate of heat transferred into ΔA_k from the reservoir is

$$\Delta\dot{Q}_k = -k_{Ts}\left(\frac{\partial T}{\partial z}\right)_{z=0}\Delta A_k \tag{11.5-4}$$

Recalling the definition of the error function from Eq. (5.5-14), we can substitute Eq. (11.5-3) into Eq. (11.5-4) and perform the differentiation to obtain

$$\Delta\dot{Q}_k = \frac{k_{Ts}\,\Delta T}{\sqrt{\pi K_{Ts}(t - t_k)}}\Delta A_k \tag{11.5-5}$$

where $\Delta T = T_J - T_I$. Equation (11.5-5) expresses the rate of heat loss to any vertical segment when $t > t_k$. If we sum all similar segments so that the largest of these

t_K is just smaller than t, we have

$$\dot{Q} = 2 \sum_{k=1}^{K} \Delta \dot{Q}_k = 2 \sum_{k=1}^{K} \frac{k_{Ts} \Delta T}{\sqrt{\pi K_{Ts}(t - t_k)}} \Delta A_k \tag{11.5-6a}$$

which in the limit of the largest ΔA_k approaching zero becomes

$$\dot{Q} = 2 \int_0^{A(t)} \frac{k_{Ts} \Delta T}{\sqrt{\pi K_{Ts}(t - u)}} dA(u) \tag{11.5-6b}$$

The 2 in these equations is to account for heat loss to both the over- and underburden. This procedure is the application of a special case of Duhumel's theorem, a form of superposition for continuously changing boundary conditions (Carslaw and Jaeger, 1959). It is convenient for later operations to convert the integration variable in Eq. (11.5-6b) to a time variable since $dA = (dA/du)\, du$

$$\dot{Q} = 2 \int_0^{t} \frac{k_{Ts} \Delta T}{\sqrt{\pi K_{Ts}(t - u)}} \frac{dA}{du} du \tag{11.5-6c}$$

Equation (11.5-6c) expresses the rate of heat loss at time t as a function of the rate of growth of the heated area. The integrand is finite because of the square root in the denominator, but it is of no use without some independent way to relate heat loss rate to time.

Overall Heat Balances

The link between $\dot{Q}$ and time comes from an overall energy balance. To simplify matters, we take T_I to be the reference temperature for the enthalpy, which means Eq. (2.5-7) applied to the reservoir now becomes

$$\dot{H}_J - \dot{Q} = \frac{d}{dt}(AH_t \rho_s U) \tag{11.5-7}$$

We have neglected the gravity term, and $\rho_s U$ is the volumetric internal energy of the over- and underburden. With the temperature reference being the original reservoir temperature, all energy terms involving the unheated or cold reservoir vanish. This simplification and the neglect of conduction imply the time derivative in Eq. (11.5-7) merely reflects the change in heated zone volume. If the reservoir thickness is constant, Eq. (11.5-7) becomes

$$\dot{H}_J = 2 \int_0^{t} \frac{k_{Ts} \Delta T}{\sqrt{\pi K_{Ts}(t - u)}} \frac{dA}{du} du + H_t M_{Tt} \Delta T \frac{dA}{dt} \tag{11.5-8}$$

where Eq. (11.5-6c) has been inserted into Eq. (11.5-7).

Equation (11.5-8) is an integral–differential equation for $A(t)$ which we solve with the initial condition $A(0) = 0$. The most direct method of solution is through

Laplace transform (Farouq Ali, 1966). The transformed solution to Eq. (11.5-8) is

$$A(\varsigma) = \frac{\dot{H}_J}{2C_1\sqrt{\pi}\varsigma^{3/2} + C\varsigma_2^2} \quad (11.5\text{-}9)$$

where ς is the Laplace transform variable, and $C_1 = k_{Ts}\Delta T/\sqrt{\pi K_{Ts}}$ and $C_2 = M_{Tt}H_t\Delta T$. The inverse transform of Eq. (11.5-10) with $\dot{H}_J$ constant (Roberts and Kaufmann, 1966) is

$$A(t_D) = \frac{\dot{H}_J H_t}{4k_{Ts}\Delta T}\left\{e^{t_D}\operatorname{erfc}(t_D^{1/2}) + \frac{2t_D^{1/2}}{\sqrt{\pi}} - 1\right\} \quad (11.5\text{-}10a)$$

where t_D is a dimensionless time defined as

$$t_D = \frac{4K_{Ts}t}{H_t^2} \quad (11.5\text{-}10b)$$

The H_t^2 in Eq. (11.5-10b) means all heat loss expressions will be especially sensitive to reservoir thickness H_t.

One important feature of the ML theory is that the final result is largely independent of shape of the heated zone. To some extent, this observation is true even if there is gravity overlay, for here the larger heat loss to the overburden is very nearly balanced by a smaller loss to the underburden. To a lesser approximation, the heated area given by Eq. (11.5-10) applies after steam reaches a producing well in steam drives if the *net* enthalpy rate (injected − produced) replaces $\dot{H}_J$.

Several immediate results follow from Eq. (11.5-10) or its time derivatives

$$\frac{dA}{dt_D} = \frac{\dot{H}_J H_t}{4k_{Ts}\Delta T}e^{t_D}\operatorname{erfc} t_D^{1/2} \quad (11.5\text{-}11a)$$

$$\frac{dA}{dt} = \frac{K_{Ts}\dot{H}_J}{H_t k_{Ts}\Delta T}e^{t_D}\operatorname{erfc} t_D^{1/2} \quad (11.5\text{-}11b)$$

Applications

Equations (11.5-7) and (11.5-11) define an expression for average heating efficiency $\overline{E}_{hs}$

$$\overline{E}_{hs} \equiv 1 - \frac{\dot{Q}}{\dot{H}_J} = e^{t_D}\operatorname{erfc} t_D^{1/2} \quad (11.5\text{-}12)$$

$\overline{E}_{hs}$ is the fraction of heat in the reservoir at time t_D expressed as a fraction of the heat entering the sand face at the injector. Figure 11-14 gives $\overline{E}_{hs}$ for a steam drive based on Eq. (11.5-12).

If we assume the displacement of a unit volume of oil from the heated zone causes the production of a unit volume of oil, the oil production rate in reservoir volumes is

$$\dot{N}_{p2} = \phi\Delta S_2 H_{NET}\frac{dA}{dt} \quad (11.5\text{-}13a)$$

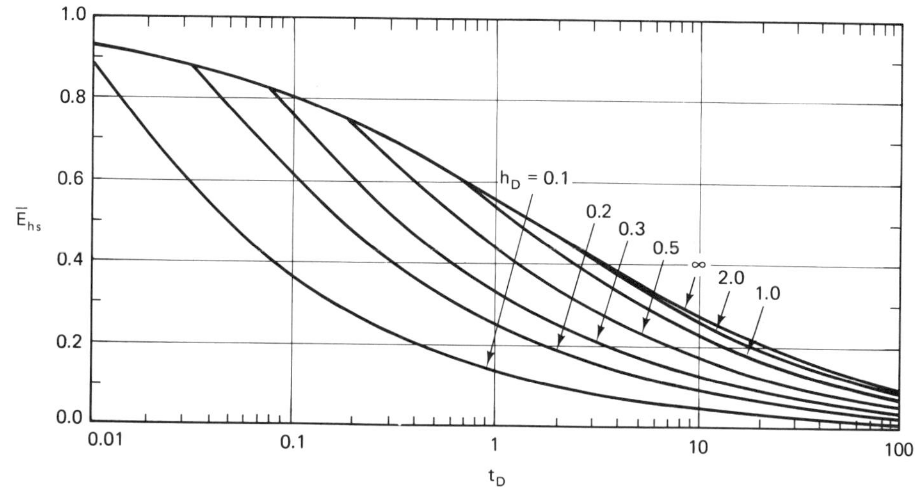

Figure 11-14 Steam zone thermal efficiency (from Myhill and Stegemeier, 1978)

from applying Eq. (2.5-1) to the oil. Using Eq. (11.5-11b), this becomes

$$\dot{N}_{p2} = \dot{H}_J\left(\frac{\phi \Delta S_2}{M_{Tt}\,\Delta T}\right)\left(\frac{H_{NET}}{H_t}\right)(e^{t_D}\ \text{erfc}\ t_D^{1/2}) \tag{11.5-13b}$$

where $\Delta S_2 = S_{2I} - S_2'$ is the oil saturation change due to heating. The equation has been slightly rearranged to include explicitly the net-to-gross thickness ratio. Correlations for the oil saturation in the steam zone S_2' are available in the literature (see Sec. 11-6).

Equation (11.5-13b) invariably gives too high an oil rate, especially after steam breakthrough in a steam drive, but it directly highlights two important parameters in thermal flooding. If the net-to-gross thickness ratio is low, the oil rate will also be proportionally low. Physically, this means a substantial amount of heat is being expended to warm up nonpay rock. The second parameter exerting a direct proportionality to the oil rate is the combination $\phi \Delta S_2$, the "delsophi," which has long been used as an indicator of thermal flooding success. Delsophi should be as large as possible for steam drive candidates. Occasionally, the $\phi \Delta S_2 H_{NET}$ from Eq. (11.5-13a) is used as a screening parameter; $\phi \Delta S_2 H_{NET}$ greater than 2 m is a good candidate.

Finally, the cumulative volume of oil displaced up to time t is

$$N_{p2} = H_{NET}\,\phi \Delta S_2 A \tag{11.5-14a}$$

And the total heat injected to time t is

$$Q = \frac{M_{Tt} H_t A\,\Delta T}{\bar{E}_{hs}} \tag{11.5-14b}$$

A unit amount of this heat resides in $(C_{p1}\,\Delta T + yL_v)$ mass of water. The volume of *cold* water required to generate Q is

$$V_1 = \frac{Q}{(C_{p1}\,\Delta T + yL_v)\rho_1} \tag{11.5-14c}$$

Equations (11.5-14a) through (11.5-14c) lead to the cumulative oil–steam ratio F_{23}

$$F_{23} = \frac{N_{p2}}{V_1} = \frac{M_{T1}}{M_{Tt}}(1 + h_D)\bar{E}_{hs}\left\{\frac{\phi \Delta S_2 H_{NET}}{H_t}\right\} \tag{11.5-14d}$$

The oil–steam ratio is a measure of economical efficiency in steam processes. The steam is expressed as cold water equivalents in Eq. (11.5-14d). Figure 11-15 plots a dimensionless oil–steam ratio

$$\left(\frac{F_{23} H_t}{\phi \Delta S_2 H_{NET}}\right)$$

based on Eq. (11.5-14d). This figure assumes a constant value for the ratio (M_{T1}/M_{Tt}).

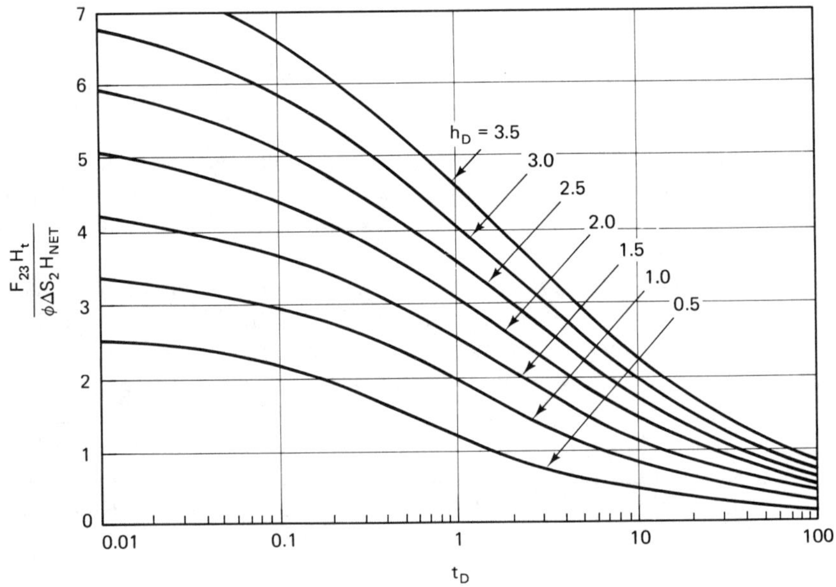

Figure 11-15 Dimensionless cumulative oil–steam ratio (from Myhill and Stegemeier, 1978)

Modifications

The ML theory has had several improvements.

Prats (1982) showed that the ML theory could accommodate different heat capacities in the over- and underburden M_{Tu} and in the reservoir M_{To}. The heated area now becomes

$$A(t) = \frac{\dot{H}_J H_t M_{To}}{4 k_{Ts} M_{Tu} \Delta T} \left\{ e^{t_D} \operatorname{erfc}(t_D^{1/2}) + \frac{2t_D^{1/2}}{\sqrt{\pi}} - 1 \right\} \tag{11.5-15a}$$

and the dimensionless time definition is

$$t_D = \frac{4 k_{Ts} M_{Tu} t}{H_t^2 M_{To}^2} \tag{11.5-15b}$$

Even though saturations may change during the process, M_{To} is approximately constant, being largely comprised of the rock's heat capacity. The quantities in Eqs. (11.5-11) through (11.5-14) still apply with this change as long as the new definitions are used.

One of the most significant advancements to the ML theory was by Mandl and Volek (1969). These workers noticed that the velocity of a steam front declines with time until it can actually propagate at a slower rate than a hot water front. After this point, the displacement forms a hot water or condensate bank that propagates ahead of the steam front from then on. The time at which this happens is called the *critical time* (see Fig. 11-16).

We can derive an expression for the critical time based on our previous equations. Let's consider a medium with a constant cross-sectional area WH_t. For this case, the steam front velocity is simply the rate of growth of the heated area dA/dt divided by the width W of the medium. The velocity of a hot water front v_{HW} has al-

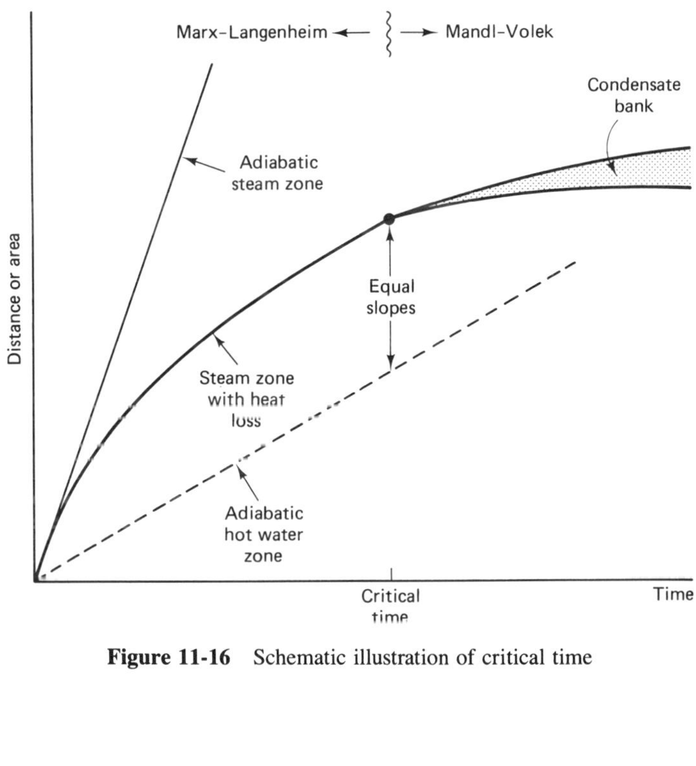

Figure 11-16 Schematic illustration of critical time

ready been given in Eq. (11.3-4). When these two velocities are equated and the area derivative eliminated by Eq. (11.5-11b), we see the critical dimensionless time t_{Dc} is the solution

$$e^{t_{Dc}} \operatorname{erfc} t_{Dc}^{1/2} = \frac{u_1 W H_t M_{Tt} \Delta T}{\phi(1 + D_{SF})\dot{H}_J} \tag{11.5-16}$$

Figure 11-16 shows the velocities of the waves involved in this determination. *Adiabatic* means no lateral heat loss.

Strictly speaking, the ML theory applies only to times less than the critical time. After this point, the more sophisticated Mandl-Volek theory or the approximate Myhill-Stegemeier (1978) theory applies. Myhill and Stegemeier included the heat of vaporization in a condensing steam drive in the ML theory by redefining the heated zone growth rate in the manner of Eq. (11.3-7). Figures 11-14 and 11-15 both include this effect through the dimensionless latent heat h_D first defined in Eq. (11.3-9). The original ML theory is the case of h_D approaching infinity. Myhill and Stegemeier correlated the cumulative oil–steam ratio of 18 steam drive projects with Fig. 11-15.

Finally, Ramey (1959) showed that the ML theory would apply to an arbitrary number of step changes in enthalpy at the injector. Applying superposition to Eq. (11.5-10), which despite everything is still linear, yields

$$A(t_D) = \frac{H_t}{4k_{Ts}\Delta T}\sum_{i=1}^{n} \dot{H}_{Ji}[\phi(t_{Di}) - \phi(t_{Di-1})] \tag{11.5-17a}$$

where

$$\phi(t_{Di}) = e^{t_{Di}} \operatorname{erfc}(t_{Di}^{1/2}) + 2\sqrt{\frac{t_{Di}}{\pi}} - 1 \tag{11.5-17b}$$

and t_{Di} in Eqs. (11.5-17) has the same form as in Eq. (11.5-10b) but with $(t - t_i)$ replacing t. The enthalpy injection rate changes from $\dot{H}_{J(i-1)}$ to $\dot{H}_{Ji}$ at t_i. Each change must maintain the same absolute value of ΔT and $t_D \geqq t_{Dn}$.

Figure 11-17 shows the results of Eq. (11.5-17) applied to a slug injection of steam. During steam injection, the heated area grows at a steadily decreasing rate because of heat losses. In fact, the difference between the indicated curve and a straight line tangent to it at the origin is the diminution of the heated area because of heat loss. At $t = 10^3$ hours, cold water at the original reservoir temperature is injected, resulting now in heat transfer from the previously heated over- and underburden into the water and a very rapid decrease in the heated zone. The diverging nature of the flow in the vicinity of the injector causes the rapid decreases where the rate of area cooled is much higher than the rate of area heated at the steam front. Calculations like those in Fig. 11-17 probably account for why thermal processes are only infrequently conducted as slugs.

With the foregoing as background, we can now address specific processes in a little more detail.

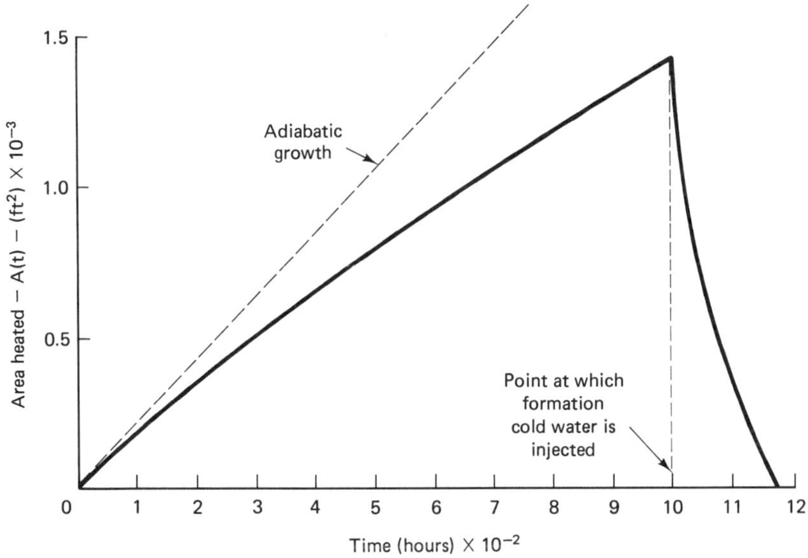

Figure 11-17 Calculated area heated from superimposed Marx-Langenheim theory

11-6 STEAM DRIVES

A steam drive process beyond the Mandl-Volek critical time consists of an unheated zone, a condensate zone, and a steam zone (Fig. 11-18). The steam zone contains a two-phase mixture of steam and water flowing with a very small amount of oil. Because steam viscosity is low, this zone is essentially at a constant pressure, which requires that it also be a constant temperature. Most flow in this zone is steam, but the steam quality is very low because of the presence of a residual water phase. The en-

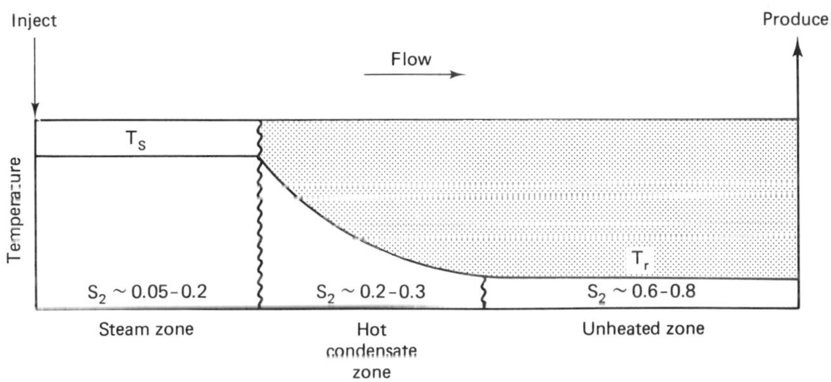

Figure 11-18 Schematic zones in a steam drive

thalpy of the steam in this zone is often neglected. The zone contains oil at a very low saturation since that remaining behind the condensate zone has been distilled. Oil saturation is also low because the wetting state of the crude is frequently altered as the steam seeks to assume the position of the most nonwetting fluid in the pores. Figure 11-19 shows a correlation of steam zone oil saturation.

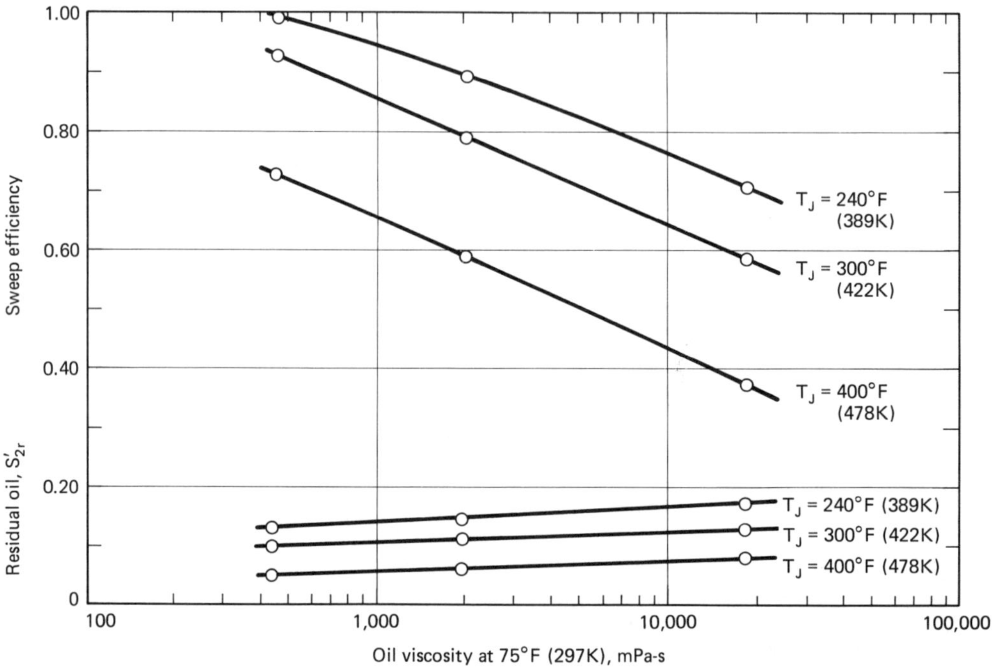

Figure 11-19 Steam zone sweep efficiency and residual oil saturation from model experiment (from Bursell and Pittman, 1975)

It is easy to see where the incremental oil comes from in such a process. If the initial oil saturation is 0.7, and the oil saturation in the steam zone is 0.1, the oil displaced is 86% of that initially in place.

We can represent the profile in Fig. 11-18 on the enthalpy–pressure plot in Fig. 11-3. The unheated zone is a point in the liquid region on a low-temperature isotherm. The condensate zone is a horizontal line segment from this isotherm to the bubble point curve, and the steam zone is a horizontal line from the bubble point curve to some small steam quality.

At a typical thermal flooding condition of 1 MPa (147 psia), the density of saturated liquid and vapor water (steam) is 885 and 5.31 kg/m^3 (55.3 and 0.33 lbm/ft^3), respectively. This pronounced difference between liquid and vapor properties is present in nearly all physical properties and contributes to several important effects in steam drives including in situ quality, viscous stability, and override.

The quality of flowing steam in the reservoir is always quite low. Suppose steam is flowing in a permeable medium in the presence of a residual water saturation. For two phases to be present, both the steam and water must be saturated. At a typical residual water saturation of 0.3 at 1 MPa, and using the above densities, the in situ quality is 1.3%. This low quality means the fluids in the pore space of the medium are just barely inside the saturated liquid line in Fig. 11-3 even though the *flowing* steam quality is nearly 100%.

A second consequence of the low steam density pertains to the issue of viscous stability. In Sec. 6-8, we said that displacing with fluid less viscous than the resident fluid in a horizontal medium inevitably leads to viscous fingering and reduced volumetric sweep efficiency. But steam displacements are quite stable for the following reasons:

1. Steam is readily converted to water. If a perturbation of the steam front were to form, it would finger into the cold zone ahead of it and immediately condense. The condensation leads to a self-stabilizing effect that suppresses fingers.
2. In a steam drive, the *kinematic* mobility ratio is usually favorable. It is more accurate for the mobility ratio to be based on kinematic viscosities for compressible flows. Mobility ratio is the ratio of pressure gradients ahead of and behind a pistonlike front in a one-dimensional displacement,

$$M_v = \frac{\left(\dfrac{dP}{dx}\right)_{\text{ahead}}}{\left(\dfrac{dP}{dx}\right)_{\text{behind}}} = \frac{\left(\dfrac{u\mu}{k}\right)_{\text{ahead}}}{\left(\dfrac{u\mu}{k}\right)_{\text{behind}}} \tag{11.6-1}$$

- If the flux u is not a function of position (fluids are incompressible), Eq. (11.6-1) reduces to that given in Sec. 5-2.
- If the mass flux ρu is not a function of position, Eq. (11.6-1) reduces to a definition of mobility ratio based on the *kinematic* viscosity (dynamic viscosity divided by density) (see Exercise 5J).

In a steam drive, neither condition is true, but the mass flux is more nearly constant. (For a more sophisticated discussion of the stability of thermal fronts, see Krueger, 1982.)

The kinematic viscosity of steam, in fact, is usually greater than that of hot water at the same temperature and pressure. Figure 11-20 shows the reciprocal kinematic mobility ratio of a steam displacement plotted against pressure. Hot waterfloods are unstable over the entire range, which partly accounts for their inferior performance compared to steam, but steamfloods are stable at all pressures less than about 1.5 MPa (220 psia). Further, superheated steam is even more stable than saturated steam. The increase in kinematic mobility ratio with pressure is the consequence of approaching the critical point of water. It further reinforces low-pressure restrictions on steam drives.

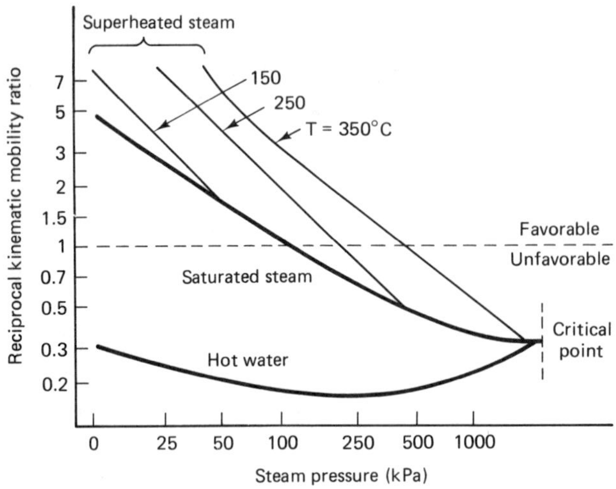

Figure 11-20 Effective mobility ratio for steam displacements (from Burger et al., 1985)

The last consequence of a low steam density is not so favorable. The mobility ratio does not strongly affect gravity segregation as does the density difference between the displacing and displaced fluids. Because of this difference, steam has a tendency to rise to the top of the reservoir causing override and reduced volumetric sweep efficiency. The gravity number N_g^0 scales the severity of the override. Figure 11-21 plots the effect of the inverse gravity number (note the altered definition from Eq. [5.2-3d] to account for compressible fluids and radial flow) on overrunning. If we review what causes N_g^0 to be large, we see many steamfloods occur under conditions that make override almost inevitable—clean sands with high horizontal and vertical permeability, small aspect ratios caused by small well spacing, and large density differences caused by heavy oils.

The gravity override has an important positive consequence. Once steam has broken through in the producing wells of a steam drive, the injection rate is usually reduced to keep more of the steam in the reservoir. At the reduced rate, the heat transfer to the cold oil remains efficient because of the large area of the now nearly horizontal interface. Oil so heated migrates to the top of the reservoir, because its density is now less than that of hot water, and subsequently flows to the producers through the steam zone. This is often called *drag* flow. If override is particularly severe, most of the oil is produced with steam through drag flow.

Two of the methods used in solvent flooding to prevent viscous fingering will mitigate override. If the reservoir has substantial dip, steam injected at the top of the reservoir will result in an interface more perpendicular to the reservoir trend. In addition, the interface can be made more vertical by adding foaming agents to the injected fluids (see Sec. 10-4). Another commonly used method is to inject near the bottom of the formation.

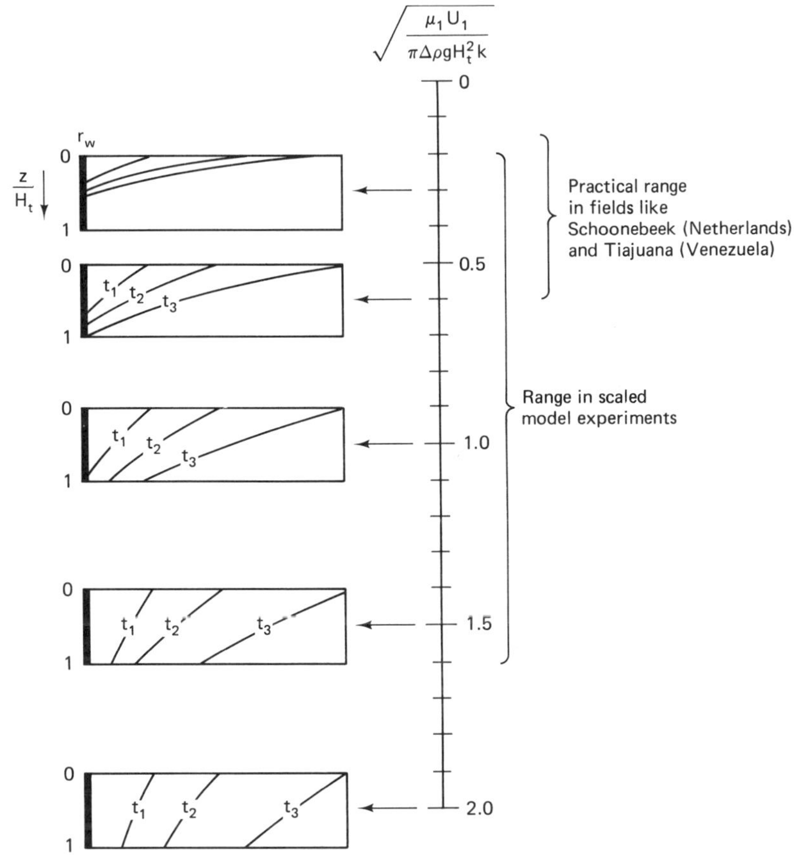

Figure 11-21 Gravity override and gravity number for steam drives (from van Lookeren, 1977)

Case History

To illustrate a steam drive, we discuss one of the phases of a highly successful project in the Kern River field, California. This large field has properties eminently typical of successful steam drives: The field is shallow, the original reservoir pressure low, the sand fairly thick, and the permeability and porosity high (Table 11-5). As we discussed before, each of these items will result in low heat losses. The cold oil viscosity is large, but not extreme and, above all, the original oil content is high in this secondary flood.

One of the projects at Kern River is the Ten-Pattern Steamflood whose well arrangement is ten seven-spots (six injectors each surrounding an injector). The high density of wells in this area is made economically possible by the shallow depths.

TABLE 11-5 SUMMARY OF RESERVOIR DATA AS OF 1968, KERN RIVER FIELD STEAMFLOOD INTERVAL (FROM BLEVINS AND BILLINGSLEY, 1975)

Depth	700–770 ft	213–235 m
Estimated original reservoir pressure	225 psig	1.53 mPa
Current reservoir pressure	60 psig	0.41 mPa
Average net sand thickness	70 ft	21 m
Reservoir temperature	80 °F	300 K
Oil viscosity at 85 °F	2,710 cp	2710 mPa − s
Oil viscosity at 350 °F	4 cp	4710 mPa − s
Average permeability to air	7,600 md	7.6 μm^2
Average porosity	35 %	35 %
Average oil content	1,437 bbl/ac-ft	0.185 m^3/m^3
Average oil saturation	52 %	52 %

Pattern size is correspondingly small. Since the productivity of the cold and heated oil is usually much less than the injectivity of steam, having more producers than injectors will better maintain fluid balance.

Figure 11-22 shows the response of the Ten-Pattern project. Steam injection began in early 1968. Oil response was immediate and very strong. The prompt response is probably due to the steam soaks that preceded the drive, but nearly all the total response is due to the drive. Oil rate peaked in late 1970 and has sustained a surprisingly gentle decline thereafter. Throughout the entire history shown, the oil rate was much greater than the estimated primary oil rate, meaning incremental oil recovery was high. The cumulative steam–oil ratio reached a minimum in early 1972 and increased thereafter as steam broke through to more and more producers.

Some of the steam breakthrough comes from gravity override. Figure 11-23

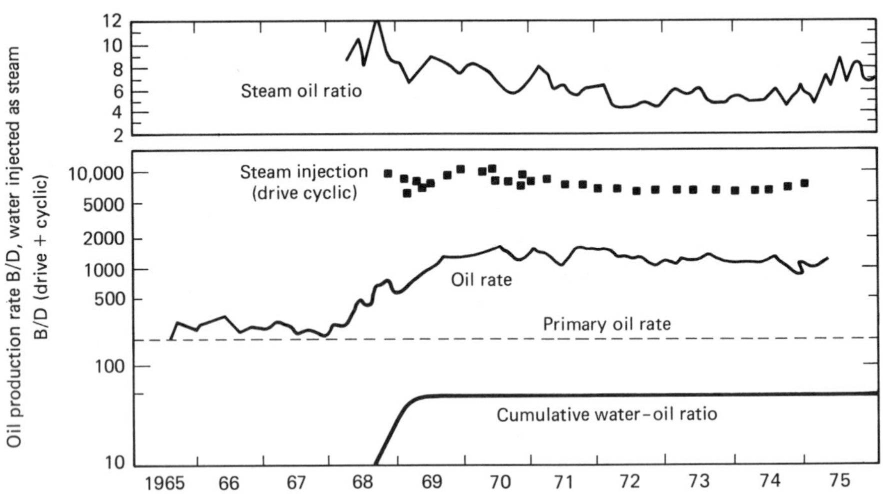

Figure 11-22 Ten-pattern performance, Kern River field (from Blevins and Billingsley, 1975)

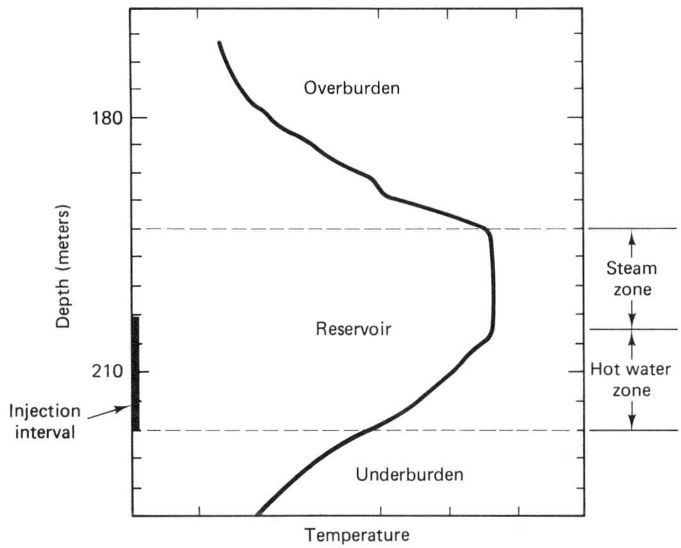

Figure 11-23 Illustration of gravity override, Kern River field (from Blevins and Billingsley, 1975)

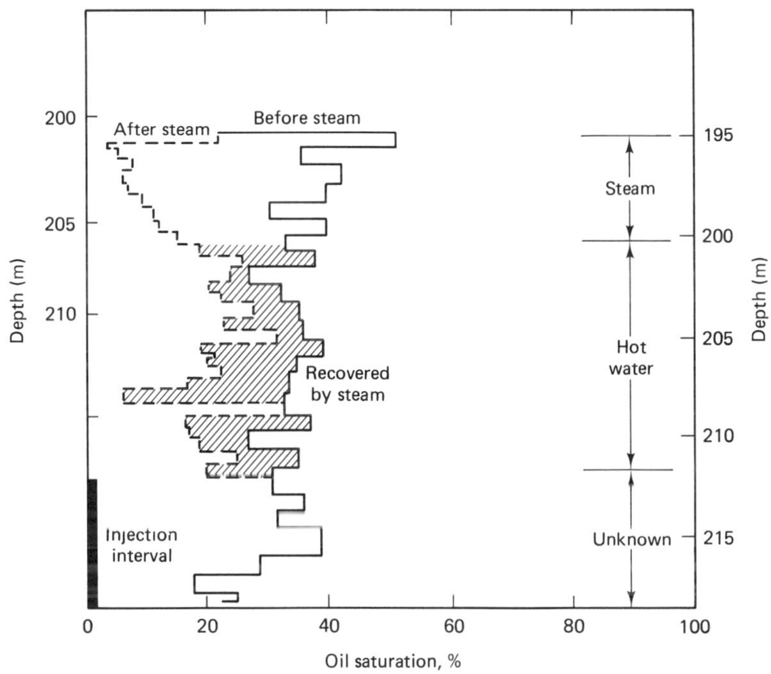

Figure 11-24 Oil saturation changes in the Kern River field (from Blevins and Billingsley, 1975)

shows a temperature survey in a nearby well compared to the injection interval in the nearest injector well. Even though these two wells are quite near each other, the steam zone (indicated by the region of constant temperature) has migrated to the top of the zone. Injecting low in the interval like this is a common way to minimize gravity segregation. The hot water zone below the steam zone shows a gradual temperature decrease that is uninterrupted at the bottom of the zone. The temperature gradients here and at the top of the zone are manifestations of heat losses to the adjacent strata.

Oil saturation changes, as observed in another nearby well (Fig. 11-24) are larger in the steamswept zone but are not insignificant in the hot water zone. The gravity segregation is more pronounced in this well even though it is only a little farther from the injector than was the well in Fig. 11-23.

11-7 STEAM SOAK

How steam soak works is counterintuitive. Evidently, the injected steam displaces relatively little of the oil from near the well. Instead, it channels through the oil to provide good thermal coverage once conduction takes place. The process produces heated oil through several mechanisms: elevated pressure, solution gas drive, thermal expansion, and gravity drainage. Even if the oil is not heated efficiently at all, increased production can result through the removal of skin damage and cleansing of the tubing string. Enough of the oil is removed near the wellbore so that subsequent injectivity improves. Thus steam soaks are frequently used as precursors to steam drives.

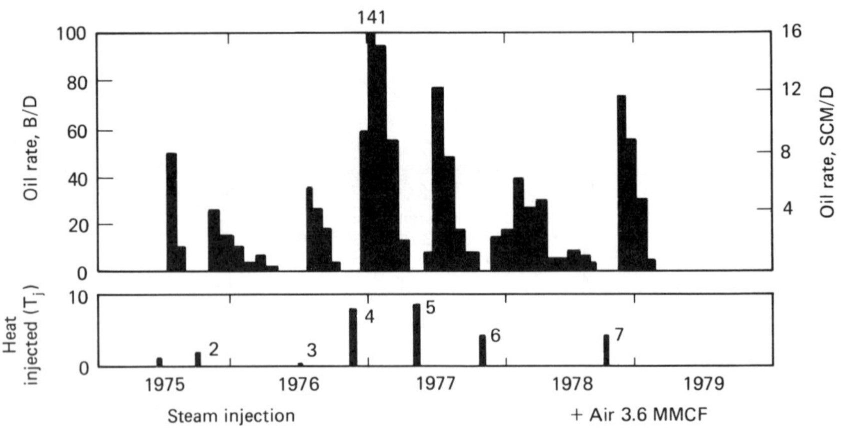

Figure 11-25 Steam soak response, Paris Valley field (from Meldau et al., 1981)

Case History

Figure 11-25 shows the response of a well in the Paris Valley field to several steam soak cycles. There were roughly two soaks per year from 1975 through 1978, each less than a month long. The cumulative oil produced after each cycle is roughly proportional to the amount of heat injected in each preceding soak. Within each cycle, the oil rate rapidly peaks and then falls in a near-exponential decline. Since a similar decline is not so evident in the oil cut, the performance suggests the total fluid rate is also declining within each cycle. The decline suggests the reservoir pressure is falling and, for this reason, the operators mixed some air with the steam in the last cycle. For a given amount of heat injected, the peak rate in each cycle should decline as the heated zone will contain successively smaller amounts of oil.

11-8 IN SITU COMBUSTION

Figure 11-26 is a schematic of in situ combustion. Usually, some form of oxidant (air or pure oxygen) is introduced into the formation, and the mixture is ignited (spontaneously or externally). Subsequent injection propagates the burning front through the reservoir. The burn front is very small (about a meter), but it generates very high temperatures. These temperatures vaporize connate water and a portion of the crude. Both of these are responsible for oil displacement. The vaporized connate water forms a steam zone ahead of the burn front that operates very much like a steam drive. The vaporized oil consists mainly of light components that form a miscible displacement. The reaction products of a high-temperature combustion can also form an in situ CO_2 flood.

If reservoir pressure or depth is too large for steam methods to work, in situ combustion might be a good alternative. In this method, burning a portion of the crude in the formation generates thermal energy. Theoretically, the portion being burned is the coke or asphaltene portion of the crude, but the issue is far less clear-cut in practice. This complex process—the most complex of the EOR processes—involves heat and mass transfer along with kinetic phenomena.

Figure 11-27 is a plot from a laboratory experiment of a differential thermal analysis (DTA) of a crude. DTA consists of heating the crude in a preprogrammed fashion, usually linear in time and measuring the rate of reactant consumption and the contents of the reaction products. We see two general points from this figure. First, oxygen is consumed in two peaks: a low-temperature oxidation at about 572 K (570°F) and a high-temperature oxidation at about 672 K (750°F). In the low-temperature oxidation, the crude is being converted to alcohols, ketones, and aldehydes. Second, in the high-temperature oxidation, the combustion proceeds entirely to carbon dioxide and carbon monoxide. Figure 11-27 shows this progression where the production of these two components is larger at high temperature. Further, high temperatures oxidize many of the minerals in the permeable media, particularly the

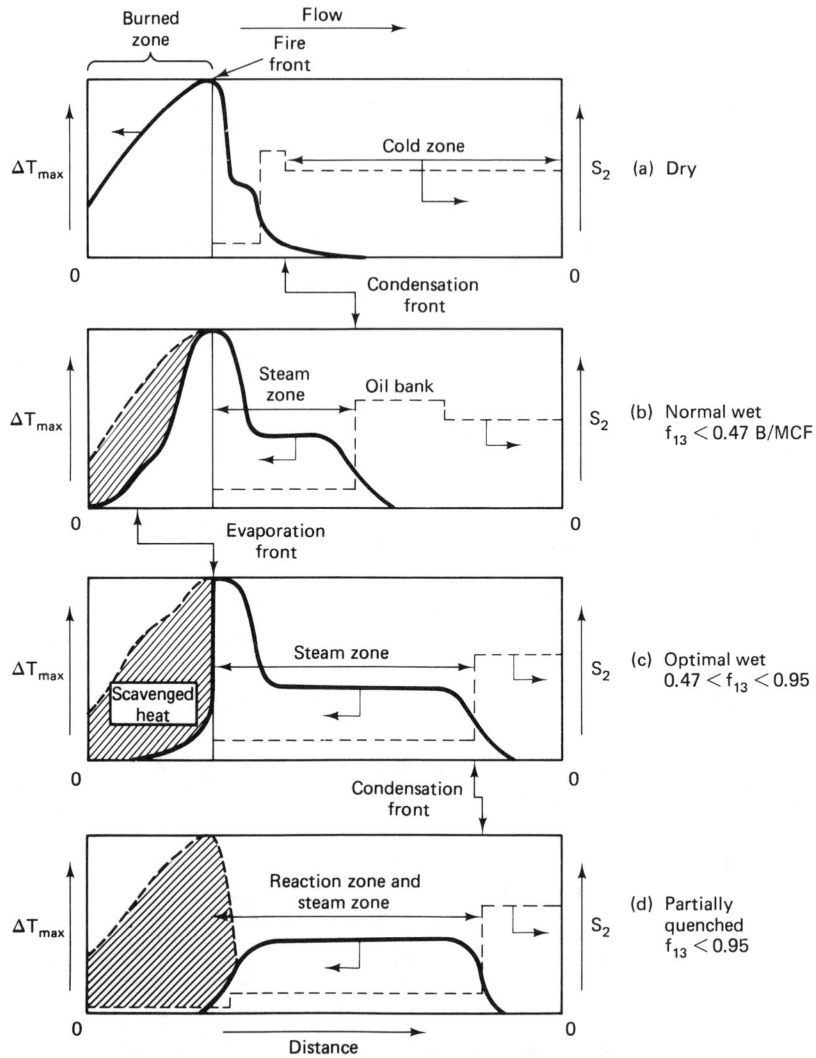

Figure 11-26 In situ combustion schematic (from Prats, 1982)

clays (these may also exert a catalytic effect) and pyrite. The high-temperature oxidation is better because it heats the oil more efficiently.

11-9 CONCLUDING REMARKS

Our discussion here, and indeed throughout the text, has been from a reservoir engineering viewpoint. For thermal methods, in particular, much of the success is due to advancements in mechanical, completion, and production technology.

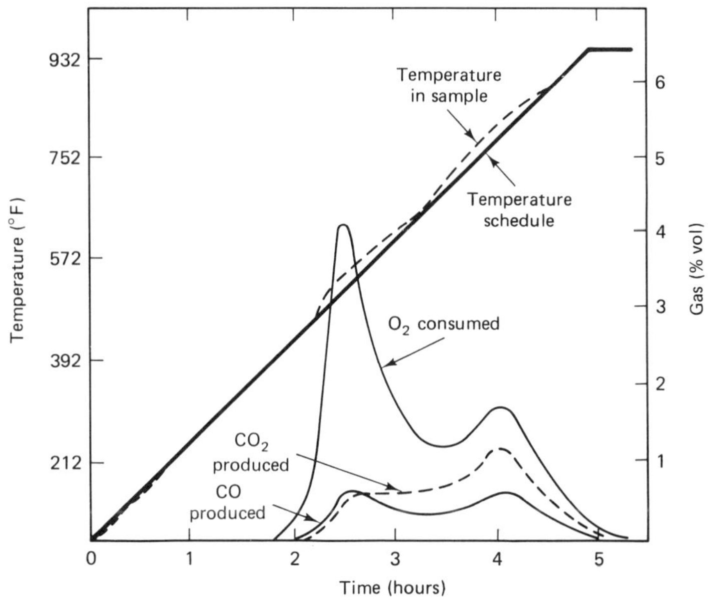

Figure 11-27 Differential thermal analysis of a crude oil (from Burger and Sahuquet, 1972)

Surface steam generation, a simple concept in principle, is not easy under field conditions. For most cases, the waters available are brines of highly variable salinity. Such water cannot be used to generate 100% quality steam because of scaling. To the contrary, most boilers generate about 80% quality steam for this reason. Moreover, in most cases, the fuel used in surface generators is the produced crude. Since this usually heavy crude tends to be especially rich in components that cause air pollution when burned, surface steam generation can represent an environmental hazard. The expense of cleaning the boiler waste gases must be borne by the entire project.

Difficulties in completing the wells plagued early steam operations, particularly the injectors. Thermal expansion of downhole equipment exacerbated the failures of the existing cementing techniques. Many of the latter difficulties have been remedied by using prestressed tubular goods in the wells. Current cement bonding techniques and the development of thermal packers have drastically reduced failure rates.

Undoubtedly, the future of thermal recovery rests on these and other technological advancements. These advancements include the cogeneration of electric power from steam boilers and the use of downhole steam generators, foams for mobility control, diluents in steam injection, and oxygen for in situ combustion. Each of these extends the range of thermal methods to heavier or lighter crudes, deeper formations, or higher-pressure reservoirs. When this extension becomes a reality, thermal methods, already proven worldwide, will directly contend with other techniques for the EOR target oil.

EXERCISES

11A. *Effect of Temperature on Productivity Improvement*. Steam soak is far from incompressible steady-state flow. However, rough estimates of productivity may be obtained by assuming both. The formula for the volumetric production rate q of a well draining two concentric cylindrical volumes is

$$q = \frac{2\pi k H_{NET}(P_e - P_{wf})}{\mu_{2h} \ln\left(\frac{R_h}{R_w}\right) + \mu_{2c} \ln\left(\frac{R_e}{R_h}\right)} \tag{11A-1}$$

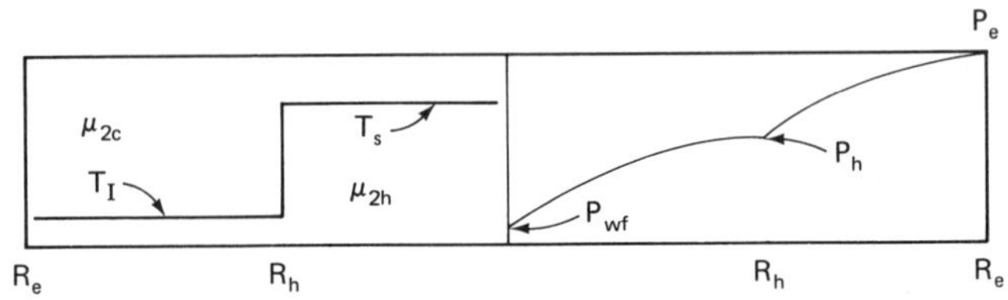

Take the inner cylinder to be the heated volume after a steam soak.

(a) Derive an expression for productivity index (PI) for this case where

$$(PI) \equiv \frac{q}{P_e - P_{wf}} \tag{11A-2}$$

Also derive an expression for the PI improvement

$$(PI)_{\text{improvement}} = \frac{(PI)_{\text{stimulated}}}{(PI)_{\text{unstimulated}}} \tag{11A-3}$$

(b) Estimate the PI improvement for a single steam cycle using the following data:

Reservoir temperature = 320 K
Heated zone temperature = 480 K
Cold oil density = 0.9 g/cm^3
Hot oil density = 0.8 g/cm^3
Drainage radius = 116 m
Heated radius = 20 m
Well radius = 7 cm
API = 20°

Use the viscosity data in Figure 11-1 for the hot μ_{2h} and cold μ_{2c} oil viscosities. Note 1 cs = 1 mm^2/s.

(c) Make subjective judgments about the effects of the following quantities on PI improvement: number of cycles, steam volume injected, cold oil viscosity, permeability, and skin factor.

11B. *Estimating Generator Performance*. Water at an initial temperature of 294 K (70°F) is being pumped through a steam generator at a rate of 15.9 m^3/day (100 bbls/day) into an injection well. The well head temperature is 533 K (500°F). What is the well head steam quality? The generator burns 1,000 SCM/day of gas that has a heating value of 300 kJ/SCM. The generator efficiency is 80%.

11C. *Boiler Scaling*. Scale (solids precipitation) will form in steam boilers if the generated steam quality is too high.

(a) Estimate the equilibrium constants and solubility products for the following intraaqueous and precipitation reactions at 350 K:

$$HCO_3^- \rightarrow H^+ + CO_3^{2-}$$
$$CaCO_3 \rightarrow Ca^{2+} + CO_3^{2-}$$
$$CaHCO_3^- \rightarrow Ca^{2+} + H^+ + CO_3^{2-}$$
$$NaCl \rightarrow Na^+ + Cl^-$$

Use the techniques described in Sec. 3-5 and the data in Table 3-7. $CaCO_3$ and NaCl are the only possible solids.

(b) Estimate the maximum steam quality that may be generated at the above temperature from water having the following overall composition:

Species	Total concentration (g/m^3)
Ca^{2+}	10
HCO_3^-	3
Na^+	352
Cl^-	451

Assume ideal solution behavior, and take the water pH to be 6.0 at these conditions.

11D. *Alternate Derivation of Thermal Velocity in Hot Water Flood*. Equation (11.3-15) can be derived in a fashion reminiscent of the composition path constructions of Sec. 7-7. The coherence constraint for Eqs. (11.3-12a) and (11.3-12c) is

$$\frac{d(\rho_1 f_1 H_1 + \rho_2 f_2 H_2)}{d\left(\rho_1 S_1 H_1 + \rho_2 S_2 H_2 + \dfrac{(1-\phi)}{\phi}\rho_s H_s\right)} = \frac{\rho_1 H_1 df_1}{\rho_1 H_1 dS_1} \tag{11D-1}$$

The density–enthalpy product has been added to the numerator and denominator of the right side to ensure units consistency below.

(a) By expansion of the numerator and denominator on the left side of Eq. (11D-1), show that lines of constant temperature satisfy the coherence condition. These will represent the saturation change at the leading edge of the cold oil bank.

(b) We know that Eq. (11D-1) will be satisfied if the numerators and denominators are identically equal. Show that equating the denominators yields an ordinary differential equation whose solution is

$$T = I_1\left[S_1(M_{T1} - M_{T2}) + \left(M_{T2} + \frac{(1-\phi)}{\phi}M_{Ts}\right)\right]^{M_{T2}/(M_{T1}-M_{T2})} \tag{11D-2}$$

where I_1 is an integration constant. Recall that $\rho_j dH_j = M_{Tj}\, dT$. Equation (11D-2) along with the T = constant lines form the composition path grid in S_1-T space. Sketch a few lines in this grid.

(c) Equate the numerators of Eq. (11D-1), and perform the analogous operation to give

$$T = I_2[\, f_1(M_{T1} - M_{T2}) + M_{T2}]^{M_{T2}/(M_{T1}-M_{T2})} \tag{11D-3}$$

where I_2 is a second integration constant. Equation (11D-3) and the T = constant lines are the composition path grid in f_1-T space.

(d) Eliminate temperature between Eqs. (11D-2) and (11D-3) to show that the temperature-varying paths in composition space follow

$$f_1 + \frac{M_{T2}}{M_{T1} - M_{T2}} = I_3\left(S_1 + \frac{M_{T2} + \dfrac{(1-\phi)}{\phi} M_{Ts}}{M_{T1} - M_{T2}} \right) \tag{11D-4}$$

where I_3 is another integration constant.

(e) Equation (11D-4) suggests $df_1 = I_3\, dS_1$. Use the similar differential forms of Eqs. (11D-3) and (11D-4) for dT to show that Eq. (11D-1) will yield $I_3 = df_1/dS_1$. Putting this back into Eq. (11D-4) gives Eq. (11.3-15).

11E. *Fractional Flow for Hot Water Floods.* The following problem is intended to reinforce the fractional flow construction in Fig. 11-6 and give practice in estimating thermal properties.

(a) We are to do a hot water flood consisting of saturated liquid water at 1 MPa pressure. Estimate the hot water temperature, the hot oil viscosity, and the volumetric heat capacities for water, oil, and the solid phase. Additional data are as follows:

	Initial (cold)	Injected (hot)
Temperature, K	300	
Water viscosity, mPa-s	1.0	0.5
Water density, g/cm^3	1.0	1.0
Oil viscosity, mPa-s	700	
Oil density, g/cm^3	0.9	0.9

Use the data or correlations in Tables 11-2 and 11-3 (the properties of water-saturated sandstone most nearly approximate the present case), Eq. (11.2-3) and Fig. 11-5. The porosity is 0.2.

(b) The exponential relative permeability curves apply to this horizontal reservoir with the following parameters:

$S_{1r} = 0.2 \qquad k_{r1}^0 = 0.3 \qquad n_1 = 2$

$S_{2r} = 0.2 \qquad k_{r2}^0 = 0.8 \qquad n_2 = 2$

You may assume these functions are independent of temperature. Using this data and that of part (a), calculate and plot the hot and cold water fractional flow curves.

(c) Calculate and plot the one-dimensional effluent history of oil and temperature based on the information given above. The initial water cut is 0.1.

11F. *Dimensional Analysis of Heat Transfer from Tubing*. In this exercise, we develop the dimensional argument for Eq. (11.4-8), the heat transfer coefficient correlation for heat flow from the tubing. Figure 11-8(a) shows the approximate velocity and temperature profiles. If the fluid flow is steady-state, laminar, Newtonian, and incompressible, the velocity profile in the tubing becomes

$$v = v_{\max}\left(1 - \left(\frac{r}{R_{ti}}\right)^2\right) \tag{11F-1}$$

(a) The energy balance of Eq. (2.3-14) applies to the flowing fluid if the porosity is set to 1. If the energy balance retains only radial conduction and axial convection, show that applied to the fluid in the tubing it reduces to

$$\rho_f C_{pf} v_{\max}\left(1 - \left(\frac{r}{R_{ti}}\right)^2\right)\frac{\partial T}{\partial z} = k_{Tf}\frac{1}{r}\frac{\partial}{\partial r}\left(r\frac{\partial T}{\partial r}\right) \tag{11F-2}$$

Equation (11F-2) assumes constant thermal conductivity k_{Tf} and viscosity μ_f of the flowing fluid. The boundary conditions on this equation are

$$\left(\frac{\partial T}{\partial r}\right)_{r=0} = 0, \qquad T(R_{ti}, z) = T_{ti}, \qquad T(r, 0) = T_f \tag{11F-3}$$

where T_{ti} is constant.

(b) Introduce the following dimensionless variables:

$$r_D = \frac{r}{R_{ti}}, \qquad z_D = \frac{K_{Tf} z}{v_{\max} R_{ti}^2}, \qquad T_D = \frac{T - T_{ti}}{T_f - T_{ti}}$$

into Eqs. (11F-2) and (11F-3), and show that they reduce to

$$(1 - r_D^2)\frac{\partial T_D}{\partial z_D} = \frac{1}{r_D}\frac{\partial}{\partial r_D}\left(r_D\frac{\partial T_D}{\partial r_D}\right)$$

$$\left(\frac{\partial T_D}{\partial r_D}\right)_{r_D=0} = 0, \qquad T_D(1, z_D) = 0, \qquad T_D(r_D, 0) = 1 \tag{11F-4}$$

The dimensionless temperature must therefore be a function only of r_D and z_D.

(c) The heat transfer rate $\dot{Q}$ from the tubing is

$$\dot{Q} = -2\pi\int_0^L \left(r k_{Tf}\frac{\partial T}{\partial r}\right)_{r=R_{ti}} dz \tag{11F-5}$$

Show that the dimensionless form of this equation is

$$\frac{\dot{Q}}{2\pi L k_{Tf} r_D (T_f - T_{ti})} = -\frac{2\pi}{Z_{DL}}\int_0^{Z_{DL}}\left(r_D\frac{\partial T}{\partial r_D}\right)_{r_D=1} d\xi \tag{11F-6}$$

where the additional term in Eq.(11F-6) is

$$Z_{DL} = \frac{K_{Tf} L}{v_{\max} R_{ti}^2} \tag{11F-7}$$

and L is the length of the heated tubing. Because of the evaluation at $r_D = 1$ and the integration between known limits, the integral is a function of Z_{DL} only.

(d) Define an average heat transfer h_{Tf} coefficient as

$$\dot{Q} \equiv \pi R_{ti} L (T_f - T_{ti}) h_{Tf} \tag{11F-8}$$

Eliminate $\dot{Q}$ between Eqs. (11F-6) and (11F-8) to show, after rearranging,

$$\frac{R_{ti} h_{Tf}}{k_{Tf}} = N_{\text{Nu}} = f(Z_{DL}) \tag{11F-9}$$

Equation (11.4-8) follows from this since Z_{DL} decomposes into

$$\frac{K_{Tf} L}{v_{\max} R_{ti}^2} = \frac{k_{Tf} L}{\rho_f C_{pf} v_{\max} R_{ti}^2} = \frac{k_{Tf}}{\mu_f C_{pf}} \cdot \frac{\mu_f}{\rho_f v_{\max} R_{ti}} \cdot \frac{L}{R_{ti}}$$
$$= \frac{1}{N_{\text{Pr}}} \cdot \frac{1}{N_{\text{Re}}} \cdot \frac{L}{R_{ti}} \tag{11F-10}$$

The Brinkman number is absent from Eq. (11F-10) since the original equation did not include viscous heating.

11G. *Calculating Heat Losses*. For steam drives, the rate of heat loss to the over- and under-burden is frequently so significant that it alone can furnish a good measure of success. In this exercise, you use theoretical relations to estimate measures of the success of a steam drive. Use the following quantities in this exercise:

$T_I = 317$ K	$k_{Ts} = 2.1$ J/s-m-K
$H_t = H_{NET} = 11$ m	$M_{To} = 2.3$ MJ/M^3
$\phi = 0.3$	$M_{Tu} = 2.8$ MJ/m^3-K
$\Delta S_2 = 0.31$	$t = 4.5$ yr
$H_3 = 44.4$ MJ/kg	

(a) Estimate the steam zone temperature. P_I, the initial reservoir pressure, is 2.72 MPa.

(b) Calculate the dimensionless time and dimensionless latent heat from Eq. (11.3-9). The steam quality y is 0.7.

(c) From the Myhill-Stegemeier charts (Figs. 11-14 and 11-15), estimate the useful heat fraction $\bar{E}_{hs}$ and dimensionless oil–steam ratio.

(d) From the results of part (c), calculate the oil–steam ratio F_{23} and the energy efficiency. The latter is defined as

$$\eta_E = \frac{\text{Oil heating value}}{\text{Heat requirement to produce steam}}$$

and given by

$$\eta_E = \frac{F_{23} \gamma_2 \eta_B H_3}{C_{p1} \Delta T (1 + h_D)} \tag{11G-1}$$

where η_B is the boiler efficiency, and γ_2 is the specific gravity of the oil. Take $\gamma_2 = 0.94$ and $\eta_B = 0.8$.

Nomenclature*

Normal

A	Area (usually cross sectional) [=] L^2
A_H, B_H, E_H, F_H	Parameters in Hand representation of ternary phase behavior
A_p	Pattern area [=] L^2
a_i	Activity of species i
a_i, b_i	Parameters in Langmuir isotherm
a_T	Geothermal temperature gradient [=] T/L
a_v	Specific surface area [=] L^{-1}
B_j	Formation volume factor for phase j [=] L^3/standard L^3
CDC	Capillary desaturation curve
C_n	Cumulative storage capacity up to layer n
C_i	Concentration of species i [=] amount/volume (definition of amount depends on the species)
$\lfloor C_i \rfloor$	Concentration in molal units [=] amount/kg solution
C_i^o	Tie line convergence point on ternary diagram [=] consistent with concentration
C_{ij}	Volume fraction of component i in phase j

*[=] means has units of, L is a length unit, F is force, m mass, t time, T temperature, and amount is moles.

C_{2pL}, C_{2pR}	Refers to left [type II(+)] and right [type II(−)] oil coordinates of the plait point
C_{pj}	Heat capacity of phase j [=] $F - L$/amount $- T$
ΔC°_{pr}	Standard heat capacity change of reaction r [=] $F - L$/amount $- T$
C_s	Salinity
c	Compressibility [=] L^2/F
D_i	Frontal advance lag or retardation factor for species i
D_{ij}	Effective binary diffusion coefficient of species i in phase [=] L^2/t
D_p	Particle or sphere diameter [=] L
D_z	Elevation or depth from a reference datum [=] L
d_p	Effective diameter of a polymer molecule [=] L
E	Effective viscosity ratio in Koval theory
$\vec{E}$	Energy flux [=] $F/L - t$
E_A	Areal sweep efficiency (fraction)
E_D	Displacement or local sweep efficiency (fraction)
E_I	Vertical sweep efficiency (fraction)
E_{MB}	Mobility buffer efficiency (fraction)
EOR	Enhanced oil recovery
E_R	Recovery efficiency (fraction)
E_V	Volumetric sweep efficiency (fraction)
F_n	Cumulative flow capacity up to layer n
F_i	Fractional flux of component i
F_{23}	Oil–steam ratio
f_a	Fraction of total pore space available to flow
f_j	Fractional flow of phase j
$\vec{g}$	Gravitation acceleration vector [=] L/t^2 (magnitude: g)
H	Enthalpy [=] $F - L$/amount
H°_r	Standard enthalpy of formation for reaction r [=] $F - L$/amount
H_K	Effective heterogeneity factor in Koval theory
H_{NET}	Net thickness [=] L
H_t	Total thickness [=] L
h_T	Heat transfer coefficient [=] $F/L^2 - t$

I	Injectivity [=] $L^5/F - t$
i	Injection rate [=] L^3/t
IFT	Interfacial tension [=] F/L
IOR	Incremental oil recovered [=] standard L^3
IR	Initial − residual
j	Leverett j-function
K_i	Equilibrium flash vaporization ratio
$\vec{\vec{K}}_{ij}$	Dispersion tensor for species i in phase j [=] L^2/t
K_ℓ	Longitudinal dispersion coefficient [=] L^2/t
K_N	Selectivity coefficient for cation exchange
K_{pl}	Power-law coefficient
K_r	Equilibrium constant for reaction r
K_T	Thermal diffusion coefficient [=] L^2/t
k	Permeability [=] L^2
k_j	Permeability to phase j [=] L^2
k_m	Mass transfer coefficient [=] t^{-1}
k^o_{rj}	Endpoint relative permeability to phase j
k_T	Thermal conductivity [=] $F/L - t$
L	Length [=] L
L_c	Lorenz coefficient (fraction)
L_v	Heat of vaporization [=] $F - L$/amount
ln	Natural logarithm
log	Base 10 logarithm
M	Mobility ratio
MP	Micellar-polymer
M_{Sh}	Shock mobility ratio
M_T	Volumetric heat capacity [=] $F/L^3 - t$
M_v	Kinematic mobility ratio
M_w	Molecular weight [=] mass/amount
M^o	Endpoint mobility ratio
$\dot{m}$	Mass flow rate [=] mass/t
N_B	Bond number
N_{Br}	Brinkman number
N_C	Total number of components

N_D	Number of spatial dimensions
N_{Da}	Damkohler number
N_{Deb}	Deborah number
N_F	Number of degrees of freedom
N_{Gr}	Grashof number
$\vec{N}_i$	Mass flux of species i [=] mass of $i/L^2 - t$
$\vec{N}_{ij}$	Mass flux of component i in phase j [=] mass i/L^2 phase $j - t$
N_L	Total number of layers
N_{Nu}	Nusselt number
N_P	Total number of phases
N_p	Cumulative mass produced [=] mass
N_{Pe}	Peclet number
N_{Pr}	Prandtl number
N_R	Total number of chemical reactions
N_{Re}	Reynolds number
N_{RL}	Rapoport and Leas number
N_{vc}	Local viscous-capillary number
n	Number of moles [=] amount
$\vec{n}$	Unit outward normal vector
n_j	Exponent on analytic relative permeability functions
n_L, n_V	Relative amounts of liquid and vapor (fraction)
n_M	Parameter in Meter model
n_{pl}	Power-law exponent
OOIP	Original oil in place [=] standard L^3
P	Pressure [=] F/L^2
P_{cjk}	Capillary pressure between phases j and k [=] F/L^2
P_v	Vapor pressure [=] F/L^2
$\dot{Q}$	Heat transfer rate [=] $F - L/t$
Q_v	Cation exchange capacity [=] equivalent/mass − substrate
q	Volumetric flow rate [=] L^3/t
$\vec{q}_c$	Conductive energy flux [=] $F/L - t$
R	Radius [=] L; Ideal gas constant [=] $F - L$/amount $- T$
R_b	Pore body radius [=] L
R_f	Resistance factor
R_h	Hydraulic radius [=] L

R_i	Insulation radius $[=] L$
R_k	Permeability reduction factor
R_n	Pore neck or entry radius $[=] L$
R_p	Polymer radius $[=] L$
R_{rf}	Residual resistance factor
R_s	Solution gas–oil ratio $[=]$ standard L^3 dissolved species/standard L^3 liquid
R_w	Well radius $[=] L$
REV	Representative elementary volume
r_i	Total reaction rate of species i $[=]$ mass i/total $L^3 - t$
r_{ij}	Homogeneous reaction rate $[=]$ mass i/L^3 phase $j - t$
r_m	Mass transfer rate $[=]$ mass$/L^3 - t$
S_F	Screen factor
S_j	Saturation of phase j
S_{jk}	Solubilization parameters between phases j and k
SCM	Standard cubic meter
SRD	Salinity requirement diagram
s	Skin factor
ς	Laplace transform variable
T	Temperature $[=] T$
t	Time $[=] t$
t_{DS}	Slug size (fraction)
t_{MB}	Mobility buffer size (fraction)
U	Internal energy $[=] F - L$/amount
U_j	Internal energy of phase j $[=] F - L$/mass
U_T	Overall heat transfer coefficient $[=] F/L - t - T$
$\vec{u}$	Superficial velocity $[=] L/t$ (magnitude: u)
V_b	Bulk volume $[=] L^3$
V_{DP}	Dykstra-Parsons coefficient (fraction)
$\bar{V}_M$	Specific molar volume $[=] L^3$/amount
V_p	Pore volume $[=] L^3$
VE	Vertical equilibrium
VFD	Volume fraction diagram
$\vec{v}$	Interstitial velocity $[=] L/t$ (magnitude: v)
v_{C_i}	Specific velocity of concentration C_i

$v_{\Delta C_i}$	Specific velocity of shock concentration change ΔC_i
W	Medium width [=] L
$\dot{W}$	Rate of work per unit volume [=] $F/L^2 - t$
W_i	Overall concentration of species i [=] amount or mass of i/L^3
W_R	Water-alternating-gas ratio
WAG	Water alternating gas
WOR	Water- (or brine-) oil ratio [=] L^3/L^3
x	Position [=] L
x_i, y_i	Mole fraction of i in vapor and liquid phases
y	Steam quality
Z_v	Cation exchange capacity [=] equivalents/L^3 of pore volume
z	Compressibility factor
z_i	Overall mole fraction of component i [=] amount i/total amount

Greek

α_ℓ, α_t	Longitudinal and transverse dispersivities [=] L
α	Dip angle
β	Heterogeneity factor (Chap. 3) or interface tilt angle (Chap. 6)
β_T	Thermal expansion coefficient [=] T^{-1}
δ_{ij}	Binary interaction coefficient between species i and j
η	Slope of tie lines
η_B	Steam boiler efficiency
Λ	Cumulative frequency
γ	Specific gravity
Δ	Operator that refers to a discrete change
$\dot{\gamma}$	Shear rate [=] t^{-1}
$\vec{\nabla}$	Gradient operator [=] L^{-1}
λ_c	Critical wavelength [=] L
λ_j	Mobility of phase j [=] $F - t$
$[\mu]$	Intrinsic viscosity [=] L^2/mass
μ	Viscosity [=] $F - t/L^2$
Φ_j	Fluid potential [=] F/L^2

$$= P_j + \rho_j g D_z \text{ for incompressible phase } j$$

$$= \rho_j \int_{P_0}^{P_j} \left(\frac{P}{\rho_j} + g D_z \right) dP \quad \text{otherwise}$$

ϕ	Porosity [=] fraction
ϕ_i	Fugacity coefficient of component i
ρ	Density [=] mass/L^3
ρ_i^o	Pure component density of species i [=] mass/L^3
ρ_j	Density of phase j [=] mass of j/L^3
ρ_j^s	Standard density of phase j [=] mass/standard L^3
ρ_M	Molar density [=] amount/L^3
σ_{ik}	Interfacial tension between phases j and k [=] F/L
τ	Shear stress [=] F/L^2; Tortuosity
θ	Contact angle
ν	Kinematic viscosity [=] L^2/t
ν_{LN}	Variance of lognormal distribution
ω_i	Overall mass fraction [=] mass i/total mass
ω_{ij}	Mass fraction of species i in phase j

Superscript

^	Quality corrected for heterogeneity
—	Average or specific
~	Pseudo
∞	Ultimate or large-time value
*	Refers to intersection between chords and tangents
′	Denotes a quantity modified by an EOR fluid (low IFT or polymer enhanced, for example)
o	Breakthrough quantities
SP	Solubility product
s, p	Refer to fractional flow curves modified by surfactant, solvent, or polymer
·	Indicates a rate
$\rightarrow$	Vector
$\rightrightarrows$	Tensor

Subscript

A, R, E	Advancing, receding, and intrinsic
c	Critical
cem	Cement

ci	Inside casing
co	Outside casing
D	Denotes a dimensionless quantity
d	Drill hole
e	Effective
eq	Equivalent
f	Floodable or flowing
I, J, K	Refer to initial, injected (slug), and chase fluids
i	Species index (first position on composition variables) 1 = water (Chaps. 3, 5, 6); heavy hydrocarbon (Chap. 7) 2 = oil (Chaps. 3, 5, 6, 8, 9); intermediate hydrocarbon (Chap. 7) 3 = displacing agent (surfactant, Chap. 9; solvent, Chap. 7; gas, Chap. 5) 4 = polymer 5 = anion 6 = divalents 7 = cosurfactant 8 = monovalents
j	phase index (second position on composition variables) 1 = water-rich (Chaps. 3, 5, 6, 8, 9); heavy hydrocarbon rich (Chap. 7) 2 = oil-rich (Chaps. 3, 5, 6, 8, 9); solvent rich (Chap. 2) 3 = microemulsion s = solid
ℓ	Layer number index ($\ell = 1, \ldots, n, \ldots, N_L$)
OPT	Optimal
R	Remaining
r	Residual (second subscript position) or relative (first position)
T	Thermal property
t	Total
ti	Inside tubing
to	Outside tubing
u, l	Upper and lower effective salinities
w, nw	Wetting, nonwetting
x, z, r	Refers to coordinate directions x, z and r

References

ABBOTT, MICHAEL M., "Cubic Equations of State," *American Institute of Chemical Engineers Journal,* 19 (May 1973), 596–601.

———, "Cubic Equations of State: An Interpretive Review," in *Equations of State in Engineering and Research,* Advances in Chemistry Series No. 182, Washington, D.C.: American Chemical Society, 1978.

ABBOTT, MICHAEL M. and H.C. VAN NESS, *Thermodynamics,* Schaum's Outline Series in Engineering, New York: McGraw-Hill, 1972.

ABRAMS, A., "The Influence of Fluid Viscosity, Interfacial Tension and Flow Velocity on Residual Oil Saturation Left by Waterflood," *Society of Petroleum Engineers Journal,* 15 (1975), 437–447.

ADER, JOSEPH C. and MICHAEL H. STEIN, "Slaughter Estate Unit CO_2 Pilot Reservoir Description via a Black Oil Model Waterflood History Match," SPE/DOE 10727, presented at the 1982 SPE/DOE Third Joint Symposium on Enhanced Oil Recovery, Tulsa, Oklahoma, April 4–7.

AHMED, T., H. CHRICHLOW, and D. MENZIE, "Preliminary Experimental Results of High Pressure Nitrogen Injection for EOR Systems," SPE 10273, presented at the 56th Annual Fall Technical Conference of the Society of Petroleum Engineers, San Antonio, Texas, 1981.

AHO, G.E. and J. BUSH, "Results of the Bell Creek Unit 'A' Micellar-Polymer Pilot," SPE 11195, presented at the 57th Annual Fall Technical Conference and Exhibition of the Society of Petroleum Engineers, New Orleans, Louisiana, 1982.

AITHISON, J. and J. A. C. BROWN, *The Lognormal Distribution,* New York: Cambridge University Press, 1957.

AKSTINAT, M. H., "Surfactants for EOR Process in High-Salinity Systems: Product Selection and Evaluation," in *Enhanced Oil Recovery,* F. J. Fayers (ed.) New York: Elsevier, 1981.

AMOTT, EARL, "Observations Relating to the Wettability of Porous Rock," *Transactions of the American Institute of Mining, Metallurgical, and Petroleum Engineers,* 216 (1959), 156–162.

ANDERSON, D. F., M. S. BIDNER, H. T. DAVIS, C. D. MANNING, and L. E. SCRIVEN, "Interfacial Tension and Phase Behavior in Surfactant-Brine-Oil Systems," SPE 5811, presented at the Fourth Symposium on Improved Oil Recovery of the Society of Petroleum Engineers, Tulsa, Oklahoma, 1976.

ANDERSON, W. G., "Wettability Literature Survey—Part 2: Wettability Measurement," *Journal of Petroleum Technology* (November 1986), 1246–1262.

ANDRADE, E. N. DA C., "The Viscosity of Liquids," *Nature* (March 1, 1930), 309–310.

ARRIOLA, ALFREDO, PAUL G. WILLHITE, and DON W. GREEN, "Trapping of Oil Drops in a Noncircular Pore Throat," SPE 9404, presented at the 55th Annual Fall Technical Conference and Exhibition of the Society of Petroleum Engineers of American Institute of Mining and Metallurgical Engineers, Dallas, Texas, 1980.

ARYA, ATUL, T. A. HEWETT, R. G. LARSON, and L. W. LAKE, "Dispersion and Reservoir Heterogeneity," Society of Petroleum Engineers. *Reservoir Engineering*, Feb. 1988, 139–148.

ARYA, A., "Dispersion and Reservoir Heterogeneity," Ph. D. dissertation, The University of Texas at Austin, 1986.

AUXIETTE, GERARD and ISABELLE CHAPERON, "Linear Gas Drives in High-Pressure Oil Reservoirs, Compositional Simulation and Experimental Analysis," SPE 10271, presented at the 56th Annual Fall Technical Conference and Exhibition of the Society of Petroleum Engineers, San Antonio, Texas, 1981.

AYDELOTTE, S. R. and G. A. POPE, "A Simplified Predictive Model for Steamdrive Performance," *Journal of Petroleum Technology* (May 1983), 991–1002.

BAIL, P. T. and S. S. MARSDEN, "Saturation Distribution in a Linear System During Oil Displacement," *Producers Monthly,* 21:8 (June 1957) 22–32.

BARAKAT, Y., L. N. FORTNEY, R. S. SCHECHTER, W. W. WADE, S. H. YIV, and A. GRACIAA, "Criteria for Structuring Surfactants to Maximize Solubilization of Oil and Water," *Journal of Colloid and Interface Science,* 92 (1983), 561–574.

Basic Petroleum Data Book, Dallas Petroleum Engineering Co., annually since 1947.

Basic Petroleum Data Book: Petroleum Industry Statistics, Washington, D.C.: American Petroleum Institute, January 1991.

BAVIERE, M., R. S. SCHECHTER, and W. WADE, "The Influence of Alcohols on Microemulsion Composition," *Journal of Colloid and Interface Science,* 9 (May 1981), 266–279.

BEAR, JACOB, *Dynamics of Fluids in Porous Media,* New York: Elsevier, 1972.

BENHAM, A. L., W. E. DOWDEN, and W. J. KUNZMAN, "Miscible Fluid Displacement—Prediction of Miscibility," *Transactions of the Society of Petroleum Engineers of the AIME,* 219 (1961), 229–237.

BENNETT, K. E., C. H. K. PHELPS, H. T. DAVIS, and L. E. SCRIVEN, "Microemulsion Phase Behavior: Observations, Thermodynamic Essentials, Mathematical Simulation," *Society of Petroleum Engineers Journal,* 21 (1981), 747–762.

BERNARD, GEORGE G. and L. W. HOLM, "Effect of Foam on Permeability of Porous Media to Gas," SPE 983, presented at the Fall Technical Conference and Exhibition of the Society of Petroleum Engineers, Houston, Texas, 1964.

BERNARD, GEORGE G., L. W. HOLM, and L. JACOBS, "Effect of Foam on Trapped Gas Saturation and on Permeability of Porous Media to Water," SPE 1204, presented at the Fall Technical Conference and Exhibition of the Society of Petroleum Engineers, Denver, Colorado, 1965.

BHUYAN, D., "The Effect of Wettability on the Capillary Desaturation Curve," M. S. thesis, The University of Texas at Austin, 1986.

BIRD, R. B., W. E. STEWART, and E. N. LIGHTFOOT, *Transport Phenomena,* New York: Wiley, 1960.

BLEAKLEY, W. B., "The How and Why of Steam, Part 5," *Oil and Gas Journal* (February 15, 1965), 121–122.

BLEVINS, T. R. and R. H. BILLINGSLEY, "The Ten-Pattern Steamflood, Kern River Field, California," *Journal of Petroleum Technology* (December 1975), 1505–1514.

BONDOR, P. L., G. J. HIRASAKI, and M. J. THAM, "Mathematical Simulation of Polymer Flooding in Complex Reservoirs," *Society of Petroleum Engineers Journal,* 12 (October 1972), 369–382.

BOURREL, M., A. M. LIPOW, W. W. WADE, R. S. SCHECHTER, and J. L. SALAGER, "Properties of Amphile/Oil/Water Systems at an Optimum Formulation of Phase Behavior," SPE 7450, presented at the 53rd Annual Fall Technical Conference and Exhibition of the Society of Petroleum Engineers, Houston, Texas, 1978.

BRAGG, J. R., W. W. GALE, W. A. MCELHANNON, and O. W. DAVENPORT, "Loudon Surfactant Flood Pilot Test," SPE 10862, presented at the Third Symposium on Enhanced Oil Recovery of the Society of Petroleum Engineers, Tulsa, Oklahoma, 1982.

BROWN, W. O., "The Mobility of Connate Water During a Waterflood," *Transactions of the American Institute of Mining, Metallurgical, and Petroleum Engineers,* 210 (1957), 190–195.

BRUGGENWERT, M. G. M. and A. KAMPHORST, "Survey of Experimental Information on Cation Exchange in Soil Systems," in *Soil Chemistry B. Physico Chemical Methods,* G. H. Bolt (ed.), New York: Elsevier, 1979.

BRYANT, STEVEN L., R. S. SCHECHTER, and L. W. LAKE, "Interaction of Precipitation/Dissolution Waves and Ion Exchange in Flow Through Permeable Media," *American Institute of Chemical Engineers Journal,* 32 (May 1986) 751–764.

BUCKLEY, S. E. and M. C. LEVERETT, "Mechanisms of Fluid Displacement in Sands," *Transactions American Institute of Mining and Metallurgical Engineers,* 146 (1942), 107–116.

BUNGE, A. L. and C. J. RADKE, "Migration of Alkaline Pulses in Reservoir Sands," *Society of Petroleum Engineers Journal,* 22 (December 1982), 998–1012.

———, "The Origin of Reversible Hydroxide Uptake on Reservoir Rock," *Society of Petroleum Engineers Journal,* 25 (October 1985), 711–718.

BURGER, J., P. SOURIEAU and M. COMBARNOUS, *Thermal Methods of Oil Recovery,* Paris: Editions Technip and Houston: Gulf Publishing Company, 1985.

BURGER, J. G. and B. C. SAHUQUET, "Chemical Aspects of In Situ Combustion—Heat of Combustion and Kinetics," *Transactions of the Society of Petroleum Engineers of the AIME,* 253 (October 1972), 410–422.

BURSELL, C. G. and G. M. PITTMAN, "Performance of Steam Displacement in the Kern River Field," *Journal of Petroleum Technology* (August 1975), 997–1004.

CAMILLERI, DOMINIC, "Micellar/Polymer Flooding Experiments and Comparison with an Improved 1-D Simulator," M.S. thesis, The University of Texas at Austin, 1983.

CAMILLERI, DOMINIC, A. FIL, G. A. POPE, B. ROUSE, and K. SEPEHRNOORI, "Improvements in Physical Property Models Used in Micellar/Polymer Flooding," Society of Petroleum Engineers, *Reservoir Engineering,* Nov. 1987, 441–451.

CAMPBELL, R. A. et al., "The Metric System of Units and SPE's Tentative Metric Standard," *Journal of Petroleum Technology* (December 1977), 1575.

CARSLAW, H. W. and J. C. JAEGER, *Conduction of Heat in Solids,* 2d ed., Oxford: Clarendon Press, 1959.

CASH, R. C., J. L. CAYIAS, R. G. FOURNIER, J. K. JACOBSON, T. SCHARES, R. S. SCHECHTER, and W. H. WADE, "Modelling Crude Oils for Low Interfacial Tension," *Society of Petroleum Engineers Journal,* 16 (1976), 351–357.

CAUDLE, B. H. and M. D. WITTE, "Production Potential Changes During Sweep-Out in a Five-Spot System," *Transactions of the American Institute of Mining and Metallurgical Engineers,* 216 (1959), 446–448.

CAUDLE, B. H., *Fundamentals of Reservoir Engineering, Part II,* Dallas: Society of Petroleum Engineers, 1968.

——— and A.B. DYES, "Improving Miscible Displacement by Gas-Water Injection," *Transactions of the American Institute of Mining, Metallurgical, and Petroleum Engineers,* 213 (1958), 281–284.

CAYIAS, J. L., R. S. SCHECHTER, and W. H. WADE, "The Measurement of Low Interfacial Tension Via the Spinning Drop Technique, Adsorption at Interfaces," L. K. Mittal (ed.), American Chemical Society Symposium Series, No. 8 (1975), 234.

CERE, A. and F. ZANOTTI, "Sharpening Behaviour and Dispersion in a Chemical Flooding," Proceedings of European Symposium on Enhanced Oil Recovery, Rome, Italy, 1985.

CHANG, H. L., H. M. AL-RIKABI, and W. H. PUSCH, "Determination of Oil/Water Bank Mobility in Micellar-Polymer Flooding," *Journal of Petroleum Technology,* 30 (July 1978), 1055–1060.

CHATZIS, J., and N. R. MORROW, "Measurement and Conditions for Entrapment and Mobilization of Residual," U. S. Department of Energy Final Report DOE/BETC/3251-12, October 1981.

CHATZIS, J., N. R. MORROW, and H. T. LIM, "Magnitude and Detailed Structure of Residual Oil Saturation," *Society of Petroleum Engineers Journal,* 23 (March–April 1983), 2.

CHOUKE, R. L., P. VAN MEURS, and C. VAN DER POEL, "The Instability of Slow, Immiscible Viscous Liquid-Liquid Displacements in Permeable Media," *Transactions of the American Institute of Mining, Metallurgical and Petroleum Engineers,* 216 (1959), 188–194.

CLAMPITT, R. L. and T. B. REID, "An Economic Polymerflood in the North Burbank Unit, Osage Country, Oklahoma," SPE 5552, presented at the 50th Annual Fall Technical Conference and Exhibition of the Society of Petroleum Engineers, Dallas, Texas, 1975.

CLARIDGE, E. L., "Prediction of Recovery in Unstable Miscible Flooding," *Society of Petroleum Engineers Journal,* 12 (April 1972), 143–155.

———, "A Method for Designing Graded Viscosity Banks," *Society of Petroleum Engineers Journal,* 18 (October 1978), 315–324.

———, "Design of Graded Viscosity Banks for Enhanced Oil Recovery Processes," Ph.D dissertation, University of Houston, 1980.

———, "Prospects for Enhanced Recovery of Oil in U.S.," Paper #829168, presented at th

17th Entersociety Energy Conversion Engineering Conference, Los Angeles, California (August 1982), 981–987.

COATS, K. H. and B. D. SMITH, "Dead-End Pore Volume and Dispersion in Porous Media," *Society of Petroleum Engineers Journal,* 4 (March 1964), 73–84.

COATS, K. H., J. R. DEMPSEY, and J. H. HENDERSON, "The Use of Vertical Equilibrium in Two-Dimensional Simulation of Three-Dimensional Reservoir Performance," *Society of Petroleum Engineers Journal,* 11 (March 1971), 63–71.

———, "An Equation of State Compositional Model," *Society of Petroleum Engineers Journal,* 20 (October 1980), 363–376.

COLLINS, R. E., *Flow of Fluids Through Porous Media,* Tulsa, Okla.: Petroleum Publishing, 1976.

COOKE, C. E., JR., R. D. WILLIAMS, and P. A. KOLODZIE, "Use of Centrifugal Measurements of Wettability to Predict Oil Recovery," U.S. Department of the Interior, Report of Investigations 7873, 1974.

COURANT, R. and K. O. FRIEDRICHS, *Supersonic Flow and Shock Waves,* New York: Springer-Verlag, 1948.

CRAIG, F. F., JR., *The Reservoir Engineering Aspects of Waterflooding,* Society of Petroleum Engineers Monograph, Dallas, 1971.

CRANE, F. E., H. A. KENDALL, and G. H. F. GARDNER, "Some Experiments on the Flow of Miscible Fluids of Unequal Density Through Porous Media," *Society Petroleum Engineers Journal,* 3 (1963), 277–280.

CRAWFORD, H. R., G. H. NEILL, B. J. LUCY, and P. B. CRAWFORD, "Carbon Dioxide—A Multipurpose Additive for Effective Well Stimulation," *Journal of Petroleum Technology,* 15 (March 1963), 237–242.

CRICHLOW, HENRY B., *A Simulation Approach,* Englewood Cliffs, N.J.: Prentice-Hall, 1977.

CROCKER, M. W., E. C. DONALDSON, and L. M. MARCHIN, "Comparison and Analysis of Reservoir Rocks and Related Clays," SPE 11973, presented at the 58th Annual Fall Technical Conference and Exhibition of the Society of Petroleum Engineers, San Francisco, California, 1983.

CULBERSON, O. L. and J. J. MCKETTA, "Phase Equilibria in Hydrocarbon-Water Systems III–The Solubility of Methane in Water at Pressures to 10,000 psia," *Transactions of the American Institute of Mining and Metallurgical Engineers,* 192 (1951), 223–226.

DAGAN, G., "Solute Transport in Heterogeneous Porous Formation," *Journal of Fluid Mechanics,* 145 (1984), 151–177.

DAKE, L. P., *Fundamentals of Reservoir Engineering,* New York: Elsevier, 1978.

DAVID, AMIEL and SULLIVAN S. MARSDEN, JR., "The Rheology of Foam," SPE 2544, Presented at the 44th Annual Fall Technical Conference and Exhibition of the Society of Petroleum Engineers, Denver, Colorado, 1969.

DEANS, H. A., "A Mathematical Model for Dispersion in the Direction of Flow," *Society of Petroleum Engineers Journal,* 3 (March 1963), 49–52.

DEGENS, E. T., *Geochemistry of Sediments* (*A Brief Survey*), Englewood Cliffs, N.J.: Prentice-Hall, 1965.

DELSHAD, MOJDEH, MOHAMMAD DELSHAD, G. A. POPE, and L. W. LAKE, "Two- and Three-Phase Relative Permeabilities of Micellar Fluids," SPE Formation Evaluation, Sept. 1987, 327–337.

———, "Measurement of Relative Permeability and Dispersion for Micellar Fluid in Berea Rock," M.S. thesis, The University of Texas at Austin, 1981.

DENBIGH, KENNETH, *The Principles of Chemical Equilibrium,* 2d ed., Oxford, England: Cambridge University Press, 1968.

DE NEVERS, N., "A Calculation Method for Carbonated Waterflooding," *Society of Petroleum Engineers Journal,* 4:1 (March 1964), 9–20.

DE ZABALA, E. F., J. M. VISLOCKY, E. RUBIN, and C. J. RADKE, "A Chemical Theory for Linear Alkaline Flooding," *Society of Petroleum Engineers Journal,* 12 (April 1982), 245–258.

DICHARRY, ROY M., T. L. PERRYMANN, and J. D. RONQUILLE, "Evaluation and Design of a CO_2 Miscible Flood Project—SACROC Unit, Kelly-Snyder Field," *Journal of Petroleum Technology,* 25 (November 1972), 1309–1318.

DIETZ, D. N., "A Theoretical Approach to the Problem of Encroaching and By-Passing Edge Water," Akad. van Wetenschappen, Amsterdam. Proc. V. 56-B (1953), 83.

DOMBROWSKI, H. S. and L. E. BROWNELL, "Residual Equilibrium Saturation of Porous Media," *Industrial and Engineering Chemistry,* 46 (1954), 1207.

DONALDSON, ERLE C., REX D. THOMAS, and PHILIP B. LORENZ, "Wettability Determination and its Effect on Recovery Efficiency," *Society of Petroleum Engineers Journal,* 9 (March 1969), 13–20.

DOUGHERTY, E. L., "Mathematical Model of an Unstable Miscible Displacement in Heterogeneous Media," *Society of Petroleum Engineers Journal,* 3 (June 1963), 155–163.

DOUGLAS, J., JR., P. M. BLAIR, and R. J. WAGNER, "Calculation of Linear Waterflood Behavior Including the Effects of Capillary Pressure," *Transactions of the American Institute of Mining, Metallurgical, and Petroleum Engineers,* 213 (1958), 96–102.

DRIA, M. A., R. S. SCHECHTER, and L. W. LAKE, "An Analysis of Reservoir Chemical Treatments," SPE Production Engineering, Feb. 1988, 52–62.

DUDA, J. L., E. E. KLAUS, and S. K. FAN, "Influence of Polymer Molecule-Wall Interactions on Mobility Control," *Society of Petroleum Engineers Journal,* 2 (1981), 613–622.

DULLIEN, F. A. L., *Fluid Transport and Pore Structure,* New York: Academic Press, 1979.

DU PREY, E. J. LEFEBVRE, "Factors Affecting Liquid-Liquid Relative Permeabilities of a Consolidated Porous Media," *Society of Petroleum Engineers Journal,* 13 (1973), 39–47.

DYES, A. B., B. H. CAUDLE, and R. A. ERICKSON, "Oil Production After Breakthrough—As Influenced by Mobility Ratio," *Transactions of the American Institute of Mining and Metallurgical Engineers,* 201 (1954), 81–86.

DYKSTRA, H. and R. L. PARSONS, "The Prediction of Oil Recovery by Waterflood," *Secondary Recovery of Oil in the United States, Principles and Practice,* 2d ed., American Petroleum Institute (1950), 160–174.

EARLOUGHER, ROBERT C., JR., *Advances in Welltest Analysis,* Society of Petroleum Engineers Monograph, Dallas, 1977.

EHRLICH, R., H. H. HASIBA, and P. RAIMONDI, "Alkaline Waterflooding for Wettability Alteration Evaluating a Potential Field Application, *Journal of Petroleum Technology* (September 1974), 1335–1352.

El Dorado Micellar-Polymer Demonstration Project, Third Annual Report, U.S. Department of Energy BERC/TPR-77/12 (1977).

ENGLESEN, SVEN ROAR, "Micellar/Polymer Flooding Simulation—Improvements in Modelling and Matching of Core Floods," M.S. thesis, The University of Texas at Austin, 1981.

Enhanced Oil Recovery—An Analysis of Potential for Enhanced Oil Recovery from Known Fields in the United States—1976 to 2000, National Petroleum Council (December 1976 and June 1984).

ERSHAGHI, I. and O. OMOREGEI, "A Method for Extrapolation of Cut vs. Recovery Curves," *Journal of Petroleum Technology* (February 1978), 203–204.

FAROUQ ALI, S. M., "Steam Injection," in *Secondary and Tertiary Oil Recovery Processes,* Oklahoma City: Interstate Oil Compact Commission, 1974.

———, "Marx and Langenheim's Model of Steam Injection," *Producers Monthly* (November 1966), 2-8.

FATT, I. and H. DYKSTRA, "Relative Permeability Studies," *Transactions of the American Institute of Mining and Metallurgical Engineers,* 192 (1951), 249.

FAYERS, F. J., "An Approximate Model with Physically Interpretable Parameters for Representing Viscous Fingering," SPE 13166, presented at 59th Annual Fall Technical Conference and Exhibition of the Society of Petroleum Engineers, Houston, Texas, 1984.

FEDKIW, P., "Mass Transfer Controlled Reactions in Packed Beds of Low Reynolds Numbers," Ph.D. thesis, University of California, 1979.

FERNANDEZ, MIGUEL ENRIQUE, "Adsorption of Sulfonates from Aqueous Solutions onto Mineral Surfaces," M.S. thesis, The University of Texas, 1978.

FLEMING, PAUL D. III, CHARLES P. THOMAS, and WILLIAM K. WINTER, "Formulation of a General Multiphase, Multicomponent Chemical Flood Model," *Society of Petroleum Engineers Journal,* 21 (1981), 63.

FLORY, PAUL J., *Principles of Polymer Chemistry,* Ithaca, New York: Cornell University Press, 1953.

FOSHEE, W. C., R. R. JENNINGS, and T. J. WEST, "Preparation and Testing of Partially Hydrolyzed Polyacrylamide Solutions," SPE 6202, presented at 51st Annual Fall Technical Conference and Exhibition of the Society of Petroleum Engineers, New Orleans, Louisiana, October 3–6, 1976.

FOSTER, W. R., "A Low Tension Waterflooding Process Employing Petroleum Sulfonate Inorganic Salts and a Biopolymer," *Journal of Petroleum Technology,* and McMillen and Foster, personal communication to R. G. Larson (February 1973).

FRANCIS, ALFRED W., *Liquid-Liquid Equilibriums,* New York: Interscience, 1963.

FRIED, A. N., "The Foam-Drive Process for Increasing the Recovery of Oil," U.S. Department of the Interior, Report of Investigations 5866, 1960.

FUSSELL, L. T. and F. D. FUSSELL, "An Iterative Technique for Compositional Reservoir Models," *Society of Petroleum Engineers Journal,* 19 (August 1979), 211–220.

GAMBILL, W. R., "You Can Predict Heat Capacity," *Chemical Engineering* (June 1957), 243–248.

GARDNER, J. W., F. M. ORR, and P. D. PATEL, "The Effect of Phase Behavior on CO_2-Flood Displacement Efficiency," *Journal of Petroleum Technology,* 33 (November 1981), 2067–2081.

GARDNER, J. W. and J. G. J. YPMA, "An Investigation of Phase Behavior-Macroscopic Bypassing Interaction in CO_2 Flooding," SPE/DOE 10686, presented at the Third Joint SPE/DOE Symposium on Enhanced Oil Recovery, Tulsa, Oklahoma, 1982.

GARRELS, ROBERT M. and CHARLES L. CHRIST, *Solutions, Minerals, and Equilibria,* San Francisco: Freeman, Cooper, 1965.

Gas Processors Suppliers Association, *Engineering Data Book,* Tulsa, 1983.

GASH, B. H., T. D. GRIFFITH, and A. F. CHAN, "Phase Behavior Effects on the Oil Displacement Mechanisms of Broad Equivalent Weight Surfactant Systems," SPE 9812, presented at the Second Joint SPE/DOE Symposium on Enhanced Oil Recovery, Tulsa, Oklahoma, 1981.

GATES, C. F. and I. SKLAR, "Combustion as a Primary Recovery Process," *Journal of Petroleum Technology* (April 1971), 981–986.

GEFFEN, TED, "Improved Oil Recovery Could Help Ease Energy Shortage," *World Oil,* 177 (October 1973), 84–88.

GELHAR, L., A. L. GUTHJAR, and R. L. NAFF, "Stochastic Analysis of Macrodispersion in a Stratified Aquifer," *Water Resources Research,* 15 (1979), 1387–1397.

GELHAR, L., A. MANTOGLOU, C. WELTY, and K. R. REHFELDT, "A Review of Field-Scale Physical Solute Transport Processes in Saturated and Unsaturated Porous Media," in Electric Power Resources Institute Final Report, EA-4190, Project RP2485-5, August 1985.

GIBBS, CHARLES W. (ed.), *Compressed Air and Gas Data,* 2d ed., Woodcliff Lake, N.J.: Ingersoll-Rand, 1971.

GILLILAND, H. E. and F. R. CONLEY, "Surfactant Waterflooding," presented at the Symposium on Hydrocarbon Exploration, Drilling, and Production, Paris, France, 1975.

GIORDANO, R. M. and S. J. SALTER, "Comparison of Simulation and Experiments for Compositionally Well-defined Corefloods," SPE 12697, presented at the Fourth Joint SPE/DOE Symposium on Enhanced Oil Recovery, Tulsa, Oklahoma, 1984.

GLINSMANN, G. R., "Surfactant Flooding with Microemulsions Formed In-situ—Effect of Oil Characteristics," SPE 8326, presented at the 54th Annual Technical Conference and Exhibition of the Society of Petroleum Engineers, Las Vegas, Nevada, 1979.

GLOVER, C. J., M. C. PUERTO, J. M. MAERKER, and E. I. SANDVIK, "Surfactant Phase Behavior and Retention in Porous Media," *Society of Petroleum Engineers Journal,* 19 (1979), 183–193.

GOGARTY, W. B., H. P. MEABON, and H. W. MILTON, JR., "Mobility Control Design for Miscible-type Waterfloods Using Micellar Solutions," *Journal of Petroleum Technology,* 22 (1970), 141–147.

GOLDBURG, A., H. PRICE, and G. W. PAUL, "Selection of Reservoirs Amenable to Micellar Flooding," U. S. Department of Energy, DOE/BC/00048 and 00051-29, August 1985.

GOODRICH, JOHN H., "Target Reservoirs for CO_2 Miscible Flooding," Final Report, U.S. Department of Energy, DOE/MC/08341-17 (October 1980).

GRACIAA, A., L. N. FORTNEY, R. S. SCHECHTER, W. H. WADE, and S. YIV, "Criteria for Structuring Surfactant to Maximize Solubilization of Oil and Water I: Commercial Nonionics," SPE 9815, presented at the Second Joint SPE/DOE Symposium on Enhanced Oil Recovery, Tulsa, Oklahoma, 1981.

GRAUE, D. J. and C. E. JOHNSON, "Field Trial of Caustic Flooding Process," *Journal of Petroleum Technology* (December 1974), 1353–1358.

GRAY, WILLIAM G., "A Derivation of the Equations for Multi-phase Transport," *Chemical Engineering Science,* 30 (1975), 229.

GREENKORN, R. A. and D. P. KESSLER, "Dispersion in Heterogeneous Nonuniform Anisotropic Porous Media," *Industrial and Engineering Chemistry,* 61 (1969) 33.

GRIM, RALPH E., *Clay Mineralogy,* New York: McGraw-Hill, 1968.

GUPTA, S. P., "Compositional Effects on Displacement Mechanisims of the Micellar Fluid Injected in the Sloss Field Test," SPE 8827, presented at the First Joint SPE/DOE Symposium on Enhanced Oil Recovery, Tulsa, Oklahoma, 1980.

GUPTA, S. P. and S. P. TRUSHENSKI, "Micellar Flooding-Compositional Effects on Oil Displacement," *Society of Petroleum Engineers Journal,* 19 (1978), 116–128.

HABERMANN, B., "The Efficiencies of Miscible Displacement as a Function of Mobility Ratio," *Transactions of the AIME*, 219 (1960), 264.

HALL, H. N. and T. M. GEFFEN, "A Laboratory Study of Solvent Flooding," in *Miscible Processes,* SPE Reprint Series No. 8, Dallas, Texas: Society of Petroleum Engineers, 1965.

HAND, D. B., "Dineric Distribution: I. The Distribution of a Consolute Liquid between Two Immiscible Liquids," *Journal of Physics and Chemistry,* 34 (1939), 1961–2000.

HARWELL, J. H., "Surfactant Adsorption and Chromatographic Movement with Application in Enhanced Oil Recovery," Ph.D. dissertation, The University of Texas at Austin, 1983.

HAWTHORNE, R. G., "Two-Phase Flow in Two Dimensional Systems—Effects of Rate, Viscosity and Density on Fluid Displacement in Porous Media," *Transactions of the American Institute of Mining, Metallurgical, and Petroleum Engineers,* 219 (1960), 81–87.

HEALY, R. N. and R. L. REED, "Physicochemical Aspects of Microemulsion Flooding," *Society of Petroleum Engineers Journal,* 14 (1974), 491–501.

HEALY, R. N., R. L. REED, and D. G. STENMARK, "Multiphase Microemulsion Systems," *Society of Petroleum Engineers Journal,* 16 (1976), 147–160.

HEARN, C. L., "Simulation of Stratified Waterflooding by Pseudo-Relative Permeability Curves," *Journal of Petroleum Technology* (July 1971), 805–813.

HEKIM, YUSUF, H. SCOTT FOGLER, and CONWELL C. MCCUNE, "The Radial Movement of Permeability Fronts and Multiple Reaction Zones in Porous Media," *Society of Petroleum Engineers Journal,* 22 (February 1982), 99–107.

HELFFERICH, F. G. and G. KLEIN, *Multicomponent Chromatography,* New York: Marcel Dekker, 1970.

———, "Theory of Multicomponent, Multiphase Displacement in Porous Media," *Society of Petroleum Engineers Journal,* 21 (February 1981), 51–62.

HELLER, JOHN P., CHENG LI LIEN, and MURTY S. KUNTAMUKKULA, "Foam-Like Dispersions for Mobility Control in CO_2 Floods," SPE 11233, presented at the 57th Annual Fall Technical Conference and Exhibition of the Society of Petroleum Engineers of the AIME, New Orleans, Louisiana, 1982.

HILL, H. J. and L. W. LAKE, "Cation Exchange in Chemical Flooding: Part 3—Experimental," *Society of Petroleum Engineers Journal,* 18 (1978), 445–456.

HILL, S., "Channelling in Packed Columns," *Chemical Engineering Science,* 1 (1952), 247–253.

HIMMELBLAU, DAVID M., *Basic Principles and Calculations in Chemical Engineering,* 4th ed., Englewood Cliffs, N.J.: Prentice-Hall, 1982.

HIRASAKI, G. J., "Application of the Theory of Multicomponent, Multiphase Displacement to Three-Component, Two-Phase Surfactant Flooding," *Society of Petroleum Engineers Journal,* 21 (April 1981), 191–204.

———, "Interpretation of the Change in Optimal Salinity with Overall Surfactant Concentration," *Society of Petroleum Engineers Journal,* 22 (1982), 971–982.

HIRASAKI, G. J. and J. B. LAWSON, "Mechanisms of Foam Flow in Porous Media—Apparent Viscosity in Smooth Capillaries," SPE 12129 presented at the 58th Annual Meeting of the Society of Petroleum Engineers, San Francisco, California, 1983.

HIRASAKI, G. J. and G. A. POPE, "Analysis of Factors Influencing Mobility and Adsorption in Flow of Polymer Solution Through Porous Media," *Society of Petroleum Engineers Journal,* 14 (1974), 337–346.

HIRASAKI, G. J., H. R. VAN DOMSELAAR, and R. C. NELSON, "Evaluation of the Salinity Gradient Concept in Surfactant Flooding," *Society of Petroleum Engineers Journal,* 23 (1983), 486–500.

HIRSHBERG, ABRAHAM, L. N. J. DE JONG, B. A. SCHIPPER, and J. G. MEYERS, "Influence of Temperature and Pressure on Asphaltenes," SPE 11202, presented at the 57th Annual Fall Technical Conference and Exhibition of the Society of Petroleum Engineers, New Orleans, Louisiana, 1982.

HOLBROOK, STAN T., JOHN T. PATTON, and WENDAR HSU, "Rheology of Mobility Control Foams," SPE/DOE paper 9809, presented at the 1981 Second Joint SPE/DOE Symposium on Enhanced Oil Recovery, Tulsa, Oklahoma, 1981.

HOLM, L. W., "A Comparison of Propane and CO_2 Solvent Flooding Processes," *American Institute of Chemical Engineers Journal* (June 1961), 179–184.

———, "Design, Performance and Evaluation of the Uniflood Micellar-Polymer Process—Bell Creek Field," SPE 11196, presented at the 57th Annual Fall Technical Conference and Exhibition of the Society of Petroleum Engineers, New Orleans, Louisiana, 1982.

HOLM, L. W. and A. K. CSASZAR, "Oil Recovery by Solvents Mutually Soluble in Oil and Water," Society of Petroleum Engineers Reprint Series No. 8, *Miscible Processes* (1965), 31–38.

HOLM, L. W. and V. A. JOSENDAL, "Mechanisms of Oil Displacement by Carbon Dioxide," *Journal of Petroleum Technology,* 26 (December 1974), 1427–1438.

———, "Effect of Oil Composition on Miscible-Type Displacement by Carbon Dioxide," *Society of Petroleum Engineers Journal,* 22 (1982), 87–98.

HOMSY, G. M., "Viscous Fingering in Porous Media," *Annual Review of Fluid Mechanics,* 19 (1987), 271–311.

HONARPOUR, MEHDI, L. F. KOEDERITZ, and HERBERT A. HARVEY, "Empirical Equations for Estimating Two-Phase Relative Permeability in Consolidated Rock," *Journal of Petroleum Technology,* 34 (December 1982), 2905–2908.

HONG, K. C., "Lumped-Component Characterization of Crude Oils for Compositional Simulation," SPE 10691, presented at the Third Joint SPE/DOE Symposium on Enhanced Oil Recovery, Tulsa, Oklahoma, 1982.

HOUGEN, OLAF A., KENNETH M. WATSON, and ROLAND A. RAGATZ, *Chemical Process Principles, Part II Thermodynamics,* New York: Wiley, 1966.

HUBBERT, M. KING, "Darcy's Law and the Field Equations of the Flow of Underground Fluids," *Transactions of the American Institute of Mining, Metallurgical, and Petroleum Engineers,* 207 (1956), 222.

HUH, C., "Interfacial Tensions and Solubilizing Ability of a Microemulsion Phase that Coexists with Oil and Brine," *Journal of Colloid and Interface Science,* 71 (1979), 408–426.

HUTCHINSON, C. A., JR. and P. H. BRAUN, "Phase Relationships of Miscible Displacement in Oil Recovery," *American Institute of Chemical Engineers Journal,* 7 (1961), 65.

JAHNKE, E. and F. EMDE, *Tables of Functions,* New York: Dover, 1945.

JEFFREY, A. and T. TANIUTI, *Nonlinear Wave Propagation with Applications to Physics and Magnetohydrodynamics,* New York: Academic Press, 1964.

JENNINGS, AARON and DAVID J. KIRKNER, "Instantaneous Equilibrium Approximation Analysis," *Journal of Hydraulic Engineering,* 110 (December 1984), 1700–1716.

JENNINGS, H. Y., JR., C. E. JOHNSON, JR., and C. E. MCAULIFFE, "A Caustic Waterflooding Process for Heavy Oils," *Journal of Petroleum Technology,* 26 (December 1974), 1344–1352.

JENNINGS, R. R., J. H. ROGERS, and T. J. WEST, "Factors Influencing Mobility Control by Polymer Solutions," *Journal of Petroleum Technology,* 23 (1971), 391–401.

JENSEN, JERRY L. and L.W. LAKE, "The Influence of Sample Size and Permeability Distribution Upon Heterogeneity Measures," SPE 15434, presented at the 61st Annual Technical Conference and Exhibition of the Society of Petroleum Engineers, New Orleans, Louisiana, 1986.

JENSEN, JERRY L., LARRY W. LAKE, and DAVID V. HINKLEY, "A Statistical Study of Reservoir Permeability: Distributions, Correlations, and Averages," *Society of Petroleum Engineers Formation Evaluation,* 2 (December 1987).

JOHNSON, C. E., JR., "Prediction of Oil Recovery by Water Flood—A Simplified Graphical Treatment of the Dykstra-Parsons Method," *Transactions of the American Institute of Mining, Metallurgical, and Petroleum Engineers,* 207 (1956), 745–746.

JOHNSON, JAMES P. and JAMES S. POLLIN, "Measurement and Correlation of CO_2 Miscibility Pressures," SPE/DOE 9790, presented at the 1981 Joint SPE/DOE Symposium on Enhanced Oil Recovery, Tulsa, Oklahoma, 1981.

JONES, KIM, "Rheology of Viscoelastic Fluids for Oil Recovery," M.S. thesis, The University of Texas at Austin, 1981.

JONES, R. S., JR., G. A. POPE, H. J. FORD, and L. W. LAKE, "A Predictive Model for Water and Polymer Flooding," SPE/DOE 12653, presented at the SPE/DOE Fourth Symposium on Improved Oil Recovery, Tulsa, Oklahoma, 1984.

JONES, STANLEY C., "Finding the Most Profitable Slug Size," *Journal of Petroleum Technology* (August 1972), 993.

JONES, STANLEY C. and W. O. ROSZELLE, "Graphical Technique for Determining Relative Permeability from Displacement Experiments," *Journal of Petroleum Technology* (May 1978), 807–817.

KEENAN, J. H., F. G. KEYS, P. G. HILL, and J. G. MOORE, *Steam Tables, Thermodynamic Properties of Water, Including Vapor, Liquid and Solid Phases,* New York: Wiley, 1969.

KHAN, SUHAIL A., "The Flow of Foam Through Porous Media," M.S. thesis, Stanford University, 1965.

KHILAR, K. C. and H. S. FOGLER, "Permeability Reduction in Water Sensitive Sandstones," in *Surface Phenomena in Enhanced Oil Recovery,* Dinesh O. Shah (ed.), New York: Plenum, 1981.

KOCH, H. A., JR. and R. L. SLOBOD, "Miscible Slug Process," *Transactions American Institute of Mining and Metallurgical Engineers,* 210 (1956), 40–47.

KOVAL, E. J., "A Method for Predicting the Performance of Unstable Miscible Displacement in Heterogeneous Media," *Society of Petroleum Engineers Journal,* 3 (June 1963), 145–154.

KRUEGER, DAVID A., "Stability of Piston-Like Displacements of Water by Steam and Nitrogen in Porous Media," *Society of Petroleum Engineers Journal,* 22 (October 1982), 625–634.

KYTE, J. R. and L. A. RAPOPORT, "Linear Waterflood Behavior and End Effects in Water-Wet Porous Media," *Transactions of the American Institute of Mining, Metallurgical and Petroleum Engineers,* 213 (1958), 423–426.

LAKE, L. W., "A Technical Survey of Micellar Polymer Flooding," presented at Enhanced Oil Recovery, A Symposium for the Independent Producer, Southern Methodist University, Dallas, Texas, November 1984.

LAKE, L. W. and F. HELFFERICH, "Cation Exchange in Chemical Flooding Part II—The Effect of Dispersion, Cation Exchange, and Polymer/Surfactant Adsorption on Chemical Flood Environment," *Society of Petroleum Engineers Journal,* 18 (1978), 435–444.

LAKE, L. W. and G. A. POPE, "Status of Micellar-Polymer Field Tests," *Petroleum Engineers International,* 51 (1978), 38–60.

LAKE, L. W. and GEORGE J. HIRASAKI, "Taylor's Dispersion in Stratified Porous Media," *Society of Petroleum Engineers Journal,* 21 (August 1981), 459–468.

LAKE, L. W., G. A. POPE, G. F. CAREY, and K. SEPEHRNOORI, "Isothermal, Multiphase, Multicomponent Fluid-Flow in Permeable Media," *In Situ,* 8 (March 1984), 1–40.

LAKE, L. W. and VITO J. ZAPATA, "Estimating the Vertical Flaring in In Situ Leaching," *In Situ,* 11 (March 1987), 39–62.

LALLEMAND-BARRES, A., and P. PEAUDECERF, "Recherche des Relations entre la Valeur de la Dispersivite Macroscopique d'un Aquifere, ses Autres Caracteristiques et les Conditions de Mesure," *Bulletin du BRGM,* 2e Serie, Section III, no. 4, 1978.

LAMBERT, MARY ELLEN, "A Statistical Study of Reservoir Heterogeneity," M.S. thesis, The University of Texas at Austin, 1981.

LARSON, R. G., "Percolation in Porous Media with Application to Enhanced Oil Recovery," M.S. thesis, University of Minnesota, 1977.

———, "A Novel Method for Controlling Numerical Dispersion in Finite Difference Simulation of Flow in Porous Media," SPE-8027, Society of Petroleum Engineers Preprint, 1979.

———, "From Molecules to Reservoirs: Problems in Enhanced Oil Recovery," Ph.D. dissertation, University of Minnesota, 1980.

LARSON, R. G., H. T. DAVIS, and L. E. SCRIVEN, "Elementary Mechanisms of Oil Recovery by Chemical Methods," *Journal of Petroleum Technology,* 34 (February 1982), 243–258.

LARSON, R. G., L. E. SCRIVEN, and H. T. DAVIS, "Percolation Theory of Residual Phases in Porous Media," *Nature,* 268 (1977), 409–413.

LEE, ANTHONY, MARIO H. GONZALEZ, and BERTRAM E. EAKIN, "The Viscosity of Natural Gases," *Journal of Petroleum Technology,* 18:8 (August 1966), 997–1000.

LEVENSPEIL, OCTAVE, *Chemical Reaction Engineering,* New York: Wiley, 1962.

LEVERETT, M. C., "Capillary Behavior in Porous Solids," *Petroleum Transactions of the American Institute of Mining and Metallurgical Engineers,* 142 (1941), 152.

LEWIS, W. K. and L. SQUIRES, "The Mechanism of Oil Viscosity as Related to the Structure of Liquids," *Oil and Gas Journal* (November 15, 1934), 92–96.

LORENZ, P. B., E. C. DONALDSON, and R. D. THOMAS, "Use of Centrifugal Measurements of Wettability to Predict Oil Recovery," U.S. Department of the Interior, Report of Investigations 7873, 1974.

MACALLISTER, D. J., "Measurement of Two- and Three-Phase Relative Permeability and Dispersion of Micellar Fluids in Unconsolidated Sand," M.S. thesis, The University of Texas at Austin, 1982.

MAERKER, J. M., "Mechanical Degradations of Partially Hydrolyzed Polyacrylamide Solutions in Unconsolidated Porous Media," *Society of Petroleum Engineers Journal,* 16 (1976), 172–174.

MANDL, G. and C. W. VOLEK, "Heat and Mass Transport in Steam-Drive Processes," *Journal of Petroleum Technology* (March 1969), 59–79.

MANNING, R. K., G. A. POPE, and L. W. LAKE, "A Technical Survey of Polymer Flooding Projects," U.S. Department of Energy DOE/BETC/10327-19, Bartlesville, Oklahoma, 1983.

MARLE, C. M., *Multiphase Flow in Porous Media,* Houston: Gulf Publishing, 1981.

MARTIN, F. DAVID, GUY L. DONARUMA, and MELVIN J. HATCH, "Development of Improved Mobility Control Agents for Surfactant/Polymer Flooding," 2nd Annual Report, U.S. Department of Energy DOE/BC/00047-13, 1981.

MARTIN, JOHN C., "Some Mathematical Aspects of Two-Phase Flow with Applications to Flooding and Gravity Segregation Problems," *Producers Monthly* (April 1958), 22.

MARTIN, JOSEPH J., "Cubic Equations of State—Which?" *Industrial and Engineering Chemistry Fundamentals,* 18 (1979), 81–89.

MARTIN, JOSEPH J. and Y. C. HOU, "Development of an Equation of State for Gases," *American Institute of Chemical Engineers Journal,* 1 (1955), 142.

MARTIN, J. W., "Additional Oil Production Through Flooding with Carbonated Water," *Producers Monthly,* 15 (July 1951), 18–22.

MARX, J. W. and R. H. LANGENHEIM, "Reservoir Heating by Hot Fluid Injection," *Transactions of the American Institute of Mining, Metallurgical, and Petroleum Engineers,* 216 (1959), 312–315.

MAXWELL, J. B., *Data Book on Hydrocarbon,* New York: D. von Nostrand, 1950.

MAYER, E. H., R. L. BERG, J. D. CARMICHAEL, and R. M. WEINBRANDT, "Alkaline Injection for Enhanced Oil Recovery—a Status Report," SPE 8848, presented at the First Joint SPE/DOE Symposium on Enhanced Oil Recovery, Tulsa, Oklahoma, 1980.

MCAULIFFE, C. E., "Crude-Oil-in-Water Emulsions to Improve Fluid Flow in an Oil Reservoir," *Journal of Petroleum Technology* (June 1973), 721–726.

MCCAIN, WILLIAM D., JR., *The Properties of Petroleum Fluids,* Tulsa, Okla.: Petroleum Publishing, 1973.

MCMILLIAN, J. M. and FOSTER, W. R., personal communication in Larson, R. G. (1977).

MCREE, B. C., "CO_2: How it Works, Where it Works," *Petroleum Engineers International,* 49 (November 1977), 52–63.

MEHRA, RAKESH K., ROBERT A. HEIDEMANN, and KHALID AZIZ, "Computation of Multiphase Equilibrium for Compositional Simulation," SPE 9232, presented at the 55th Annual Fall Technical Conference and Exhibition of the Society of Petroleum Engineers, Dallas, Texas, 1980.

MELDAU, ROBERT F., ROBERT G. SHIPLEY, and KEITH H. COATS, "Cyclic Gas/Steam Stimulation of Heavy-Oil Wells," *Journal of Petroleum Technology* (October 1981), 1990–1998.

MELROSE, J. C., "Interpretation of Mixed Wettability States in Reservoir Rocks," *Society of Petroleum Engineers of the AIME,* Technical Paper No. 10971 (1982), 2.

MELROSE, J. C. and C. F. BRANDNER, "Role of Capillary Forces in Determining Microscopic Displacement Efficiency for Oil Recovery by Water Flooding," *Journal of Canadian Petroleum Technology,* 13 (October–December 1974), 54–62.

METCALFE, R. S., "Effects of Impurities on Minimum Miscibility Pressures and Minimum Enrichment Levels for CO_2 and Rich-Gas Displacements," *Society of Petroleum Engineers Journal,* 22 (1981), 219–225.

METCALFE, R. S. and LYMAN YARBOROUGH, "The Effect of Phase Equilibria on the CO_2 Displacement Mechanism," *Society of Petroleum Engineers Journal,* 19 (August 1978), 242–252.

METER, D. M. and R. B. BIRD, "Tube Flow of Non-Newtonian Polymer Solutions, Parts 1 and 2—Laminar Flow and Rheological Models," *American Institute of Chemical Engineers Journal* (November 1964), 878–881, 1143–1150.

MINSSIEUX, L., "Waterflood Improvement by Means of Alkaline Water," in *Enhanced Oil Recovery by Displacement with Saline Solutions,* Houston: Gulf Publishing, a Symposium held May 20, 1976, at Britannic House, London, England.

MOHANTY, K., and S. J. SALTER, "Mulitphase Flow in Porous Media II: Pore Level Modelling," SPE 11018, presented at the 57th Annual Technical Conference and Exhibition of the Society of Petroleum Engineers, New Orleans, Louisiana, 1982.

MONGER, T. G., "The Impact of Oil Aromaticity on CO_2 Flooding," *Society of Petroleum Engineers Journal,* 25 (December 1985), 865–874.

MONGER, T. G. AND J. M. COMA, "A Laboratory and Field Evaluation of the CO_2 Huff 'n' Puff Process for Light Oil Recovery," SPE 15501, presented at the 61st Annual Fall Technical Conference and Exhibition of the Society of Petroleum Engineers of the AIME, New Orleans, 1986.

MOORE, T. F. and R. L. SLOBOD, "The Effect of Viscosity and Capillarity on the Displacement of Oil by Water," *Producers Monthly,* 20 (August 1956), 20–30.

MOREL-SEYTOUX, H. J., "Unit Mobility Ratio Displacement Calculations for Pattern Floods in Homogeneous Medium," *Society of Petroleum Engineers Journal,* 6 (September 1966), 217–227.

MORROW, NORMAN R. and IOANNIS CHATZIS, "Measurements and Correlation of Conditions for Entrapment and Mobilization of Residual Oil," U.S. Department of Energy, DOE/BETC/3251-12, 1981.

———, "The Effect of Surface Roughness on Contact Angles," Preprint-35, 48th National Colloid Symposium of the American Chemical Society, Austin, Texas, 1974.

———, "Capillary Pressure Correlations for Uniformly Wetted Porous Media," *Journal of Canadian Petroleum Technology* (October–December 1976), 49–69.

MUNGAN, N., "Programmed Mobility Control in Polymer Floods," Preprint A-1, presented at the 3rd Assoc. Rech. Tech. Forages and Prod. (ARTFP) Symposium, Pau, France, 1968.

MYHILL, N. A. and G. L. STEGEMEIER, "Steam-Drive Correlation and Prediction," *Journal of Petroleum Technology,* February 1978, 173–182.

NEASHAM, J. W., "The Morphology of Dispersed Clay in Sandstone Reservoirs and its Effect on Sandstone Shaliness," SPE 6858, presented at the 52nd Annual Fall Exhibition of the Society of Petroleum Engineers, Denver, Colorado, 1977.

NEEDHAM, R. B., C. B. THRELKELD, and J. W. GALL, "Control of Water Mobility Using Polymers and Multivalent Cations," SPE 4747, presented at the Improved Oil Recovery Symposium of the Society of Petroleum Engineers, Tulsa, Oklahoma, 1974.

NELSON, R. C., J. B. LAWSON, D. R. THIGPEN, and G. L. STEGEMEIER, "Cosurfactant-Enhanced Alkaline Flooding," SPE/DOE 12672, presented at the Fourth Joint SPE/DOE Symposium on Enhanced Oil Recovery, Tulsa, Oklahoma, 1984.

——— and G. A. POPE, "Phase Relationships in Chemical Flooding," *Society of Petroleum Engineers Journal,* 18 (1978), 325–338.

———, "The Salinity Requirement Diagram—A Useful Tool in Chemical Flooding Research and Development," *Society of Petroleum Engineers Journal,* 22 (1982), 259–270.

———, "Effect of Live Crude on Phase Behavior and Oil-Recovery Efficiency of Surfactant Flooding Systems," *Society of Petroleum Engineers Journal,* 23 (1983), 501–520.

NG, K. M., H. T. DAVIS, and L. E. SCRIVEN, "Visualization of Blob Mechanics in Flow Through Porous Media," *Chemical Engineering Science,* 33 (1978), 1009–1017.

NGHIEM, LONG X. and KHALID AZIZ, "A Robust Iterative Method of Flash Calculations Using the Soave-Redlich-Kwong or the Peng-Robinson Equation of State," SPE 8285, presented at the 54th Annual Fall Technical Conference and Exhibition of the Society of Petroleum Engineers, Las Vegas, Nevada, 1979.

OH, S. G. and J. SLATTERY, "Interfacial Tension Required for Significant Displacement of Residual Oil," Paper D-2, presented at the Second ERDA Symposium on Enhanced Oil and Gas Recovery, Tulsa, Oklahoma, 1976.

Oil and Gas Journal Biennial Survey Since 1972, Tulsa, Oklahoma: PennWell. May 3, 1971, pp. 87–91; March 25, 1974, pp. 70–75; April 5, 1976, pp. 107–138; March 27, 1978, pp. 113–140; March 31, 1980, pp. 79–124; April 5, 1982, pp. 139–159; April 2, 1984, pp. 83–105; April 14, 1986, pp. 71–101; April 18, 1988, pp. 18–73; April 23, 1990, pp. 49–82.

ORR, F. M., JR. and C. M. JENSEN, "Interpretation of Pressure-Composition Phase Diagrams for CO_2-Crude Oil Systems," SPE 11125, presented at the 57th Annual Fall Technical Conference and Exhibition of the Society of Petroleum Engineers of AIME, New Orleans, Louisiana, 1982.

ORR, F. M., JR., C. M. JENSEN, and M.K. SILVA, "Effect of Solution Gas on the Phase Behavior of CO_2-Crude Oil Mixtures," presented at the International Energy Agency Workshop on Enhanced Oil Recovery, Winfuth, United Kingdom, 1981.

ORR, F. M., JR. and M. K. SILVA, "Equilibrium Phase Compositions of CO_2 Hydrocarbon Mixtures: Measurement by a Continuous Multiple Contact Experiment," SPE/DOE 10726, presented at the Third Joint SPE/DOE Symposium on Enhanced Oil Recovery, Tulsa, Oklahoma, 1982.

ORR, F. M., JR., M. K. SILVA, and M. T. PELLETIER, "Laboratory Experiments to Evaluate Field Prospects for CO_2 Flooding," *Journal of Petroleum Technology,* 34 (1982), 888–903.

OVERBEEK, J. TH. G., *Colloids and Surface Chemistry, A Self-Study Subject, Study Guide, PART 2, Lyophobic Colloids,* Cambridge, Mass.: Center for Advanced Engineering Study and Department of Chemical Engineering, Massachusetts Institute of Technology, 1972.

OWENS, W. W. and D. L. ARCHER, "The Effect of Rock Wettability on Oil-Water Relative Permeability Relationships," *Journal of Petroleum Technology* (July 1971), 873–878.

PARKINSON, W. J. and N. DE NEVERS, "Partial Molal Volume of Carbon Dioxide in Water Solutions," *Industrial and Engineering Chemistry Fundamentals,* 8 (November 1969), 709–713.

PATTON, JOHN T., S. HOLBROOK, M. KUNTAMUKKULA, and R. L. LONG, "Enhanced Oil Recovery by CO_2 Foam Flooding, Second Annual Report," U.S. Department of Energy, DOE/MC/03259-15, 1981.

PAUL, G. W. and H. R. FRONING, "Salinity Effects of Micellar Flooding," *Journal of Petroleum Technology,* 25 (1973), 957–958.

PAUL, G. W., L. W. LAKE, G. A. POPE, and G. B. YOUNG, "A Simplified Predictive Model for Micellar/Polymer Flooding," SPE 10733, presented at the California Regional Meeting of the Society of Petroleum Engineers, San Francisco, California, 1982.

PAYATAKES, A. C., R. W. FLUMERFELT, and K. M. NG, "On the Dynamics of Oil Ganglia Populations During Immiscible Displacement," American Institute of Chemical Engineers 84th National Meeting, Atlanta, Georgia, 1978.

PEACEMAN, DON W., *Fundamentals of Numerical Reservoir Simulation,* New York: Elsevier, 1977.

PENG, DING-YU and DONALD B. ROBINSON, "A New Two-Constant Equation of State," *Industrial and Engineering Chemistry Fundamentals,* 15 (1976), 59–64.

———, "Two and Three Phase Equilibrium Calculations for Systems Containing Water," *Canadian Journal of Chemical Engineers,* 54 (December 1976), 595–599.

PERKINS, T. K. and O. C. JOHNSTON, "A Review of Diffusion and Dispersion in Porous Media," *Society of Petroleum Engineers Journal,* 3 (March 1963), 70–80.

PERRY, G. E., "First Annual Report, Weeks Island 'S' Sand Reservoir B Gravity Stable, Miscible CO_2 Displacement, Iberia Parish, Louisiana," U.S. Department of Energy, METC/CR-78/13, Morgantown, West Virginia (June 1978).

PETERS, EKWERE J., "Stability Theory and Viscous Fingering in Porous Media," Ph.D. dissertation, University of Alberta, 1979.

PICKENS, J. F. and G. E. GRISAK, "Scale-Dependent Dispersion in a Stratified Granular Aquifer," *Water Resources Research,* 17 (1981), 1191–1211.

PIRSON, SYLVAIN J., *Geologic Well Log Analysis,* 3rd. edition Houston: Gulf Publishing, 1983.

POPE, G. A., "The Application of Fractional Flow Theory to Enhanced Oil Recovery," *Society of Petroleum Engineers Journal* (June 1980), 191–205.

POPE, G. A., L. W. LAKE, and F. G. HELFFERICH, "Cation Exchange in Chemical Flooding, Part 1: Basic Theory Without Dispersion," *Society of Petroleum Engineers Journal,* 18 (1978), 418–434.

POPE, G. A. and R. C. NELSON, "A Chemical Flooding Compositional Simulator," *Society of Petroleum Engineers Journal,* 18 (October 1978), 339–354.

POPE, G. A., BEN WANG, and KERMING TSAUR, "A Sensitivity Study of Micellar-Polymer Flooding," *Society of Petroleum Engineers Journal,* 19 (December 1979), 357–368.

POZZI, A. L. and R. J. BLACKWELL, "Design of Laboratory Models for Study of Miscible Displacement," *Society of Petroleum Engineers Journal,* 3 (March 1963), 28–40.

PRATS, M., *Thermal Recovery,* Dallas: Society of Petroleum Engineers of American Institute of Mining and Metallurgical Engineers, Henry L. Doherty Series, Monograph 7, 1982.

PRAUSNITZ, J. M., T. F. ANDERSON, E. A. GRENS, C. A. ECKERT, R. HSIEH, and J. P. O'CONNELL, *Multicomponent Vapor–Liquid and Liquid–Liquid Equilibria,* Englewood Cliffs, New Jersey: Prentice-Hall, Inc., 1980.

PROUVOST, LAURENT P., "Research Report on Phase Behavior of Micellar Systems," Center for Enhanced Oil and Gas Recovery Research Report No. 84-2, The University of Texas at Austin, 1984.

Puerto, M. C. and R. L. Reed, "Three-Parameter Representation of Surfactant/Oil/Brine Interaction," *Society of Petroleum Engineers Journal,* 23 (1982), 669–683.

Raimondi, Pietro and Michael A. Torcaso, "Distribution of the Oil Phase Obtained Upon Imbibition of Water," *Transactions of the American Institute of Mining and Metallurgical Engineers,* 231 (1964), 49–55.

Ramakrishnan, T. S. and D. T. Wassan, "Alkaline Waterflooding—A Model for Interfacial Activity of Acidic Crude/Caustic Systems," SPE 10716, presented at the Third Symposium on Enhanced Oil Recovery of the Society of Petroleum Engineers, Tulsa, Oklahoma, 1982.

Ramey, H. J., Jr., Discussion of "Reservoir Heating by Hot Fluid Injection," *Transactions of the American Institute of Mining and Metallurgical Engineers,* 216 (1959), 364–365.

———, "Wellbore Heat Transmission," *Journal of Petroleum Technology,* 14 (April 1962), 427–435.

———, "How to Calculate Heat Transmission in Hot Fluids," *Petroleum Engineer* (November 1964), 110–120.

Rapoport, L. A. and W. J. Leas, "Properties of Linear Waterfloods," *Transactions of the American Institute of Mining and Metallurgical Engineers,* 198 (1953), 139–148.

Reed, R. L. and R. N. Healy, "Some Physico-Chemical Aspects of Microemulsion Flooding: A Review," *Improved Oil Recovery by Surfactant and Polymer Flooding,* D. O. Shah and R. S. Schecter (eds.) New York: Academic Press, 1977.

———, "Contact Angles for Equilibrated Microemulsion Systems," SPE 8262, presented at the 54th Annual Fall Technical Conference and Exhibition of the Society of Petroleum Engineers, Houston, Texas, 1979.

Reid, Robert C., John M. Prausnitz, and Thomas K. Sherwood, *The Properties of Gases and Liquids,* 3d ed., St. Louis: McGraw-Hill, 1977.

Rieke, Herman H., George V. Chilingarian, and Paul Vorabutr, "Clays," in Chapter 5 of *Drilling and Drilling Fluids,* G. V. Chilingarian and P. Vorabutr (eds.), New York: Elsevier, 1983.

Risnes, R. and Vilgeir Dalen, "Equilibrium Calculations of Coexisting Liquid Phases," SPE 11126, presented at the 57th Annual Fall Technical Conference and Exhibition of the Society of Petroleum Engineers, New Orleans, Louisiana, 1982.

Roberts, G. E. and H. Kaufman, *Table of Laplace Transforms,* Philadelphia: Saunders, 1966.

Rowe, A. M., Jr., "The Critical Composition Method—A New Convergence Pressure Method," *Society of Petroleum Engineers Journal,* 7 (March 1967), 54–60.

Sage, Bruce H. and William N. Lacey, *Volumetric and Phase Behavior of Hydrocarbons,* Houston: Gulf Publishing, 1939.

Salathiel, R. A., "Oil Recovery by Surface Film Drainage in Mixed-Wettability Rocks," *Journal of Petroleum Technology* (October 1973), 1216–1224.

Salter, S. J., "The Influence of Type and Amount of Alcohol on Surfactant-Oil-Brine Phase Behavior and Properties," SPE 6843, presented at the 52nd Annual Technical Conference and Exhibition of the Society of Petroleum Engineers, Denver, Colorado, 1977.

———, "Selection of Pseudo-Components in Surfactant-Oil-Brine Alcohol Systems," SPE 7056, presented at the Fifth Symposium on Improved Methods for Oil Recovery of the Society of Petroleum Engineers, Tulsa, Oklahoma, 1978.

———, "Optimizing Surfactant Molecular Weight Distribution I. Sulfonate Phase Behavior and Physical Properties," SPE 12036, presented at the 58th Annual Technical Conference and Exhibition of the Society of Petroleum Engineers, San Francisco, California, 1983.

——— and K. K. MOHANTY, "Multiphase Flow in Porous Media: Macroscopic Observations and Modeling," SPE 11017, presented at the 57th Annual Technical Conference and Exhibition of the Society of Petroleum Engineers, New Orleans, Louisiana, 1982.

SAMIZO, N., "Numerical Simulation of Two-Phase Compressible Flow by the Moving Point Method," M.S. thesis, The University of Texas at Austin, 1982.

SATTER, ABDUS, "Heat Losses During Flow of Steam Down a Wellbore," *Journal of Petroleum Technology* (July 1965), 845–851.

SAVINS, J. G., "Non-Newtonian Flow Through Porous Materials," *Industrial and Engineering Chemistry,* 61 (1969), 19.

SAZEGH, SELIM G., "Phase Behavior Measurements of Pembina Cardium Stock Tank Oil-Methane-Carbon Dioxide Mixtures at 45°C and 13.8 MPa," Petroleum Recovery Institute Research Report 81-3, Calgary, Alberta, Canada, 1981.

SCHNEIDER, F. N. and W. W. OWENS, "Steady-State Measurements of Relative Permeability for Polymer/Oil Systems," *Society of Petroleum Engineers Journal,* 22 (February 1982), 79–86.

SCHNEIDER, G. M., "Phase Equilibria in Fluid Mixtures at High Pressures," *Advances in Chemical Physics,* I. Prigogine and Stuart A. Rice (eds.), New York: Interscience, 1970.

SCRIVEN, L. E., "Equilibrium Bi-Continuous Structures," in *Micellization, Solubilization, and Microemulsions,"* K. L. Mittal (ed.), New York: Plenum, 1976.

SEBASTIAN, H. M., R. S. WINGER, and T. A. RENNER, "Correlation of Minimum Miscibility Pressure for Impure CO_2 Streams," SPE/DOE 12648, presented at the Fourth Joint SPE/DOE Symposium on Enhanced Oil Recovery, Tulsa, Oklahoma, 1984.

SERIGHT, R. S., "The Effects of Mechanical Degradation and Viscoelastic Behavior on Injectivity of Polyacrylamide Solutions," *Society of Petroleum Engineers Journal,* 23 (June 1983), 475–485.

SHELTON, J. L. and F. N. SCHNEIDER, "The Effects of Water Injection on Miscible Flooding Methods Using Hydrocarbons and Carbon Dioxide," *Society of Petroleum Engineers Journal,* 15 (June 1975), 217–226.

SHINODA, KOZO and HIRONOBU KUNIEDA, "Phase Properties of Emulsions," in *Emulsions, Microemulsions, and Solubilization,* New York: Wiley, 1979.

SHUPE, R. D., "Chemical Stability of Polyacrylamide Polymers," *Journal of Petroleum Technology,* 33 (August 1981), 1513–1529.

SHUTLER, N. D. and T. C. BOBERG, "A One-Dimensional Analytical Technique for Predicting Oil Recovery by Steamflooding," *Society of Petroleum Engineers Journal* (October 1972), 489–498.

SIMLOTE, V. M., V. J. ZAPATA, and L. A. BELVEAL, "A Study of Caustic Consumption in a High-Temperature Reservoir," SPE 12086, presented at the 58th Annual Fall Meeting of the Society of Petroleum Engineers, San Francisco, California, 1983.

SIMON, R. and D. J. GRAUE, "Generalized Correlation for Predicting Solubility, Swelling, and Viscosity Behavior for CO_2-Crude Oil Systems," *Transactions of the American Institute of Mining and Metallurgical Engineers,* 234 (1965), 102–130.

SLATTERY, JOHN C., *Momentum, Energy, and Mass Transfer in Continua,* New York: McGraw-Hill, 1972.

SMITH, J. M. and H. C. VAN NESS, *Introduction to Chemical Engineering Thermodynamics,* 2d ed., New York: McGraw-Hill, 1975.

SOAVE, GIORGIO, "Equilibrium Constants from a Modified Redlich-Kwong Equation of State," *Chemical Engineering Science,* 27 (June 1972), 1197–1203.

SOMASUNDARAN, M. C., A. GOYAL, and E. MANEV, "The Role of Surfactant Precipitation and Redissolution in the Adsorption of Sulfonate on Minerals," SPE 8263, presented at the 54th Annual Technical Conference and Exhibition of the Society of Petroleum Engineers, Las Vegas, Nevada, 1979.

SOMERTON, W. H., "Thermal Properties of Hydrocarbon Bearing Rocks at High Temperatures and Pressures," API Research Project 117, Final Report, College of Engineering, University of California, Berkeley, 1973.

SPENCE, ANDREW P., JR. and ROBERT W. WATKINS, "The Effect of Microscopic Core Heterogeneity on Miscible Flood Residual Oil Saturation," SPE 9229, presented at the 55th Annual Fall Technical Conference and Exhibition of the Society of Petroleum Engineers, Dallas, Texas, 1980.

STALKUP, FRED I., *Miscible Flooding Fundamentals,* Society of Petroleum Engineers Monograph Series, 1983.

———, "Displacement of Oil by Solvent at High Water Saturation," *Society of Petroleum Engineers Journal,* 10 (December 1970), 337–348.

STANDING, M. B., *Volumetric and Phase Behavior of Oil Field Hydrocarbon Systems,* Dallas: Society of Petroleum Engineers of the American Institute of Mining and Metallurgical Engineers, 1977.

STEGEMEIER, G. L., "Relationship of Trapped Oil Saturation to Petrophysical Properties of Porous Media," SPE 4754, presented at the Improved Oil Recovery Symposium of the Society of Petroleum Engineers, Tulsa, Oklahoma, 1974.

———, "Mechanisms of Entrapment and Mobilization of Oil in Porous Media," Paper 13c, presented at the 81st National Meeting of the American Institute of Chemical Engineers, Kansas City, Missouri, 1976.

STEGEMEIER, G. L., D. D. LAUMBACH, and C. W. VOLEK, "Representing Steam Processes with Vacuum Models," SPE 6787, presented at the 52nd Annual Fall Technical Conference and Exhibition of the Society of Petroleum Engineers, Denver, Colorado, 1977.

STONE, H. L., "Probability Model for Estimating Three Phase Relative Permeability and Residual Oil Saturation," *Journal of Petroleum Technology* (1970), 214–218.

SYNDANSK, R. D., "Elevated-Temperature Caustic/Sandstone Interaction: Implications for Improving Oil Recovery," *Society of Petroleum Engineers Journal* (August 1983), 453–462.

TABER, J. J., I. S. K. KAMATH, and R. L. REED, "Mechanism of Alcohol Displacement of Oil from Porous Media," *Miscible Processes,* SPE Reprint Series No. 8 (1965), 39–56.

——— and W. K. MEYER, "Investigations of Miscible Displacements of Aqueous and Oleic Phases from Porous Media," *Miscible Processes,* SPE Reprint Series No. 8 (1965), 57–68.

———, "Dynamic and Static Forces Required to Remove a Discontinuous Oil Phase from Porous Media Containing Both Oil and Water," *Society of Petroleum Engineers Journal,* 9 (1969), 3–12.

——— and F. D. MARTIN, "Technical Screening Guides for the Enhanced Recovery of Oil," SPE 12069, presented at the 58th Annual Conference and Exhibition of the Society of Petroleum Engineers, San Francisco, California, 1983.

TAYLOR, G. I., "The Dispersion of Matter in Solvent Flowing Slowly Through a Tube," *Proceeding of Royal Society London,* Series A, 219, 1953, 189–203.

TIFFIN, DAVID L. and WILLIAM F. YELLIG, "Effects of Mobile Water on Multiple Contact Miscible Gas Displacements," SPE/DOE 10687, presented at the 1982 Third Joint SPE/DOE Symposium on Enhanced Oil Recovery, Tulsa, Oklahoma, 1982.

TINKER, G. E., R. W. BOWMAN, and G. A. POPE, "Determination of In Situ Mobility and Wellbore Impairment from Polymer Injectivity Data," *Journal of Petroleum Technology* (May 1976), 586–596.

TODD M. R., and C. A. CHASE, "A Numerical Simulator for Predicting Chemical Flood Performance," SPE-7689, presented at the 54th Annual Fall Technical Conference and Exhibition of the Society of Petroleum Engineers, Houston, Texas, 1979.

TODD, M. R. and W. J. LONGSTAFF, "The Development, Testing, and Application of a Numerical Simulator for Predicting Miscible Flood Performance," *Journal of Petroleum Technology,* 7 (July 1972), 874–882.

TRIEBER, L. E., DUANE L. ARCHER, and W. W. OWENS, "A Laboratory Evaluation of the Wettability of Fifty Oil-Producing Reservoirs," *Society of Petroleum Engineers Journal,* 12 (December 1971), 531–540.

TRUSHENSKI, S. P., "Micellar Flooding: Sulfonate-Polymer Interaction," in *Improved Oil Recovery by Surfactant and Polymer Flooding,* D. O. Shah and R. S. Schecter (eds.) New York: Academic Press, 1977.

TSAUR, K., "A Study of Polymer/Surfactant Interactions for Micellar/Polymer Flooding Applications," M.S. thesis, The University of Texas at Austin, 1978.

TUMASYAN, A. B., V. G. PANTELEEV, and G. P. MEINSTER, "The Effect of Carbon Dioxide Gas on the Physical Properties of Crude Oil and Water," Nauk.-Tekh. Sb Ser. Neftepromyslovoe Delo #2, 20–30, 1969.

TUREK, EDWARD A., ROBERT S. METCALFE, LYMAN YARBOROUGH, and ROBERT L. ROBINSON, "Phase Equilibrium in Carbon Dioxide-Multicomponent Hydrocarbon Systems: Experimental Data and Improved Prediction Technique," SPE 9231, presented at the 55th Annual Fall Technical Conference and Exhibition of the Society of Petroleum Engineers, Dallas, Texas, 1980.

VAN LOOKEREN, J., "Calculation Methods for Linear and Radial Steam Flow in Oil Reserves," SPE 6788, presented at the 52nd Annual Technical Conference and Exhibition of the Society of Petroleum Engineers, Denver, Colorado, 1977.

VOGEL, J. L. and L. YARBOROUGH, "The Effect of Nitrogen on the Phase Behavior and Physical Properties of Reservoir Fluids," SPE 8815, presented at the First Joint SPE/DOE Symposium on Enhanced Oil Recovery, Tulsa, Oklahoma, 1980.

VUKALOVICH, M. P. and V. V. ALTUNIN, *Thermodynamic Properties of Carbon Dioxide,* London: Collet's Publishers, 1968.

WAGGONER, JOHN R. and LARRY W. LAKE, "A Detailed Look at Simple Viscous Fingering," presented at the Second International Symposium on Enhanced Oil Recovery, Maracaibo, Venezuela, 1987.

WAGNER, O. R. and R. O. LEACH, "Improving Oil Displacement by Wettability Adjustment," *Transactions of the American Institute of Mining and Metallurgical Engineers,* 216 (1959), 65–72.

WATSON, K. M., E. F. NELSON, and GEORGE B. MURPHY, "Characterization of Petroleum Fractions," *Industrial and Engineering Chemistry,* 27 (1935).

WEAVER, C. E. and L. D. POLLARD, *The Chemistry of Clay Minerals,* New York: Elsevier, 1973.

WELCH, BRANDON, "A Numerical Investigation of the Effects of Various Parameters in CO_2 Flooding," M.S. thesis, The University of Texas at Austin, 1982.

WELGE, H. J., "A Simplified Method for Computing Oil Recovery by Gas or Water Drive," *Transactions of the American Institute for Mining and Metallurgical Engineers,* 195 (1952), 91–98.

WELLINGTON, S. L., "Biopolymer Solution Viscosity Stabilization—Polymer Degradation and Antioxidant Use," SPE 9296, presented at the 55th Annual Fall Technical Conference and Exhibition of the Society of Petroleum Engineers, Dallas, Texas, 1980.

WHITE, PHILLIP D. and JON T. MOSS, *Thermal Recovery Methods,* Tulsa, Oklahoma: PennWell, 1983.

WHITEHEAD, W. R., O. K. KIMBLER, W. R. HOLDEN, and A. T. BOURGOYNE, "Investigations of Enhanced Oil Recovery Through Use of Carbon Dioxide," Final Report, DOE/MC/03103-6, U.S. Department of Energy, 1981.

WHITEHEAD, W. R., O. K. KIMBLER, R. M. HOSHMAN, and J. R. HERVEY, "Investigations of Enhanced Oil Recovery Through Use of Carbon Dioxide," U.S. Department of Energy Progress Report, DOE/MC/03103-2, U.S. Department of Energy, June 1980.

WHITSON, CURTIS H., "Effect of Physical Properties Estimation on Equation-of-State Predictions," SPE 11200, presented at the 57th Annual Fall Technical Conference and Exhibition of the Society of Petroleum Engineers of AIME, New Orleans, Louisiana, 1982.

WILLHITE, G. P. and J. G. DOMINGUEZ, "Mechanisms of Polymer Retention in Porous Media," in *Improved Oil Recovery by Surfactant and Polymer Flooding,* D. O. Shah and R. S. Schechter (eds.), New York: Academic Press, 1977.

———, "Over-all Heat Transfer Coefficients in Steam and Hot Water Injection Wells," *Journal of Petroleum Technology* (May 1967), 607–615.

WILLIAMS, CRAIG A., ERDINC N. ZANA, and GEORGE HUMPHREYS, "Use of the Peng-Robinson Equation of State to Predict Hydrocarbon Phase Behavior and Miscibility for Fluid Displacement," SPE 8817, presented at the First Joint SPE/DOE Symposium on Enhanced Oil Recovery, Tulsa, Oklahoma, 1980.

WINSOR, P. A., *Solvent Properties of Amphiphilic Compounds,* London: Butterworths, 1954.

YARBOROUGH, LYMAN, "Application of a Generalized Equation of State to Petroleum Reservoir Fluids," in *Equations of State in Engineering and Research,* Washington, D.C.: American Chemical Society, 1978.

YARBOROUGH, LYMAN and L. R. SMITH, "Solvent and Driving Gas Compositions for Miscible Slug Displacement," *Society of Petroleum Engineers Journal,* 10 (October 1970), 298–310.

YELLIG, W. F. and R. S. METCALFE, "Determination and Prediction of CO_2 Minimum Miscibility Pressures," *Journal of Petroleum Technology* (January 1980), 160–168.

YOKOYAMA, YOSHIO, "The Effects of Capillary Pressure on Displacements in Stratified Porous Media," M.S. thesis, The University of Texas at Austin, 1981.

YOKOYAMA, YOSHIO and LARRY W. LAKE, "The Effect of Capillary Pressure on Immiscible Displacements in Stratified Porous Media," SPE-10109, presented at the 56th Annual Fall Technical Conference and Exhibition of the Society of Petroleum Engineers, San Antonio, Texas, 1981.

YORTSOS, Y. C., "Discussion of the Radial Movement of Permeability Fronts and Multiple Reaction Zones in Porous Media," *Society of Petroleum Engineers Journal,* 22 (December 1982), 903–905.

YOUNG, LARRY C. and ROBERT E. STEPHENSON, "A Generalized Compositional Approach for Reservoir Simulation," SPE 10516, presented at the Sixth Symposium on Reservoir Simulation of the Society of Petroleum Engineers, New Orleans, Louisiana, 1982.

———, "The Use of Dispersion Relationships to Model Adverse Mobility Ratio Miscible Displacements," SPE/DOE 14899, presented at the Fifth Joint SPE/DOE Symposium on Improved Oil Recovery, Tulsa, Oklahoma, 1986.

ZAPATA, VITO J. and LARRY W. LAKE, "A Theoretical Analysis of Viscous Crossflow," SPE 10111, presented at the 56th Annual Fall Technical Conference and Exhibition of the Society of Petroleum Engineers, San Antonio, Texas, 1981.

ZUDKEVITCH, DAVID and JOSEPH JOFFE, "Correlation and Prediction of Vapor-Liquid Equilibria with the Redlich-Kwong Equation of State," *American Institute of Chemical Engineers Journal,* 16 (1970), 112–199.

Index